Mechanisms and Machine Science

Volume 120

This book series establishes a well-defined forum for monographs, edited Books, and proceedings on mechanical engineering with particular emphasis on MMS (Mechanism and Machine Science). The final goal is the publication of research that shows the development of mechanical engineering and particularly MMS in all technical aspects, even in very recent assessments. Published works share an approach by which technical details and formulation are discussed, and discuss modern formalisms with the aim to circulate research and technical achievements for use in professional, research, academic, and teaching activities.

This technical approach is an essential characteristic of the series. By discussing technical details and formulations in terms of modern formalisms, the possibility is created not only to show technical developments but also to explain achievements for technical teaching and research activity today and for the future.

The book series is intended to collect technical views on developments of the broad field of MMS in a unique frame that can be seen in its totality as an Encyclopaedia of MMS but with the additional purpose of archiving and teaching MMS achievements. Therefore, the book series will be of use not only for researchers and teachers in Mechanical Engineering but also for professionals and students for their formation and future work.

The series is promoted under the auspices of International Federation for the Promotion of Mechanism and Machine Science (IFToMM).

Prospective authors and editors can contact Mr. Pierpaolo Riva (publishing editor, Springer) at: pierpaolo.riva@springer.com

Indexed by SCOPUS and Google Scholar.

More information about this series at https://link.springer.com/bookseries/8779

Andreas Müller · Mathias Brandstötter
Editors

Advances in Service and Industrial Robotics

RAAD 2022

Springer

Editors
Andreas Müller
Institute of Robotics
Johannes Kepler University Linz
Linz, Austria

Mathias Brandstötter
Institute for Robotics and Mechatronics
Joanneum Research
Klagenfurt, Austria

ISSN 2211-0984 ISSN 2211-0992 (electronic)
Mechanisms and Machine Science
ISBN 978-3-031-04869-2 ISBN 978-3-031-04870-8 (eBook)
https://doi.org/10.1007/978-3-031-04870-8

This Springer imprint is published by the registered company Springer Nature Switzerland AG
The registered company address is: Gewerbestrasse 11, 6330 Cham, Switzerland

Preface

The 31st International Conference on Robotics in Alpe-Adria-Danube Region, RAAD 2022, was held at the Lakeside Science & Technology Park in Klagenfurt am Wörthersee, Austria, on June 8–10, 2022. The conference brought together robotics researchers in academy and industry from 29 countries, mainly in the Alpe-Adria-Danube Region, but also from other countries worldwide.

This book comprises 71 contributed papers that were selected for inclusion after a thorough review process. The conference covers all relevant topics and ongoing trends in robotics research. It is organized in eight tracks, which correspond to the following chapters of this proceedings book:

I. Robot Modeling and Identification
II. Perception and Learning
III. Motion Planning and Control
IV. Robot Design and Synthesis
V. Assistive and Medical Robots
VI. Autonomous Mobile Manipulation
VII. Humanoids, Legged, and Bioinspired Robots
VIII. Collaborative Robotics.

Discussion of current trends and exchange of recent results in robotics research in an open and collegial atmosphere is a key aspect of the RAAD conference, and besides the technical sessions, plenary talks by renowned experts are a key component of the conference. We are thankful and honored that we had the following keynote speakers sharing their ideas on current topics:

- Arne Wahrburg (Principal Scientist at ABB, Darmstadt, Hesse, Germany) on "Trash or treasure? On AI in industrial robotics (and why we still need mechatronics)"
- Bruno Siciliano (Director of ICAROS Center, Università degli Studi di Napoli Federico II) on "Robot Manipulation and Control"
- Mike Mayer (Global Key Account Manager at SCHUNK GmbH & Co.KG) on "What it means to build smart grasping services".

The 31st RAAD conference is special in that it is the first conference taking place in presence, after it had to be organized as an online event in 2020 and 2021 due to the COVID-19 pandemic. Moreover, the 2021 RAAD conference should have been the occasion for celebration and reflection on three decades of RAAD. Finally, this year's conference was the venue for celebrating the 31st anniversary of RAAD.

The first event in this series was organized during 1991 in Portorož, Slovenia, by the Jožef Stefan Institute as the 1st International Meeting on Robotics in Alpe-Adria Region. In the following years, its name was changed to International Workshop on Robotics in Alpe-Adria Region (RAA) until 1996, when it became the International Conference on Robotics in Alpe-Adria-Danube Region (RAAD).

It is our great pleasure that the RAAD founder, Jadran Lenarčič from the Jožef Stefan Institute, Ljubljana, Slovenia, gave a welcome address during the opening ceremony and has contributed a welcome message to this book.

Organizing scientific events relies on the dedication of authors and reviewers, but also on the kind support of our sponsors. We thank the International Federation for the Promotion of Mechanism and Machine Science (IFToMM) and the Austrian Section of Institute of Electrical and Electronics Engineers (IEEE) for allowing us to organize this conference under their institutional umbrella. We are particularly grateful for the financial support of the following companies

- SCHUNK Intec GmbH, Austria (Gold Sponsor)
- ABB AG, Austria (Silver Sponsor)
- Universal Robots GmbH, Germany (Silver Sponsor).

Special thanks go to the RAAD Scientific Conference Committee for support and advertisement of the conference.

June 2022

Andreas Müller
Mathias Brandstötter

Welcome Address from the Founder of RAAD, Jadran Lenarčič

When I entered robotics as a young researcher in the late 1970s, I made my first acquaintances and connections in the USA and Japan. It is absurd that these were mostly colleagues from Europe, and we Europeans met and got to know each other outside Europe. As a young researcher, I felt this lack of cooperation in Europe, and I looked with admiration at my American colleagues, who seemed to me to come from the same university or institute. Things began to improve gradually with the first European projects and also with networking supported by the European Science Foundation.

So I made many acquaintances among colleagues, especially in Northern Italy and Austria. After the independence of Slovenia in 1991, my idea of scientific collaboration in Europe turned into a necessity. Both from a scientific and especially cultural point of view, I especially felt the need to establish closer ties in the Alps-Adriatic Region. This was like a mission for me. Although many of us already knew each other well, we did not have a common place to meet, so I came up with the idea to organize the first conference on robotics in Alpe-Adria Region. Organized by the Jožef Stefan Institute, it was held during 1991 in Portorož on the Slovenian coast, and to my delight, it was attended by everyone I invited. That is how the RAA came about.

A few years later, mainly due to the activities of our Austrian colleagues, we added the Danube Region, and we also began to invite colleagues from all over Italy. This is the current RAAD as we know it today. After more than ten years of cooperation, I moved away from the conference due to my new position at the Jožef Stefan Institute, so I watched with even greater joy and pride how RAAD is developing, how generations are changing, and how researchers are making contacts and new friends, exchanging ideas, and proposing new projects.

The coincidence that RAAD 2022 is taking place in Klagenfurt particularly excites me. This is the birthplace of the world-famous physicist Jožef Stefan, whom both Slovenes and Austrians share as their own. Namely Stefan was a Slovene who worked in Vienna all his career. After him the larger Slovenian research center is named, where I worked for 42 years and where I was also the director for almost 16 years. Dear colleagues, I wish you many more professional successes and, above all, personal connections, which are the bases for any progress in science.

June 2022 Jadran Lenarčič

Contents

Motion Planning and Control

Robot Design and Synthesis

Robot Modeling and Identification

Geometric Identification of Denavit-Hartenberg Parameters with Optical Measuring System

Leon Žlajpah(✉) and Tadej Petrič

Department for Automation, Biocybernetics and Robotics, Jožef Stefan Institute, Ljubljana, Slovenia
{leon.zlajpah,tadej.petric}@ijs.si

Abstract. The absolute accuracy of robots plays an essential role in modern robotics applications that go beyond the typical industrial environment. This paper proposes a method to improve absolute robot accuracy by using an optical measurement system. The method is based on geometric operations used to calculate the rotation axes of each joint. Once all the rotation axes are identified, the complete kinematic model is calculated using the Denavit-Hartenberg (DH) parameters. The effectiveness of the proposed method was validated on a 7-DoF Franka Emika Panda robot. The results show a significant improvement in absolute accuracy. Thanks to the optical measurement system, the proposed method can be easily automated to improve collaborative robot accuracy, especially in applications that require high positioning accuracy.

Keywords: Serial robot manipulators · Kinematic parameter identification · Optical tracking system

1 Introduction

In high-volume production, robots have long been established as an effective programmable tool. Here, operators typically program the robot to move to the desired position, which is recorded and then repeated to complete the task. However, programming is usually time-consuming and not economically justified in low-volume manufacturing. This can be partially circumvented by using off-line programming strategies, which on the other hand, require absolute robot accuracy to complete the task on a real robot. Therefore, the knowledge of robot kinematics is essential. Robot manufacturers usually provide a nominal kinematic model, e.g., defining the Denavit-Hartenberg (DH) parameters. Unfortunately, the nominal model usually differs from the actual one due to errors in robot assembly, manufacturing, etc. In addition, the robot model parameters are generally not accessible and are managed internally by the robot controller. Consequently, the use of nominal DH parameters increases the error of the robot end-effector and limits the practical use of robots in many low-volume applications.

To overcome the mentioned problem, several online and offline methods have been described in the literature to improve the absolute accuracy of the robot [6, 10]. For example, open-loop calibration of DH parameters with an external measurement system

This work was supported by Slovenian Research Agency grant P2-0076.

A. Müller and M. Brandstötter (Eds.): RAAD 2022, MMS 120, pp. 3–10, 2022.
https://doi.org/10.1007/978-3-031-04870-8_1

using a laser tracker can achieve high absolute accuracy, but it is costly and not flexible [9,11]. On the other hand, closed-loop calibration with measurement equipment is cheap but requires manual teaching and is challenging to automate, which affects the overall accuracy of the calibration [2,3]. In addition, these approaches usually do not consider the influence of the accuracy of tool positioning. With the increasing number of robots, better accessibility, and their introduction into low-volume manufacturing areas, improving absolute accuracy remains a challenge for robotics to this day.

To solve this problem, we propose in this work an approach where the DH parameters are identified by using geometric operations like in [4] but by measuring robot base and end-effector positions with an optical tracking system, similar to the one described in [8], except that we were using passive markers. First, we give the kinematic modelling of a robot and algorithms to calculate DH parameters. Then we describe our experimental setup and measurement procedure. Finally, we present the experimental results, emphasizing position accuracy improvement due to DH parameter correction.

2 Kinematic Parameters

A serial robot manipulator is composed of links connected with joints. The joints can be of different types, having one or more degrees-of-freedom (DOF). However, as most modern robots use only revolute joints, we will assume in the following that the robot has only single DOF revolute joints.

For kinematic analysis we attach a coordinate frame to each link. In particular, frame $\mathcal{F}_i$ with origin $\boldsymbol{p}_i$ and axes $(\boldsymbol{x}_i, \boldsymbol{y}_i, \boldsymbol{z}_i)$ is attached to link i and joint i connects frames $\mathcal{F}_{i-1}$ and $\mathcal{F}_i$. Note that $\mathcal{F}_0$ is representing the robot base frame and $\mathcal{F}_n$ is attached to the robot end-effector. The transformation between frames $\mathcal{F}_{i-1}$ and $\mathcal{F}_i$ is expressed with homogeneous transformation matrix ${}^{i-1}\mathbf{T}_i(q_i)$, where q_i is a joint variable (position). Hence, the pose (position and orientation) of the robot end-effector is given as

$$ {}^0\mathbf{T}_E = {}^0\mathbf{T}_n(\boldsymbol{q})^n\mathbf{T}_E = \left(\prod_{i=1}^{n} {}^{i-1}\mathbf{T}_i(q_i)\right) {}^n\mathbf{T}_E \tag{1} $$

where ${}^n\mathbf{T}_E$ is fixed transformation matrix between the robot flange and the end-effector.

Denavit-Hartenberg Parameters: Various models have been proposed to calculate the direct kinematic equation (1). However, the most commonly used approach is the Denavit-Hartenberg (DH) convention [12], which provides a minimal representation and an intuitive method for determining its parameters using linear algebra.

Following the DH convention, z-axes of frames $\mathcal{F}_i$ are aligned with corresponding joints and the x-axes are parallel to the common normal of consecutive joint axes as shown on Fig. 1. Then, a set of four parameters a, α, d, and θ is used to define the transformation ${}^{i-1}\mathbf{T}_i$

$$ {}^{i-1}\mathbf{T}_i = \mathrm{Rot}_{\boldsymbol{z}_{i-1}}(\theta_i) \cdot \mathrm{Trans}_{\boldsymbol{z}_{i-1}}(d_i) \cdot \mathrm{Trans}_{\boldsymbol{x}_i}(r_i) \cdot \mathrm{Rot}_{\boldsymbol{x}_i}(\alpha_i) \tag{2} $$

where $\mathrm{Trans}_{\boldsymbol{a}}(\cdot)$ and $\mathrm{Rot}_{\boldsymbol{a}}(\cdot)$ represent the translation and rotation along vector $\boldsymbol{a}$, respectively.

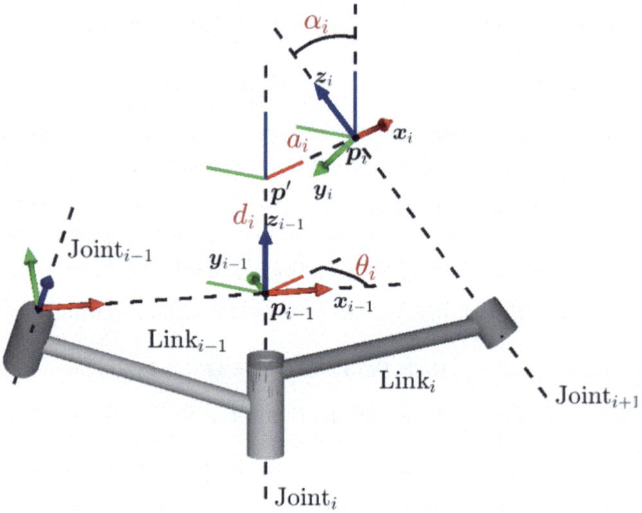

Fig. 1. DH parameters for serial robot manipulators

Calculation of DH Parameters: The manufacturer of a robot usually provides the nominal values of DH parameters or some other data, which can be used to calculate DH parameters. However, the actual DH parameters of a robot may differ from the nominal values due to manufacturing or assembly errors. Therefore, to reduce the positioning errors of the end-effector, it is necessary to identify the actual kinematic parameters of a robot, in our case, the exact DH parameters.

A straightforward method for calculating DH parameters is a geometric approach based on known joint axes and linear algebra. Assuming that joint axes $\boldsymbol{z}_i, i = 1, \ldots, n$ are known, the parameters can be determined in two steps: determination of relative frames for each joint and calculation of DH parameters using these two frames. The procedure is recursive, starting from the robot base and proceeding toward the end-effector. The location and orientation of the frame $\mathcal{F}_i$ is determined by the intersection or the closest points of lines defined by joint axes $\boldsymbol{z}_{i-1}$ and $\boldsymbol{z}_i$.

In general, two lines in 3D space intersect, are parallel, collinear, or skewed to each other. There are several methods to determine their relationship and the intersection or nearest points [4,7,10]. One of the proposed methods is used to determine the nearest points on axis lines, i.e., the intersections with a common normal, are obtained and used in subsequent steps.

Knowing the frame $\mathcal{F}_{i-1}$ and the line on which the axis of the next joint ($\boldsymbol{z}_i$) lies, we first calculate the intersection or closest points of lines defined by $\boldsymbol{z}_{i-1}$ and $\boldsymbol{z}_i$. Then, the location and orientation of the frame $\mathcal{F}_i$ is determined in the following way (see also Fig. 1):

1. Per definition, $\boldsymbol{z}_{i-1}$ represents the axis of the i-th joint.
2. The frame origins $\boldsymbol{p}'$ and $\boldsymbol{p}_i$ are located at the closest points on the lines defined by $\boldsymbol{z}_{i-1}$ and $\boldsymbol{z}_i$, respectively.

3. Axis $\boldsymbol{x}_i$ can be determined from the common normal to both joint axes, i.e., it is in the direction from $\boldsymbol{p}'$ to $\boldsymbol{p}_i$.
4. Axis $\boldsymbol{y}_i$ is directly determined once the $\boldsymbol{z}_i$ and $\boldsymbol{x}_i$ are known, $\boldsymbol{y}_i = \boldsymbol{z}_i \times \boldsymbol{x}_i$.

After calculating the coordinate frames $\mathcal{F}_{i-1}$ and $\mathcal{F}_i$, the DH parameters used to describe each transformation ${}^{i-1}\mathbf{T}_i$ are calculated as follows:

– angle θ_i between x-axes $\boldsymbol{x}_{i-1}$ and $\boldsymbol{x}_i$ around z-axis $\boldsymbol{z}_{i-1}$

$$\theta_i = \text{atan2}((\boldsymbol{x}_{i-1} \times \boldsymbol{x}_i) \cdot \boldsymbol{z}_{i-1}, \boldsymbol{x}_{i-1} \cdot \boldsymbol{x}_i) \tag{3}$$

– distance d_i between the origins $\boldsymbol{p}_{i-1}$ to $\boldsymbol{p}'$ (to the common normal)

$$d_i = (\boldsymbol{p}_i - \boldsymbol{p}_{i-1}) \cdot \boldsymbol{z}_{i-1} = (\boldsymbol{p}' - \boldsymbol{p}_{i-1}) \tag{4}$$

– distance a_i between the origins $\boldsymbol{p}'$ and $\boldsymbol{p}_i$ (length of the common normal)

$$a_i = (\boldsymbol{p}_i - \boldsymbol{p}_{i-1}) \cdot \boldsymbol{x}_i = (\boldsymbol{p}_i - \boldsymbol{p}') \tag{5}$$

– angle α_i between z-axes $\boldsymbol{z}_{i-1}$ and $\boldsymbol{z}_i$ around x-axis $\boldsymbol{x}_i$ (the common normal)

$$\alpha_i = \text{atan2}((\boldsymbol{z}_{i-1} \times \boldsymbol{z}_i) \cdot \boldsymbol{x}_i), \boldsymbol{z}_{i-1} \cdot \boldsymbol{z}_i) \tag{6}$$

Using Eqs. (3–6) we can calculate only the DH parameters sets $(a_i, \alpha_i, d_i, \theta_i), i = 1, ..., n-1$. The last set of parameters $(a_n, \alpha_n, d_n, \theta_n)$ can be calculated using (1). Namely, it reasonably be assumed to know ${}^n\mathbf{T}_E$. Assuming that the location of the end-effector frame $\mathcal{F}_E$ described by ${}^0\mathbf{T}_E$ is also known and using the previously obtained transformation matrices ${}^{i-1}\mathbf{T}_i$ yields

$${}^{n-1}\mathbf{T}_n(a_n, \alpha_n, d_n, \theta_n) = \left(\prod_{i=1}^{n} {}^{i-1}\mathbf{T}_i(q_i)\right)^{-1} {}^0\mathbf{T}_E \left({}^n\mathbf{T}_E\right)^{-1} \tag{7}$$

As ${}^{n-1}\mathbf{T}_n$ has the form given in (2), the n-th DH parameter set can be calculated directly form ${}^{n-1}\mathbf{T}_n$.

Determination of Joint Axis: As proposed by several authors [1,4,5], for revolute joints, the rotation axis can be easily obtained by rotating only one joint at a time. By measuring the 3D spatial position of a point, we obtain a set of 3D points in space. The obtained dataset of points is then fitted to a circle (see Fig. 2). Among different available methods [4,5,13], we used a common approach based on singular value decomposition (SVD) to obtain the normal $\boldsymbol{n}$ to the best fitting plane and the circle center $\boldsymbol{p}_c$, i.e., the intersection of normal $\boldsymbol{n}$ with the plane $\mathcal{P}$. Point $\boldsymbol{p}_c$ and normal $\boldsymbol{n}$ define the location and the direction of the joint axis, respectively.

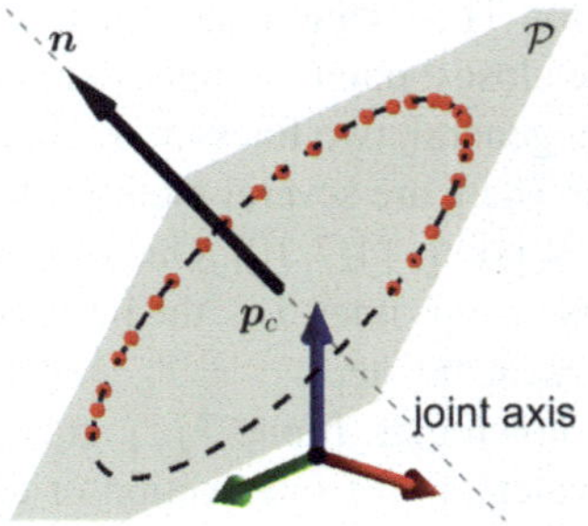

Fig. 2. Captured 3D points (red dots) defining a circle, center point and normal

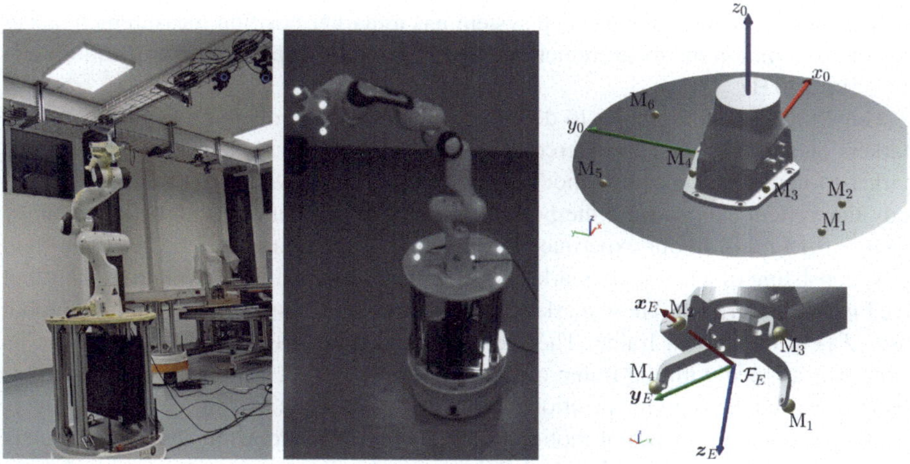

Fig. 3. MoCap scene: Franka Emika Panda with marker tool (left), IR image captured by MoCap camera (middle), markers and optical rigid body frame on robot base (upper right), and markers and optical rigid body frame on tool (lower right)

3 Experimental Identification of DH Parameter

The goal was to verify that the DH parameters of the Franka Emika Panda collaborative robot were consistent with specifications. To estimate the kinematic model parameters of the robot, the joint positions and the poses in the task space have to be measured. Then, the parameters can be calculated using the kinematic model. We used the method described in the previous section. As explained earlier, to identify the DH parameters, the axes of rotation of all joints must be determined. It is not difficult to see that all the rotational axes of a robot can be identified by measuring only the position of an appropriate point on the end effector while one joint is moving and all other joints are fixed in any position. Note that it is not even necessary to know the exact position of this point on the robot body. The only requirement is that the measurement point is not on the axis of rotation. An exception is the last set of DH parameters. To identify this set, it is necessary to measure the position and orientation of the end effector.

3.1 Experimental Setup

Our experimental setup consists of a Franka Emika Panda robot mounted on a plate with markers, a tool with markers attached to the end effector of the robot, and an optical motion capture system (MoCap) as shown in Fig. 3. The robot is controlled by a control system based on Matlab and ROS, which allows easy motion programming and acquisition of robot states.

The actual pose of the tool was determined by the MoCap system. More specifically, we used a passive marker-based MoCap system that can locate the marker positions and to track their movement in space in real-time. Our MoCap system consists of 16 IR cameras and has a measurement range larger than the robot's working range.

With proper calibration, the MoCap system has a marker position measurement accuracy of ±0.3 mm in the measurement volume. It can also capture marker positions at a rate of up 120 Hz.

Our MoCap system allows to define a rigid body as a collection of three or more markers on an object that are interconnected to each other with an assumption that the marker-to-marker distance does not deviate beyond the allowable tolerance. For a rigid body the system can measure the pose (position and orientation) with better accuracy (up to ±0.15 mm). In our experiments, we defined two virtual rigid bodies. One rigid body, consisting of a set of six markers, was used to define the base frame of the robot (see Fig. 3, upper right). These markers were used to determine the location of the robot base ($\mathcal{F}_0$) in the optical frame. The second virtual rigid body was defined so that its frame was aligned with the frame of the robot tool $\mathcal{F}_E$ (see Fig. 3, lower right). This object was used measure the position of the robot end-effector.

We generated two types of motion trajectories: the first to estimate DH parameters and the second to verify the identified DH parameters.

First, we generated trajectories where only one joint moved at a time, resulting in a circular trajectory of the tracked point. The joint movement was at low speed (∼10% of the robot's maximum speed) and the range of motion included almost the entire available joint range. During this motion, the other joints were at fixed positions in an optimal configuration with respect to external influences on measurement accuracy.

The second type of movement was used for verification. In this case, the robot moved between random configurations throughout the robot's workspace.

3.2 Results

From the circular trajectories, we extracted points needed to estimate DH parameters. The obtained circles for each joint are shown in Fig. 4. To minimize the influence of measurement errors of the MoCap system, we measured 3000–14000 points for each joint, depending on the range of each joint. In addition, the estimation procedure was repeated 20× and the final corrected values of DH parameters are calculated as the average of all runs. The nominal and the corrected DH parameters are given in Table 1.

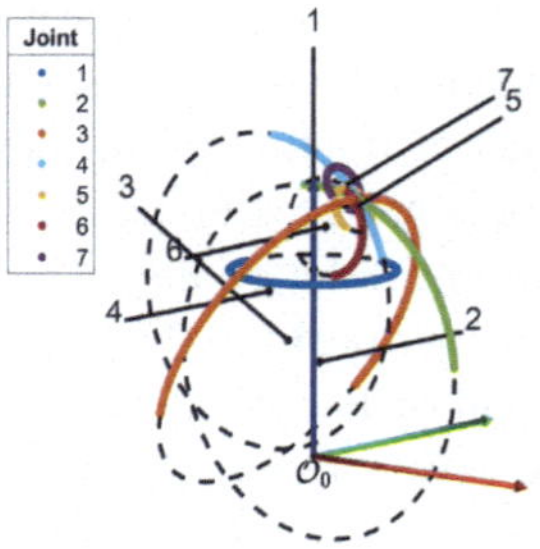

Fig. 4. Captured circles for all joints

The improvement in position accuracy due to the corrected DH parameters was verified on a set of ∼300000 acquired points taken during a random verification move. The mean position error for all acquired points using the nominal DH parameters was 7.9 ± 3.0 mm, and the maximum position error was 14.2 mm. Using the corrected DH parameters, the position error decreased significantly: the mean position error was 1.7 ± 0.6 mm and the maximum error was 4.2 mm. The standard deviation of the position error for the corrected DH parameters is almost within the range of the accuracy of the MoCap system. This improvement is clearly shown in the error histograms in Fig. 5. Moreover, the 3D representation in Fig. 5 shows the distribution of position errors in the robot's workspace. As we can see, there are some

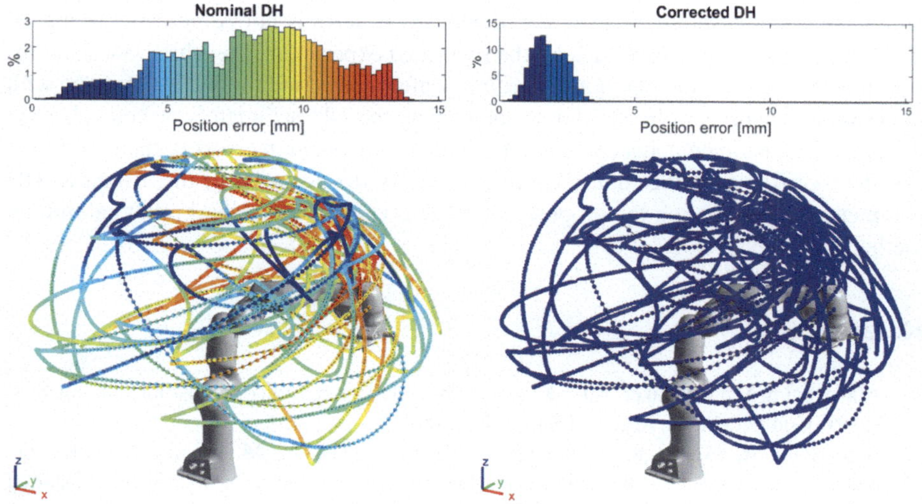

Fig. 5. Position errors with nominal and corrected DH parameters for motion between random configurations: error histograms for ~300000 captured points (upper figures) and position error of a subset of captured points in 3D task space - color indicates the position error (lower figures).

motion parts for the original DH parameters where the error is significantly larger. These parts do not depend on the position in the workspace, but on the joint configurations. By using the corrected DH parameters, the error is improved in all regions.

Table 1. Nominal and corrected DH parameters of Franka Emika Panda robot

	Nominal				Corrected			
Joint	a [mm]	d [mm]	α [°]	θ [°]	a [mm]	d [mm]	α [°]	θ [°]
1	0	333	-90	0	−0.1	332.7	−90.05	0.28
2	0	0	90	0	0.2	−0.5	90.02	−0.36
3	82.5	316	90	0	81.9	316.4	89.93	0.28
4	−82.5	0	-90	0	−82.4	0.3	−89.98	−0.17
5	0	384	90	0	0.1	385.4	89.72	0.74
6	88	0	90	0	88.0	−0.9	90.12	−0.40
7	0	107	0	0	0.1	106.8	−0.40	0.37

4 Conclusion

In this paper, we present how to improve the positioning accuracy of a robot by using a common method to determine the DH parameters based on a MoCap system. Compared to other calibration strategies such as using laser tracker-based systems or mechanical measurement systems, using a MoCap system is more cost-effective and flexible.

Of course, the measurement accuracy of MoCap systems is not comparable to that of laser tracking systems. Nevertheless, the presented experimental results show a significant improvement of the positioning accuracy after correcting the DH parameters with the presented approach. In addition to identifying the DH parameters, we can also use the optical measurement system to compensate for the position errors of the base frame and the tool frame. In the future, we plan to verify the accuracy of the estimated DH parameters by comparing the proposed approach with some standard robot calibration methods.

References

1. Abderrahim, M., Whittaker, A.R.: Kinematic model identification of industrial manipulators. Robot. Comput.-Integr. Manuf. **16**(1), 1–8 (2000)
2. Brau-Avila, A., Acero, R., Santolaria, J., Valenzuela-Galvan, M., Icasio-Hernández, O.: Kinematic parameter identification procedure of an articulated arm coordinate measuring machine based on a metrology platform. Int. J. Adv. Manuf. Technol. **104**(1–4), 1027–1040 (2019)
3. Cheng, L., Wang, W., Weng, Y., Shi, G., Yang, H., Lu, K.: A novel kinematic parameters identification method for articulated arm coordinate measuring machines using repeatability and scaling factor. Math. Probl. Eng. **2018** (2018)
4. Faria, C., Vilaça, J.L., Monteiro, S., Erlhagen, W., Bicho, E.: Automatic denavit-hartenberg parameter identification for serial manipulators. In: IECON 2019 - 45th Annual Conference of the IEEE Industrial Electronics Society, vol. 1, pp. 610–617 (2019). https://doi.org/10.1109/IECON.2019.8927455
5. Hayat, A.A., Chittawadigi, R.G., Udai, A.D., Saha, S.K.: Identification of denavit-hartenberg parameters of an industrial robot. In: AIR 2013, pp. 1–6. Association for Computing Machinery, New York (2013)
6. He, R., Zhao, Y., Yang, S., Yang, S.: Kinematic-parameter identification for serial-robot calibration based on POE formula. IEEE Trans. Rob. **26**(3), 411–423 (2010)
7. Ketchel, J.S., Larochelle, P.M.: Collision detection of cylindrical rigid bodies using line geometry. In: Proceedings of the ASME International Design Engineering Technical Conferences and Computers and Information in Engineering Conference - DETC 2005, vol. 7 B, pp. 811–825 (2005)
8. Liu, Y., Li, Y., Zhuang, Z., Song, T.: Improvement of robot accuracy with an optical tracking system. Sensors **20**(21), 1–26 (2020)
9. Majarena, A.C., Conte, J., Santolaria, J., Acero, R.: A new methodology for kinematic parameter identification in laser trackers. In: Kinematics. InTech, December 2017. https://doi.org/10.5772/intechopen.71444
10. Rajeevlochana, C., Subir, S., Shivesh, K.: Automatic extraction of DH parameters of serial manipulators using line geometry. In: The 2nd Joint International Conference on Multibody System Dynamics (May), pp. 1–9 (2012)
11. Shen, H., Meng, Q., Li, J., Deng, J., Wu, G.: Kinematic sensitivity, parameter identification and calibration of a non-fully symmetric parallel Delta robot. Mech. Mach. Theory **161**, 104311 (2021)
12. Siciliano, B., Sciavicco, L., Villani, L., Oriolo, G.: Robotics - Modelling, Planning and Control. Springer, London (2009). https://doi.org/10.1007/978-1-84628-642-1
13. Umbach, D., Jones, K.N.: A few methods for fitting circles to data. IEEE Trans. Instrum. Meas. **52**(6), 1881–1885 (2003)

Kinematics of the "Ai-Gerim" Robot Arm

Zhumadil Baigunchekov[1,3](✉), Giuseppe Carbone[2], Zhadyra Zhumasheva[1], Bekzat Amanov[1], Yernar Zholdassov[1], Serik Tolenov[1], and Alibek Tleukhanov[1]

[1] Al-Farabi Kazakh National University, Almaty 050040, Republic of Kazakhstan
bzh47@mail.ru
[2] University of Calabria, Rende, Italy
[3] Satbayev University, Almaty 050013, Republic of Kazakhstan

Abstract. In this paper, the direct and inverse kinematics of the "Ai-Gerim" humanoid robot arm are solved. This robot is a remotely controlled social robot. For the study of kinematics of the robot arm, the Denavit – Hartenberg transformation matrices are derived. The pose of the end-effector (hand) in the direct kinematics is determined by multiplying these matrices. For the study of the inverse kinematics of the robot arm, a reverse decoupling method is used to analytically determine the joint angles.

Keywords: Humanoid robot arm · Direct and inverse kinematics

1 Introduction

Industrial robots have been used for a long time in many industries: in mechanical engineering, military-space industry, agriculture, etc. Since the 1990s, the development of service robots began, which do useful work for people, with the exception of industrial automation tasks, for example, robots - cleaners, robots - guides, robots - consultants. The development of service robots, in turn, led to the creation of social robots capable of social interactions with humans [1–9], for example, assistant - robots, companion - robots, robots for rehabilitation. Robot - assistants, in addition to performing basic functions, carry out social interactions with a person to provide assistance. For example, the humanoid robot "AVA" of the company "i-robot" [10], except for cleaning the house, collects information about the medical condition of its owner and transmits this data to the attending physician, robot - warden of Asian Forum for Corrections group [11], except for observation of convicts, mediates between prisoners and wardens, the humanoid robot Octavia [12] interacts with firefighters when extinguishing a fire, robot Robonaut 2 [13] works on the NASA international space station, interacting with astronauts. Robots for rehabilitation support the mental health of people, especially the elderly, through communication, and also involve people with disabilities in public life (work, study, communication). Widely known is the robot - the baby seal Paro [14], which perceives people and the environment, demonstrates joy when it is ironed. This causes a calming effect and positive emotions in older adults and hospital patients.

According to the international work organization, about 650 million people on our planet are with disabilities. In some countries, up to 80% of people with disabilities are

A. Müller and M. Brandstötter (Eds.): RAAD 2022, MMS 120, pp. 11–18, 2022.
https://doi.org/10.1007/978-3-031-04870-8_2

unemployed. This is a big problem, since this category of people breaks away from social life.One of the directions to solve this problem is to attract people with disabilities to work and public life with the help of remotely controlled social robots. Remote control of such social robots by operators - people with disabilities, being at home, on the one hand attracts them to work and public life, and on the other hand, reduces the cost of robots. It is known that the bulk of the cost of robots operating offline is their control systems.

The "Ai–Gerim" robot is a remotely controlled social humanoid robot used to provide employment for people with disabilities, as well as to provide contactless medical and other services [15]. In this paper, the direct and inverse problems of kinematics of the robot's arm are solved.

2 Direct Kinematics

Figure 1 shows 3D CAD models of the general view, construction and kinematic diagram of the "Ai–Gerim" humanoid robot. The robot "Ai-Gerim" has 20 degrees of freedom (DOF) [16], including a head has 2 DOF, a body has 3 DOF, a movement mechanism has 3 DOF, and each arm has 6 DOF. Each 6 DOF arm has the 2 DOF shoulder joints, the 2 DOF elbow joints, and the 2 DOF wrist joints. Figure 2 shows the kinematic diagram of the robot's right arm with coordinate frames.

Coordinate frames are set as follows. The base coordinate frame, $O_{B1}X_{B1}Y_{B1}Z_{B1}$ (as reference coordinate frame) is established at the center of the neck and is the reference coordinate frame for the head and the arms. Following the Denavit–Hartenberg's (D-H) convention [17], the Z_i axes of the $X_iY_iZ_i$ coordinate systems (i = 1, 2, ..., 6) are directed along the rotation axes of revolute joints, the axes X_i are directed along the common normals between the revolute axes of adjacent revolute joints. D–H parameters are a set of four variables that define the spatial relationship between two neighbouring coordinate frames. These variables are d_i (translation along the axis Z_i), θ_i (rotation about the axis Z_i), a (translation along the axis X_i) and α_i (rotation about the axis X_i). Table 1 shows the D-H parameters of the right hand of the "Ai-Gerim" robot.

The pose (position and orientation) of frame i with respect to its nearest neighbour frame i − 1 is described by the following homogeneous transformation matrix [17]:

$$
{}^{i-1}\mathbf{A}_i = \begin{bmatrix} c\theta_i & -c\alpha_i \cdot s\theta_i & s\alpha_i \cdot s\theta_i & \alpha_i \cdot c\theta_i \\ s\theta_i & c\alpha_i \cdot c\theta_i & -s\alpha_i \cdot c\theta_i & \alpha_i \cdot s\theta_i \\ 0 & s\alpha_i & c\alpha_i & d_i \\ 0 & 0 & 0 & 1 \end{bmatrix} = \begin{bmatrix} {}^{\mathbf{i}}\mathbf{R}_{\mathbf{i}-1} & \vdots\ {}^{\mathbf{i}}\mathbf{p}_{\mathbf{i}-1} \\ \ldots\ldots & \ldots\vdots\ldots \\ 0\ 0\ \ 0 & \vdots\ 1 \end{bmatrix}, \quad (1)
$$

where the following abbreviations were used: $c\theta_i = cos\theta_i$, $s\theta_i = sin\theta_i$, $c\alpha_i = cos\alpha_i$, $s\alpha_i = sin\alpha_i$. The upper left 3 × 3 matrix is the rotation matrix ${}^{\mathbf{i}}\mathbf{R}_{\mathbf{i}-1}$ giving the orientation of frame i with respect to frame i − 1, expressed in i − 1 coordinates. The upper right 3 × 1 vector ${}^{\mathbf{i}}\mathbf{p}_{\mathbf{i}-1}$ is the position vector from the origin of i − 1 to the origin of i, expressed in i − 1 coordinates. Substituting the D-H parameters from Table 1 into the transformation matrix (1), we obtain the six matrices ${}^{0}\mathbf{A}_1$, ${}^{1}\mathbf{A}_2$, ${}^{2}\mathbf{A}_3$, ${}^{3}\mathbf{A}_4$, ${}^{4}\mathbf{A}_5$, ${}^{5}\mathbf{A}_6$.

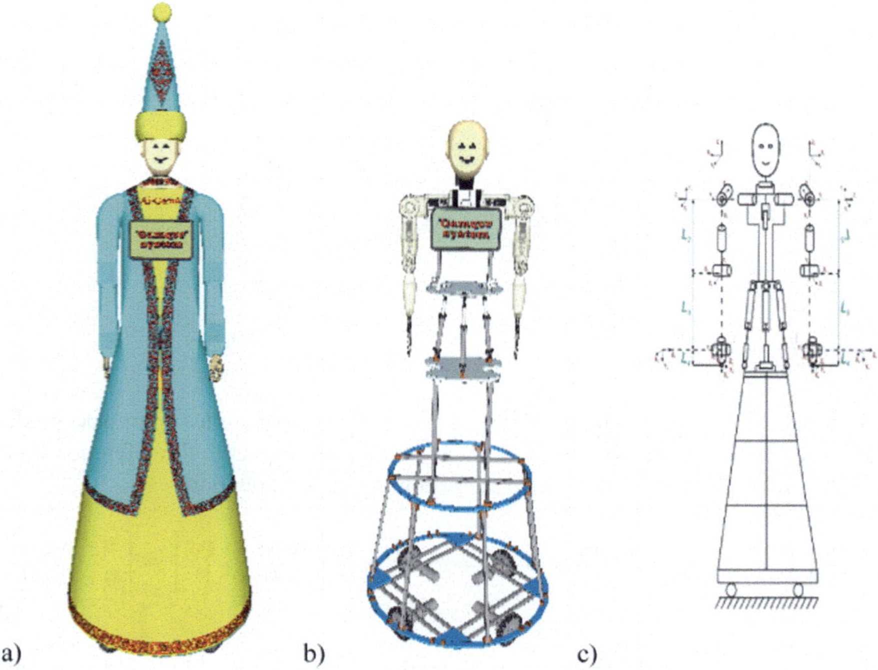

Fig. 1. 3D CAD models of the "Ai - Gerim" robot.

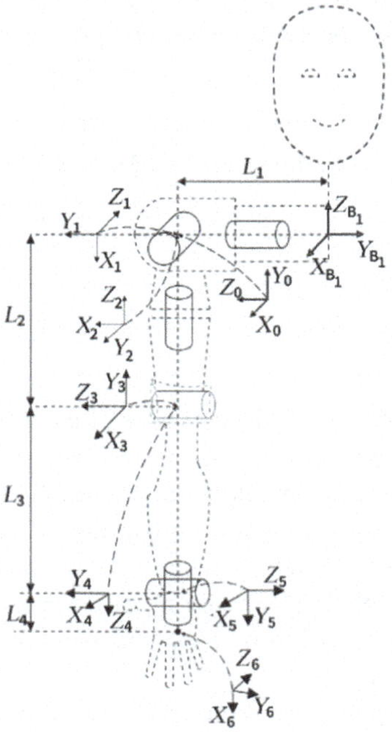

Fig. 2. Kinematic diagram of the right arm.

Table 1. D-H parameters for right arm.

i	θ_i	α_i	a_i	d_i
1	$-90+\theta_1$	90	0	0
2	θ_2+90	−90	0	0
3	θ_3+90	90	0	$-L_2$
4	θ_4	90	0	0
5	θ_5	90	0	L_3
6	$90+\theta_6$	0	L_4	0

In the direct kinematics of the robot "Ai-Gerim" six joint angles θ_i are specified, it is necessary to define the pose of the end - effector with respect to the reference frame $O_{B1}X_{B1}Y_{B1}Z_{B1}$. To do this, we derive the following matrix equation:

$$\mathbf{A}_6 = \prod_{\mathbf{i}=1}^{6} {}^{\mathbf{i}-1}\mathbf{A}_\mathbf{i} = {}^0\mathbf{A}_1 \cdot {}^1\mathbf{A}_2 \cdot {}^2\mathbf{A}_3 \cdot {}^3\mathbf{A}_4 \cdot {}^4\mathbf{A}_5 \cdot {}^5\mathbf{A}_6 = \begin{bmatrix} \mathbf{x}_6 & \mathbf{y}_6 & \mathbf{z}_6 & \mathbf{p}_6 \\ 0 & 0 & 0 & 1 \end{bmatrix} = \begin{bmatrix} \mathbf{n} & \mathbf{s} & \mathbf{a} & \mathbf{p} \\ 0 & 0 & 0 & 1 \end{bmatrix}, \tag{2}$$

where $\mathbf{x}_6$, $\mathbf{y}_6$ and $\mathbf{z}_6$ are the unit vectors along the principal axes of the hand frame and $\mathbf{p}_6$ is the position vector describing the location of the hand relative to the shoulder. These three unit vectors describe the orientation of the hand coordinate frame relative to the reference coordinate frame $O_{B1}X_{B1}Y_{B1}Z_{B1}$. The vectors **n**, **s**, **a** and **p** represent the normal vector, sliding vector, approach vector, and position vector, respectively [18]:

$${}^0\mathbf{A}_6 = {}^0\mathbf{A}_1 \cdot {}^1\mathbf{A}_2 \cdot {}^2\mathbf{A}_3 \cdot {}^3\mathbf{A}_4 \cdot {}^4\mathbf{A}_5 \cdot {}^5\mathbf{A}_6 = \begin{bmatrix} n_x & s_x & a_x & p_x \\ n_y & s_y & a_y & p_y \\ n_z & s_z & a_z & p_z \\ 0 & 0 & 0 & 1 \end{bmatrix}, \tag{3}$$

3 Inverse Kinematics

In the inverse kinematic problem, the desired pose of the arm end-effector is given with respect to the reference frame $O_{B1}X_{B1}Y_{B1}Z_{B1}$, need to determine the joint angles. When solving the inverse kinematics of a manipulator, a closed – form solution exists if three consecutive joint axes of the manipulator are parallel to one another, or intersect at a single point [19, 20]. A joint solution is said to be "closed - form" if the unknown joint angles can be solved for symbolically in terms of the arc-tangent function. The three shoulder joined axes of the "Ai-Gerim" robot intersect at a single point, therefore a closed-form kinematic solution exists.

Pieper [21] indicates, the position vector **p** of the wrist coordinate frame decouples into positioning and orientation subsystems if the last three joints intersect at a point. The last three joint angles are determined from [**n**, **s**, **a**] and the solved first three joint angles. Using this knowledge, the arm can be viewed in reverse so that the last three joints make up the shoulder, thus the position and orientation of the shoulder frame can be described relative to the hand frame. This new position vector $\mathbf{p}'$ is only a function of θ_4, θ_5 and θ_6, and thus decouples the arm kinematics into position and orientation components. The inverse kinematic problem of the "Ai-Gerim" robot is solved in this reverse method by taking the inverse both sides of Eq. (2):

$$^{0}\mathbf{A}_6^{-1} = \begin{bmatrix} \mathbf{n}\ \mathbf{s}\ \mathbf{a}\ \mathbf{p} \\ 0\,0\,0\,1 \end{bmatrix}^{-1} = \begin{bmatrix} \mathbf{n}'\ \mathbf{s}'\ \mathbf{a}'\ \mathbf{p}' \\ 0\ \ 0\ \ 0\ \ 1 \end{bmatrix} = {}^{6}\mathbf{A}_5 \cdot {}^{5}\mathbf{A}_4 \cdot {}^{4}\mathbf{A}_3 \cdot {}^{3}\mathbf{A}_2 \cdot {}^{2}\mathbf{A}_1 \cdot {}^{1}\mathbf{A}_0 = {}^{6}\mathbf{A}_0, \quad (4)$$

Using the inverse matrices $^{\mathbf{i}-1}\mathbf{A}_\mathbf{i}$, Eq. (4) can be written as:

$$\begin{bmatrix} n'_x & s'_x & a'_x & p'_x \\ n'_y & s'_y & a'_y & p'_y \\ n'_z & s'_z & a'_z & p'_z \\ 0 & 0 & 0 & 1 \end{bmatrix} = {}^{6}\mathbf{A}_0 = {}^{6}\mathbf{A}_5 \cdot {}^{5}\mathbf{A}_4 \cdot {}^{4}\mathbf{A}_3 \cdot {}^{3}\mathbf{A}_2 \cdot {}^{2}\mathbf{A}_1 \cdot {}^{1}\mathbf{A}_0 \quad (5)$$

We can use the inverse transform method to solve for the last three joint angles by multiplying both sides of Eq. (4) by matrix $^{5}\mathbf{A}_6$:

$$^{5}\mathbf{A}_6 \cdot {}^{0}\mathbf{A}_6^{-1} = {}^{5}\mathbf{A}_6 \cdot {}^{6}\mathbf{A}_5 \cdot {}^{5}\mathbf{A}_4 \cdot {}^{4}\mathbf{A}_3 \cdot {}^{3}\mathbf{A}_2 \cdot {}^{2}\mathbf{A}_1 \cdot {}^{1}\mathbf{A}_0. \quad (6)$$

or

$$^{5}\mathbf{A}_6 \cdot \begin{bmatrix} \mathbf{n}'\ \mathbf{s}'\ \mathbf{a}'\ \mathbf{p}' \\ 0\ \ 0\ \ 0\ \ 1 \end{bmatrix} = {}^{5}\mathbf{A}_0. \quad (7)$$

The left hand side of Eq. (7) is

$$\mathbf{A}^{LHS} = {}^{5}\mathbf{A}_6 \cdot \begin{bmatrix} \mathbf{n}'\ \mathbf{s}'\ \mathbf{a}'\ \mathbf{p}' \\ 0\ \ 0\ \ 0\ \ 1 \end{bmatrix} = \begin{bmatrix} C_6 & -S_6 & 0 & L_4C_6 \\ S_6 & C_6 & 0 & L_4S_6 \\ 0 & 0 & 1 & 0 \\ 0 & 0 & 0 & 1 \end{bmatrix} \cdot \begin{bmatrix} n'_x & s'_x & a'_x & p'_x \\ n'_y & s'_y & a'_y & p'_y \\ n'_z & s'_z & a'_z & p'_z \\ 0 & 0 & 0 & 1 \end{bmatrix} \quad (8)$$

The right hand side of Eq. (7) is

$$\mathbf{A}^{RHS} = {}^{5}\mathbf{A}_0 = {}^{5}\mathbf{A}_4 \cdot {}^{4}\mathbf{A}_3 \cdot {}^{3}\mathbf{A}_2 \cdot {}^{2}\mathbf{A}_1 \cdot {}^{1}\mathbf{A}_0 = \begin{bmatrix} a_{511} & a_{512} & a_{513} & a_{514} \\ a_{521} & a_{522} & a_{523} & a_{524} \\ a_{531} & a_{532} & a_{533} & a_{534} \\ 0 & 0 & 0 & 1 \end{bmatrix}, \quad (9)$$

By equating the last columns (position terms) of Eqs. (8) and (9), we can solve for the last three joint angles

$$\theta_4 = atan2(\pm\sqrt{1 - C_4^2}, C_4). \quad (10)$$

$$\theta_5 = atan2(S_5, \pm\sqrt{1 - S_5}). \quad (11)$$

$$\theta_6 = atan2(-C_4L_2 - L_3, S_4C_3L_2) - \psi. \tag{12}$$

where

$$C_4 = \frac{(p_x^{'} + L_4)^2 + (p_y^{'})^2 + (p_x^{'})^2 - L_2^2 - L_3^2}{2L_2L_3}, \tag{13}$$

$$S_5 = \frac{p_x^{'}}{S_4L_2} = \pm\frac{p_x^{'}}{\sqrt{1 - C_4^2L_2}}, \tag{14}$$

$$\psi = atan2(p_y^{'}, p_x^{'} + L_4), \tag{15}$$

To solve the joint angles θ_1, θ_2 and θ_3, multiplying both sides of Eq. (5) by matrices ${}^3\mathbf{A}_4, {}^4\mathbf{A}_5, {}^5\mathbf{A}_6$. This results in an equation where the left side of the equation is:

$$\mathbf{A}^{LHS} = {}^3\mathbf{A}_4 \cdot {}^4\mathbf{A}_5 \cdot {}^5\mathbf{A}_6 \cdot \begin{bmatrix} n_x^{'} & s_x^{'} & a_x^{'} & p_x^{'} \\ n_y^{'} & s_y^{'} & a_y^{'} & p_y^{'} \\ n_z^{'} & s_z^{'} & a_z^{'} & p_z^{'} \\ 0 & 0 & 0 & 1 \end{bmatrix} = \begin{bmatrix} a_{411} & a_{412} & a_{413} & a_{414} \\ a_{421} & a_{422} & a_{423} & a_{424} \\ a_{431} & a_{432} & a_{433} & a_{434} \\ 0 & 0 & 0 & 1 \end{bmatrix}, \tag{16}$$

and the right side of the equation is:

$$\mathbf{A}^{RHS} = {}^3\mathbf{A}_2 \cdot {}^2\mathbf{A}_1 \cdot {}^1\mathbf{A}_0 = {}^3\mathbf{A}_0 = \begin{bmatrix} C_1C_2C_3 - S_1S_3 & C_1S_3 + C_2S_3S & C_3S_2 & 0 \\ -C_1S_2 & -S_1S_2 & C_2 & L_2 \\ C_3S_1 + C_1C_2S_3 & C_2S_1S_3 - C_1C_3 & S_2S_3 & 0 \\ 0 & 0 & 0 & 1 \end{bmatrix}. \tag{17}$$

By equating the element (2,3) of Eqs. (16) and (17), we obtain θ_2:

$$C_2 = a_x^{'}(S_4C_5C_6 - C_4S_6) - a_y^{'}(C_4C_6 + S_4C_56S_6) + a_z^{'}S_4S_5, \tag{18}$$

$$\theta_2 = atan2(\pm\sqrt{1 - C_2^2}, C_2). \tag{19}$$

By comparing the elements (1,3) and (3,3), of the left and right side matrices, we can determine the joint solution θ_3:

$$\frac{S_3}{C_3} = \frac{a_x^{''}S_5C_6 - a_y^{''}S_5S_6 - a_z^{''}C_5}{a_x^{''}(S_4S_6 + C_4C_5C_6) + a_y^{''}(S_4C_6 - C_4C_5S_6) + a_z^{'}C_4S_5}, \tag{20}$$

$$\theta_3 = atan2(g_{433}, g_{431}). \tag{21}$$

By comparing the elements (2,1) and (2,2) of the left and right side matrices, we can determine the joint solution θ_1

$$\frac{-S_1}{-C_1} = \frac{s_x^{'}(S_4C_5C_6 - C_4S_6) - s_y^{'}(C_4C_6 + S_4C_56S_6) + s_z^{'}S_4S_5}{n_x^{'}S_4C_5C_6 - C_4S_6) - n_y^{'}(C_4C_6 + S_4C_56S_6) + n_z^{'}S_4S_5}, \tag{22}$$

$$\theta_1 = atan2(-g_{422}, -g_{412}). \tag{23}$$

4 Conclusion

Methods for solving the direct and inverse problems of the kinematics of the social humanoid robot "Ai - Gerim" arm have been developed. This robot is remotely controlled and is used to perform contactless medical and other services. Each robot arm has six DOF. On the basis of the D-H method, the transformation matrices of adjacent joints are derived. In the direct kinematics problem, the position of the end – effector (hand) is determined from the given values of the joint angles by successive multiplication of the D-H matrices. Solving the inverse problem of kinematics of the robot arm, i.e. determining the joint angles from the specified end- effector pose is more complex. To solve the inverse kinematics problem, the reverse decoupling method is used. In this case, the axes of rotation of the last three joints intersect at one point and the solution of the inverse kinematics problem is carried out analytically in closed-form.

Acknowledgments. This research is founded by the Science Committee of the Ministry of Education and Science of the Republic of Kazakhstan (Grant No AP08857522).

References

1. Hegel, F., et al.: Understanding social robots. In: Second International Conference on Advances in Computer – Human Interactions, pp. 169–174. IEEE (2009)
2. Duffy, B.R., Rooney, C., O'Hare, G.M., O'Donoghe, R.: What is a social robot? In: 10th Irish Conference on Artificial Intelligence amp; Cognitive Science. University College Cork, Irland, 1–3 September, pp. 96–118 (1999)
3. Young, J.E., et al.: Evaluating human – robot interacting. Int. J. Soc. Robots **3**(1), 53–67 (2011)
4. Breazeal, C.: Social interactions in HRI: the robot view. IEEE Trans. Syst. Man Cybern. Part C (Appl. Rev.) **34**(2), 181–186 (2004). https://doi.org/10.1109/TSMCC.2004.826268
5. Fong, I., Nourbakhsh, I., Dautenkahn, K.: A survey of socially interactive robots. Robot. Autom. Syst. **42**(3), 143–166 (2003)
6. Brazeal, C.: Toward sociable robots. Robot. Autom. Syst. **42**(3), 167–175 (2003)
7. Graaf, M.M.A., Ben Allouch, S., Dijk, J.A.G.M.: What makes robots social?: a user's perspective on characteristics for social human-robot interaction. In: Tapus, A., André, E., Martin, J.C., Ferland, F., Ammi, M. (eds.) Social Robotics. ICSR 2015. Lecture Notes in Computer Science (Lecture Notes in Artificial Intelligence), vol. 9388. Springer, Cham (2015). https://doi.org/10.1007/978-3-319-25554-5_19
8. Engelhardt, K.G., Edwards, R.A.: Humanrobot Integration for Service Robotics, pp. 315–346. Human- Robot Interaction. London (1992)
9. Yang, S., et al.: Experiences developing socially acceptable interactions for a robotic trash Barrel. In: 24-th IEEE International Symposium on Robot and Human Interactive Communication (ROMAN), pp. 277–284 (2015)
10. AVA Overview Irobot. http://www.irobot.com/ava
11. Bloss, R.: Robots go to prison – as guards. Int. J. Indust. Robot. **39**, 148–158 (2012)
12. Martinson, E., et al.: Fighting fires with human robot teams. In: IEEE/RSJ International Conference on Intelligent Robots and Systems, pp. 2682–2683 (2012)
13. Diftler, M.A., et al.: Robonaunt2 – the first humanoid robot in space. In: IEEE International Conference on Robotics and Automation (ICRA), pp. 2178–2183 (2011)

14. Sabanovis, S., et al.: Paro robot affects divers interaction modalities in group sentory therapy for older adult with dementia. In: IEEE International Conference on Rehabilitation Robotics (ICORR), pp. 1–6 (2013)
15. Baigunchekov, Z., Zhumasheva, Z., Amanov, B., Zholdasov, E., Sagitzhanov, B., Tleukhanov, A.: Robotic system for employment of people with disabilities. Int. J. Adv. Robot. Mech. Eng. **2** (2020). https://doi.org/10.32474/ARME.2020.02.000144
16. Baigunchekov, Z., Sarosh, P., et al.: Social Humanoid robot "Ai-Gerim". In: Proceedings of the 7th IEEE International Conference on Control Automation and Robotics. 23–26 April 2021, Singapore (ICCAR 2021) (2021)
17. Denavit, J., Hartenberg, R.S.: A kinematic notation for lower-pair mechanisms based on matrices. Trans. ASME J. April Mech. **23**, 215–221 (1955)
18. Fu, K.S., Gonzalez, R.S., Lee, C.S.G.: Robotics: Control, Sensing, Vision and Intelligence. McGraw-Hill Book company (1988)
19. Park, H.A., Ali, M.A., Lee, C.S.G.: Closed-form inverse kinematic position solution for humanoid robots. Int. J. Hum. Robot. **9**(3), 1250022 (28 p) (2012)
20. Fabrice Inverse Kinematics for a Humanoid Robot: A Mix Between Closed Form and Geometric Solutions. Technical Report (2017)
21. Pieper, D.L.: The Kinematics of Manipulators under computer control. PhD dissertation. Stanford University (1968)

Singularity Robust Inverse Kinematics of Serial Manipulators by Means of a Joint Arc Length Parameterization

Tobias Marauli(✉), Hubert Gattringer, and Andreas Müller

Institute of Robotics, Johannes Kepler University Linz, Altenberger Strasse 69, 4040 Linz, Austria
{tobias.marauli,hubert.gattringer,a.mueller}@jku.at

Abstract. The inverse kinematics problem is ill-posed near kinematic singularities leading to extremely large joint velocities. This classical problem is addressed from a different perspective; instead of prescribing the end-effector (EE) motion in terms of a path parameter, the motion is parameterized in terms of the arc length in joint space. This allows tackling the inverse kinematics problem when approaching kinematic singularities. Thereupon a singularity consistent sampling scheme can easily be devised. A sampling of the arc length gives direct control over the joint motion for prescribed EE motion. The arc length parameterization can be used for solving the (time) optimal path following problem, avoiding the ill-conditioning upfront.

Keywords: Serial robots · Inverse kinematics · Singularities · Path planning · Arc length parameterization · Non-equidistant discretization

1 Introduction

A solution of the inverse kinematics problem is a relation of the EE-motion and the joint motion. Near a kinematic singularity, small EE-velocities lead to extremely large joint velocities, which is a well-known problem for the robot control. To mitigate this problem, path planning strategies for avoiding singularities exist [9]. The problem of large velocities (causing wear of mechanical components and possibly reaching the rated actuator limits) remains, however, even when the robot is only near a singularity. Moreover, often, a prescribed EE-path is deduced from process requirement (e.g. welding, polishing) without taking singularities into account. The velocity inverse kinematics map (defined by the generalized inverse of the forward kinematics Jacobian) transforms EE-velocities to joint velocities, and its metric properties change drastically near singularities. The EE-motion along a prescribed path is described by a path parameter σ. An equidistant sampling of the EE-path is translated to extreme changes in the sampling intervals in joint space. The length of the corresponding path in joint space is described by the arc length s. Now an obvious approach to

A. Müller and M. Brandstötter (Eds.): RAAD 2022, MMS 120, pp. 19–27, 2022.
https://doi.org/10.1007/978-3-031-04870-8_3

avoid high joint velocities, i.e. large changes of s within a time step, is to adaptively reduce the EE-velocity, i.e. the change of σ, when traversing the path. Moreover, the EE-motion could be described in terms of s, and thus be sampled according to prescribed (e.g. equidistant) increments of s, allowing to prescribe the joint velocity while following a given EE-path.

In this paper, the inverse kinematics problem is addressed for the 1) parameterization of an EE-path $\mathbf{z}(\sigma)$ in terms of the path parameter σ [3,7], and 2) the parameterization of the joint motion $\mathbf{q}(s)$ in terms of the (kinetic) arc length s. Each of these parameterizations gives rise to a particular inverse kinematics problem. Both are derived in this paper, and computational methods for their solution are proposed. The two parameterizations refer exclusively to the geometry of motion, and are independent of the temporal execution of the motion. Clearly, a smooth motion in terms of joint arc length s is desirable (rather than in the path parameter σ, as it is the standard approach). It thus provides an avenue for smooth singularity-robust optimal motion planning.

2 Kinematics in Terms of Path Parameter

2.1 Forward Kinematics

Denote with $\mathbf{q} \in \mathbb{R}^n$ the vector of joint variables, and with $\mathbf{z} \in SE(3)$ the EE-pose. Throughout the paper, the EE-path $\mathbf{z}(\sigma)$ is assumed to be given, e.g. from a path planning, and is parameterized in terms of a path parameter $\sigma(t) \in [0,1]$. Since the joint motion is determined from the EE-motion, this induces a parameterization of joint coordinates in terms of σ. The latter is usually used for motion planning and control. Notice, that the relation of path parameter in work space and joint space is generally not unique. With slight abuse of notation, the same symbol is used throughout the paper. The geometric forward kinematics map

$$\mathbf{z} = f(\mathbf{q}) \tag{1}$$

determines the EE-pose for prescribed joint coordinates $\mathbf{q}$ [14]. The EE-twist, containing the translation and angular velocity, is determined by the velocity forward kinematics

$$\mathbf{V}(\mathbf{q}, \dot{\mathbf{q}}) = \begin{pmatrix} \mathbf{v} \\ \omega \end{pmatrix} = \mathbf{J}(\mathbf{q})\,\dot{\mathbf{q}} \tag{2}$$

where $\mathbf{J}(\mathbf{q}) = \partial \mathbf{V}/\partial \dot{\mathbf{q}}$ is the geometric Jacobian. As the EE-twist depends linearly on the joint rates, and since $\dot{\mathbf{q}} = \mathbf{q}'\dot{\sigma}$, the velocity forward kinematics (2) can be expressed in terms of $\dot{\sigma}$ as

$$\mathbf{V} = \frac{\partial \mathbf{V}}{\partial \dot{\sigma}}\dot{\sigma} = \mathbf{J}\mathbf{q}'\dot{\sigma}. \tag{3}$$

This yields $\mathbf{V}(\sigma, \dot{\sigma}) = \mathbf{h}(\sigma)\dot{\sigma}$ and the relation

$$\mathbf{h}(\sigma) = \mathbf{J}(\mathbf{q}(\sigma))\mathbf{q}'(\sigma) \tag{4}$$

where $\mathbf{h}(\sigma) := \partial \mathbf{V}/\partial \dot{\sigma}$ is the *partial path-velocity.* The relation (4) is referred to as the *velocity forward path-kinematics problem*, which will be central for the singularity-robust adaptive sampling. Here $\mathbf{z}(\sigma(t)), \mathbf{h}(\sigma), \dot{\sigma}$, and thus $\mathbf{V}(\sigma, \dot{\sigma})$ are given from the trajectory planning (which is not the topic of this paper).

2.2 Inverse Kinematics

The standard inverse kinematics problem consists in finding the joint coordinates for given an EE-pose $\mathbf{z}$. This can be solved analytically for particular robots, such as wrist-partitioned serial manipulators [1]. In general, however, it can only be solved numerically. Various different computational methods were discussed in [1,6,11,15], including nonlinear root-finding, fixed point iteration, integration of the joint velocity. In this paper, the numerical integration of the joint velocity is pursued. This amounts to solving the relation (2) for $\mathbf{q}(t)$ with given EE-twist $\mathbf{V}(t)$. When the path is parameterized with $\sigma(t)$ the EE-twist is determined by the partial path-velocity via (3). Then the curve, and thus $\mathbf{q}$, is uniquely determined by (4), where the partial path velocity $\mathbf{h}(\sigma)$ describes the motion along the path.

The *velocity inverse path-kinematics problem* is to determine $\mathbf{q}'$ for given partial path velocity $\mathbf{h}$. A solution to the velocity inverse path-kinematics is

$$\mathbf{q}' = \mathbf{J}(\mathbf{q})^{+}\mathbf{h} \tag{5}$$

where $\mathbf{J}(\mathbf{q})^{+}$ is a generalized pseudo inverse [2,10]. The *geometric inverse path-kinematics problem* is to determine the joint motion $\mathbf{q}(\sigma)$ in terms of the path parameter σ. The solution of the inverse path-kinematics problem is the solution of the initial value problem (IVP) (5) with $\mathbf{q}(0) = \mathbf{q}_0$ as initial value. When numerically solving the IVP (5), numerical drifts are unavoidable. In order to reduce the drift, the partial path-velocity error is introduced as

$$\mathbf{e}' = \mathbf{h}_{\mathrm{d}} - \mathbf{J}(\mathbf{q})\,\mathbf{q}' \tag{6}$$

with $\mathbf{h}_{\mathrm{d}}$ denoting the prescribed partial path-velocity. Therewith, the system (5) is amended as

$$\mathbf{q}' = \mathbf{J}(\mathbf{q})^{+}(\mathbf{h}_{\mathrm{d}} + \mathbf{K}\mathbf{e}) \tag{7a}$$

$$\mathbf{0} = \mathbf{e}' + \mathbf{K}\mathbf{e} \tag{7b}$$

where (7b) governs the error dynamics. The latter asymptotically converges to zero for $\mathbf{K} > 0$ [4]. The geometric error $\mathbf{e}$ in (7b) comprises the position error and orientation error, and can be expressed as

$$\mathbf{e} = \begin{pmatrix} \mathbf{e}_{\mathrm{p}} \\ \mathbf{e}_{\mathrm{o}} \end{pmatrix} = \begin{pmatrix} \mathbf{r}_{\mathrm{d}} - \mathbf{r}(\mathbf{q}) \\ \frac{1}{2}(\tilde{\mathbf{n}}(\mathbf{q})\mathbf{n}_{\mathrm{d}} + \tilde{\mathbf{s}}(\mathbf{q})\mathbf{s}_{\mathrm{d}} + \tilde{\mathbf{a}}(\mathbf{q})\mathbf{a}_{\mathrm{d}}) \end{pmatrix} \tag{8}$$

where $\mathbf{r}_{\mathrm{d}}$ is the prescribed EE-position, and $\mathbf{R}_{\mathrm{d}} = [\mathbf{n}_{\mathrm{d}}\ \mathbf{s}_{\mathrm{d}}\ \mathbf{a}_{\mathrm{d}}]$ the presdribed EE-orientation, with columns $\mathbf{n}, \mathbf{s}, \mathbf{a}$ [13], and $\widetilde{\mathbf{x}} \in \mathbb{R}^{3,3}$ is the skew symmetric (cross product) matrix associated to $\mathbf{x} \in \mathbb{R}^3$. With these definitions and a proper choice of $\mathbf{K}$, the velocity inverse path-kinematics problem can be solved numerically stable using time integration schemes. In summary, the solution of (5) yields a parameterization of the joint coordinates $\mathbf{q}(\sigma)$ in terms of the parameter σ.

3 Kinematics in Terms of Arc Length

3.1 Arc Length Parameterization

The inverse path-kinematics problem is ill-posed near singularities which not only leads to numerical problems, but also causes extremely fast and undesirable joint motions. Following a path with constant speed $\dot{\sigma}_0$, for instance, leads to an EE-motion with a well-defined bounded EE-velocity. The corresponding inverse kinematics solution $\mathbf{q}(t)$ will not have well-defined bounds, especially at forward kinematics singularities, and the joint velocity $\dot{\mathbf{q}}(t)$ tends to infinity. As a consequence, when prescribing the EE-motion by an equidistant sampling of the path parameter σ, the 'distance' of joint coordinates $\mathbf{q}$ corresponding to two consecutive values of σ tend to vary drastically. A measure of 'distance' of two joint coordinate vectors is the arc length s defined by the line element $\mathrm{d}s = \sqrt{\mathrm{d}\mathbf{q}^T\mathbf{M}\mathrm{d}\mathbf{q}}$ with metric $\mathbf{M}$ (using the generalized mass matrix $\mathbf{M}$ of the robot would yield the kinetic line element and ensure homogeneity). The arc length gives rise to a well-defined discretization of the curve in joint space in terms of the step size. For the rest of this paper $\mathbf{M}$ is chosen as identity matrix.

The arc length path-kinematics problem is to determine the EE-motion $\mathbf{z}(s)$ as a function of s, and thus to derive parameterizations $\sigma = \sigma(s)$ and $\mathbf{q} = \mathbf{q}(s)$. It is known from the theory of metric curves that there is no such explicit relation in general. Therefore, the problem is formulated as an ordinary differential equation (ODE). With the path-parameterization $\mathbf{q}(\sigma)$, the arc length in joint space is [5]

$$s = \int_0^\sigma \|\mathbf{q}'\|_{\mathbf{M}} \mathrm{d}\xi \tag{9}$$

with $\|\mathbf{q}'\|_{\mathbf{M}} := \sqrt{\mathbf{q}'^T\mathbf{M}\mathbf{q}'}$ denoting the *joint path speed*. This leads to $\partial s/\partial\sigma = \|\mathbf{q}'\|_{\mathbf{M}}$, and with (4) follows

$$\frac{\partial\sigma}{\partial s} = \frac{1}{\|\mathbf{J}^+\mathbf{h}\|_{\mathbf{M}}} = \frac{1}{\|\mathbf{q}'\|_{\mathbf{M}}}, \quad \frac{\partial\mathbf{q}}{\partial s} = \frac{\mathbf{J}^+\mathbf{h}}{\|\mathbf{J}^+\mathbf{h}\|_{\mathbf{M}}} = \frac{\mathbf{q}'}{\|\mathbf{q}'\|_{\mathbf{M}}} \tag{10}$$

with $\mathbf{q}'$ as (5). This ODE system can be solved with given initial conditions σ_0 and $\mathbf{q}_0$ for $s = 0$.

3.2 Numerical Solution of the Initial Value Problem

The system (10) can be solved for a given parameter intervall $s \in [0, s_\mathrm{T}]$. This necessitates knowledge of the terminal arc length s_T, which is unknown, however. The condition on the terminal arc length is that

$$\sigma(s_\mathrm{T}) = 1. \tag{11}$$

This serves as indicator function to terminate the integration. That is, the system (10) is solved numerically, until $\sigma(s) = 1$. This can be carried out numerically stable using integration schemes with zero-crossing event detection using advanced root finding methods. For practical applications, the integration can be stopped simply when $\sigma > 1$. Notice that the solution $\sigma(s)$ merely serves for this detection of terminal arc length.

As for the standard geometric inverse path-kinematics problem, the numerical drift of the arc length parameterized inverse path-kinematics subsystem (10) must be stabilized. This is again achieved by introducing the path-velocity error

$$\frac{\partial \mathbf{e}}{\partial s} = \frac{\partial \mathbf{h}_{\mathrm{d}}}{\partial s} - \mathbf{J}\frac{\partial \mathbf{q}}{\partial s}, \tag{12}$$

with $\partial \mathbf{h}_{\mathrm{d}}/\partial s = \mathbf{h}'_{\mathrm{d}}/\|\mathbf{q}'\|_{\mathbf{M}}$. The system (10) is amended as

$$\frac{\partial \mathbf{q}}{\partial s} = \frac{\mathbf{J}^{+}\mathbf{h}}{\|\mathbf{J}^{+}\mathbf{h}\|_{\mathbf{M}}} + \mathbf{J}^{+}\mathbf{K}\mathbf{e} \tag{13a}$$

$$\mathbf{0} = \frac{\partial \mathbf{e}}{\partial s} + \mathbf{K}\mathbf{e} \tag{13b}$$

where (13b) governs the error dynamics. Asymptotically convergence to zero is ensured for $\mathbf{K} > 0$.

The numerical solution must take into account situations where $\mathbf{q}' = \mathbf{0}$, due to the division by $\|\mathbf{q}'\|_{\mathbf{M}}$. This can happen for two reasons. The first case, when $\mathbf{h} = \mathbf{0}$ is avoided by assuming a regular EE-path so that always $\mathbf{h} \neq \mathbf{0}$. Secondly, the partial path-velocity $\mathbf{h}$ is an element of the null space $\mathbf{h} \in \ker \mathbf{J}^{+}$. As a consequence, the corresponding EE movement has no effect on any joint coordinates. This, then is a contradiction to the forward kinematics.

4 A Singularity-Consistent Sampling Scheme

The adaptive sampling strategy is based on the velocity inverse path-kinematics problem. It is assumed that the EE-path using σ as parameter is equivalent to the curve parameterized with the arc length s. As mentioned earlier, prescribing the EE-motion by an equidistant sampling $\Delta\sigma$ inevitably causes drastically varying joint coordinates $\mathbf{q}$, of two consecutive samples, when solving the inverse path kinematics problem (5). The 'distance' between consecutive values of $\mathbf{q}$ is measured by the arc length (9). The latter is a nonlinear relation of s and σ. For illustration purpose, Fig. 1a shows the mapping $s(\sigma)$ for a 6-DOF *Comau Racer3* robot following a straight line in the work space with constant orientation and speed $\dot{\sigma}_0$, Fig. 1b. It can be seen that the manipulator passes near a singularity at $\sigma \approx 0.15 \ldots 0.25$, indicated by the drastically increase of the joint arc length.

By parameterizing the prescribed path using the joint arc length s, Sect. 3, the evolution of the joint coordinates $\mathbf{q}$ and the relation between s and σ is maintained. Following the trajectory then with constant *arc length speed* $\dot{s}_0$ and

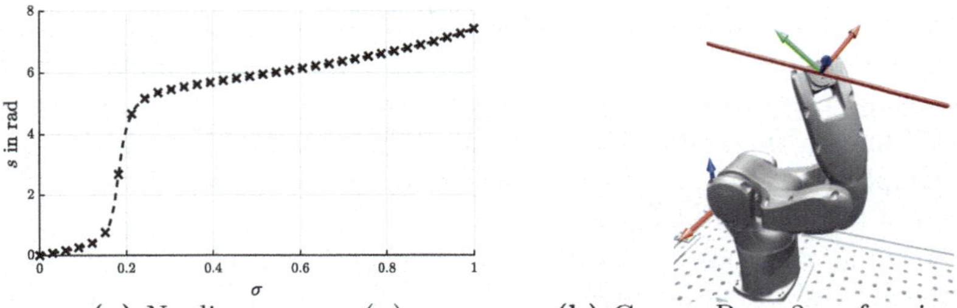

(a) Nonlinear map $s(\sigma)$ (b) Comau Racer3 performing the task

Fig. 1. 6-DOF Comau Racer3 following a straight line in task-space with constant orientation and nonlinear relation between s and σ.

equidistant sampling for s, a varying speed $\dot{\sigma}$ due to the nonlinear mapping (9) is induced. As a consequence of the relation, smaller speeds $\dot{\sigma}$ are achieved at higher rates $\partial s/\partial\sigma$ leading to more sampling points in σ (locally). Since $\dot{\mathbf{q}} = \mathbf{q}'\dot{\sigma}$, lower joint velocities are expected in such regions. Moreover, this can be used as an adaptive sampling strategy for σ due to the nonlinear mapping. This discretization scheme, then can be applied for solving time optimal path following problems, e.g. [8,12,16].

5 Simulation Results

Again considering the 6-DOF *Comau Racer3* robotic arm manipulator following a straight line with constant orientation, Fig. 1. In order to get comparable results, the EE follows the prescribed path with 1) constant speed $\dot{\sigma}_0$ and equidistant sampling in σ when using the velocity inverse path-kinematics (5), and 2) constant arc length speed $\dot{s}_0$ with equidistant discretization in s for the arc length parameterized velocity inverse path-kinematics (10). As shown in Fig. 2, the same joint coordinate samples are obtained with both strategies. The arc length parameterized solution in Fig. 1a shows an accumulation of sampling points at $\sigma \in [0.15, 0.25]$, which indicates the passing near a singularity (Sect. 4).

As previously discussed, for fixed step size Δs (which means fixed time step size due to constant $\dot{s}_0$), higher sampling rates of the joint coordinates in this area are induced and hence lower joint velocities $\dot{\mathbf{q}} = \mathbf{q}'\dot{\sigma}$ as Fig. 3 shows. This is particularly pronounced for $\dot{q}_4$. The arc length parameterized velocity inverse path-kinematic solution yields a joint velocity $\dot{q}_4$ more than 10 times lower than the solution obtained with (5).

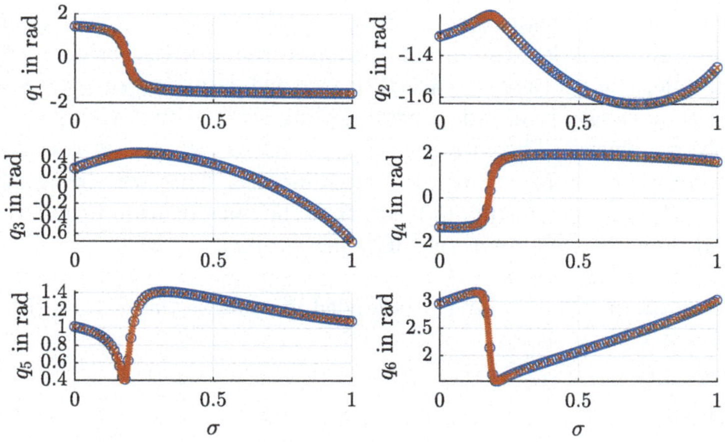

Fig. 2. Samples of the joint coordinates $\mathbf{q}$. Blue circles: using solution (5) for the inverse path kinematic with constant speed $\dot{\sigma}_0$. Orange crosses: using solution (10) for arc length parameterized inverse path kinematics with constant arc length speed $\dot{s}_0$

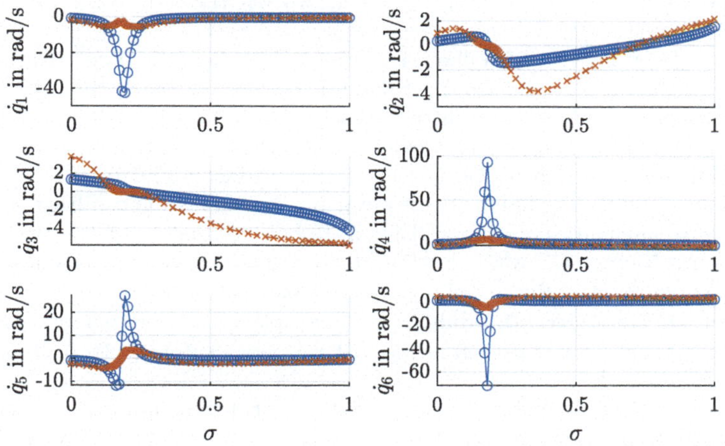

Fig. 3. Resulting joint velocities $\dot{\mathbf{q}}$. Blue circles: using (5) for the inverse path kinematics with constant speed $\dot{\sigma}_0$. Orange crosses: using (10) for the inverse path kinematics with constant speed $\dot{s}_0$.

6 Conclusion and Outlook

The inverse kinematics problem was expressed in terms of the arc length s in joint space when the EE-motion is parameterized by a path parameter σ. This leads to an ODE system for $\mathbf{q}(s)$ and $\sigma(s)$ which is solved for $s \in [0, s_{\mathrm{T}}]$. It is shown that using the arc length as independent parameter (instead of σ) allows for coping with inverse kinematic singularities. Moreover, the joint motion can be fully controlled, instead of controlling the EE-speed. Instead of large joint motions near singularities, the method leads to reduced EE-motions. As shown in the example, the arc length parameterized velocity inverse path-kinematics solution (10) gives rise to an adaptive sampling scheme that yields significantly smaller joint velocities. The introduced approach is applicable to time-optimal path following that is robust against singularities. This will be addressed in future work. Preliminary results indicate that the optimization problem is better conditioned and excessive joint velocities are avoided.

Acknowledgement. This research was funded in whole, or in part, by the Austrian Science Fund (FWF) [I 4452-N].

References

1. Aristidou, A., Lasenby, J.: Inverse kinematics: a review of existing techniques and introduction of a new fast iterative solver (2009)
2. Ben-Israel, A., Greville, T.N.: Generalized Inverses: Theory and Applications, vol. 15. Springer, Heidelberg (2003). https://doi.org/10.1007/b97366
3. Breivik, M., Fossen, T.: Path following for marine surface vessels. In: Oceans 2004 MTS/IEEE Techno-Ocean 2004 (IEEE Cat. No. 04CH37600), vol. 4, pp. 2282–2289 (2004)
4. Siciliano, B., Sciavicco, L., Villani, L., Oriolo, G.: Differential kinematics and statics. In: Siciliano, B., Sciavicco, L., Villani, L., Oriolo, G. (eds.) Robotics, pp. 105–160. Springer, London (2009). https://doi.org/10.1007/978-1-84628-642-1_3
5. Do Carmo, M.P.: Differential Geometry of Curves and Surfaces: Revised and Updated, 2nd edn. Courier Dover Publications, Mineola (2016)
6. Goldenberg, A.A., Benhabib, B., Fenton, R.: A complete generalized solution to the inverse kinematics of robots. IEEE J. Robot. Autom. **1**, 14–20 (1985)
7. Hladio, A., Nielsen, C., Wang, D.: Path following for a class of mechanical systems. IEEE Trans. Control Syst. Technol. **21**(6), 2380–2390 (2013)
8. Kaserer, D., Gattringer, H., Müller, A.: Time optimal motion planning and admittance control for cooperative grasping. IEEE Robot. Autom. Lett. **5**(2), 2216–2223 (2020)
9. Marani, G., Kim, J., Yuh, J., Chung, W.K.: A real-time approach for singularity avoidance in resolved motion rate control of robotic manipulators. In: Proceedings 2002 IEEE International Conference on Robotics and Automation (Cat. No. 02CH37292), vol. 2, pp. 1973–1978 (2002)
10. Penrose, R.: A generalized inverse for matrices. In: Mathematical Proceedings of the Cambridge Philosophical Society, pp. 406–413 (1955)
11. Reiter, A., Mueller, A., Gattringer, H.: On higher-order inverse kinematics methods in time-optimal trajectory planning for kinematically redundant manipulators. IEEE Trans. Industr. Inform. **14**, 1681–1690 (2018)
12. Shin, K., McKay, N.: A dynamic programming approach to trajectory planning of robotic manipulators. IEEE Trans. Autom. Control **31**(6), 491–500 (1986)
13. Siciliano, B.: A closed-loop inverse kinematic scheme for on-line joint-based robot control. Robotica **8**(3), 231–243 (1990)
14. Waldron, K.J., Schmiedeler, J.: Kinematics. In: Siciliano, B., Khatib, O. (eds.) Springer Handbook of Robotics, pp. 11–36. Springer, Cham (2016). https://doi.org/10.1007/978-3-319-32552-1_2

15. Thomopoulos, S.C., Tam, R.Y.: An iterative solution to the inverse kinematics of robotic manipulators. Mech. Mach. Theory **26**(4), 359–373 (1991)
16. Verscheure, D., Demeulenaere, B., Swevers, J., De Schutter, J., Diehl, M.: Time-optimal path tracking for robots: a convex optimization approach. IEEE Trans. Autom. Control **54**(10), 2318–2327 (2009)

Simulation of the Effects of Backlash on the Performance of a Collaborative Robot: A Preliminary Case Study

Roberto Guida(✉), Andrea Raviola, Domenico Fabio Migliore, Andrea De Martin, Stefano Mauro, and Massimo Sorli

Department of Mechanical and Aerospace Engineering, Politecnico di Torino, Corso Duca degli Abruzzi 24, 10129 Turin, Italy
roberto.guida@polito.it

Abstract. The life cycle of industrial manipulators is heavily affected by wear due to continuous and repetitive movements the joints make. The consequent increase of backlash can jeopardize the correct execution of a task and, eventually, cause unexpected downtimes and economical losses. The present paper proposes a preliminary analysis of the effects of the presence of backlash between the motor and the joint shafts in a collaborative robot. To do so, different levels and locations of such fault have been simulated in a pick-and-place application using a dynamic model of a UR5 manipulator. The simulation results show how the deviations of the position of the tool center point from the desired trajectory highly depend on which joint presents a non-nominal behavior. The manuscript also introduces a methodology to detect the presence of backlash in industrial manipulators without any need for additional sensors. Besides the differences in the angular position measured by the motor (input) and the joint (output) encoders, spikes, whose magnitude depends on the backlash severity, arise in the motor currents. From these signals, a pool of health features candidates can be then extracted and considered for future applications in diagnostics and prognostics routines.

Keywords: Industrial robots · Collaborative robotics · High-fidelity modelling · Backlash · Health management · Diagnostics

1 Introduction

The level of automation within industrial companies has grown significantly in the last decade, driven in no small part by the increasingly widespread use of collaborative robots (cobots) [1]. Since any critical malfunctioning of these machines could lead to production wastes, unexpected downtimes, and economical losses, their correct health assessment is of primary importance. To do this, Data-Driven Models (DDMs) are used to extract health features from signals coming from the robot and other additional sensors to detect a fault at its early stage and optimize the maintenance procedures on the machine [2, 3]. The drawback of this approach consists in its need for large datasets of both healthy and degraded robots to properly detect and classify a faulty unit. However, since a robot

A. Müller and M. Brandstötter (Eds.): RAAD 2022, MMS 120, pp. 28–35, 2022.
https://doi.org/10.1007/978-3-031-04870-8_4

mean time between failures is in the order of tens of thousands of hours [4], there is a shortage of data coming from robots operating in non-nominal conditions [5]. To overcome this problem, a possible solution is represented by the development of a High-Fidelity (HF) model of the manipulator in which to inject simulated faults and failures of different types and magnitudes. Signals generated by the robot simulation model would be representative of its behavior both in nominal and degraded operating conditions and could be fed to DDMs for health features selection. The feasibility of this approach has been proven in previous studies like [6], where a properly validated mathematical model of a robotic roller hemming device has been used to inform the machine learning routines used both for diagnostics and prognostics purposes. To do so, proper knowledge of the possible failure modes affecting both industrial and collaborative manipulators, and the effects on their performances, is crucial. In general, a robot is made up of three main macro-elements: the control unit, the teach pendant and the robotic arm. Except for accidental damages, the first two components are only subjected to electrical failures, which can be easily recognized by the built-in control logic. On the contrary, mechanical faults to the joints gearboxes, besides representing the most common failure mode in industrial robots [7], are less likely to be autonomously detected by the robot [8].

In this context, the present study provides a first insight into the effects of the presence of backlash between the input and the output shaft in a joint of a collaborative robot. Using a simulation model of a UR5 manipulator from Universal Robots, different levels and locations of backlash have been simulated and their effects on the Tool Center Point (TCP) position and motor currents have been quantified. The research campaign is a continuation of the work done in [9], where a UR5 multibody model has been developed in a MATLAB/Simulink environment and validated with experimental results.

2 Dynamic Model of a UR5 Collaborative Robot

To build a simulation model able to replicate the behavior of the real robot, it has been necessary to specify both the kinematic and dynamic properties of the manipulator and to properly tune the Proportional-Integrative (PI) parameters with which the joints control logic have been modeled. The Denavit-Hartenberg coefficients, used to build the robot kinematic chain, together with the values of mass and position of the center of mass of each joint/link, are provided by the manufacturer [10]. On the other hand, inertia tensors and joints friction have been identified according to the methodologies provided in [11] and [12], respectively. However, since not all the robot inertia parameters are identifiable, a mismatch among simulated and measured joints currents still persists. This issue will be addressed in future studies where the UR5 inertia tensors will be fully identified according to the methodology described in [13].

As for the real robot, the only input provided to the UR5 simulation model is the set of joints angular positions (q_{set}) required to perform the desired trajectory. All the other quantities (i.e., motor voltages, currents, joints angular velocities, and torques) are internally calculated, for each one of the six joints of the manipulator, according to the joint scheme reported in Fig. 1.

The control electronics has been structured in three nested control loops. The outermost ones are used to compensate the errors related to joints angular positions (e_{pos}) and

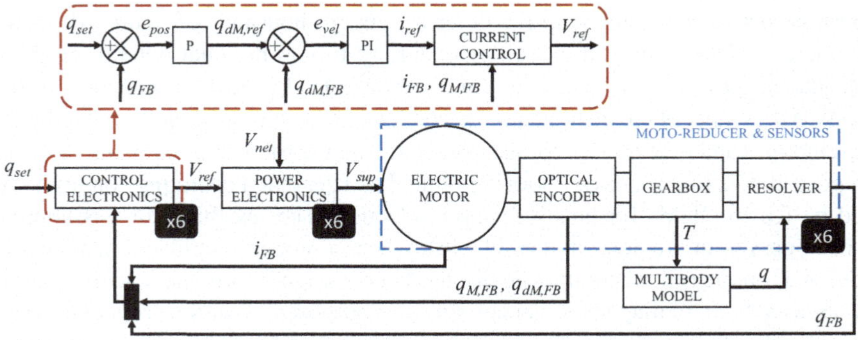

Fig. 1. Schematic representation of the dynamic model of the UR5 collaborative robot.

the motor angular velocities (e_{vel}). On the other hand, the current control logic evaluates the reference voltage (V_{ref}) as a function of the reference (i_{ref}) and the feedback (i_{FB}) motor currents, and the motor angular position ($q_{M,FB}$). This information is sent to the power electronics that modulates the network voltage (V_{net}) to provide the electric motor the supply voltage (V_{sup}). Then, the gearbox model estimates the joint torque (T) taking into account friction losses. Torques are then given as input to the robot multibody model from which, through direct dynamics, joints angular positions (q) are determined and sent to the resolvers from which the feedback signals q_{FB} are obtained.

In case of a fault (i.e., backlash), the control logic will try to compensate the off-nominal behavior of the robot by minimizing the deviation of the actual joints positions from the nominal ones. As it will be better highlighted in Sect. 3, the macroscopic effects of this drop in the manipulator performance will affect the TCP pose which is crucial, for the end user, for a proper competition of the assigned tasks.

2.1 Moto-Reducers

To build a high-fidelity simulation model of a collaborative robot, detailed models of all its components should be developed. However, this would lead to high simulation times, so simplifications should be adopted. In the present research, only the shoulder joint has been modeled using a three-phase AC motor, while, for the other ones, simplified models of a DC brushless motor have been adopted. This choice has been driven by the fact that, according to [14], this is the joint with the highest stopping time.

The reduction of the motor speed, and the consequent increase in the joint torque, is achieved through a gearbox with a reduction ratio of 101:1.

The mechanical transmission is described through a damping-stiffness (c-k) model [15] so that the ideal output torque (T_{id}) is defined as:

$$T_{id} = \begin{cases} k\theta_{rel} + c\theta_{d,rel} & if\,\alpha = 0 (nobacklash) \\ k(\theta_{rel} - \alpha) + c\theta_{d,rel} & if\,\theta_{rel} > \alpha \wedge k(\theta_{rel} - \alpha) + c\theta_{d,rel} > 0 \\ 0 & if\,\theta_{rel} > \alpha \wedge k(\theta_{rel} - \alpha) + c\theta_{d,rel} < 0 \\ k\theta_{rel} + c\theta_{d,rel} & if\,\theta_{rel} < 0 \wedge k\theta_{rel} + c\theta_{d,rel} < 0 \\ 0 & if\,\theta_{rel} < 0 \wedge k\theta_{rel} + c\theta_{d,rel} > 0 \\ 0 & otherwise (nocontact) \end{cases} \tag{1}$$

being θ_{rel} and $\theta_{d,rel}$, respectively, the relative angular position and velocity of the gear teeth, and α the maximum clearance, expressed in radians, between them. However, due to the presence of friction within the joint, a friction torque (T_f) must be subtracted to T_{id} to obtain the actual torque (T) at the output shaft. This value is determined by an approximation of the Stribeck curve:

$$T_f = \left(T_{f,s} - T_{f,c}\right) \cdot [1 - tanh(10\omega)] + T_{f,c} + f_v \cdot \omega, \quad (2)$$

being $T_{f,s}$ and $T_{f,c}$, respectively, the static and the Coulomb friction torques, f_v the viscous friction coefficient, and ω the joint angular velocity at the output shaft. In the left image of Fig. 2, are depicted, for all the robot joints, the values of these contributions with respect to both the joint and motor angular velocities. For an accurate description of this phenomenon, the viscous friction coefficients are determined as a function of the joints temperature according to the findings reported in [16]. These are used to determine the experimental trends of the friction torques for the UR5 shoulder joint, reported in the right graph of Fig. 2, in a range of temperature likely to be encountered by a robot in an industrial environment.

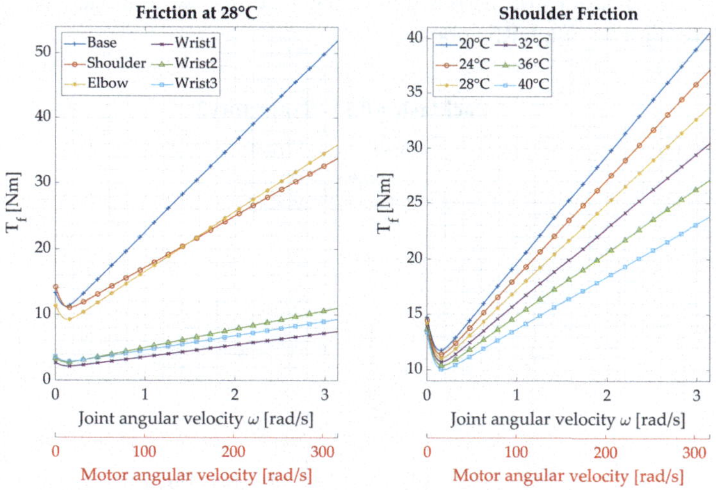

Fig. 2. Friction torques of the UR5 joints at 28 °C (left); Friction torque of the UR5 shoulder joint at different operating conditions (right).

2.2 Sensors

Before being used to close the relative control loops, the signals of the joint angular position, motor angular velocity, and current coming from the moto-reducers respectively pass through blocks used to simulate the presence of three sensors:

1. Resolver: a magnetic encoder, integral with the joint shaft, which measures the feedback angular position (q_{FB}) used to close the position control loop;

2. Optical encoder: mounted on the motor shaft to measure its angular position ($q_{M,FB}$) and velocity ($q_{dM,FB}$) used to close the motor speed control loop.
3. Current sensor: it measures the motor current (i_{FB}) used to close the relative loop.

First-order transfer functions have been adopted to simulate the current sensors, while second-order ones have been implemented for the two encoders. In addition, to replicate a real device, white noise, with the same standard deviation as the one of the signals measured from the UR5, has been added to the simulated quantities which have been then digitalized using a Sample and Hold circuit and a quantizer [9].

3 Simulations Results

Since, according to [17], robots are mainly used for handling tasks, it has been decided to study the effect of backlash on the UR5 TCP position during the pick-and-place of a 2 kg mass. The trajectory taken as a reference in this work is the second of the five ones described in [9] which have been designed to represent the movements usually performed by an anthropomorphic manipulator during such applications. As an example, in Fig. 3, are reported the deviations of the position of the TCP in case of a backlash α equal to 0.1° in different joints of a UR5 cobot.

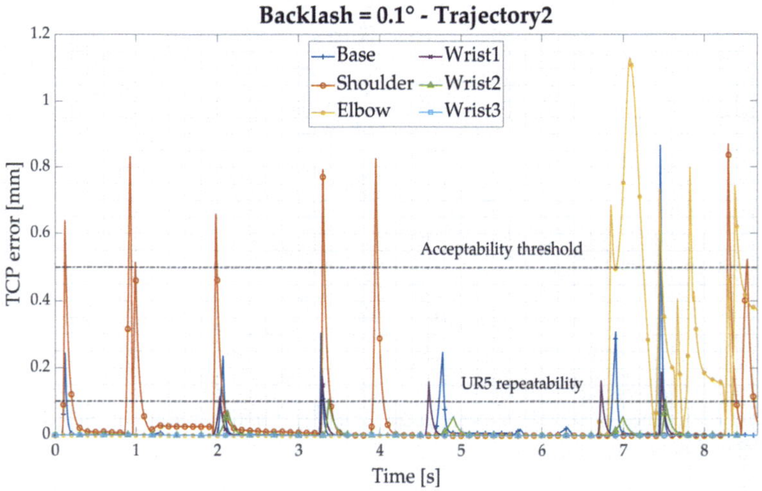

Fig. 3. Effects of backlash in different joints on the robot TCP position over the same pick-and-place trajectory.

Since the value of the backlash α over which Universal Robots manipulators autonomously detect a failure is not known, a trial-and-error approach has been used to detect such a threshold. Pick-and-place applications do not often require high accuracy in the positioning of an object. So, a maximum deviation of 0.5 mm, with respect to the nominal TCP position, has been labeled as not acceptable for the correct execution of the task. On the contrary, for other uses such as assembly or welding, which demand high precision movements, lower acceptability thresholds could be chosen. This information allows identifying the corresponding maximum acceptable level of backlash α, equal to 0.1°, which, as shown in Fig. 3, would lead to a deviation of the TCP position from the desired path higher than the above-mentioned limit.

From the simulations, it is possible to see that, for the proposed case study, the most critical joints are the ones of the base, the shoulder, and the elbow. This can be attributed to the fact that in general, for industrial manipulators, the wrist group is quite compact, while longer links connect the first three joints of a robot arm. So, each joint has a different impact on the proper position and orientation of the TCP, although similar results have been obtained in [9] for other trajectories.

Nevertheless, since the TCP pose derives from the forward kinematics applied to the joints angular positions, unless it is measured by external devices, it cannot be directly used to detect the presence of backlash in a robot arm. Instead, this can be done by comparing the measurements coming from the two encoders, mounted before and after the gearbox, integral with the motor and the joint shaft respectively. In the presence of backlash, there would be a mismatch between the two signals which will not be scaled only by the gear ratio. In addition, as reported in Fig. 4, current signals are also affected by the presence of such a fault and can be then used to detect an anomaly in the robot behavior.

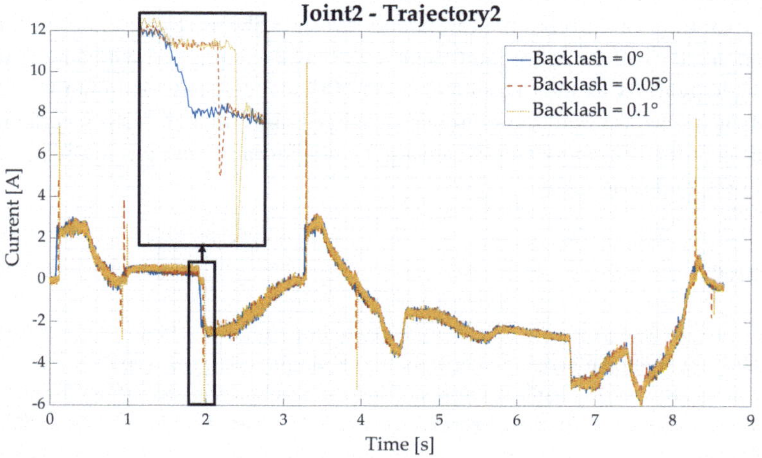

Fig. 4. Comparison of the effects of different levels of backlash on the motor current of the shoulder joint during a pick-and-place trajectory.

In the proposed case study, the analysis of the motor currents has been performed only in the case of a backlash on the shoulder joint. This choice has been driven by the fact that, as already mentioned, this component is the only one in which the motor is modeled as a three-phase, as it actually is in the robot. In Fig. 4, it can be observed that, due to the presence of the backlash, whenever the sign of the quadrature current undergoes a variation, the transmitted torque is null for the time required to recover the clearance. This is due to the loss of contact among the gear teeth that can be caused by both a commanded inversion of the joint motion or by a change in the direction of the combined action of gravity and inertia forces. This leads to a mismatch between the reference and the feedback signals of the joint angular velocity which makes the current control loop command an increment, in absolute value, of the motor voltage to compensate such error, represented in Fig. 4 by the current spikes whose magnitude is correlated to the level of backlash.

4 Conclusions

The present paper provides a first insight into the macroscopic effects that the presence of backlash inside a joint of a collaborative manipulator can have on its overall performance. To do this, a mathematical model of a UR5 collaborative robot has been developed to study the deviations of the TCP positions and the motor currents from their nominal operating conditions in a pick-and-place application. Other typical robot tasks, such as welding, polishing, and drilling will be an object of future research.

For the proposed case study, simulation results suggest that the mismatch between joint angular positions measured by the two encoders and the spikes in the motor currents could be used as possible features candidates for studies aimed to assess the health status of an industrial robot.

Future work will be focused on the validation of the UR5 simulation model in non-nominal operating conditions. For more accurate results, the three-phase AC motor model will be extended to all the robot joints which will be also equipped with a more detailed model of the gearbox as the one introduced in [8]. An extensive simulation campaign will be then run to simulate the presence of other failure modes, both mechanical and electrical, in order to support data-driven algorithms for faults and failures identification, classification, and prognosis.

References

1. International Federation of Robotics: The Impact of Robots on Productivity, Employment and Jobs, International Federation of Robotics, Frankfurt, Germany (2017)
2. Jaber, A.A., Bicker, R.: Fault diagnosis of industrial robot gears based on discrete wavelet transform and artificial neural network. Insight Non-Destruct. Test. Cond. Monit. **58**(4), 179–186 (2016)
3. Jaber, A.A., Bicker, R.: Industrial robot backlash fault diagnosis based on discrete wavelet transform and artificial neural network. Int. J. Progn. Heal. Manag. **7**(2), 21–31 (2016)
4. Majid, M.A.A., Fudzin, F.: Study on robots failures in automotive painting line. ARPN J. Eng. Appl. Sci. **12**(1), 62–67 (2017)
5. Qiao, G., Weiss, B.A.: Accuracy degradation analysis for industrial robot systems. In: ASME 2017 12th International Manufacturing Science and Engineering Conference, MSEC 2017 collocated with the JSME/ASME 2017 6th International Conference on Materials and Processing, vol. 3, p. 9 (2017)
6. Grosso, L.A., De Martin, A., Jacazio, G., Sorli, M.: Development of data-driven PHM solutions for robot hemming in automotive production lines. Int. J. Progn. Heal. Manag. **11**(January), 2022 (2020)

7. Zhou, Q., Wang, Y., Jianming, X.: A Summary of health prognostics methods for industrial robots. In: 2019 Prognostics & System Health Management Conference - Qingdao, pp. 5–10 (2019)
8. Raviola, A., De Martin, A., Guida, R. , Jacazio, G., Mauro, S., Sorli, M.: Harmonic drive gear failures in industrial robots applications: an overview, pp. 350–360 (2021)
9. Migliore, D. F.: Modellazione e fault detection di giochi in robot collaborativi. Master's thesis, Politecnico di Torino, Torino, Italy (2021)
10. Universal Robots A/S, Universal Robots - DH Parameters for calculations of kinematics and dynamics. https://www.universal-robots.com/articles/ur/application-installation/dh-parameters-for-calculations-of-kinematics-and-dynamics/. Accessed 18 Jan 2022
11. Kufieta, K.: Force Estimation in Robotic Manipulators: Modeling, Simulation and Experiments, Norwegian University of Science and Technology. Master's thesis, Trondheim, Norway (2014)
12. Raviola, A., De Martin, A., Guida, R., Pastorelli, S., Mauro, S., Sorli, M.: Identification of a UR5 collaborative robot dynamic parameters. In: Zeghloul, S., Laribi, M.A., Sandoval, J. (eds.) RAAD 2021. MMS, vol. 102, pp. 69–77. Springer, Cham (2021). https://doi.org/10.1007/978-3-030-75259-0_8
13. Bahloul, A., Tliba, S., Chitour, Y.: Dynamic parameters identification of an industrial robot: a constrained nonlinear WLS approach, 26th Mediterr. Conf. Control Autom. **51**(15), 443–448 (2018)
14. Universal Robots A/S, Universal Robots - What is CB3 Stop performance categories? (2018). https://www.universal-robots.com/articles/ur/safety/what-is-cb3-stop-performance-categories/. Accessed 18 Jan 2022
15. Akoto, C.L., Spangenberg, H.: Modeling of backlash in drivetrains. In: 4th CEAS Air and Space Conference, September 2013 (2013)
16. Raviola, A., Guida, R., De Martin, A., Pastorelli, S., Mauro, S., Sorli, M.: Effects of temperature and mounting configuration on the dynamic parameters identification of industrial robots. Robotics **10**(3), 83 (2021)
17. International Federation of Robotics: World Robotics Press Conference 2021, International Federation of Robotics, Frankfurt, Germany (2021)

Modeling and Analysis of an S-Shape Link for an Anthropomorphic Robotic Arm

Qun Xing Xue(✉) and Fengfeng Xi

Ryerson University, Toronto, ON M5B 2K3, Canada
{qunxing.xue,fengxi}@ryerson.ca

Abstract. In this paper, an S-shape link is presented for a compact anthropomorphic robot for narrow space applications and modeled for stiffness analysis. A finite element method is adopted to discretize this link and establish a compliance matrix that relates the link tip deflections to the link tip loads. It is shown that there are eight extra entries in the compliance matrix of the S-shape link compared with the one of a straight link. These extra entries are termed the secondary deflections caused by three phenomena associated with the S-shape link. The first is the secondary moment caused by the misalignment of the two link ends. The second is the secondary twisting and the third is the secondary compression along the link length, both caused by the bending forces and moments, which are otherwise neglected in a straight link. Furthermore, the S-shape link is parameterized in terms of the length and the thickness, and an optimization method is applied to determine these parameters leading to a minimum link tip deflection. A case study is provided to demonstrate how to determine this compliance matrix numerically.

Keywords: S-shape link · Finite element method · Compliance matrix · Secondary deflections · Compact robot arm

1 Introduction

Nowadays automation is being applied commonly in almost every corner of our society and the demand for more flexible and compact robot arms arises when facing narrow and enclosed workspace. A bio-inspired hyper-redundant robot such as a snake robot could be functional when dealing with a light payload, but the numerous freedoms would cause a high level of complexity in control [1]. Other solutions, such as continuum robots, have the same limitation on payloads [2]. A movable anthropomorphic robot mimicking human arm operation with a load-bearing structure was originally designed for wing-box fastening [3] and this robot is applicable to a wide range of enclosed area applications. In order to make a robot arm as compact as a human arm as possible, an S-shape link was proposed for the compact robot to reduce the width when the adjacent link pairs were joined together, which motivated the study of this paper.

In this paper, we focus on the geometric structure of the S-shape link and the deflection of a robot arm using three S-shape links when a payload is applied on the end-effector. Compared to the conventional robot links which are of a straight shape, the S-shape link

A. Müller and M. Brandstötter (Eds.): RAAD 2022, MMS 120, pp. 36–43, 2022.
https://doi.org/10.1007/978-3-031-04870-8_5

has the two ends not aligned, causing a secondary moment and twisting, which need to be addressed. For this reason, a finite element method is applied to model the S-link and an augmented compliance matrix is obtained. The S-shape link is parameterized, and an optimization method is used to find an optimal shape with the minimum deflection of a three-link robot arm. In what follows, the details of this study are provided.

2 Problem Formulation

Normally, when two links are connected by a revolute joint, they stack up in the direction of the joint's axis. When the workspace is narrow and enclosed, the link pairs are required to be as compact as possible to occupy the least space. For better load-bearing capacity, the stiffness of the link becomes essential. The S-shape link was designed as a module to expand a planar robot arm by increasing the number of modules when the robot is required to reach into an enclosed workspace through a narrow path. Figure 1 shows our original design of an anthropomorphic robot, which is made of a planar robot of three S-shape links that can rotate about the horizontal support beam and slide along the vertical column. The entire robot system is mounted on a mobile platform, and it can be readily calibrated against a targeted application area using a vision system [4].

In Fig. 1, the stack-up width is indicated at a joint where two adjacent links are connected. As stated in [3], this width should be designed equally to the average width of a human elbow in order to allow the replacement of a human arm for a narrow area application. A comprehensive study was conducted and reported in [5]. A complete kinematic modeling was developed to deal with a multi-DOF planar (mimicking a human arm) that can translate vertically (adjusting the arm's height) and rotate horizontally (adjusting the arm's orientation) yet mobile (adjusting the arm's global position). Newton-Euler methods were used to calculate the link tip loads from the robot payload and link weights as a function of joint angles. Based on the required load conditions, the S-shape deflections were analyzed using a commercial FEM, and a prototype system was fabricated for experimental validation.

The purpose of this work is to further our study by developing a link-based stiffness modeling for analysis of the robot system stiffness. As shown in Fig. 4 that depicts the skeleton of an S-shape link, its two ends are not aligned along the link length, and this misalignment causes a secondary comment for an axial load, which normally does not occur in a straight link. As shown in Fig. 3, the link is of a thin plate shape, i.e., its width larger than its thickness, leading to high stiffness along with the former and low stiffness along the latter. This rectangle cross-section would induce twisting when a bending load or moment is applied at the link tip, which does not occur in a straight link. All these phenomena require a new study.

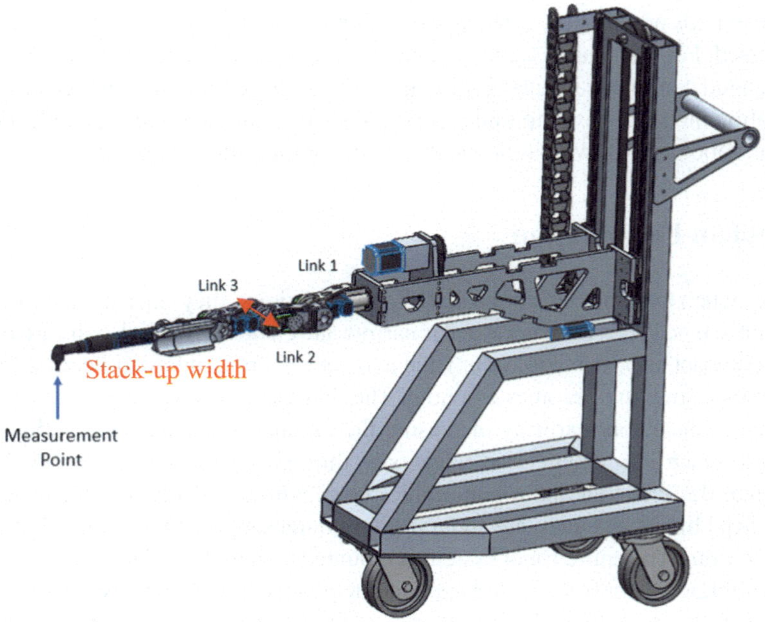

Fig. 1. Structure of the anthropomorphic robot and the S-shape link

3 S-Shape Link Design and Analysis

The general structure of the S-shape link is shown in Fig. 2. The total length of the link is given by *L*, the total height is given by *H*, and the total width is given by *W*. The well-shaped cavities on both ends are designed for the motor to be installed. Due to this feature, the motor has to have a size that fits the link. The total height distributes to the two ends with $H = H_1 + H_2$ which H_1 and H_2 are the heights on the left end and right end respectively. H_1 is one of the parameters that will be used to determine the optimized stiffness and once H_1 is chosen, H_2 can be obtained. A reasonable range of H_1 is picked for it to vary from 1/3 of the total height to 2/3 of the total height. The other parameter is *D*, which is how long are the end parts, varying from half of *W* to $(L - W)/3$. For simplicity, the two ends are made to have the same length. The semi-circle part on both ends with a radius of $0.5W$. To avoid colliding when rotating, the parameter *D* has to be larger than $0.5W$. The length, height, and width are comparable to a human forearm since it is designed for an anthropomorphic robot.

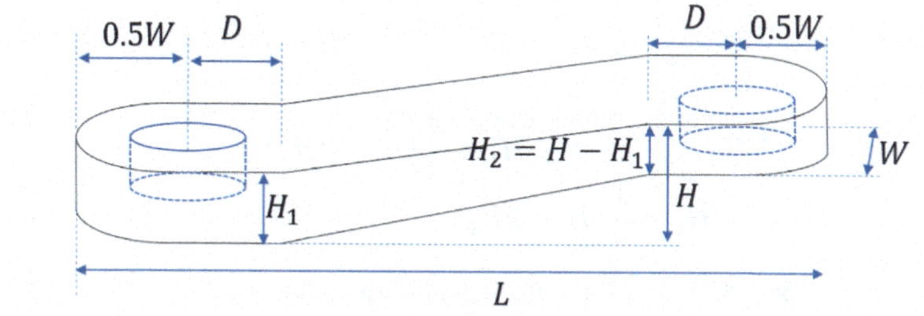

Fig. 2. Structure of the S-shape link

3.1 Discretized Modelling

A long cantilever beam shown in Fig. 3 is to be used to model the S-shape link. The load is considered as applying at the center of the well hence the semi-circle parts on both ends are neglected for simplicity.

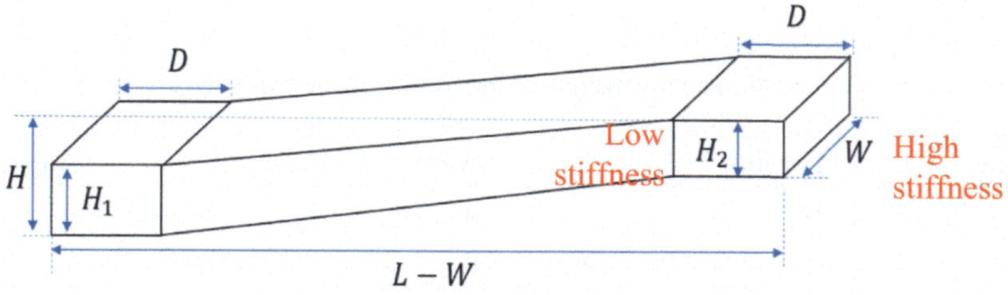

Fig. 3. Analytical model of the S-shape link

By applying a finite element method, the link is discretized as five cantilever beam elements with rectangular cross-sections connected together resulting in six nodes shown in Fig. 4. The beam elements on the two ends have uniform cross-sections, while the middle three beam elements are treated with equal length but the heights are sloped. Therefore, for the non-uniform height elements in the middle, a trapezoid is adopted for modeling. Dividing it into three pieces with equal length will give us a height distribution as shown in Fig. 5. For the finite element analysis, the height is to be chosen as the average of the top and bottom sides of each discretized trapezoid.

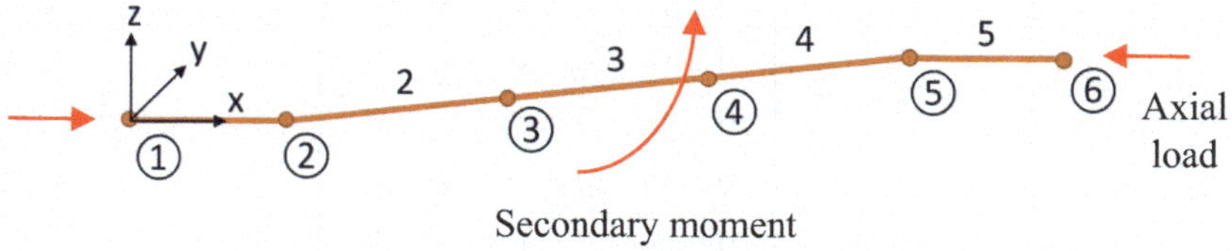

Fig. 4. Element layout of the S-shape link for finite element analysis

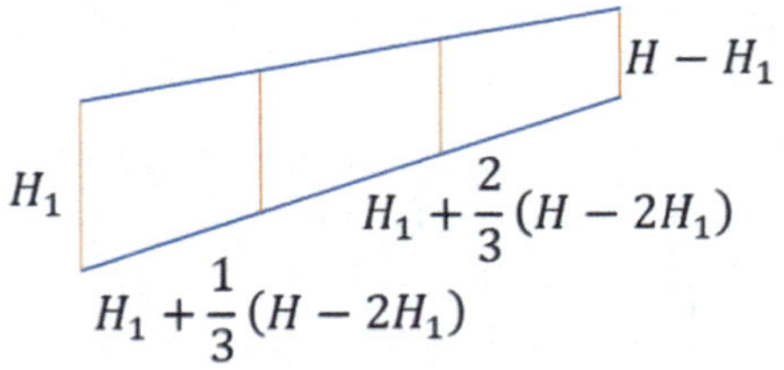

Fig. 5. Side view of the non-uniform middle part

The vertical distance between the two centroid neutral axes of the first element and last element is half of the total height since together they make up the total height. Meanwhile, the horizontal distance between the elements on both ends is the total length subtract the semi-circles, together with length *W*, and twice the length *D*. These can be observed from the tilted angle equation Eq. (1). In conclusion, we have Table 1 with the information of each element. Note, the tilted angle of the middle part is given by:

$$\theta_t = tan^{-1} 0.5H/(L - W - 2D) \tag{1}$$

Table 1. Dimensions of the five beam elements.

Element nodes	Length	Width	Height	Angle
①②	D	W	H_1	0
②③	$(L - W - 2D)/3cos\theta_t$	W	$(4H_1 + H)/6$	θ_t
③④	$(L - W - 2D)/3cos\theta_t$	W	$H/2$	θ_t
④⑤	$(L - W - 2D)/3cos\theta_t$	W	$(5H - 4H_1)/6$	θ_t
⑤⑥	D	W	$H - H_1$	0

If following the procedure for beam element analysis, we would obtain a total stiffness matrix with a dimension of 36 by 36 [6], corresponding to 6 degrees of freedom, i.e., 3 linear deflections and 3 torsions, for each of six nodes. Since we are only concerned about the link tip deflection relative to the fixed link base, the link stiffness will be expressed similar to that of a cantilever beam of straight shape as given below.

$$C_{straight} = \begin{bmatrix} \frac{L}{AE} & 0 & 0 & 0 & 0 & 0 \\ 0 & \frac{L^3}{3EI_z} & 0 & 0 & 0 & \frac{-L^2}{2EI_z} \\ 0 & 0 & \frac{L^3}{3EI_y} & 0 & \frac{L^2}{2EI_y} & 0 \\ 0 & 0 & 0 & \frac{L}{GJ} & 0 & 0 \\ 0 & 0 & \frac{L^2}{2EI_y} & 0 & \frac{L}{EI_y} & 0 \\ 0 & \frac{-L^2}{2EI_z} & 0 & 0 & 0 & \frac{L}{EI_z} \end{bmatrix} \tag{2}$$

In fact, Eq. (2) is a compliance matrix relating the link tip forces to the link tip deflection. We will be augmenting this matrix for the S-shape link by considering the secondary moment and twisting effect.

3.2 Augmentation of Compliance Matrix

The compliance matrix given in Eq. (2) will be augmented numerically using the FEM model described in Sect. 3.1. In addition to the non-zero entries in Eq. (2), there will be more non-zero entries for the S-shape link. For performing the FEM analysis, the same dimensions as the original paper [3] are used for all the numerical analyses. The total length $L = 170$ mm, the width $W = 50$ mm which means the effective length $L - W = 120$ mm, and the total height is $H = 88$ mm. Assuming the link is made of aluminum alloy 6061-T6, we have the modulus of elasticity $E = 68.9$ GPa and shear modulus $G = 26$ GPa [7]. For obtaining a compliance matrix, the unit forces $f_x = f_y = f_z = 1$ N and the unit moment $m_x = m_y = m_z = 1$ Nm are used to compute the link deflections and angular deflections. In this simulation, we vary parameters H_1 and D in an attempt to find out the optimal values with minimal tip deflections. Table 2 lists the secondary deflections which do not occur in a straight beam. The axial force f_x would cause a secondary moment inducing an angular deflection about the y-axis and a deflection along the z-axis. The bending force f_y would cause a twisting about the x-axis. The bending force f_z would cause a deflection in the x-axis which is usually omitted, and it appears significant due to the low stiffness in the thickness direction. The torsion m_x would induce a twisting about the z-axis and a deflection along the y-axis. The bending motion m_y would produce a deflection in the x-axis which is usually omitted, and it appears significant due to the low stiffness in the thickness direction. The bending moment m_z would generate a twisting about the x-axis. In total, there will be eight more non-zero entries in the augmented compliance matrix for the S-shape link.

Table 2. Table of each secondary deflection due to unit forces and moments

Force/moment	Secondary deflection	
f_x	*y*-axis curvature	*z*-axis deflection
f_y	*x*-axis twisting	
f_z	*x*-axis deflection	
m_x	*y*-axis deflection	*z*-axis twisting
m_y	*x*-axis deflection	
m_z	*x*-axis twisting	

In general, the compliance matrix varies with the robot's configuration. Due to the space limitation, only one case is discussed just to prove a point. The robot configuration under consideration is shown in Fig. 1, where the motor axes (z-axis) of the three S-shape links are placed horizontally. In this case, the motor weights will add to the force f_z to make it more significant. Hence, we use the values leading to the least linear deflection in the *z*-direction, $H_1 = H/3 = 29.33$ mm, $D = W/2 = 25$ mm for the compliance matrix. The straight link with the same width and length and height of $H/2$ would be used for comparison.

With the parameters given, we can compute the deflections due to each of the unit forces and each of the moments. Recall that node 6 is fixed and loads are applied on node 1, we can numerically solve the endpoint deflections, leading to the compliance matrix C on the endpoint with stronger stiffness on z-direction as below.

$$C = 10^{-6}\begin{bmatrix} 0.0032 & 0 & -0.0059 & 0 & -0.0840 & 0 \\ 0 & 0.0199 & 0 & 0.0659 & 0 & -0.2423 \\ -0.0059 & 0 & 0.0176 & 0 & 0.2964 & 0 \\ 0 & 0.0659 & 0 & 6.7848 & 0 & 0.6744 \\ -0.0840 & 0 & 0.2964 & 0 & 8.0210 & 0 \\ 0 & -0.2423 & 0 & 0.6744 & 0 & 4.9140 \end{bmatrix} \mathrm{m/N} \quad (3)$$

$$C_{straight} = 10^{-6}\begin{bmatrix} 0.0008 & 0 & 0 & 0 & 0 & 0 \\ 0 & 0.0182 & 0 & 0 & 0 & -0.2280 \\ 0 & 0 & 0.0236 & 0 & 0.2944 & 0 \\ 0 & 0 & 0 & 5.6751 & 0 & 0 \\ 0 & 0 & 0.2944 & 0 & 4.9070 & 0 \\ 0 & -0.2280 & 0 & 0 & 0 & 3.8000 \end{bmatrix} \mathrm{m/N} \quad (4)$$

Comparing C in Eq. (3) with $C_{straight}$ in Eq. (4), we can see that there are eight more non-zero entries belonging to the secondary deflections for the S-shape link as explained in Table 2. A way to identify them is to associate each column to its corresponding unit force or moment. Let us start with the first column, i.e., a unit force in the x-direction. In addition to the primary axial deflection of 0.0032e−6 mm in the x-axis, it causes the secondary deflection of −0.0059e−6 mm in the z-axis (third entry) and a −0.0840e−6 rad curvature deflection about the y-axis (fifth entry), both due to the secondary moment as mentioned before. In fact, the secondary deflection in the z-direction is larger than the primary one in the x-direction, which may be easily understood since the link is relatively stiffer in the x-axis than in the z-axis. Next, the unit force in the y-axis (the second column) induces a secondary twisting of 0.0659e−6 rad about the x-direction (the fourth entry). The unit force in the z-axis (the third column) induces a secondary compression of −0.0059 mm in the x-direction (the first entry).

Now let us look at moments. A unit moment of the x-axis triggers a secondary deflection of 0.0659e−6 mm in the y-axis as well as a secondary twisting of 0.6744e−6 rad about the z-axis. A unit moment of the y-axis prompts a secondary compression of −0.0840e−6 mm in the x-axis. A unit moment of the z-axis provokes a secondary twisting of 0.6744e−6 rad about the x-axis.

Now for a given load f' on the endpoint in the local frame of the S-shape link, we can calculate the endpoint deflection d' in the local frame given by:

$$d' = Cf' \quad (6)$$

The deflection of the robot's end-effector can be expressed as.

$$d_{\mathrm{robot}} = J_G d' \quad (7)$$

where J_G is the generalized Jacobian matrix considering the infinitesimal motions of the link tips in three directions for both rotation and translation. Due to the space limitation, the robot deflection analysis is not provided here.

4 Conclusion

In this paper, an S-shape link is modeled as a cantilever beam using a finite element approach as a procedure of finding a compliance matrix when a load is applied at the end of the link. It is shown that the compliance matrix of a straight link should be augmented with eight extra entries for an S-shape link. It is explained that these extra entries, termed as the secondary deflections, are caused by three phenomena associated with the S-shape link. The first is the secondary moment produced by the misalignment of the two link ends. The second is the secondary twisting and the third is the compression along the link length, both caused by the bending forces and moments, which are neglected in a straight link. Furthermore, the S-shape link is parameterized in terms of the length and the thickness, and an optimization method is applied to determine these parameters leading to a minimum link tip deflection. A case study is provided to demonstrate how to determine this compliance matrix numerically. In the end, it is shown that the deflections of the robot's end-effector can be determined following a conventional method with the only difference being to use the compliance matrix developed in this paper.

References s

1. Transeth, A.A., Pettersen, K., Liljeback, P.: A survey on snake robot modeling and locomotion. Robotica **27**, 999–1015 (2009)
2. Yang, Z., Zhao, B., Bo, L., Zhu, X., Xu, K.: CurviPicker: a continuum robot for pick-and-place tasks. Assembly Autom. (2018)
3. Jiang, J., Xi, F., You, J., Xue, Q.: Mechanical design of a new anthropomorphic robot for fastening in wing-box. In: ASME 2021 International Design Engineering Technical Conferences and Computers and Information in Engineering Conference, Volume 8B: 45th Mechanisms and Robotics Conference (MR), Virtual Online (2021)
4. Soria Olmedo, V.: Localization of a Mobile Fastening Robot. Master thesis, Ryerson University (2020)
5. Balsara, F.: Robot Design and Modeling for Enclosed Area Assembly. Master thesis, Ryerson University (2022)
6. Logan, D.: A First Course in the Finite Element Method. 4th Edition. By Nelson. A Division of Thomson Canada Limited (2007)
7. 6061-T6 Aluminium – The Ultimate Guide – Matmatch. https://matmatch.com/learn/material/6061-t6-aluminium
8. Mohamed, R., Xi, F., Finistauri, A.: Module-based static structural design of a modular reconfigurable robot. ASME **132**(1), 014501 (2010)

Parallelized Forward Kinematics Using Product of Exponentials in PyTorch

Hristo Vrigazov(✉) and Kaloyan Yovchev

Faculty of Mathematics and Informatics, Sofia University "St. Kliment Ohridski", Sofia, Bulgaria

hvrigazov@uni-sofia.bg, k.yovchev@fmi.uni-sofia.bg

Abstract. In open-chain robots, the forward and inverse kinematics problems are foundational for manipulating the robot. The forward kinematics takes as input the position of each joint of the robot and computes the position and orientation of the robot's end-effector. Commonly used methods for solving the forward kinematics problem include the product of exponentials formula and the Denavit-Hartenberg (D-H) representation. In this paper, we focus on a specific Python implementation of the product of exponentials (POE) formula (modern_robotics package), which is one of the most popular implementations used in production, with nearly 1000 stars on Github, and has been downloaded over 5000 times in the last year. This POE implementation does not natively support computation in parallel for multiple inputs. Also, the product of exponentials implementation cannot be used in machine learning models that use automatic differentiation, such as neural networks. This paper contributes a parallelized, automatically differentiable implementation of the modern_robotics Python implementation using PyTorch, one of the most popular deep learning frameworks. We achieve this by introducing a natural extension using PyTorch tensors to the special orthogonal group so(3) and the special euclidean group se(3). The newly introduced extensions form a group themselves, and all operations needed for the POE formula can be implemented in them in parallel, leading to significant speedup on our hardware - both on CPU (over 200 times) and especially on GPU (over 1600 times), compared to a naive sequential implementation in Python.

Keywords: Forward kinematics · Product of exponentials · Pytorch · Parallelized

1 Introduction

In a robot arm consisting of links and joints, forward kinematics refers to the computation performed to obtain the coordinates of the end effector, given the values of its joints. There are several methods for solving the forward kinematics problem. While Denavit-Hartenberg (D-H) representation is the most popular

A. Müller and M. Brandstötter (Eds.): RAAD 2022, MMS 120, pp. 44–51, 2022.
https://doi.org/10.1007/978-3-031-04870-8_6

notation used for solving the forward kinematics, the product of exponentials formula (POE) [1] has been gaining popularity. This paper focuses on the POE and on a particular implementation in Python, which is available online and is one of the most popular open-source implementations of the POE formula. When computing the forward kinematics for a significant number of input configurations, this implementation takes advantage of neither vectorized CPU instructions nor GPU devices available. A vectorized implementation of the POE formula would provide several benefits, including improved initialization for inverse kinematics methods based on initial guess and gradual improvements in real-time computation of possible collisions in motion planning.

There has been an increasing number of machine learning papers in many areas of robotics (e.g., [2]). Many of these papers use gradient descent methods and their derivatives and are typically implemented using frameworks supporting automatic differentiation like PyTorch [3] or Tensorflow [4]. Using an automatic differentiation framework allows automatic computation of the gradient given a loss function, as opposed to symbolic derivation, and finds application in the computation of the inverse kinematics, or when training an end-to-end machine learning model. For example, [5] trains a machine learning model that predicts best suction point and best grasp proposal in the image space. A forward kinematics implementation that uses an automatic differentiation framework would enable a similar work, but predict directly the grasp and suction point in world coordinates (combined with a 3D sensor). However, in most cases, the implementation of forward kinematics does not include automatic differentiation frameworks, making it challenging to have this part in the training process. The Python implementation on which we focus (the modern_robotics packages) [6] uses Numpy, but only for a single input matrix, not a multi-dimensional input. Both Numpy and PyTorch are array programming (vectorized) solutions [7], which allow the application of operations to multiple values at once. Array programming solutions allow implicit parallelization, enabling the compiler to create a parallelized solution based on the program, written in a vectorized, multi-dimensional array way. For this reason, array programming solutions do not need language tools related to parallel execution, as opposed to explicit parallelization solutions, which make use of abstractions such as threads and processes, and their spawning and management is left to the programmer.

PyTorch has the additional advantage of support for automatic differentiation, and for this reason, in this paper, we aim to create a vectorized implementation of the modern_robotics package in PyTorch. The main abstraction of both PyTorch and Tensorflow is a tensor, which represents a multi-dimensional array with automatic differentiation support.

To implement the operations needed for forward kinematics in PyTorch, we provide a simple representation of the input tensor and the associated PyTorch operations to achieve identical behavior to a simple, sequential implementation (code is included in the repository [8]). We ensure that the outputs of the two approaches match and present a benchmark, which demonstrates the dramatic decrease in latency of the proposed approach. Lastly, we briefly possible applications of the novel approach, which include: faster collision detection during planning, improved initialization in inverse kinematics methods such as Newton-Raphson, and the usage of automatically differentiable methods.

2 Forward Kinematics Using Product of Exponentials

In this section, we briefly introduce the main parts of the POE formula for forward kinematics.

The forward kinematics problem consists of calculating the robot's end-effector frame's position and orientation from its joint coordinates θ. To solve the forward kinematics problems, the POE approach uses exponential coordinates.

We will use a notation, consistent with [1], and use $SO(3)$ to denote the group of the 3×3 rotational matrices, also known as the special orthogonal group.

Let $R \in SO(3)$, and $p \in \mathbb{R}^3$. We will call all matrices T of the form:

$$T = \begin{bmatrix} R & p \\ 0 & 1 \end{bmatrix} \tag{1}$$

homogeneous transformation matrices and we will denote their group by $SE(3)$, which stands for special euclidean group.

Let T_{ij} be a homogeneous transformation matrix from joint i to joint j. Then, to solve the forward kinematics problem for a robot with n joints, we have to solve the following product:

$$T_{0,n} = \prod_{i=0}^{n-1} T_{i,i+1} \tag{2}$$

Let $M \in SO(3)$ be the initial configuration, and let θ_i denote the orientations of each joint of the robot, which is given when solving the forward kinematics problem. Then we can use the POE rule [1]:

$$\prod_{i=0}^{n-1} T_{i,i+1} = \prod_{i=0}^{n-1} e^{[S_i]\theta_i} M \tag{3}$$

S_i where $i \in [0, n)$ denotes the screw axis of joint i as expressed in the fixed base frame and $S_i = (\omega_i, v_i)$, where ω_i is unit vector in the positive direction of joint axis i, and $v_i = -\omega_i \times q_i$ any arbitrary point on joint axis i as written in coordinates in the fixed base frame.

The $se(3)$ Lie algebra consists of matrices which use the notation $[S]$:

$$[S] = \begin{bmatrix} [\omega] & v \\ 0 & 0 \end{bmatrix} \tag{4}$$

For $[\omega]$, where $\omega = [\omega_0 \quad \omega_1 \quad \omega_2]^T$, the same notations means the skew-symmetric representation of:

$$[\omega] = \begin{bmatrix} 0 & -\omega_3 & \omega_2 \\ \omega_3 & 0 & -\omega_1 \\ -\omega_2 & \omega_1 & 0 \end{bmatrix} \tag{5}$$

The set of skew-symmetric matrices forms a group called $so(3)$.

The next building block needed to compute the result is the conversion of a $so(3)$ matrix to a rotation matrix $R \in SO(3)$ (see Eq. (6)).

$$R = \mathrm{Rot}(\omega, \theta) = e^{[\omega]\theta} = I + \sin\theta[\omega] + (1 - \cos\theta)[\omega][\omega]^T \in SO(3) \tag{6}$$

Once this product is computed, we can use it to compute the final homogeneous transformation matrix $T \in SE(3)$ (see Eq. (7)).

$$e^{[S]\theta} = \begin{bmatrix} e^{[\omega]\theta} & \left(I\theta + (1 - \cos\theta)[\omega] + (\theta - \sin\theta)[\omega]^2\right) v \\ 0 & 1 \end{bmatrix} \tag{7}$$

A possible improvement in terms of computational effort could be done using more efficient forms of Eq. 6 and Eq. 7, such as [10], which would be subject of future work.

3 Parallelization Approach

The input to the forward kinematics function is represented by a PyTorch tensor of shape `(N, C)`, where N is the number of inputs, and C is the degrees of freedom of the robot.

3.1 Vector to $so(3)$

To perform parallelized conversion to $so(3)$, we separate the process into two steps: first, we initialize the resulting tensor with zeros, with the correct output shape (we know that this shape is `(N, 3, 3)`). We follow this step by 6 single instruction, multiple data (SIMD) instructions, which set the correct values on the respective positions along the last two dimensions (see the function `VecToso3` in the supplementary materials[1]), corresponding to the 6 nonzero items in the definition of $so(3)$ (see Eq. (5)). The newly defined extension of $se(3)$ consists of tensors with 3 dimensions, the dimension zero being the number of inputs, and the last two dimensions corresponding to the dimensions of the $so(3)$ group. It is also obvious by clear expansion that the extension of $se(3)$ is itself a group, which keeps its important properties for further parallelization operating on items of the group.

3.2 $so(3)$ to Vector

The inverse conversion to the previous subsection is straightforward - we just have to select all of the relevant 3 items with a single SIMD instruction, and then stack them using PyTorch's built-in function `stack`.

[1] https://github.com/hristo-vrigazov/modern_robotics_torch.

3.3 Vector to $se(3)$

Similar logic can be used for the conversion from a vector to $se(3)$ (see Eq. (4)). The initialization shape is `(N, 4, 4)`, with the last row being zeros, i.e. there is no need for an SIMD instruction for the last row. We also need two more SIMD instructions - first, we write $[\omega]$ into the result matrix, taking advantage of PyTorch's vectorized slicing along the last axis, and then we slice the second half of the input vector and write it following the defined notation in Eq. (4) (see the function `VecTose3` in the application [8]). Similar to $so(3)$, is clear that the newly defined extension is itself a group.

3.4 MatrixExp3

Once the product has been computed, the next step is to compute the matrix exponentiation of the previously computed product (see Eq. (3) and 6). The input to this function (called `MatrixExp3` in the application [8]) is a member of the extension of the $so(3)$ group. While this implementation also closely follows Eq. (6), it is slightly more challenging compared to the previous subsections, because of the fact that some inputs may have a norm close to zero, which would lead to a division by zero error. However, PyTorch already supports vectorized implementation of computing the norm along a given dimension. We use this feature to create a mask tensor, which indicates whether a given input matrix has a zero norm, and combine it with PyTorch's broadcasting feature, to write to the outputs for these particular input matrices in a single instruction. After that, the implementation closely follows Eq. (6).

3.5 MatrixExp6

Once we have converted a vector into $se(3)$, we need to be able to further convert it into a the homogeneous transformation matrix $T \in SE(3)$ corresponding to the matrix exponential of a $se(3)$ (see Eq. (7)). We use the same technique described in the previous section to handle elements with zero norm, with the output tensor being of shape `(N, 4, 4)`, and follow Eq. (7) (see the function `MatrixExp6` in the application [8]).

3.6 Full Product Implementation

Finally, after we can compute the homogeneous transformation matrices $T \in SE(3)$ for every step of the computation, we can compute the final product (Eq. (3)) using simple `matmul` operation from PyTorch, which is vectorized along the last dimension.

4 Results

Once we have described the vectorized implementation of the forward kinematics, next we focus on testing and benchmarking the novel implementation.

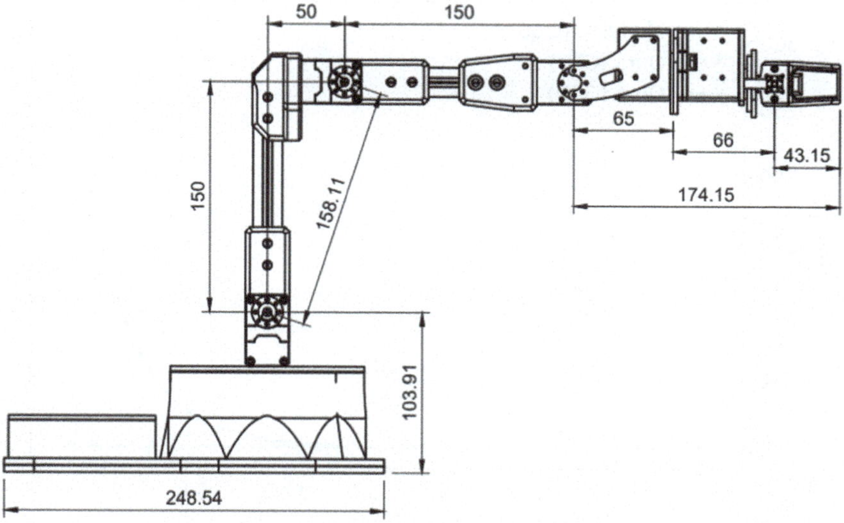

Fig. 1. The sizes in mm of the links of the robot. Source: [9]

We perform tests and benchmarks on a PincherX 150 Robot Arm, which has 5-DOF (see Fig. 1 to see the lengths of the links). We test our implementation by comparing the results to the original sequential approach, and we see that both implementations return the same results (see `test_core.py` in the application [8]). To compare the performance of the implementations, we measure the time it takes to compute the result for an exponentially increasing number of inputs (see Fig. 2). The PyTorch implementation is shown in two flavours: executed on a CPU or a GPU (in our case a Nvidia Geforce 1080Ti). We see that there is a dramatic performance improvement, with the GPU implementation being over 1000 times faster than the sequential implementation, while the CPU vectorized implementation is over 200 times faster, demonstrating the feasibility of the proposed approach (Fig. 3 and Table 1).

Table 1. Time taken for computation (seconds), columns - number of inputs

	1	32	1024	32768	524288
Baseline [6]	0.00237	0.01939	0.60888	19.48991	314.24987
Ours (CPU)	0.00294	0.00260	0.00670	0.09150	1.18750
Ours (GPU)	0.00618	0.00619	0.00663	0.02024	0.22349

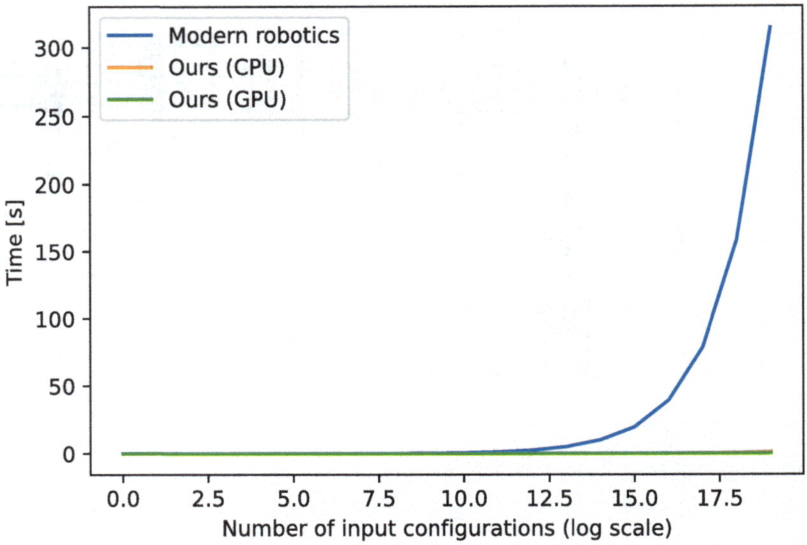

Fig. 2. Comparison of the time it takes to compute the result for exponentially increasing number of input configurations. "Modern robotics" refers to the baseline implementation [6]

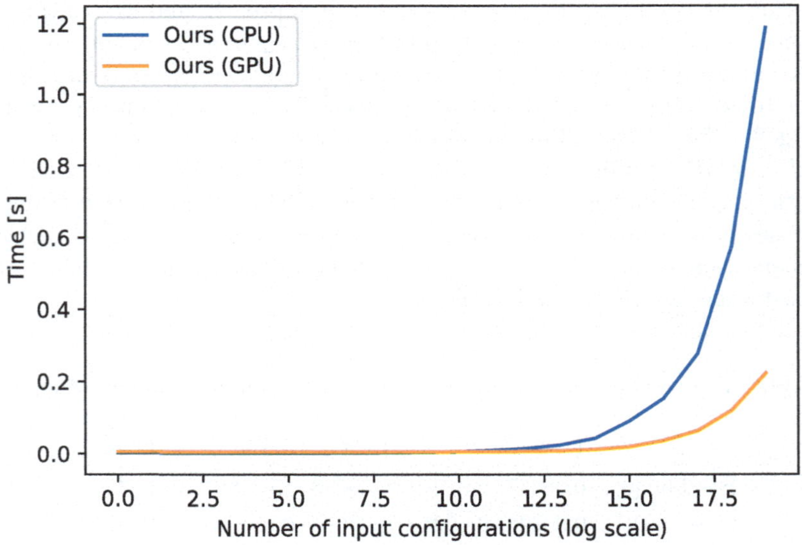

Fig. 3. Comparison of executing the novel implementation on CPU vs a GPU

5 Conclusion

Forward kinematics, the computation of the position and orientation of the robot's end-effector given its joint configurations, is a foundational problem with many available approaches. One of the popular approaches is the POE formula, which has many implementations available online. However, to our knowledge, none of the implementations support parallel computation for a large number of input configurations, or automatic differentiation. In this paper, we propose a novel implementation of the POE formula in PyTorch, which enables vectorizated, parallel computation both on CPU on GPU. We test that our solution returns the same results as the sequential implementation, and we also show that it leads to significant performance improvement (over 200 times on the CPU and over 1600 times on GPU) over the original Python implementation of one of the most popular implementations of POE formula.

Acknowledgement. The research presented in this paper was supported by the Bulgarian National Science Fund under the contract # KP-06-M47/5 - 27.11.2020.

References

1. Brockett, R.W.: Robotic manipulators and the product of exponentials formula. In: Fuhrmann, P.A. (ed.) Mathematical Theory of Networks and Systems, pp. 120–129. Springer, Heidelberg (1984). https://doi.org/10.1007/BFb0031048
2. OpenAI, Akkaya, I., et al.: Solving rubik's cube with a robot hand (2019)
3. Paszke, A., et al.: Pytorch: an imperative style, high-performance deep learning library. In: Advances in Neural Information Processing Systems 32, pp. 8026–8037 (2019)
4. Abadi, M.: TensorFlow: learning functions at scale. In: Proceedings of the 21st ACM SIGPLAN International Conference on Functional Programming (2016)
5. Zeng, A., et al.: Robotic pick-and-place of novel objects in clutter with multi-affordance grasping and cross-domain image matching. In: 2018 IEEE International Conference on Robotics and Automation (ICRA). IEEE (2018)
6. Baseline implementation repository. https://github.com/NxRLab/ModernRobotics. Accessed 6 Mar 2022
7. Harris, C.R., et al.: Array programming with NumPy. Nature **585**(7825), 357–362 (2020)
8. Code repository. https://github.com/hristo-vrigazov/modern_robotics_torch. Accessed 6 Feb 2022
9. Technical drawing of Pincher 150x. https://www.trossenrobotics.com/Shared/Xseries/PXA-150-MDrawing.pdf. Accessed 6 Feb 2022
10. Müller, A.: Review of the exponential and Cayley map on SE (3) as relevant for Lie group integration of the generalized Poisson equation and flexible multibody systems. Proc. Roy. Soc. A **477**(2253), 20210303 (2021)

Simultaneous Calibration and Stiffness Identification of Flexible Link Robots Using Lumped Parameter Model

Stefan Gadringer[1(✉)], Patrick Klement[1], Hubert Gattringer[1], Andreas Mueller[1], and Ronald Naderer[2]

[1] Institute of Robotics, Johannes Kepler University Linz, Linz, Austria
{stefan.gadringer,patrick.klement,hubert.gattringer,a.mueller}@jku.at
[2] FerRobotics Compliant Robot Technology GmbH, Linz, Austria
ronald.naderer@ferrobotics.at

Abstract. Elastic joints and/or links greatly influence the position and orientation accuracy of robots. Hence, standard geometric calibration might be insufficient for elastic robots and stiffness identification is also required for improved accuracy. This paper describes the simultaneous identification of geometric as well elasticity parameters of a 3-DoF flexible link robot using a lumped parameter model. These parameters were identified via the usage of 100 optimal poses and thoroughly verified with more than 9500 poses, which were distributed over the whole workspace and measured via a laser tracker. Experimental results for a standard geometric calibration as well as a combined geometric and stiffness identification approach are presented. The simultaneous identification approach leads to a maximum position error of 1.50 mm compared to 5.71 mm for geometric calibration only. The determined parameters as well as dependencies are shown and discussed for physical interpretability.

Keywords: Robot calibration · Lumped parameter model · Stiffness identification · Position accuracy · Orientation accuracy · Laser tracker

1 Introduction

Accurate position controlled robots become more and more important in the industry and hence robot calibration is essential. The standard approach is to consider robots as rigid and perform geometric calibration, which determines deviations in the geometric parameters of the kinematic model [9]. However, the consideration of elasticities could further improve the accuracy. Neubauer et al. [5] identified such non-geometric errors at first and then combined the elastic deflections with geometric calibration in the second step. Gear elasticities, modeled as torsional springs, were also considered in [2,6] for enhanced accuracy.

In contrast to conventional bulky robots, elasticity is much more significant in lightweight robots. Advantages of such robots are lower cost, higher payload-to-robot-weight ratio, reduced inertia and hence safer operation [4].

A. Müller and M. Brandstötter (Eds.): RAAD 2022, MMS 120, pp. 52–59, 2022.
https://doi.org/10.1007/978-3-031-04870-8_7

However, this requires proper elasticity modeling and control for accurate positioning. Like described in [8], the lumped parameter model is one of the main finite-dimensional models for elastic robots. It simply considers a spring-mass system and is similar to rigid robots with elastic joints. This allows to use calibration methods as described above even for elastic robots.

2 Lumped Parameter Model

Figure 1 shows an elastic 3-DOF-robot, modeled as lumped parameter model (LPM), with arm coordinates $\mathbf{q}_A = (q_{A1}, q_{A2}, q_{A3})^\top$ and corresponding body-fixed frames $\mathcal{F}_1$ to $\mathcal{F}_3$ at its rigid arms. For collision avoidance, the respective limits are $-75° \leq q_{A1} \leq 75°$, $-20° \leq q_{A2} \leq 75°$ and $-90° \leq q_{A3} \leq 90°$. The inertial frame $\mathcal{F}_I$ serves as general reference point. For the robot calibration, a Leica laser tracker with reference frame $\mathcal{F}_L$ measures the position of three Super CatEye reflectors, mounted on a calibration tool with end effector frame $\mathcal{F}_E$.

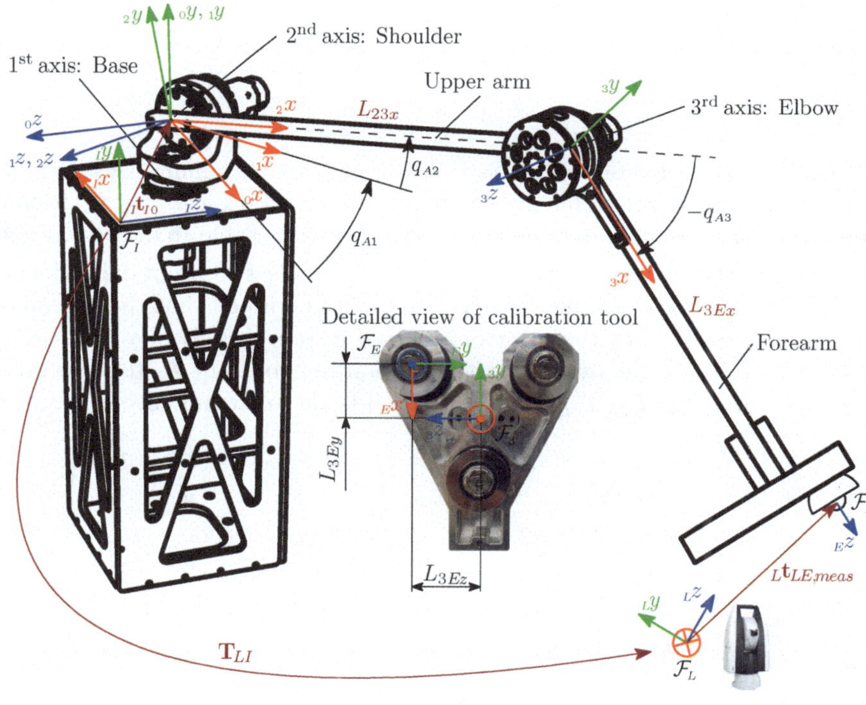

Fig. 1. Elastic robot with defined frames

2.1 Kinematic Model

The kinematics is modelled via homogeneous transformation matrices using the same notation as in [1]. Similarly, rotation matrices $\mathbf{R}(\alpha, \beta, \gamma) \in SO(3)$ utilize

Table 1. Forward kinematics with geometric parameters and arm coordinates

(a) Nominal transformations

$\mathbf{T}$	$\mathbf{p}_{geo}$		$\mathbf{q}_A$
	$\overline{\varphi}_{Cardan}$ in rad	$\overline{\mathbf{t}}$ in mm	
$\mathbf{T}_{I0}$	0	$L_{I0x} = 167.5$	
	π	$L_{I0y} = 120.0$	
	0	$L_{I0z} = 167.5$	
$\mathbf{T}_{01}$	0	0	
	0	0	q_{A1}
	0	0	
$\mathbf{T}_{12}$	0	0	
	0	0	
	0	0	q_{A2}
$\mathbf{T}_{23}$	0	$L_{23x} = 959.5$	
	0	0	
	0	0	q_{A3}
$\mathbf{T}_{3E}$	$-\pi/2$	$L_{3Ex} = 1040.4$	
	$\pi/2$	$L_{3Ey} = 17.0$	
	0	$L_{3Ez} = 39.0$	

(b) Geometric error parameters of the stiff robot

$\hat{\mathbf{T}}$	$\mathbf{p}_S$	
	$\mathbf{p}_L$ in m	$\mathbf{p}_R$ in rad
$\hat{\mathbf{T}}_{I0}$	$p_{L\,I0x}$	$p_{R\,I0x}$
	$p_{L\,I0y}$	$p_{R\,I0y}$
	$p_{L\,I0z}$	$p_{R\,I0z}$
$\hat{\mathbf{T}}_{01}$		
$\hat{\mathbf{T}}_{12}$	$p_{L\,12x}$	$p_{R\,12x}$
	$p_{L\,12y}$	$p_{R\,12y}$
	$p_{L\,12z}$	
$\hat{\mathbf{T}}_{23}$	$p_{L\,23x}$	$p_{R\,23x}$
	$p_{L\,23y}$	$p_{R\,23y}$
	$p_{L\,23z}$	
$\hat{\mathbf{T}}_{3E}$	$p_{L\,3Ex}$	$p_{R\,3Ex}$
	$p_{L\,3Ey}$	$p_{R\,3Ey}$
	$p_{L\,3Ez}$	$p_{R\,3Ez}$

Cardan angles as parameters for $x-y-z$ rotations. Table 1a summarizes the nominal kinematics of the zero reference configuration (orientation $\overline{\varphi}_{Cardan}$, position $\overline{\mathbf{t}}$) of the robot using nominal parameters $\mathbf{p}_{geo}$. By contrast, Table 1b introduces geometric error parameters $\mathbf{p}_S = \left(\mathbf{p}_L^\top, \mathbf{p}_R^\top\right)^\top \in \mathbb{R}^{n_S=22}$ of the stiff robot due to length deviations $\mathbf{p}_L$ and axes misalignment $\mathbf{p}_R$. It should be noted that the kinematic model can be described by the usage of arm coordinates $\mathbf{q}_A$ only.

Like shown in [1], the overall transformation matrix from $\mathcal{F}_E$ to $\mathcal{F}_I$ is given by $\mathbf{T}_{IE} = \mathbf{T}_{I0}\,\mathbf{T}_{01}\,\mathbf{T}_{12}\,\mathbf{T}_{23}\,\mathbf{T}_{3E}$, which in detail leads to an end effector position

$$
\begin{aligned}
{}_I\mathbf{t}_{IE} = L_{23x}\begin{pmatrix} -\cos(q_{A1})\cos(q_{A2}) \\ \sin(q_{A2}) \\ \sin(q_{A1})\cos(q_{A2}) \end{pmatrix} + L_{3Ex}\begin{pmatrix} -\cos(q_{A1})\cos(q_{A2}+q_{A3}) \\ \sin(q_{A2}+q_{A3}) \\ \sin(q_{A1})\cos(q_{A2}+q_{A3}) \end{pmatrix} \\
+ L_{3Ey}\begin{pmatrix} \cos(q_{A1})\sin(q_{A2}+q_{A3}) \\ \cos(q_{A2}+q_{A3}) \\ -\sin(q_{A1})\sin(q_{A2}+q_{A3}) \end{pmatrix} - L_{3Ez}\begin{pmatrix} \sin(q_{A1}) \\ 0 \\ \cos(q_{A1}) \end{pmatrix} + \begin{pmatrix} L_{I0x} \\ L_{I0y} \\ L_{I0z} \end{pmatrix}.
\end{aligned} \tag{1}
$$

Considering $n_S = 22$ geometric error parameters, as shown in Table 1b, the transformation matrix from $\mathcal{F}_E$ to $\mathcal{F}_I$ changes to

$$
\mathbf{T}_{IE} = \hat{\mathbf{T}}_{I0}\,\hat{\mathbf{T}}_{01}\,\hat{\mathbf{T}}_{12}\,\hat{\mathbf{T}}_{23}\,\hat{\mathbf{T}}_{3E} = \begin{bmatrix} \hat{\mathbf{R}}_{IE} & {}_I\hat{\mathbf{t}}_{IE} \\ 0 & 1 \end{bmatrix}, \tag{2}
$$

with position ${}_I\hat{\mathbf{t}}_{IE}$, represented in $\mathcal{F}_I$ and rotation matrix $\hat{\mathbf{R}}_{IE}$, describing the orientation of $\mathcal{F}_E$ relative to $\mathcal{F}_I$.

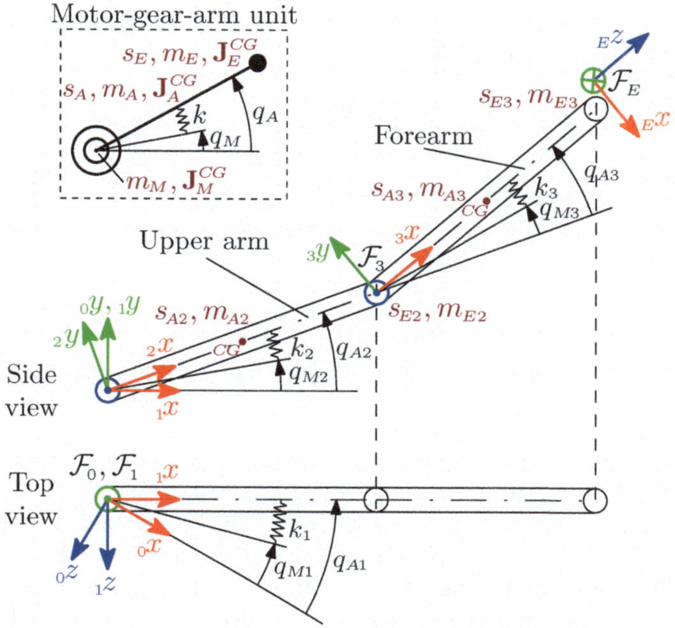

Fig. 2. Lumped parameter model of elastic robot [10]

2.2 Dynamic Model

Figure 2 shows the lumped element model of the robot, consisting of three motor-gear-arm units. As mentioned in [7], the motor dynamics is described by

$$\mathbf{M}_M \ddot{\mathbf{q}}_M + \boldsymbol{\tau}_F(\dot{\mathbf{q}}_M) + \mathbf{K}\left(\mathbf{q}_M - \mathbf{q}_A\right) = \boldsymbol{\tau}_M, \tag{3}$$

with motor inertia $\mathbf{M}_M$, friction torques $\boldsymbol{\tau}_F(\dot{\mathbf{q}}_M)$ and motor torques $\boldsymbol{\tau}_M$. The motor and arm coordinates are coupled with the diagonal stiffness matrix $\mathbf{K}$, which comprises the lumped stiffness of the individual joints. The dynamics of the rigid arms (neglecting external disturbances) is given by

$$\mathbf{M}(\mathbf{q}_A)\,\ddot{\mathbf{q}}_A + \mathbf{C}(\mathbf{q}_A, \dot{\mathbf{q}}_A)\dot{\mathbf{q}}_A + \mathbf{G}(\mathbf{q}_A) - \mathbf{K}\left(\mathbf{q}_M - \mathbf{q}_A\right) = \mathbf{0}, \tag{4}$$

with inertia matrix $\mathbf{M}(\mathbf{q}_A)$, centrifugal and Coriolis terms $\mathbf{C}(\mathbf{q}_A, \dot{\mathbf{q}}_A)\dot{\mathbf{q}}_A$ as well as generalized gravity forces $\mathbf{G}(\mathbf{q}_A)$. Rewriting (4) leads to the motor positions

$$\mathbf{q}_M = \mathbf{q}_A + \mathbf{K}^{-1}\left(\mathbf{M}(\mathbf{q}_A)\,\ddot{\mathbf{q}}_A + \mathbf{C}(\mathbf{q}_A, \dot{\mathbf{q}}_A)\dot{\mathbf{q}}_A + \mathbf{G}(\mathbf{q}_A)\right) \tag{5}$$

or for the static case

$$\mathbf{q}_{M,0} = \mathbf{q}_{A,0} + \mathbf{K}^{-1}\mathbf{G}(\mathbf{q}_{A,0}). \tag{6}$$

Considering the lumped parameter model of Fig. 2 as well as additional motor and gear error parameters $\mathbf{p}_M = \left(\mathbf{p}_Z^\top, \mathbf{p}_K^\top\right)^\top \in \mathbb{R}^{n_M=6}$, consisting of zero motor position errors $\mathbf{p}_Z = (p_{qM1}, p_{qM2}, p_{qM3})^\top$ and gear elasticity errors $\mathbf{p}_K = (p_{K1}, p_{K2}, p_{K3})^\top$, then (6) changes to

$$\mathbf{q}_{M,0} = \mathbf{q}_{A,0} - \mathbf{p}_Z + \hat{\mathbf{K}}^{-1}\mathbf{G}(\mathbf{q}_{A,0}) = \begin{pmatrix} q_{A1,0} - p_{qM1} \\ q_{A2,0} - p_{qM2} + \frac{a_1 g \cos(q_{A2,0}+q_{A3,0})+a_2 g \cos(q_{A2,0})}{k_2(1+p_{K2})} \\ q_{A3,0} - p_{qM3} + \frac{a_1 g \cos(q_{A2,0}+q_{A3,0})}{k_3(1+p_{K3})} \end{pmatrix} \tag{7}$$

with

$$\begin{aligned} a_1 &= m_{E3}\, s_{E3} + m_{A3}\, s_{A3} \\ a_2 &= (m_{E2} + m_{E3} + m_{A3}) \underbrace{(L_{23x} + p_{L\,23x})}_{s_{E2}} + m_{A2}\, s_{A2}. \end{aligned} \tag{8}$$

The corresponding parameters are $m_{A2} = 2.25\,\text{kg}$, $m_{E2} = 6.30\,\text{kg}$ (mass of motor 3 is included), $m_{A3} = 1.40\,\text{kg}$, $m_{E3} = 1.73\,\text{kg}$, $s_{A2} = 475\,\text{mm}$, $s_{E2} = L_{23x} + p_{L\,23x}$, $s_{A3} = 505\,\text{mm}$, $s_{E3} = 894\,\text{mm}$, $k_2 = 13\,500\,\text{N m/rad}$, $k_3 = 7500\,\text{N m/rad}$.

Equation (8) allows to calculate $\mathbf{q}_{M,0}$ from given arm coordinates. However, for robot calibration the inverse relationship $\mathbf{q}_{A,0} = \mathrm{f}(\mathbf{q}_{M,0}, \mathbf{p}_M)$ is also required. An accurate solution could be derived by a detailed case-by-case analysis. Alternatively, it is easier to get an approximation of $\mathbf{q}_{A,0}$ via replacing it in the gravitational forces with $\mathbf{q}_{M,0} + \mathbf{p}_Z$, i.e. $\mathbf{G}(\mathbf{q}_{A,0}) \approx \mathbf{G}(\mathbf{q}_{M,0} + \mathbf{p}_Z)$, which leads to

$$\hat{\mathbf{q}}_{A,0} = \mathbf{q}_{M,0} + \mathbf{p}_Z - \hat{\mathbf{K}}^{-1}\mathbf{G}(\mathbf{q}_{M,0} + \mathbf{p}_Z) = \begin{pmatrix} q_{M1,0} + p_{qM1} \\ q_{M2,0} + p_{qM2} - \frac{a_1 g \cos(q_{M2,0}+q_{M3,0}+p_{qM2}+p_{qM3})+a_2 g \cos(q_{M2,0}+p_{qM2})}{k_2(1+p_{K2})} \\ q_{M3,0} + p_{qM3} - \frac{a_1 g \cos(q_{M2,0}+q_{M3,0}+p_{qM2}+p_{qM3})}{k_3(1+p_{K3})} \end{pmatrix}. \tag{9}$$

Furthermore, the usage of this simple approximation inside the cosine allows the derivation of the even better approximation

$$\hat{\hat{\mathbf{q}}}_{A,0} = \mathbf{q}_{M,0} + \mathbf{p}_Z - \hat{\mathbf{K}}^{-1}\mathbf{G}(\hat{\mathbf{q}}_{A,0}) = \begin{pmatrix} q_{M1,0} + p_{qM1} \\ q_{M2,0} + p_{qM2} - \frac{a_1 g \cos(\hat{q}_{A2,0}+\hat{q}_{A3,0})+a_2 g \cos(\hat{q}_{A2,0})}{k_2(1+p_{K2})} \\ q_{M3,0} + p_{qM3} - \frac{a_1 g \cos(\hat{q}_{A2,0}+\hat{q}_{A3,0})}{k_3(1+p_{K3})} \end{pmatrix}. \tag{10}$$

This approximation is sufficient for robot calibration because the resulting error due to composition of (8) and (10) is negligible. Alternatively, a simpler but less accurate model can be derived by the negligence of $\mathbf{p}_Z$ in the gravitational forces of (9), i.e. $\mathbf{G}(\mathbf{q}_{A,0}) \approx \mathbf{G}(\mathbf{q}_{M,0})$. The consequence will be apparent in Sect. 4.

3 Parameter Identification

The model used for robot calibration combines the kinematic and the static gravity model, leading to $n_e = 28$ error parameters $\mathbf{p}_e = \left(\mathbf{p}_S^\top, \mathbf{p}_M^\top\right)^\top \in \mathbb{R}^{n_e}$

in total. The combination of (2) and (10) allows the calculation of the end effector pose using measured motor positions. The corresponding end effector error $\Delta\mathbf{z} = \left(\Delta_I\mathbf{t}_{IE}^\top, \Delta\boldsymbol{\varphi}_{IE}^\top\right)^\top$ consists of position error $\Delta_I\mathbf{t}_{IE} \in \mathbb{R}^3$ and orientation error $\Delta\boldsymbol{\varphi}_{IE} \in \mathbb{R}^3$ using the Axis–Angle representation. As described in [1], $\Delta\mathbf{z} = \mathbf{0}$ leads to a non-linear optimization problem with six independent equations. The error parameters $\mathbf{p}_e$ can be found via a Taylor series of $\Delta\mathbf{z} = \Delta\mathbf{z}(\mathbf{z}_{meas}, \mathbf{q}_M, \mathbf{p}_e)$ at the start parameters $\mathbf{p}_e^{(0)}$, i.e. $\Delta\mathbf{z}(\mathbf{z}_{meas,j}, \mathbf{q}_j, \mathbf{p}_e^{(0)}) + \boldsymbol{\Theta}_j\,\Delta\mathbf{p}_e = \mathbf{0}$ with $\boldsymbol{\Theta}_j = \partial\Delta\mathbf{z}/\partial\mathbf{p}_e|_{\mathbf{p}_e^{(0)}}$. For the evaluation of $j = 1\ldots m$ measurements, this leads to

$$\underbrace{\begin{pmatrix} \Delta\mathbf{z}(\mathbf{z}_{meas,1}, \mathbf{q}_{A,1}, \mathbf{p}_e^{(0)}) \\ \vdots \\ \Delta\mathbf{z}(\mathbf{z}_{meas,m}, \mathbf{q}_{A,m}, \mathbf{p}_e^{(0)}) \end{pmatrix}}_{\mathbf{Q}} + \underbrace{\begin{pmatrix} \boldsymbol{\Theta}_1 \\ \vdots \\ \boldsymbol{\Theta}_m \end{pmatrix}}_{\boldsymbol{\Theta}} \Delta\mathbf{p}_e = \mathbf{0}, \tag{11}$$

with regressor matrix $\boldsymbol{\Theta} \in \mathbb{R}^{n,n_e}$ considering $n = 6m$ rows. However, for the determination of $\overline{n}_e$ identifiable parameters or rather base parameters $\overline{\mathbf{p}}_e \in \mathbb{R}^{\overline{n}_e}$, a QR decomposition of $\boldsymbol{\Theta}$ is necessary. Following [3] yields

$$\overline{\mathbf{Q}} + \overline{\boldsymbol{\Theta}}\,\Delta\overline{\mathbf{p}}_e = \mathbf{0} \qquad \text{with} \qquad \Delta\overline{\mathbf{p}}_e = -[\overline{\boldsymbol{\Theta}}^\top\overline{\boldsymbol{\Theta}}]^{-1}\overline{\boldsymbol{\Theta}}^\top\overline{\mathbf{Q}} \tag{12}$$

as solution for the parameters deviations $\Delta\overline{\mathbf{p}}_e \in \mathbb{R}^{\overline{n}_e}$ via well known method of least squares and usage of regressor matrix $\overline{\boldsymbol{\Theta}} \in \mathbb{R}^{6m,\overline{n}_e}$ as well as $\overline{\mathbf{Q}} \in \mathbb{R}^{6m,1}$. The first solution of the error parameters is determined by $\overline{\mathbf{p}}_e^{(1)} = \overline{\mathbf{p}}_e^{(0)} + \Delta\overline{\mathbf{p}}_e$. For a more accurate solution of the non-linear model, more iterations are necessary.

4 Results

The robot base parameters $\mathbf{p}_{base} = \overline{\mathbf{p}}_e^{(20)}$ are determined using 100 optimal poses. Depending on the chosen robot model $\overline{n}_e = 22$ (LPM), $\overline{n}_e = 20$ (LPM with $\mathbf{G}(\mathbf{q}_{A,0}) \approx \mathbf{G}(\mathbf{q}_{M,0})$) or just $\overline{n}_e = 18$ (rigid robot model) parameters are identifiable. Table 2 provides a statistical error analysis of 9501 verification poses.

Table 2. Position and orientation error of 9501 verification poses

Error in mm or mrad	Uncalibrated rigid model	Uncalibrated LPM	Calibrated $\overline{\mathbf{p}}_e \in \mathbb{R}^{18}$	Calibrated $\overline{\mathbf{p}}_e \in \mathbb{R}^{20}$	Calibrated $\overline{\mathbf{p}}_e \in \mathbb{R}^{22}$
Max $\lVert\Delta_I\mathbf{t}_{IE}\rVert$	68.685	62.219	5.713	1.872	1.503
Mean $\lVert\Delta_I\mathbf{t}_{IE}\rVert$	56.860	46.018	2.584	0.706	0.577
$Q_{0.9}$ $\lVert\Delta_I\mathbf{t}_{IE}\rVert$	66.589	60.816	4.106	1.227	0.919
Max $\lVert\Delta\boldsymbol{\varphi}_{IE}\rVert$	68.597	65.954	6.671	2.354	2.320
Mean $\lVert\Delta\boldsymbol{\varphi}_{IE}\rVert$	58.676	53.956	2.869	0.716	0.649
$Q_{0.9}$ $\lVert\Delta\boldsymbol{\varphi}_{IE}\rVert$	63.293	59.257	5.077	1.042	0.915

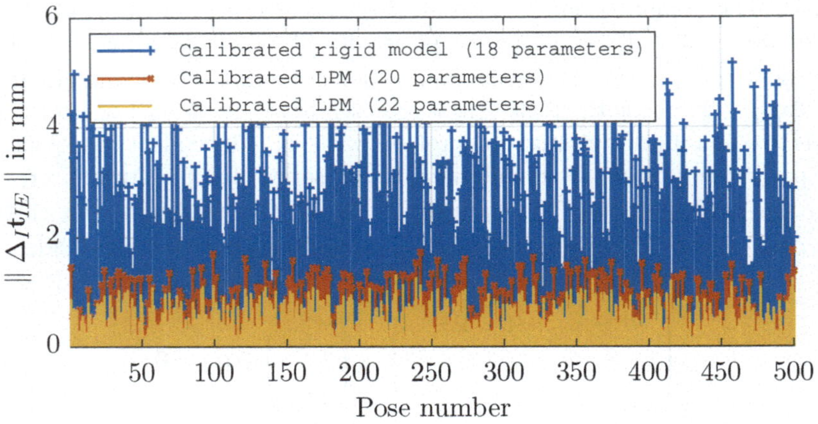

Fig. 3. Position error of 500 random verification poses

Table 3. Identifiable error parameters of lumped parameter models

(a) LPM with $\mathbf{G}(\mathbf{q}_{A,0}) \approx \mathbf{G}(\mathbf{q}_{M,0})$

#	Identifiable Parameters	#	
1	$p_{L\,I0x}$	12	$p_{R\,I0x}$
2	$p_{L\,I0y} + p_{L\,12y}$	13	$p_{R\,I0z}$
3	$p_{L\,I0z}$	14	$p_{R\,12x}$
4	$p_{L\,12x}$	15	$p_{R\,23x}$
5	$p_{L\,12z} + p_{L\,23z} + p_{L\,3Ez}$	16	$p_{R\,23y}$
6	$p_{L\,23x}$	17	$p_{R\,3Ex}$
7	$p_{L\,23y} + L_{23x}(p_{R\,3Ey} - p_{qM3})$	18	$p_{R\,3Ez}$
8	$p_{L\,3Ex} - L_{3Ey}\,p_{R\,3Ey}$	19	p_{K2}
9	$p_{L\,3Ey} + L_{3Ex}\,p_{R\,3Ey}$	20	p_{K3}
10	$p_{qM1} + p_{R\,I0y} + p_{R\,12y}$		
11	$p_{qM2} + p_{qM3} - p_{R\,3Ey}$		

(b) Determined LPM parameters $\overline{\mathbf{p}}_e \in \mathbb{R}^{20}$

#	$\mathbf{p}_{base}$	SI	#	$\mathbf{p}_{base}$	SI
1	-7.434×10^{-4}	m	12	3.327×10^{-4}	rad
2	-1.586×10^{-3}	m	13	-2.577×10^{-4}	rad
3	-3.119×10^{-4}	m	14	8.912×10^{-4}	rad
4	1.205×10^{-3}	m	15	-5.136×10^{-2}	rad
5	-2.917×10^{-3}	m	16	-2.836×10^{-2}	rad
6	3.181×10^{-3}	m	17	-3.160×10^{-3}	rad
7	1.197×10^{-2}	m	18	1.267×10^{-2}	rad
8	-1.121×10^{-3}	m	19	-1.775×10^{-1}	
9	5.944×10^{-3}	m	20	-5.078×10^{-1}	
10	1.281×10^{-3}	rad			
11	-2.015×10^{-2}	rad			

(c) LPM with $\mathbf{G}(\mathbf{q}_{A,0}) \approx \mathbf{G}(\mathbf{q}_{M,0} + \mathbf{p}_Z)$

#	Identifiable Parameters	#	
1	$p_{L\,I0x}$	12	p_{qM3}
2	$p_{L\,I0y} + p_{L\,12y}$	13	$p_{R\,I0x}$
3	$p_{L\,I0z}$	14	$p_{R\,I0z}$
4	$p_{L\,12x}$	15	$p_{R\,12x}$
5	$p_{L\,12z} + p_{L\,23z} + p_{L\,3Ez}$	16	$p_{R\,23x}$
6	$p_{L\,23x}$	17	$p_{R\,23y}$
7	$p_{L\,23y}$	18	$p_{R\,3Ex}$
8	$p_{L\,3Ex}$	19	$p_{R\,3Ey}$
9	$p_{L\,3Ey}$	20	$p_{R\,3Ez}$
10	$p_{qM1} + p_{R\,I0y} + p_{R\,12y}$	21	p_{K2}
11	p_{qM2}	22	p_{K3}

(d) Determined LPM parameters $\overline{\mathbf{p}}_e \in \mathbb{R}^{22}$

#	$\mathbf{p}_{base}$	SI	#	$\mathbf{p}_{base}$	SI
1	-7.273×10^{-4}	m	12	-1.71×10^{-1}	rad
2	-1.468×10^{-3}	m	13	2.832×10^{-4}	rad
3	-3.538×10^{-4}	m	14	-2.804×10^{-4}	rad
4	1.145×10^{-3}	m	15	9.296×10^{-4}	rad
5	-2.970×10^{-3}	m	16	-5.466×10^{-2}	rad
6	-4.616×10^{-3}	m	17	-2.120×10^{-2}	rad
7	-1.170×10^{-1}	m	18	-3.144×10^{-3}	rad
8	-1.972×10^{-3}	m	19	-3.686×10^{-2}	rad
9	4.430×10^{-2}	m	20	1.253×10^{-2}	rad
10	1.274×10^{-3}	rad	21	-1.075×10^{-1}	
11	1.133×10^{-1}	rad	22	-5.215×10^{-1}	

The best results can be achieved with the calibrated LPMs. Compared to the uncalibrated models, also the geometrically calibrated model vastly improves both position and orientation accuracy. Figure 3 shows the position error of 500 randomly selected verification poses of the calibrated models. Table 3 summarizes the corresponding determined base parameters $\mathbf{p}_{base}$ considering full pose measurement. Table 3b shows realistic parameter range but this is not the case in Table 3d. The parameters related to the gravitational direction (e.g. $p_{L\,23_y}$) seem to be independently identifiable but have nonsensical magnitude.

5 Conclusion

We proposed a parameter identification procedure of a flexible link robot using a simple lumped parameter model. Depending on the model complexity, up to $\overline{n}_e = 22$ parameters are identifiable and in the whole workspace, a maximum position error of 1.50 mm is achievable. However, the physical parameter interpretability is adversely affected by the error parameters in the gravitational forces. If this is undesired but higher accuracy is required, then error compensation via neural network would be a promising alternative. This will be tested in future work.

Acknowledgments. Supported by the "LCM-K2 Center for Symbiotic Mechatronics" within the framework of the Austrian COMET-K2 program.

References

1. Gadringer, S., Gattringer, H., Müller, A., Naderer, R.: Robot calibration combining kinematic model and neural network for enhanced positioning and orientation accuracy. IFAC-PapersOnLine **53**(2), 8432–8437 (2020)
2. Joubair, A., Bonev, I.A.: Non-kinematic calibration of a six-axis serial robot using planar constraints. Precis. Eng. **40**, 325–333 (2015)
3. Khalil, W., Gautier, M.: Calculation of the identifiable parameters for robots calibration. IFAC Proc. Vol. **24**(3), 687–691 (1991)
4. Kiang, C.T., Spowage, A., Yoong, C.K.: Review of control and sensor system of flexible manipulator. J. Intell. Robot. Syst. **77**(1), 187–213 (2015)
5. Neubauer, M., Gattringer, H., Müller, A., Steinhauser, A., Höbarth, W.: A two-stage calibration method for industrial robots with joint and drive flexibilities. Mech. Sci. **6**(2), 191–201 (2015)
6. Nubiola, A., Slamani, M., Joubair, A., Bonev, I.A.: Comparison of two calibration methods for a small industrial robot based on an optical CMM and a laser tracker. Robotica **32**(3), 447–466 (2014)
7. Pucher, F., Gattringer, H., Müller, A.: Collision detection for flexible link robots using accelerometers. IFAC-PapersOnLine **52**(16), 514–519 (2019)
8. Rahimi, H., Nazemizadeh, M.: Dynamic analysis and intelligent control techniques for flexible manipulators: a review. Adv. Robot. **28**(2), 63–76 (2014)
9. Roth, Z., Mooring, B., Ravani, B.: An overview of robot calibration. IEEE J. Robot. Autom. **3**(5), 377–385 (1987)
10. Staufer, P., Gattringer, H.: State estimation on flexible robots using accelerometers and angular rate sensors. Mechatronics **22**(8), 1043–1049 (2012)

Analysis of the Singularities Influence on the Forward Kinematics Solution and the Geometry of the Workspace of the Gough-Stewart Platform

D. I. Malyshev(✉), L. A. Rybak, A. S. Pisarenko, and V. V. Cherkasov

Belgorod State Technological University n.a. V.G. Shukhov, Kostyukova 46, 308012 Belgorod, Russia

rl_bgtu@intbel.ru

Abstract. The article considers the problem of analyzing the influence of singularities on the forward kinematics of the 3/6 Gough - Stewart platform. A numerical algorithm for solving the forward kinematics of platform based on deterministic optimization methods and interval analysis has been developed. An analysis of the change in the number of forward kinematics solutions near the singularity is performed. An algorithm has been synthesized that makes it possible to determine a singularity-free workspace for given ranges of change in platform orientation angles. An analysis of the change in the volume of the workspace depending on the range of change in the angles of the platform orientation was carried out. The simulation was performed using parallel computing and the procedure for exporting 3D models of the platform positions and workspace to the STL model. The results of mathematical modeling are presented.

Keywords: Gough-Stewart platform · Forward kinematics · Workspace · Singularity

1 Introduction

Robots based on parallel mechanisms have a number of advantages over serial mechanisms [1, 2]. The disadvantage of such structures is the appearance of singularities. The robot loses its degree of freedom or loses controllability in such condition. In this regard, the task of analyzing the workspace arises, taking into account the condition for excluding singularities. Singularity are of two types: boundary singularities and internal singularities [3]. One of the methods for singularity analysis is based on the determinant of the Jacobi matrix obtained as a result of differentiation of the dependence functions of input and output coordinates. C. Gosselin and J. Angeles in their work [4] described the conditions for the occurrence of singularities when the Jacobi matrix of input or output coordinates is degenerate. The complexity of calculating the Jacobi matrix increases significantly when solving problems with more than three degrees of freedom, due to the complexity of finding the determinant of the matrix. C. Gosselin and J. Wang [5]

A. Müller and M. Brandstötter (Eds.): RAAD 2022, MMS 120, pp. 60–67, 2022.
https://doi.org/10.1007/978-3-031-04870-8_8

used numerical methods to solve the inverse problem of kinematics and find all singularities of the mechanism inside the workspace. K.H. Hunt in his work [6] for the first time pointed out a singularity that occurs when all lines passing through the axis of rectilinear drives cross a common line. Then E. F. Fichter [7] showed that a singularity is achieved when the platform rotates about an axis orthogonal to the base plane through an angle of 90°. J. P. Merlet in his works [8, 9] proposed an approach for determining singular positions based on the use of Grassmannian geometry. J. P. Merlet noted that a singularity for the Gough -Stewart platform occurs when the axis of any of the rods hits the plane of the movable platform. Using Grassmannian geometry, it is easy to determine the geometric conditions that lead to a particular configuration, but it is difficult to express them analytically. In [10] M. Slavutin gives an approach for determining the singularities of various configurations of platforms 6/6, the basis of the method includes the application of the Kennedy theorem. Investigation of existing studies has shown a significant backlog in the field of analysis of singularities for the Gough-Stewart platform. However, analytical methods for determining singularities are difficult to implement; high-performance computing systems are required. The existing works do not consider the task of investigating the workspace of the Gough-Stewart platform for various ranges of change in orientation angles, considering singularities based on numerical methods. In the framework of this work, numerical algorithms for solving the forward kinematics and analysis of the workspace are proposed based on deterministic methods of global optimization.

2 Mathematical Model of the Platform

The Gough-Stewart platform (Fig. 1) has 6 kinematic chains that connect the fixed base to the movable platform by means of linear drives (L_i) and universal joints at the bases (A_i и B_i). The mobile platform is an equilateral triangle.

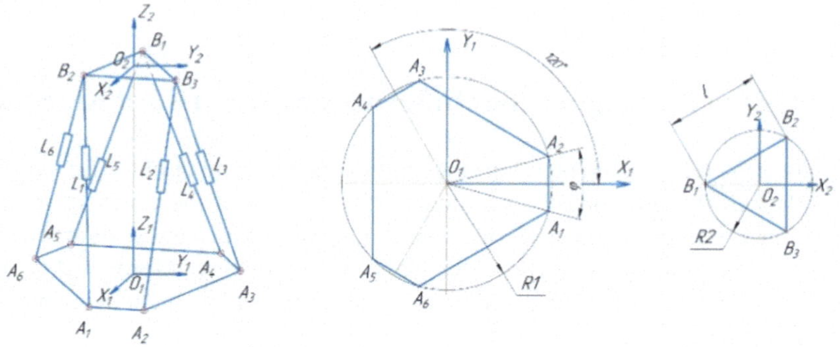

Fig. 1. The structure of the Gough-Stewart platform.

The inverse kinematics of the Gough - Stuart platform for the input coordinate L_1 can be determined

$$L_1 = \left((A_{1x} - B_{2x})^2 + \left(A_{1y} - B_{2_y}\right)^2 + \left(A_{1z} - B_{2_z}\right)^2\right)^{1/2} \quad (1)$$

where B_{2x}, B_{2y}, B_{2z} is the point coordinate B_2 in the $X_1Y_1Z_1$ fixed platform coordinate system. The values of other input coordinates are determined similarly. B_i defined as.

$$\boldsymbol{B}_i = \left(B_{i_x}\ B_{i_y}\ B_{i_z}\ 1 \right)^T = \boldsymbol{M} \cdot \left(B_{i_x}^{(2)}\ B_{i_y}^{(2)}\ B_{i_z}^{(2)}\ 1 \right)^T, \tag{2}$$

where $B_{ix}^{(2)}, B_{iy}^{(2)}, B_{iz}^{(2)}$ are the coordinates of the points B_i in the $X_2Y_2Z_2$ coordinate system, $\boldsymbol{M}$ is the transformation matrix from $X_2Y_2Z_2$ to $X_1Y_1Z_1$. We use Euler angles (α, β, γ) to compile the matrix $\boldsymbol{M}$. Let us introduce the following notation: $c_\alpha = \cos(\alpha)$, $s_\alpha = sin(\alpha)$, $c_\beta = \cos(\beta)$, $s_\beta = sin(\beta)$, $c_\gamma = \cos(\gamma)$, $s_\gamma = sin(\gamma)$. In this case

$$M = \begin{pmatrix} i_x\ j_x\ k_x\ O_{2_x} \\ i_y\ j_y\ k_y\ O_{2_y} \\ i_z\ j_z\ k_z\ O_{2_z} \\ 0\ 0\ 0\ 1 \end{pmatrix} = \begin{pmatrix} c_\alpha c_\gamma - c_\beta s_\alpha s_\gamma & -c_\gamma s_\alpha - c_\alpha c_\beta s_\gamma & s_\beta s_\gamma & O_{2_x} \\ c_\beta c_\gamma s_\alpha + c_\alpha s_\gamma & c_\alpha c_\beta c_\gamma - s_\alpha s_\gamma & -c_\gamma s_\gamma & O_{2_y} \\ s_\alpha s_\beta & c_\alpha s_\beta & c_\beta & O_{2_z} \\ 0 & 0 & 0 & 1 \end{pmatrix} \tag{3}$$

where $O_{2_x}, O_{2_y}, O_{2_z}$ are the coordinates of the O_2 moving platform center in the coordinate system $X_1Y_1Z_1$.

J. P. Merlet in his works [8, 9] noted that a singularity for the Gough -Stewart platform occurs when the axis of any of the drive rods hits the plane of the movable platform. It follows from this that a singularity arises when at least one of the points A_i belongs to the plane of the moving platform. The condition for the occurrence of a singularity can be written in the following form

$$aA_{ix} + bA_{iy} + cA_{iz} + d = 0, \tag{4}$$

where a, b, c, d are the coefficients of the equation of the plane of the moving platform:

$$a = k_x, b = k_y, c = k_z, d = k_x O_{2_x} + k_y O_{2_y} + k_z O_{2_z},$$

3 Analysis of the Singularities Influence on the Forward Kinematics

Let's analyze the relationship between singularities and the number of the forward kinematics solutions. Determining the position and orientation of the center of the moving platform O_2 is the main goal of solving a forward kinematics. Its coordinates $O_{2_x}, O_{2_y}, O_{2_z}$ and orientation angles of the plane of the moving platform α, β, γ (Euler angles) are determined using the coordinates of the vertices of B_i the moving platform, which is a regular triangle

$$O_{2_x} = \frac{\sum B_{ix}}{3}, O_{2_y} = \frac{\sum B_{iy}}{3}, O_{2_z} = \frac{\sum B_{iz}}{3} \tag{5}$$

Euler angles, given (3), can be defined as

$$\beta = \arccos(k_z), \alpha = \arccos\left(\frac{j_z}{\sin(\beta)}\right), \gamma = \arccos\left(\frac{-k_y}{\sin(\beta)}\right), \tag{6}$$

The values j_z k_y, k_z we define using the coordinates of the points B_i. Since it B_1 lies on the O_2X_2 axis (Fig. 1), one can find the vector I collinear to the transformed axis O_2X_2

$$\overrightarrow{I} = (I_x, I_y, I_z) = -1 \cdot (O_{2_x} - B_{1x}, O_{2_y} - B_{1y}, O_{2_z} - B_{1z}),$$

Convert vector I to unit vector $\overrightarrow{i}$

$$\overrightarrow{i} = \left(I_x/\left(I_x{}^2 + I_y{}^2 + I_z{}^2\right)^{1/2}, I_y/\left(I_x{}^2 + I_y{}^2 + I_z{}^2\right)^{1/2}, I_z/\left(I_x{}^2 + I_y{}^2 + I_z{}^2\right)^{1/2}\right).$$

Let's find the axis vector O_2Z_2 and define it as $\overrightarrow{k} = \overrightarrow{i} \times \overrightarrow{m}$, where $\overrightarrow{m}$ is the unit vector O_2B_3. The unit vector $\overrightarrow{j}$ is defined as $\overrightarrow{j} = \overrightarrow{i} \times \overrightarrow{k}$. Thus, considering formulas (5), (6), the solution of the forward kinematics is reduced to determining the position of points B_i for given rod lengths. We compose a system of equations:

$$\begin{cases} g_1(x) = 0, \\ \dots \\ g_9(x) = 0, \\ a_i \le x_i \le b_i, \; i = 1, \; \dots, \; 9. \end{cases} \tag{7}$$

where $g_j(x)$ are the functions of the rod lengths based on the transformation of Eq. (1) between the joints of the movable and fixed platforms, as well as the joints of the movable platform between themselves, that is $g_1(x) = (A_{1_x} - B_{2_x})^2 + (A_{1_y} - B_{2_y})^2 + (A_{1_z} - B_{2_z})^2 - L_1^2, \ldots, g_9(x) = (B_{3_x} - B_{1_x})^2 + (B_{3_y} - B_{1_y})^2 + (B_{3_z} - B_{1_z})^2 - l^2$. An algorithm (Fig. 2) based on interval analysis methods [11, 12] considered earlier in [13] can be used to approximate the solution set of this system.

We use parallel computing to speed up its execution. These cycles are highlighted in Fig. 2 by dotted lines. Based on the proposed algorithm, we will analyze the influence of singularities on the solution of the forward kinematics. The analysis consists of two stages. The first stage consists in solving the kinematics problem using formulas (1), (2) for the position and orientation of the platform, at which a singularity arises in accordance with condition (4). The second stage consists in solving of forward kinematics for the case of a singularity and the case near a singularity. As a result of solving of forward kinematics for the case of a singularity, five possible positions were identified. The Table 1 presents center coordinates O_2 and platform orientation angles α, β, γ for five solutions. The first solution according to Table 1 (Fig. 3) corresponds to a singularity.

Analyzing the projections of the position of the platform on the plane, we can conclude that the axes of the rods L_4 and L_5 belong to the plane of the movable platform $B_1B_2B_3$. When changing the lengths of the links L_4 and L_5 as a result of solving forward kinematics, 2 possible achievable positions of the movable platform near the singularity were revealed (Fig. 4). Analyzing the obtained results, we can conclude that when passing a singularity, the number of solutions to forward kinematics changes, which can lead to jamming and malfunction. Accordingly, this should be taken into account in the design.

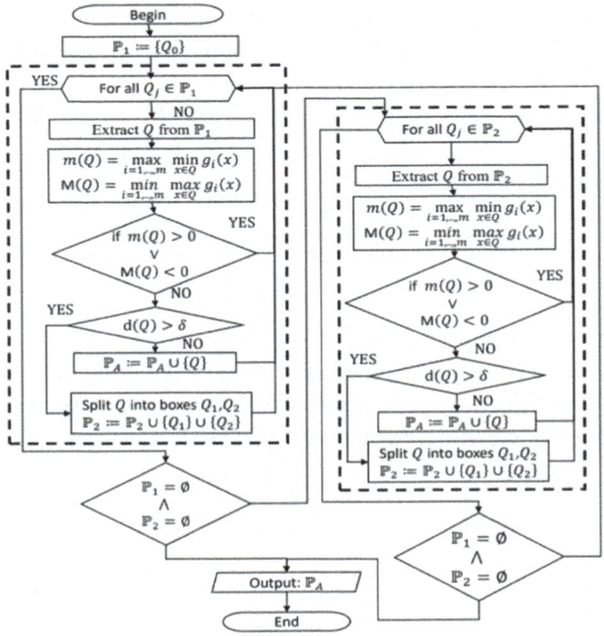

Fig. 2. Algorithm for forward kinematics solving.

Table 1. Solutions of forward kinematics.

No.	O_X	O_y	O_Z	α	β	γ
1	7.25	0	797.02	90	−70.05	−90
2	−38.77	101.57	727.01	135.96	110.7	43.97
3	−191.45	−1.41	730.95	94.77	172.37	95.27
4	−58.01	−0.42	694.61	89.57	37.92	90.5
5	−38.83	−101.67	727.04	44.08	110.77	136.04

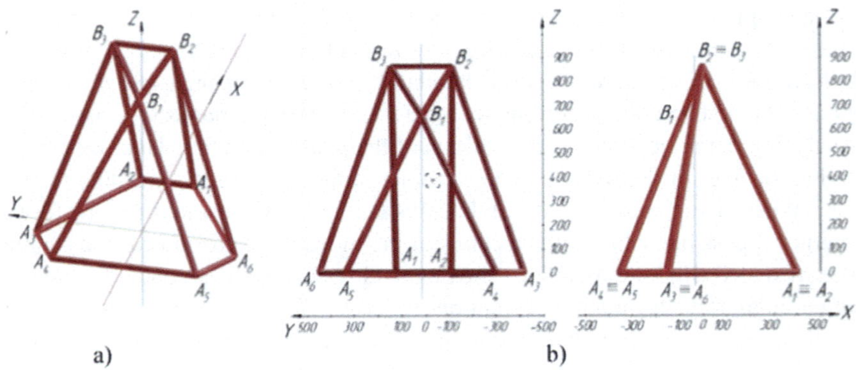

Fig. 3. Platform configuration Gough-Stewart in a singularity: a) isometry b) projections on a plane

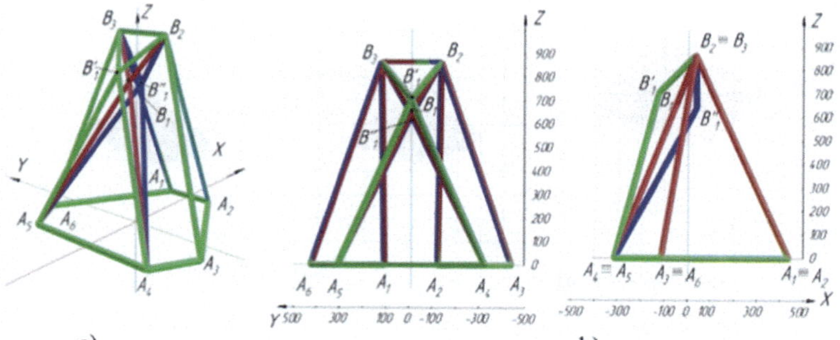

a) b)

Fig. 4. Comparison of the results of solving the forward kinematics near a singular position: a) isometry b) projections on the plane

4 Analysis of the Influence of Singularities on the Geometry of the Workspace

Consider the problem of determining the workspace, within which any position of the platform is achievable for any orientation of the platform within the given ranges of Euler angles, provided there are no singularities. To do this, we introduce the following restrictions: for rod lengths L : $L_i \in [L_{min}; L_{max}]$; on the position and orientation of the platform to avoid singularities. It can be taken into account by adding the condition for the positivity of the left side of Eq. (4).

The algorithm for determining the workspace consists in iterative enumeration of platform coordinates O_2 with a discrete step δ that may belong to the workspace. Coordinate values are added to the coverage of the workspace if, for any angle values from the ranges of orientation angles, the conditions are met in accordance with the restrictions on the lengths of the rods and on the position and orientation of the platform to exclude singularities. To do this, the lengths of the rods are calculated using the formulas (1)–(2) and the value of the function of the left side of Eq. (4) for each of the rods. Adding to the coverage of the workspace involves introducing integers into an ordered set to reduce the amount of data while maintaining the accuracy [14]. The considered algorithm is implemented in the C++ programming language using parallel computing and exporting to a three-dimensional STL model.

Let's perform the simulation for the following parameters: $L_{min} = 600\,\text{mm}$, $L_{max} = 1000\,\text{mm}$, $\varphi = 60°$, $R_1 = 450\,\text{mm}$, $R_2 = 150\,\text{mm}$, $\delta = 10\,\text{mm}$. Figure 5 shows the results of modeling in isometry and projections on a plane.

Let's analyze the dependence of the volume of the workspace on the range of Euler angles for the following restrictions on the length of the rods: $L_{min} = 500\,\text{mm}$, $L_{max} = 1000\,\text{mm}$. During the simulation, the range of Euler angles was consistently increased for all angles from ±0° to ±90° with step 1°. A graph of the dependence of the volume of the workspace on the range of angles is shown in Fig. 6. The orange line shows the change in the volume of the workspace with a limitation only on the lengths of the rods, the dotted black line shows the volume of the singularity-free workspace.

It can be seen from the graph that singularities within the workspace occur at a range of Euler angles ±15°, however, the percentage of the workspace that cannot be

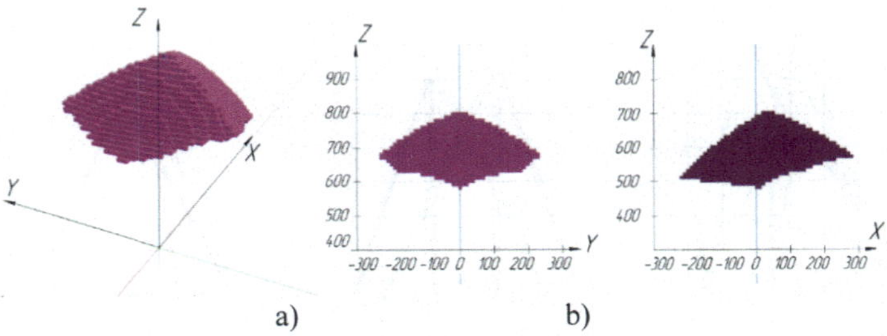

Fig. 5. Workspace: a) isometry b) projections on the plane.

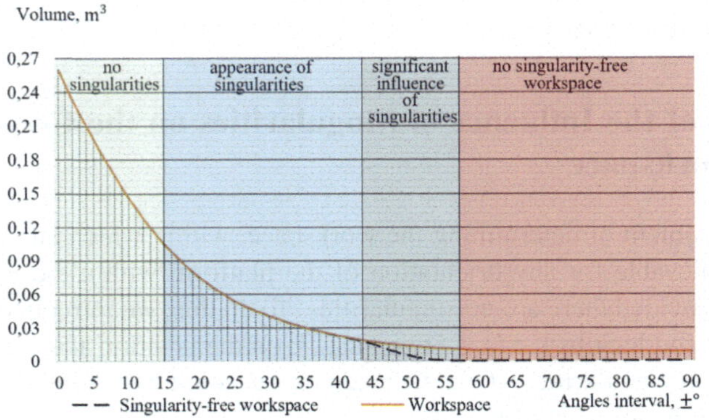

Fig. 6. Dependence of the volume of the workspace on the range of Euler angles.

reached without passing singularities is less than 5% up to the range ±43°. With a further increase in the range of angles, singularities have a significant impact on the volume of the workspace. For a range ±58°, there is no single position at which all platform orientations are achievable without passing singularities.

5 Conclusion

The analysis of the singularities influence on the forward kinematics solution for the Gough - Stewart platform is carried out. It is revealed that the number of solutions of the forward kinematics changes when it enters the singularity zone. When the drive coordinate values are changed, the platform may move to one of two possible positions, which may cause jamming and malfunction. The influence of singularities on the geometry of the workspace is also studied. An algorithm has been synthesized that makes it possible to determine the singularity-free workspace for given ranges of change in platform orientation angles. It is revealed that the singularity zones inside the workspace arise in the range of Euler angles of ±15°. For angle ranges greater than ±43°, singularities have a significant effect on the volume of the workspace. For a range of ±58°, there is no position at which all platform orientations are achievable without going through singularities. The results obtained can be used for a rational choice of platform parameters, taking into account the presence of singularity zones, as well as for the design of a control system. Control of the robot in the resulting workspace can be performed without uncertainty and does not require checking whether the end-effector is in a singular position. However, this reduces the maneuverability of the end-effector due to the limitation of orientation angles.

Acknowledgments. This work was supported by the Ministry of Science and Higher Education of the Russian Federation, the agreement number 075-11-2021-060 of June 24, 2021 "Development of a high-tech production of robotic 6DOF mobility platforms for trainers and simulators for the aviation and rocket-space industry" (unique number 000000S407521QLB0002).

References

1. Merlet, J.-P.: Parallel Robots. Kluwer Academic Publishers, Netherlands (2000)
2. Kong, X., Gosselin, C.M.: A class of 3-dof translational parallel manipulator with inear input/output equations. In: Workshop on Fundamental Issues and Future Research for Parallel Mechanisms and Manipulators, pp. 25–32. Quebec. Quebec City, Canada (2002)
3. Craig, J.J.: Introduction to Robotics: Mechanics and Control, pp. 105–108. Stanford University, Stanford, CA, USA, (2018)
4. Gosselin, C.M., Angeles, J.: Singularity analysis of closed-loop kinematic chains. IEEE. Trans. Robot. Autom. **6**(3), 281–290 (1990)
5. Gosselin, C.M., Wang, J.: Singularity loci of planar parallel manipulators. In: Ninth World Congress on the Theory of Machines and Mechanisms, Milano, Italy (1995)
6. Hunt, K.H.: Kinematic Geometry of Mechanisms. Oxford University Press, Oxford (1990)
7. Fichter, E.F.: A stewart platform-based manipulator: General theory and practical construction. Int. J. Robot. Res. **5**(2), 157–182 (1978)
8. Merlet, J.-P.: Parallel manipulators part 2: singular configurations and Gassmann geometry. Technical report, INRIA, Sophia Antipolis, France (1988)
9. Merlet, J.-P.: Singular configurations of parallel manipulators and Grassmann geometry. Int. J. Robot. Res. **8**(5), 45–56 (1989)
10. Slavutin, M., Sheffer, A., Shai, O.: A complete geometric singular characterization of the 6/6 Stewart platform. J. Mech. Robot. **10**(4), 041012 (2016)
11. Merlet, J.P.: Interval Analysis and Robotics Tracts in Advanced Robotics, vol. 66, pp. 147–156 (2010)
12. Malyshev, D., Posypkin, M., Rybak, L., Usov, A.: Approaches to the determination of the working area of parallel robots and the analysis of their geometric characteristics. Eng. Trans. **67**(3), 333–345 (2019)
13. Rybak, L.A., Gaponenko, E.V., Malyshev, D.I., Virabyan, L.G.: The algorithm for planning the trajectory of the 3-RPR robot, taking into account the singularity zones based on the method of non-uniform covering. IOP Conf. Ser. Mater. Sci. Eng. **489**(1), 012060 (2019)
14. Rybak, L., Malyshev, D., Gaponenko, E.: Optimization algorithm for approximating the solutions set of nonlinear inequalities systems in the problem of determining the robot workspace. Commun. Comput. Inf. Sci. **1340**, 27–37 (2020)

Perception and Learning

Structure Synthesis for Extended Robot State Automata

Lukas Sauer(✉) and Dominik Henrich

Universität Bayreuth, Universitätsstr. 30, 95447 Bayreuth, Germany
{lukas.sauer,dominik.henrich}@uni-bayreuth.de

Abstract. Opening up untapped potential in smaller enterprises requires methods for robot programming usable by non-experts. In a previous work on an automata-based programming approach without a graphical interface, linking to previous states to close loop structures was identified as challenging for users. In this paper, we propose an approach to generate such structures automatically from small overlap within the programmed instructions. To this end, a marking function for state pairs is calculated, which evaluates whether the structures starting from two states are conflicting or in agreement (and in the latter case, how much). Algorithms for calculating and utilizing this marking function are presented. Our experiments on an example task confirm the effectiveness of the approach.

Keywords: Robot programming · Automata · Automatic synthesis

1 Introduction

While robots have a long and successful history in large scale industry, there is both increasing use and still untapped potential in small and medium-sized enterprises or workshops [1], where only domain experts are readily available, but no expert robot programmers. This necessitates more intuitive forms of robot programming. Automata-based robot programs are promising in this regard, because they can express control flow in a conceptually easy way: in a given situation, specify what happens next. In our research [2,3], we attempt to gauge the viability of an automata-based approach which does not use a graphical editor. This can reduce the amount of hardware required in the workspace and the number of changes in input devices during programming. Potentially, all interactions necessary to program the robot can be done via the robot in this system.

The overall approach from our previous work is as follows: In each step, the user specifies the system's next action, which in the automaton representation generates a new state and a corresponding transition. Actions can be e.g.

This work has partly been supported by the Deutsche Forschungsgemeinschaft (DFG) under grant agreement He2696/15 INTROP.

A. Müller and M. Brandstötter (Eds.): RAAD 2022, MMS 120, pp. 71–79, 2022.
https://doi.org/10.1007/978-3-031-04870-8_9

movement to some target robot pose or connecting the current state to another existing state (called *linking*). Actions can either be defined as spontaneous (i.e. in execution, they will trigger automatically) or requiring a specific input or perception.

Closing loops within the program (by linking) without a graphical representation proved to be conceptually challenging in the experiments for our previous work [3]. The user needs a good understanding of how to identify states within the automaton, and where to link, to create specific control flow. The approach proposed in this paper attempts to alleviate this by letting the system detect automatically when parts of the programmed structure start to overlap. When that happens, the system can simplify the structure by itself. The detection of overlap is based on a marking function on state pairs which describes how well the structures starting from these states fit onto each other, or whether there is some contradiction that distinguishes them.

In the following, we give a brief overview over works related to the approach. In Sect. 3, the marking function is formally defined. Algorithms for generating and utilizing the marking function are presented in Sect. 4. In Sect. 5, we describe our experiments. Finally, Sect. 6 contains our conclusion.

2 Related Work

Automata as a basic concept are used in robotics in varying contexts and abstraction levels [4–9]. There is also a number of approaches where users can explicitly generate and edit robot programs in the form of (different variants of) automata [10–12]. However, all of these use graphical editors. Our research, in contrast, is trying to gauge the validity of an approach without a GUI, as motivated in Sect. 1.

Because the formalism of *Extended Robot State Automata* (ERSA) is unique to our research, there is no previous work on automatic simplification of structures within them. Some existing approaches for structural synthesis in a more general context use Version Spaces [13–15], for *Programming by Demonstration* (PbD) with applications from text editing macros to robot programs. Other works use planning systems for this goal [16,17]. There are also classical PbD approaches that, if they synthesize structure, usually do so from repeated demonstrations [8,18,19]. In contrast, the proposed approach operates directly on the automata, and requires only a small amount of overlap with an existing part of a program to close a loop (as opposed to multiple complete demonstrations). Finally, there is a number of works on structured robot programs (including loops) that employ visual programming languages, which we explicitly want to avoid for our system [20–23].

3 Marking Formalism

The goal of our method is to simplify programs when they start *overlapping* (i.e. when the user starts to program a part of the task that has been programmed before) and generate more complex structure (e.g. closed loops) within the program from that. To do this, we track whether state pairs are *conflicting*, meaning there is some information that lets us rule out that two states would represent a shared state in a more concise representation of the task. The opposite term is *unifiable*, i.e. states that could possibly represent a shared logical task state and to which unification might be applied in the future. For unifiable state pairs, the system also calculates how much overlap the two states provide in the structure of their successors. This idea of marking state pairs according to unifiability is directly inspired by unification of states in finite state automata, used for automata minimization (see e.g. [24,25]).

In [3] we defined the model of ERSA, a variant of finite state automata. An ERSA is given as a tuple $M = (Q, \Sigma, P, D, \delta, u, q_s)$, where (with the set of states Q, input alphabet Σ, transition function δ, and initial state q_s as usual) the additional components are the space of poses P, the space of pose variables $D = P^n$, and the update function u. This u is the general update function, which represents the action executed by the robot when the system progresses along a transition as defined by $\delta(q, \sigma) = q'$. We defined the restricted update functions $u^{q,\sigma}$ (to a specific q and σ) as satisfying one of a small number of possible forms. This formulation ensures that the semantics of a logical state of the system do not differ e.g. when the robot is at two different poses in its workspace. The definition, for completeness, is $u^{q,\sigma} : P \times D \times P \mapsto P \times D$, where

$$u^{q,\sigma} \in \left\{ \begin{array}{l} (p, d, p') \mapsto (\bar{p}, d), \\ (p, d, p') \mapsto (p \cdot \bar{p}, d), \\ (p, d, p') \mapsto (p' \cdot \bar{p}, d), \\ (p, d, p') \mapsto (p, d_1, \ldots, d_{k-1}, p, d_{k+1}, \ldots, d_n), \\ (p, d, p') \mapsto (d_k, d) \end{array} \right\}, 1 \leq k \leq n, \bar{p} \in P \qquad (1)$$

For more details on ERSA, we refer to [3].

In marking state pairs, we need to consider the updates along pairs of outgoing transitions from the two states. If the successor states (the targets of the two transitions) are not conflicting, but a different update is performed (e.g. movement in one case and saving the current pose, or a different movement, in the other), the preceding states themselves cannot be considered unifiable. Substituting one by the other would lead to a different action being performed.

Formally: There is a function

$$M_{trans} : \{\{q, q'\} \mid q \in Q, q' \in Q, q \neq q'\} \times (\Sigma \cup \{\varepsilon\}) \mapsto \{-1, 0\},$$

$$M_{trans}(\{q, q'\}, \sigma) = \begin{cases} -1 \text{ if } u^{q,\sigma} \text{ and } u^{q',\sigma} \text{ are from different cases in (1),} \\ \quad \text{or the parameters } k \text{ or } \bar{p} \text{ are different,} \\ 0 \text{ otherwise} \end{cases}$$

We will use this in the definition of the marking function. A final piece of notation: we can express $Q = Q_b \uplus Q_v \uplus Q_t$ as a disjoint partition, where

- $Q_b = \{q \in Q \mid \exists \sigma \in \Sigma : \delta(q, \sigma) \text{ defined}\}$ are *branching states* (states with at least one transition that requires a specific input)
- $Q_v = \{q \in Q \mid \delta(q, \varepsilon) \text{ defined}\}$ are *via states* (states with a spontaneous transition)
- $Q_t = Q \setminus \{Q_b \cup Q_v\}$ are states for which no transitions are defined so far

In branching states, the automaton can also branch into several possible transitions. A branching state and a via state are automatically conflicting, because unifying them would create a state with non-deterministic behaviour (following the spontaneous transition or gathering an input).
The *marking function* for state pairs $\{q, q'\}$ is then defined as follows:

$$M : \{\{q, q'\} \mid q \in Q, q' \in Q, q \neq q'\} \mapsto \mathbb{N} \cup \{-1\},$$

$$M(\{q, q'\}) = \begin{cases} -1 \quad \text{if } ((\text{w.l.o.g. } q \in Q_b \wedge q' \in Q_v) \\ \qquad \vee (\exists \sigma \in \Sigma \cup \{\varepsilon\} : M(\{\delta(q, \sigma), \delta(q', \sigma)\}) = -1) \\ \qquad \vee (\exists \sigma \in \Sigma \cup \{\varepsilon\} : M_{trans}(\{q, q'\}, \sigma) = -1)), \\ \min\big(\max\limits_{\sigma \in \Sigma \cup \{\varepsilon\}} \{M(\{\delta(q, \sigma), \delta(q', \sigma)\}) + 1\}, |Q|\big) \text{ otherwise} \end{cases}$$

I.e., $M(\{q, q'\}) = -1$ iff the two states are of conflicting type, or there is a conflicting successor pair under some input σ, or the outgoing transitions under some input σ are conflicting. Otherwise, it is one more than the highest marking of a successor pair (under any input σ), defaulting to zero. We will call a non-conflicting marking of value $n \in \mathbb{N}$ *supporting* (to depth n).

4 Algorithms

The table-filling algorithm used to determine the correct markings is presented in pseudocode in Fig. 1. Note that calculating values of M_{trans} does not require any iteration or recursion, only simple lookups within the automaton.

```
Data: Table M for {{q, q'} | q ∈ Q, q' ∈ Q, q ≠ q'} (e.g. lower diagonal matrix)
if M was newly generated then
    for {{q, q'} | q ∈ Q, q' ∈ Q, q ≠ q'} do
        if w.l.o.g. q ∈ Q_b ∧ q' ∈ Q_v then
            M({q, q'}) ← −1
        else
            M({q, q'}) ← 0
else if a new state q* was just generated then
    for {q ∈ Q | q ≠ q*} do
        M({q, q*}) ← 0
repeat
    for {{q, q'} | q ∈ Q, q' ∈ Q, q ≠ q'} do
        if ∃σ ∈ Σ ∪ {ε} : M({δ(q, σ), δ(q', σ)}) = −1 ∨ M_trans({q, q'}, σ) = −1 then
            M({q, q'}) ← −1
        else
            M({q, q'}) ← min( max_{σ∈Σ∪{ε}} {M({δ(q, σ), δ(q', σ)}) + 1}, |Q|)
until no entry of M changed in this iteration;
```

Fig. 1. Table-filling algorithm to generate correct markings

If values of M higher than a threshold of m are calculated for any pair of states, this pair is subjected to *unification*, which denotes merging the two states into a single state. The unification procedure is presented in Fig. 2. It makes use of a disjoint-set (or union-find) data structure [26,27] for tracking the central information: which sets of states will be unified into which *representative*.

```
Data: Disjoint-set data structure I (e.g. disjoint-set forest) for states
for {{q, q'} | q ∈ Q, q' ∈ Q, q ≠ q', M({q, q'}) ≥ m} do
    I.union(q, q')
repeat
    for {q ∈ Q | I.find(q) ≠ q} do
        for {σ ∈ Σ ∪ {ε}} do
            I.union(δ(q, σ), δ(I.find(q), σ))
until no previously distinct state sets were unified in this iteration;
for {q ∈ Q} do
    for {σ ∈ Σ ∪ {ε}} do
        δ(q, σ) ← I.find(δ(q, σ))
Set initial and current state to their representatives; Remove states no longer reachable
 from initial or current state
```

Fig. 2. Unification procedure when sufficient support has been detected.

We first note the pairs with sufficient support, and set one of both states as the representative. Then, we unify states with their representatives, i.e. we note their successor pairs for unification. This may entail further unification in transitive successors. When there are no more updates, we set all transition targets as well as the initial and current state to their correct representatives. Finally, we remove any states no longer reachable from either the initial or the current state; this can be done by simple graph reachability tests.

5 Experiments

The above methods for calculating markings and unifying state pairs were implemented into the system described in our previous work [3]. In this system, the user kinesthetically guides a lightweight robot arm with a mounted gripper and camera. New states are generated by different haptic interactions (specific movements the user executes with the robot). Each new state is connected to the automaton by a new transition. The update along this transition represents the programmed action (e.g., movement to a target pose to which the robot was guided). The table-filling algorithm for calculating the markings is run after each action. When sufficient support (overlap) is detected, unification takes place.

Experiments consisted of the same task from [3], programmed here without any explicit linking. Briefly summarized, it includes all parts of the ERSA formalism: Distinguishing two kinds of objects in an outer loop with two branches, where in the first branch objects of one kind are picked up and placed again after some movement. In the second branch, objects of the other kind are placed in a row by iterating over target positions in a second small loop. The overall setup for this task is depicted in Fig. 3 (left). Here, unification was applied for closing the outer loop (of identifying an arriving object), as well as the inner loop (of iterating over target positions).

The parameter m was set to 3 for the experiments. This is the lowest appropriate value: The task involves overlap of two states after the initial branch-

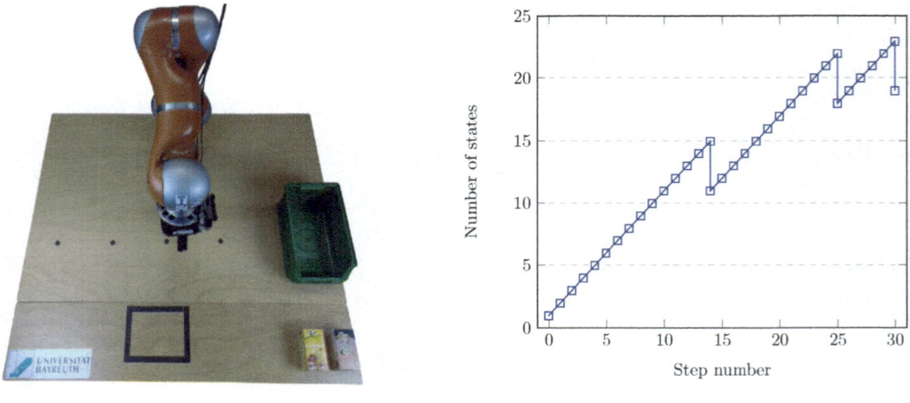

Fig. 3. Left: setup, reproduced from [3]. Objects arrive in the black rectangle and have to be dipped into the container and put back, or placed at the first free spot marked with a black dot. Right: number of states over number of programmed steps, in one run. Each step adds a new state. On steps 14, 25 and 30, a new state is generated, but unification afterwards reduces the number of states.

ing state (moving to a grasp pose above the object, and grasping it, in both branches).

Figure 4 shows an automaton before and after unification of the first branch into a closed loop. It illustrates how only partial overlap of the two instances of the loop body is sufficient for the system to suggest unification.

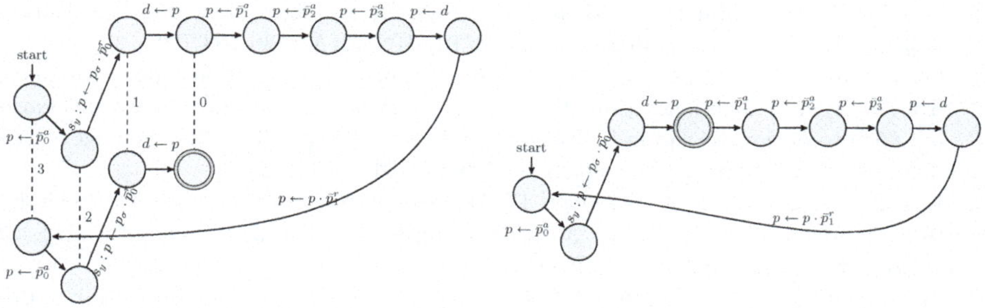

(a) Before unification. Support depth is printed on the dashed edges.

(b) After unification.

Fig. 4. Partial automaton before and after the first unification. The leftmost state pair has a support depth of 3. From there, successor states are unified. Current state after unification is the unified version of the previously current state. (Current states marked by double boundary.)

The approach of marking and unification was verified on the task by application by one of the authors as well as two non-expert participants. These followed the same procedure as the study depicted in [3]. In all three cases, unification was applied successfully. As Fig. 3 (right) shows for a single run, the number of states progresses linearly up to detection of sufficient support, at which point the number of states drops by unification.

6 Conclusion

In this paper we proposed an approach to automatically close structures within ERSA from overlap in the programmed state chains. We presented the formalism of the underlying marking function on state pairs, and the algorithms for calculating and utilizing the marking function. We also outlined our experimental verification. In conclusion, the approach generates the intended structures without any explicit linking operation required from the user. This facilitates the programming of structured tasks by non-experts.

References

1. Perzylo, A., et al.: SMErobotics: smart robots for flexible manufacturing. IEEE Robot. Autom. Mag. **26**, 78–90 (2019). https://doi.org/10.1109/MRA.2018.2879747

2. Sauer, L., Henrich, D., Martens, W.: Towards intuitive robot programming using finite state automata. In: Benzmüller, C., Stuckenschmidt, H. (eds.) KI 2019. LNCS (LNAI), vol. 11793, pp. 290–298. Springer, Cham (2019). https://doi.org/10.1007/978-3-030-30179-8_25
3. Sauer, L., Henrich, D.: Extended state automata for intuitive robot programming. In: Zeghloul, S., Laribi, M.A., Sandoval, J. (eds.) RAAD 2021. MMS, vol. 102, pp. 61–68. Springer, Cham (2021). https://doi.org/10.1007/978-3-030-75259-0_7
4. Brooks, R.A.: A robust layered control system for a mobile robot. IEEE J. Robot. Autom. **2**, 14–23 (1986). https://doi.org/10.1109/JRA.1986.1087032
5. Grollman, D.H., Jenkins, O.C.: Can we learn finite state machine robot controllers from interactive demonstration? In: Sigaud, O., Peters, J. (eds.) From Motor Learning to Interaction Learning in Robots. SCI, vol. 264, pp. 407–430. Springer, Heidelberg (2010). https://doi.org/10.1007/978-3-642-05181-4_17
6. König, L., Mostaghim, S., Schmeck, H.: Decentralized evolution of robotic behavior using finite state machines. Int. J. Intell. Comput. Cybern. **2**, 1–33 (2009). https://doi.org/10.1108/17563780911005845
7. Marino, A., Parker, L., Antonelli, G., Caccavale, F.: Behavioral control for multi-robot perimeter patrol: a finite state automata approach. In: Proceedings - IEEE International Conference on Robotics and Automation, pp. 831–836 (2009). https://doi.org/10.1109/ROBOT.2009.5152710
8. Orendt, E.M., Henrich, D.: Control flow for robust one-shot robot programming using entity-based resources. In: 18th International Conference on Advanced Robotics, pp. 68–74 (2017). https://doi.org/10.1109/ICAR.2017.8023498
9. Riano, L., Mcginnity, T.M.: Automatically composing and parameterizing skills by evolving finite state automata. Robot. Auton. Syst. **60**(4), 639–650 (2012). https://doi.org/10.1016/j.robot.2012.01.002
10. Brunner, S.G., Steinmetz, F., Belder, R., Dömel, A.: RAFCON: a graphical tool for engineering complex, robotic tasks. In: IEEE International Conference on Intelligent Robots and Systems, pp. 3283–3290 (2016). https://doi.org/10.1109/IROS.2016.7759506
11. Steinmetz, F., Wollschlager, A., Weitschat, R.: RAZER-A HRI for visual task-level programming and intuitive skill parameterization. IEEE Robot. Autom. Lett. **3**, 1362–1369 (2018). https://doi.org/10.1109/LRA.2018.2798300
12. Thomas, U., Hirzinger, G., Rumpe, B., Schulze, C., Wortmann, A.: A new skill based robot programming language using UML/P statecharts. In: Proceedings - IEEE International Conference on Robotics and Automation, pp. 461–466 (2013). https://doi.org/10.1109/ICRA.2013.6630615
13. Lau, T.A., Weld, D.S.: Programming by demonstration using version space algebra. Mach. Learn. **53**, 111–156 (2003). https://doi.org/10.1023/A:1025671410623
14. Lau, T., Domingos, P., Weld, D.S.: Learning programs from traces using version space algebra. In: Proceedings of the 2nd International Conference on Knowledge Capture, pp. 36–43 (2003). https://doi.org/10.1145/945645.945654
15. Pardowitz, M., Glaser, B., Dillmann, R.: Learning repetitive robot programs from demonstrations using version space algebra. In: Proceedings of the 13th IASTED International Conference on Robotics and Applications, RA 2007, Proceedings of the IASTED International Conference on Telematics, pp. 394–399 (2007)
16. Hu, Y., Levesque, H.: Planning with loops: some new results. In: ICAPS Workshop on Generalized Planning, pp. 35–42 (2009)
17. Levesque, H.J.: Planning with loops. In: IJCAI - International Joint Conferences on Artificial Intelligence, pp. 509–515 (2005)

18. Eker, S., Lee, T.J., Gervasio, M.: Iteration learning by demonstration. In: AAAI Spring Symposium - Technical report SS-09-01, pp. 40–47 (2009)
19. Friedrich, H., Dillmann, R.: Robot programming based on a single demonstration and user intentions. In: Proceedings 3rd European Workshop on Learning Robots (1995)
20. Alexandrova, S., Cakmak, M., Hsiao, K., Takayama, L.: Robot programming by demonstration with interactive action visualizations. In: Robotics: Science and Systems Foundation (2015). https://doi.org/10.15607/rss.2014.x.048
21. Alexandrova, S., Tatlock, Z., Cakmak, M.: RoboFlow: a flow-based visual programming language for mobile manipulation tasks. In: Proceedings IEEE International Conference on Robotics and Automation, pp. 5537–5544 (2015). https://doi.org/10.1109/ICRA.2015.7139973
22. Huang, J., Lau, T., Cakmak, M.: Design and evaluation of a rapid programming system for service robots. In: ACM/IEEE International Conference on Human-Robot Interaction, pp. 295–302 (2016). https://doi.org/10.1109/HRI.2016.7451765
23. Sefidgar, Y.S., Agarwal, P., Cakmak, M.: Situated tangible robot programming. In: ACM/IEEE International Conference on Human-Robot Interaction, Part F1271, pp. 473–482 (2017). https://doi.org/10.1145/2909824.3020240
24. Hopcroft, J.E., Motwani, R., Ullman, J.D.: Introduction to Automata Theory, Languages, and Computation. Addison-Wesley Publishing Company (1979)
25. Watson, B.: A taxonomy of finite automata minimization algorithms. Eindhoven University of Technology, Department of Mathematics and Computing Science, Computing Science Section (1993)
26. Hopcroft, J.E., Ullman, J.D.: Set merging algorithms. SIAM J. Comput. **2**, 294–303 (1973). https://doi.org/10.1137/0202024
27. Galil, Z., Italiano, G.F.: Data structures and algorithms for disjoint set union problems. ACM Comput. Surv. **23**, 319–344 (1991). https://doi.org/10.1145/116873.116878

Intuitive Optimization of Kinesthetic Programmed Trajectories for Fiber Spraying

Edgar Schmidt(✉) and Dominik Henrich

University of Bayreuth, Universitätsstr. 30, 95447 Bayreuth, Germany
{edgar.schmidt,dominik.henrich}@uni-bayreuth.de

Abstract. The optimization of robot programs before their execution enables to correct inaccurate and erroneous programming. In this work, an extension to kinesthetic programming is presented, with the aim to optimize robot trajectories in their geometrical and temporal course. For this, we define a set of optimization methods, which can be used quickly and without geometrical information about the used workpiece. The optimization of a robot trajectory can be performed via a simple user interface. We evaluate our approach in terms of intuitiveness and required amount of time for the optimization of robot trajectories in the application fiber spraying processes. We show the preferred order of optimization steps for the application fiber spraying processes. The resulting extension enables the elimination of a major disadvantage of kinesthetic programming.

Keywords: Industrial robots · Flexible manufacturing · Small batch size · Kinesthetic programming · Fiber spraying

1 Introduction

Robot-based automation is still not distributed in small and medium-sized enterprises, even though a low cost and high quality production is possible with the use of robots [1]. This low spread is largely due to the production of small batches, which requires frequent reconfiguration of the robot system [2]. A fast reconfiguration can be achieved through a speed-up of robot programming time with intuitive programming approaches like *kinesthetic programming* [3,4]. This approaches aim to program robot systems fast by experts in the application (*domain experts*), who have little to no programming expertise.

For the manufacturing and inspection of near-contour components, as for fiber spraying processes required [5], the robot path programming must be as accurate as possible. However, inaccuracies during kinesthetic programming are hardly avoidable, as the accuracy of manual guidance depends heavily on the operator. Therefore, a kinesthetic programming framework has to be extended to correct erroneous robot trajectories after demonstration. Our underlying system can perform adjustments to the program structure and thus significantly simplify programming [6]. Therefore, we extend our framework with the aim to optimize

A. Müller and M. Brandstötter (Eds.): RAAD 2022, MMS 120, pp. 80–88, 2022.
https://doi.org/10.1007/978-3-031-04870-8_10

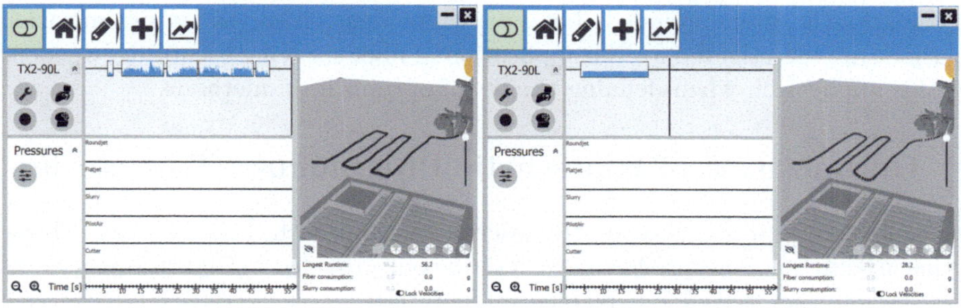

Fig. 1. An domain expert should correct and adapt a kinesthetic programmed trajectory (left) within the programming framework for an efficient execution (right).

the geometrical and temporal course of a robot trajectory. Besides correcting erroneous movements, we also want to adapt trajectories so that they can be reused for new tasks, as shown in Fig. 1, in which a program for the coating of a plate is adapted to coat a shaft. Further, the resulting extension should be simple and quick to use by domain experts and should not require additional information of the used workpiece (such as a CAD model).

Section 2 gives a overview of trajectory optimization. In Sect. 3, we describe our concept for an intuitive optimization of trajectories. We evaluate our approach in terms of intuitiveness and required amount of time for an optimization in Sect. 4. Section 5 summarizes and concludes the paper.

2 Related Work

The optimization of robot paths can be divided into two main approaches. The first main approach is direct optimization, in which the entire movement or user-defined sections are directly adapted by the user [7]. For this, intermediate positions on the trajectory are select-able and adjustable. The neighboring intermediate points are then weighted and shifted using, for example, a Gaussian kernel [8]. In the second main approach, the programming system analyzes the robot program and proposes the optimized alternative directly. These methods can be further classified according to the optimization criteria used. For example, the minimization of the execution time reduces the production costs, or the minimization of the energy results in smooth robot paths [9].

Most of the optimization approaches are based on a simulation of the process, as seen in case of a uniform coating thickness in spraying processes [10] or thermal spraying [11]. Such a simulation requires geometrical information of the given workpiece in the process. Further, this geometrical information can also be used for programming and direct optimization of the geometrical course of a robot trajectory. For example, based on the robot path, the corresponding path on the workpiece can be calculated. The robot path is then validated and edited through adaption of the path on the workpiece [7].

In this work, we focus on the optimization of robot trajectories without the use of additional geometrical information of a used workpiece, since this information is often provided, but not self-evident. Also, as no optimization methods are investigated for the application of fiber spraying according, we focus on this application when defining the set of optimization methods.

3 Optimization of Kinesthetic Programmed Trajectories

In the following subsections we will describe our approach. First, we define basic terms used in our concept. Then, we describe our provided optimization methods and the interface design is explained in detail last.

3.1 Definitions

Kinesthetic Programming describes a programming approach which consists of a programming phase and a playback phase. In the programming phase, the robot is guided kinesthetically or by a teach-pendant. While guiding, the programming framework records the robot joints motions and stores them. In the playback phase, the stored motions are played back like they were recorded. We call the concatenation of motions in a playback program a *sequence*. A sequence of joint configurations with timestamps for each joint configuration is called a *trajectory*.

3.2 Optimization Methods

The aim of our concept is to achieve a minimal number of methods, with which it is possible to adjust trajectories for most application cases. For this, we first examine optimization methods by interview with experts of the fiber spraying process. The fiber spraying process is to be classified as a primary forming process [5] and so the material application on the negative mold requires methods which achieve a uniform thickness. A uniform material application can be achieved by adjustment of the geometrical course of the trajectory and the adjustment of the robot's velocity. Further, methods to reduce downtime and execution time are required, since the process is costly in execution.

The adjustment of the geometrical course can be performed in various ways. Direct adjustment of configurations is very exact, but exhausting when many adjustments are needed. Thus, extended approaches are required: Equalizing a sequence to a defined geometrical form, for example a line, a plane or a circle defined by a minimum of points, create a specific geometrical course. Additionally, a automatic segmentation of a sequence into geometrical forms like lines and circles is provided. The use of smoothing approaches is dependent on the goal of the optimization. Smoothing a sequence to line segments allows to generate a meander-like path from a erroneous demonstration while the smoothing in general avoids jerky changes of direction and thus spares the mechanics of the robot. Also, the scaling of a trajectory around a specific center point is needed. The set of the provided optimization methods is shown in Table 1. These methods require only few parameters and are calculated in less then a second. Additional safety checks as well as an insertion of transition configurations after optimization guarantee the executability of the resulting robot program.

Table 1. Provided functions for the optimization of robot trajectories.

Icon	Description	Icon	Description
	Align sequence to a line		Smoothing sequence to line segments
	Scale sequence with respect to a center point	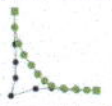	Blending of corners in the geometrical course
	Align sequence to a circle		Smoothing sequence
	Thin out configurations with same position coordinates		Delete downtime of the robot
	Segmentation into sections	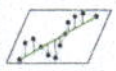	Align configurations of sequence onto a plane
$\vec{V}$	Adjust velocity of trajectory		Direct adjustment of configurations

3.3 User Interface

Our interface for optimization is divided into three parts as shown in Fig. 2 left. On the top, buttons for the optimization methods are placed in a scroll area. The different methods are distinguishable by icons shown in Table 1. In the center right, the parameters of the current selected method are displayed. In the left center, general settings of the optimization are placed. The first setting defines how the area for the optimization is chosen (whole trajectory, selection from timeline display, direct selection of configurations in the 3D simulation window). The second setting defines which function is used to insert transitions for the avoidance of gaps in the robot path. At the bottom, buttons for the control logic are placed: Display preview, discard preview, accept preview as new robot program, exit dialog (and discard preview).

To perform a optimization, a method and the associated parameters, as well as the area on which the optimization is executed are set. After that, the result of the optimization is calculated and a preview in form of a before/after comparison in the timeline representation and also in the 3D simulation window is displayed,

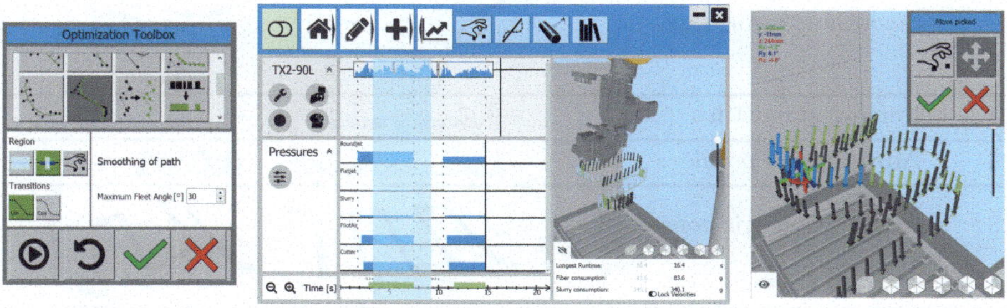

Fig. 2. After setting the optimization method and the associated parameters in the optimization dialog (left), a preview is displayed with characteristic values (center). Direct editing of the geometrical course is also possible (right).

as shown in Fig. 2 center. Additionally, changes in characteristic values of the program (e.g. execution time and material consumption) are displayed to provide more information about the resulting robot program.

4 Experiments

In this section, we first explain our experiment setup, in which we measure the intuitiveness and required amount of time for the optimization of trajectories. After this, we describe the outcomes of our study. Finally, we discuss specific results of our experiment in detail.

4.1 Experiment Setup

We evaluate the intuitiveness of our approach in a user study with nine participants (average: 26.6 years, 24 years youngest and 28 years oldest participant). Four of the participants have experience with fiber spraying processes and can be called domain experts. All of the domain experts have little to no programming knowledge. The participants use a tablet to operate our programming framework, in which a 3D model of a robot cell for fiber spraying [12] is displayed.

We rate the intuitiveness in the use of our programming framework with the MINERIC toolkit [13]. The participants must optimize robot programs in various ways. For this, a complete optimization task for fiber spraying was separated into five sub-tasks. In the last and sixth task, the first five tasks are performed in combination, resulting in the optimization of a real-world production task.

Table 2. Tasks of the user study with initial robot program and one possible solution. The images for task 3 and task 4 show the timeline representation of the robot trajectory. The other images show the robot path in the 3D simulation window.

Task	Description	Initial Program	Possible Solution
1	Straightening path		
2	Smoothing path edges		
3	Remove standstill		
4	Adjust speed profile		
5	Adapt geometric course		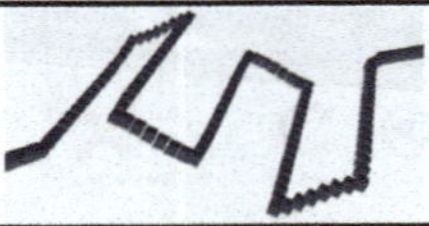
6	Optimization for fiber spraying production task		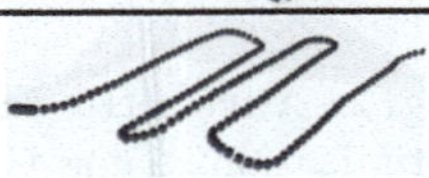

An overview of the tasks is given in Table 2. The participants are introduced shortly into the programming framework with an example task presented by the interviewer. The required amount of time for solving the tasks is also measured. The time measurements start when the participants accept to work on the task after reading the task description and end with accepting the resulting robot program as a solution for the task. For the sixth task, the participants are also asked to define a sequence of the optimization steps required to solve the task.

4.2 Experiment Results

The QUESI score and its sub-scores are shown in Fig. 3 left. The QUESI score is measured with five sub-scores on a scale from 1 to 5, where 5 means that the participants are fully satisfied with the framework and 1 means that they are not satisfied [14]. The QUESI score of 3.17 describes that participants were rather satisfied when using the framework, but there can be further improvements. The score of 2.6 for perceived error rate indicates, that in the process of using the framework, some errors were made. Compared to our work in [15], the use of the framework received good ratings although the tasks are more complex.

The effectivity score, the efficiency score and the measured required amount of time are shown in the table of Fig. 3 right. The effectivity score is measured in percent where a higher score indicates an effective use. The overall rating is 81%, which indicates a rather effective framework to use. The efficiency score ranges from 0 to 220, where lower scores indicate a better efficiency. The overall rating of 44.5 is translatable to little effort when using the framework. The average amount of time to solve the third and fourth tasks is lower then 180 seconds. The amount of time for performing optimizations on the geometrical course rises up to 490 seconds for the fifth task, in which most participants required more manual steps to solve, since the direct adjustment of configurations was used.

The order of the optimizations steps for fiber spraying are commonly preferred with first deleting downtime in the robot trajectory (task 3). Thereafter, the geometrical course is optimized (with the partial steps of task 1, 2 and 5 in any order) and as final step, the adjustment of the velocity is mentioned.

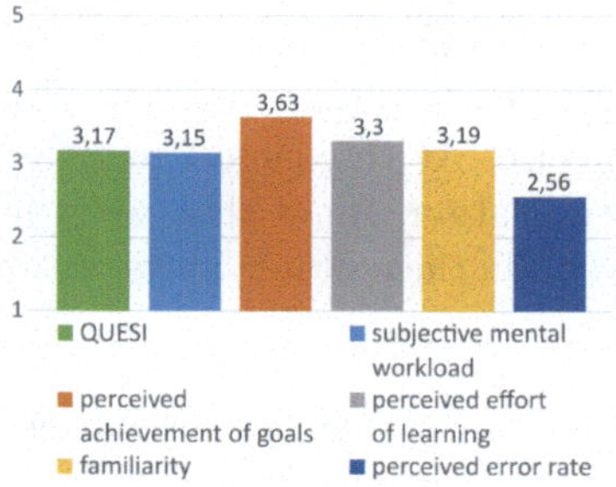

Task	1	2	3	4	5	6	Overall
Effectivity [%]	80	64	93	84	82	84	81
Efficiency	41.1	53.3	22.2	30.6	41.7	76	44.2
Required Time [sec]	422	342	158	174	490	668	–

Fig. 3. Results of our experiments. Left: QUESI score with sub-scores. Right: Effectivity score, efficiency score and required amount of time for solving the tasks.

4.3 Discussion

Different methods can be used to solve the second task. The first method is blending of corners, where the corners are defined by the user. The second method is smoothing of the path with a maximum fleet angle. The third method is equalizing configurations to a circle. Most participants choose the blending of corner as it is listed in the interface first. This method has two parameters, whose description is difficult to understand for the participants and thus confuse them and thus result in a low effectivity rating and the high efficiency score.

The summed efficiency score of the first five tasks is 188.9 (very high effort). For the sixth task, the efficiency was rated 76 (reasonably effort). The summed amount of time for solving the first five tasks is 1586 seconds, while solving the sixth task required 668 seconds. This observations may indicate, that the users are quickly getting familiar with our programming framework. This conclusion is supported by the observed effectivity of 84%, which is slightly higher as the effectivity over all tasks although the second task (effectivity score 64%) has to be performed as part of the sixth task.

5 Conclusion

In this work, an extension to the editable kinesthetic programming is proposed, with the aim to optimize already programmed robot trajectories before their execution. For this, we choose optimization methods which can be used to adapt the geometrical and temporal course of a robot trajectory. The methods can be calculated quickly without the use of geometrical information of the used workpiece. We evaluate our extension to the framework and show, that the extension is still intuitive in use even if the operations are more complex. The required amount of time to complete a optimization for a fiber spraying task is around eleven minutes. The proposed extension allows to eliminate a major disadvantage of kinesthetic programming and facilitates the programming of trajectories for complex tasks. The preferred order of optimization steps for fiber spraying trajectories are first the reduction of downtime, thereafter optimizing the geometrical course and last the adaption of the traversal speed.

However, an accurate programming of complex trajectories is still strenuous via kinesthetic programming, even through our extension for optimization. In industrial production, a CAD model of the workpiece is often provided and can be used to review robot trajectories more precise. At the same time, the CAD model can be used for a more accurate optimization. As of this, we will investigate the possibility to include CAD information in the editable kinesthetic programming and its applicability in future work.

Acknowledgments. The authors acknowledge Prof. Dr.-Ing. Stefan Schafföner, Georg Puchas and Jonas Winkelbauer from the Chair for Ceramic Materials Engineering (University of Bayreuth) for the excellent cooperation.

References

1. Perzylo, A., et al.: SMErobotics: smart robots for flexible manufacturing. IEEE Robot. Autom. Mag. **26**(1), 78–90 (2019). https://doi.org/10.1109/MRA.2018.2879747
2. Dietz, T., et al.: Programming system for efficient use of industrial robots for deburring in SME environments. In: ROBOTIK 2012, VDE, pp. 1–6 (2012)
3. Müller, R., et al.: Handbuch Mensch-Roboter-Kollaboration, pp. 243–270. Hanser-Fachbuch (2019)
4. Schraft, R.D., Meyer, C.: An intuitive teaching method for small and medium enterprises. In: Intelligent Production Machines and Systems, pp. 568–571 (2006). https://doi.org/10.1016/B978-008045157-2/50099-7
5. Winkelbauer, J., Puchas, G., Schafföner, S., Krenkel, W.: Short fiber-reinforced oxide fiber composites. Int. J. Appl. Ceram. Technol. **19**(2), 1136–1147 (2021). https://doi.org/10.1111/ijac.13931
6. Riedl, M., Henrich, D.: A fast robot playback programming system using video editing concepts. In: Schüppstuhl, T., Tracht, K., Roßmann, J. (eds.) Tagungsband des 4. Kongresses Montage Handhabung Industrieroboter, pp. 259–268. Springer, Heidelberg (2019). https://doi.org/10.1007/978-3-662-59317-2_26
7. Meyer, C., et al.: Programming by demonstration for assistive systems - intuitive programming of welding and gluing trajectories. IT-Inf. Technol. **49**(4), 238–246 (2007). https://doi.org/10.1524/itit.2007.49.4.238
8. Wölfel, K., Müller, J., Henrich, D.: ToolBot: robotically reproducing handicraft. In: Ardito, C., et al. (eds.) INTERACT 2021. LNCS, vol. 12934, pp. 470–489. Springer, Cham (2021). https://doi.org/10.1007/978-3-030-85613-7_32
9. Ratiu, M., Prichici, M.A.: Industrial robot trajectory optimization - a review (2017). https://doi.org/10.1051/matecconf/201712602005
10. Gadow, R., Candel, A., Floristán, M.: Optimized robot trajectory generation for thermal spraying operations and high-quality coatings on free-form surfaces. Surf. Coat. Technol. **205**(4), 1074–1079 (2010). https://doi.org/10.1016/j.surfcoat.2010.08.121
11. Hegels, D.: Optimierung thermischer Verhältnisse bei der Bahnplanung für das thermische Spritzen mit Industrierobotern. Dissertation (2017). https://doi.org/10.17877/DE290R-18069
12. Schmidt, E., Winkelbauer, J., Puchas, G., Henrich, D., Krenkel, W.: Robot-Based Fiber Spray Process for Small Batch Production. In: Schüppstuhl, T., Tracht, K., Henrich, D. (eds.) Annals of Scientific Society for Assembly, Handling and Industrial Robotics, pp. 295–305. Springer, Heidelberg (2020). https://doi.org/10.1007/978-3-662-61755-7_27
13. Orendt, E.M., Fichtner, M., Henrich, D.: MINERIC toolkit: measuring instruments to evaluate robustness and intuitiveness of robot programming concepts. In: 26th IEEE International Symposium on Robot and Human Interactive Communication (RO-MAN), pp. 1379–1386 (2017). https://doi.org/10.1109/ROMAN.2017.8172484

14. Naumann, A., Hurtienne, J.: Benchmarks for intuitive interaction with mobile devices. In: Proceedings of the 12th International Conference on Human Computer Interaction with Mobile Devices and Services (MobileHCI 2010), pp. 401–402 (2010). https://doi.org/10.1145/1851600.1851685
15. Schmidt, E., Henrich, D.: Playback robot programming framework for fiber spraying processes. In: Schüppstuhl, T., Tracht, K., Raatz, A. (eds.) Annals of Scientific Society for Assembly, Handling and Industrial Robotics, pp. 351–362. Springer, Cham (2021). https://doi.org/10.1007/978-3-030-74032-0_29

Towards Dynamic Obstacle Avoidance for Robot Manipulators with Deep Reinforcement Learning

Friedemann Zindler(✉), Matteo Lucchi, Lucas Wohlhart, Horst Pichler, and Michael Hofbaur

JOANNEUM RESEARCH - ROBOTICS, Klagenfurt am Wörthersee, Klagenfurt, Austria
{friedemann.zindler,matteo.lucchi,lucas.wohlhart,horst.pichler, michael.hofbaur}@joanneum.at
https://www.joanneum.at/en/robotics

Abstract. We use reinforcement learning (RL) to demonstrate an easily reproducible setup to learn dynamic obstacle avoidance for a robotic arm based on sensory input as it follows a pre-planned trajectory. Training takes place exclusively in a simulation environment with random obstacle movements around the robot. We show that training dynamic obstacle avoidance in simulation translates well to the real environment with a UR5 manipulator and yields similar performance and success without further tuning of the learned policy. This is a step towards learning general skills needed to enable robots to operate in dynamic environments shared with humans. Source code, data and application videos are available at: https://www.robogym.net.

Keywords: Machine learning for robot control · Collision avoidance · Transfer learning

1 Introduction

Typical applications for industrial robots are pick-and-place operations, palletizing and assembly. Traditionally, these operations are performed in closed cells limiting human access, and the robot programs consist of standardized, repetitive operations without regard to external environmental variations. Current efforts to perform operations in open, collaborative scenarios, however, require the ability to adapt robot programs to dynamic changes in the workspace in real-time.

To achieve *dynamic obstacle avoidance* several solutions have been proposed. Each of them requires specialized expertise, experience, the local environment of the application, as well as robot-specific motion primitive programming, hardware and software.

F. Zindler and M. Lucchi—These authors contributed equally.

A. Müller and M. Brandstötter (Eds.): RAAD 2022, MMS 120, pp. 89–96, 2022.
https://doi.org/10.1007/978-3-031-04870-8_11

In contrast to that, we envision a system that does not focus on a particular use case, but allows developers to model more generic robotic skills. These skills are trained in simulation, by defining metrics upon which to optimize the performance and behavior using data-driven AI approaches, to be transferred to real robots with very little effort.

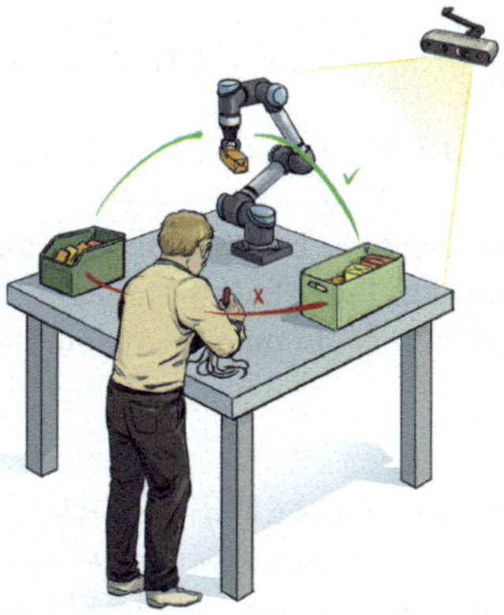

Fig. 1. Avoiding collision with a human collaborator during pick and place.

As a first step in this direction, we introduced robo-gym [11], an open-source framework that enables reproducible RL experiments on simulated and real robots using a common interface based on ROS. Now we extend the skills of the robotic arm by introducing an environment for dynamic obstacle avoidance (see illustration in Fig. 1).

The main contributions presented in this work are 1) modeling a parametrizable robot manipulator obstacle avoidance task as a RL problem, resulting non-erratic motions that stay close to a planned trajectory, and 2) developing a reproducible, open source environment to evaluate the performance of different algorithms and strategies in simulation and, after transfer, in real environments.

2 Related Work

In motion planning methods, the strategy is to compute the path to be followed by the robot before execution, the reaction to an unexpected event such as the presence of an obstacle on the planned path generally requires re-planning. Developments in this area allow complex robotic tasks to be modeled and solved, e.g., [6,9]. However, these methods typically tend to be computationally intensive and in highly dynamic environments the performance decreases significantly. Approximations and heuristics have been introduced to mitigate the problem [5]. An alternative solution involves the adoption of custom designed chips that allow for real-time application of aforementioned algorithms [12]. Reactive policy approaches focus on generating prompt feedback control based on the present state of the robot and sensors data [3,7,8]. Deep RL has been successfully applied to solve multiple complex robotic tasks [10,13]. Its application to achieve collision avoidance for robot manipulators has been explored as well [4,15]. Given the data sampling requirements of AI methods it is often very convenient to exploit the flexibility and power of simulations to collect experience. Different works studied the transfer of policies trained in simulation to applications in the real world [13,14].

3 Problem Description

In this paper, we address the problem of adapting a planned trajectory in real-time to move a robot manipulator from an arbitrary start position to an arbitrary goal position to avoid dynamic obstacles in the robot's workspace. The robotic arm used in this work is the UR5 collaborative robot from Universal Robots. The robot is controlled by means of joint position inputs and the interface is enacted by the ROS Driver[1]. Although we present a solution for obstacle avoidance for this particular robot model, the solution is general enough to be quickly adapted to other robotic arms with similar control interfaces. The basic goal of the robot is to avoid dynamic obstacles while following a pre-planned trajectory, which we programmed with the UR's teach pendant. This trajectory is sampled at a frequency 20 Hz, which is the frequency used by the Deep RL module to control the robot. The pre-planned path is executed continuously, but stops at certain waypoints, also predefined, which the robot must reach in the right order.

The idea behind this setup is to model a pick-and-place operation where the hard constraints are the position, i.e., the waypoints from which the object is to be picked up and at which the object is to be placed, and there are no hard constraints on the path between the two waypoints as long as collision-free motion is achieved.

In this work, we model an obstacle as a point. The goal is to maintain a minimum distance $d_{min} = 45$ cm between the obstacle and two points on the robot: the end effector and the elbow joint. The object can move freely throughout the robot's workspace without constraints. When feasible the robot should find a way around the obstacle and continue to the goal joint configuration, otherwise wait for the obstacle to move away before proceeding.

The detection of the obstacle position is shown with two different setups: (1) OptiTrack Motion Capture System for showing obstacle avoidance with a reliable and precise reference system (2) Skeleton pose tracking with RGB-D camera for demonstrating our approach for human robot collaboration.

4 Method

We exploit the flexibility and power of simulation to recreate the task and to collect experience in parallel instances that do not require human supervision or safety measures. To enable researchers to test their algorithms on the same problem, we implemented an OpenAI Gym [2] environment interface, the de facto

[1] https://github.com/UniversalRobots/Universal_Robots_ROS_Driver.

standard for RL environments. In addition, we use the robo-gym framework to simultaneously provide the same interface for the physical robot and to be able to model the control problem as a Markov Decision Process thanks to the Command Handler component of robo-gym [11]. Through a process of interaction with the simulated environment, we use Deep RL to find a policy that solves the given task of avoidance.

4.1 Simulation

The simulation of the environment is implemented in Gazebo and includes the robot itself and a simulated obstacle modeled as a moving point in space. Since randomization in the simulation can help to achieve a more robust and general policy [14,16] by exposing the learning agent to a variety of situations, we generate random trajectories for the moving obstacle each time the environment is reset, eliminating the possibility of the learning agent remembering the obstacle's movements.

4.2 Gym Environment

The observation of the environment $\boldsymbol{s} = (\boldsymbol{P_{ee}}, \boldsymbol{P_{fa}}, \boldsymbol{q_r}, \boldsymbol{q_t}, f, \boldsymbol{\delta})$ is described by: The spherical coordinates of the obstacle calculated with respect to the end effector frame $\boldsymbol{P_{ee}} = (r_{ee}, \theta_{ee}, \varphi_{ee})$ and with respect to the forearm link frame that is centered in the elbow joint $\boldsymbol{P_{fa}} = (r_{fa}, \theta_{fa}, \varphi_{fa})$; the current joint positions of the robot $\boldsymbol{q_r} \in \mathbb{R}^6$; the position on the trajectory where the robot currently should be $\boldsymbol{q_t} \in \mathbb{R}^6$; a flag f indicating if $\boldsymbol{q_t}$ is a waypoint; and a delta that describes the distances between the current robot joint positions and the desired trajectory joint positions $\boldsymbol{\delta} = \boldsymbol{q_t} - \boldsymbol{q_r}$. All joint positions are normalized with respect to the joint position limits.

An action $\boldsymbol{a} \in \mathbb{R}^5$ describes the relative variation of joint positions for the first five joints, with respect to the joint positions of the pre-planned trajectory point.

In this example, we formulated a reward based on our assessment of how important certain aspects should be. The reward function is formulated as the sum of several weighted reward components where each one provides feedback for a specific desired behavior that we want the policy to achieve. In particular the reward used is

$$R_{w_r} = \boldsymbol{w_r}^T [r_{dist}, r_\delta, r_a, r_{\dot{a}}, r_{waypoint}, r_{collision}] \tag{1}$$

where: r_{dist} is a binary signal equal to 1 if the distance between the obstacle and the end effector or the forearm is less than a predefined d_{min}; r_δ is proportional to the absolute value of the δ of each joint if they are below a certain threshold; r_a is formulated as the mean of the squared actions; $r_{\dot{a}} = \sum_{i=1}^{5} r_{\dot{a}_i}$ where for each joint $r_{\dot{a}_i}$ is 1 if the difference between the current action and the one taken in the previous step is higher than a certain threshold; $r_{waypoint}$ is 1 if the current $\boldsymbol{q_r}$ is a waypoint and $r_{collision}$ is 1 if the robot is in collision with the obstacle or the environment.

In the reward composition provided in Table 1 the most important part is keeping a certain predefined distance to an obstacle w_{dist}. There is an incentive to stay close to the original trajectory w_δ and for reaching the predefined waypoints $w_{waypoint}$. The reward parts w_a and $w_{\dot{a}}$ do not target the basic functionality but the agent's movement behavior, so that it has an incentive to deviate from the original trajectory as little as possible from w_a and avoid high jerks in robot joints $w_{\dot{a}}$. Finally, collisions with the environment or with the robot itself are punished by $w_{collision}$.

Table 1. Reward composition

Weight	Value	Description
w_{dist}	-2×10^{-3}	Negative signal if obstacle is too close
w_δ	1×10^{-3}	Reward for staying close to pre-planned trajectory
w_a	-1.5×10^{-3}	Reward for using small action for more stable behavior
$w_{\dot{a}}$	-2×10^{-4}	Punishment for rapid changes in action taken
$w_{waypoint}$	0.05	Reward for reaching one of the predefined waypoints in the pre-planned trajectory
$w_{collision}$	-0.1	Terminal state with big negative signal

As mentioned in Sect. 3, the task of the robot is to reach a number of selected joint configurations in an ordered manner. At each step taken in the environment, an internal counter progresses and the subsequent step of the pre-planned robot trajectory is included in the observation returned to the agent. The progression continues until a waypoint of the pre-planned trajectory is reached. Once this happens, the observation will include the waypoint joint positions as pre-planned trajectory joint positions until the given configuration is reached by the robot. Once the robot succeeds to reach the waypoint joint configuration, the state machine transitions to the successive state and updates the pre-planned trajectory segment to be followed.

4.3 Policy Training

To learn the control policy, we use the same setup as in [11], a standard implementation of Distributed Distributional Deep Deterministic Policy Gradients (D4PG) [1]. For a list of the used hyperparameters see the web page of the framework. Training was performed exclusively in simulation, using 16 instances in parallel for a total of 2 million environment steps.

5 Experiments

To evaluate the performance of the learned policy, we deployed the model trained in simulation to perform a set of experiments on the real and simulated application. For all the experiments the pre-planned trajectory is the same, whereas the trajectory of the obstacle changes for each test.

(a) Robot passing above the hand to keep minimum distance to it

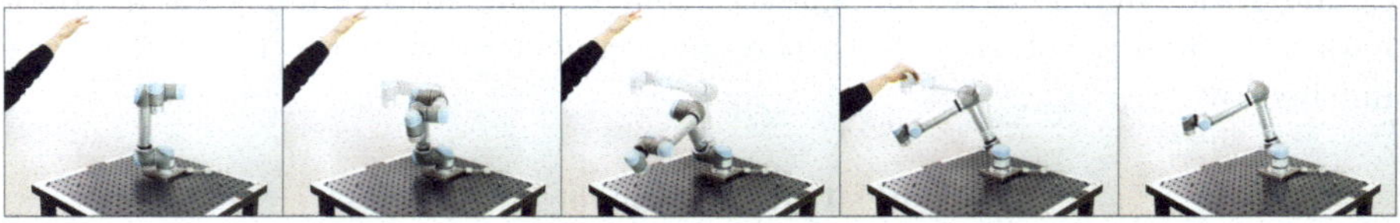

(b) Robot passing below the hand to keep minimum distance to it

Fig. 2. Real world experiments with avoidance of human hand. The faded robot represents the position of the robot if the obstacle avoidance algorithm is not applied.

The effectiveness of a policy is measured with custom metrics that indicate whether the policy models the desired behavior. First, we monitor if the robot collides with the object or the table (C). Second, we check the minimum distance (d_{min}) between the object and the closest of the two selected points of the robot for each step of an episode. Based on d_{min} (see Sect. 3), we define four different conditions:

- No Violation: $distance \geq d_{min}$
- Soft Violation: $d_{min} > distance \geq \frac{2}{3}\, d_{min}$
- Hard Violation: $\frac{2}{3}\, d_{min} > distance \geq \frac{1}{3}\, d_{min}$
- Critical Violation: $\frac{1}{3}\, d_{min} > distance > 0$

An important note here is that the obstacle could still collide with the base of the robot or the robot could hit the table without violating the distance metric. We report the percentage of episode steps in each of the four categories, calculated over the total number of steps. Finally, we measure $\overline{W}$, the average number of waypoints reached throughout an episode with a possible maximum value of 5.

Three different types of tests were performed, one in simulation and the other two testing the performance on the robot using different sensors to detect the obstacle. The results obtained are showed in Table 2.

For the test run in simulation, 50 randomly generated splines were generated (Setup i). To put the performance metrics from the simulation into context, we also measure these metrics with the same obstacle splines while only following the pre-planned trajectory (Setup i).

The second set of experiments (Setup iii) was conducted on the physical robot and using the OptiTrack Motion Capture System to detect the obstacle's position. The obstacle in this experiment is a cube-shaped object on a stick, equipped with markers, which is then maneuvered by a human operator. Although the trajectories performed by the operator differ significantly from the randomly generated splines, the results are still very comparable. Finally, we

Table 2. (i) Simulation without collision avoidance (ii) Simulation with collision avoidance (iii) Test on the robot using OptiTrack (iv) Test on the robot using the skeleton tracker

Setup	Collision free (C)	$\overline{W}$	No V.	Soft V.	Hard V.	Critical V.
i	48%	3.52	80.16%	14.17%	5.43%	0.24%
ii	94%	4.88	98.06%	1.74%	0.2%	0%
iii	100%	4.55	95.45%	4.42%	0.11%	0.01%
iv	95%	4.35	96.51%	3.00%	0.45%	0.04%

ran 20 episodes on the robot using the skeleton tracker to monitor the position of a human's right hand (Setup iv), see Fig. 2 for two sequences extracted from these experiments. Even with this less stable pose estimator compared to the OptiTrack system, the results change only slightly.

Overall the trained agent showed comparable performance in all the different setups. We noticed that even in the case in which the robot occluded the hand, causing the skeleton tracker to temporarily lose the detection, the agent produced coherent behavior. Even though the agent has been trained purely in simulation on continuous obstacle trajectories, it also performed well in real test cases where sensors introduce discontinuities, errors in obstacle position measurements.

6 Conclusion

In this paper, we present a method for setting up a dynamic collision avoidance task for robot manipulators using Deep RL. Our approach dynamically generates joint trajectories based on pre-planned trajectories and sensor inputs that provide the position of an obstacle. The experiments show good performance in transferring policies learned in simulation to the real world with a variety of sensors used. On the other hand, this approach requires manual tuning of the reward weights for a specific use case. Further investigation is required to improve the overall performance of the method and to generalize the solution for the application in different scenarios, as well as the learning of additional robot skills that can be combined to a broad variety of reactive manipulation tasks.

Acknowledgements. This research has received funding from the Austrian Ministry for Transport, Innovation and Technology (bmvit) within the project "Credible & Safe Robot Systems (CredRoS)", from the "Kärntner Wirtschaftsförderung Fonds" (KWF) and the "European Regional Development Fund" (EFRE) within the PATTERN-Skin project 26616/34294/49769.

References

1. Barth-Maron, G., et al.: Distributed distributional deterministic policy gradients. In: 6th International Conference on Learning Representations, ICLR 2018 - Conference Track Proceedings, pp. 1–16 (2018)
2. Brockman, G., et al.: OpenAI Gym, June 2016. http://arxiv.org/abs/1606.01540
3. Cheng, C.A., et al.: RMPflow: a geometric framework for generation of multi-task motion policies. Technical report (2020)
4. Duguleana, M., Barbuceanu, F.G., Teirelbar, A., Mogan, G.: Obstacle avoidance of redundant manipulators using neural networks based reinforcement learning. Robot. Comput. Integr. Manuf. **28**(2), 132–146 (2012)
5. Hauer, F., Tsiotras, P.: Deformable rapidly-exploring random trees. In: Robotics: Science and Systems (2017)
6. Kavraki, L.E., Švestka, P., Latombe, J.C., Overmars, M.H.: Probabilistic roadmaps for path planning in high-dimensional configuration spaces. IEEE Trans. Robot. Autom. **12**(4), 566–580 (1996)
7. Khatib, O.: A unified approach for motion and force control of robot manipulators: the operational space formulation. IEEE J. Robot. Autom. **3**(1), 43–53 (1987)
8. Kroeger, T.: Opening the door to new sensor-based robot applications - The Reflexxes Motion Libraries. In: Proceedings - IEEE International Conference on Robotics and Automation (2011)
9. LaValle, S.M.: Planning Algorithms. Cambridge University Press, Cambridge (2006)
10. Levine, S., Finn, C., Darrell, T., Abbeel, P.: End-to-end training of deep visuomotor policies. J. Mach. Learn. Res. **17**(1), 1334–1373 (2016)
11. Lucchi, M., Zindler, F., Mühlbacher-Karrer, S., Pichler, H.: robo-gym – an open source toolkit for distributed deep reinforcement learning on real and simulated robots. In: 2020 IEEE/RSJ International Conference on Intelligent Robots and Systems (IROS), pp. 5364–5371 (2020)
12. Murray, S., Floyd-Jones, W., Konidaris, G., Sorin, D.J.: A programmable architecture for robot motion planning acceleration. In: Proceedings of the International Conference on Application-Specific Systems, Architectures and Processors, July 2019, pp. 185–188 (2019)
13. OpenAI, et al.: Solving rubik's cube with a robot hand (2019). http://arxiv.org/abs/1910.07113
14. Peng, X.B., Andrychowicz, M., Zaremba, W., Abbeel, P.: Sim-to-real transfer of robotic control with dynamics randomization. In: Proceedings - IEEE International Conference on Robotics and Automation, pp. 3803–3810 (2018)
15. Sangiovanni, B., Rendiniello, A., Incremona, G.P., Ferrara, A., Piastra, M.: Deep reinforcement learning for collision avoidance of robotic manipulators. In: 2018 European Control Conference, ECC 2018, pp. 2063–2068 (2018)
16. Tobin, J., Fong, R., Ray, A., Schneider, J., Zaremba, W., Abbeel, P.: Domain randomization for transferring deep neural networks from simulation to the real world. In: 2017 IEEE/RSJ International Conference on Intelligent Robots and Systems (IROS), pp. 23–30 (2017)

Cable-Driven Parallel Robot Accuracy Improving Using Visual Servoing

Ferdaws Ennaiem[1,2(✉)], Abdelbadiâ Chaker[2], Juan Sandoval[1], Sami Bennour[2], Abdelfattah Mlika[2], Lotfi Romdhane[2,3], Saïd Zeghloul[1], and Med Amine Laribi[1]

[1] Department of GMSC, Pprime Institute CNRS, ENSMA, University of Poitiers, UPR 3346, Poitiers, France
ferdaws.ennaiem@univ-poitiers.fr

[2] Mechanical Laboratory of Sousse (LMS), National Engineering School of Sousse, University of Sousse, 4000 Sousse, Tunisia

[3] American University of Sharjah, P.O. Box 26666, Sharjah, United Arab Emirates

Abstract. A visual based control method based on coupling a camera with a cable-driven parallel robot (CDPR) is proposed in this paper. This work is motivated by the need of overcoming the robot uncertainties resulting from cable vibrations or inaccurate initial configurations. The poses taken by the end-effector are recorded in real time and used as feedback for the control loop to enhance the robot accuracy. Experimental tests were conducted on a planar CDPR dedicated to rehabilitation and allowing to assist the patient's upper limb along a preselected path. The reference path of the prescribed exercise was recorded, for a healthy volunteer, using a motion capture system. The implementation results show a remarkable increase in the robot accuracy. In addition, thanks to the proposed correction, the initial pose of the end-effector has no longer an impact on the robot behavior. The prescribed task can be performed independently of the starting pose.

Keywords: Visual based control · Rehabilitation · Cable-driven parallel robot (CDPR) · Accuracy · Starting pose

1 Introduction

Cable-driven parallel robots (CDPRs) are a particular type of parallel robots where the rigid links are replaced by cables coiled on actuated pulleys [1]. By adjusting the length of each cable, the poses of the end-effector can be controlled. Despite the multiple advantages offered by the flexible nature of the cables, such as extending the workspace and lightening the weight of the moving parts [2], this flexibility affects considerably the control and the position accuracy of the robot. Several other factors may undesirably affect the robot accuracy such as inaccurate quantification of the initial lengths of cables or inexact positions of the anchor and exit points (points connecting the cables to the end-effector and to the robot fixed frame, respectively) [3].

A. Müller and M. Brandstötter (Eds.): RAAD 2022, MMS 120, pp. 97–105, 2022.
https://doi.org/10.1007/978-3-031-04870-8_12

Open-loop control is generally adopted for CDPRs actuation, where the input is the desired angular positions of the actuators. The resulting end-effector displacements are estimated using the Forward Kinematic Model (FKM). However, the FKM is generally difficult to solve due to the system redundancy [4]. Therefore, the real pose of the end-effector cannot be computed and thus compared to the input order.

As a solution to this issue, external sensors can be added to the robot control loop in order to measure the end-effector poses. In [5], the authors proposed to introduce cable angular position and length sensors. The use of a force sensor is discussed in [6] to assess the cable tensions and thus to control the mobile platform accurately and guarantee its stability. In [7], the combination of three laser distance sensors and one camera were employed to track the actual positions and orientations of the end-effector.

This paper proposes the integration of an exteroceptive sensor in the feedback control loop of a planar CDPR. This sensor is formed by an OptiTrack camera and a set of passive reflective markers. The proposed robot was developed to assist patients with upper limb motor disability during rehabilitation therapies. The performed experiment consists of a patient holding a handle attached to the end-effector. This latter is instructed to follow a prescribed path having an "8" shape. Using the data collected by the camera, the end-effector poses are continuously corrected to minimize the deviation from the desired path.

The paper is structured as follows: Sect. 2 details the experimental protocol followed to identify the task's path as well as the robot prototype design. Section 3 presents the proposed closed-loop control model to improve the CDPR accuracy and shows the obtained results. The last section concludes the paper.

2 Experimental Setup

2.1 Rehabilitation Task Analysis

Robotic rehabilitation is gaining a wide interest compared to traditional reeducation techniques thanks to its capability of performing repetitive and intensive exercises [8]. They allow assisting the affected limb along preprogrammed tasks in the aim of recovering its functional abilities. In this paper, the selected exercise consists of generating an 8 shaped path with a variable rotation angle φ around the z-axis normal to the trajectory plane.

In this context, a healthy person was asked to perform the desired task. A motion capture system, using four Qualisys cameras and a set of reflective markers attached to the volunteer's hand, was set up to record the latter's motion (see Figs. 1a and 1b). These measurements, illustrated in Figs. 1c and 1d will be used as the reference path to control a planar CDPR for upper limb rehabilitation.

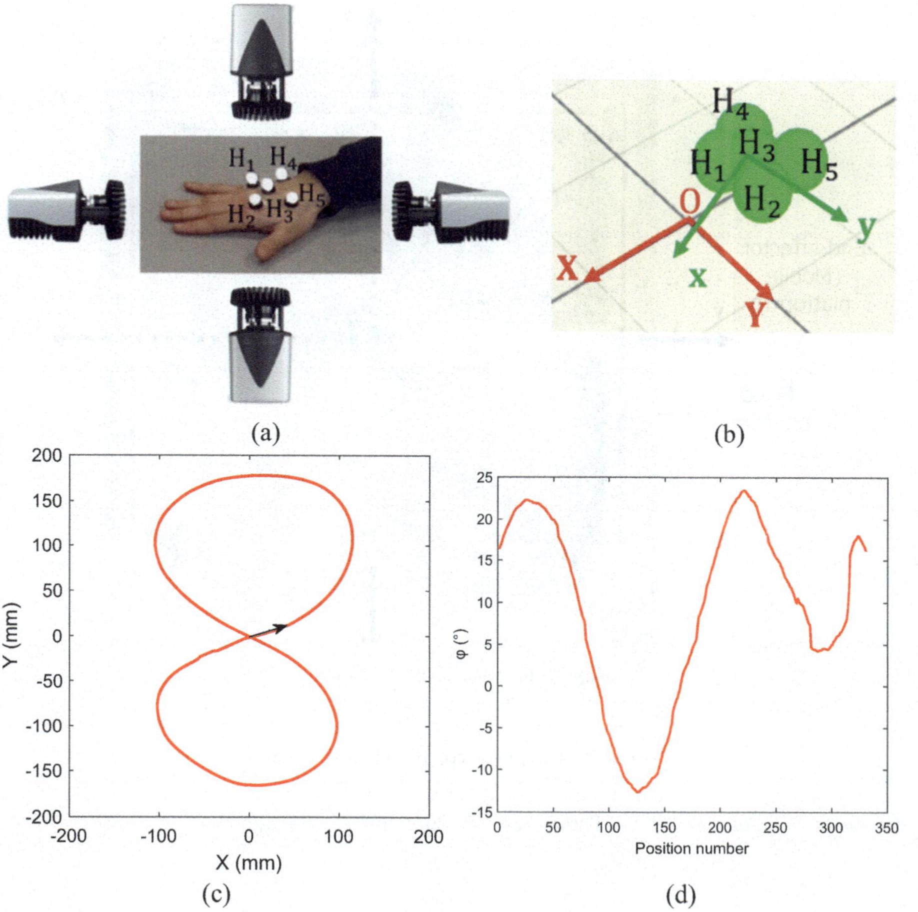

Fig. 1. Extraction of the reference path: (a) Motion capture setup, (b) Markers labeling, (c) "8" shaped path identified by tracking the marker (H_3), and (d) the corresponding orientations.

2.2 CDPR Prototype

The robot under study is a planar CDPR which its prototype is shown in Fig. 2. The CDPR is formed by four cables, wrapped around four motorized pulleys, and a square shape end-effector. The position of the actuators, as well as the size of the mobile platform, have been optimized in a previous work [9], where the minimization of the tension in each cable and the maximization of the elastic stiffness and the dexterity were the selected criteria. Four servomotors of type Dynamixel MX 106T were controlled using Robot Operating System (ROS), an open-source robotics middleware. The geometric parameters characterizing the CDPR prototype are summarized in Table 1.

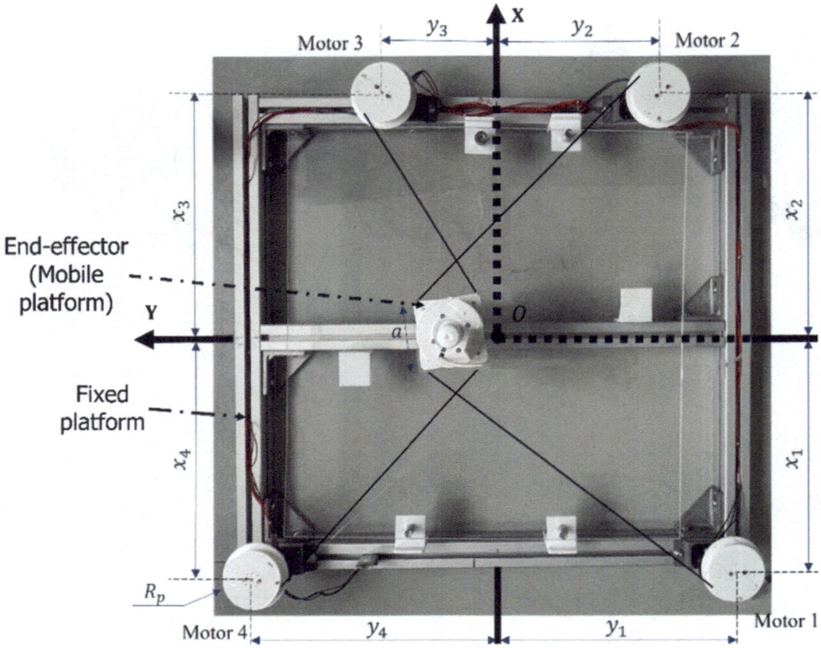

Fig. 2. Planar CDPR prototype.

Table 1. The prototype parameters values.

Parameter	Value [mm]	Parameter	Value [mm]	Parameter	Value [mm]
x_1	−375	x_3	375	a	100
y_1	−385	y_3	180	R_p	37
x_2	375	x_4	−375		
y_2	−260	y_4	385		

3 Visual-Based Control Technique

3.1 Experimental Setup and Visual Servoing

A V120:Trio tracking device, from OptiTrack, was used to measure the end-effector poses. It is composed of a set of 3 aligned cameras mounted on a support and connected to a computer. The data acquisition was performed via an OptiTrack software called "Motive". This camera sends continuously at a rate of 120 Hz three coordinates and four elements of rotation (quaternions), in its frame, of a rigid body representing its positions and the orientations. The rigid body is composed of at least three markers using a predefined package called "mocap OptiTrack".

The ROS environment of the proposed solution is given in Fig. 3 and the exchanging data between nodes are detailed in Table 2. In order to measure the poses of the end-effector with respect to the robot fixed frame $\mathcal{R}_0(O, \mathbf{X}, \mathbf{Y})$, two rigid bodies were created, as shown in Fig. 3. One (MP) is associated with the mobile platform and the second (FP) with the base of the robot. A first node was created under ROS to compute the transformation matrices between the different coordinate systems.

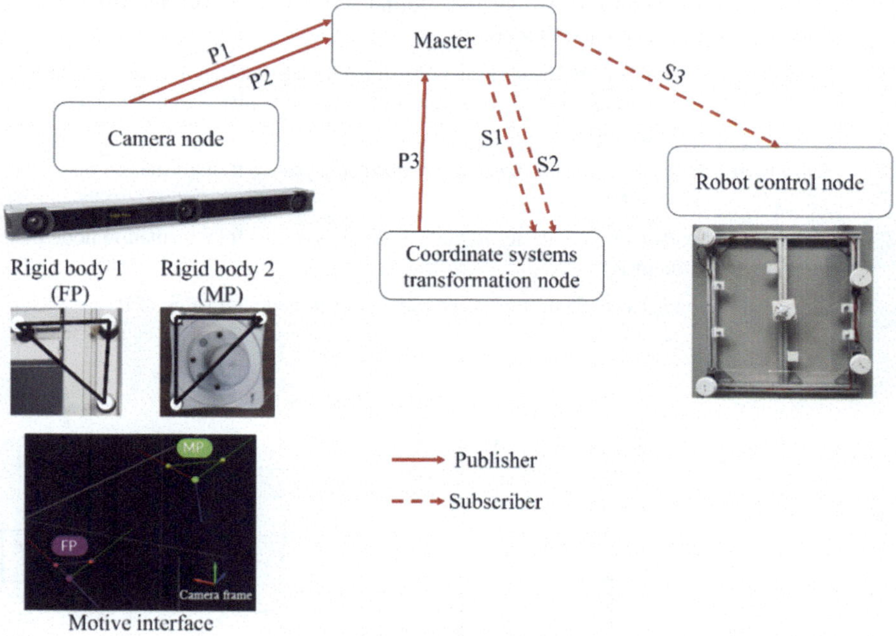

Fig. 3. ROS environment of the proposed method.

A second node controlling the robot was also constructed based on the control law illustrated in Fig. 4. The function named "Angular position computation" takes as an input the current pose (χ_i) and the next pose (χ_{i+1}) of the mobile platform and returns as output the needed rotation angles of the four actuators allowing them to reach the target point. The desired path, presented in Fig. 1a, is segmented into n points and for each point, the position error measured between the desired and the real poses is computed and corrected at the following point. The definition of the different parameters used in Fig. 4 are listed in Table 3.

3.2 Results

The control law detailed above has been experimentally validated and its robustness is confirmed by Fig. 5, where the desired end-effector poses as well as the real ones measured by the camera for the closed-loop control match closely, which is not the case with the open-loop control (Fig. 5).

Table 2. Topics details.

Topic	Type	Function
P1	Publisher	Publishes from the camera node to the master the positions and the orientations of the rigid body forming the fixed platform (FP) with respect to the camera frame
P2	Publisher	Publishes from the camera node to the master the positions and the orientations of the rigid body forming the mobile platform (MP) with respect to the camera frame
P3	Publisher	Publishes from the coordinate systems transformation node to the master the positions and the orientations of the mobile platform in the fixed platform frame
S1	Subscriber	Sends from the master to the coordinate systems transformation node the value of P1
S2	Subscriber	Sends from the master to the coordinate systems transformation node the value of P2
S3	Subscriber	Sends from the master to the robot control node the value of P3

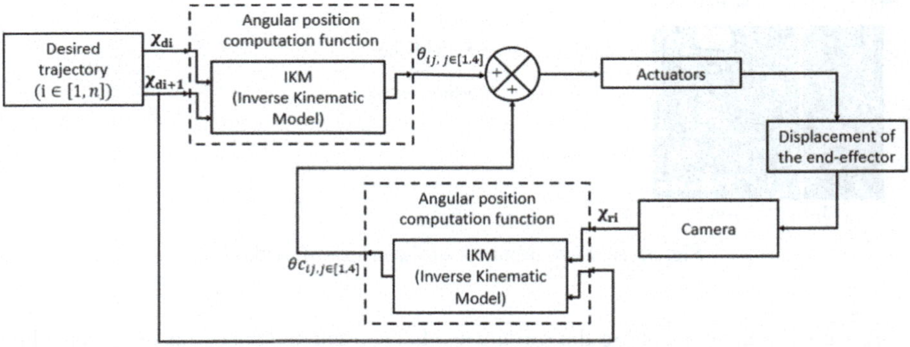

Fig. 4. The proposed control schema.

Table 3. The signification of the control schema parameters.

Designation	Definition
χ_{di}	The ith desired end-effector pose
θ_{ij}	The ith angular position of the actuator j
θc_{ij}	The correction value of the angular position added to the actuator j at the ith pose
χ_{ri}	The ith real end-effector pose
n	The number of points forming the desired path

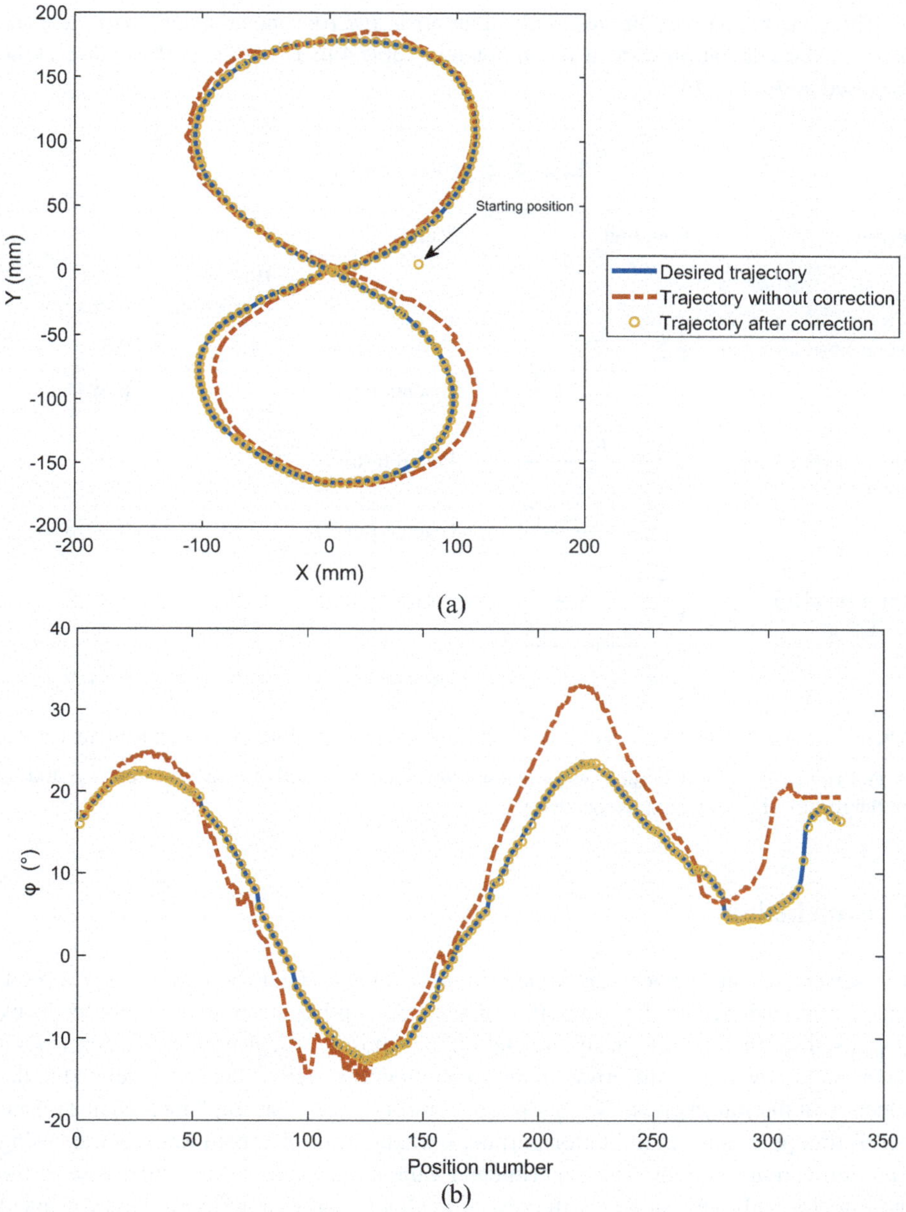

Fig. 5. Obtained results, comparison between the desired and the real (a) trajectories and (b) orientations without and with visual technique correction.

For the open-loop control, the end-effector was positioned manually, as close as possible to its initial pose. However, for the closed-loop control, the end-effector was positioned initially at an arbitrary pose, which will be corrected automatically thanks to the proposed closed-loop approach.

Three errors, namely the mean absolute error, the root-mean-square error, and the error standard deviation were also computed in Table 4 to assess the performance of the proposed control method.

Table 4. Computed errors.

Error	Equation	Value	Before correction	After correction
Mean absolute error	$\frac{1}{n}\sum_{i=1}^{n}\lvert\mathcal{P}_{di}-\mathcal{P}_{ir}\rvert$	Position (mm)	11.3	0.5
	$\frac{1}{n}\sum_{i=1}^{n}\lvert\omega_{di}-\omega_{ir}\rvert$	Orientation (°)	3.9	0.0841
Root-mean-square error	$\sqrt{\frac{\sum_{i=i}^{n}(\mathcal{P}_{di}-\mathcal{P}_{ir})^2}{n}}$	Position (mm)	18.7	1.07
	$\sqrt{\frac{\sum_{i=i}^{n}(\omega_{di}-\omega_{ir})^2}{n}}$	Orientation (°)	5.01	0.1428
Error standard deviation	$\sqrt{\frac{\sum_{i=i}^{n}(\varepsilon_{\mathcal{P}i}-\bar{\varepsilon}_{\mathcal{P}})^2}{n-1}}$	Position (mm)	17.6	1.07
	$\sqrt{\frac{\sum_{i=i}^{n}(\varepsilon_{\omega i}-\bar{\varepsilon}_{\omega})^2}{n-1}}$	Orientation (°)	4.5	0.1429

Where $\mathcal{P}_{di}$, ω_{di}, $\mathcal{P}_{ir}$, and ω_{ir} refer to the ith desired and real positions and rotation angle, respectively. $\varepsilon_{\mathcal{P}i} = \mathcal{P}_{di} - \mathcal{P}_{ir}$ and $\varepsilon_{\omega i} = \omega_{di} - \omega_{ir}$ are the ith position and orientation errors. $\bar{\varepsilon}_{\mathcal{P}}$ and $\bar{\varepsilon}_{\omega}$ are the means of $\varepsilon_{\mathcal{P}i}$ and $\varepsilon_{\omega i}$, respectively.

4 Conclusion

In this paper, a visual servoing approach aiming to enhance CDPRs accuracy is proposed. This method is based on the integration of an exteroceptive sensor in the robot feedback loop, used to track continuously the end-effector poses. A real-time regulation is then performed to overcome the errors in the robot model as well as the issues related to the selection of the starting pose. The proposed control method was implemented in a planar CDPR for upper limb rehabilitation purposes, where the desired path was recorded using a motion capture system. The experimental validation, namely the comparison of the end-effector path without and with correction clearly shows a noticeable improvement in the robot accuracy in position and orientation.

Acknowledgments. This work was financially supported by the "PHC Utique" program of the French Ministry of Foreign Affairs and Ministry of higher education, research and innovation and the Tunisian Ministry of higher education and scientific research in the CMCU project number 19G1121.

References

1. Qian, S., Zi, B., Shang, W.W., Xu, Q.S.: A review on cable-driven parallel robots. Chinese J. Mech. Eng. **31**(1), 1–11 (2018)
2. Xiong, H., Diao, X.: A review of cable-driven rehabilitation devices. Disabil. Rehabil. Assist. Technol. **15**(8), 885–897 (2020)
3. Wang, B., Caro, S.: Exit point, initial length and pose self-calibration method for cable-driven parallel robots. In: Zeghloul, S., Laribi, M.A., Arsicault, M. (eds.) MEDER 2021. MMS, vol. 103, pp. 90–101. Springer, Cham (2021). https://doi.org/10.1007/978-3-030-75271-2_10
4. Santos, J.C., Gouttefarde, M.: A real-time capable forward kinematics algorithm for cable-driven parallel robots considering pulley kinematics. In: Lenarčič, J., Siciliano, B. (eds.) ARK 2020. SPAR, vol. 15, pp. 199–208. Springer, Cham (2021). https://doi.org/10.1007/978-3-030-50975-0_25
5. Fortin-Côté, A., Cardou, P., Campeau-Lecours, A.: Improving cable driven parallel robot accuracy through angular position sensors. In: 2016 IEEE/RSJ International Conference on Intelligent Robots and Systems (IROS), pp. 4350–4355. IEEE (2016)
6. Picard, E., Caro, S., Claveau, F., Plestan, F.: Pulleys and force sensors influence on payload estimation of cable-driven parallel robots. In: 2018 IEEE/RSJ International Conference on Intelligent Robots and Systems (IROS), pp. 1429–1436. IEEE (2018)
7. Korayem, M.H., et al.: A novel method for recording the position and orientation of the end effector of a spatial cable-suspended robot and using for closed-loop control. Int. J. Adv. Manuf. Technol. **72**(5–8), 739–755 (2014). https://doi.org/10.1007/s00170-014-5681-2
8. Maciejasz, P., Eschweiler, J., Gerlach-Hahn, K., Jansen-Troy, A., Leonhardt, S.: A survey on robotic devices for upper limb rehabilitation. J. Neuroeng. Rehabil. **11**(1), 1–29 (2014)
9. Ennaiem, F., et al.: Task-Based design approach: development of a planar cable-driven parallel robot for upper limb rehabilitation. Appl. Sci. **11**(12), 5635 (2021)

Explaining Local Path Plans Using LIME

Amar Halilovic(✉) and Felix Lindner

Ulm University, Ulm, Germany
{amar.halilovic,felix.lindner}@uni-ulm.de

Abstract. As robots are becoming a more significant part of humans' daily life, there is a challenge to bridge the gap between robots' actions and humans' understanding of what robots are doing and how they make their decisions. We present an approach to local navigation explanation based on Local Interpretable Model-agnostic Explanations (LIME), a popular approach from the Explainable Artificial Intelligence (XAI) community for explaining individual predictions of black-box models. We show how LIME can be applied to a robot's local path planner. We experimentally evaluate the explanation method's runtime, quality, and robustness, and discuss implications for the robotic domain.

Keywords: Explainable artificial intelligence · Robot navigation · Obstacle avoidance

1 Introduction

Robots in social environments raise the requirement for explainability of robot's behavior [1]. As the tendency of robots' presence in society grows [2], this requirement becomes more pronounced. According to the IEEE Guideline for Ethically Aligned Design [3], robots' decisions should be transparent and backed up by clear justifications. The need for accountability of autonomous intelligent robots and transparency of their actions underlines the human right to explanation in face of machines making decisions that affect humans. Current decision-making methods in robotics largely lack explainability [4], and thus limit the faster adoption of robots in important tasks [5]. A lack of explainability also is a safety issue when robots behave in unexpected ways putting humans in highly sensitive environments at risk.

We address explainability in robotics by focusing on explainable robot navigation in social environments [6,7]: Imagine a robot navigating in a known environment with the possibility to encounter humans and obstacles. Local path planners allow robots to follow a global path plan while dynamically reacting to unexpected occurrences. Some of the robot's decisions may require abrupt stops or changes of direction and path deviations thus surprising people in the neighborhood or even scaring them. This can lead to trust loss, which needs to get mitigated [8]. One mitigation strategy is explanation. Using Local Interpretable Model-agnostic Explanations (LIME) [9], an established method from

A. Müller and M. Brandstötter (Eds.): RAAD 2022, MMS 120, pp. 106–113, 2022.
https://doi.org/10.1007/978-3-031-04870-8_13

Explainable Artificial Intelligence (XAI) [10], we demonstrate how a robot can generate visual explanations of its local decision-making in path planning and obstacle avoidance.

The paper is structured as follows: We first review related work on explainable robot navigation. Section 3 introduces LIME as a standard XAI method. In Section 4, we show how LIME can be used in tandem with a local path planner (we are using the Timed Elastic Band (TEB) planner [11]) to generate explanations. We present results from a lab experiment. Finally, we discuss limitations and implications of our work.

2 Related work

Recently, explainable robot navigation has gained considerable interest. Rosenthal and colleagues [12] and Perera and colleagues [13] have worked on verbal explanations of robot navigation. The authors propose a method for automatically generating summaries of whole trajectories. The work by Stein [14] is also targeted at generating natural language explanations of navigational choices. It focuses on generating explanations from interpretable-by-design navigation algorithms. Contrary to the mentioned approaches, we aim at explaining local deviations from the original path plan instead of whole narrations of trajectories. Moreover, we focus on visual explanations rather than natural-language explanations.

He and colleagues [15] use visual explanations but present them as numerical data or refer to a specific neural network that is still far from a simple enough explanation for laymen. Bautista-Montesano and colleagues [16] present a navigation explanation approach that is limited to only fuzzy reinforcement learning navigation agents. Thus, these approaches make strong assumptions about the underlying algorithm to be explained, as their explanation methods require looking into the black box.

Brandao and colleagues [17] focus on gloabl path optimality for explanation by presenting users with an alternative global path and comparing it with the optimal one. We do not focus on optimal path explanation, but rather on explanations of local deviations from the initial path. The work by Brandao et al. [18] is focused on explanations of planning failures, which we also encompass as situations when a local planner cannot find a traversable (joint) local plan.

In summary, we observe that current state-of-the-art approaches are constrained to model-dependent methods, and to explanations in terms of static objects known before navigation, or to a whole trajectory. Instead, we apply LIME as an out-of-the-box model-agnostic explanation method to obtain visual explanations of a robot's decision-making in local navigation tasks.

3 Technical Background

LIME [9] is a model-agnostic local explainable AI (XAI) technique that explains predictions of a black-box model by learning an interpretable model around the instance of interest. We focus on visual explanations, so we use an image (viz., the local costmap, see below) as the instance of interest. LIME for images[1] takes the input image and partitions it into segments—called superpixels—thereby creating interpretable features. Then it perturbs interpretable features by turning them off to generate new input samples (perturbations) in the neighborhood of the instance of interest. The number of perturbations is a user-defined parameter. There can be a maximum of 2^n unique perturbations, where n is the number of interpretable features. For every perturbation, LIME queries the black-box model and thereby generates a local data set of (perturbed) neighbor images and the respective black-box model's predictions. On this new dataset LIME trains an interpretable model, viz., a weighted linear regression model. The explanation is obtained by interpreting the coefficients of the trained linear model: The importance of each segment in the image for the behavior of the black-box model is represented by one coefficient in the linear model. Depending on the sign of the coefficient, the interpretable feature (viz., the segment in the image) positively or negatively affects the black-box model's prediction. That said, to apply LIME to visually explaining local navigation, we need to provide a suitable method for computing a segmentation of a local costmap (viz., the interpretable features). Moreover, the output of the local planner has to get interpreted as the prediction of some black-box model.

4 Explaining Local Paths with LIME

4.1 Technical Set-up

Our set-up is situated in the context of the ROS navigation stack.[2] That is, a global path planner has generated a global path plan for the robot to navigate to a specified goal position. For path following and obstacle avoidance, a local path planner takes the local costmap and the global path as input and outputs a local path (in terms of a velocity vector) for the robot to execute. For LIME to be applicable, the black-box behavior needs to be deterministic. Therefore, we do not employ sampling-based planners, such as DWA or RRT, but instead, employ the TEB planner [11].

To use LIME for generating visual explanations of local path plans in terms of obstacles in the local neighborhood of the robot, we use the local costmap as an instance of interest, and the TEB planner as the black box that takes that costmap as input and outputs some local path. LIME first segments the local costmap into obstacles as interpretable features. The size of interpretable feature space is set to 8, as it covered all cases in our lab environment. The SLIC [19] segmentation algorithm is used to get obstacle segments.[3] If there are less than 8 obstacles, obstacle segments are divided, starting from the biggest one,

[1] https://github.com/marcotcr/lime.

[2] http://wiki.ros.org/navigation.

[3] We used the following SLIC parametrization: *n_segments* = 10, *compactness* = 100.0, *max_iter* = 1000, *min_size_factor* = 0.01, *max_size_factor* = 10.

until the number of 8 segments is reached. As the second step, LIME obtains perturbations of the segmented costmap by turning-off segments by replacing them with free space. The perturbed local costmap together with the global plan, the robot's footprint, and its current velocities form the input to the local path planner, which then outputs a local plan for the perturbation at hand. The deviation of the so-calculated local plan from the global plan is taken as a target for the interpretable model. The deviation is calculated as a sum of the minimal point-to-point L2 differences between the local and the global plan. The other important LIME parameters—cosine distance metric and Ridge linear regressor—are kept as in the LIME standard implementation.

The coefficients of the linear regression model output by LIME are the primary representation of the explanation. We turn these explanations into some visually more appealing and accessible representation. By coloring segments based on their coefficients, we get an explanation image for each local navigation decision. The sign of the coefficient dictates the color: Positive-weighted segments are colored green and negative-weighted segments are colored red. A green-colored segment indicates *"without that segment, the local plan would deviate less from the global plan"* and a red-colored segment means *"without that segment, the local plan would deviate more from the global plan"*. Color intensity is set proportional to the coefficient with intensities in the range $[0, 255]$ in RGB color space, where 255 corresponds to the coefficient with the highest absolute value.

4.2 Results

Qualitative Results. Figure 1a, 1d and 1g show three characteristic local navigation cases (C1, C2, and C3) in our lab. The robot (a TIAGo from PAL robotics) tries to follow the global plan that leads it through the doorway. Figures 1b, 1e, and 1h show the local costmaps for three local navigation cases with black obstacles, the robot's location (white), and free space (grey). In C1, the local plan (yellow dots) mostly coincides with the global plan (blue dots), while in C2 starting point and endpoint of the local plan could not be connected into a joint trajectory. In C3, a local plan is found, which however clearly deviates from the global plan. With LIME we explain how obstacles and/or parts of obstacles contribute to the deviation. From the explanation images 1c, 1f, and 1i we can derive the following explanations:

- C1: "The right (green) wall segment increases deviation, while the left (red) wall segments decreases it, squeezing it to the doorway."
- C2: "The (green) obstacle increases the deviation because if it was not there the robot would follow the global plan. If the wall (red) was not there, the local path planner would be able to create the connected local plan and deviate from the global plan, thus increasing deviation distance."
- C3: "Both obstacles increase the deviation, but the round one does so more significantly. If it were not there, the robot would follow the global plan without deviating from the global path. If the rectangular obstacle was not there, the robot would still deviate, but with a smaller deviation."

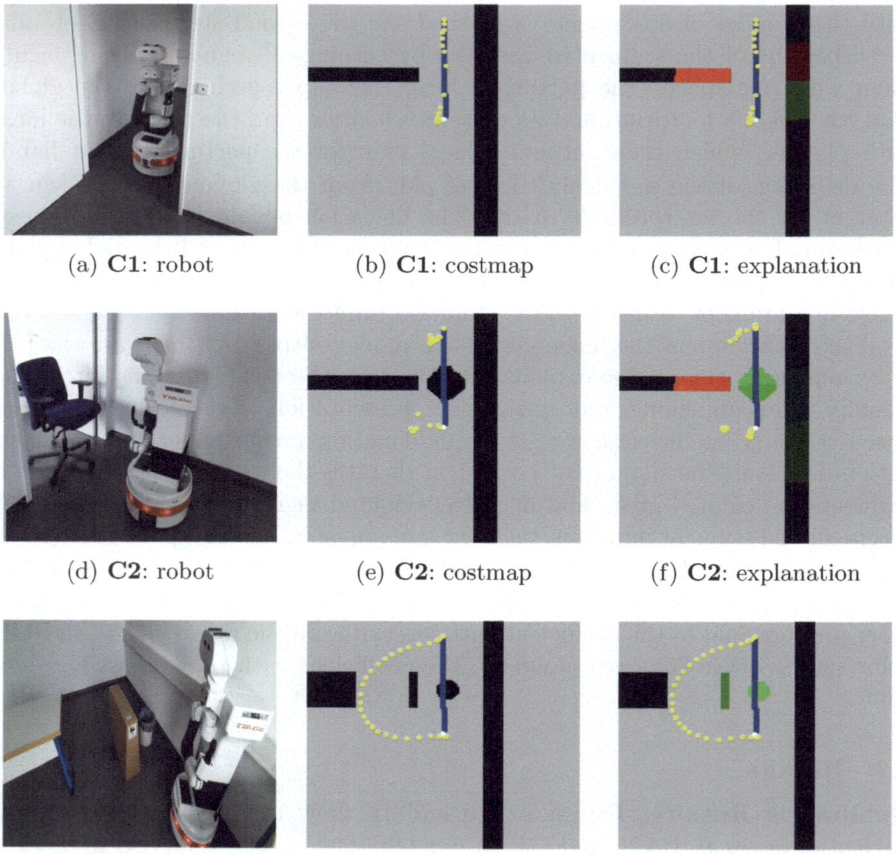

(a) **C1**: robot (b) **C1**: costmap (c) **C1**: explanation

(d) **C2**: robot (e) **C2**: costmap (f) **C2**: explanation

(g) **C3**: robot (h) **C3**: costmap (i) **C3**: explanation

Fig. 1. C1: A free doorway allows TIAGo to follow the initial trajectory and move through the doorway. Because TIAGo is too close to the right wall it has to adjust its position and then proceed through the doorway. **C2**: The same doorway is blocked with a chair, so TIAGo cannot progress through the doorway. It stops and rotates in place to try to go left. **C3**: A table, box, and a wall form two doorways where the right doorway through which TIAGo should go is suddenly blocked with a trash can. The robot must deviate from the initial trajectory traversing through the free doorway.

Quantitative Results. We measure *objective explanation quality* as the difference of variance and mean of absolute values of the coefficients in a given LIME explanation. The higher this difference is the bigger the separation between the importance of the segments. The bigger the separation is, the more important segments are more prominent, and it is clearer which segments contribute to the local path planner's decision. If the separation is small, it may not be clear which segments contribute and which do not, and explanation may not be useful. Figure 2a shows that this difference increases in relation to the increase of the

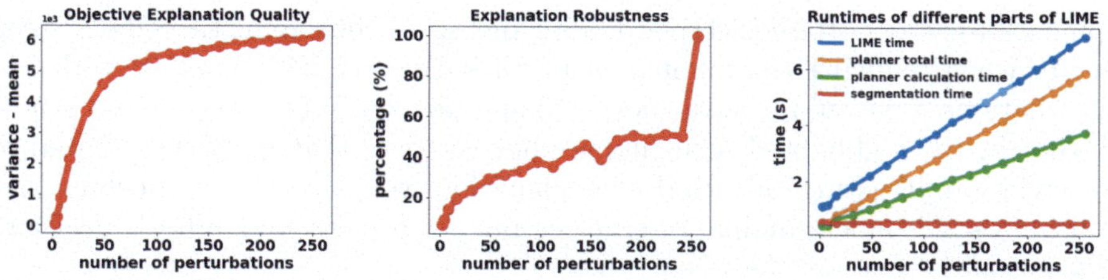

(a) Quality (b) Robustness (c) Runtime

Fig. 2. a) The difference of variance and mean of absolute values of the explanations' coefficients. The higher the better. b) Percentage of the order of coefficients preserved in relation to the decrease of the number of perturbations. The higher the better. c) Total runtimes of different parts of the explanation method.

number of perturbations. As could be expected, with more perturbations LIME outputs better explanations.

For *explanation robustness*, we measure to what extent the relative segment importance is affected by reducing the number of perturbations. Figure 2b shows the percentage of the order of coefficients preserved in relation to the decrease of the number of perturbations. After reducing the number of samples by 50%, only around 40% of the original order of coefficients is preserved. If the reduction continues, under 30% of the original order of coefficients is preserved.

Finally, we analyse *explanation runtime*.[4] LIME's runtime is generally high and increases linearly in the number of perturbations as shown in Fig. 2c. The runtimes of the most important parts of the LIME procedure are plotted. *Planner total time* (orange) takes the biggest part of the total explanation runtime and includes the preparation of input data for the planner (TEB), planner calculation time of all the paths for each perturbation, and the collection of the planner's outputs. Figure 2c also shows that the *planner calculation time* (green) takes the biggest part of the *planner total time*. Both *planner total time* and *planner calculation time* increase relative to the increase of the number of perturbations. As segmentation only needs to be done once for each explanation, its runtime (red) is not affected by the number of perturbations.

5 Discussion

The results show that LIME generates better and more robust visual explanations of local path plans with more perturbations. This translates to having a better distinction of more important segments and better stability of their importance as the number of perturbations grows. The objective explanation quality measure declines slowly until the 100 perturbations and then more rapidly. We could thus reduce the number of perturbations to 100 and still get good

[4] All reported runtime experiments run on a Lenovo Thinkpad E14 Gen2 with AMD Ryzen 5 4500U 2.3 GHz and 16 GB of RAM.

explanations (w.r.t. the objective quality measure), but robustness falls steep already when reducing the number of perturbations to 240. That is, with 240 perturbations, we would receive good explanations which however are not truthful, that is, the highlighted segments are not the ones that truly contributed to the deviation. Trusting this kind of explanation would result in a feeling that we have a solid understanding of true reasons for a particular robot's decision, while actually, they are the wrong ones.

On the other side of the coin, using more perturbations means slower explanation generation. The LIME runtime of around 8 s with 256 perturbations might be good enough for the post-hoc generation of explanations of recorded decisions. This could be helpful in situations where fast robot explanation feedback is not required, such as in a lab, remote testing grounds, and environments where people are informed of the robot's intentions and can anticipate a certain set of robot behaviors. But LIME has clear limitations in that this method cannot, in its current form, be used for real-time explanations. Fast-changing and socially-complex environments like streets or places with people might require real-time explanations. The runtime measures indicate that only with less than 50 perturbations, we do have a runtime of around 2 s (which could be interpreted as fast in some contexts). But this comes with a big decrease in explanation quality and robustness. Achieving good and robust explanations in near real-time can be achieved only when a very small number of obstacles are considered as potential explanations. But even with a time frame of 2 s, not every TEB call (every 200 ms) can be explained in real-time.

6 Conclusions

To address the explainability of robot navigation, we have presented an explanation method based on LIME which generates visual explanations of the robot's local path-following behavior. The results from our lab experiments showed that to achieve high explanation quality and robustness, LIME needs to be configured to sample exponentially many perturbations (in the number of segments). Runtime thus grows exponentially in the number of obstacles considered as potential parts of an explanation. This limits the usefulness for real-time explanation generation. Consequently, our future work will focus on speeding up explanation generation towards real-time performance. Moreover, we will integrate semantic information to improve the segmentation algorithm and to enhance the explanations by natural-language annotations.

References

1. Lomas, M., Chevalier, R., Cross, E.V., Garrett, R.C., Hoare, J., Kopack, M.: Explaining robot actions. In: Proceedings of the Seventh Annual ACM/IEEE International Conference on Human-Robot Interaction, pp. 187–188 (2012)
2. Lim, V., Rooksby, M., Cross, E.S.: Social robots on a global stage: establishing a role for culture during human-robot interaction. Int. J. Soc. Robot. **13**(6), 1307–1333 (2021)

3. IEEE. Ethically aligned design, first edition overview – a vision for prioritizing human well-being with autonomous and intelligent systems. In: Technical Report (2019)
4. Wachter, S., Mittelstadt, B. and Floridi, L.: Transparent, explainable, and accountable AI for robotics. Sci. Robot. **2**(6), eaan6080 (2017)
5. Edmonds, M., et al.: A tale of two explanations: enhancing human trust by explaining robot behavior. Sci. Robot. **4**(7), eaay4663 (2019)
6. Lindner, F.: Towards a formalization of explanations for robots' actions and beliefs. In: JOWO 2020 Proceedings of the FOIS Workshop Ontologies for Autonomous Robotics (ROBONTICS 2020) (2020)
7. Karalus, J., Halilovic, A., Lindner, F.: Explanations in, explanations out: human-in-the-loop social navigation learning. In: ICDL Workshop on Human aligned Reinforcement Learning for Autonomous Agents and Robots (2021)
8. Tolmeijer, S.: Taxonomy of trust-relevant failures and mitigation strategies. In: Proceedings of HRI 2020 (2020)
9. Ribeiro, M.T., Singh, S. and Guestrin, C.: Why should I trust you?" explaining the predictions of any classifier. In: Proceedings of the 22nd ACM SIGKDD International Conference on Knowledge Discovery and Data Mining, pp. 1135–1144 (2016)
10. Miller, T.: Explanation in artificial intelligence: Insights from the social sciences. Artif. Intell. **267**, 1–38 (2019)
11. Rösmann, C., Feiten, W., Wösch, T., Hoffmann, F., Bertram, T.: Trajectory modification considering dynamic constraints of autonomous robots. In: ROBOTIK 2012; 7th German Conference on Robotics, VDE, pp. 1–6 (2012)
12. Rosenthal, S., Selvaraj, S.P., Veloso, M.M.: Verbalization: narration of autonomous robot experience. In: IJCAI, vol. 16, pp. 862–868 (2016)
13. Perera, V., Selveraj, S.P., Rosenthal, S., Veloso, M.: Dynamic generation and refinement of robot verbalization. In: 2016 25th IEEE International Symposium on Robot and Human Interactive Communication (RO-MAN), pp. 212–218 (2016)
14. Stein, G.: Generating high-quality explanations for navigation in partially-revealed environments. In: Advances in Neural Information Processing Systems, vol. 34 (2021)
15. He, L., Aouf, N., Song, B.: Explainable deep reinforcement learning for UAV autonomous path planning. Aerospace Sci. Technol. **118**, 107052 (2021)
16. Bautista-Montesano, R., Bustamante-Bello, R., Ramirez-Mendoza, R.A.: Explainable navigation system using fuzzy reinforcement learning. Int. J. Interact. Des. Manuf. (IJIDeM) **14**(4), 1411–1428 (2020)
17. Brandao, M., Coles, A., Magazzeni, D.: Explaining path plan optimality: fast explanation methods for navigation meshes using full and incremental inverse optimization. In: Proceedings of the International Conference on Automated Planning and Scheduling, vol. 31, pp. 56–64 (2021)
18. Brandao, M., Canal, G., Krivić, S., Magazzeni, D.: Towards providing explanations for robot motion planning. In: 2021 IEEE International Conference on Robotics and Automation (ICRA), pp. 3927-3933. IEEE (2021)
19. Achanta, R., Shaji, A., Smith, K., Lucchi, A., Fua, P., Süsstrunk, S.: Slic superpixels compared to state-of-the-art superpixel methods. IEEE Trans. Patt. Anal. Mach. Intell. **34**(11), 2274–2282 (2012)

A Fast Method for Explanations of Failures in Optimization-Based Robot Motion Planning

Matthias Eder(✉) and Gerald Steinbauer-Wagner

Graz University of Technology, 8010 Graz, Austria
{matthias.eder,steinbauer}@ist.tugraz.at

Abstract. The transparent interaction between an operator and a robot system is essential for successful task completion. This requires a mutual understanding of decisions and processes in order to provide accurate diagnoses and troubleshooting alternatives in the event of a failure. Due to inaccuracies in the environmental perception or planner configuration, errors can occur in robot motion planning that are hard to understand by the operator. In this work we present a method that is able to provide explanations for motion planning failures quickly. In the context of optimization-based planners, failures origin from planning constraints can be identified using an adaption of the FastDiag algorithm. It is able to provide one preferred minimal diagnosis in logarithmic time, also for large constraint sets. To evaluate the applicability of the proposed method, experiments are conducted that compare the computational performance to an existing method while considering different parameters such as number of constraints and requested diagnoses.

Keywords: Explainable artificial intelligence · XAI · Human-robot interaction · Motion planning · FastDiag · Diagnosis

1 Introduction

The research field of robot motion planning nowadays has its focus on the efficient computation of decent trajectories which are applicable in various scenarios. Therefore, the consideration of not only environmental constraints but the robot's kinematics and dynamics in planning is a crucial step to compute motions that are executable by the robotic system [9]. Common motion planning algorithms are sample-based planners (e.g. [13]), optimization-based trajectory planners (e.g. [12]), and probabilistic planners using motion primitives (e.g. [16]). A comprehensive review on motion planning techniques can be found in [9]. While these frameworks aim to increase the autonomy of a robot by providing navigable trajectories they are prone to difficulties arising from the robot's environment, its (noisy) perception, and the planner configuration. This might

This work was partially supported by the Austrian Research Promotion Agency (FFG) with the project Easier.

A. Müller and M. Brandstötter (Eds.): RAAD 2022, MMS 120, pp. 114–121, 2022.
https://doi.org/10.1007/978-3-031-04870-8_14

lead to failures in path planning that have to be resolved manually by the operator. To do so, the operator must be able to reason about the robot's failure and operate the robot into a state in which it is able to compute valid plans again [5]. In this work, a failed plan is defined as not finding a traversable path for motion execution. However, the reasoning about a robot's state is a complex task, as the robotic system is often presented as a black-box to the operator. The field of explainable artificial intelligence (XAI) aims to open the black box of intelligent systems and to understand its internal procedures [1] better. Thereby, XAI deals with two major topics. On the one hand it deals with the explanation of data-driven approaches like deep learning techniques [4] while on the other hand it aims to shed light on goal-driven approaches by focusing on the explainability of an agent's behavior, such as motion planning [15]. While recent work already focuses on the provision of explanations for motion planning failures, the approach uses inefficient algorithms which work on small constraint sets only [2].

This work aims to overcome this problem and presents a fast explanation method for failures in robot motion planning. In detail, we focus on explaining efficiently failures of optimization-based motion planners that are caused by inconsistencies in large constraint sets. The constraint set represents the plan space which includes e.g. the environment, the planner configuration, and the robot's dynamics. By comparing and evaluating the proposed method to HSDAG [18], a commonly used algorithm for identifying minimal conflict sets, we aim to get insights into its performance and applicability. The experiments are conducted using the motion planner STOMP [12] which provides feedback on the validity of a plan that can be used to generate explanations for the operator.

The rest of this paper discusses related work (Sect. 2), presents the methodology to efficiently compute diagnoses (Sect. 3) and its evaluation (Sect. 4).

2 Related Work

This chapter discusses research which is related to this work. First, minimal hitting set algorithms, their application and performance for computing diagnoses are discussed. Then, recent research on explaining and narrating robot motion plans and the provision of alternatives in case of failures is presented.

2.1 Diagnosis for Over-Constraint Problems

According to Reiter [18], in order to diagnose a faulty system minimal hitting sets (MHS) are essential as they can represent a set of constraints that cause the system's behavior to be inconsistent. In his early work he proposed a diagnosis theory and defined diagnoses as the minimal hitting sets of conflicts within a system. He also presents an algorithm for computing MHS which was improved by Greiner et al. [10] using directed acyclic graphs instead of a tree, leading to the Hitting Set Directed Acyclic Graph (HSDAG). Other concepts were published as well [7,14], but the HSDAG stayed the most common basic concept. As a starting point, the algorithms need a conflict detector that provides minimal conflicts (set of assumptions or constraint causing an unwanted behavior). For this purpose, algorithms such as simple conflict detection [6] or QuickXPlain can be used [11]. A performance comparison of different minimal hitting set algorithms for diagnosis can be found in [17]. As these algorithms are based on the standard hitting set algorithm from Reiter [18] whose problem is NP-hard in the general case, an exponential runtime for finding all minimal hitting sets is to be assumed. Approaches to find a single minimal cardinality diagnosis typically take less time, but these diagnoses are in many settings not the preferred ones [7]. The major goal of model-based diagnosis tasks is the identification of preferred diagnoses, which are not necessarily the ones with minimal cardinality [7]. The authors in [8] present an efficient diagnosis algorithm for inconsistent constraint sets, called FastDiag, which uses the characterization of a preferred diagnosis as the sorting of the constraints by their importance. This allows the calculation of one preferred diagnosis at a time with logarithmic complexity. Due to its efficient runtime behavior, this algorithm can be used to provide a fast computation of explanations in case of a failure in robot motion planning.

2.2 Explanations in Robot Motion Planning

In [19], the authors present an approach to describe a planned path through an indoor environment in natural language. This verbalization is defined "as the process by which an autonomous robot converts its own experience into language" [19]. By representing the environment as a graph, containing additional information such as floor level, building, room numbers, and distances, simple explanations for the path can be generated in natural language. To adapt the explanations based on user preferences, the concept of verbalization space is introduced which represents variations of possible descriptions based on the level of abstraction, the specificity, and locality. While the presented concept verbally describes the planned path in different levels of abstraction, an explanation for failures in planning or execution and a provision of alternatives is missing.

A recent study on path explanations presents a method to explain the optimality of a path to improve the user's understanding [3]. Using this approach, a user can present an alternative path and ask why it is not the optimal path. By adapting the navigation map until the user's path is optimal, changes can be identified and used as explanation. The proposed method allows for a fast reasoning on why a given alternative was not considered.

The authors of [2] present a concept for explainable motion planners which are capable of providing explanations and alternatives in case of planning failures. They introduce the taxonomy for motion planning explanations and separate them into a set of explanation types such as constraint-based explanations, cost-based explanations, design-based explanations, and environment-based explanations. Using this separation, they present an optimization-based planner as well as a sampling-based planner which are capable of explaining motion planning failures in natural language. In their experiments, they focus on constraint-based failures and offer alternatives by searching and removing conflicting constraints. The authors have shown valuable work in providing natural explanations in robot motion planning [2]. However, the presented work optimizes the constraint set by searching through a powerset which shows exponential runtime. As a solution, constraints are merged into four groups, which are used to generate the explanations. Thus, this approach cannot identify single constraints and limits the granularity of the explanation. In our work, a method for efficiently providing explanations on a failed motion plan is presented to overcome the performance issue when increasing the possible number of constraints for motion planning and to receive more fine-grained explanations.

3 A Fast Explanation Method for Failed Motion Planning

This section presents a fast explanation method for robot motion planning based on [2] which efficiently computes an explanation of a failed planning attempt by identifying one preferred diagnosis as presented in [8]. While the explanation framework in [2] is defined in general terms such that different planning variants are applicable, we focus on optimization-based motion planning, since failures in sample-based approaches can also be caused by randomization or heuristics and thus are not applicable for the presented diagnosis algorithm.

A trajectory ξ generated by a motion planner can be represented as a function $\xi(t) : \mathcal{T} \rightarrow \mathcal{Q}$ which maps time $t \in \mathcal{T}$ into a robot configuration $q \in \mathcal{Q}$ which typically includes a combination of torques, contact forces, link poses, and joint angles [2]. Depending on the application, time can be discrete and represented as an integer $\mathcal{T} = \mathbb{Z}$ or continuous $\mathcal{T} = \mathbb{R}$. To obtain a kinematically feasible trajectory $\xi(t)$ between two given robot configurations $q_s, q_g \in \mathcal{Q}$, a motion planning problem is formulated which aims to minimize a cost function $F[\xi] : \Xi \rightarrow \mathbb{R}^+$ while respecting a set of constraints $\mathcal{H}$ such that $H_i[\xi] \leq 0$. Constraints in planning can be e.g. obstacles or kinematic limitations of the robot. The motion planner then solves the optimization problem by minimizing costs w.r.t. the constraints for planning the trajectory and can thus be formulated as $argmin_\xi \ F[\xi]$ s.t. $\mathcal{H}$. In case of failure, an explanation involves the identification of a minimal diagnosis $\mathcal{D}$ which holds a set of constraints that lead to a failed planning if included in $\mathcal{H}$. A diagnosis is called minimal, if there exists no diagnosis $\mathcal{D}' \subset \mathcal{D}$ s.t. $\mathcal{H} \setminus \mathcal{D}'$ that makes planning consistent [8]. To obtain a preferred diagnosis, a ordering relation '$<$' is to be applied on $\mathcal{H}$, enumerating the constraints $\mathcal{H}$ by their importance in increasing order $H_1..H_n$, starting with the least important constraint such that $H_i < H_j \Rightarrow i < j$. Thus, based on the definition of a total lexicographical order [8], a diagnosis $\mathcal{D}$ is called preferred, iff there exists no other minimal diagnosis $\mathcal{D}'$ with $\mathcal{D}' >_{lex} \mathcal{D}$. Based on these definitions, an explanation for a failed robot motion planning can be computed efficiently as proposed in Algorithm 1. The method takes the set of motion planning constraints as input, which also includes the start and goal configuration q_s, q_g of the desired motion plan. If a valid solution is found no alternative explanation needs to be provided (line 3). Otherwise, a minimal diagnosis for a valid motion plan has to be calculated (line 9) using FastAlt, which is based on FastDiag [8]. Following a divide-and-conquer strategy, the total constraint set C is divided into two sets C_1, C_2, to check if $C' - C_1$ is valid. If this is the case, C_2 can be omitted since one minimal diagnosis is already identified in C_1. If one element is remaining in C and the C' is still invalid, the element in C is part of the diagnosis and C_2 is further divided into subsets to be checked for validity. If it is desired to calculate more than one preferred diagnosis, an adoption of the HSDAG algorithm [10] can be done by substituting the resolution of minimal conflicts with the deletion of single diagnosis elements, identified by FastAlt. A more detailed discussion on the algorithm can be found in [8].

Algorithm 1. Motion Planning Explanation

function EXPLAIN($\mathcal{H}$) : $\mathcal{E}$ ▷ Where $\mathcal{H}$ - constraint list, including q_s, q_g; $\mathcal{E}$ - explanation for user
    if valid($argmin_\xi$ $F[\xi]$ s.t. $\mathcal{H}$) **then**
        return "Found valid path ξ, no alternatives needed"
    else
        $\mathcal{D} = \text{FastAlt}(\emptyset, \mathcal{H}, \mathcal{H})$
        return "No valid path found. The plan would be feasible if $\mathcal{D}$ is dropped."
    end if
end function
function FASTALT($D, C = \{c_1..c_n\}, C'$) : $\mathcal{D}$ ▷ Where D, C, C' - constraint (sub)sets; $\mathcal{D}$ - preferred diagnosis (set of conflicting constraints)
    if $D \neq 0$ and valid($argmin_\xi$ $F[\xi]$ s.t. C') **then** return $\emptyset$
    else if $|C| = 1$ **then** return C
    else
        $k = \lfloor \frac{|C|}{2} \rfloor$
        $C_1 = \{c_1...c_k\}, C_2 = \{c_{k+1}..c_n\}$
        $D_1 = \text{FastAlt}(C_1, C_2, C' - C_1)$
        $D_2 = \text{FastAlt}(D_1, C_1, C' - D_1)$
        return $D_1 \cup D_2$
    end if
end function

4 Experiments and Results

To investigate the performance of the proposed method, an experimental setup in a simulation was created. As simulation and planning environment, Gazebo is used together with MoveIt. A simulation of a Panda manipulator within a simple environment comprising various obstacles such as walls, desks, or lamps is designed and used for evaluation. For motion planning, the optimization-based planner STOMP [12] was used. The framework was also integrated into ROS. Different constraint types were implemented for $\mathcal{H}$, defining constraints representing the robots kinematic capabilities, obstacles within the environment, or other user-defined motion constraints such as orientation constraints. Dynamic constraints such as acceleration and velocities of joints were not considered, as MoveIt provides kinematically feasible plans only. The ordering of the constraints for the algorithm was defined such that $H_{goal} < H_{start} < H_{kinematics} <$

$H_{collision}$, putting the highest importance on obstacles within the environment. The obstacle constraints were additionally sorted in ascending order based on the Euclidean distance between the starting point and the center of the obstacle. Conflicts to enforce failed motion plans were constructed manually using a combination of goal and collision constraints.

The proposed algorithm was evaluated in terms of performance considering various configurational settings that influence the runtime and required validation checks of a diagnostic algorithm. Therefore, parameters such as the number of conflicts $|S|$, the number of constraints $|\mathcal{H}|$ and the number of desired diagnoses N were explored. As a baseline, the HSDAG algorithm [10] using QuickXPlain [11] for minimal conflict detection was used. Figure 1 shows the influence of the two parameters $|S|$ and $|\mathcal{H}|$ on a logarithmic scale. While the HSDAG algorithm increases exponentially with increasing number of conflicts (Fig. 1 left), FastAlt provides a guaranteed logarithmic solution for one preferred diagnosis as proven in [8]. When evaluating the influence of the constraint set size $|\mathcal{H}|$ (Fig. 1 right), only little influence on the runtime can be seen. While FastAlt provides a diagnosis within 1–2 s, HSDAG takes at least 5 times longer, also depending on the size of the conflict set $|S|$. However, the exponential downside of FastAlt can be seen in Fig. 2 which shows the run time for a different number of preferred diagnoses N. Due to the algorithmic design, FastAlt offers a preferred solution in logarithmic time, while the HSDAG already has an exponential factor in it to find the first diagnosis. As soon as the first diagnosis has been found, computing the next diagnosis in HSDAG takes only little time, leading to a nearly constant runtime over N. For FastAlt however, the performance gets significantly worse with increasing number of diagnoses. After computing 29 diagnoses, it becomes slower than the baseline algorithm.

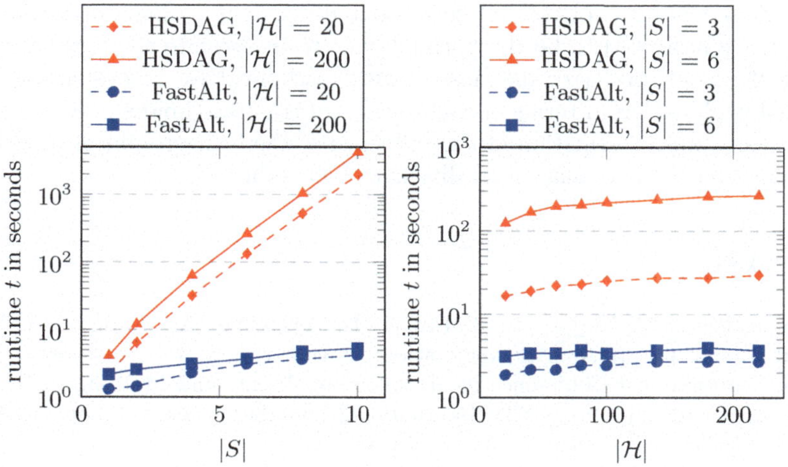

Fig. 1. Runtime comparison for providing an explanation ($N = 1$) with HSDAG and FastAlt on a different number of conflicts $|S|$ with a constraint size $|\mathcal{H}|$ of 20 and 200 (left), and on a different size of constraints $|\mathcal{H}|$ with a conflict set size of 3 and 6 (right).

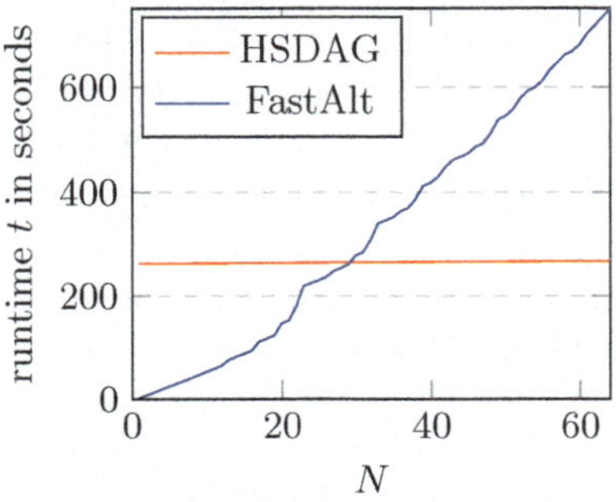

Fig. 2. Runtime comparison of HSDAG and FastAlt on the number of requested diagnoses N with $|\mathcal{H}| = 200$, $|S| = 6$. While HSDAG stays constant over N, FastAlt gets significantly worse until, after 29 diagnoses, it gets slower than HSDAG.

5 Conclusion

This work presents a method to efficiently compute explanations in robot motion planning. By using the FastDiag algorithm for computing a preferred diagnosis with logarithmic complexity, an explanation of a failed motion plan can be provided in short time. From the experiments it can be concluded, that the size of the constraint and conflict set does not influence the performance of the proposed method. Thus, it improves the approach presented in [2] by efficiently computing explanations, also for larger constraint sets which makes the collection of constraints in subsets for performance optimization obsolete. However, it is also shown that the proposed method is only efficient if only one preferred diagnosis is needed. But to provide a possible explanation for the failed motion plan, it is sufficient to compute only one preferred diagnosis. Still, a single preferred solution is not guaranteed to be the optimal one. To investigate the optimality and quality of a single preferred diagnosis, future work involves the evaluation of the provided explanation in terms of usefulness and trustworthiness in a user study. Additionally, an investigation of the influence of the lexicographical ordering of constraints on the explanation quality is to be conducted.

References

1. Anjomshoae, S., Najjar, A., Calvaresi, D., Främling, K.: Explainable agents and robots: results from a systematic literature review. In: Proceedings of the 18th International Conference on Autonomous Agents and Multi-agent Systems, AAMAS 2019, pp. 1078–1088. International Foundation for Autonomous Agents and Multiagent Systems, Richland (2019)
2. Brandao, M., Canal, G., Krivic, S., Magazzeni, D.: Towards providing explanations for robot motion planning. In: 2021 IEEE International Conference on Robotics and Automation (ICRA), pp. 3927–3933. IEEE (2021). https://doi.org/10.1109/ICRA48506.2021.9562003

3. Brandão, M., Coles, A., Magazzeni, D.: Explaining path plan optimality: fast explanation methods for navigation meshes using full and incremental inverse optimization. Proc. Int. Conf. Autom. Plan. Sched. **31**(1), 56–64 (2021). https://ojs.aaai.org/index.php/ICAPS/article/view/15947
4. Das, A., Rad, P.: Opportunities and challenges in explainable artificial intelligence (xai): a survey (2020), https://arxiv.org/pdf/2006.11371
5. Devin, S., Alami, R.: An implemented theory of mind to improve human-robot shared plans execution. In: 2016 11th ACM/IEEE International Conference on Human-Robot Interaction (HRI), pp. 319–326. IEEE (2016). https://doi.org/10.1109/HRI.2016.7451768
6. Felfernig, A. (ed.): Knowledge-based configuration: from research to business cases. Elsevier, Amsterdam and Heidelberg (2014)
7. Felfernig, A., Schubert, M., Reiterer, S.: Personalized diagnosis for over-constrained problems. In: International Joint Conferences on Artificial Intelligence (IJCAI). AAAI Press, Menlo Park (2013)
8. Felfernig, A., Schubert, M., Zehentner, C.: An efficient diagnosis algorithm for inconsistent constraint sets (2021). https://arxiv.org/pdf/2102.09005
9. Gonzalez, D., Perez, J., Milanes, V., Nashashibi, F.: A review of motion planning techniques for automated vehicles. IEEE Trans. Intell. Transp. Syst. **17**(4), 1135–1145 (2016). https://doi.org/10.1109/TITS.2015.2498841
10. Greiner, R., Smith, B.A., Wilkerson, R.W.: A correction to the algorithm in Reiter's theory of diagnosis. Artif. Intell. **41**(1), 79–88 (1989). https://doi.org/10.1016/0004-3702(89)90079-9
11. Junker, U.: QUICKXPLAIN: preferred explanations and relaxations for over-constrained problems. In: Proceedings of the 19th National Conference on Artifical Intelligence, AAAI 2004, pp. 167–172. AAAI Press (2004)
12. Kalakrishnan, M., Chitta, S., Theodorou, E., Pastor, P., Schaal, S.: Stomp: stochastic trajectory optimization for motion planning. In: Bicchi, A. (ed.) International Conference on Robotics and Automation, pp. 4569–4574. IEEE, Piscataway (2011). https://doi.org/10.1109/ICRA.2011.5980280
13. Karaman, S., Walter, M.R., Perez, A., Frazzoli, E., Teller, S.: Anytime motion planning using the RRT*. In: Bicchi, A. (ed.) International Conference on Robotics and Automation, pp. 1478–1483. IEEE, Piscataway (2011). https://doi.org/10.1109/ICRA.2011.5980479
14. de Kleer, J.: Hitting set algorithms for model-based diagnosis. 22nd International Workshop on the Principles of Diagnosis, pp. 100–105 (2011)
15. Korpan, R., Epstein, S.L.: Toward natural explanations for a robot's navigation plans. Notes from the Explainable Robotic Systems Worshop, HRI (2018)
16. Loew, T., Bandyopadhyay, T., Williams, J., Borges, P.: Prompt: probabilistic motion primitives based trajectory planning. In: Robotics: Science and Systems XVII. Robotics: Science and Systems Foundation (2021). https://doi.org/10.15607/RSS.2021.XVII.058
17. Pill, I., Quaritsch, T., Wotawa, F.: On the practical performance of minimal hitting set algorithms from a diagnostic perspective. Int. J. Prognost. Health Manag. **7**(2) (2016). https://doi.org/10.36001/ijphm.2016.v7i2.2363
18. Reiter, R.: A theory of diagnosis from first principles. Artif. Intell. **32**(1), 57–95 (1987). https://doi.org/10.1016/0004-3702(87)90062-2
19. Rosenthal, S., Selvaraj, S.P., Veloso, M.: Verbalization: narration of autonomous robot experience. In: Proceedings of the Twenty-Fifth International Joint Conference on Artificial Intelligence, IJCAI 2016, pp. 862–868. AAAI Press (2016)

Intelligent Robotic Knowledge-Supported Visual Recognition of Handled Objects in Condictions of Acquiring Incomplete Information

Jovan Šumarac[1], Uroš Ilić[2], Aleksandar Rodić[1(✉)], and Xiangrong Xu[3]

[1] Institute Mihajlo Pupin, University of Belgrade, Belgrade, Serbia
aleksandar.rodic@pupin.rs
[2] Faculty of Mechanical Engineering, University of Belgrade, Belgrade, Serbia
[3] Anhui University of Technology, Ma'anshan, China

Abstract. The paper presents a knowledge-supported approach to the synthesis of visual-perceptual system of industrial bi-manual robot for collaborative work with human, intended for performing service tasks in random task space in conditions of collecting incomplete data set on processing objects within technological operation. This system is intended for Industry 4.0, i.e. smart industrial production, that connects (integrates) cyber-physical systems with modern information and communication technologies. In Industry 4.0, machine vision systems and knowledge-based perceptions involving artificial intelligence are of great importance. The new principle of visual perception, described in this paper, will be used by the robot controller for faster and more accurate planning of the movement and manipulative actions of the robot in unstructured conditions, less tidy workspace and normal ambient lighting. The visual perception system uses the methods of classical 3D machine vision, a structured database containing CAD models of objects, technical drawings and photographs of bodies made from different angles of observation. The database also stores useful information on the technological stages of the implementation of operations on read objects using a syntax that is understandable to a robotic controller. The controller uses recorded images of objects and/or video streams from the scene or from the task workspace. In doing so, the robot uses artificial intelligence algorithms to reconstruct the scene from an incomplete set of object geometry data and determine unknown objects and their positions, and then create a strategy for capturing and manipulating bodies in accordance with the requirements of the technological process. The experimental system of a bi-manual collaborative robot, for the verification of the concept of knowledge-supported visual perception, is in the phase of physical integration of the system, so simulation results will be demonstrated here. For this purpose, the Robotic Vision Toolbox for the Mathworks' Matlab/Simulink was used, which is connected to one Microsoft RealSense RGB-D camera as a video sensor via the appropriate software interface. The results of the experimental verification are presented in the corresponding figures in the paper with expert comments attached. In conclusion, future directions of development related to the inclusion of complementary, tactile perception on the robot gripper are given.

Keywords: Industry 4.0 · Bi-manual collaborative robot · Industrial humanoid · Knowledge-supported perception · Intelligent grasping

A. Müller and M. Brandstötter (Eds.): RAAD 2022, MMS 120, pp. 122–132, 2022.
https://doi.org/10.1007/978-3-031-04870-8_15

1 Introduction

Human skills, both physical and cognitive, have always been of interest to researchers in various fields of science - biomechanics, medicine, psychology, sports, arts, etc. In industry, well-trained (trained) workers perform their tasks better and faster than less well-trained counterparts. However, even in the case of maximally trained workers, there are fine differences in the skills of performing the set work tasks that make a qualitative difference between individuals. In this paper, which deals with the synthesis of knowledge-based intelligent visual perception in industrial robots, we have developed a methodology to ensure that human 3D perception and spatial reasoning skills are effectively mapped (implemented) to robot machine skills. Besides, an additional quality is that the technologist does not require special preparation of the task workspace (strict arrangement of the scene) and special intensity and structure of lighting, but that daylight or normal ambient lighting in which the man-worker operates is sufficient. The aim of the research in this paper is to ultimately enable the robot to intelligently, reliably capture objects using heterogeneous information and artificial intelligence algorithms. The example in this paper presents 3D puzzle parts but this approach could be used in different real life scenarios. For instance, some palletizing tasks can require more complex object placement, or placement in an order, or fitting the objects in a certain way. Electronic waste sorting can also use such a system to recognize the size and shape of different parts such as plastic, batteries, boards etc.

The work is organized in several sections. Section 1 provides some introductory remarks related to the research objectives in this paper. Section 2 provides a brief overview of conventional methods and technologies applied in machine vision systems in industry. Section 3 briefly presents a way to describe the geometry of a 3D body suitable for implementation in robotic controllers. Section 4 explains the syntax for conveniently describing the stages of a technological process in a way understandable to a robot. Section 5 presents a new method, the so-called know-supported visual perceptions in robots and explains the process of forming an image database. Finally, Sect. 6 presents a case study in which the methodology presented in the paper is explained and experimentally verified.

2 Robotic Visions – Conventional Practice

Robotic vision is the technology and methods used to provide imaging-based guidance of robots usually in industry. Machine vision [1, 2] applications utilizing 3D imaging are a growing niche within the industry [3, 4]. The most commonly used method for 3D imaging is scanning based triangulation which utilizes motion of the product or image during the imaging process. A laser is projected onto the surfaces of an object. In machine vision this is accomplished with a scanning motion, either by moving the workpiece, or by moving the camera & laser imaging system. The line is viewed by a camera from a different angle; the deviation of the line represents shape variations. Lines from multiple scans are assembled into a depth map or point cloud [5]. Stereoscopic vision is used in special cases involving unique features present in both views of a pair of cameras [5]. Other 3D methods used for machine vision are time of flight and grid based [3, 5]. One method is grid array based systems using pseudorandom structured light system as employed by the Microsoft Kinect system [6, 7]. After an image is acquired, it is processed [9]. Central processing functions are generally done by a CPU, GPU, FPGA or a combination of these [9]. Deep learning training and inference impose higher processing performance requirements [9]. Multiple stages of processing are generally used in a sequence that ends up as a desired result. A typical sequence might start with tools such as filters which modify the image, followed by extraction of objects, then extraction (e.g. measurements, reading of codes) of data from those objects, followed by communicating that data, or comparing it against target values to create and communicate "pass/fail" results. Machine vision image processing methods include: a) stitching/registration, b) filtering, c) thresholding, d) pixel counting, e) segmentation, f) edge detection, g) color analysis, h) blob detection and extraction, i) deep learning or machine learning processing, j) pattern recognition, k) gauging/metrology, and l) comparison against target values to determine a "pass or fail" or "go/no go" result. In recent years traditional (non-learning) computer vision methods have included various corner detection methods and combined them with image feature descriptors in order to recognize different objects [14].

The approach promoted in this paper differs from standard machine vision methods in the following aspects: i) does not require a large number of recordings and/or video streams because the images are not used for neural network training and knowledge generalization but only for 3D peak detection and edge of the body using an RGB-D Microsoft RealSense camera (Kinect sensor). Additionally, as support, data taken from a previously prepared database, which contain CAD models of the body, i.e. mash models of objects determined by their points and the lengths that connect them.

3 Description of the Object Geometry Understandable to Robot

Most robotic systems (computer systems in general) are unable to comprehend 3D models generated with CAD software. Only numerical type of model's data is supposed to be submitted to the robot's control unit. Image and 3D models are unusable in general. It is necessary to create such a file that contains every geometrical data of the model. The conversion of STL files to Microsoft Excel spreadsheets was chosen as the optimal solution in our previous work [11]. At this point of research, this way of storing converted data was found to be convenient because it is also easily understandable and readable by humans. Within the spreadsheet there are three types of data: (i) normal vectors of triangles created by triangulation of object's surfaces, (ii) x-y-z coordinates of vertices of these triangles, and (iii) connectivity list, i.e. the way the triangles are formed from extracted points. The robot control unit/computer pulls this data and, with appropriate homogeneous transformations and kinematic algorithms, can compute the desired trajectory of the end-effector.

4 Syntax for Describing Technological Operations Clear to Robots

When defining the technological process (assembly, welding, pick-and-place, etc.), it is first necessary to define the technological bases on the workpieces. The base is a geometric characteristic of a part of an object (point, line or plane) that is important for a technological process. Depending on which activity the technological bases (TBs) are related to, they can be divided into: (i) TB manipulations, (ii) TB supports and (iii) TB body fusions. Manipulation bases are those surfaces of parts of objects that are in contact with the executive body of the manipulator. The technological bases of the connection are the vectors of the normal surface of the different parts in contact. The goal of assembling the parts is to bring the mentioned two vectors to the same support, with different signs. TB supports are mostly related to production processes and concern support works in work accessories or on the work surface within the stage. These can be the edges and vertices of the parts, if some negative of the pyramid is envisaged in which the parts should be inserted, as will be described in the example below. Figure 1 shows an example of five parts of a 3D pyramid (in the form of a 3D puzzle) marked with numbers from one to five. If all the parts are laid with their largest side turned down and rotated so that the longest edge is towards us, the shown layout is obtained. The parts under ordinal numbers 4 and 5 have marked diagonals in one of the rhombuses, which represent the pyramid seen from above. In other words, only these two works, in this position, have one theme above all others. Also, since the selected surfaces of the parts are also the largest, they are located in the folded pyramid on the outside. If the robot had a vacuum catcher, it would be ideal to adopt these surfaces as manipulation bases for the technological operation of connecting 3D puzzle.

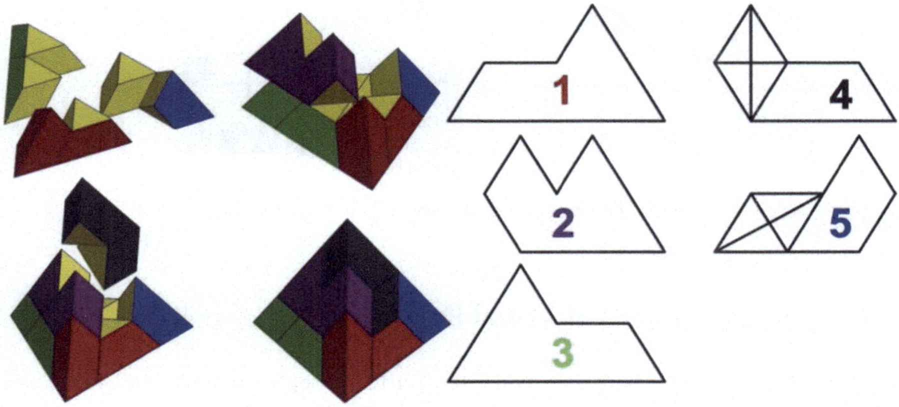

Fig. 1. Phases of the puzzle assembling (left side). The nomenclature of the parts that make up the puzzle according to [11] (right side).

After defining contact surfaces and manipulation bases, it is necessary to define the so-called Bourjault diagram [13] of assembly formation. Figure 2 gives one example of a handout. It is necessary to form a complete technological operation of assembling the ABCD assembly (top left), where the connections between the parts are marked with numbers 1–4. For example, if parts A and D were to touch by some chance, there would be a fifth connection between nodes A and D in the graph at the bottom left.

Finally, an algorithm for merging or filling connections is formed. Generally speaking, there are several ways to form the resulting assembly (puzzle assemblies). With this systematization (like the one in Fig. 2) it is possible to define finally many puzzle assembly processes. The robot would, with the help of the application of artificial intelligence, select those parts that it first recognizes on the table. This would gradually eliminate the parts in the selection and converge to the final solution (assembled puzzle), i.e. accurate recognition of the (last) part. Joining the parts of the pyramid shown in Fig. 1, can be defined by a similar Bourjault diagram and the merge sequences themselves are stored in a readable form in a database (e.g. one Excel spreadsheet). Also, it is necessary to define illegal operations, which can occur by applying this method of assembly. Let's say that when connecting the last part of the 3D puzzle, in the matrix of complex parts (the body of the pyramid is composed of parts in Fig. 1), it is necessary to invest a certain physical effort (insertion force) and angle to insert the last part into the assembly. This data is also entered into a database so that the technological operation is fully defined, unambiguous and "understandable" for use by a robotic controller.

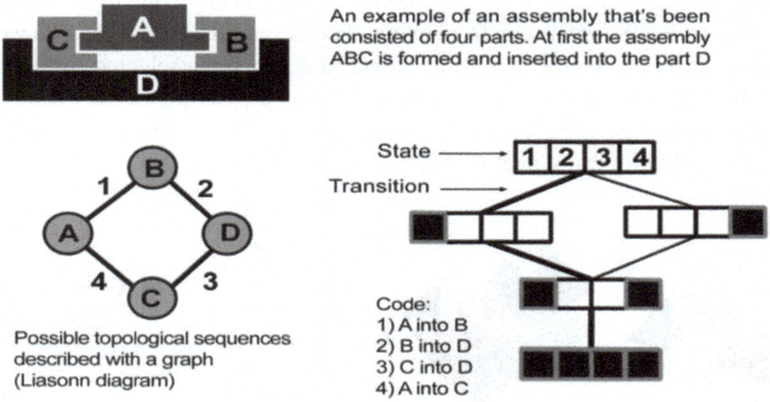

Fig. 2. The process of forming a Bourjault diagram.

5 Knowledge-Supported Visual Perception

The robot vision system hardware consists of a RGB depth camera, Intel RealSense D435i equipped with an inertial measurement unit (IMU) used for detection of its movements and rotations. Its' ideal range is 0.3 to 3 m which is suitable for a typical robot workspace.

The primary software package used for image processing is the Machine Vision Toolbox (MVTB) for MATLAB, version 4.3. The task for this vision system is to recognize parts of the 3D puzzle in robot's workspace, obtain their coordinates, correctly identify each part, and enumerate them in the order the robot needs to manipulate them. In order to do this a database of 3D numerical models of each puzzle part is used. These models provide detailed information about each part that will later be used in end effector trajectory and grasping planning. They also provide the correct order in which the objects should be grasped. For this reason, it is first necessary to create a database of images of each individual part and match them with the 3D models. Each part has the same color, and their geometric configuration is such that they can look very similar from certain points of view (POVs). The goal for this paper was also to find a computationally less expensive way of identifying objects. This means that taking a large number of images of each part from different POVs should be avoided. For this to work a simpler unique feature for each part must be identified. An analysis of the parts and their 3D models shows that the distinctive feature of a part is its largest side (by area). The idea is to take images of just the largest sides for each part. After processing the vertices should be identified as well as the length of each side edge. The coordinates of the vertices and the edge lengths can be saved and compared and matched with the adequate 3D model. This means that a relatively small number of images would be used to create an image database. Another feature of the parts is that their edges only have 3 distinct lengths, small (median value 25 mm), medium (54 mm) and large (81 mm). This information will be used in object matching.

To create image database each part is photographed so that the object is positioned with its largest side facing the camera directly. No particular lighting system is used, and each object is illuminated by diffuse light coming from office lighting and desk lamp. The camera is about 30 cm away from the object. An example image of a part is given in Fig. 3a).

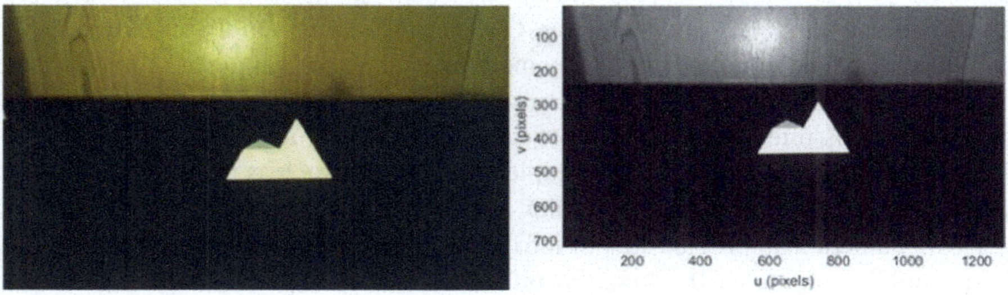

Fig. 3. a) Color image of a 3D puzzle part against a dark background. b) The image, loaded in MATLAB as a black and white photo, with horizontal and vertical axes in pixels

The post processing begins with image segmentation. The image is loaded in MATLAB as a black and white image where each pixel has its intensity (0–256) as shown in Fig. 3b). This form is quite suitable for many post processing operations. Since the background is dark and the largest side of the object is the brightest it is suitable to perform a binary segmentation. The goal is to segment the image into just two classes, object and background. As can be seen in Fig. 3b) in some cases, another side of the object will be partially visible. In this case it is the small, triangular upper side. It is less bright and an appropriate threshold should be set so that it is classified as background. Also, in Fig. 3b) there is a noted reflection from the table lamp in the background. Since the region of interest is the middle region of the image, where the object is, these pixels are set to background value.

After the segmentation a morphological operation of closing is performed. It functions as a hole filling filter to fill possible areas within the object that could be wrongly classified as background. The binary, segmented image is shown in Fig. 4a). This image is used in the next step which is corner detection. A Harris corner detector is used to identify corners in the image, which in fact identifies side vertices. The binary image is very suitable for such a detector and it is often used in literature [12]. After detection the vertices of this side are identified and shown in Fig. 4b).

Fig. 4. a) Binary image of a 3D puzzle part. b) The zoomed image, with corners marked with red squares.

In comparison with other work that uses corner detection for object recognition this is a much simpler approach. Authors have combined Harris or the similar ShiTomasi corner detectors [12] with other algorithms to describe image features such as combined SIFT and SURF [14] algorithms. This is a necessity when dealing with drastically different objects in terms of size, shape and color. Our approach differs in that it only uses the Harris detector, and combines it with camera depth info to reconstruct the side edge length. While it's simpler, this approach is effective when dealing with same-colored objects, which is often the case in practical tasks. Also, while similar work [15] has explored using this detector to detect simple object shapes such as triangles, circles and squares, the shapes here are more complex as shown in Fig. 4.

Depth camera also provides depth info which is the distance of every point in the image from the camera itself. Using this info and the knowledge about camera intrinsic parameters (focal length, and the principal point of projection) the pixel coordinates (u,v) of each vertex can be translated into 3D coordinates x, y and z in the coordinate frame of the camera [12]. When the coordinates of each vertex are known the length of the edges can be calculated. Since there are only three classes of edge lengths each length is coded as short (1), medium (2) or long (3). The largest side can then be coded by the edge lengths. Starting from the left most vertex in Fig. 4b) the edges go as follows: short, short, short, medium, long (code 11123).

The 3D numerical model describes the object in a complete way. Therefore, it is important to match info from object images with info contained in the matrices of vertices, triangles, and normal vectors. The triangles in the 3D model are grouped based on common normal vectors and potential body sides are defined. The sides with the largest number of triangles are singled out and the number of unique vertices on each side is checked. The model is rotated so that only the largest side is visible. Eliminating the non-edge points leaves only the vertices set. From there it is simple to calculate the edge lengths and to code them in the same way as above. Simple circularly shifted vector comparison yields a match for each side.

The goal for the vision system is to be able to recognize any puzzle part in the robot workspace in situations when all the parts are in the workspace at the same time. This is a complex task as all the objects are the same color and similarly shaped. For each individual object a mapping of its largest side (or sides, in case there are two) has been done. This was done with a minimum of images as the idea is to only take multiple images of the group scene with all the parts. The goal is to recognize all the parts using different POVs of the scene to identify largest sides for each object.

6 The Object Identification Algorithm with an Experimental Example

The experimental setup consists of five objects in the robot workspace ('the scene'). For this experiment a single camera was physically moved to five different marked workspace locations, placed on a tripod, and the scene was photographed. This is repeated five times, once from above, and once from each side of the scene, meaning the camera has been shifted 90° for each new image. A LED lamp has been mounted on the camera so that there is a direct source of light from every POV. In future experiments the camera will be mounted at the robot end effector and will move with the robot.

The object configuration should reflect a typical example of object positions in robot workspace. A trivial configuration would be one where each object's largest side is clearly visible, and all largest sides are facing the same direction. An extreme situation would be one where each object is lying on its largest side. The example in this experiment is a typical situation where some objects are on their largest sides and for others the largest sides are visible but are not facing the same direction (Fig. 5a).

The group images are segmented and filtered so that they become binary images. Each binary image is segmented in particular regions which correspond to a different object. The position of each object in the camera's frame is determined. Since the camera position in the global frame is known, transformation matrices are used to get the objects' coordinates in global frame. This way each object's position is unique regardless of the POV of the camera. The corners with most strength are then found using the Harris corner detector. Groups of corners are formed depending on the region in the image they belong to. Each object in the binary image now has 'its own' vertices (Fig. 5b); they are correctly arranged and the edge lengths are calculated.

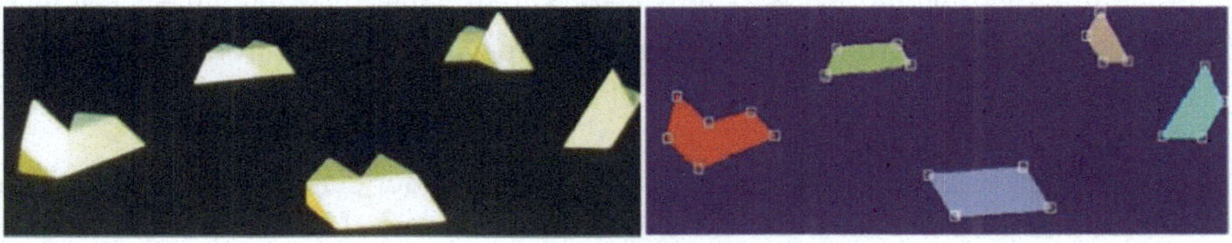

Fig. 5. a) An image of the scene with objects. b) A labeled image where each region is distinguished, here shown in different colors. The white squares represent the detected corners.

The view of all the objects is complex, and not all largest sides are easily visible. Some knowledge about the objects is used in the decision-making process:

1. Each object has either 5 or 6 vertices and 5 or 6 edges, on its largest side.
2. The objects can be classified in two classes depending on the length of their largest edge. Two parts have the longest edge of about 81 mm (class 1) while for the remaining three the longest edge is around 54mm (class 2).
3. The perimeter of each object is between 170 and 200 mm.

With this in mind the main procedure of the object detection algorithm is as follows:

1. The program examines the saved info and looks for objects with 5 or 6 edges.
2. Their perimeter is checked and if it falls in between 150 and 220 mm they become candidates for recognition.
3. Their edge lengths are examined. If they are roughly close to the defined lengths of 'small', 'medium' or 'large' (±5 mm tolerance from the median values) they are coded as 1, 2 and 3 respectively. Otherwise they are coded 0. A vector of coded values is formed (for instance 11231 or 12001).
4. If one of the edge lengths is 'large' (code 3) an indicator is added to the object info. This indicator marks the object as a class 1 object.
5. Next, the program looks for objects with 4 or 7 edges only to check if they contain the largest edge. If they do a class 1 indicator is added to their info.
6. The objects with less than 4 and more than 7 identified vertices are too uncertain to provide reliable information and are discarded.
7. The program then goes through the info with the coded edge lengths from group scene images ('group object info'). Those who contain no uncertainties (no 0 code values) are then compared with all the information vectors obtained from the single object images ('single object info'). If there is a match that object becomes a very strong candidate for detection. The group info also provides object global position so that particular object is checked in all other POVs to see in any other information is available (for instance if it is class 1 object). Each single object info is checked to see if it has already been matched with another group object. If no other contradictory information is obtained the match is made. If there is contradictory information the program looks for the strongest information. For instance, if one object has been indicated as class 1 from two different POV images then this information is more reliable than if another object has been classified based on information from a single image.
8. The program examines other group info. It first looks for class 1 indicators in other objects that have not been identified. If for instance one class 1 object has already been detected, it is enough for the program to find just a reliable class 1 indicator in another object to identify it.
9. If up to this point four objects are identified the fifth one is automatically detected. If there is more than one unidentified object an option is left to the user to take more photos from new points of view or to use the incomplete codes (with zeroes) to find the best possible matching and identify objects.

For this example, five images were enough for all the objects in group scenes to be identified and matched with the information from single object images.

7 Conclusions

The paper presents a new, knowledge-supported system of visual perception of robots for intelligent capture, based on complementary information on the geometry of work objects, technological operations on them and archived photographs of the body. The advantages of the new approach are that: (i) It makes processing data about objects on the robotic scene far simpler and faster for microprocessor processing than processing the huge number of pixels that make up a digitized image of geometric bodies; ii) The intensity and character of lighting is not a crucial factor for successful identification. In most cases, normal daylight or standard ambient light are sufficient to form the image database, while a camera mounted LED lamp is adequate for group scene images; iii) Since object and shape recognition is not based on common deep learning conducted over a large set of data (data-set) obtained by image processing or video-stream, this method is significantly more suitable for real-time applications in industry; it even utilizes less data than other machine learning methods such as support vector machine (iv) The proposed method does not use colors as attributes of objects because processing objects in industry are often monochrome. If the objects are polychromatic, the colors can be additionally used for shape recognition and body detection, which makes it easier for the robotic controller to plan movements. In future research, work will be done on the development of a tactile robotic preceptor that would include the mapping of human tactile skills to the robot's capture process. Also, one of the key future goals is to perform comprehensive method verification on a number of different scene configurations to examine its success rate.

Acknowledgements. Development of the robot prototype is the result of the bilateral R&D project entitled "Development and experimental verification of performance of mobile dual-arm robot for cooperation with humans" from the "Science and Technology Development Program - Funding of Joint Research and Development Projects of the R. Serbia and the P. R. China", 2017–2019.

References

1. Steger, C., Ulrich, M., Wiedemann, C.: Machine Vision Algorithms and Applications, 2nd edn, p. 1. Wiley-VCH, Weinheim (2018). ISBN 978-3-527-41365-2
2. Beyerer, J., Fernando, P.L., Frese, C.: Machine Vision - Automated Visual Inspection: Theory, Practice and Applications. Springer, Berlin (2016). https://doi.org/10.1007/978-3-662-47794-6
3. Murray, C.J.: 3D Machine Vision Comes into Focus. Design News. Archived from the original on 2012–06–05, February 2012. Accessed 12 May 2012
4. Davies, E.R.: Computer and Machine Vision: Theory, Algorithms, Practicalities, 4th edn, pp. 410–411. Academic Press (2012). ISBN 9780123869081. Accessed 13 May 2012
5. Turek, F., Jackson, K.: 3-D imaging: a practical overview for machine vision. Qual. Mag. **53**(3), 6–8 (2014)

6. Zhang, Y., Xiong, Z., Wu, F.: Hybrid structured light for scalable depth sensing. University of Science and Technology of China, Hefei, China Microsoft Research Asia, Beijing, China. http://research.microsoft.com/en-us/people/fengwu/depth-icip-12.pdf
7. Morano, R., Ozturk, C., Conn, R., Dubin, S., Zietz, S., Nissano, J.: Structured light using pseudorandom codes. IEEE Trans. Pattern Anal. Mach. Intell. **20**(3), 322–327 (1998)
8. Rice, D., Thousand, A.: Critical considerations for embedded vision design. Photonics Spectra magazine published by Laurin Publishing Co., pp. 60–64, July 2019
9. Fussell, M.: Finding the optimal hardware for deep learning inference in machine vision. Vision Systems Design Magazine, pp. 8–9, September 2019
10. Rodić, A., Stevanović, I., Jovanović, M.: Smart cyber-physical system to enhance flexibility of production and improve collaborative robot capabilities – mechanical design and control concept. In: Aspragathos, N.A., Koustoumpardis, P.N., Moulianitis, V.C. (eds.) RAAD 2018. MMS, vol. 67, pp. 627–639. Springer, Cham (2019). https://doi.org/10.1007/978-3-030-00232-9_66
11. Ilić, U., Šumarac, J., Stevanović, I., Rodić, A.: Interface development dedicated to connecting CAD tools for 3D modeling of complex objects and industrial robot's controller. ROI1.2 In: Proceedings of the IcETRAN2021, Banja Vrućica, Bosnia and Herzegovina, September 2021
12. Corke, P.: Robotics, Vision and Control, 2nd edn. Springer, Berlin (2017). https://doi.org/10.1007/978-3-319-54413-7
13. Bourjault, A.: Contribution à une approche méthodologique de l'assemblage automatisé: Elaboration automatique des séquences opératiores, Ph.D. thesis, L'Université de Franche Comté (1984)
14. Bansal, M., Kumar, M., Kumar, M., Kumar, K.: An efficient technique for object recognition using Shi-Tomasi corner detection algorithm. Soft. Comput. **25**(6), 4423–4432 (2020). https://doi.org/10.1007/s00500-020-05453-y
15. Kitti, T., et al.: An object recognition and identification system using the Harris corner detection method. Int. J. Mach. Learn. Comput. 462–465 (2012)

A Brief Survey of Sim2Real Methods for Robot Learning

Konstantinos Dimitropoulos, Ioannis Hatzilygeroudis(✉), and Konstantinos Chatzilygeroudis

Computer Engineering and Informatics Department, University of Patras, Patras, Greece
{kdimitrop,ihatz,costashatz}@upatras.gr

Abstract. Simulation has been crucial for robotics research development almost from the beginning of its existence. While simulation has been widely used for education, testing, and prototyping, only very recently the robotics community has attempted transferring behaviors learned in simulation to the real world (this process is usually referred to as Sim2Real). Those attempts have opened-up a novel research direction that has produced some exciting results that were previously thought impossible to achieve. In this paper, we attempt to give a quick overview of the most promising Simulation-To-Reality (Sim2Real) methods, results and directions.

Keywords: Sim-to-real · Robot learning · Domain randomization · Machine Learning

1 Introduction

Traditional control methods have been used in a wide variety of successful robotics applications [25]. Analytic control methods, however, have two main disadvantages: (a) they require robotics experts to design, develop and maintain solutions/control architectures, and (b) cannot take advantage of data-driven methods to continuously improve and adapt to the circumstances at hand [9]; this means that the designer needs to anticipate beforehand for all possible situations [8,15].

Artificial Intelligence (AI) methods, especially Machine Learning (ML) ones, on the other hand are able to produce systems that can effectively learn complex behaviors without the need for specific and advanced modeling [9]. In this regime, Reinforcement Learning (RL) [26] has produced learning methods that allow robots to learn and adapt by trial and error. For example, using RL algorithms a humanoid agent was able to learn how to walk successfully in a wide variety of environments [13].

Testing, however, RL methods in real robots is often dangerous, since those methods often attempt dangerous behaviors that can damage the robot. For this reason, most successful RL methods for robotics applications are mainly demonstrated in simulation environments [13,16]. A simulated environment though can

A. Müller and M. Brandstötter (Eds.): RAAD 2022, MMS 120, pp. 133–140, 2022.
https://doi.org/10.1007/978-3-031-04870-8_16

never replicate the real world, and methods attempting to transfer learned behaviors from simulation to reality will have to deal with, what is called, the *"reality gap"* [21]. The reality gap refers to the discrepancies between the simulated and the real, physical world. For example, we can have some un-modeled dynamics (e.g. friction or soft-bodies are hard to model accurately) or situations that we have never encountered in simulation (e.g. a bird randomly passing through our field of view).

The Sim2Real transfer concerns the task of transferring robot skills acquired in a simulated environment to a physical setting. There are numerous works that try to close the *reality gap* with the use of ML techniques. There are efforts that rely only on the simulations [3,17,20] and others that additionally use a small amount of real-world data [5,28]. Also, there are techniques that try to face the *reality gap* by exposing the algorithm to a variety of random environments and different types of randomization [10,16,19,22,24] and "hope" that the real world will be just another variation. Other methods take advantage of domain-invariant representations so that behaviors learned in simulation can better generalize to the real system [11,27]. Approaches that utilize a hybrid simulator [1,7,23] have also flourished. The aim of this paper is not the presentation of all efforts related to the sim2real transfer problem [29], but a short review of some of the fundamental recent efforts in this domain.

In this view, we categorize the sim2real methods in to two main categories: (a) simulation-only methods, and (b) adaptation methods. In the first category, the main assumption behind the approaches is that we can introduce enough variations in simulation in order to learn a robust behavior/model/policy that will be able to "see" the real world as another variation and perform adequately there. In the second category, methods attempt to utilize the simulator to identify a good initial point or to perform pre-computation so that the online learning can be done efficiently.

Lastly, we discuss separately *Imitation Learning (IL)* methods that have been a core part in many Sim2Real methods.

2 Simulation-Only Methods

The methods in this category attempt to perform what is usually called, one-shot transfer learning. The main idea behind those methods is to learn in a big set of simulated environments with the goal of learning a robust model or policy that will "see" the physical world as another variation and transfer without any adaptation. The main technique to robustify the learned policies/models is to randomize the environment the agent learns on; this process is often also referred to as "domain randomization".

2.1 Randomizing for Perception

One of the main applications of domain randomization is to learn robust perception models. Here the controller is given (or assumed trivial to learn/program) and the focus is on creating perception models that can generalize to the physical world.

In the case of ambidextrous robot grasping [20], synthetic data was generated from a set of 3D computer-aided design models, using models from geometry and mechanics. The vast amount of data consisted of point clouds, grasps and reward labels for heterogeneous grippers. The synthetic training environment used simulated grasp outcomes by evaluating rewards according to the ability of a grasp to resist forces and torques, due to gravity and other random perturbations. The sim2real policy was trained using Grasp Quality Convolutional Neural Networks, one for each gripper. The policy was implemented on a physical robot with two grippers on a bin picking task with reliability greater than 95%.

In [17], in trying to transform real-world images to canonical images, the authors used a generator that translated randomized simulation images to a chosen canonical simulation version and after that it could generalize to accept an image from the real world and produce a canonical RGB image, a segmentation mask and a depth image. That synthetic data was used to train their robot grasping agent. Finally the agent was transferred to the real world and attained 70% zero-shot success on unseen objects.

2.2 Randomizing for Control

In [16], domain randomization is used to learn an end-to-end reactive controller that is able to run in the real world. The authors randomized the initial configurations of the environment (joint positions, basket and box for a pick-and-place task), the location of the camera, the type of shadows, and other visual properties (e.g. texture and material). They also added random distractor objects. Their algorithm was able to learn a pick-and-place task that operates on the real world with a 100% success rate, while only been trained with simulated data. Similarly [28] used domain randomization to learn a robotic reaching task, and chose to randomize the following: number, shape, pose, color of distractors, camera field of view, color of the table, floor and target object. All this randomization was tuned manually to simulate variations in the real scene.

In [5], the authors attempt to identify the crucial factors of virtual scene randomization for robotic grasping. They create different datasets with various kinds of randomization: (a) no randomization, (b) visual randomization, (c) dynamics randomization and (d) visual and dynamics randomization. In the first one, only the camera pose and bin location were varied, whereas in the second one, things like the texture and color of the object along with lighting direction. In the third one, the object mass and object rolling and spinning friction coefficients were varied. In the final one, a mix of dynamics and visual randomization was used. The approach was evaluated with over 25000 physical test grasps and they managed to reduce the number of real-worlds samples needed by up to 50 times when used in conjuction with online adaptation (see Sect. 3).

Selecting the right parameters to tune and the appropriate ranges for randomizing is a challenging problem by itself. This can be tackled by using an automated curriculum [2,18]. In [18], the authors use an adaptive terrain curriculum in order to find an effective policy for blind quadrupedal locomotion that generalizes to different environments. The method realizes a training curriculum that gradually synthesizes terrains based on the controller's performance. As the neural network learns, the distribution of terrain parameters become more challenging, leading to a terrain of greater difficulty.

3 Adaptation Techniques

These techniques make the assumption that simulation and domain randomizing is not enough by themselves, and take advantage of real data.

3.1 Combining Simulated and Real Data

In [5], the authors used a single real-world and several simulated datasets. For their approach, domain adaptation was used to improve the efficiency of deep robotic grasping. They combined feature level adaptation with their own pixel level adaptation technique. In feature level adaptation, domain adversarial neural networks were trained to extract domain invariant features, while in pixel level adaptation a convolutional neural network, given original synthetic images, could produce adapted images that looked more realistic. The real-world dataset was collected by progressively better grasp prediction networks. Real datasets were also used for a robotic reaching task in [28]. The adversarial discriminative approach that was introduced achieved a similar success rate by using less than half of the labeled data that other similar approaches used. The model was pre-trained using labeled simulated data as input in convolutional neural networks that were part of the perception module of the architecture and then it was adapted using both simulated and real data. Finally, the real images that were used, were collected by a real Baxter robot.

In [10], the goal was to find a distribution of simulation parameters that brings observations induced by the policy, trained under this distribution, closer to the observations of the real world. After the policy was trained on current distribution, it was sampled both in the real world and in simulation. The discrepancy between the simulated and real samples was, then, used to update the simulation parameter distribution in their algorithm. The difference in this effort is that the simulation parameters were updated in each real-world iteration, in contrast to standard domain randomization, where the initial distribution of the parameters remains intact. The robotic tasks that evaluated the transfer were swing-peg-in-hole and opening a cabinet drawer.

3.2 Hybrid Models

Physics simulators are important for robot control, because they allow the robot to plan its actions and estimate the effects of these actions in the environment. However, simulators can never be perfect and thus there is risk that the resulting behavior will greatly diverge from the original plan. On the other hand, data-driven methods are able to learn directly from experience and improve over time. They need, however, many samples to be able to learn a reliable and effective model; something that is usually not possible to do with a physical robot. As such, many techniques have been developed that attempt to combine simulated priors with data-driven methods.

A physics simulator augmented with interaction networks [4] was used for a complex planar pushing task, while the interaction networks played the role of data-driven model. The physics simulator was combined with an object-based neural network that increased the prediction accuracy of the simulator [1]. Augmented simulation was also used in [23]. The full physics engine augmented with a machine learning model was used for predictive control. After the initialization of the physics engine with parameters of the real system and the correction of the simulation errors, the residual was modeled, and the final hybrid model was able to guide the agent to drive a marble to the center of a maze environment.

In [6], the authors propose a new GP learning scheme that combines model identification and nonparametric model learning (called GP-MI) and, then, performs policy with the Black-DROPS algorithm [7]. The main idea of GP-MI is to use simulators with tunable parameters (e.g., different lengths of links). Searching for the parameters that best matches the observations can be seen as a model identification procedure, which could be solved via minimizing the mean squared error; nevertheless, the authors formulate it in a way so that they can exploit the GP framework to jointly optimize for the kernel hyperparameters and the mean parameters, which allows the modeling procedure to balance between nonparametric and parametric modeling. Black-DROPS with GP-MI was able to robustly learn controllers for a physical damaged hexapod in less than 30 seconds of interaction time.

4 Imitation Learning

Imitation Learning (IL) deals with the problem of learning a behavior policy by "imitating" a set of demonstrations. Sim2Real methods have been using IL techniques to boost their performance (e.g. privileged learning [18]), and thus we briefly discuss about IL methods that could potentially be used in conjunction with Sim2Real methods.

In [30], the authors combine an IL approach [14] with a model-free deep reinforcement learning method in order to develop a unified training framework for solving robotic manipulation tasks. High-throughput reinforcement learning algorithms were used along with a physics engine to simulate data of a robot performing physical interactions. An IL method, that made use of a network policy and a discriminator network, used demonstration data, by allowing the agent to interact with the environment and learn from its experiences.

A method for policy learning that combines imitation and reinforcement learning is presented in [12]. Relay policy learning is a method that first uses imitation learning to produce goal-conditioned hierarchical policies and then with the help of reinforcement learning these policies are fine tuned to be efficient for different tasks. In the imitation learning phase, hierarchical polices are pre-trained using demonstrations of meaningful activities provided by the user, but without assuming that those demonstrations can accomplish any specific goal from their own. However, the test-time goals should be from the same distribution as the goals accomplished from the demonstrations. This allows the method to be flexible enough and easily generalized for various long-horizon problems.

5 Conclusion

In this short survey, we attempted to give a brief overview of the latest Sim2Real methods for robot control. We identified two main directions for effective Sim2Real transfer: (a) environment randomization, and (b) online adaptation. In the first case, the methods attempt to use randomized versions of the simulated world in order to learn a robust policy that can work without any adaptation on the physical system. In the second case, methods assume that learning in simulation only is not enough and further adaptation on the real system is needed. In the following we make an effort to give a few guidelines for effective Sim2Real algorithms:

- If we face a perception problem (Sect. 2.1), then the usage of static datasets yields very good results. We can use sets of diverse synthetic images (with 3D rendering), and real-world images that are collected effortlessly (e.g. by moving a camera around). Moreover, using a simulator we can generate artificial situations that can make the model more robust to outliers or novel situations.
- If the problem at hand is that of learning a controller (Sect. 2.2), then the Sim2Real transfer is more difficult to achieve since the controller basically defines the robot behavior. In those cases, we can use diverse sets of randomized simulations to learn a robust initial set of controller parameters, but we surely need some online adaptation technique.
- Lastly, imitation learning can be beneficial for both perception and control problems. Demonstrations can come from expert or non-expert users. Usually, systems that take advantage of imitation learning for policy learning scenarios, combine imitation with other reinforcement learning methods and algorithms that finetune these policies for task performance. An effective way to combine IL with RL methods is privileged learning, where the ground truth information (easily available in simulators) is given to an RL algorithm to find optimal controllers, while IL methods are used to "diffuse" this controller into another one with realistic inputs (e.g. RGB images instead of object positions).

Overall, the field of Sim2Real is rapidly growing with many exciting results and novel research questions to be solved.

References

1. Ajay, A., et al.: Combining physical simulators and object-based networks for control. In: 2019 International Conference on Robotics and Automation (ICRA), pp. 3217–3223. IEEE (2019)
2. Akkaya, I., et al.: Solving rubik's cube with a robot hand. arXiv preprint arXiv:1910.07113 (2019)
3. Alghonaim, R., Johns, E.: Benchmarking domain randomisation for visual sim-to-real transfer. In: 2021 IEEE International Conference on Robotics and Automation (ICRA), pp. 12802–12808. IEEE (2021)
4. Battaglia, P.W., Pascanu, R., Lai, M., Rezende, D., Kavukcuoglu, K.: Interaction networks for learning about objects, relations and physics. In: Conference on Neural Information Processing Systems (NeurIPS) (2016)
5. Bousmalis, K., et al.: Using simulation and domain adaptation to improve efficiency of deep robotic grasping. In: 2018 IEEE International Conference on Robotics and Automation (ICRA), pp. 4243–4250. IEEE (2018)
6. Chatzilygeroudis, K., Mouret, J.B.: Using parameterized black-box priors to scale up model-based policy search for robotics. In: 2018 IEEE International Conference on Robotics and Automation (ICRA), pp. 5121–5128. IEEE (2018)
7. Chatzilygeroudis, K., Rama, R., Kaushik, R., Goepp, D., Vassiliades, V., Mouret, J.B.: Black-box data-efficient policy search for robotics. In: 2017 IEEE/RSJ International Conference on Intelligent Robots and Systems (IROS), pp. 51–58. IEEE (2017)
8. Chatzilygeroudis, K., Vassiliades, V., Mouret, J.B.: Reset-free trial-and-error learning for robot damage recovery. Robot. Autonom. Syst. **100**, 236–250 (2018)
9. Chatzilygeroudis, K., Vassiliades, V., Stulp, F., Calinon, S., Mouret, J.B.: A survey on policy search algorithms for learning robot controllers in a handful of trials. IEEE Trans. Robot. **36**(2), 328–347 (2019)
10. Chebotar, Y., Handa, A., Makoviychuk, V., Macklin, M., Issac, J., Ratliff, N., Fox, D.: Closing the sim-to-real loop: Adapting simulation randomization with real world experience. In: 2019 International Conference on Robotics and Automation (ICRA), pp. 8973–8979. IEEE (2019)
11. Gangapurwala, S., Geisert, M., Orsolino, R., Fallon, M., Havoutis, I.: Rloc: Terrain-aware legged locomotion using reinforcement learning and optimal control. arXiv preprint arXiv:2012.03094 (2020)
12. Gupta, A., Kumar, V., Lynch, C., Levine, S., Hausman, K.: Relay policy learning: solving long horizon tasks via imitation and reinforcement learning. In: Conference on Robot Learning (CoRL) (2019)
13. Heess, N., et al.: Emergence of locomotion behaviours in rich environments. arXiv preprint arXiv:1707.02286 (2017)
14. Ho, J., Ermon, S.: Generative adversarial imitation learning. Conf. Neural Inf. Process. Syst. (NeurIPS) **29**, 4565–4573 (2016)
15. Isermann, R.: Fault-Diagnosis Systems: An Introduction From Fault Detection to Fault tolerance. Springer Science & Business Media, Cham (2005)
16. James, S., Davison, A.J., Johns, E.: Transferring end-to-end visuomotor control from simulation to real world for a multi-stage task. In: Conference on Robot Learning, pp. 334–343. PMLR (2017)
17. James, S., et al.: Sim-to-real via sim-to-sim: Data-efficient robotic grasping via randomized-to-canonical adaptation networks. In: Proceedings of the IEEE/CVF Conference on Computer Vision and Pattern Recognition, pp. 12627–12637 (2019)

18. Lee, J., Hwangbo, J., Wellhausen, L., Koltun, V., Hutter, M.: Learning quadrupedal locomotion over challenging terrain. Sci. Robot. **5**(47), eabc5986 (2020)
19. Lee, T.E., et al.: Camera-to-robot pose estimation from a single image. In: 2020 IEEE International Conference on Robotics and Automation (ICRA), pp. 9426–9432. IEEE (2020)
20. Mahler, J., et al.: Learning ambidextrous robot grasping policies. Sci. Robot.**4**(26), eaau4984 (2019)
21. Mouret, J.B., Chatzilygeroudis, K.: 20 years of reality gap: a few thoughts about simulators in evolutionary robotics. In: Proceedings of the Genetic and Evolutionary Computation Conference Companion, pp. 1121–1124 (2017)
22. Muratore, F., Treede, F., Gienger, M., Peters, J.: Domain randomization for simulation-based policy optimization with transferability assessment. In: Conference on Robot Learning, pp. 700–713. PMLR (2018)
23. Ota, K., Jha, D.K., Romeres, D., van Baar, J., Smith, K.A., Semitsu, T., Oiki, T., Sullivan, A., Nikovski, D., Tenenbaum, J.B.: Data-efficient learning for complex and real-time physical problem solving using augmented simulation. IEEE Robot. Autom. Lett. **6**(2), 4241–4248 (2021)
24. Rajeswaran, A., Ghotra, S., Ravindran, B., Levine, S.: Epopt: Learning robust neural network policies using model ensembles. In: International Conference on Learning Representations (ICLR) (2017)
25. Siciliano, B., Khatib, O.: Springer Handbook of Robotics. Springer, Berlin (2016)
26. Sutton, R.S., Barto, A.G.: Reinforcement Learning: An Introduction. MIT press, London (2018)
27. Tanwani, A.K.: Domain invariant representation learning for sim-to-real transfer. In: Conference on Robot Learning (2020)
28. Zhang, F., Leitner, J., Ge, Z., Milford, M., Corke, P.: Adversarial discriminative sim-to-real transfer of visuo-motor policies. Int. J. Robot. Res. **38**(10–11), 1229–1245 (2019)
29. Zhao, W., Queralta, J.P., Westerlund, T.: Sim-to-real transfer in deep reinforcement learning for robotics: a survey. In: 2020 IEEE Symposium Series on Computational Intelligence (SSCI), pp. 737–744. IEEE (2020)
30. Zhu, Y., et al.: Reinforcement and imitation learning for diverse visuomotor skills. In: Proceedings of Robotics: Science and Systems. Pittsburgh, Pennsylvania, June 2018. https://doi.org/10.15607/RSS.2018.XIV.009

Applying Monte Carlo Search and Monte Carlo Tree Search on Embedded Systems to Play Connect Four with a Robotic Arm

Moritz Duarte Pinheiro-Torres Vogt(✉), Jörn Fischer, and Thomas Ihme

Robotics Institute, Mannheim University of Applied Sciences,
Paul-Wittsack-Str. 10, 68163 Mannheim, Germany
m.vogt@posteiro.de, {j.fischer,t.ihme}@hs-mannheim.de

Abstract. Monte Carlo based methods have brought a remarkable improvement in performance of artificial intelligence in the realm of games in recent years, whereby human champions could be beaten in several board games of high complexity. In this work, two Monte Carlo based approaches, the Monte Carlo Search and the Monte Carlo Tree Search, are applied to the game of *Connect Four*. To test and compare these two approaches, games against each other are simulated. The better approach is then further placed against human opponents using a website (https://carloconnect.com/) and a robotic arm to evaluate the algorithm's strength. All tests are conducted on a Raspberry Pi 3 Model B to investigate the approaches' performance in an environment with limited available computing power and memory. The Experiments show that under these conditions the Monte Carlo tree search significantly outperforms the Monte Carlo search and is therefore the better solution. This approach is called CarloConnect and proves itself against human opponents by winning most of the games, thus playing at human level even while having only low computing resources.

Keywords: Artificial intelligence for games · Connect Four · Monte Carlo Search · Monte Carlo Tree Search · Raspberry Pi · Robotic arm

1 Introduction

In the last decades, artificial intelligence (AI) has become increasingly important in a multitude of domains. Tasks, that are difficult or impossible to program by hand, can be tackled by utilizing machine learning. In the domain of games, too, AI has been applied in recent years to enhance the performance of virtual players significantly – especially in games with a high degree of complexity.

Implementing such a search algorithm for games with perfect information usually requires an evaluation function based on heuristic knowledge [5], which determines the value of a game state. The more complex the game is the more difficult and time-consuming the defining of an evaluation function becomes.

A. Müller and M. Brandstötter (Eds.): RAAD 2022, MMS 120, pp. 141–148, 2022.
https://doi.org/10.1007/978-3-031-04870-8_17

There are several Monte Carlo based approaches which neither require an evaluation function, making them applicable to any game of finite length, nor need to carry out exhaustive searches through a complete search tree of states [2].

In this study, two Monte Carlo based search algorithms, the Monte Carlo Search [9] and the Monte Carlo Tree Search [6], are applied to the game of *Connect Four* to investigate their performance in games with a comparable degree of complexity. The better solution for this application is called CarloConnect. While Google is utilizing computers with extremely high computing power, it is becoming more and more relevant to implement AI's on embedded systems with limited computing power and memory, which is why all experiments of this work are carried out on a single-board computer. For further evaluation and applicability, CarloConnect is then tested by carrying out games against human opponents via an online website hosted on the single-board computer.

2 Connect Four

Connect Four is a game with perfect information in which two players can drop discs into a vertical board with 42 spaces distributed in six rows and seven columns. The players make their moves in turn, where the configuration of the board at a given turn is referred to as the state of the game, which allows for a specific set of valid actions for a player. The first player to get four discs building a line wins, whereby orientation can be either horizontal, vertical or diagonal.

The game is solved by *Allis* [1] showing that in case of perfect playing of both players, the first player always wins if he chooses the middle column as his first move and loses if he selects the four outermost columns. The selection of the other two columns leads to a draw. Furthermore, *Edelkamp et al.* [7] calculated the total amount of possible states of over $4.5 \cdot 10^{12}$ which would require nearly 43.8 terabytes of memory if the evaluation value of a position is stored in 2 bits.

3 Monte Carlo Methods

Monte Carlo methods repeat an experiment many times until the average of all results stabilizes around the expected value, which is ensured by the law of large numbers. With these experiments, a temporary heuristic game-specific knowledge for a given state can be obtained. In this work's context, the computationally obtained knowledge is an approximation for a best action to undertake for a given player in a given state of the game.

3.1 Monte Carlo Search

The Monte Carlo Search algorithm [9] is a straightforward technique where for each of the available actions of a current state, a large number of simulations are run, limited either by a specified amount of simulations or by computing time. In each simulation, a complete play through is performed. In this work,

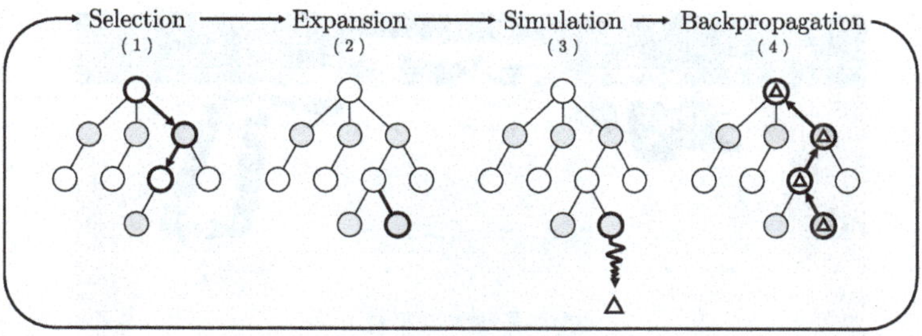

Fig. 1. The four phases of an iteration of the MCTS algorithm, colored for two-player games where each color represents a player. In the first phase the tree is traversed using a tree policy until a node with unexpanded child nodes is reached (1). Next, one of the unexpanded child nodes is created (2) and a game is simulated by selecting actions for each player at random until a final game state is reached, returning a result Δ (3). This result is then used to update all traversed nodes (4).

the term *play through* implies that the game is played virtually from a selected action by randomly selecting actions for each player until a final game state is reached, returning the winning player. As soon as all simulations are concluded or the time limit is reached, a win rate $rate_{win}$ is calculated using the win count $wins_{action}$ and the amount of simulations num_{sim}, as in *Eq.* (1). This procedure is applied to all the available actions of a current game state, eventually returning the action with the best win rate.

$$rate_{win} = \frac{wins_{action}}{num_{sim}} \tag{1}$$

3.2 Monte Carlo Tree Search

As the MCS simulations are random and thus unguided, the MCS has a lack of tactical insight. To compensate for this disadvantage, *Coulum* [6] combined the Monte Carlo Search with a search tree, thereby creating the so-called Monte Carlo Tree Search algorithm. This algorithm iteratively builds a search tree of possible future game states, node by node, where each node represents a state at a given player's turn and its estimated value. When a limiting factor, i.e. time or memory, is reached, the search iterations stop and the most promising action is returned. The larger the tree grows the higher the probability becomes to increase the value of the returned move, thus getting closer to the best possible move. The way the tree is built depends on how nodes in the tree are evaluated and selected. It should be noted that each node is an instance that requires memory, which can be limited by the embedded system that is utilized. An individual search iteration can be split into four phases [3] as presented in *Fig.* 1.

The players, which are represented by the levels in the tree, are identified by different colors – here shown for a two-player game. When the limiting factor is reached and consequently the iterations are stopped, the most promising child node of the root node, representing an action, is returned. For the final action selection, there are five common policies [4, 6].

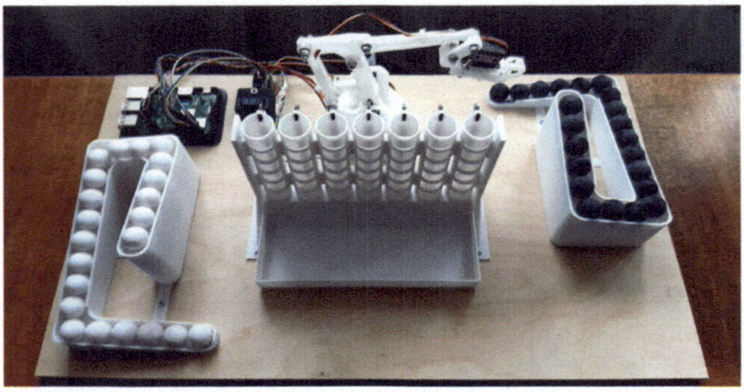

Fig. 2. 3D constructed *Connect Four* game board with a robotic arm as opponent. The robotic arm is connected via a Raspberry Pi to play autonomously against human players whereby human plays are detected with micro switches at the top of the game board's tubes.

4 Materials and Methods

For the experiments, the two Monte Carlo based search algorithms described in *Sect.* 3, MCS and MCTS, are applied to the game of *Connect Four* – implemented in C++. As the final selection criterion in accordance with the studies of *Chaslot* [4], the so-called *Robust Max child* is utilized. In order to obtain reliable results and to ensure consistent behavior of both algorithms, their limiting condition has to be equal. Therefore, a fixed amount of computing time to calculate a best move is used. It is notable, that the computing time a simulation takes varies depending on the game's progress. The more progressed a game is, the less moves and thus less time is needed to reach a final game state. This results in a decrease of the computing time of a simulation and hence in an increase of the algorithms' performance – they become stronger in the late game. Additionally, time is an overarching variable that can be set independently of the used embedded system, which is why it is utilized as the limiting factor for the two approaches.

Both search algorithms are tested by carrying out games where they play against each other. The strength of both algorithms can thereby be evaluated by calculating the win rate of the respective search algorithms. Furthermore, the simulation speeds of both search algorithms, representing how many simulations of a search for the first move of a game are conducted within an equal given time, are compared. Therefore, for the tests, the defined first player (the search algorithm making the first move) as well as the limitation factor (a computing time between 100 and 5000 ms) are varied. Each of the in total 12 tests consists of 100 simulated games and is run five times in order to quantify the fluctuations of the performance. The stronger algorithm, CarloConnect, is then placed against human opponents to further evaluate the AI's strength. For this purpose, a website (https://carloconnect.com/) is set up and a robotic arm (http://eezyrobots.it/eba_mk1.html) is 3D printed. The fully mounted physical game board is shown in *Fig.* 2.

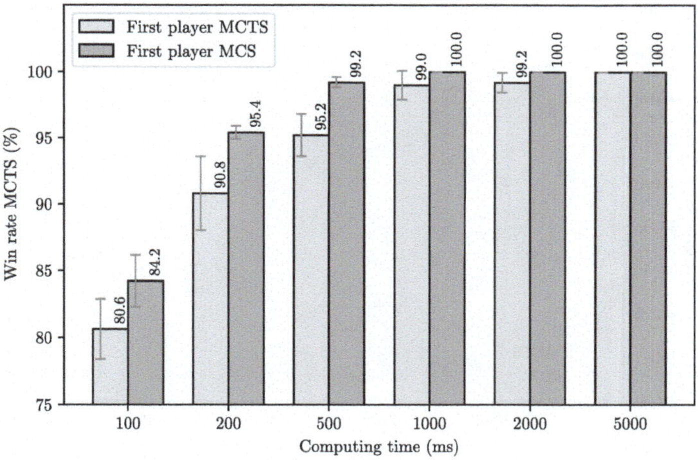

Fig. 3. Results of performance tests of the MCTS algorithm playing *Connect Four* against the MCS algorithm. Tests with different conditions are conducted where the starting player and the amount of available computing time as limitation for the algorithms are varied. Five times 100 games are performed for each test.

All tests are performed on a Raspberry Pi 3 Model B with a 64-bit 1.2 GHz quad-core processor and 1 gigabyte of memory to ensure sufficient memory for the MCTS, running on Ubuntu 16.04.6 LTS.

5 Results and Discussion

The results of comparing the presented search algorithms and testing the stronger one against humans are described hereinafter.

MCS versus MCTS – Win Rate. The results, shown in *Fig.* 3, prove that the MCTS clearly outperforms the MCS. The more computing time is given, the stronger the MCTS comparatively becomes until it wins all games. Due to the lack of tactical insight of the MCS algorithm, the opponent's behavior does not receive sufficient consideration, which is why the potential of MCS's performance is limited. With a rising degree of complexity of a game, the importance of the tactical insight grows. Therefore, the MCS does not suffice anymore and the MCTS begins to lead instead. A similar applicability is also described in [8].

In addition, the results show that for each of the differently set computing times the win rate of the MCTS is consistently lower if the MCTS starts first, even when the standard error is taken into account. Investigations of the first moves of the search algorithms prove that in all of the tests, both algorithms always start with a move in the middle column, which provides the potential of a safe win (see *Sect.* 2). Yet, the quantitatively stronger algorithm, the MCTS, wins less if it starts, which contradicts the conclusion *Allis* [1] that the win chances are the highest with a start in the middle column. According to the results, the

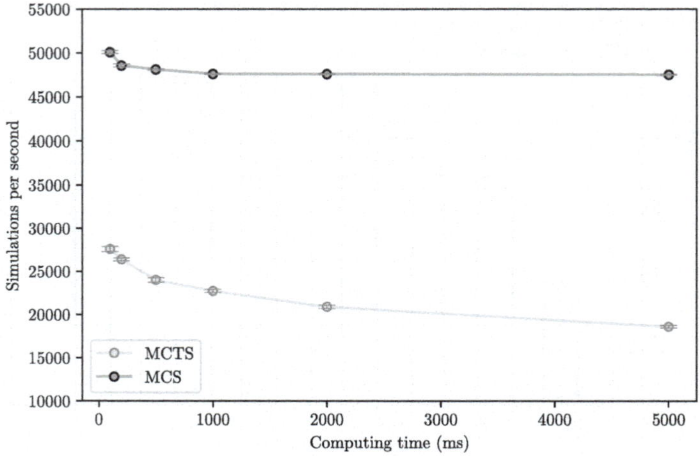

Fig. 4. Comparison of the amount of simulations the MCTS and the MCS conducted per second during performance tests. For evaluation, the amount of simulations conducted for a first move is presented as *simulations per second*. A characteristic curve is plotted with the values of each test (dots in the graph).

opposite occurs as the MCTS wins more often as second player after the MCS made its first move selecting the middle column. One might expect that while the algorithms do not play perfectly, as required in the theorem, the approach that overall performs better still should tend to perform better especially when it gets the first move, which is not the case. This behavior could be caused by the impact of the early and the late game performances of the approaches.

MCS versus MCTS – Simulation Speed. In order to compare the simulation speed in consideration of the different set computing time limits, the simulation speed is specified in the unit *simulations per second*. This unit is calculated by dividing the set computing time limit by its respective achieved amount of simulations (see *Fig.* 4).

This resultant graph shows that the MCTS performs approximately two times less simulations than the MCS, which is due to its iterative and complex tree building process. Both slopes reveal a drop of simulations at very low computation times, which are caused by initializing operations. Their positive effect on the simulation rate averages out with more performed simulations. Moreover, it is noticeable that compared to MCS' slope the slope of the MCTS continuously decreases with increasing computing time which can be explained by the fact that the longer the MCTS algorithm computes, the bigger the search tree grows. Thus, the branches get longer and so does the time that it takes to reach a leaf node where a play through is conducted – every simulation becomes more complex with increasing computing time. Still, also the slope of MCTS tend to flatten with higher computing time which can be attributed to the finite nature of *Connect Four*. When a branch reaches its maximum length, which is defined through the maximum amount of turns – 42 in this case – it cannot be expanded

Table 1. Win and tie rates of CarloConnect playing against human opponents. For three different computing times, games are played via a website, whereby in every played game, as an advantage for the human player, the human player starts first. The resulting win and tie rates are presented in this Table.

Computing time [ms]	Total games played	Win rate [%], human player	Win rate [%], CarloConnect	Tie [%]
15	505	21.1	78.9	0.0
60	235	10.9	89.1	0.0
1500	536	5.2	94.4	0.4

any further as a final game state is reached at this point. Consequently, the time it takes to traverse the tree up to a leaf node cannot exceed the time for a full traverse.

CarloConnect versus Humans. In this evaluation phase, CarloConnect is placed against human opponents via the website and using the robotic arm. Three different difficulties are thereby tested to investigate a wide spectrum of the AI's strength. The win rates of the human player and CarloConnect as well as the tie rate are calculated. The results including these rates of the games that are played on the website are presented in *Table* 1.

With a computing time of only 15 ms (difficulty *Easy*), CarloConnect already wins about 79% of the games. At the difficulty *Moderate*, which provides 60 ms for CarloConnect to search for a move, the win rate for the human players almost halves from 21.1% (difficulty *Easy*) to 10.9%. From the difficulty *Moderate* to *Hard* the human players' win rate is slightly more than halved, again. Human players only win 5.2% of the games at this stage, though rarely managing to cause a tie with a rate of 0.4%. These results demonstrate that CarloConnect can outperform most humans.

6 Conclusion

In this study, two Monte Carlo based approaches, MCS and MCTS, are compared in order to investigate their performance in the game of *Connect Four*. The experiments evince that the MCTS outperforms the MCS due to a lack of tactical insight of the MCS, even though the total amount of simulations is comparatively low. Concerning the behavior in performance of the MCTS, as expected, the accuracy increases with the available computing time. Still, it is unclear why the MCTS wins more often if its weaker opponent MCS starts first as this should provide a lower win chance for the MCTS according to the game solution found in the literature.

Regarding the fact that the MCTS needs an amount of memory that depends on the number of nodes that are created during a search, the embedded system must fulfill the requirement of sufficient memory. In this work, a Raspberry Pi 3 Model B is utilized which supplies 1 gigabyte of memory, which is why the

MCTS can be used with no memory limitations and is thus the better approach to select. Common small computer platforms (like most Arduinos) often only provide little memory from 2 to 32 kilobytes, limiting the MCTS in expanding the search tree, as each node of the tree is an instance that requires memory. This leads to a small possible amount of simulations wherefore the MCS would be the better choice. Nevertheless, considering the strength of the MCTS, a computing platform that fulfils the requirement of the MCTS would be chosen, which is why the final program CarloConnect uses the MCTS based approach.

Using the website, many human players can play against CarloConnect simultaneously from anywhere. The robotic arm, in turn, provides an authentic game feeling. Overall, the games against human opponents show that Carlo-Connect performs strong enough to reach a win rate of 94.4 against human players having 1500 ms computing time.

References

1. Allis, V.: A knowledge-based approach of connect-four. ICGA J. **11**(4), 165 (1988). https://doi.org/10.3233/ICG-1988-11410
2. Browne, C.B., et al.: A Survey of monte Carlo tree search methods. IEEE Trans. Comput. Intell. AI in Games **4**(1), 1–43 (2012). https://doi.org/10.1109/TCIAIG.2012.2186810
3. Chaslot, G., et al.: Monte-Carlo tree search: a new framework for game AI. In: Proceedings of the AAAI Conference on Artificial Intelligence and Interactive Digital Entertainment, California (2008). https://doi.org/10.5555/3022539.3022579
4. Chaslot, G.M.J.B., et al.: Progressive strategies for Monte-Carlo tree search. New Math. Nat. Comput. **4**(3), 343–357 (2008). https://doi.org/10.1142/S1793005708001094
5. Clune, J.: Heuristic evaluation functions for general game playing. In: AAAI, pp. 1134–1139 (2007). https://doi.org/10.5555/1619797.1619828
6. Coulom, R.: Efficient selectivity and backup operators in Monte-Carlo tree search. In: van den Herik, H.J., Ciancarini, P., Donkers, H.H.L.M.J. (eds.) CG 2006. LNCS, vol. 4630, pp. 72–83. Springer, Heidelberg (2007). https://doi.org/10.1007/978-3-540-75538-8_7
7. Edelkamp, S., Kissmann, P.: Symbolic classification of general two-player games. In: Dengel, A.R., Berns, K., Breuel, T.M., Bomarius, F., Roth-Berghofer, T.R. (eds.) KI 2008. LNCS (LNAI), vol. 5243, pp. 185–192. Springer, Heidelberg (2008). https://doi.org/10.1007/978-3-540-85845-4_23
8. Silver, D., et al.: Mastering the game of Go with deep neural networks and tree search. Nature **529**(7587), 484–489 (2016). https://doi.org/10.1038/nature16961
9. Tesauro, G., Galperin, G.R.: On-line policy improvement using monte-carlo search. In: The MIT Press: Advances in Neural Information Processing, pp. 1068–1074 (1996). https://doi.org/10.5555/2998981.2999131

Improving Kinesthetic Teaching of Fine Tasks Using a Teaching Agent

Aljaž Baumkircher(✉), Marko Munih, and Matjaž Mihelj

Laboratory of Robotics, Faculty of Electrical Engineering, University of Ljubljana, Tržaška cesta 25, Ljubljana, Slovenia
aljaz.baumkircher@fe.uni-lj.si

Abstract. In this study, programming by demonstration (PbD) teaching agent was proposed. PbD of fine movements proves to be difficult since sub-millimetre positioning accuracy is required. The agent presented in this study provided users with the visual enhancement feature in order to reduce the positioning error of demonstrations. Furthermore, an optimization of Gaussian mixture models (GMM) components positioning is presented, which allows for better detection and learning of relevant sub-millimetre movements within a larger movement.

Keywords: Kinesthetic teaching · Sub-millimetre accuracy · Teaching agent · Visual enhancement · Frequency-based component distribution

1 Introduction

Programming by demonstration (PbD) is a concept of robot task learning, where the task is learned through a human demonstration [1,12]. The often-used method to perform a demonstration is kinesthetic teaching, which allows users to teach robots in ways that are usually faster and more intuitive than conventional robot programming [6]. It does not require an input device since the robot movement is controlled directly by the users grabbing and moving individual segments [3,9].

While trajectory demonstrations are intuitive for naïve users and allow complicated skills to be taught, it might be hard for users to manipulate a high degree-of-freedom (DoF) robot while sustaining a smooth trajectory. This is especially the case with fine movements that require sub-millimetre accuracy. Visual enhancement might be necessary to improve accuracy since it would increase the user's sensitivity.

Furthermore, to generalize the demonstrations for later adaptations to different start and end conditions, so-called movement primitives can be used [4,7,8,11]. Gaussian mixture models (GMM) is a commonly-used probabilistic movement primitive that encodes the task from a set of demonstrations by fitting multiple components (clusters) on the given data points through expectation maximization (EM) algorithm [5]. The Gaussian mixture regression (GMR)

Supported by Slovenian Research Agency (research core funding No. P2-0228).

A. Müller and M. Brandstötter (Eds.): RAAD 2022, MMS 120, pp. 149–156, 2022.
https://doi.org/10.1007/978-3-031-04870-8_18

is then used to reconstruct the sequence (i.e., a movement trajectory) from the mixture model; however, relevant sub-millimetre information might be lost if the mixture model components are not optimally placed. An additional algorithm that would optimize the initial placement of mixture model components might be necessary for better detection.

Both the visual enhancement feature and use of movement primitives can be combined in a so-called teaching agent to improve the learning of fine tasks. In our study, we used kinesthetic teaching to demonstrate a fine task by following a reference trajectory. A teaching agent has been used to provide the user with the visual enhancement feature and the learning feature that generalized the demonstrated trajectories using a GMM method. The method was modified by optimizing the mixture model component placement using frequency analysis to improve the generalization of relevant parts of the trajectory.

2 Materials and Methods

2.1 Experimental Setup

The experimental setup consisted of robot manipulator Panda (Franka Emika, GmbH) and a horizontally placed LCD monitor (Fig. 1, left). The Panda robot was used as an input device during the kinesthetic teaching, while the horizontally placed LCD monitor served as a user interface, as it rendered the target trajectory and the visual enhancement feature (Fig. 1, middle). We designed a custom end-effector with a narrow tip that enabled the user to assume the end-effector's position with sub-millimetre accuracy.

The robot's end-effector orientation was constant throughout the experiment, and the movement was limited to a horizontal plane. The constraints were done to simplify the comparison between demonstrations with and without the visual enhancement feature. Recorded data contained all relevant sensor information from the Panda robot (end-effector position and orientation, joint positions, joint velocities, etc.) that was acquired with 1 kHz sampling frequency and sent to MATLAB (The MathWorks, Inc.) in real-time using UDP communication.

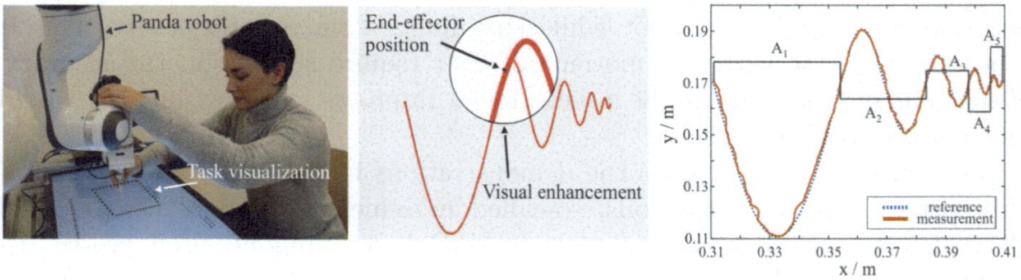

Fig. 1. An experimental setup overview (left), the reference trajectory while using the visual enhancement feature (middle), and a demonstrated trajectory along the reference (right).

A crucial issue that was also overcome was the robot's kinematic model accuracy. Due to tolerances in mechanical construction and assembly, small errors in end-effector pose can occur. Errors distributed along the arm are amplified due to the robot's open chain kinematic structure. During testing, it turned out that end-effector position errors as large as 5 mm occurred when moving the robot to the same desired end-effector position under different joint configurations. Therefore a calibration of the kinematic model was required. In our case, after the calibration, the average error of the robot's end-effector position was reduced to 0.1 mm.

2.2 Trajectory Demonstration

In order to test the influence of the target trajectory's spatial dimensions on the accuracy of a demonstration, we designed a reference trajectory with gradually smaller spatial dimensions (Fig. 1, right). It was based on a sine wave, whose amplitude decreases with each period:

$$y_i = A_i sin(\omega_i x_i), \quad \mathbf{A} = \begin{pmatrix} 0.06 \\ 0.02 \\ 0.01 \\ 0.005 \\ 0.0025 \end{pmatrix} m. \tag{1}$$

In order to ensure a smooth trajectory, the ω_i had to be scaled accordingly:

$$\omega_i = \frac{A_{i-1}}{A_i}\omega_{i-1}, \quad \omega_0 = 0.01256. \tag{2}$$

During this task, the user had to move the robot's end-effector along the target trajectory while following the line as accurately as possible. A gradual decrease in amplitude allowed for a smooth transition from coarse to fine movements. The target trajectory represented a fine movement during the last two periods (A_4, A_5). We will refer to parts of the trajectory with the same A_i as trajectory segments (i.e. trajectory segment A_3 represents the part of a trajectory where $A_i = A_3$).

A visual enhancement feature was implemented as part of the teaching agent (Fig. 1, middle). Its purpose was to visually enlarge the area under the Panda robot's end-effector (3× magnification) to increase the user's positioning error detection and, therefore, influence a demonstration's accuracy. It was created using the Unity game engine (Unity Software Inc.).

The tasks were performed by 31 participants (20 male, 11 female; 11 robotic experts, 20 robotic non-experts; average age 32 ± 9.8 years). Before the experiment started, each had a short training session, testing the kinesthetic teaching and the visual enhancement feature. Preliminary studies showed that amount of training needed to minimize the learning effect during the experiment varied from participant to participant based on their motoric skills and previous experience. Therefore enough time was given to each participant in order for them to feel comfortable with controls. A set of small tests helped us indicate whether participants are still improving or started to stagnate and are therefore ready to start with the experiment. After the training, the user had to follow the reference trajectory four times with the visual enhancement turned off and four times with the visual enhancement turned on. We then analyzed the demonstrations by determining the positioning errors and movement velocities.

2.3 Frequency-Based GMM Component Density Modification

In order to learn from the demonstrated trajectories, we used a GMM method as part of the teaching agent. GMM generalizes the acquired data points with a set of Gaussian components placed along the trajectories. For our purposes, the trajectories have been pre-processed in order to contain a predetermined number of data points, which are also time-aligned for a correct comparison by using the dynamic time warping (DTW) method [2]. Therefore each trajectory can be described as a function of normalized time s.

During the GMM initialization procedure, the initial component positions are usually placed uniformly along the independent variable, which is, in our case, normalized time s. This is fine since their position is then further modified through the expectation-maximization algorithm that fits the components on the data points. However, since each component results from many data points, finer details may be lost in the overall trajectory, which is a problem when fine details represent relevant information. One way to tackle this issue is by increasing the number of components. Consequently, each component has fewer data points to adjust to, and finer details have a more significant impact. However, increasing the number of components significantly increases the model's complexity and the required computational resources. Therefore we propose that instead of changing the number of components, we modify the component's initial positioning so that their positioning density is higher in areas with relevant information and lower elsewhere. In our case, the relevant information from the demonstrated trajectory are segments A_4 and A_5 since they represent a fine movement.

In order to achieve that, we have to determine the dominant movement frequencies along the normalized time s since higher-frequency regions represent finer movements and, therefore, relevant information. Each trajectory's dimensions are first reduced to one dimension, where instead of being defined with $x(s)$ and $y(s)$ cartesian coordinates, it is defined with the polar coordinate $\varphi(s)$. Then the newly defined trajectory $\varphi(s)$ is processed with a low-pass filter (10 Hz cutoff frequency) in order to remove signal noise. Frequency analysis is then performed on the filtered trajectory using Welch's method which returns a frequency spectrum for each s (Fig. 2, left) [13]. The amplitudes of each frequency spectrum are then normalized based on their maximum value. To determine the dominant frequency, we establish multiple time-frequency sectors by defining time windows and spectrum bands. The median value of normalized frequency spectra in the sector is determined for each time-frequency sector. Finally, for each time window, the time-frequency sector with the highest value is selected, as it represents the dominant signal frequency band in that time window. Since each trajectory produces slightly different results, the overall function of the dominant frequency band along the normalized time $f_D(s)$ is the median result from all trajectories.

In our case, higher frequency represents the relevant information, and therefore the mixture model components should have a higher density around these areas. To achieve that, we calculate the total area under the function $f_D(s)$. Then each mixture model component gets its s value so that the area under the function $f_D(s)$ between two consecutive components is the same. This way, the components have a different initial position s as if they would have been uniformly distributed. Consequently, the resulting mixture-model after the expectation-maximization algorithm might be different as well (Fig. 2, right).

Frequency-based GMM component density modification algorithm was compared to uniform and linearly-increasing component density. Latter was chosen since it is a reasonable assumption of the required distribution based on the trajectory's prior knowledge. The comparison has been made on ten different subsets of 20 randomly-chosen demonstrations. In all cases, GMM used 20 components to learn. This number was determined through BIC and AIC criteria [10].

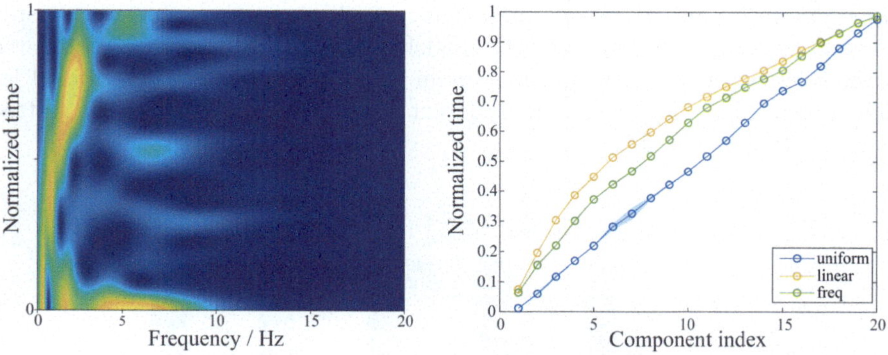

Fig. 2. A result of frequency analysis using Welch's transformation (left) and distribution of mixture-model components along the trajectory (right).

3 Results

In this section, we separately present the results of the trajectory demonstration experiment, and the influence of mixture model component distribution on the quality of the GMM generated trajectory.

3.1 Trajectory Demonstration

For our experiment, we acquired 248 demonstrations. The results are plotted using box and whiskers plots and present the demonstrated movement positioning error and movement velocity (Fig. 3). They are plotted separately for each of the last three trajectory segments (A_3, A_4, A_5). Furthermore, the figure also depicts the influence of the visual enhancement feature.

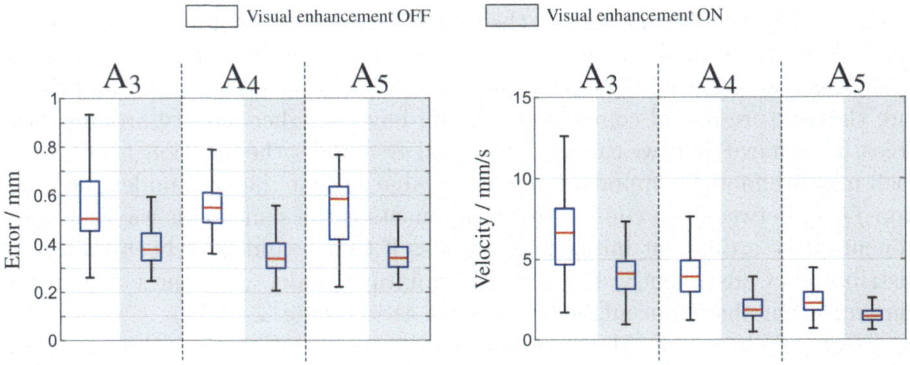

Fig. 3. Influence of the visual enhancement feature on the positioning error (left) and movement velocity (right) in trajectory segments A_3, A_4 and A_5.

3.2 Frequency-Based GMM Component Density Modification

Three different mixture-model component distribution approaches were tested in ten iterations. Each iteration, the model-generated trajectory was compared to the reference trajectory, and maximum position errors were determined for each segment. In order to compare different approaches, the results were plotted using a variance plot (Fig. 4).

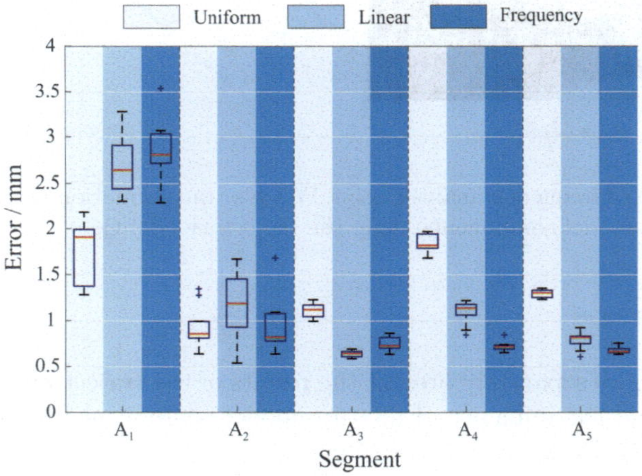

Fig. 4. Influence of the initial component positioning on the maximum error of the model-generated trajectory.

4 Discussion

In order to determine the influence of the teaching agent on the kinesthetic teaching of fine movements, it is necessary to determine the influence of each of its features - visual enhancement and frequency-based GMM component density modification.

During the trajectory demonstration, sub-millimetre positioning error was consistently achieved in all three segments when visual enhancement feature was not used (Fig. 3, left). However, an additional significant improvement in the positioning error was achieved when the visual enhancement feature was used. The improvement can be attributed to users' greater sensitivity due to the visual enhancement. The movement velocity drops significantly from one segment to the next (Fig. 3, right), which can be attributed to the fact that each segment required a movement smaller in spatial dimensions, which was more challenging for users to generate. Consequently, the movement velocity dropped significantly from one segment to the next to achieve similar accuracy. Additionally, visual enhancement reduced the velocity in each segment, which is most evident in segment A_3. Due to greater sensitivity, users could achieve higher accuracy and consequently reduce the movement velocity. Therefore, based on the results, we can say that the visual enhancement feature benefits the teaching agent since the positioning error of the kinesthetic teaching is significantly reduced.

For the second feature of the teaching agent, it is evident that the distribution of mixture-model components is significantly different between the uniform distribution and the frequency-based distribution, which interestingly enough more closely resembles the linear distribution (Fig. 2, right). This indicates that the proposed algorithm successfully detects a shift to higher frequencies, although the latter signal amplitudes were significantly smaller and harder to detect than the low-frequency ones. Furthermore, it can be seen that with the uniform distribution, the maximum positioning error remained relatively consistent throughout the trajectory (Fig. 4). However, frequency-based distribution significantly reduces the maximum positioning error in the last three segments, which can be attributed to higher mixture-model component density and consequently a better description of fine movement. As a compromise, the positioning error in the first segment is higher compared to the uniform distribution due to the lower mixture-model component density; however, this is acceptable since it does not represent relevant information. Therefore we can say that the frequency-based mixture-model distribution benefits the teaching agent and is an interesting option when detection of finer movements is desired.

5 Conclusion

In this study, we analyzed whether a teaching agent improves kinesthetic teaching of fine movements. The agent provided users with the visual enhancement feature, which significantly reduced the positioning error of demonstrations. Furthermore, the agent used an algorithm to modify the Gaussian mixture-model components placement in order to better learn fine movements. The proposed frequency-based algorithm improves the detection of fine movements since the maximum positioning error is significantly reduced. However, it achieves that by sacrificing the description of other trajectory segments. Therefore we believe that the teaching agent is a relevant addition when users teach fine movements using kinesthetic teaching.

References

1. Argall, B.D., Chernova, S., Veloso, M., Browning, B.: A survey of robot learning from demonstration. Robot. Autonom. Syst. **57**(5), 469–483 (2009)
2. Berndt, D.J., Clifford, J.: Using dynamic time warping to find patterns in time series. In: KDD Workshop, Seattle, vol. 10, pp. 359–370 (1994)
3. Billard, A., Calinon, S., Dillmann, R., Schaal, S.: Robot programming by demonstration. In: Siciliano, B., Khatib, O. (eds.) Springer Handbook of Robotics. Springer, Berlin (2008). https://doi.org/10.1007/978-3-540-30301-5_60
4. Calinon, S., Guenter, F., Billard, A.: On learning, representing, and generalizing a task in a humanoid robot. IEEE Trans. Syst. Man Cybern. Part B (Cybernetics) **37**(2), 286–298 (2007)
5. Dempster, A.P., Laird, N.M., Rubin, D.B.: Maximum likelihood from incomplete data via the EM algorithm. J. R. Stat. Soc. Ser. B (Methodological) **39**(1), 1–22 (1977)
6. Fischer, K., et al.: A comparison of types of robot control for programming by demonstration. In: 2016 11th ACM/IEEE International Conference on Human-Robot Interaction (HRI), pp. 213–220. IEEE (2016)
7. Huang, Y., Rozo, L., Silvério, J., Caldwell, D.G.: Kernelized movement primitives. Int. J. Robot. Res. **38**(7), 833–852 (2019)
8. Ijspeert, A.J., Nakanishi, J., Hoffmann, H., Pastor, P., Schaal, S.: Dynamical movement primitives: learning attractor models for motor behaviors. Neural Comput. **25**(2), 328–373 (2013)
9. Kronander, K., Billard, A.: Learning compliant manipulation through kinesthetic and tactile human-robot interaction. IEEE Trans. Hapt. **7**(3), 367–380 (2013)
10. Kuha, J.: AIC and BIC: comparisons of assumptions and performance. Sociol. Methods Res. **33**(2), 188–229 (2004)
11. Paraschos, A., Daniel, C., Peters, J., Neumann, G., et al.: Probabilistic movement primitives. In: Advances in Neural Information Processing Systems (2013)
12. Ravichandar, H., Polydoros, A.S., Chernova, S., Billard, A.: Recent advances in robot learning from demonstration. Ann. Rev. Control Robot. Autonom. Syst. **3** (2020)
13. Welch, P.: The use of fast Fourier transform for the estimation of power spectra: a method based on time averaging over short, modified periodograms. IEEE Trans. Audio Electroacoust. **15**(2), 70–73 (1967)

Motion Planning and Control

Projecting Robot Dynamics onto Trajectories

Friedrich Pfeiffer(✉)

Lehrstuhl fuer Angewandte Mechanik, TU-Muenchen, Boltzmannstrasse 15, 85748 Garching, Germany

friedrich.pfeiffer@tum.de

Abstract. The projection of arbitrary many robot degrees of freedom on a given path results in a set of nonlinear equations with one degree of freedom only, the path coordinate s. Application of this idea can be used to construct a mobility space $(\ddot{s}, \dot{s}, s)$ for any combination of coordinates and constraints. The paper presents a corresponding approach for n-link robots by applying multibody system theory. Practical aspects are discussed, and an example is given.

Keywords: Robot dynamics · Multibody systems · Projections · Quadratic forms

1 Introduction

The problem of trajectory planning for robotic systems is well known from the beginning of robotic research during the seventies and eighties. Especially one approach has been particularly successful at that time concerning robots with three up to ten revolute joints. It represented the prescribed path by one degree of freedom, namely the curvilinear coordinate s along that path. All robot degrees of freedom are then projected to this single path coordinate coming out with a linear system of ordinary differential equations of motion in the square of the path velocity. Many large applications and comparisons with measurements were carried through at that time, [1,3,7,8]. By analyzing various types of such examples, we find an astonishingly large number being defined by one trajectory only, from manufacturing to transport technologies. Therefore the one-dimensional robot example represents a certain amount of generality.

2 General Concept

2.1 Motion and Projections

Usually kinematics and possibly some forces of a robot-process are known, and the dynamics is unknown. This requires a multibody dynamic model of the robot with the necessity, that all components follow the prescribed trajectory. A classical set of equations of motion with constraints are given by many textbooks [2,5,8,10]. Jourdain's principle writes

$$\sum_{i=1}^{n} \underbrace{\delta \begin{pmatrix} \boldsymbol{v}_{O'} \\ \boldsymbol{\omega} \end{pmatrix}^T}_{\dot{\boldsymbol{z}}^T} \left\{ \underbrace{\begin{pmatrix} m\boldsymbol{I} & m\tilde{\boldsymbol{r}}_{O'S}^T \\ m\tilde{\boldsymbol{r}}_{O'S} & \boldsymbol{\Theta}_{O'} \end{pmatrix}}_{\boldsymbol{M}} \underbrace{\begin{pmatrix} \boldsymbol{a}_{O'} \\ \dot{\boldsymbol{\omega}} \end{pmatrix}}_{\ddot{\boldsymbol{z}}} + \underbrace{\begin{pmatrix} m\tilde{\boldsymbol{\omega}}\tilde{\boldsymbol{\omega}}\boldsymbol{r}_{O'S} \\ \tilde{\boldsymbol{\omega}}\boldsymbol{\Theta}_{O'}\boldsymbol{\omega} \end{pmatrix}}_{\boldsymbol{f}^g} - \underbrace{\begin{pmatrix} \boldsymbol{F} \\ \boldsymbol{M} \end{pmatrix}}_{\boldsymbol{f}^e+\boldsymbol{f}^a+\boldsymbol{f}^p} - \underbrace{\begin{pmatrix} \boldsymbol{F}^c \\ \boldsymbol{M}_{O'}^c \end{pmatrix}}_{\boldsymbol{f}^c} \right\}_i = 0, \tag{1}$$

A. Müller and M. Brandstötter (Eds.): RAAD 2022, MMS 120, pp. 159–166, 2022.
https://doi.org/10.1007/978-3-031-04870-8_19

($\boldsymbol{v}_{O'}, \boldsymbol{\omega}$ translational and rotational velocities, $m, \boldsymbol{I}, \boldsymbol{\Theta}_{O'}$ masses and mass moments of inertia, O', S reference point and center of gravity, $\boldsymbol{f}^g, \boldsymbol{f}^e, \boldsymbol{f}^a, \boldsymbol{f}^p, \boldsymbol{f}^c$ gyroscopic, applied, driving, process and constraint forces and torques). The virtual velocities $\delta\dot{\boldsymbol{z}}$ cannot be chosen arbitrarily, but they must watch side conditions, which might be considered by a process called Lagrange I [8]. The corresponding equations write

$$\boldsymbol{M}\ddot{\boldsymbol{z}} + \boldsymbol{f} - \boldsymbol{f}^c = \boldsymbol{0}, \qquad \boldsymbol{f} = \boldsymbol{f}^g - \boldsymbol{f}^e - \boldsymbol{f}^a - \boldsymbol{f}^p, \qquad \boldsymbol{f}^c = -\boldsymbol{W}(\boldsymbol{z}, t)\boldsymbol{\lambda},$$
$$\dot{\boldsymbol{\Phi}} = \boldsymbol{W}^T\dot{\boldsymbol{z}} + \bar{\boldsymbol{w}} = \boldsymbol{0}, \quad \ddot{\boldsymbol{\Phi}} = \boldsymbol{W}^T\ddot{\boldsymbol{z}} + [(\frac{d\boldsymbol{W}^T}{dt})\dot{\boldsymbol{z}} + (\frac{d\bar{\boldsymbol{w}}}{dt})] = \boldsymbol{W}^T\ddot{\boldsymbol{z}} + \hat{\boldsymbol{w}} = \boldsymbol{0}, \qquad (2)$$

The mass matrix is positive definite and symmetric. The constraints $\dot{\boldsymbol{\Phi}}$ include robot and path constraints $\boldsymbol{\Phi}$. The non-minimal coordinates depend on the minimal ones $[\boldsymbol{z} = \boldsymbol{z}(\boldsymbol{q})]$ and from this $[\dot{\boldsymbol{\Phi}} = \boldsymbol{W}^T(\frac{\partial \boldsymbol{z}}{\partial \boldsymbol{q}})\dot{\boldsymbol{q}} + \bar{\boldsymbol{w}} = \boldsymbol{0}, (\frac{\partial \dot{\boldsymbol{\Phi}}}{\partial \dot{\boldsymbol{q}}}) = \boldsymbol{W}^T(\frac{\partial \boldsymbol{z}}{\partial \boldsymbol{q}}) = (\frac{\partial \boldsymbol{z}}{\partial \boldsymbol{q}})^T\boldsymbol{W} = \boldsymbol{0}]$. Applying that to (2) gives ($\boldsymbol{M}_q = [(\frac{\partial \boldsymbol{z}}{\partial \boldsymbol{q}})^T\boldsymbol{M}(\frac{\partial \boldsymbol{z}}{\partial \boldsymbol{q}})]$)

$$\boldsymbol{M}_q\ddot{\boldsymbol{q}} + (\frac{\partial \boldsymbol{z}}{\partial \boldsymbol{q}})^T\boldsymbol{M}\left[\frac{\partial}{\partial \boldsymbol{q}}(\frac{\partial \boldsymbol{z}}{\partial \boldsymbol{q}})\dot{\boldsymbol{q}}\right]\dot{\boldsymbol{q}} + (\frac{\partial \boldsymbol{z}}{\partial \boldsymbol{q}})^T(\boldsymbol{f}^g - \boldsymbol{f}^e - \boldsymbol{f}^a - \boldsymbol{f}^p) = \boldsymbol{0}, \qquad (3)$$

Projecting (3) in direction of a path coordinate s, watching $\frac{dq}{dt} = \frac{dq}{ds}\dot{s}$, $\frac{dq}{ds} = \boldsymbol{q}'$ and second derivatives we come to

$$\boldsymbol{A}(s)(\dot{s}^2)' + \boldsymbol{B}(s)(\dot{s}^2) + \boldsymbol{C}(s) = \boldsymbol{T}(s), \qquad (\boldsymbol{A}, \boldsymbol{B}, \boldsymbol{C}, \boldsymbol{T}) \in \mathbb{R}^{n_q},$$
$$\boldsymbol{B}(s) = \boldsymbol{M}_q\boldsymbol{q}'' + (\frac{\partial \boldsymbol{z}}{\partial \boldsymbol{q}})^T\boldsymbol{M}[\frac{\partial}{\partial \boldsymbol{q}}(\frac{\partial \boldsymbol{z}}{\partial \boldsymbol{q}})\boldsymbol{q}']\boldsymbol{q}',$$
$$\boldsymbol{A}(s) = \frac{1}{2}\boldsymbol{M}_q\boldsymbol{q}', \qquad \boldsymbol{C}(s) = \boldsymbol{f}_q^e, \qquad \boldsymbol{T}(s) = \boldsymbol{f}_q^a,$$
$$\boldsymbol{f}_q = (\frac{\partial \boldsymbol{z}}{\partial \boldsymbol{q}})^T(\boldsymbol{f}^g - \boldsymbol{f}^e - \boldsymbol{f}^a - \boldsymbol{f}^p)_q, \qquad \boldsymbol{f}_q^g = (\frac{\partial \boldsymbol{z}}{\partial \boldsymbol{q}})^T\boldsymbol{M}[\frac{\partial}{\partial \boldsymbol{q}}(\frac{\partial \boldsymbol{z}}{\partial \boldsymbol{q}})\dot{\boldsymbol{q}}]\dot{\boldsymbol{q}}, \qquad (4)$$

Note, that $\boldsymbol{f}_q^g$ is already part of the coefficient $\boldsymbol{B}(s)$. Equation (4) represents a quadratic form being linear in $(\dot{s}^2)$.

2.2 Mobility Space by Ruled Surfaces

Considering for each drive a maximum and minimum torque, Eq. (4) gives

$$T_{i,min} \leq [A_i(s)(\dot{s}^2)' + B_i(s)(\dot{s}^2) + C_i(s)] \leq T_{i,max},$$
$$(\dot{s}^2)' = \frac{-B_i(s)(\dot{s}^2) - C_i(s) + \begin{pmatrix} +T_{i,max} \\ -T_{i,min} \end{pmatrix}}{A_i(s)}, \quad (\dot{s}^2)_{A=0} = \frac{-C_i(s) + \begin{pmatrix} +T_{i,max} \\ -T_{i,min} \end{pmatrix}}{B_i(s)}, \qquad (5)$$

Further possible constraints are the following: The angular or translational velocities of constraining elements may be limited due to some maximum speeds of the drive train components ($\dot{q}_{i,min} \leq [q_i'\dot{s}] \leq \dot{q}_{i,max}$), and the path velocity itself might become constrained by some process ($-v_{max} \leq |\mathbf{r}'|\dot{s} \leq +v_{max}$) with the vector $\mathbf{r}$ from the constrained element under consideration to a path point. These relations define together a maximum velocity $\dot{s}_L$ along the path which must not be exceeded:

$$0 \leq (\dot{s}^2) \leq (\dot{s}^2)_L, \quad \text{with} \quad (\dot{s}^2)_L = \min\left[(\frac{\dot{q}_{i,max}}{q_i'})^2, \quad (\frac{v_{max}}{|\mathbf{r}'|})^2, \quad (\dot{s}^2)_{A=0}\right]. \quad (6)$$

A first step to the solution of the mobility problem may be done in the following way (Fig. 1):

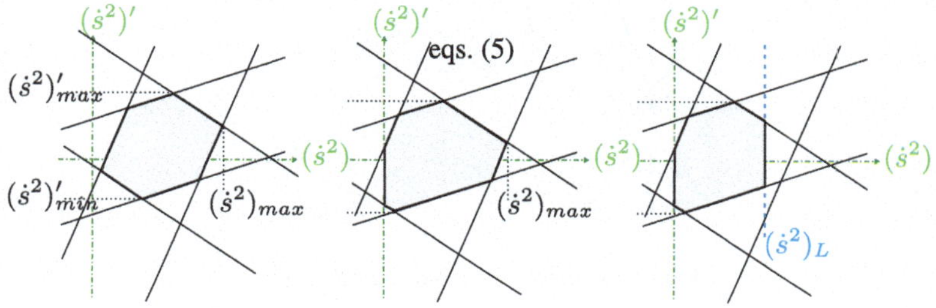

Fig. 1. Polygons of possible motion for s = constant (see also [7])

Equations (5) for the limiting cases $(T_{i,min}, T_{i,max})$ state, that maximum mobility can only be achieved by applying in a maximum number of joints the limiting torques or forces. In operating at the power limits Eq. (5) define two straight lines in the $[(\dot{s}^2)', (\dot{s}^2)]$-plane. The altogether $(2f_D)$ straight lines $(f_D \leq f)$ form a polygon confined at the left side by the axis $(\dot{s}^2) = 0$ and at the right side by the $(T_{i,min}, T_{i,max})$-straight-lines or by the constraint $(\dot{s}^2)_L$ given by Eq. (6). Without violating the constraints, which may be easily extended, motion can take place only within or on the polygons, which contain the following information: the maximum possible velocity $\dot{s}^2_{max}$ and for each point $(\dot{s}^2)$ a maximum and a minimum acceleration or deceleration $[(\dot{s}^2)'_{max}, \ (\dot{s}^2)'_{min}]$. For evaluating a time-optimal solution we have to find a trajectory on the ruled surface of the motion space, which connects start and end in a continuous way. It has to be on the surface, because all polygon lines defining this surface include the extreme torques. For finding these solutions we have to go along the assemblage of these polygons and generate what we call extremals.

Combining all polygons for all path points s we obtain a constrained phase space bounded by ruled surfaces due to the straight line characteristics of the polygons [4]. We call this motion space or mobility space. These polygons appear as plain cuts perpendicular to the s-axis, Fig. 5. The evaluation of these polygons with the help of Eq. (4) is difficult. From many trials a numerical procedure is chosen, being the most efficient at the time being.

3 Example Multi-DOF-Robot with Revolute Joints

According to Eq. (1) we need for the **equations of motion** kinematic and kinetic magnitudes. A robot with n_q revolute joints can be described by Cartesian z-coordinates and by angular $\boldsymbol{q}$-coordinates. The projections ($\boldsymbol{z} \to \boldsymbol{q} \to s$) have been discussed before. With respect to the $\boldsymbol{q}$-coordinates two version are needed (Fig. 2).

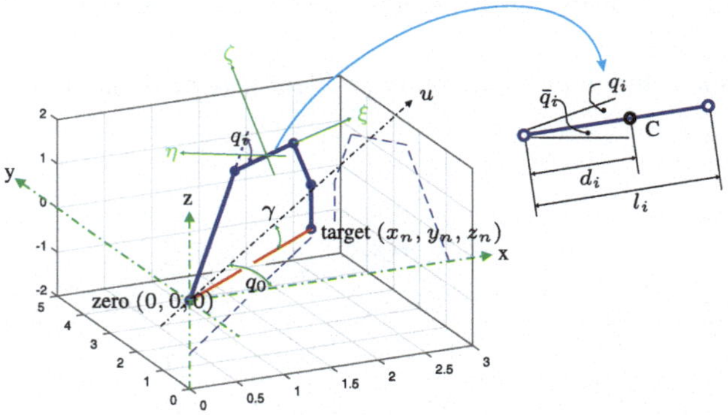

Fig. 2. Geometry of a multi-DOF robot

$$\bar{q}_i = \frac{\pi}{2} - \sum_{j=1}^{i} q_j = \bar{q}_{i-1} - q_i, \qquad q_i = \bar{q}_{i-1} - \bar{q}_i, \qquad \bar{q}_0 = \frac{\Pi}{2}. \tag{7}$$

With these definitions the joint vectors $\boldsymbol{z}_{Ji}$ and the center of mass vectors $\boldsymbol{z}_{Ci}$ possess the following elements

$$\boldsymbol{r}_{Ji} = \sum_{j=1}^{i-1} l_j \begin{pmatrix} \cos\bar{q}_j \cos q_0 \\ \cos\bar{q}_j \sin q_0 \\ \sin\bar{q}_j \end{pmatrix}, \quad \boldsymbol{r}_{Ci} = \sum_{j=1}^{i-1} l_j \begin{pmatrix} \cos\bar{q}_j \cos q_0 \\ \cos\bar{q}_j \sin q_0 \\ \sin\bar{q}_j \end{pmatrix} + d_i \begin{pmatrix} \cos\bar{q}_i \cos q_0 \\ \cos\bar{q}_i \sin q_0 \\ \sin\bar{q}_i \end{pmatrix} \tag{8}$$

It should be noted, that for the rotation around the vertical inertial z_1-axis (Fig. 2) the above vectors are zero, $\boldsymbol{r}_{Ji} = 0$, $\boldsymbol{r}_{Ci} = 0$. The angular velocities $\boldsymbol{\omega}_i$ are determined by the well-known relations ${}_I\tilde{\boldsymbol{\omega}}_i = \dot{\boldsymbol{A}}_{IB}\boldsymbol{A}_{BI} = \dot{\boldsymbol{A}}_{IB}\boldsymbol{A}_{IB}^T$ with the result [8] (I=inertial, B-body–)

$${}_I\boldsymbol{\omega}_i = \begin{pmatrix} -\dot{\bar{q}}_i \sin q_0 \\ +\dot{\bar{q}}_i \cos q_0 \\ \dot{q}_0 \end{pmatrix}, \quad {}_B\boldsymbol{\omega}_i = \begin{pmatrix} -\dot{q}_0 \sin\bar{q}_i \\ \dot{\bar{q}}_i \\ +\dot{q}_0 \cos\bar{q}_i \end{pmatrix}, \quad \boldsymbol{A}_{IB} = \begin{pmatrix} \cos q_0 cos\bar{q}_i & -\sin q_0 & \cos q_0 \sin\bar{q}_i \\ \sin q_0 cos\bar{q}_i & +\cos q_0 & \sin q_0 \sin\bar{q}_i \\ -\sin\bar{q}_i & 0 & \cos\bar{q}_i \end{pmatrix} \tag{9}$$

The matrix $\boldsymbol{A}_{IB}$ follows from two rotations, first around the inertial z-axis, second around the body-fixed y-axis. For i=1 (first degree of freedom with q_0) the angular velocity is ${}_I\boldsymbol{\omega}_1 = {}_B\boldsymbol{\omega}_1 = (0 \ 0 \ \dot{q}_0)^T$.

For positioning the robot end effector at the target point we apply formula (8) and require (Fig. 2)

$$\boldsymbol{r}_{Jn} = \boldsymbol{r}(\boldsymbol{q}(s)) = \sum_{j=1}^{n} l_j \begin{pmatrix} \cos\bar{q}_j \cos q_0 \\ \cos\bar{q}_j \sin q_0 \\ \sin\bar{q}_j \end{pmatrix} = \begin{pmatrix} x \\ y \\ z \end{pmatrix}_{trajectory} . \tag{10}$$

This represents a nonlinear set of equations for the determination of robot coordinates $\boldsymbol{q}$.

From the definition of the vectors to the joints and the centers of mass all necessary derivations can be obtained in a straightforward manner. The projections of $\boldsymbol{q}$ with respect to s are evaluated numerically by

$$\boldsymbol{q}' = \frac{\boldsymbol{q}(s_{i+1}) - \boldsymbol{q}(s_{i-1})}{s_{i+1} - s_{i-1}}, \qquad \boldsymbol{q}'' = \frac{\boldsymbol{q}(s_{i+1}) - 2\boldsymbol{q}(s_i) + \boldsymbol{q}(s_{i-1})}{[\frac{1}{2}(s_{i+1} - s_{i-1})]^2}. \tag{11}$$

To determine the same information for the angle $\boldsymbol{q}_0$ we start with the (x,y,z)-position of each trajectory point and calculate q_0 and its derivatives (Fig. 2)

$$\tan q_0 = (\frac{y}{x})_{path}, \quad q_0' = \frac{y'x - yx'}{x^2 + y^2},$$
$$q_0'' = \left[\frac{y''x - yx''}{x^2 + y^2} + 2\frac{x'y'(y^2 - x^2) - xy((y')^2 - (x')^2)}{(x^2 + y^2)^2}\right] \tag{12}$$

For our specific case of a n-link robot following a trajectory we determine $\boldsymbol{q} = \boldsymbol{q}(\boldsymbol{r}(s))$ by an optimization algorithm.

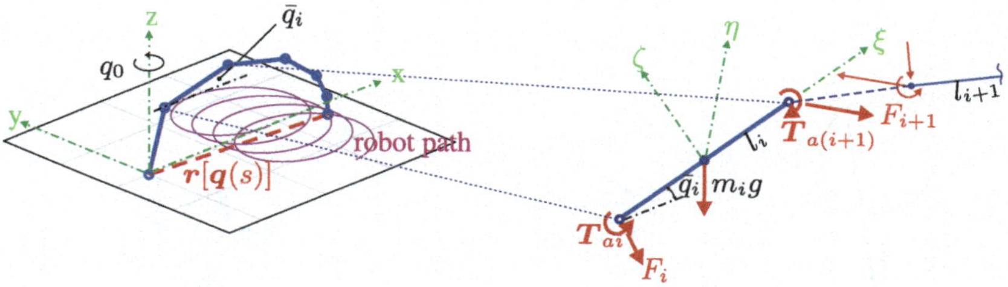

Fig. 3. Trajectory, torques and forces at link i

Following Eq. (1) the mass matrix and its derivatives write

$$\boldsymbol{M}_i = \begin{pmatrix} m\boldsymbol{E} & m\tilde{\boldsymbol{r}}^T_{O'S} \\ m\tilde{\boldsymbol{r}}_{O'S} & \boldsymbol{\Theta}_{O'} \end{pmatrix}_i \in \mathbb{R}^{n_m, n_m}, \quad \text{with} \quad \boldsymbol{\Theta}_{O'i} = -\sum_{\Delta m}[\tilde{\boldsymbol{r}}_{O'S}\tilde{\boldsymbol{r}}_{O'S}\Delta m]_i$$

$$\frac{d\boldsymbol{M}_i}{dt} = \begin{pmatrix} \boldsymbol{0} & m\dot{\tilde{\boldsymbol{r}}}^T_{O'S} \\ m\dot{\tilde{\boldsymbol{r}}}_{O'S} & \dot{\boldsymbol{\Theta}}_{O'} \end{pmatrix}_i, \quad \frac{d\boldsymbol{\Theta}_{O'i}}{dt} = -\sum_{\Delta m}[\dot{\tilde{\boldsymbol{r}}}_{O'S}\tilde{\boldsymbol{r}}_{O'S} + \tilde{\boldsymbol{r}}_{O'S}\dot{\tilde{\boldsymbol{r}}}_{O'S}]_i \Delta m_i \tag{13}$$

For some link i we choose as a reference point the center of mass S, which makes $\boldsymbol{r}_{O'S} = 0$ and $\dot{\boldsymbol{r}}_{O'S} = 0$. A special case is the first coordinate q_0, where the reference point O' is the coordinate origin with $(\boldsymbol{r}_{O'S})_i \neq 0$ being the distance to the mass centers of each link i. For this case and according to the Eq. (8) we have $(\boldsymbol{r}_{O'S})_i = \boldsymbol{r}_{Ci}$. The moment if inertia for the bar-like links alone is $\dot{\boldsymbol{\Theta}}_{Si} = \frac{1}{3}m_i l_i^2[1 - 3(\frac{d}{l}) + 3(\frac{d}{l})^2]$.

As a further step we have to evaluate the forces and torques considered in an inertial system. The external and applied forces and torques (Fig. 3) are:

$${}_I\boldsymbol{f}^e_i = [0\ 0\ (-m_i g)\ 0\ 0\ 0]^T, \quad {}_I\boldsymbol{f}^e_1 = \boldsymbol{0}, \quad {}_I\boldsymbol{f}^a_1 = [0\ 0\ 0\ 0\ 0\ (T_1)]^T,$$
$${}_I\boldsymbol{f}^a_{i>1} = [0\ 0\ 0\ (-(T_{a(i+1)} - T_{ai})\sin q_0)\ (+(T_{a(i+1)} - T_{ai})\cos q_0)\ 0]^T, \tag{14}$$

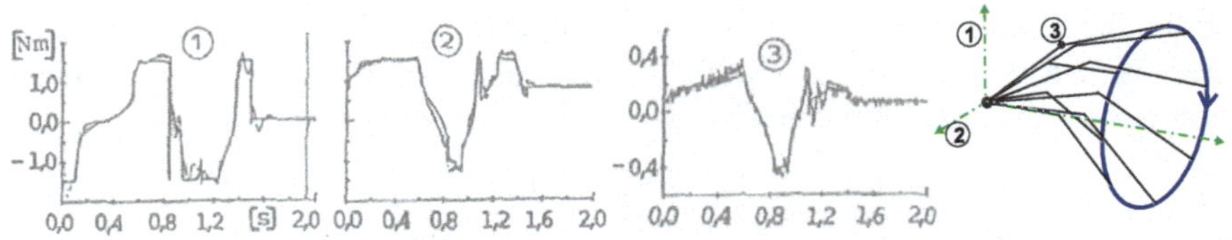

Fig. 4. Comparison theory/measurements for a two-link robot performing a circular vertical trajectory, (1, 2, 3) are the joints, ordinate joint torque, abscissa path coordinate [6,7]

The indicated theory has been realized by a computer code, operating with a time scale of some seconds. In the following we present **some results**. An illustrative example is a two-link robot performing a circular vertical path, for which we give as a first step some old comparisons with measurements of the time-optimal path torques along a vertical circle, Fig. 4. Theory and measurements compare well, the high frequency oscillations of the measured data are due to the robot measurement system, details are described in [6,8,9]. The method has been used to optimize robots and trajectories for various manufacturing problems, like pick and place or peg-in-hole problems.

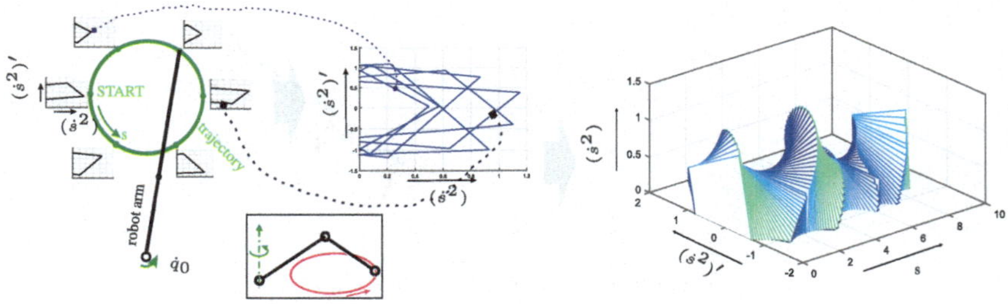

Fig. 5. Motion areas (polygons) along a horizontal circular trajectory, **left**:configuration with selected polygons, **right**: arrangement of all polygons along the circular path

In the following we consider horizontal circular path configurations or moving circles similar to polishing trajectories. Figure 5 illustrates the solution-procedure by depicting polygons of six points along a circular path, which are collected in one graph (third picture of Fig. 5). Each polygon presents the kinematics of a point s. Arranging that along the path with all points s, we come out with a three-dimensional structure of ruled surfaces (right picture of Fig. 5). Motion can only take place on these ruled surfaces originating from the extremum drive torques or forces, or it can take place within the space bounded by these ruled surfaces.

The motion space of Fig. 5 represents a convenient tool for studying various solution and motion concepts, for example any prescribed motion within the polygon-space or a time-optimal solution on the ruled surface of that space. The last problem needs the introduction of extremals. From Eq. (5) these extremals are constructed on both sides of the motion space, the side with maximum accelerations $(\dot{s}^2)'_{max}$ containing the max-extremals and the side with minimum accelerations $(\dot{s}^2)'_{min}$ containing the min-extremals. Figure 6 illustrates the projections of these extremals together with the maximum velocities from motion space onto the $[\dot{s}^2 - s]$ - plane.

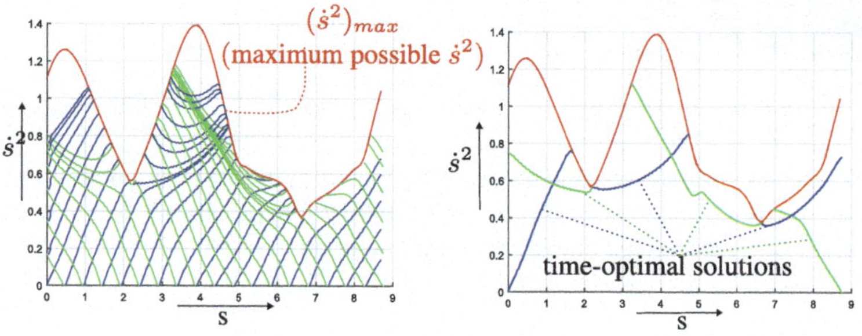

Fig. 6. Field of extremals (blue lines max-extremal, green lines min-extremals) and an extremal selection forming the time-optimal solution

A time-optimal solution for example must be composed by several of these extremals for following reasons. Starting with $\dot{s}^2 = 0$ at s=0 and $s = s_F$ requires to follow a certain combination of extremals for arriving at the final point. Such a procedure needs an integration of the equations of motion along the ruled surfaces and thus from polygon at s to the next one at $(s + \Delta s)$. Integration on polygons excludes normally a progress along the rim of the motion space, exceptions are possible.

Figure 6 illustrates that procedure. The blue extremal at the beginning connects s=0 with the maximum velocity curve at $s \approx 1.7$, and the green extremal at the end connects $s = s_F \approx 8.8$ with $s \approx 6.9$. Note, that the blue extremals are on the motion space side with maximum accelerations and the green on the side with minimum accelerations. As a further step we construct all extremals originating from the saddle points (minimum points) at $s \approx 2.2$ and at $s \approx 6.7$. These are points, where at least one coefficient $A_i \in \boldsymbol{A}$ of the Eq. (4) vanishes, thus allowing a smooth step from one side of the motion space to the other one. The extremals emerging from these points intersect the extremals coming from the beginning and from the end (right graph of Fig. 6). The resulting ensemble of six curves form the time-optimal solution for the motion under consideration, avoiding any limits and integration problems.

For considering the driving torques along a time-optimal solution we select a spiral trajectory as shown by Fig. 7. The torques for following six cycles and one cycle are shown. The result confirms, that for time-optimal solutions at least one driving torque has to operate at its limits. In addition the left graph indicates an useful property, namely a tunnel-like free area, where any kind of motion might take place without touching limiting constraints.

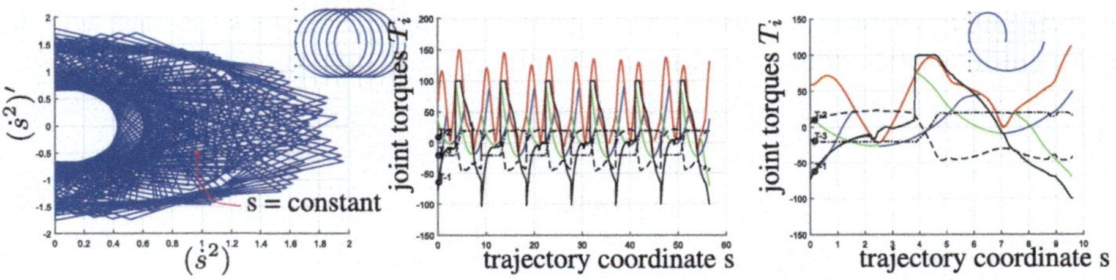

Fig. 7. left:collection of polygons and trajectory for polishing (robot position as in Fig. 5), **middle**, **right**: joint torques for a two-link robot, black solid T_1, black dashed T_2, black dash-dotted T_3, red line $(\dot{s}^2)_{max}$, blue-green lines time-optimal solution

4 Conclusion

Starting with a formulation according to Lagrange I (Jacobi) the degrees of freedom of any multibody system are projected to one prescribed trajectory of arbitrary shape resulting in relations following one degree of freedom only. The projecting tools are derivatives, which are evaluated from the geometry and constraints of the multibody structure together with the given trajectory. The emerging relations represent a system of linear differential equations in $(\dot{s})^2$ with highly nonlinear coefficients including the complete MBS (robot) - configuration. They can be used to build a mobility space, where the motion takes place. This space represents a useful tool for improving and for facilitating design of mechanical systems in general and of robotic systems in particular. It will be a matter of ongoing research.

References

1. Bobrow, J.E., Dubowsky, S., Gibson, J.S.: Time optimal control of robotic manipulators along specified paths. Int. J. Robot. Res. **4**, 3–17 (1985)
2. Bremer, H.: Elastic Multibody Dynamics. Springer Science+Business Media, Switzerland (2008)
3. Dubowsky, S., Shiller, Z.: Optimal dynamic trajectories for robotic manipulators. In: Proceedings of 5th Symposium on Theory and Practice of Robotics and Manipulators (1985)
4. Hoschek, J.: Liniengeometrie. B. I. Hochschulskripten, Bibliographisches Institut Zürich (1971)
5. Jacobi, C.G.J.: Vorlesungen über Dynamik. Edited by Clebsch, A. Georg Reimer, Berlin (1866)
6. Johanni, R.: Optimale Bahnplanung bei Robotern, Fortschritt-Berchte VDI, Reihe 18, Nr. 51, VDI-Verlag Düsseldorf (1988)
7. Pfeiffer, F.: Optimal Trajectory Planning for Manipulators. Pergamon Press, Systems and Control Encyclopedia, Oxford, New York (1990)
8. Pfeiffer, F.: Mechanical System Dynamics. Springer, Heidelberg (2008). https://doi.org/10.1007/978-3-540-79436-3
9. Pfeiffer, F.: Motion spaces of machine-process combinations. Arch. Appl. Mech. **89**(10), 2115–2132 (2019). https://doi.org/10.1007/s00419-019-01564-7
10. Woernle, C.: Mehrkörpersysteme. Springer, Berlin (2016). https://doi.org/10.1007/978-3-662-46687-2

Fitting Constrained Trajectory with High Variability into Redundant Robot Workspace

Zvezdan Lončarević[1,2(✉)], Tadej Petrič[1,2], and Andrej Gams[1,2]

[1] Jožef Stefan Institute, Jamova cesta 39, 1000 Ljubljana, Slovenia
[2] Jožef Stefan International Postgraduate School, Jamova cesta 39, 1000 Ljubljana, Slovenia
zvezdan.loncarevic@ijs.si
https://abr.ijs.si/

Abstract. One of the important aspects of the robotic tasks is their variability. Most of the nowadays tasks are not pre-programmed and because of increased usage of reinforcement learning and other autonomous learning techniques, robotic trajectories have to be fitted in the robot workspace so that they have as much space for learning as possible, i.e., as high variability as possible. In this paper, we present and compare our methods for fitting a trajectory in a 2-D sagittal plane in the robot workspace. The method is shown on the example of robotic throwing action with the TALOS humanoid robot.

Keywords: Robot workspace · Manipulability · Optimization · Neural networks · Movement primitives · Generalization

1 Introduction

Reinforcement learning (RL) has recently made a huge success in a whole variety of tasks. One of the biggest problems when applying it in robotics is that a lot of trajectories generated by RL algorithms are simply not executable due to physical limitations of the robot or due to problems with calculating inverse kinematics. One of the most widely applied methods for learning is combination of learning by demonstration (LbD) and RL. The initial trajectory, that is somewhat similar to the desired one, is shown to the robot and then refined using RL algorithms [1]. When using the robots with narrow joint limits and small manipulability measure, the process of refining this trajectory may lead to unexecutable trajectories. Our goal is to develop a method that would allow us to position the trajectory in the constrained space of the robot so that there is a highest number of executable trajectories similar to the initial (demonstrated) one, thus allowing for successful RL.

Most of the work presented in this direction of research considers positioning of mobile robots and determining initial position so that the robot has highest manipulability and distance from joint limits but for only pre-programmed trajectories [2,3]. Also, by exploiting the robot's redundancy, configurations with

A. Müller and M. Brandstötter (Eds.): RAAD 2022, MMS 120, pp. 167–175, 2022.
https://doi.org/10.1007/978-3-031-04870-8_20

higher manipulability can be favored in order to allow better adaptation during execution [4]. However, in the case of learning, robots cannot be re-positioned and the configuration cannot be essentially different because it needs to stay similar to the trajectory obtained by LbD. That is why in the scope of this paper, we examine the positioning of the trajectory so that it has the biggest possible search space for the RL algorithm without considering reconfiguration of the robot.

In order to operate in various environments, robot's ability to supervise their own movements is essential. A measure that gives the information about robot's ability to maneuver in the workspace is a manipulability measure introduced in [5]. In addition to manipulability analysis, joint limits also play an important role in the robot's ability to move in the workspace [6]. In order to address this problems, penalization terms can be used [7] or Jacobian matrix can be augmented and analyzed [8]. In our work, we consider both manipulability and the distance from joint limits as a direct measure of how good is the configuration in order to position the trajectory in the most suitable way.

The paper is structured as follows. In the next Section we present experimental environment and explain the problem. Sect. 3 explains data acquisition using Dynamic Movement Primitives (DMPs). Used methods are presented in Sect. 4, followed by the results in Sect. 5 and Conclusion in Sect. 6.

2 Robot and Definition of the Problem

As mentioned before, our approach is tested on the robotic throwing problem. Since solving this problem requires collecting a database, the experiments were performed in simulated environment - MuJoCo [9]. We used the left arm of humanoid robot TALOS [10] equipped with a spoon in order to extend the throwing range as shown in Fig. 1 - left. Throwing is performed in the vertical plane as shown in Fig. 1 - right. Initial throw was given as a part of a parabola that the robot was able to execute in the plane parallel to xz plane. If we would start modifying this initial throw (for instance using RL), we would face the problems of reaching joint limits or getting into the singular configurations. In order to overcome this problems, we need to position the trajectory differently. This is why we decided to try rotating it around its initial point (Fig. 1 - right) and finding at which angle there are the highest number of feasible trajectories. The initial position could be rotated for $2°$ counter-clockwise and for $11°$ clockwise without robot reaching any joint limits, so this was taken as the range in which the optimal orientation needs to be found.

3 Dynamic Movement Primitives and Database Acquisition

In order to check which trajectories are executable and which not, our first step was to rotate the initial trajectory. We generated 53 trajectories in the range

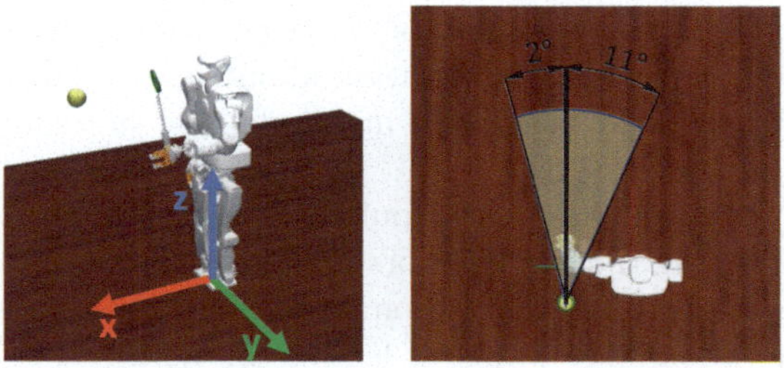

Fig. 1. Left image shows the throwing action and coordinate system of the robot and right image shows the feasible range of angles with the origin of the arc put into the initial position of the throw.

$[2°, -11°]$ with rotating initial trajectory for up to 2° counter-clockwise and up to 11° clockwise with the step of 0.25°. In further text and images, we will use positive sign for counter-clockwise direction and negative for the clockwise. This set of trajectories was then encoded in a form of Cartesian Dynamic Movement Primitives (CDMPs). For the clarity of the paper, we briefly explain parameters of CDMPs that are important for our application but full description and auxiliary math are given in [11].

3.1 Cartesian Dynamic Movement Primitives (CDMPs)

Cartesian Movement Primitives are parametric representation of the trajectory that is able to present the trajectory in fixed and reduced number of parameters:

$$\mathbf{W} = [\boldsymbol{w}_k^p, \mathbf{g}_k^p, \mathbf{y}_{0,k}^p, \boldsymbol{w}_k^o, \mathbf{g}_k^o, \mathbf{y}_{0,k}^o, \tau], \tag{1}$$

where $\boldsymbol{w}_k^p$ defines a shape of the position part of trajectory, $\mathbf{g}_k \in \mathbb{R}^3$ defines the final position of trajectory and $\mathbf{y}_{0,k}^p \in \mathbb{R}^3$ defines the initial position of the trajectory. Subscript k together with superscript p refers to position trajectory components in x, y and z directions. Similarly w_k^o describes a shape of the orientation part of trajectory. Final and starting orientations are presented with quaternions and therefore they are in four dimensional space: $\mathbf{g}_k^o \in \mathbb{R}^4$ and $\mathbf{y}_{0,k}^o \in \mathbb{R}^4$. Subscript k together with superscript o refers to orientation trajectory components about x, y and z axes. Both $\boldsymbol{w}_k^p$ and $\boldsymbol{w}_k^o$ were chosen to be vectors of size $N = 20$ for all the components. Parameter τ is time scaling factor.

3.2 Database Acquisition

When having the initial trajectory rotated in different angles with respect to initial position and encoded in CDMPs, we were able to start generating the database of trajectories. In order to do that, we needed to modify forcing terms for position and orientation. Parameters that describe the forcing term are weights $\boldsymbol{w}_k^p$, $\boldsymbol{w}_k^o$. Since we want trajectories to stay in the same plane in order for the direction of the throw not to be changed, we modify only the weights defining the position in z-axis and rotation about y-axis. For all the reachable points of this trajectories we calculated manipulability measure and distance to joint limits.

Manipulability is a measure of manipulating ability of robotic mechanisms in positioning and orienting the end-effectors [5]. This number describes the robot's ability to change its end-effector position given the current joint configuration. The greater it gets, the greater range of feasible motions can be performed from that configuration. If this number is zero, it means that the robot is in singular configuration and it cannot generate velocity in one of the Cartesian directions. This measure is defined for each k-th point in the dataset as:

$$\mu_k = \sqrt{det(\mathbf{J}\mathbf{J}^T)}, \tag{2}$$

In our case of a redundant robotic arm with 7-DoF, the Jacobian matrix is of size 6 × 7. Since throwing is performed only in one plane, we are not interested in ability to move in the direction of y-axis and in rotations about x and z axes. This is why we used reduced Jacobian of size 3 × 7 where we left out corresponding rows of the matrix. Normalized values of manipulability index ($\hat{\mu}_k = \frac{\mu_k}{max(\mu_k)_{k=1}^K}$) for all the executed points in the database are shown in Fig. 2.

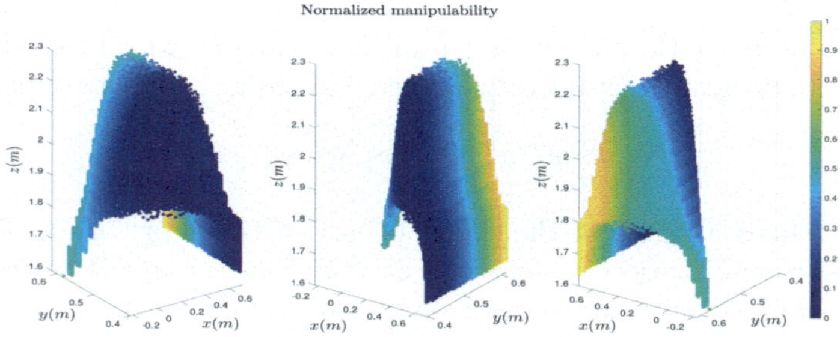

Fig. 2. Normalized manipulability measure for the reachable points

Distance from the joint limits is the second but equally important parameter for successfully executing the trajectory. As a measure of distance from the joint limits, we used the smallest normalized distance to the joint limit from all the 7 joints of the robot arm. For all K points in dataset it was calculated as:

$$\begin{aligned} \phi_k &= min(j_i)_{i=1}^7, \\ j_i &= \frac{min(l_{i,1}, l_{i,2})}{|\theta_i^{max} - \theta_i^{min}|}, \\ l_{i,1} &= |\theta_i^{max} - \theta_i|, \\ l_{i,2} &= |\theta_i - \theta_i^{min}|, \end{aligned} \tag{3}$$

where θ_i^{min}, θ_i^{max} and θ_i are lower, upper limit and the current state of the i-th joint respectively. The values are presented with a color-bar in Fig. 3.

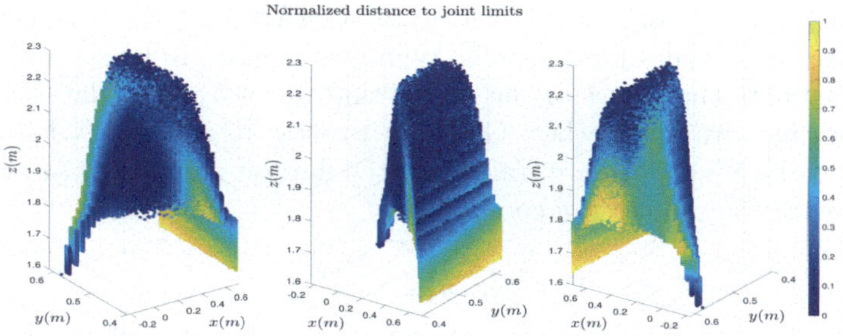

Fig. 3. Normalized minimal distance to the joint limit for the reachable points

Once we calculated all the manipulability measures μ_k and joint limit distance ϕ_k, we also calculated crossection of this two values:

$$c_k = \frac{\hat{\mu}_k + \phi_k}{2}, \tag{4}$$

and this number was used as criteria of how appositely it is for a robot to be in that state and its values are given in Fig. 4.

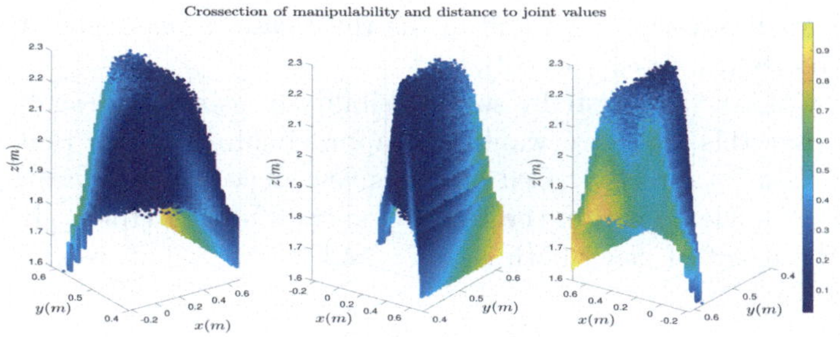

Fig. 4. Crossection of the normalized manipulability measure and minimal distance to the joint limit for the reachable points

Since the problem is planar and we can change only the angle of the throw, our next step was to project the mean value of crossection for all the points in dataset that have unique x and y coordinates to xy-plane as shown in Fig. 5 and Fig. 6.

4 Generalization and Evolutionary Algorithm

With having the database of points, we needed to approximate which angle is the best for the throw. We present three different approaches for solving this problem. We apply evolutionary algorithm (CMA-ES [12]), use generalization and compare it to brute-force search. Whatever approach we choose, first step is to approximate the values of the points that are not in the dataset. For this purpose, we use two approaches: Gaussian Process Regression (GPR [13]) and neural network (NN). Both are trained on the dataset containing M crossection c_m values with the unique x, y coordinates:

$$\mathcal{D} = \left\{\{x, y\}_m, c_m\right\}_{m=1}^{M}. \tag{5}$$

With both approaches trained, we could approximate values of all the N points of trajectory in one line of the throw, rotated for a certain degree using GPR:

$$G(\mathcal{D}) : \{x, y\}_n \rightarrow c_n, \quad n = 1...N \tag{6}$$

or NN:

$$N(\mathcal{D}) : \{x, y\}_n \rightarrow c_n', \quad n = 1...N. \tag{7}$$

When using CMA-ES, our goal was to find the desired angle and the reward was determined by the minimal crossection value in the certain line of the throw:

$$R = min(\hat{c}_n)_{n=1}^{N}. \tag{8}$$

We decided for this approach and not using mean value because we wanted to avoid having any bottlenecks with small variability. We tested 3 different rewards. In the first case, crossection was calculated with GPR ($\hat{c}_n = c_n$), in the second one with NN ($\hat{c}_n = c_n'$) and in the third case we used mixed reward of these two values ($\hat{c}_n = (c_n + c_n')/2$).

Except CMA-ES, we tested also a possibility of using GPR again. The difference is that this time, we wanted to approximate function that describes the reward (Eq. 8) directly based on the given angle. Rewards were given for P equidistant angles ρ within the range of successful trajectories. Reward was based on the result of using both $\hat{c}_n = c_n$ and $\hat{c}_n = c_n'$. This sets our training dataset to:

$$\mathcal{D} = \left\{\rho_p, R_p\right\}_{p=1}^{P}. \tag{9}$$

With having trained GPR, we could approximate rewards for much more angles in-between this points. We did it for L points and chosen q_l that had maximal approximated reward R_l:

$$G(\mathcal{D}) : \rho_l \rightarrow R_l, \quad l = 1...L. \tag{10}$$

In order to confirm the efficiency of the approaches, we compared them to the brute-force method. In this third approach, we split the range of possible angles to Q equidistant angles and simply tested which one of them has the highest reward. Same as with previous method, we used two rewards where one was determined by GPR approximated values $\hat{c}_n = c_n$ and the other one where we used NN approximated values $\hat{c}_n = c_n'$.

5 Results

Color bars in Fig. 5-left and Fig. 6 show the crossection values for all $M = 1856$ points with unique x, y-coordinates. Results of learning appropriate angle using CMA-ES is given in Fig 5. Learning with GPR approximated reward is presented with green line, with NN approximated reward with red line and with mixed reward with blue line in Fig. 5-left. All the rewards were determined as minimal crossection value among $N = 3000$ points. Right side of Fig. 5 shows angles that CMA-ES needed to approximate reward in its 10 iterations with population size of 40 in order to converge. Results show that all the reward systems had similar and stable convergence to the final solutions of $-3.2719°$ when using GPR, $-3.1127°$ when using NN and $-3.1244°$ when using mixed reward.

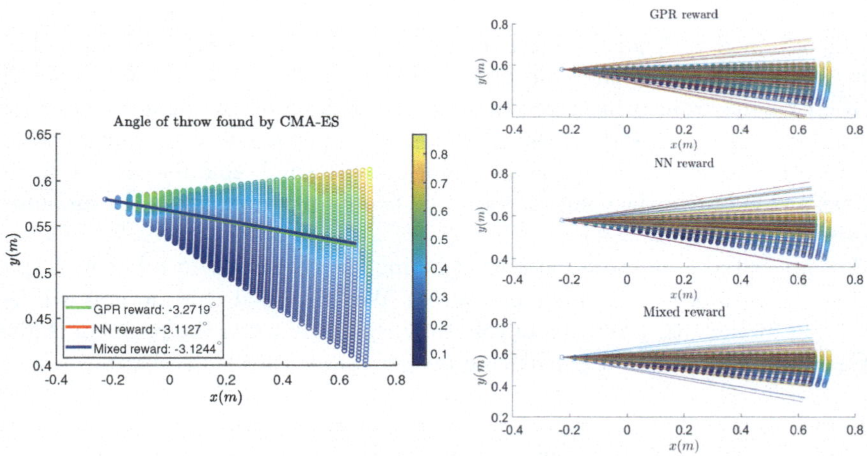

Fig. 5. Angle found by CMA-ES (left) and tested angles until convergence (right)

Figure 6-left shows the results when using GPR on $L = 100$ points for two cases of using GPR and NN to approximate the reward. The approach was tested also with higher amount of points (up to 800) but there was no significant change in the outcome. Final solution of this approach were angles of $-3.1922°$ for the case of GPR reward and $-3.036°$ for the case of NN reward.

In order to compare the results between GPR and CMA-ES, we performed brute-force search on 1000 different angles and determined that best angles for performing the trajectory are $-3.2703°$ for the case of GPR approximated reward and $-3.1271°$ for the case of NN approximated reward.

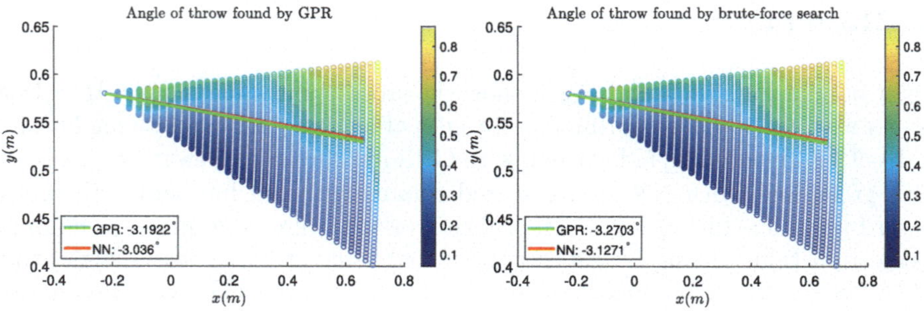

Fig. 6. Angle found by GPR (left) and by brute-force method (right)

6 Conclusion

Result obtained with all the methods produced similar results and showed efficient way of fitting planar trajectory into the robot workspace. As a reference we used results from brute-force method. CMA-ES was the fastest and easiest method to implement and it reached almost the same result as the reference one. On the other hand, GPR found close solution with small amount of data but did not get any closer to the solution found by brute-force even when the amount of data was increased.

We presented a methods capable of fitting the trajectory in robot workspace so that the variability is as high as possible. We think that this method will help in solving further RL problems in robotics, especially when the problem is given in Cartesian space so that calculating inverse kinematics is necessary.

Acknowledgement. This research was funded program group Automation, robotics, and Biocybernetics (P2-0076) supported by the Slovenian Research Agency.

References

1. Lončarević, Z., Pahič, R., Ude, A., Gams, A.: Generalization-based acquisition of training data for motor primitive learning by neural networks. Appl. Sci. **11**(3), 1013 (2021)
2. Kapusta, A., Kemp, C.C.: Task-centric optimization of configurations for assistive robots. Autonom. Robot. **43**(8), 2033–2054 (2019). https://doi.org/10.1007/s10514-019-09847-2
3. Zacharias, F., Borst, C., Beetz, M., Hirzinger, G.: Positioning mobile manipulators to perform constrained linear trajectories. In: International Conference on Intelligent Robots and Systems, pp. 2578–2584 (2008)
4. Vahrenkamp, N., Asfour, T.: Representing the robot's workspace through constrained manipulability analysis. Autonom. Robot. **38**, 1–14 (2014)
5. Yoshikawa, T.: Manipulability of robotic mechanisms. Int. J. Robot. Res. **4**, 3–9 (1985)
6. Flacco, F., De Luca, A. and Khatib, O.: Prioritized multi-task motion control of redundant robots under hard joint constraints. In: Proceedings of the International Conference on Intelligent Robots and Systems, pp. 3970–3977 (2012)

7. Tsai, M.-J.: Workspace Geometric Characterization and Manipulability of Industrial Robots. PhD thesis, The Ohio State University (1986)
8. Abdel-Malek, K., Yu, W., Yang, J.: Placement of robot manipulators to maximize dexterity. Int. J. Robotics Autom. **19**(1), 6–14 (2004)
9. Todorov, E., Erez, T., Tassa, Y.: Mujoco: a physics engine for model-based control. In: International Conference on Intelligent Robots and Systems, pp. 5026–5033 (2012)
10. Stasse, O., et al.: TALOS: a new humanoid research platform targeted for industrial applications. In: 17th International Conference on Humanoid Robotics, pp. 689–695 (2017)
11. Ude, A., Nemec, B., Petric, T., Morimoto, J.: Orientation in cartesian space dynamic movement primitives. In: International Conference on Robotics and Automation, pp. 2997–3004 (2014)
12. Hansen, N.: The CMA evolution strategy: a tutorial (2016)
13. Williams, C., Rasmussen, C.: Gaussian processes for regression. In: Touretzky, D., Mozer, M.C., Hasselmo, M. (eds.) Advances in Neural Information Processing Systems, vol. 8, MIT Press (1996)

Experimental Validation of a Variable Stiffness Joint Based on Antagonistic Principle

María Guadalupe Contreras-Calderón[1(✉)], Juan Sandoval[2], Eduardo Castillo-Castañeda[1], and Med Amine Laribi[2]

[1] Instituto Politécnico Nacional, Mexico City, México
ma.gpe.contreras.c@gmail.com
[2] Department of GMSC, Pprime Institute CNRS, ENSMA, UPR 3346, University of Poitiers, Poitiers, France

Abstract. In human-robot interaction environments, flexible mechanisms have been implemented to relate force with movement to make interaction safer. Mechanism flexibility can be established through the use of variable stiffness joints which allow simultaneous change of the position and stiffness in joints. This paper presents a Variable Stiffness Joint (VSJ) based on the antagonistic principle with two springs. The proposed VSJ can provide constant or variable torque, depending on the task requirements. It includes a motor, associated with a set of gears and racks, to vary the lengths of the springs. The stiffness behavior is controlled with a position control law of the motor. This work details the joint mathematical model and its validation by simulation. For this purpose, a specific experimental setup is developed for torque measurement. The design of a new proposal VSJ is a compact version that maintains the same antagonistic configuration of springs consisting of two stages is presented.

Keywords: Variable stiffness joint · Antagonistic springs · Antagonist principle · Human-robot interaction · Variable torque

1 Introduction

Several recent innovations in robotics are oriented toward human interaction and primarily on improving safety during the interaction. Traditional robots have been considered as rigid bodies, and new technologies are oriented towards being adaptable to daily activities, flexible, and safe. At present, the interaction between robots and humans is more frequent, both in service robots and in industrial robots. That is why it is important to include flexible elements, with the possibility of controlling their stiffness and, to achieve a safe environment during the interaction. Some advantages of using flexible elements in mechanisms are reducing high-impact collisions. In addition, they improve inertia, damping, accuracy, energy efficiency [1], and the accuracy of force [2]. There are two methods for controlling the stiffness in a mechanical system. The first method uses torque feedback concerning the difference between the input and the output position, as does impedance control. The second method is to control the stiffness of the robot's

A. Müller and M. Brandstötter (Eds.): RAAD 2022, MMS 120, pp. 176–183, 2022.
https://doi.org/10.1007/978-3-031-04870-8_21

joints [3], including elastic elements such as springs, beams, cables, among others. Variable Stiffness Joints (VSJ) have been developed as an alternative to include flexibility in a mechanism since its stiffness can be controlled with mechanical settings [4]. These joints can have a series of antagonistic configurations and include flexible elements, among other materials [5]. For example, the AMASC joint at [6] features a spring and cable configuration. The joints in [2, 7] are based on the principle of antagonistic springs, whose function is similar to that of the arm muscles, biceps, and triceps. Joints [8–10, 15] are based on the lever principle, where the fulcrum slides through the bar in order to vary stiffness. Some such as [11, 12] vary their stiffness with spring pretension. The stiffness-self-adjustable module [13] uses the lever principle and springs. Some of the above have a combination of elastic elements.

This paper presents the experimental validation of a VSJ based on the principle of antagonistic springs [6]. The adjustment is carried out through a motor that controls the pretension of the springs. This joint provides variable or constant stiffness determined at the output. The adjustment is carried out through a motor that controls the pretension of the springs. The paper is organized as follows. Section 2 describes the concept of the VSJ as well as the mathematical model based on Hooke's law. Section 3 presents the validation on simulation of the VSJ mathematical model. Section 4 deals with the experimental setup and presents the results. Section 5 describes a new VSJ version. Finally, Sect. 6 presents our conclusions.

2 Concept and Mathematical Model

The proposed joint is based on the principle of antagonistic springs, as reported in (Fig. 1). The VSJ consists of a link with a pinion P_e (with radius r) in the rotational joint. At the gear P_e are attached two springs (R_1 and R_2) through two racks, where one of the ends of the springs is fixed to racks (C_3 and C_4) and are meshed to the gear P_e. The other ends of the springs are fixed to racks C_1 and C_2, which are meshed to a set of gears ($P_{1,2,3}$). Moreover, a motor is coupled to gear P_3 that allows the motion in the set.

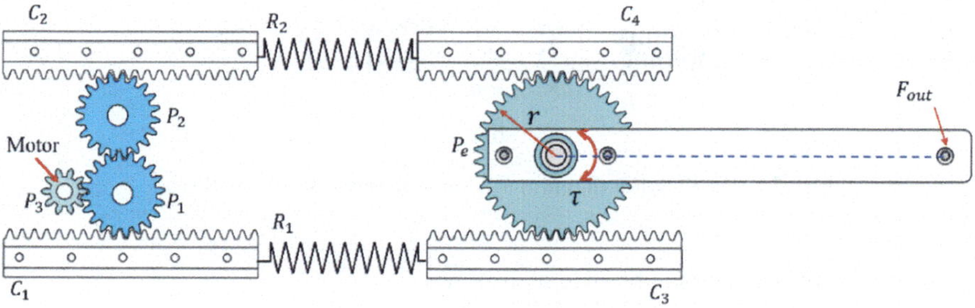

Fig. 1. VSJ diagram

The antagonistic function of the springs is due to the movement and direction of joint rotation. When the link rotates counterclockwise, the spring R_1 is extended, while the spring R_2 is compressed (see Fig. 2a); when it rotates clockwise, the spring R_1 is compressed and R_2 extended (see Fig. 2b). The characteristics of the spring are used in order to generate the joint torque and stiffness, in this way, obtaining a variable force at the end of the link, that is, at a greater rotational, the length deformed due to rotation is defined as angle (θ), higher torque, and stiffness. The length of springs R_1 and R_2, are l_1 and l_2 respectively $x_m = \theta r$. In order to obtain a constant torque, a motor is coupled to gear P_3 allowing the springs to extend or compress (depending on the rotation direction of P_3), the length deformed through the set of gears and racks is x_{li}; thus increasing the spring effort. Then, the gear set has two functions: the first is to set the length x_{li} at a specific value to obtain different ranges of variation in the stiffness; the greater is x_{li} greater the torque achieved. The second allows to control the length x_{li}, to provide a constant stiffness by extending or compressing the springs during the joint rotation. The second one provides constant torque. The total deformation of the spring length is $X_T = x_{li} \pm x_m$.

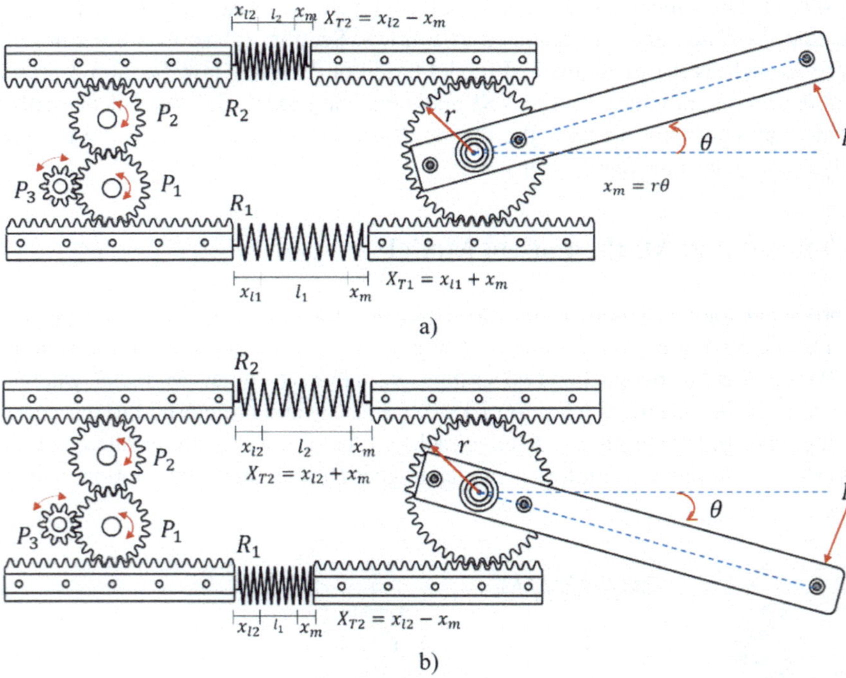

Fig. 2. VSJ Position diagram a) Counterclockwise b) Clockwise

The torque of the joint is obtained as follows:

$$\tau = rk(x_{l1} - x_{l2}) + 2kr^2\theta \tag{1}$$

where $x_{li} = \left[\frac{F}{k}\right] \pm x_m$ $i = 1, 2$ based on Hook's law. Furthermore, the stiffness is obtained as:

$$K = \tau/\theta \tag{2}$$

It is necessary to control the motor in P_3 to adjust the length x_{li} in order to obtain a constant Torque. Therefore, the relation between the length of the spring and the motor rotation is determined as:

$$\theta_{P3} = \frac{x_{li} Z_{P1,2}}{r P_{1,2} Z_{P3}} \tag{3}$$

where $Z_{P1,2}$ is the teeth number of the gears, P_1 and P_2 and Z_{P3} of the gear P_3. The pitch of the racks and gears is $P_{1,2}$.

3 Simulation of VSJ Torque

The mathematical model is simulated in Matlab to obtain the torque behavior in the joint during rotation. The rotation angle θ is performed from $-30°$ to $30°$, as shown in (Fig. 3a). The position of P_3 is considered fixed at $182.84°$ for $x_{li} = 0.02$ m and $275°$ for $x_{li} = 0.03$ m. The length of the springs used is $l_i = 00384$ m, with a constant $k = 510$ N/m. The pitch module of the racks and gears is $\text{Mpi} = 1.25$, the teeth number of the gears $Z_{P1,2} = 20$ and $Z_{P3} = 10$. The links length is $l_{Pe} = 0.15$ m. Figure 3b shows the torque behavior.

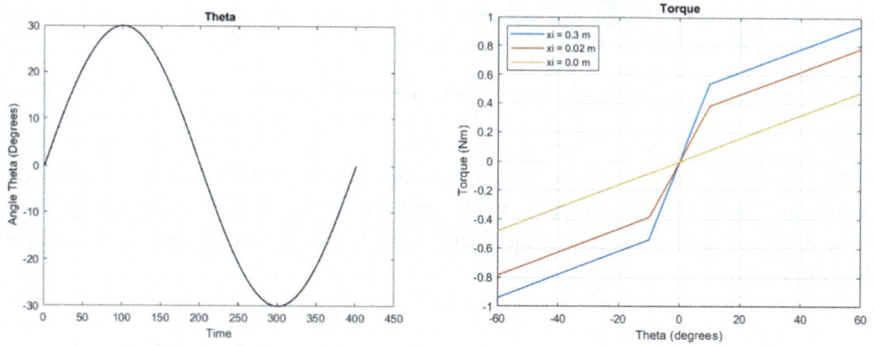

a) b)

Fig. 3. a) Joint angular displacement b) VSJ torque behavior with fixed P_3.

It can be seen that the torque increases as θ increases in both directions. In the equilibrium position, the torque is zero, whereas for the angle $\theta = \pm 30°$. Also, the toque increases when increase x_{li} as shown in Fig. 3b, the torque corresponding to $x_{li} = 0.02$ m is ± 0.78 Nm (red line); the torque corresponding to $x_{li} = 0.03$ m is ± 0.93 Nm (blue line) and the torque conforming $x_{li} = 0.0$ m is ± 0.48 Nm (yellow line). It can be observed that the torque starts with value, this is because when increasing a value x_{li} the effort in springs increases, starting with a previous effort, the yellow line corresponds to $x_{li} = 0.0$ m and no presents preload.

4 Experimental Test

The VSJ experimental validation measures the torque to analyze the behavior during the desired trajectory. Set x_{li} at different lengths and rotate the joint between $\theta = -10\text{to}10°$. The springs are of length $x_{li} = 0.039$ m with constant $k = 1180$ N/m. Torque is measured with a six-axis force sensor from Sensix company, coupled to the joint shaft, as shown in Fig. 4.

The sensor measures the force by using a strain gauge, i.e. applying force causes deformation. The force measurement in the sensor is due to the deformation between the lower part concerning the upper part that remains fixed. The sensor's lower part is fixed to the joint shaft, and the upper part is fixed to a rigid support, which is screwed to the joint plate. This allows the sensor to be fixed from the top, as shown in Fig. 4.

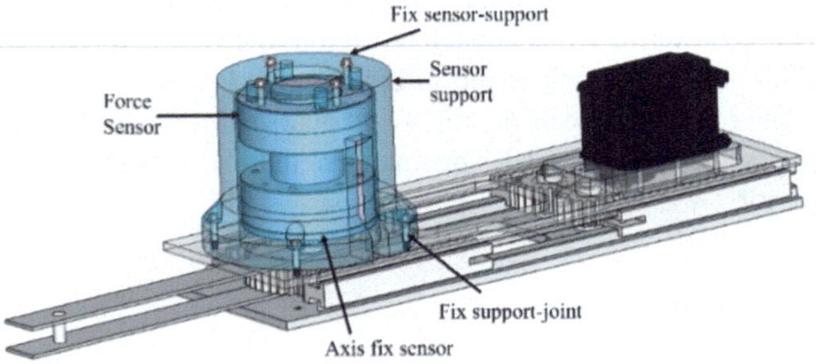

Fig. 4. Mounting the sensor on the joint

The sensor measures the torque and provides a voltage signal at the output. Data acquisition is performed with the instrumentation of Fig. 5. The force sensor output voltage is processed by a signal adapter, acquired with a NI 9181 card, and visualized with Matlab.

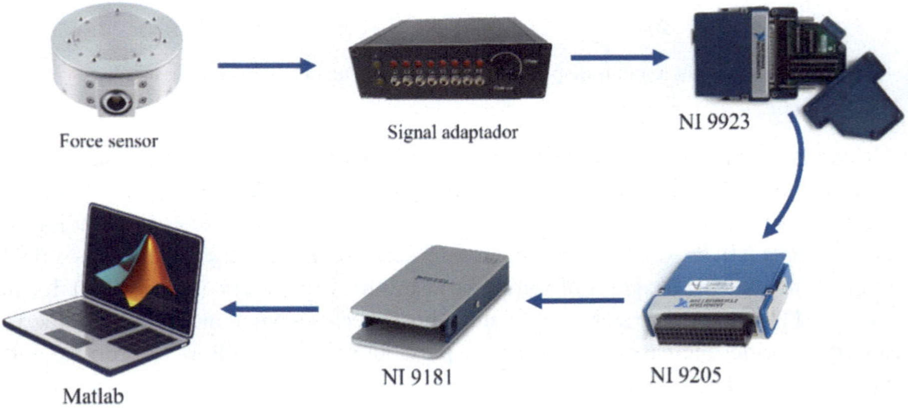

Fig. 5. Signal conditioning interface

The link rotates from the initial position (equilibrium) to $\theta = 10°$ and $-10°$; the joint torque is measured during the rotation. Two measurements were made at different lengths of $x_{li} = 0.01$ m and $x_{li} = 0.02$ m. The joint set-up for testing is as in Fig. 6, in order to increase the stiffness, it is possible to collocate more springs, in this case, we use three springs on each side.

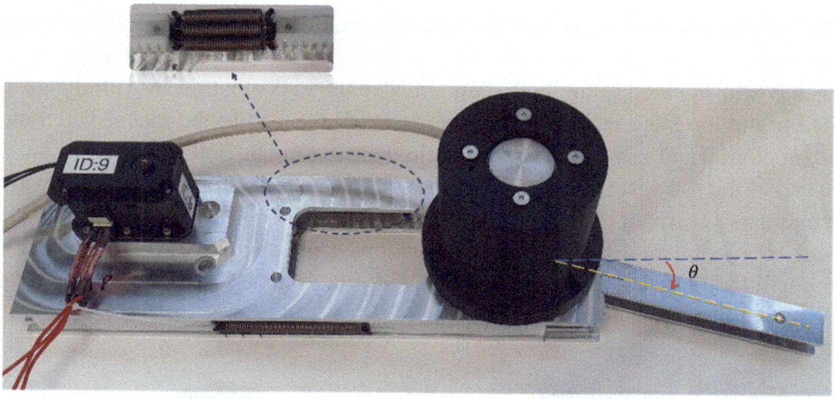

Fig. 6. Experimental setup and VSJ testing

The torque behavior of the joint during rotation can be seen in Fig. 7. It is observed that as the θ angle increases, the joint torque increase.

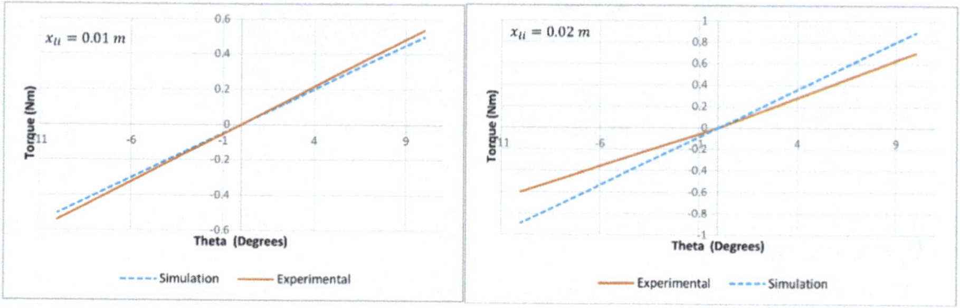

a) b)

Fig. 7. Torque vs angular displacement curve a) $x_{li} = 0.01$ m b) $x_{li} = 0.02$ m

Two increments were performed on x_{li}. The first is $x_{li} = 0.01$ m corresponding to Fig. 7a, where is obtained a range of torque of ± 0.5 Nm (orange line); the blue line represents the simulation curve. Figure 7b corresponds to the torque behavior for $x_{li} = 0.02$ m and where is obtained a torque of ± 0.7 Nm (orange line); the blue line corresponds to simulation curve. One observes that the torque increases as x_{li} increases. Therefore, it is shown that the joint torque can be modified by controlling the motor position in the gear P_3 to vary the length x_{li}. In this experimental test, the displacements for x_li son until 10° due to the constant k is not linear and need to be characterized mathematically. In addition, the friction from elements as racks is not considered in this test.

5 VSJ Compact Version

The design of the more joint compact version maintains the same antagonistic principle (see Fig. 8). The new version consists of two stages. Stage A consists of a bottom plate in it, is the gear P_e fixed to the link, there are also the racks that mesh with the gear P_e. Located in the top plate in stage B are the gears set meshed to the racks. The motor is placed on plate B and connected to gears below the plate. The bottom racks are attached to one end of the springs, while the other end is attached to the top racks. In this way, the deformation is generated by the joint rotation and the motor. The benefit in compactness is about 50% compared to the previous version. This advantage of the reduction in volume will facilitate the integration of the VSJ into systems such as the rehabilitation device based on five bar-mechanism [14].

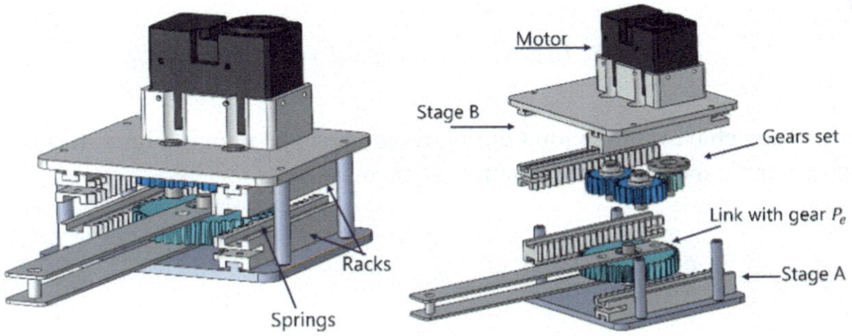

Fig. 8. VSJ compact version

6 Conclusions

This VSJ is based on antagonistic springs that provide stiffness to the joint. The VSJ consists of a set of gears and racks that allows to spring extend or compress by a motor on gear (P_3). Two racks are meshed in gear (P_e) and allow to springs to extend or compress depending on the rotational direction. This paper has presented the validation of the VSJ, verifying that is possible to obtain a variable torque with relation to the length spring. The mathematical model validation demonstrated that the joint torque increases when the angle θ increases. In the experimental test, the torque was measured with a sensor placed on the joint axis. In this experimental test, we observed the friction in racks and gears, which is not considered in the mathematical model. The torque increases as θ and x_{li} increase; Further, it is possible to control the joint torque by controlling the motor position.

The future work will be focused on the development of the new version and implementing the position control of the motor for a nonlinear variable torque/stiffness behavior. Also, the next version will be focused on solve the friction problem in actual version.

References

1. Hussain, I., et al.: Design and control of a discrete variable stiffness actuator with instant stiffness switch for safe human-roboto interaction. IEEE Access **9**, 118215–118231 (2021)
2. Jafari, A.: Coupling between the output force and stiffness in different variable stiffness actuators. MDPI **3**(3), 270–284 (2014)
3. English, C., Russell, D.: Implementation of variable joint stiffness through antagonistic actuation using rolamite springs. Mech. Mach. Theory **34**, 27–40 (1991)
4. Malosio, M., Spagnuolo, G., Prini, A., Molinari Tosatti, L., Legnani, G.: Principle of operation of RotWWC-VSA, a multi-turn rotational variable stiffness actuator. Mech. Mach. Theory **116**, 34–49 (2017)
5. Contreras-Calderón, M.G., Castillo-Castañeda, E., Laribi, M.A.: Integration of variable stiffness actuators in a five-bar device to assist in active rehabilitation of the arm. Meccanica **57**(3), 697–706 (2022)
6. Hurst, J.W., Chestnutt, J.E., Rizzi, A.: The actuator with mechanically adjustable series compliance. IEEE Trans. Rob. **26**(4), 597–606 (2010)
7. Govindan, N., Ramesh, S., Thondiyath, A.: Design of a variable stiffness joint module to quickly change the stiffness and to reduce the power consumption. IEEE Access **8**, 138318–138330 (2020)
8. Contreras-Calderón, M.G., Castillo-Castañeda, E.: Design of a variable stiffness joint for a five-bar-mechanism. In: Zeghloul, S., Laribi, M.A., Sandoval Arevalo, J.S. (eds.) RAAD 2020. MMS, vol. 84, pp. 54–63. Springer, Cham (2020). https://doi.org/10.1007/978-3-030-48989-2_7
9. Jafari, A., Tsagarakis, N.G., Caldwell, D.G.: A novel intrinsically energy efficient actuator with adjustable stiffness (AwAS). IEEE/ASME Trans. Mechatron. **18**(1), 355–365 (2013). https://doi.org/10.1109/TMECH.2011.2177098
10. Jafari, A., Tsagarakis, N.G., Sardellitti, I., Caldwell, D.G.: A new actuator with adjustable stiffness based on a variable ratio lever mechanism (AwAS-II). IEEE Trans. Mechatron. **19**(1), 55–63 (2014)
11. Wolf, S., Hirzing, G.: A new variable stiffness design: matching requirements of the next robot generation. In: IEEE International Conference in Robotics and Automation, pp. 1741–1746 (2008)
12. Visser, L.C., Carloni, R., Unal, R., Stramigioli, S.: Modeling and Design of Energy Efficient Variable Stiffness Actuators. IEEE Robotics and Automation Society Press (2010)
13. Liu, Y., Wang, D., Yang, S., Liu, J., Hao, G.: Design and experimental study of a passive power-source-free-adjustable mechanism. Front. Mech. Eng. **16**, 32–45 (2021)
14. Contreras-Calderón, M.G., Laribi, M.A., Castillo-Castañeda, E.: New variable stiffness joint (VSJ): study and simulation. In: Zeghloul, S., Laribi, M.A., Arsicault, M. (eds.) MEDER 2021. MMS, vol. 103, pp. 284–293. Springer, Cham (2021). https://doi.org/10.1007/978-3-030-75271-2_30
15. Tsagarakis, N., Sardellitti, I., Caldwell, D.: A new variable stiffness actuator (CompACT-VSA): design and modeling. In: IEEE International Conference on Intelligent Robotics Systems (IROS), pp. 378–383 (2011)

Optimal Positioning of Mobile Manipulators Using Closed Form Inverse Kinematics

Giovanni Colucci(✉), Lorenzo Baglieri, Andrea Botta, Paride Cavallone, and Giuseppe Quaglia

Politecnico di Torino, 10129 Turin, Italy
{giovanni_colucci,andrea.botta,paride.cavallone, giuseppe.quaglia}@polito.it, lorenzo.baglieri@studenti.polito.it

Abstract. Mobile manipulators have recently been subject to studies and researches thanks to their augmented mobility and interaction capability. In the precision agriculture field, the development and implementation of such systems can be advantageous in every aspect of the farm activities, e.g. harvesting, pruning, or trimming. This paper presents the implementation of a 7 degree of freedom manipulator upon a mobile rover prototype, designed for precision agriculture, in order to perform grape sampling tasks. While the redundancy of the arm is used to perform off-line collision avoidance with the environment and the mobile base itself, thanks to sampling based path planning methods, a closed form inverse kinematics solution allows to select the posture which maximizes the manipulability index of the manipulator. To do so, base mobility is used to reach the target and properly position the arm. The overall architecture was implemented on the real system and successfully validated through experimental tests.

Keywords: Mobile manipulation · Manipulability · Precision agriculture

1 Introduction

The idea and development of mobile manipulators, robotic systems that can be decomposed into a robotic arm and a mobile base, was firstly introduced at the end of the 80's for industrial purposes, and, since then, the research world has produced systems with a high level of complexity in several application fields [8]. Such general architecture was studied and implemented for precision agriculture purposes to perform common farm activities, e.g. harvesting [6], pruning [3] and spraying [9]. In all of these applications, the use of a robotic arm is fundamental to undertake tasks with a high level of dexterity, such as the selective spraying routine presented in [9].

In this paper, the authors present the implementation of a commercial assistive robotic arm upon the custom mobile base Agri.Q designed for monitoring

A. Müller and M. Brandstötter (Eds.): RAAD 2022, MMS 120, pp. 184–191, 2022.
https://doi.org/10.1007/978-3-031-04870-8_22

and sampling crops and soil. The design of the base and its related locomotion system was widely discussed in previous articles (see [2,10]) and preliminary studies about the arm installation were carried out [11]. In this work, the overall system architecture is presented, starting from the arm inverse kinematics (IK) formulation and related motion planning to the combined motion of the base and the arm to outperform sampling tasks. The final section presents some experimental results, where the real prototype is tested.

1.1 System Architecture

In Fig. 1 the Agri.Q mobile manipulator prototype, mainly composed of a custom mobile base and a commercial seven degree of freedom (dof) robotic arm, is presented. The mobile base consists of two locomotion modules, front and rear one, both provided by two driving units. Each driving unit, in turn, is composed of two wheels connected by a rocker arm. The total number of eight wheels provides a good load distribution and the capability to move and navigate also in uneven terrain.

The upper part of the body is covered by a solar panel, designed to recharge the 24 V lithium-ion battery that feeds the locomotion motors, the robotic arm and all the other on-board electronic components. The panel can be oriented thanks to a pitch mechanism, that also can move the arm, causing the increase of its total workspace.

2 Redundancy Solution and Manipulability Optimization

A representation of the commercial 7dof arm, a Kinova Jaco 2 model, is depicted in Fig. 2a. It is composed of seven consecutive revolute joints but with a significant offset between links 3 and 4 (the distance d in Fig. 2b). Since the first three

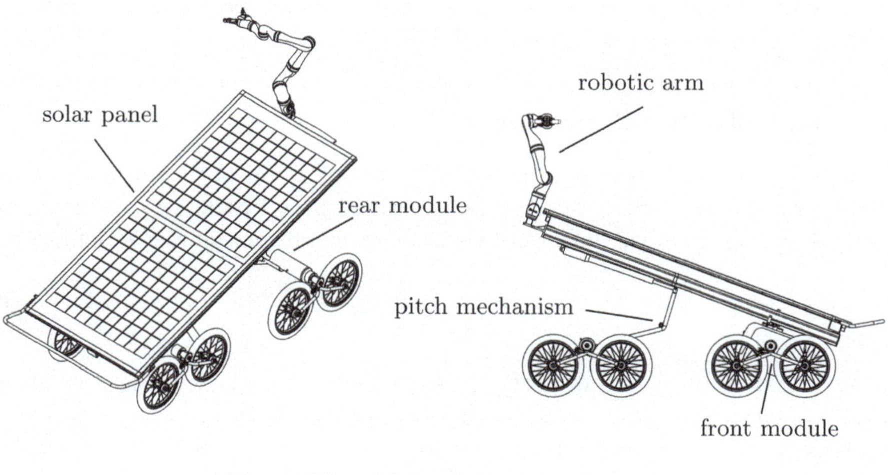

Fig. 1. The mobile manipulator Agri.Q.

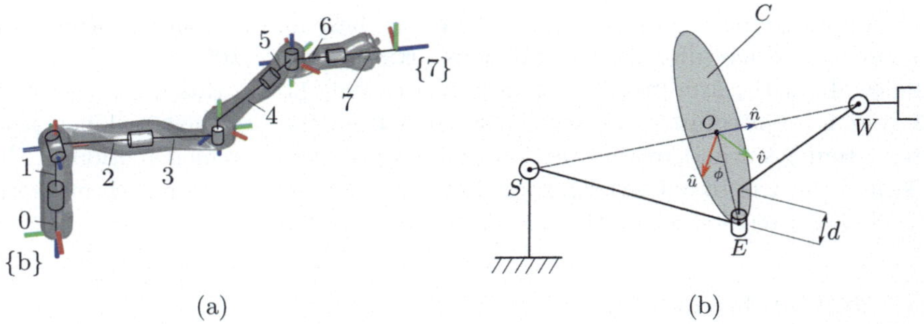

Fig. 2. (a) Kinematic model of the manipulator. (b) Definition of the swivel angle ϕ. For a given constant pose, E must lie on circumference C. Please notice that there is an offset between link 3 and 4.

joint axes intersect in a point, they can be treated as a single spherical joint, and the same goes for the last three joints. Moreover, the arm is intrinsically redundant, because its number of degrees of freedom is greater than the dimension of the operational space.

The inverse kinematics problem allows ∞^1 solutions, which can be found through analytical or numerical methods. As an example, the solution can be found using inverse differential kinematics techniques where a cost function $g(\dot{q})$ is minimized [5]. Besides, the authors present a closed form method where all possible solutions are calculated as a function of the Swivel angle ϕ [1,7]. As shown in Fig. 2b, the arm can be modelled as two spherical joints with an intermediate revolute joint which represents, in fact, the redundancy of the system. The circumference C is orthogonal to the shoulder-elbow-wrist (SEW) plane and, moving E around C, all possible IK solutions can be found. The swivel angle, or elbow angle, is defined as the angle between $\hat{u}$ and OE, where $\hat{u}$ can be arbitrarily defined, and Yan et al. proposed a method in order to avoid singularities in its definition [13]. In Fig. 3 an illustration of three different postures for the same constant pose is presented.

2.1 Manipulability Evaluation

Given that the solution is parametric in terms of ϕ, an optimization method can be introduced to find the best pose with respect to the task. A common approach is the use of the inverse of the 2-norm condition number κ defined as:

$$c = \frac{1}{\kappa(\mathbf{J})} = \frac{\lambda_{min}}{\lambda_{max}} \tag{1}$$

where λ_{min} and λ_{max} are respectively the smallest and largest eigenvalues of Jacobian matrix $\mathbf{J}(\theta)$, and its values are in the range $[0, 1]$. If c is close to 0, the arm is near to a singular configuration, if $c = 1$, the corresponding posture is called *isotropic*.

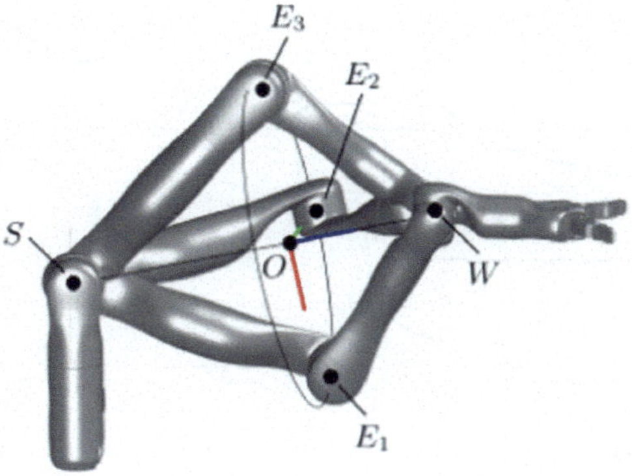

Fig. 3. Use of swivel angle to find three different postures for a given constant pose

To consider also joint limits boundaries, before the computation of c, the Jacobian matrix $\mathbf{J}$ can be modified with some penalization factor p_j, with $j = 1, \ldots, 7$, which are a quantitative evaluation of joint limits closeness of a given posture [4,12]. As firstly introduced in [4] and implemented in [5,12], a gradient function $\nabla h(\theta)_j$ can be defined as:

$$\nabla(\theta)_j = \frac{\partial h(\theta)}{\partial \theta_j} = \frac{(\theta_{j,max} - \theta_{j,min})^2 (2\theta_j - \theta_{j,max} - \theta_{j,min})^2}{\alpha(\theta_{j,max} - \theta_j)^2 (\theta_j - \theta_{j,min})^2} \tag{2}$$

where $\theta_{j,max}$, $\theta_{j,min}$ are respectively the j-th joint upper and lower limit, and α is a scale factor which relax or emphasize the closeness to joint limits. The penalization factor p_j for the j-th column of $\mathbf{J}$ can now be defined as:

$$p_j = \frac{1}{\sqrt{1 + |\nabla h(\theta)_j|}} \tag{3}$$

For our purpose, a value of $\alpha = 4$ was chosen, as [12] suggests, and, as described by Eq. (2), p_j goes from 1, when $\theta_j = (\theta_{j,max} - \theta_{j,min})/2$, to ∞, when θ_j tends to its upper or lower limit.

In Fig. 4 the joint angle curves for a given pose are illustrated as a function of the swivel angle ϕ, and a comparison between the original manipulability index c and its new version c_{mod}, modified by the joint limits penalization factors. Please note that, even if some solutions are exact, they are not feasible with the joint limits constraint (dashed line).

Starting from the total set of solutions described by the closed-form inverse kinematics formulation of above, the extracted solution is the one that maximize the value of c_{mod}, i.e. the associated modified manipulability index is maximum.

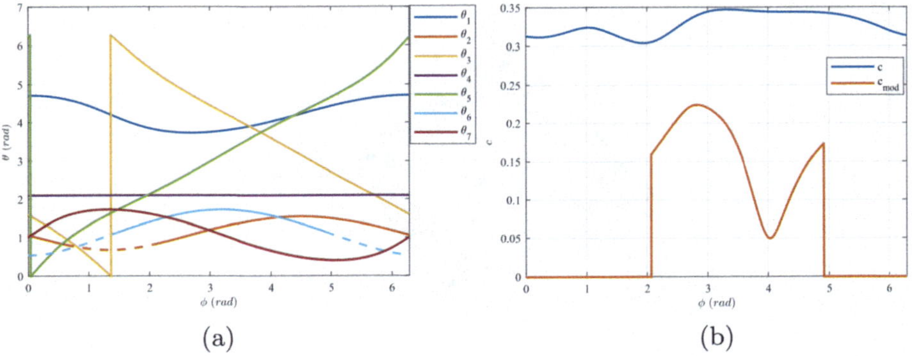

Fig. 4. (a) Joint angles θ_i $i = 1, ..., 7$ as function of swivel angle ϕ. A dashed line is used to highlight not allowed values due to joint limits. (b) Comparison between c and c_{mod}. c_{mod} is the inverse of the 2-norm condition number of the modified Jacobian matrix for the manipulator.

3 Combined Mobility and Task Execution

To properly interact with its environment, the mobile manipulator Agri.Q can use both the mobile base and manipulator mobility. Since the base mobility is significantly less accurate than the manipulator one, the authors propose a combined motion where the mobile base mobility is used to position the task object inside the manipulator workspace and then only the manipulator performs the task execution. To do so, the base can be modelled as shown in Fig. 5, where the position x and the pitch angle ψ are the two degrees of freedom.

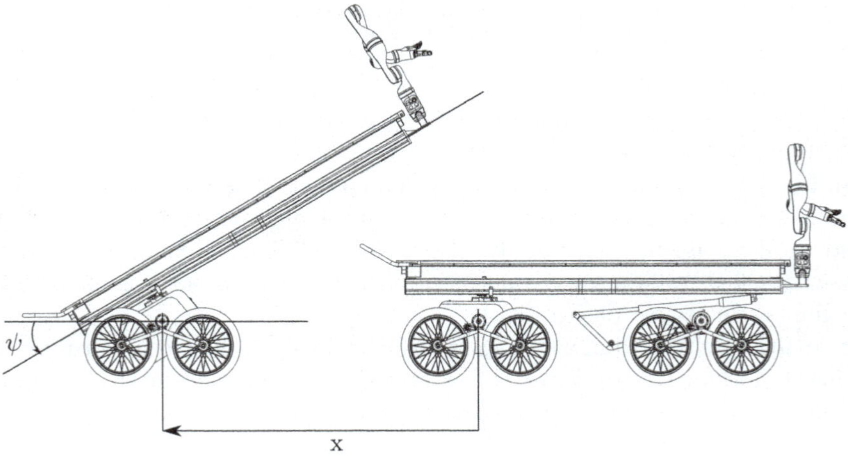

Fig. 5. x linear position and ψ pitch angle for robot arm positioning.

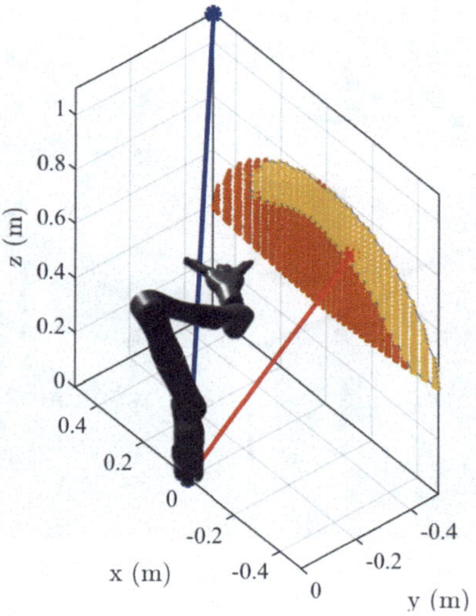

Fig. 6. Manipulability mapping of the workspace of the arm. Blue line: target point, red area: arm workspace, yellow area: best manipulability area, red line: manipulability barycenter of the best manipulability area.

In Fig. 6, the developed motion planning of the base is illustrated. For a generic pick-and-place task, if the target object (blue line) is outside the arm workspace (red and yellow area), the developed motion planning procedure moves the mobile base so that the arm can reach the target, also with a good manipulability index. To do so, the target point is moved inside the best manipulability area of the manipulator (yellow area inside the workspace), and specifically into the barycenter of the best manipulability area (red line), where the manipulability index that was used is the above defined c_{mod}. Once the motion is completed, the new target pose is calculated and a motion planning procedure, based on the sampling-based method RRT (rapidly exploring random tree) connect, that is fundamental to guarantee the collision avoidance between the arm and the solar panel of the mobile base, is carried out.

4 Experimental Tests

In Fig. 7 the experimental setup to validate the system architecture is shown. During this pick-and-place procedure, the arm grasps the target object (the blue cylinder upon the table) and then moves it to its back, where the future placing area of the rover will be placed. All the requested algorithms, e.g., IK solution, motion planning, time scaling, were done on a local pc on Matlab environment. The communication with the mobile base, fundamental to send commands to the pitch and traction actuators, was done through a serial port communication, while the communication with the arm was developed through a Matlab - ROS interface, then sending commands through ROS topics.

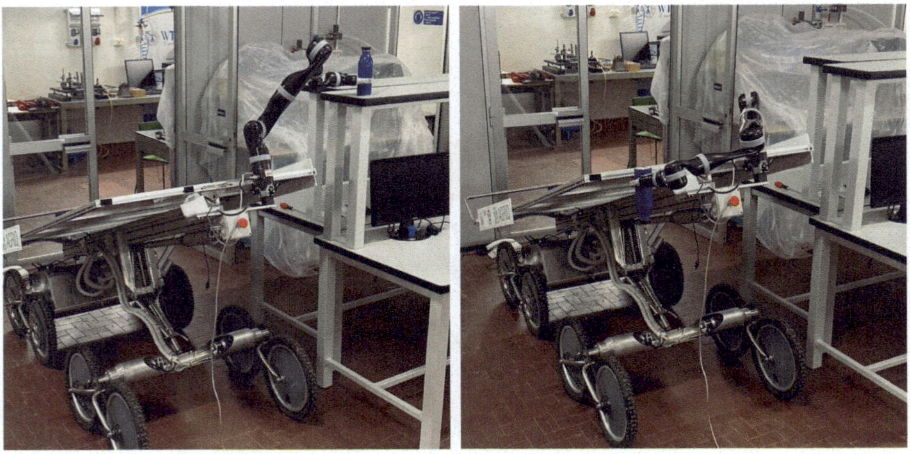

(a) (b)

Fig. 7. The prototype during an experimental test. (a) grasp posture (b) place posture

5 Conclusions

In this paper, a planning strategy based on manipulability optimization for the mobile manipulator prototype Agri.Q has been presented. The redundancy of the arm was solved thanks to a closed form formulation, and the solution is parametric in terms of the elbow angle. The mobility of the base is used to properly position the arm so that it can reach the target and manipulate it with the best possible manipulability index. According to that, a planning pipeline, that takes into account both base and arm mobility but separately, was developed in Matlab environment, with the use of RRT connect sampling based method for collision avoidance of the arm with the solar panel of the base. The overall architecture was implemented on the real prototype and validated through experimental tests.

References

1. Artemiadis, P.: Closed-form inverse kinematic solution for anthropomorphic motion in redundant robot arms. In: ICRA 2013, (2013). https://doi.org/10.4172/2168-9695.1000110
2. Botta, A., Cavallone, P.: Robotics applied to precision agriculture: the sustainable agri.q rover case study. In: Quaglia, G., Gasparetto, A., Petuya, V., Carbone, G. (eds.) I4SDG 2021. MMS, vol. 108, pp. 41–50. Springer, Cham (2022). https://doi.org/10.1007/978-3-030-87383-7_5
3. Botterill, T., et al.: A robot system for pruning grape vines. J. Field Robot. **34**(6), 1100–1122 (2017). https://doi.org/10.1002/rob.21680, https://onlinelibrary.wiley.com/doi/abs/10.1002/rob.21680

4. Chan, T.F., Dubey, R.: A weighted least-norm solution based scheme for avoiding joint limits for redundant joint manipulators. IEEE Trans. Robot. Autom. **11**(2), 286–292 (1995). https://doi.org/10.1109/70.370511
5. Chen, F., Selvaggio, M., Caldwell, D.G.: Dexterous grasping by manipulability selection for mobile manipulator with visual guidance. IEEE Trans. Ind. Inf. **15**(2), 1202–1210 (2019). https://doi.org/10.1109/TII.2018.2879426
6. De Preter, A., Anthonis, J., De Baerdemaeker, J.: Development of a robot for harvesting strawberries andreas de preter is supported by a baekeland PhD scholarship (150712) through flanders innovation and entrepreneurship (VLAIO). IFAC-PapersOnLine **51**(17), 14–19 (2018). https://doi.org/10.1016/j.ifacol.2018.08.054, https://www.sciencedirect.com/science/article/pii/S2405896318311704
7. Faria, C., Ferreira, F., Erlhagen, W., Monteiro, S., Bicho, E.: Position-based kinematics for 7-DoF serial manipulators with global configuration control, joint limit and singularity avoidance. Mech. Mach. Theory **121**, 317–334 (2018). https://doi.org/10.1016/j.mechmachtheory.2017.10.025
8. Hvilshøj, M., Bøgh, S., Skov Nielsen, O., Madsen, O.: Autonomous industrial mobile manipulation (AIMM): past, present and future. Ind. Robot Int. J. **39**(2), 120–135 (2012). https://doi.org/10.1108/01439911211201582
9. Oberti, R., et al.: Selective spraying of grapevines for disease control using a modular agricultural robot. Biosys. Eng. **146**, 203–215 (2016). https://doi.org/10.1016/j.biosystemseng.2015.12.004, https://www.sciencedirect.com/science/article/pii/S1537511015001865
10. Quaglia, G., Visconte, C., Carbonari, L., Botta, A., Cavallone, P.: Agri.q: a sustainable rover for precision agriculture. In: Visa, I., Duta, A. (eds.) Solar Energy Conversion in Communities. SPE, pp. 81–91. Springer, Cham (2020). https://doi.org/10.1007/978-3-030-55757-7_6
11. Quaglia, G., Visconte, C., Scimmi, L.S., Melchiorre, M., Cavallone, P., Pastorelli, S.: Design of a UGV powered by solar energy for precision agriculture. Robotics **9**(1), 13 (2020). https://doi.org/10.3390/robotics9010013, https://www.mdpi.com/2218-6581/9/1/13
12. Vahrenkamp, N., Asfour, T., Metta, G., Sandini, G., Dillmann, R.: Manipulability analysis. In: 2012 12th IEEE-RAS International Conference on Humanoid Robots (Humanoids 2012). pp. 568–573. IEEE (2012). https://doi.org/10.1109/HUMANOIDS.2012.6651576, http://ieeexplore.ieee.org/document/6651576/
13. Yan, L., Mu, Z., Xu, W.: Analytical inverse kinematics of a class of redundant manipulator based on dual arm-angle parameterization. In: 2014 IEEE International Conference on Systems, Man, and Cybernetics (SMC), pp. 3744–3749 (2014). https://doi.org/10.1109/SMC.2014.6974513, ISSN: 1062-922X

Multistage Approach for Solving the Optimal Control Problem for a Wheeled Inverted Pendulum with Infeasible Initial Guess

Christian Zauner(✉), Hubert Gattringer, and Andreas Müller

Institute of Robotics, Johannes Kepler University Linz, Altenbergerstraße 69, 4040 Linz, Austria
{christian.zauner,hubert.gattringer,a.mueller}@jku.at

Abstract. A robotic system constructed as a wheeled inverted pendulum (WIP) serves as an impressive demonstrator, since this kind of system is inherently non-linear, unstable and non-minimum-phase. But the same properties which make it impressive also introduce serveral difficulties, when it comes to control and trajectory planning. Therefore this paper shows a method for deriving a highly dynamic trajectory compliant with the system dynamics. The constraints inherent to the definition of this trajectory are non-convex. The constraint functions have a local minimum in an infeasible region. This poses a problem when the initial guess is within this infeasible reagion. For a WIP this happens when it has to pass (in limbo) underneath an obstacle. To overcome this, a multi-stage approach is proposed for the optimal control problem to evade this infeasible local minimum. After solving 4 stages of related optimization problems the optimal trajectory is obtained. Additionally the center of gravity of the examined WIP can be influenced slightly by the mounting point of the battery pack. By incorporating this into the optimization problem, a task dependent optimal system configuration can be found.

Keywords: Optimal control · Trajectory planning · Mobile robotic systems · Wheeled inverted pendulum · Non-minimum phase system

1 Introduction

One of the benefits of a wheeled inverted pendulum (WIP), despite its manoeuvrability, is that it is capable of following exceptional trajectories as indicated in Fig. 1. This special type of obstacle avoidance is inspired by a limbo dance and has several interesting properties. Firstly, it pushes the system to its limits, and can therefore be used to test how a path following control performs near the system limits. Furthermore it is impressive to watch, and therefore is ideal for demonstration purposes. But finally, it also has some interesting mathematical properties regarding the solvability of the optimal control problem (OCP). Opposed to [5], where obstacles on the ground are addressed, this paper addresses

A. Müller and M. Brandstötter (Eds.): RAAD 2022, MMS 120, pp. 192–200, 2022.
https://doi.org/10.1007/978-3-031-04870-8_23

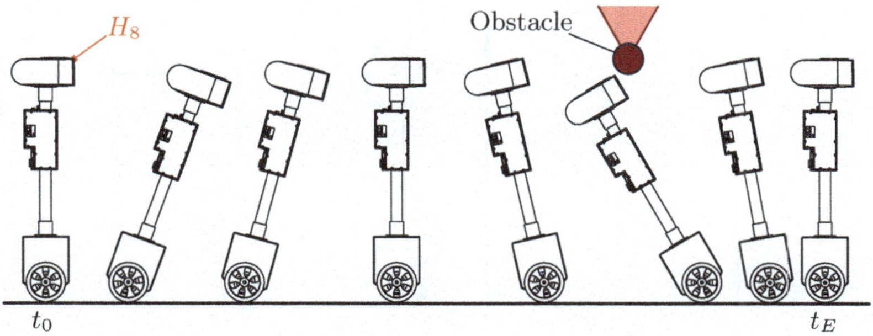

Fig. 1. Examplary trajectory with obstacle

obstacle avoidance as indicated in Fig. 1, similar to [6]. Special about that is that the height of the point H_8 starts increasing if the robot is tilted backwards until its maximum height is reached. Only then the height of H_8 decreases if the robot is tilted further. This results in non-convex constraints for the obstacle avoidance with a possibly infeasible local minimum and infeasible initial guess. This paper addresses this issue, by means of a multistage approach for solving the according optimal control problem. Despite of finding a feasible solution for the OCP, the proposed procedure allows to obtain the minimal height of the obstacle, for which a feasible solution exists, and to perform a limited form of structural optimization in order to find the optimal mounting point of the battery pack for the given task.

2 Dynamic Model

The derivation of the dynamic model of a wheeled inverted pendulum with two independently driven wheels as shown in Fig. 2 can be found in [7]. Therein movements on the two dimensional ground surface are considered, which results in non-holonomic constrains at velocity level for the dynamic model. Since in this paper the desired trajectory follows a straight line on the ground surface, a simplified model may be used. By setting the orientation angle $\gamma = 0$ and the ${}_I y$-coordinate of the position $y = 0$, the generalized coordinates $\mathbf{z} = \begin{bmatrix} x \; \theta \; \xi \; \eta \end{bmatrix}^{\mathsf{T}}$ can be used to fully describe the positions and orientations of the basis and the two wheels. The coordinate x describes the position of the ground contact point along the ${}_I x$-direction, θ is the inclination angle of the basis, and ξ and η denote the relative angles of the wheels. Since ideal rolling of the wheels is assumed, the kinematic relation $\dot{\mathbf{z}} = \mathbf{H}\dot{\mathbf{s}}$ can be introduced with the matrix

$$\mathbf{H} = \begin{bmatrix} 1 & 0 & \frac{2}{D_W} & \frac{2}{D_W} \\ 0 & 1 & -1 & -1 \end{bmatrix}^{\mathsf{T}} \tag{1}$$

the diameter of the wheels D_W and the vector of minimal velocities $\dot{\mathbf{s}} = \begin{bmatrix} \dot{x} \; \dot{\theta} \end{bmatrix}^{\mathsf{T}}$. Since the rotation angles of the wheels do not appear in equations of motion,

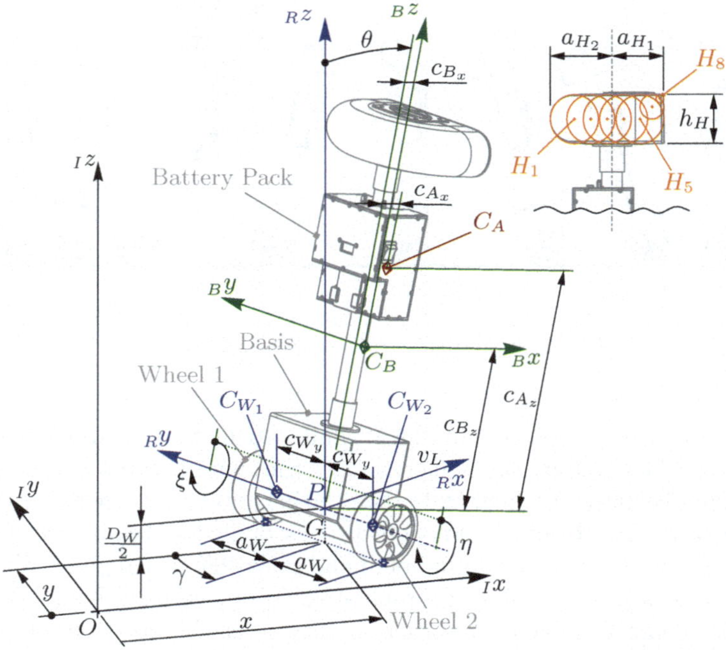

Fig. 2. Wheeled inverted pendulum on two dimensional ground surface

it is sufficient to choose the minimal coordinates as $\mathbf{q} = \begin{bmatrix} x \ \theta \end{bmatrix}^\mathsf{T}$. The minimal coordinates and the minimal velocities can be combined to the state vector $\mathbf{x} = \begin{bmatrix} \mathbf{q}^\mathsf{T} \ \dot{\mathbf{s}}^\mathsf{T} \end{bmatrix}^\mathsf{T}$. Due to the symmetric structure of the robot and the mentioned simplification of the model the driving torques have to be equal for each wheel. Therefore the vector of inputs can be stated by $\mathbf{u} = \begin{bmatrix} M \end{bmatrix}$ with M denoting the driving torque for each wheel. Since the optimization process should also yield a task dependent optimal position for the battery pack, this battery pack is modelled as independent body fixed to the body of the basis.

A detailed derivation of the dynamic model would be too extensive to be provided here. The procedure is similar to the approach in [7]. At first the velocities and angular velocities as well as vectors for the momentum and angular momentum are derived. Then synthetic methods are used as shown in [3] to deduce the equations of motion (EOM), which have the form

$$\mathbf{M}(\theta, \mathbf{p})\ddot{\mathbf{s}} + \mathbf{g}(\theta, \dot{\mathbf{s}}, \mathbf{p}) = \mathbf{B}(\mathbf{p})\mathbf{u} \tag{2}$$

with the symmetric positive definite mass matrix $\mathbf{M}(\theta, \mathbf{p})$, the vector of nonlinear terms $\mathbf{g}(\theta, \dot{\mathbf{s}}, \mathbf{p})$, the constant input matrix $\mathbf{B}(\mathbf{p})$, and the vector of parameters $\mathbf{p}$. The EOM can also be converted to the state space representation with affine-input structure, resulting in

$$\dot{\mathbf{x}} = \begin{bmatrix} \mathbf{H}(\mathbf{p})\dot{\mathbf{s}} \\ -\mathbf{M}(\theta, \mathbf{p})^{-1}\mathbf{g}(\theta, \dot{\mathbf{s}}, \mathbf{p}) \end{bmatrix} + \begin{bmatrix} \mathbf{0} \\ \mathbf{M}(\theta, \mathbf{p})^{-1}\mathbf{B}(\mathbf{p}) \end{bmatrix} \mathbf{u} = \mathbf{f}(\mathbf{x}, \mathbf{p}) + \mathbf{G}(\mathbf{x}, \mathbf{p})\mathbf{u}. \tag{3}$$

By solving the equation system $\mathbf{g}(\theta_e, \mathbf{0}, \mathbf{p}) = \mathbf{B}(\mathbf{p})\left[M_e\right]$, the inclination angle θ_e and the drive torque $M_e = 0$ of the equlibrium state can be obtained.

3 Distance to Obstacle

As can be seen in Fig. 1, it is very likely, that the robot may hit the obstacle with the upper part regarded as the 'head'. Therefore the cross section of the head is approximated, as indicated in Fig. 2, by 5 circles located at the points

$$
{}_B\mathbf{r}_{PH_i} = \begin{bmatrix} \left(a_{H_1} - \frac{h_H}{2}\right)\frac{i-1}{4} - a_{H_2}\frac{5-i}{4} \\ 0 \\ H - \frac{h_H}{2} \end{bmatrix} \tag{4}
$$

with a radius $r_{H_i} = \frac{h_H}{2}$, $i \in \{1, \ldots, 5\}$ and the height H of the WIP measured from P to the top of the head. To account for the sharp edge at the front of the head, the approximation is refined by adding two additional circles at the points ${}_B\mathbf{r}_{PH_6} = \left[a_{H_1} - \frac{h_H}{4}\ 0\ H - \frac{h_H}{4}\right]^\mathsf{T}$ and ${}_B\mathbf{r}_{PH_7} = \left[a_{H_1} - \frac{h_H}{12}\ 0\ H - \frac{h_H}{12}\right]^\mathsf{T}$ with the radii $r_{H_6} = \frac{h_H}{4}$ and $r_{H_7} = \frac{h_H}{12}$ and the point ${}_B\mathbf{r}_{PH_8} = \left[a_{H_1}\ 0\ H\right]^\mathsf{T}$ with $r_{H_8} = 0$, for consistency. With the absolute position vectors ${}_I\mathbf{r}_{OH_i} = {}_I\mathbf{r}_{OP} + \mathbf{R}_{IB}\,{}_B\mathbf{r}_{PH_i}$ with $i \in \{1, \ldots, 8\}$ and the position vector to the center of the obstacle ${}_I\mathbf{r}_{OC_O} = \left[a_{O_x}\ 0\ a_{O_z}\right]^\mathsf{T}$ the distance of each point to the center of the obstacle can be stated by $\|{}_I\mathbf{r}_{H_iC_O}\|_2$ with ${}_I\mathbf{r}_{H_iC_O} = {}_I\mathbf{r}_{OC_O} - {}_I\mathbf{r}_{OH_i}$. A non colliding trajectory always has to satisfy $\|{}_I\mathbf{r}_{H_iC_O}\|_2 \geq r_{H_i} + r_O$ with $i \in \{1, \ldots, 8\}$ and the radius of the obstacle r_O. Furthermore to ensure, that trajectories are excluded, where a test point passes above the obstacle, a test point may never lie inside the triangular shaped region above the center of the obstacle, as indicated in Fig. 1. This can be ensured by satisfying the condition

$$
\frac{{}_I\mathbf{n}_O^\mathsf{T}\,{}_I\mathbf{r}_{H_iC_O}}{\|{}_I\mathbf{r}_{H_iC_O}\|_2} \leq \cos\left(\frac{\pi}{6}\right) \text{ with } {}_I\mathbf{n}_O = \left[0\ 0\ {-1}\right]^\mathsf{T}. \tag{5}
$$

4 Optimal Control Problem

For solving the optimal control problem a direct multiple shooting approach [4] is chosen. Therefore the time span from the start $t = 0$ until the end of the trajectory $t = T_E$ is discretized with the sampling time $T_S = \frac{T_E}{N}$, $N = 1000$, resulting in the vector of time steps $\hat{\mathbf{t}} = \left[0\ T_S \cdots iT_S \cdots T_E\right]^\mathsf{T} \in \mathbb{R}^{N+1}$. Between each time step the value of the input vector $\mathbf{u}$ is assumed to be constant. This results in the matrices of discrete input values $\hat{\mathbf{u}} = \left[\hat{\mathbf{u}}_0\ \hat{\mathbf{u}}_1 \cdots \hat{\mathbf{u}}_{N-1}\right] \in \mathbb{R}^{1\times N}$ and state values $\hat{\mathbf{x}} = \left[\hat{\mathbf{x}}_0\ \hat{\mathbf{x}}_1 \cdots \hat{\mathbf{x}}_N\right] \in \mathbb{R}^{4\times(N+1)}$. As integration scheme the explicit Runge-Kutta method of fourth order RK4 is used, which is implemented explicitly in order to exploit the automatic differentiation capability of the used solver. Therefore the state of the next time step can be obtained by a function $\mathbf{f}_{\text{RK4}}(\hat{\mathbf{x}}_i, \hat{\mathbf{u}}_i, T_E, \mathbf{p})$. The optimal control problem can then be stated as

$$\min_{\hat{\mathbf{u}},\hat{\mathbf{x}},T_E,c_{A_z}} \left(\nu_t T_E + \nu_u \frac{T_E}{N} \sum_{i=0}^{N-1} \hat{\mathbf{u}}_i^\mathsf{T} \hat{\mathbf{u}}_i \right) \tag{6}$$

$$\text{s.t.} \quad \hat{\mathbf{x}}_{i+1} = \mathbf{f}_{\text{RK4}}(\hat{\mathbf{x}}_i, \hat{\mathbf{u}}_i, T_E, \mathbf{p}) \qquad i \in \{0, \ldots, N\} \tag{7}$$

$$\hat{\mathbf{x}}_0 = \mathbf{x}_0 \tag{8}$$

$$\hat{\mathbf{x}}_N = \mathbf{x}_E \tag{9}$$

$$\mathbf{x}_{\min} \leq \hat{\mathbf{x}}_i \leq \mathbf{x}_{\max} \qquad i \in \{0, \ldots, N+1\} \tag{10}$$

$$\mathbf{u}_{\min} \leq \hat{\mathbf{u}}_i \leq \mathbf{u}_{\max} \qquad i \in \{0, \ldots, N\} \tag{11}$$

$$|\omega_W(\hat{\mathbf{x}}_i, \mathbf{p})| \leq \omega_{W,\max} \qquad i \in \{0, \ldots, N+1\} \tag{12}$$

$$|P_M(\hat{\mathbf{x}}_i, \hat{\mathbf{u}}_i, \mathbf{p})| \leq P_{M,\max} \qquad i \in \{0, \ldots, N\} \tag{13}$$

$$r_{H_j} + r_O \leq \left\| {}_I\mathbf{r}_{H_j C_O}(\hat{\mathbf{x}}_i, \mathbf{p}) \right\|_2 \qquad i \in \{0, \ldots, N+1\},\ j \in \{1, \ldots, 8\} \tag{14}$$

$$\frac{{}_I\mathbf{n}_{OI}^\mathsf{T}\mathbf{r}_{H_j C_O}(\hat{\mathbf{x}}_i, \mathbf{p})}{\left\| {}_I\mathbf{r}_{H_j C_O}(\hat{\mathbf{x}}_i, \mathbf{p}) \right\|_2} \leq \cos\left(\frac{\pi}{6}\right) \qquad i \in \{0, \ldots, N+1\},\ j \in \{1, \ldots, 8\} \tag{15}$$

$$c_{A_z,\min} \leq c_{A_z} \leq c_{A_z,\max} \tag{16}$$

with the time weight ν_t, the input weight ν_u, the initial state $\mathbf{x}_0 = \left[0\ \theta_e\ 0\ 0\right]$ and the terminal state $\mathbf{x}_E = \left[x_E\ \theta_e\ 0\ 0\right]$. The vectors $\mathbf{x}_{\min}$ and $\mathbf{x}_{\max}$ form box constraints for the states and are used to restrict especially the inclination angle θ to admissible values, whereas the box constraints for the input, given by $\mathbf{u}_{\min}$ and $\mathbf{u}_{\max}$, account for the maximal permissible drive torque. The drive speed, derived by $\omega_W(\mathbf{x}, \mathbf{p}) = \frac{2\dot{x}}{D_W} - \dot{\theta}$, is limited to the maximal drive speed $\omega_{W,\max}$, the drive power $P_M(\mathbf{x}, \mathbf{u}, \mathbf{p}) = \omega_W(\mathbf{x}, \mathbf{p})M$ is constrained to the maximal drive power $P_{M,\max}$, and the constraints regarding the obstacle are according to Sect. 3. Finally, the task dependent position of the battery pack c_{A_z} is limited due to system restrictions to $c_{A_z,\min}$ and $c_{A_z,\max}$.

For solving this non-linear optimization problem, the optimization framework *CasADi* [1,2] with the solver *Ipopt* is used. In order to gain the minimum obstacle height which would still result in a feasible solution a quite common approach would be to iteratively reduce the obstacle height until no feasible solution can be found. But if this problem is tried to be solved for any obstacle height $a_{O_z} < \frac{D_W}{2} + H + r_O$ with, for example, the trivial initial guess according to

$$\hat{\mathbf{x}}_i^{\mathrm{I}_0} = \frac{N+1-i}{N+1}\mathbf{x}_0 + \frac{i}{N+1}\mathbf{x}_E, \tag{17}$$

$$\hat{\mathbf{u}}^{\mathrm{I}_0} = \mathbf{0}, \qquad T_E^{\mathrm{I}_0} = \frac{20(x_E - x_0)}{D_W \omega_{W,\max}}, \qquad c_{A_z}^{\mathrm{I}_0} = c_{A_z,\max} \tag{18}$$

with $i \in \{1, \ldots, N+1\}$ the solver tends to terminate with an error due to detection of an infeasible problem. This is because, the overall height of the robot has a local minimum at $\theta = 0$ and the constraints violation regarding the obstacle is increasing if the inclination angle increases or decreases, until the point of the maximum overall height is exceeded. Thus these constraints are non-convex with an infeasible local minimum. To overcome this, a multistage approach is used to provide an initial solution which avoids entering the region of this local minimum.

5 Multistage Approach

The first stage consists of finding a valid solution for the optimal control problem (6) without the constraints regarding the obstacle. The optimization problem thus simplifies to

$$\min_{\hat{\mathbf{u}},\hat{\mathbf{x}},T_E,c_{A_z}} \left(\nu_t T_E + \nu_u \frac{T_E}{N} \sum_{i=0}^{N-1} \hat{\mathbf{u}}_i^{\mathsf{T}} \hat{\mathbf{u}}_i \right) \tag{19}$$

$$\text{s.t.} \quad (7) \text{ to } (13) \text{ and } (16)$$

Using the trivial initial guess (17) to (18) the problem (19) can be solved and the results for the optimization variables $\hat{\mathbf{u}}$, $\hat{\mathbf{x}}$, T_E, and c_{A_z} can be used as initial guess for the next stage.

In the second stage the constraints regarding the obstacles are relaxed in such a way, that these constraints no longer have a local minimum at $\theta = 0$. To this end, the vector to the center of the head ${}_I\mathbf{r}_{OC_H} = {}_I\mathbf{r}_{OP} + \mathbf{R}_{IB\,B}\mathbf{r}_{PC_H}$ with ${}_B\mathbf{r}_{PC_H} = \left[0\;0\;H - \frac{h_H}{2}\right]^{\mathsf{T}}$ is introduced. The distance of this point to the center of the obstacle $\|{}_I\mathbf{r}_{C_H C_O}\|_2$ with ${}_I\mathbf{r}_{C_H C_O} = {}_I\mathbf{r}_{OC_O} - {}_I\mathbf{r}_{OC_H}$ now has to satisfy the relaxed relation

$$\|{}_I\mathbf{r}_{C_H C_O}\|_2 \geq r_O + \sqrt{a_{H_1}^2 + \frac{h_H^2}{4}}. \tag{20}$$

Furthermore the height of the obstacle is minimized starting at a height which is known to not interact with the initial guess obtained by the previous stage. The resulting optimization problem can be stated as

$$\min_{\hat{\mathbf{u}},\hat{\mathbf{x}},T_E,c_{A_z},a_{O_z}} \left(\nu_t T_E + \nu_u \frac{T_E}{N} \sum_{i=0}^{N-1} \hat{\mathbf{u}}_i^{\mathsf{T}} \hat{\mathbf{u}}_i + \nu_O a_{O_z} \right) \tag{21}$$

$$\text{s.t.} \quad (7) \text{ to } (13) \text{ and } (16)$$

$$r_O + \sqrt{a_{H_1}^2 + \frac{h_H^2}{4}} \leq \|{}_I\mathbf{r}_{H_c C_O}(\hat{\mathbf{x}}_i, \mathbf{p})\|_2 \qquad i \in \{0, \ldots, N+1\} \tag{22}$$

$$\frac{{}_I\mathbf{n}_O^{\mathsf{T}}\,{}_I\mathbf{r}_{H_c C_O}(\hat{\mathbf{x}}_i, \mathbf{p})}{\|{}_I\mathbf{r}_{H_c C_O}(\hat{\mathbf{x}}_i, \mathbf{p})\|_2} \leq \cos\left(\frac{\pi}{6}\right) \qquad i \in \{0, \ldots, N+1\} \tag{23}$$

$$a_{O_z,\min} \leq a_{O_z} \tag{24}$$

with the weight $\nu_O \gg \nu_t$ and $\nu_O \gg \nu_u$.

In order to push the minimum height of the obstacle to the system limits, the optimization problem (21) is solved again, but now with the original constraints regarding the obstacle. The according optimization problem is given by

$$\min_{\hat{\mathbf{u}},\hat{\mathbf{x}},T_E,c_{A_z},a_{O_z}} \left(\nu_t T_E + \nu_u \frac{T_E}{N} \sum_{i=0}^{N-1} \hat{\mathbf{u}}_i^{\mathsf{T}} \hat{\mathbf{u}}_i + \nu_O a_{O_z} \right) \tag{25}$$

$$\text{s.t.} \quad (7) \text{ to } (16) \text{ and } (24)$$

If the results of problem (25) are used as initial guess for problem (6), the solver terminates successfully and the resulting solution can be obtained.

6 Results

The result of the final optimization of (6) with the solution of (25) used as initial guess are shown in Figs. 3, 4 and 5. Figure 3 depicts the states of the WIP over time and Fig. 4 shows the drive torque and speed with respect to the maximum values. In Fig. 5 the resulting trajectory is indicated, which shows the cross section of the head during motion and the obstacle. The resulting optimal mounting point for the battery pack is at the lowest possible position, i.e. $c_{A_z,\mathrm{opt}} = c_{A_z,\mathrm{min}}$, which would be different for an optimization without an obstacle. The minimal feasible height of the center of the obstacle results in $a_{O_z,\mathrm{opt}} = 0.686\,\mathrm{m}$. Therefore, with an overall height of the WIP of 0.7m and the radius of the obstacle $r_O = 0.05\,\mathrm{m}$, the lowest point of the obstacle lies 0.064 m underneath the highest point of the WIP in the equilibrium state.

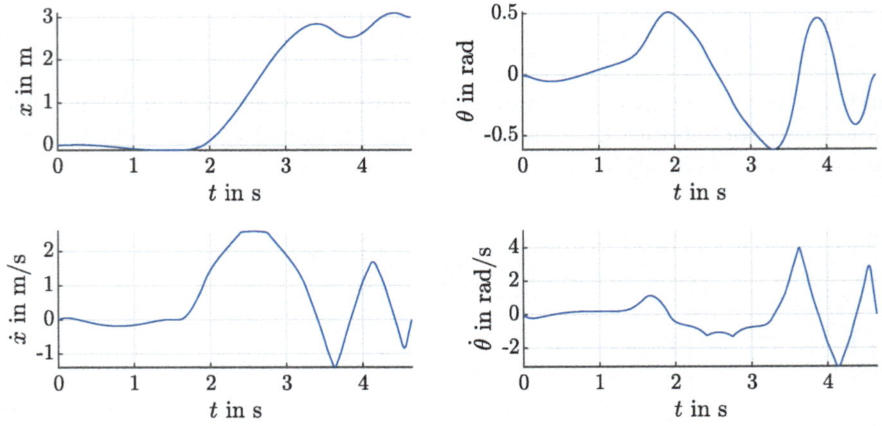

Fig. 3. Optimization result: states

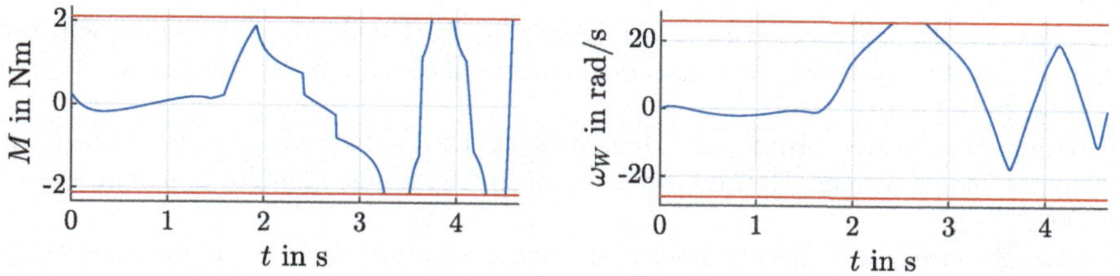

Fig. 4. Optimization result: drive torque and speed

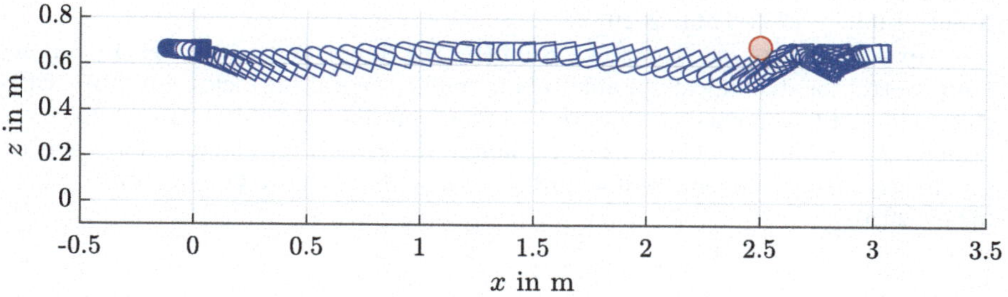

Fig. 5. Optimization result: obstacle avoiding

7 Conclusion

In this work an optimal control problem for a wheeled inverted pendulum has been derived for a desired trajectory with obstacle avoidance inspired by a limbo dance. The resulting optimization problem has non-convex constraints with an infeasible local minimum. Therefore gradient based solvers tend to terminate with an error due to infeasible problem detection, if the initial guess is within this infeasible region. In order to prevent tedious manual tuning of the initial guess, a multistage approach has been proposed to provide an initial solution which avoids entering the region of this local minimum. With this approach, involving 4 closely related optimization problems, it has been possible to reliably obtain a solution for the OCP and to find the optimal mounting point for the battery pack und the minimal obstacle height, for which a solution compliant to the system limits exists.

Acknowledgement. Supported by the "LCM - K2 Center for Symbiotic Mechatronics" within the framework of the Austrian COMET-K2 program.

References

1. Andersson, J., Åkesson, J., Diehl, M.: Dynamic optimization with CasADi. In: 2012 IEEE 51st IEEE Conference on Decision and Control (CDC), pp. 681–686. IEEE (2012)

2. Andersson, J., Gillis, J., Horn, G., Rawlings, J., Diehl, M.: CasADi: a software framework for nonlinear optimization and optimal control. Math. Program. Comput. **11**(1), 1–36 (2019)
3. Bremer, H.: Elastic Multibody Dynamics: A Direct Ritz Approach. Science and Engineering, Springer, Netherlands, Intelligent Systems, Control and Automation (2008)
4. Diehl, M., Gros, S.: Numerical optimal control. Optimization in Engineering Center (OPTEC) (2011)
5. Ning, Y., Yue, M., Yang, L., Hou, X.: A trajectory planning and tracking control approach for obstacle avoidance of wheeled inverted pendulum vehicles. Int. J. Control **93**(7), 1735–1744 (2020)
6. Teeyapan, K., Wang, J., Kunz, T., Stilman, M.: Robot limbo: optimized planning and control for dynamically stable robots under vertical obstacles. In: 2010 IEEE International Conference on Robotics and Automation, pp. 4519–4524 (2010)
7. Zauner, C., Müller, A., Gattringer, H., Jörgl, M.: Trajectory planning for a segway model exploiting inherent feedforward structure. IFAC-PapersOnLine **53**(2), 8444–8449 (2020)

Computational-Efficient Resolved Motion Rate Control with Task-Space Trajectory Tracking

Theodor Borangiu, Silviu Răileanu(✉), and Nick-Andrei Ivănescu

Department of Automation and Applied Informatics, University Politehnica of Bucharest, Bucharest, Romania
{theodor.borangiu,silviu.raileanu,nick.ivanescu}@upb.ro

Abstract. The paper presents a computational-efficient resolved motion rate control software solution for task space trajectory tracking of a 6-d.o.f. robot, with direct control of the end-effector velocity. The centralized computation of non-linear space transformations is configured to get the desired motion precision in position and velocity. A low-cost multiprocessor implementation is described.

Keywords: Robot motion · Task-space control · Resolved-motion rate control · Trajectory generator · Multiprocessor implementing · Computational efficiency

1 Introduction. Task-Space Control

In control schemes developed directly in the operational space, the robot motion is specified in terms of the *task space* which is isomorphic to the special Euclidian group $\mathbf{SE}(3) = \mathbf{R}^3 \times \mathbf{SO}(3)$ where: $\mathbf{p} \in \mathbf{R}^3$ is the position of the origin of the task frame (the end-effector frame), and φ is a minimal representation of the task's frame current orientation (e.g., based, on the Euler angles y, p, r – yaw, pitch and roll) derived from the rotation matrix $\mathbf{R}$ - member of the orthogonal group $\mathbf{SO}(3)$: $\mathbf{R}(t) = [\mathbf{n}(t)\mathbf{s}(t)\mathbf{a}(t)]$ where $\mathbf{n}, \mathbf{s}, \mathbf{a}$ are respectively the normal, sliding and approach vectors along the principal axes x_n, y_n, z_n of the task coordinate system. The currently measured joint space variables $\mathbf{q}(t), \dot{\mathbf{q}}(t)$ are transformed into the corresponding operational space variables through direct kinematics coordinate mappings like $\mathbf{dk}(\cdot)$ and $\mathbf{J}_A(\cdot)$.

The desired trajectory $\mathbf{x}_d(t)$, representing the *direct output* of the task-space interpolation made by the trajectory generator, will be compared with the reconstructed task-space signals. The resulting task-space error $\tilde{\mathbf{x}}(t)$ is taken to zero by feedback control; trajectory inversion is usually embedded with incremental coordinate transformations.

The motion control with operational-space trajectory tracking operates directly with representations of the end-effector's position and orientation. It embeds on its direct path an inverse kinematics transformation (position IK or Jacobian inverse), to access finally the joint control channels $\mathbf{u}$, and a direct kinematic computation on its feedback path, since evaluation of current operational variables is performed indirectly in most cases starting from the measured joint position $\mathbf{q}_s(t)$ and velocity $\dot{\mathbf{q}}_s(t)$.

A. Müller and M. Brandstötter (Eds.): RAAD 2022, MMS 120, pp. 201–209, 2022.
https://doi.org/10.1007/978-3-031-04870-8_24

The problem considered is to find a control providing the desired end effector motion in the manipulator's task space. There is a linear mapping between the instantaneous end-effector task-space velocity and the joint velocities given by the manipulator Jacobian which is a $r \times n$ matrix, n being the number of robot joints and r the dimension of the task-space. For $r = m = n$ the Jacobian is square and if invertible it can be used to map task-space velocity to joint-space velocity.

The ability of a robot manipulator to achieve an arbitrary end-effector velocity is a scalar function of the manipulator Jacobian, e.g., a measure of the manipulability [1]. Choosing the end-effector velocity to perform the task can be achieved using techniques such as potential fields [2], resolved-rate motion control [3], or quadratic programming [4]. These approaches can be augmented to achieve secondary tasks such as joint-limit avoidance and manipulability maximization [5]. Manipulability is maximized in [6] by projecting the gradient of the manipulability into the null space of the differential kinematics. The ability to achieve a secondary task is possible if the robot is redundant, by exploiting null-space motion; a novel approach to controlling the instantaneous velocity of a robot end-effector that can simultaneously maximize manipulability and avoid joint limits is defined in [1]. It operates on non-redundant and redundant robots, which is achieved by adding artificial redundancy in the form of controlled path deviation.

Resolved-motion rate control (RMRC) is an old but effective technique to control the velocity of a robot's end-effector in task space [7]. This control type allows directly controlling the robot's end-effector velocity, without expensive path planning. Resolved-rate is a Jacobian-based control scheme for moving the end-effector of an open chain manipulator at a desired Cartesian velocity without computing the inverse kinematics at each time step, but only the inverse of the Jacobian matrix to account for the updated joint angles. The advantage of this incremental RMRC approach of updating joint angles is that the robot moves smoothly between support points on the trajectory, without jerky movements caused by computing the IK at each sample period [8].

Applications that use RMRC techniques are robot welding and processing part borders. In surgical robots, it is often necessary for the robot arm to move in a straight line at a constant velocity as it lacerates tissues [9]. Resolved-velocity position-based visual servoing is analysed in [10]: visual servoing can be obtained by computing the imposed reference trajectory $\mathbf{q}_r(t)$ for the joint variables on the basis of visual measurements, with the operational space error going asymptotically to zero. The control problem considers the object fixed relative to the robot base frame and an eye-in-hand camera. The RMRC technique can be also applied for image-based visual servoing, when the vector of some object feature parameters should have a desired constant value corresponding to the imposed pose of the camera.

The paper presents a computational-efficient RMRC software solution with low-cost multiprocessor implementation. Section 2 defines the solution and introduces the task-space pipelined RMRC scheme. Section 3 describes the trajectory generator with hybrid task-space feedback and analyses the computational requirements to get the desired motion precision in terms of position and velocity. Section 4 defines the hierarchy of parent-child software processes and proposes a cost-efficient implementation for a 6-d.o.f. educational robot. Section 5 formulates conclusions and indicates future research.

2 RMRC with Pipelined Task-Space Trajectory Tracking

RMRC is a centralised nonlinear task-space motion control scheme that commands the end-effector to move in the desired Cartesian direction in a coordinated position and rate control. The motions of the individual joints are *combined* and *resolved* in separable end-effector motions along the world coordinate axes. This implies that several joint motors must run simultaneously at different, time-varying rates to achieve desired coordinated task-space motion along any world coordinate axis. Consequently, it will be allowed to specify the direction and speed along any arbitrary path to be followed by the manipulator.

The RMRC algorithm introduced in [10] is rearranged as:

$$\tilde{\mathbf{q}}(k+1) = \mathbf{J}_A^{-1}(\mathbf{q}(k)) \cdot \left[\mathbf{x}_d(k+1) - \mathbf{dk}(\mathbf{q}(k))\right], \; \boldsymbol{\tau}(k) = \mathbf{K}_s\tilde{q}(k)$$

where, for notation simplification, $\mathbf{q}(k) = \mathbf{q}(t_k)$, etc., with $t_{k+1} - t_k = T_s, \forall k \in \mathbf{Z}$ and T_s is the sampling period of the motion control. The block diagram of the task – space RMRC scheme based on inverse Jacobian computation is shown in Fig. 1.

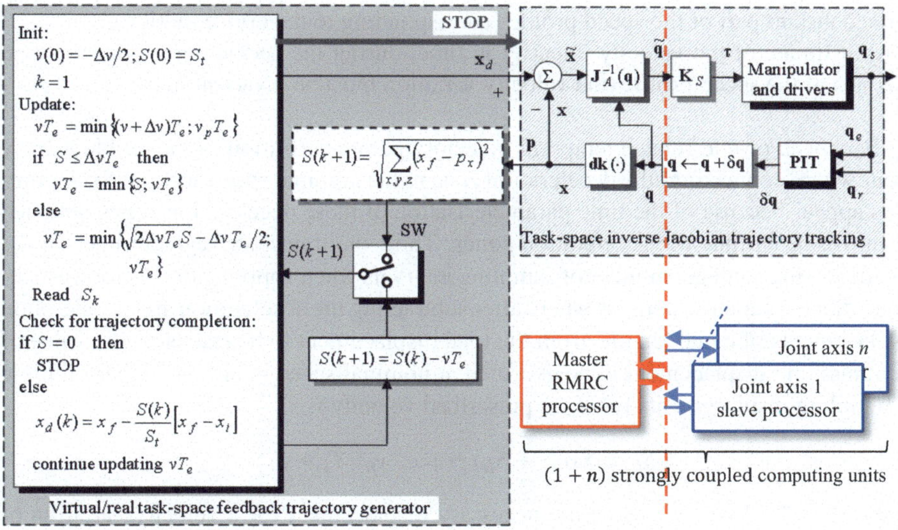

Fig. 1. Task-space pipelined RMRC scheme

Assuming that the task-space error $\tilde{\mathbf{x}}(t)$ is sufficiently small for good trajectory tracking, this deviation can be transformed into the corresponding joint space deviation $\tilde{\mathbf{q}}$ through the inverse of the analytical Jacobian, $\mathbf{J}_A^{-1}(\cdot)$. Then, the control input $\mathbf{u} = \boldsymbol{\tau}$ for the torque-controlled servo motors can be computed from $\tilde{\mathbf{q}}$ with the positive definite diagonal matrix $\mathbf{K}_s$ of feedback gains. The linear mapping $\tilde{\mathbf{q}} \to \boldsymbol{\tau}$ and the numerical integration $\mathbf{q} \leftarrow \mathbf{q} + \delta\mathbf{q}$ updating the absolute current joint positions represent decoupled processing activities; on the contrary, the nonlinear coordinate transformations $\mathbf{J}_A^{-1}(\cdot)$ and $\mathbf{dk}(\cdot)$ represent centralised computation processes that are executed by the same processing unit which is responsible for trajectory generation.

It is also considered that the manipulator has incremental position encoders can sense the angular displacements $\mathbf{q}_s$ as dual pulse trains $\mathbf{q}_l$, $\mathbf{q}_r$ (respectively for clockwise and counter clockwise rotations) and produce the pulse count $\delta\mathbf{q}$ corresponding to the relative displacement of the joint within one current sampling interval T_s with respect to the absolute position of the joint at the end of the previous sampling period.

The trajectory generation subsystem shown in Fig. 1 (left) controls the *velocity* along the task-space trajectory by computing the displacement space for each sampling period. A scalar value $S_{k+1} = S(k+1)$ is generated at the $(k+1)^{th}$ sampling period depending on $S_k = S(k)$ computed for the kth period so that the value meets the time variation imposed by the velocity profile. v_{k+1} is the average speed during period $k+1$.

$$v_{k+1} = (S_{k+1} - S_k)/T_s$$

The term 'accuracy' in generating a displacement corresponding to a sampling period is used in this context to express from a qualitative point of view the measure in which sampling displacements conform with the imposed speed profile, i.e.:

- A characterisation of the time invariance of generated sampling displacements during the constant part of the speed profile (corresponding to the cruise motion);
- An estimate of the velocity linearity in time, during the acceleration / deceleration zones of the speed profile (the velocity variation must be invariant in time: $a \cdot T_s$).

For speed profile-related temporal trajectory parameterisation, some problems may occur when 'connecting' the deceleration zone to the nominal speed section. Such problems appear because of the time parameterisation of these deceleration zones obtained by mirroring in time the acceleration zones. Thus, the trajectory segment may be not executable in an integer number of sampling intervals when applying the mirroring technique. Such a situation appears when, after subtracting the acceleration and deceleration displacements of equal lengths from the total displacement to be executed, the value of the remaining displacement to be executed at nominal speed is not an integer multiple of sampling displacements $v_p \cdot T_s$ at prescribed velocity v_p:

$$S_f = S_{ACC} + S_{DEC} + n \cdot v_p \cdot T_s + S_r$$

where: $n \in \mathbf{N}$; $S_{ACC} = S_{DEC}$: the necessary displacement for a velocity variation of v_p (from 0 to v_p respectively from v_p to 0); S_f: the total displacement to be executed; S_r: the 'rest of displacement' resulting from the above equation and from the associated parameterisation according to the mirroring technique. A trapeze profile was used.

3 Trajectory Generator with Hybrid Task-Space Feedback

A *hybrid algorithm* was chosen to control the displacement velocity along the trajectory. This algorithm has two computational sections:

- An *iterative* section (less time consuming) which checks whether the prescribed nominal velocity is attained, and hence is associated to the acceleration and constant velocity parts of the speed profile.

- An *analytic* section that monitors the approach towards the final point of the trajectory segment, being hence related to the deceleration part of the speed profile.

The sequence of steps of this trajectory generation algorithm with simultaneous control of the speed profile is given below (see Fig. 1): Step 0: Initialisation; Step 1: update $v \cdot T_s = min\{(v + \Delta v) \cdot T_s, v_p \cdot T_s\}$; Step 2: compare S with $\Delta v \cdot T_s$ and take the decision; Step 3: update $S = S - v \cdot T_s$; Step 4: check for trajectory termination ($S \dot{:} 0$); Step 5: compute command $x_d = x_f - (S/S_t) \cdot (x_f - x_i)$ and loop to Step 1.

One can observe that the iterative section of the algorithm is performed in **Step 1**. There is no numeric drift, because the threshold value $v_p \cdot T_s$ represents an invariant data for the application segment and is not obtained in previous iterations. The variables and constant parameters that appear in Step 1 have the following significance: v - current displacement velocity; T_s - sampling period for speed profile update; $\Delta v = a \cdot T_s$ - velocity variation corresponding to a sampling interval; v_p - user-prescribed cruise velocity. As the effective value of velocity is of no interest, Step 1 can be rewritten as:

$$(v \cdot T_s) \leftarrow \min\{(v \cdot T_s) + (\Delta v \cdot T_s); (v_p \cdot T_s)\}$$

The function "min" discriminates in this relationship between the acceleration section on one hand, and the cruise and deceleration sections on the other hand. The initialisation of $v \cdot T_s$ is $v \cdot T_s \leftarrow -\Delta v \cdot T_s/2$, so that after the first speed incrementing the velocity on the trajectory reaches the value $\Delta v/2$. The multiplication $v_p \cdot T_s$ is done off-line at the start of the current trajectory segment, while $\Delta v \cdot T_s$ is a constant data.

In **Step 2**, the distance between the current point on the trajectory and the final point (S) is compared with the maximum distance the execution of which is allowed during the last sampling period of the trajectory ($\Delta v \cdot T_s = a \cdot T_s^2$).

If $S \leq (\Delta v \cdot T_s)$, it is $(v \cdot T_s) \leftarrow \min\{S; v \cdot T_s\}$; using this operation, it is intended to execute the remaining space in a single sampling period if $S < v_p \cdot T_s$ (when $v_p < \Delta v$).

If $S > (\Delta v \cdot T_s)$, the space increment in the current sampling period is defined as:

$$(v \cdot T_s) \leftarrow \min\left\{\sqrt{(2\Delta v \cdot T_s)S} - \Delta v \cdot T_s/2; (v \cdot T_s)\right\}$$

where the first term is:

$$\sqrt{(2\Delta v \cdot T_s)S} - \Delta v \cdot T_s/2 = T_s(\sqrt{2aS} - \Delta v/2)$$

and represents the max. displacement increment during the sampling period in which the current point on the trajectory segment is at distance S from the final point.

The function 'min' discriminates now between the deceleration section and the acceleration and cruise sections. The value of S is initialised at the beginning of the current trajectory segment with a quantity S_t representing the distance between the initial and the final point on the trajectory. Thus, for a displacement of the end-effector along a linear path in the Cartesian space and for v_p prescribed as the travelling velocity of the origin of the task frame (end-effector frame), S_t is defined as the semi norm:

$$S_t = \sqrt{(p_{xf} - p_{xi})^2 + (p_{yf} - p_{yi})^2 + (p_{zf} - p_{zi})^2}, x = \left[\mathbf{p}^T \varphi^T\right]^T, \mathbf{p} = [p_x p_y p_z]^T$$

The square root and multiplications in one decision branch of **Step 2** will not significantly increase the computational effort. As the current displacement increment $v \cdot T_s$ has the nature of velocity (or position on the trajectory parameterised with S), its value will be computed on a number of bits that ensure the imposed relative speed precision.

Absolute precision constraints are imposed only for the velocity variation (acceleration); this aspect is really important only in points close to the final target position, where S and $(v \cdot T_s)$ are smaller and, although represented in order to satisfy the relative precision, their absolute precision will be also improved. Hence, the operands will be expressed on a number of bits imposed by the *relative velocity precision* ε_{vv} and by the *absolute acceleration precision* ε_{aa}, such that the constraints below are satisfied:

$$\frac{|(v \cdot T_s)_{est} - (v \cdot T_s)|}{v \cdot T_s} < \varepsilon_{vv}, |(v \cdot T_s)_{est} - (v \cdot T_s)| < \varepsilon_{aa} \cdot T_s^2,$$

where $(v \cdot T_s)_{est}$ represents the *estimated* displacement increment over a sampling period. From these relationships it results:

$$\frac{\varepsilon_{aa} \cdot T_s^2}{v \cdot T_s} > \frac{\varepsilon_{aa} \cdot T_s^2}{v_{max} \cdot T_s} > \varepsilon_{vv}$$

which in general means that, satisfying the relative velocity precision is sufficient to provide the required performances of the speed profile.

Step 3 updates the scalar value of the control space. The updating equation of **Step 3** represents however a virtual, software feedback for the closed-loop trajectory generation only. The update of $S(k)$ at each sampling period $k \in \mathbf{Z}$ leads to an open-loop, pipelined RMRC in the task space, because the currently measured position feedback closes at trajectory tracking level only. This pipelined motion control scheme implies that the upward entry of the software switch SW will be collected as virtual feedback data from the trajectory generator in the scheme represented in Fig. 1. Closing the overall control system composed by the trajectory generator and trajectory tracking modules occurs when the downward entry of the switch SW is collected, which corresponds to:

- Measuring the joint positions $\mathbf{q}_s$ from the dual (left - right direction) pulse trains of each joint encoder ($\mathbf{q}_l, \mathbf{q}_r$).
- Updating the current joint configuration $\mathbf{q}$ at each sampling period.
- Computing the current end-effector pose $\mathbf{x}$ through a nonlinear coordinate transformation $\mathbf{dk}(\cdot)$– direct kinematics. x_f, y_f, z_f are the end-tip frame coordinates.
- Extracting the current task-frame position $\mathbf{p}$ and updating the control space S:

$$S(k) = \sqrt{\left(x_f - p_x(k)\right)^2 + \left(y_f - p_y(k)\right)^2 + \left(z_f - p_z(k)\right)^2}$$

The test for trajectory termination is executed in **Step 4** by comparing S with zero, and a corresponding *centralized* command STOP is issued.

If the trajectory segment is not terminated, i.e., $S > 0$, **Step 5** computes the reference signal $\mathbf{x}_d(k)$ for the trajectory tracking subsystem as $\mathbf{x}_d(k) = \mathbf{x}_f - S(k) \cdot \mathbf{V}$. By defining the unit vector $\mathbf{V}$ of the tangent vector to the linear desired trajectory in the sense of semi norms defined for each type of trajectory (in this case a linear trajectory), it is:

$$\mathbf{V} = \frac{[\mathbf{x}_f - \mathbf{x}_i]}{S_t} = \frac{1}{S_t}[p_{xf} - p_{xi}, p_{yf} - p_{yi}, p_{zf} - p_{zi}, y_f - y_i, p_f - p_i, r_f - r_i]$$

where $\mathbf{p} = [p_x p_y p_z]^T$ and $\varphi = [ypr]^T$ and the indices f, i are respectively associated to the final and initial points on the trajectory segment. **Step 5** must be executed with sufficiently high absolute precision, so that the computed reference points $\mathbf{x}_d(k)$ lie at a distance (in sense Cebîshev) smaller than a maximal allowed one (this is a controller-dependent constant data) relative to the ideal trajectory.

Thus, $|\hat{x}_{dj} - x_{dj}| < \varepsilon_{aj}$, $1 \leq j \leq n$, where ε_{aj} is the absolute deviation error with respect to the linear trajectory along axis j. The components of vector $\mathbf{V}$ are off-line computed with high accuracy. The maximum error in the computation of $\mathbf{x}_d$ points along the trajectory segment is in the initial point (when $S = S_t$). For example, considering that ε_{rx} is the relative projection error of the unit vector $\mathbf{V}$ onto the Cartesian axis X, the following condition must be fulfilled:

$$\varepsilon_{ax} > \varepsilon_{rx} \cdot |x_f - p_x(k)| = \textit{abs.position error on the trajectory}$$

where ε_{ax} is the maximum admissible error on axis X. It results hence that the mathematical library to be used for the computation of $\mathbf{x}_d$ must have the representation of operand mantissa on a sufficiently large number of bits, such that:

$$\varepsilon_{av} > \varepsilon_{rv} \cdot \max_{(taskspace)} |x_f - x_i|, \text{ where:}$$

- ε_{av} is the max. absolute admissible error of the point's deviation along one axis;
- ε_{rv} is the relative error (off-line evaluated), when computing the projection of the unit vector $\mathbf{V}$ onto the respective coordinate axis;
- $\max|x_f - x_i|$ is the max. movement along the *leading* task space coordinate *axis*.

As suggested in the control scheme in Fig. 1, $\mathbf{J}_A^{-1}(q)$, $\mathbf{dk}(\cdot)$ and $S(k+1)$ are real time centralized computation tasks that require increased processing power.

4 A Low-Cost Multiprocessor Implementation Solution

The software implementation for the pipelined task space RMRC scheme consists into creating hierarchies of *parent* and *multiple child* software processes running simultaneously on strongly coupled processors (Fig. 2).

The centralized computation of $S(k)$ closes the motion control loop in task space; *dragging axes* (joint axes for which the tracking error $\tilde{q}_i$ exceeds a limit value) are identified and reported instantly to the parent process that initiates an emergency procedure.

The implementing solution, generically described in the lower right part of Fig. 1, is composed of a central node with Raspberry Pi Single Board Computer (SBC) in charge of the computation for trajectory generation and inverse Jacobian trajectory tracking,

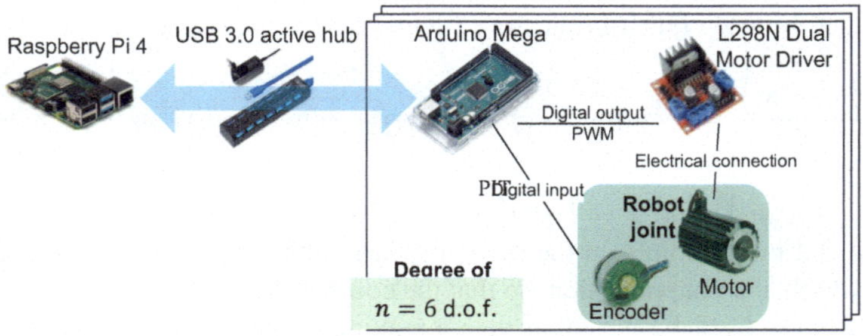

Fig. 2. Low-cost implementation of RMRC for the 6-d.o.f. educational robot

and a set of 6 Arduino-based controllers for joint d.c. motor control and encoder reading - subordinated to the central node. The output τ_i, $1 \leq i \leq 6$ of each Arduino Mega 2560 controller is amplified using a L298N Dual Full-Bridge Driver.

The Raspberry Pi ver. 4 SBC uses a 64-bit quad-core ARM Cortex-A72 processor with computing performance about 1 GFLOPS, multitasking and USB3.0 connection ports with up to 4.8 Gbit/s communication speed. Arduino Mega SBC has USB serial connection of max. communicating speed of 2mil bauds, processing power of 16MIPS at 16 MHz (ATmega2560) and I/O system operating in PWM mode (output for motor control) and PIT mode (input from encoder) with sampling speed up to 16 MHz.

5 Conclusion

The paper presents a computational-efficient RMRC software solution for task space trajectory tracking of a 6-d.o.f. robot, with direct control of the end-effector velocity. For joint speeds of about 300°/s, a relative velocity precision of 1% was imposed; the incremental computations at sampling periods $T_s = 16$ ms with 10-bit operands leads to an absolute position error on the trajectory of 40 μm, confirmed by simulations.

A strongly coupled multiprocessor structure based on low-cost single-board computing units implements the robot motion controller. The elements are chosen considering their real-time capabilities, high-speed communication and affordable costs for educational robotics purpose. The basic software for position and velocity kinematics models, trajectory generation and task-space tracking with RMRC and is currently tested; it will be then integrated on the robot controller by the educational system manufacturer.

Future research will be directed towards applying RMRC for the tracking of object feature parameters extracted from images acquired by an arm-mounted camera in image-based visual servoing of the robot working in unstructured environments.

References

1. Haviland, J., Corke, P.: Maximising Manipulability During Resolved-Rate Motion Control, arXiv:2002.11901 [cs.RO] (2020)

2. Khatib, O.: Real-time obstacle avoidance for manipulators and mobile robots. Int. J. Robot. Res. **5**(1), 90–98 (1986). https://doi.org/10.1177/027836498600500106
3. Whitney, D.E.: Resolved motion rate control of manipulators and human prostheses. IEEE Trans. Man-Mach. Syst. **10**(2), 47–53 (1969)
4. Guo, D., Zhang, Y.: Acceleration-level inequality-based MAN scheme for obstacle avoidance of redundant robot manipulators. IEEE Trans. Ind. Electron. **61**(12), 6903–6914 (2014)
5. Zhang, Y., Yan, X., Chen, D., Guo, D., Li, W.: QP-based refined manipulability-maximizing scheme for coordinated motion planning and control of physically constrained wheeled mobile redundant manipulators. Nonlinear Dyn. **85**(1), 245–261 (2016). https://doi.org/10.1007/s11071-016-2681-9
6. Park, J., Wangkyun, C., Youngil, Y.: Computation of gradient of manipulability for kinematically redundant manipulators including dual manipulators system. Trans. Control Autom. Syst. Eng. **1**(1), 8–15 (1999)
7. Jaquier, N., Rozo, L., Caldwell, D.G., Calinon, S.: Geometry aware manipulability learning, tracking and transfer, arXiv preprint arXiv:1811.11050 (2018)
8. Yao-An, L., Tang, K., Wang, C.-Y.: Collision-free and smooth joint motion planning for six-axis industrial robots by redundancy optimization. Robotics CIM **68**, 102091 (2021)
9. Zevallos, N., et al.: A Real-time Augmented Reality Surgical System for Overlaying Stiffness Information (2018). https://doi.org/10.13140/RG.2.2.17472.64005
10. Siciliano, B., Sciavicco, L., Villani, L., Oriolo, G.: Robotics. Modelling, Planning and Control. Springer, London (2010). https://doi.org/10.1007/978-1-84628-642-1. ISBN 978-1-84628-642-1

Fractional-Order $PII^{1/2}DD^{1/2}$ Control of a Mechatronic Axis: Influence of the Discrete-Time Approximation of the Half-Order Derivative and Integral

Luca Bruzzone(✉) and Shahab E. Nodehi

University of Genoa, 16145 Genoa, GE, Italy
luca.bruzzone@unige.it, shahabedin.nodehi@edu.unige.it

Abstract. The proposed work investigates the influence of the discrete-time approximation in the digital implementation of the $PII^{1/2}DD^{1/2}$ control of a mechatronic axis. The most widespread approach to apply Fractional Calculus to position control is the $PI^{\lambda}D^{\mu}$ scheme, in which the integral and the derivative of the error have respectively non-integer orders λ and μ. The $PII^{1/2}DD^{1/2}$ scheme is an alternative approach in which an integral and a derivative of order 1/2 are added to the first-order ones instead of replacing it. The exact calculation of derivatives and integrals of non-integer order requires to consider the whole time history of the signal, which is unfeasible for real-time computation. Consequently, a discrete-time approximation comprising a digital filter with finite order, sampling time and time window needs to be adopted. This paper discusses the influence of the main digital filter parameters on the dynamic performance of the $PII^{1/2}DD^{1/2}$ control of a rotary mechatronic axis.

Keywords: Fractional-order control · Distributed-order control · Position control

1 Introduction

Fractional Calculus (FC) is the extension of derivatives and integrals to non-integer orders [1]. Its most widespread application to control system design is the $PI^{\lambda}D^{\mu}$ scheme, which exploits an integral term and a derivative term of orders λ and μ [2]. Design techniques [3, 4], optimization methods [5, 6] and practical applications [7, 8] of $PI^{\lambda}D^{\mu}$ controllers have been extensively discussed in the scientific literature.

An alternative way to apply FC to control systems is to add fractional-order terms to the PID scheme, as proposed in [9] with the $PDD^{1/2}$ scheme, in which a half-derivative term (derivative of order 1/2) is added to the PD controller. This scheme has been investigated in [10] adopting a nondimensional approach for a SISO purely inertial system. Experimental validations of the performance improvement of the $PDD^{1/2}$ control over the classical PD in the transient behavior are reported in [11], with reference to a linear motor, and in [12], with reference to a rotary axis. The PD, PD^{μ} and $PDD^{1/2}$

A. Müller and M. Brandstötter (Eds.): RAAD 2022, MMS 120, pp. 210–217, 2022.
https://doi.org/10.1007/978-3-031-04870-8_25

schemes have been compared by simulation in the position control of mechatronic systems, showing that both the fractional-order controllers overperforms the integer-order one, with the PDD$^{1/2}$ having slightly better readiness but slightly higher overshoot [13]; this comparison is experimentally validated in [14] in real working conditions.

Besides applications to Single Input – Single Output systems, the adoption of the PDD$^{1/2}$ scheme for position control of a Multi Input – Multi Output system (a parallel robot) is discussed in [15]; remaining in the area of robotics, the addition of the half-derivative term can bring to alternative fractional-order evolutions of impedance control [16] of industrial robots, a field which is beginning to be explored [17, 18].

Considering also the integral action for steady-state accuracy, the PID controller can be extended to the PII$^{1/2}$DD$^{1/2}$ scheme, adding a half-integral term besides the half-derivative one. A Bode plot based tuning method for the PII$^{1/2}$DD$^{1/2}$ scheme has been proposed and validated in simulation and experimentally in [19], with application to a mechatronic rotary axis.

The exact calculation of derivatives and integral of non-integer order requires to consider the whole time history of the signal [20], which is unfeasible for real-time computation; as a consequence, for the digital implementation of fractional-order controllers it is necessary to adopt a discrete-time approximation comprising a finite-order filter. In the following of the paper, the same case study discussed in [19] is considered, comparing the performances of the PID and PII$^{1/2}$DD$^{1/2}$ schemes for position control of a rotary axis, but investigating the influence of the main filter parameters: sampling time and filter order.

2 Discrete Time Implementation of Fractional-Order Derivatives and Integrals

In FC both integration and differentiation are represented by the same integro-differential operator D_t^α, in which α is the order, which can be even complex; in the following, we will consider only real values of α. Adopting the Grünwald-Letnikov definition, which is suitable for robust discrete-time implementation [21], the differentiation of fractional order α (if $\alpha > 0$) or the integration of fractional order $-\alpha$ (if $\alpha < 0$) of a function of time $x(t)$ is defined as:

$$D_t^\alpha x(t) = \lim_{h \to 0} \left[\frac{1}{h^\alpha} \sum_{k=0}^{\left[\frac{t}{h}\right]} (-1)^k \frac{\Gamma(\alpha+1)}{\Gamma(k+1)\Gamma(\alpha-k+1)} x(t-kh) \right] \tag{1}$$

where h is the time increment, Γ is the Gamma function [20] and $[y]$ is the integer part of y. In Eq. (1) the number of terms of the sum tends to infinite, since h tends to zero; therefore, for a numerical computation, Eq. (1) must be rewritten using a small but finite sampling time T_s [21]:

$$D_t^\alpha x(t) \cong D_t^\alpha x(kT_s) \cong \frac{1}{T_s^\alpha} \left(\sum_{j=0}^{k} w_j^\alpha x((k-j)T_s) \right), k = [t/T_s] \tag{2}$$

where:

$$w_0^\alpha = 1,\ w_j^\alpha = \left(1 - \frac{\alpha+1}{j}\right) w_{j-1}^\alpha, j = 1, 2, \ldots \tag{3}$$

The calculation of Eq. (2) involves all the $k + 1$ sampled values of the x, consequently, even with finite sampling time, the continuous increase of the number of addends makes unfeasible the real-time implementation of Eq. (2), unless only a limited number of n previous steps is considered. This corresponds to adopting a nth order digital filter with fixed memory length $L = nT_s$. This approximation, which considers only the recent past of the function $x(t)$, does not yield relevant approximations in the evaluation of the fractional-order derivatives and integrals, according to the *short-memory principle* [20].

3 $\text{PII}^{1/2}\text{DD}^{1/2}$ Control of a Rotary Axis

In the following of the paper, the PID and $\text{PII}^{1/2}\text{DD}^{1/2}$ controllers will be compared with reference to position control of a rotary axis (Fig. 1a), modelled as a second-order linear system with overall inertia $J = 1.04 \times 10^{-3}$ kgm^2 (motor rotor included) and friction which is linearized by a viscous coefficient $B = 1.45 \times 10^{-3}$ Nms/rad; consequently, the open-loop transfer function is the following:

$$J\frac{d^2}{dt^2}\theta + B\frac{d}{dt}\theta = M(e_\theta) \tag{4}$$

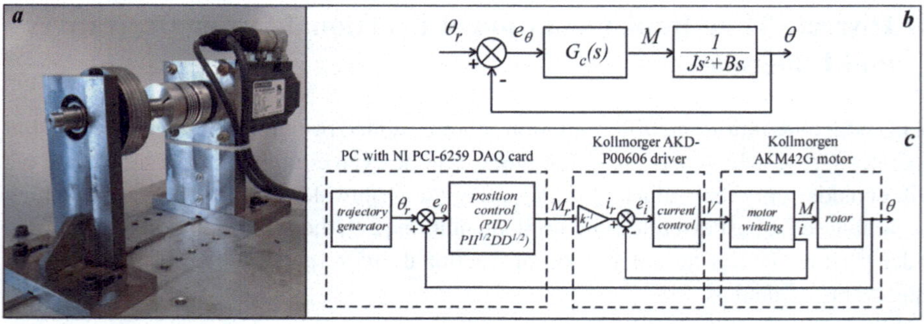

Fig. 1. Considered case study: rotary axis (a), theoretical closed-loop position control scheme (b), controller implementation scheme (c).

According to the position control scheme of Fig. 1b, the torque applied to the rotor is calculated as a function of the error e_θ, which is the difference between the set-point θ_r and the current angle θ. In case of PID scheme, the control law is:

$$M(e_\theta) = \left(K_p + K_i D^{-1} + K_d D^1\right) e_\theta \tag{5}$$

where K_p, K_i, K_d are respectively the proportional, integral and derivative gains. In case of $\text{PII}^{1/2}\text{DD}^{1/2}$ scheme, the control law is:

$$M(e_\theta) = \left(K_p + K_i D^{-1} + K_{hi} D^{-1/2} + K_d D^1 + K_{hd} D^{1/2}\right) e_\theta \tag{6}$$

where K_{hd} and K_{hi} are respectively the *half-derivative gain* and the *half-integral gain* [19].

Applying the Laplace transform to Eqs. (5) and (6), the two transfer functions of the controllers can be obtained:

$$G_{c,PID}(s) = K_p + \frac{K_i}{s} + K_d s = K_i \frac{(1 + (K_p/K_i)s + (K_d/K_i)s^2)}{s} \quad (7)$$

$$\begin{aligned} G_{c,PII^{1/2}DD^{1/2}}(s) &= K_p + \frac{K_i}{s} + \frac{K_{hi}}{s^{1/2}} + K_d s + K_{hd} s^{1/2} \\ &= K_i \frac{1 + (K_{hi}/K_i)s^{1/2} + (K_p/K_i)s + (K_{hd}/K_i)s^{3/2} + (K_d/K_i)s^2}{s} \end{aligned} \quad (8)$$

As discussed in [19], the asymptotic Bode plot of the $PII^{1/2}DD^{1/2}$ controller is characterized by an initial magnitude slope of −20 dB/dec and an initial phase of $-\pi/2$ rad at low frequencies; there are four corner frequencies ($\omega'_{c,i}$, $i = 1 \ldots 4$); after each corner frequency the magnitude slope increases of 10 dB/dec, and the phase increases of $\pi/4$; consequently, at high frequencies the magnitude slope and the phase tend respectively to + 20 dB/dec and to + $\pi/2$. In [19] two Bode plot based tuning methods for the $PII^{1/2}DD^{1/2}$ controller have been proposed, which derive the control gains from a given PID controller. The first method ($PII^{1/2}DD^{1/2}{}_H$, red plot in in Fig. 2) imposes the symmetry of the magnitude plot with respect to the frequency ω_{min}, and the coincidence of the initial and final asymptotes and of the central zone with the magnitude Bode plot of the reference PID (blue plot in Fig. 2). The second proposed tuning ($PII^{1/2}DD^{1/2}{}_L$, green plot in Fig. 2) is obtained from the first multiplying all the control gains for the ratio min_{PID}/min_{PIIDD} (Fig. 2) in order to obtain the same minimum magnitude of the reference PID controller not for the asymptotic Bode plots, but for the exact Bode plots (Fig. 2, blue and green dashed lines).

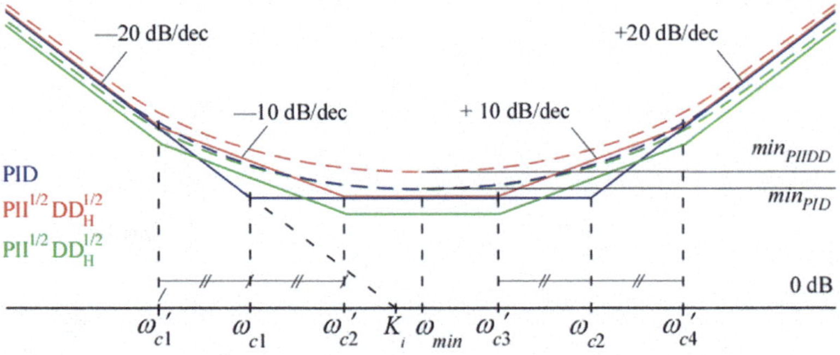

Fig. 2. Bode plot based $PII^{1/2}DD^{1/2}$ tuning. Reference PID: blue; $PII^{1/2}DD^{1/2}{}_H$, with equal minimum magnitude of the asymptotic Bode plot: red; $PII^{1/2}DD^{1/2}{}_L$, with equal minimum magnitude of the exact Bode plot: green (continuous lines: asymptotic magnitude Bode plots; dashed lines: exact magnitude Bode plots).

The two corner frequencies of the PID controller are ω_{c1} and ω_{c2}. Once decided the ratio $\rho = \omega_{c1}/\omega'_{c1} = \omega'_{c2}/\omega_{c1} = \omega_{c2}/\omega'_{c3} = \omega'_{c4}/\omega_{c2}$, all the $PII^{1/2}DD^{1/2}$ gains can be derived from the reference PID gains, using equations which are demonstrated in [19] and that are not recalled here for reasons of space. In the following of the paper, the PID / PII$^{1/2}$DD$^{1/2}$ comparison considers the same case study discussed in [19], with the control parameters collected in Table 1, and the investigation is focussed on the influence of sampling time and order of the digital filters used to calculate the half-derivative and the half-integral by means of Eq. (2).

Table 1. Control parameters for the PID/PII$^{1/2}$DD$^{1/2}$ comparison.

Controller	K_p [Nm/rad]	K_i [Nm/rad·s]	K_{hi} [Nm/rad·s$^{1/2}$]	K_d [Nms/rad]	K_{hd} [Nms$^{1/2}$/rad]
PID	2.5×10^{-1}	5.0×10^{-3}	–	3.5×10^{-2}	–
PII$^{1/2}$DD$^{1/2}{}_\mathrm{H}$	3.3×10^{-1}	5.0×10^{-3}	9.3×10^{-2}	3.5×10^{-2}	2.5×10^{-1}
PII$^{1/2}$DD$^{1/2}{}_\mathrm{L}$	1.5×10^{-1}	2.3×10^{-3}	4.2×10^{-2}	1.6×10^{-2}	1.1×10^{-1}

The comparison has been performed by simulation using Simulink, considering the discrete-time implementation of the digital filters, but the model has been experimentally validated by means of the control layout depicted in Fig. 1c: the position loop is implemented in Simulink Desktop Real Time running on a PC; a National Instrument PCI-6259 DAQ card reads the encoder signal θ and generates the reference torque signal M_r, which is sent to the driver (Kollmorgen AKD P00606) and then imposed to the motor (Kollmorgen AKM42G) by means of the current loop.

Figure 3 shows the good agreement between simulation and experimental test for a setpoint with trapezoidal speed law, characterized by a first phase of 0.2 s with acceleration of 500 rad/s^2, a second phase of 0.6 s with constant speed of 100 rad/s, and a third phase of 0.2 s with deceleration of -500 rad/s^2; subsequently, the setpoint varies from 0 to 80 rad in 1 s. The results in Fig. 3 are referred to a sampling time $T_s = 0.006$ s and a filter order $n = 6$.

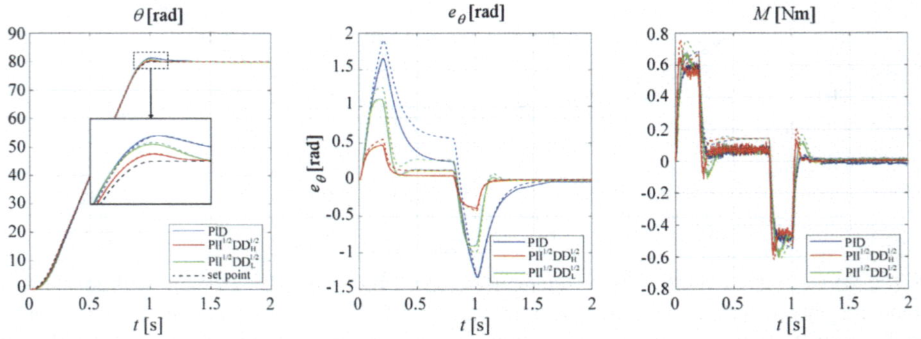

Fig. 3. Trapezoidal speed law response, comparison of experimental tests (continuous line) and discrete-time simulation (dashed line): angle θ (left), error e_θ (center) and motor torque M (right); PID: blue; PII$^{1/2}$DD$^{1/2}{}_H$: red; PII$^{1/2}$DD$^{1/2}{}_L$: green; set-point: black.

4 Analysis of the Influence of Sampling Time and Filter Order

Once validated the model, the influence of the sampling time and of the filter order has been investigated by simulation, without being bounded by the computational power provided by the currently adopted hardware and software implementation.

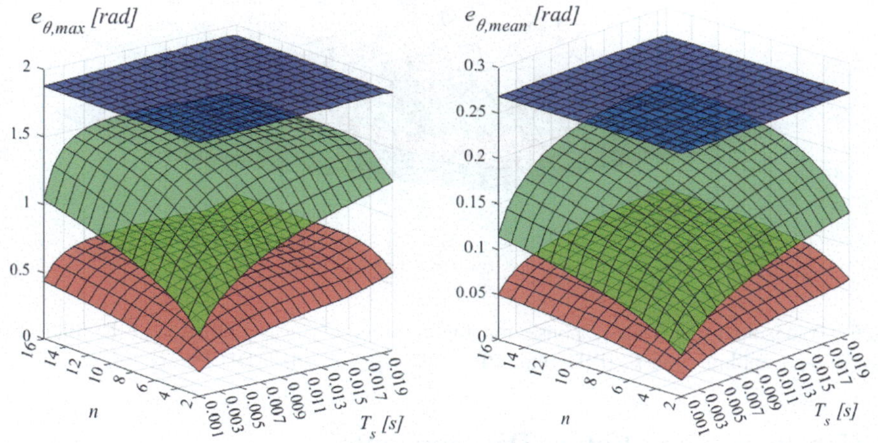

Fig. 4. Trapezoidal speed law response: maximum error $e_{\theta,max}$ (right) and mean error $e_{\theta,mean}$ (left) as a function of the filter order n and of the sampling time T_s (PID: blue; PII$^{1/2}$DD$^{1/2}$$_H$: red; PII$^{1/2}DD^{1/2}$$_L$: green).

The 3D graphs of Figs. 4 and 5 show the results in terms of maximum error $e_{\theta,max}$, mean error $e_{\theta,mean}$ and maximum torque M_{max} as a function of n (ranging from 2 to 16) and T_s (ranging from 1 to 20 ms). The blue surfaces representing the PID and are not influenced by n, since there is no use of Eq. (2), and slightly influenced by T_s. The following conclusion can be summarized:

- a smaller sampling time reduces the maximum and mean errors for all the controllers, as expected;
- the PII$^{1/2}$DD$^{1/2}$ controllers have always lower maximum and mean errors, but with comparable maximum control torque only if T_s is not too high ($<6 \div 7$ ms); in particular, approximately with $n > 6 \div 7$ and $T_s < 5$ ms there is a zone in which the PII$^{1/2}$DD$^{1/2}$$_H$ has lower maximum torque than the PID;
- on the other hand, for higher T_s the PID has lower maximum torque, and the negative effect of the sampling time increase is larger for the PII$^{1/2}$DD$^{1/2}$$_H$ than for the PII$^{1/2}$DD$^{1/2}$$_L$;
- a higher filter order reduces the maximum torque of both the PII$^{1/2}$DD$^{1/2}$ controllers, but slightly increases the maximum and mean errors; this effect is higher for small sampling times, for which the application of the fractional-order controllers is more convenient.

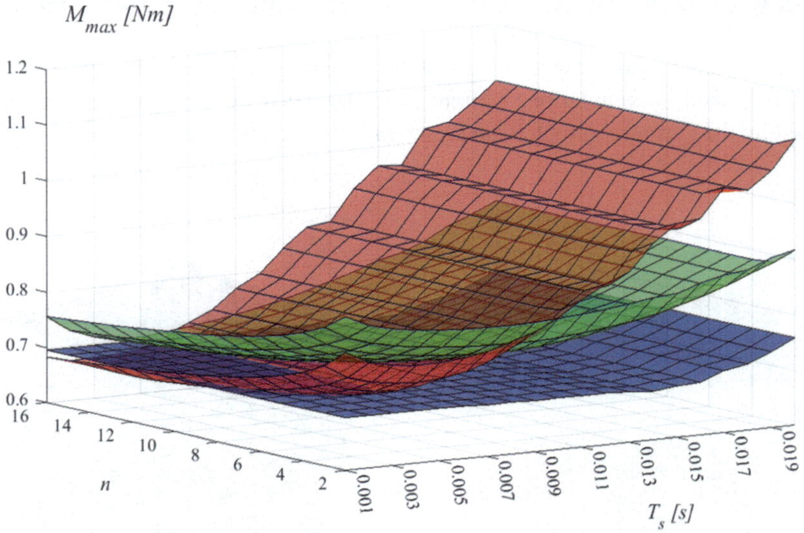

Fig. 5. Trapezoidal speed law response: maximum torque M_{max} as a function of the filter order n and of the sampling time T_s (PID: blue; $PII^{1/2}DD^{1/2}{}_H$: red; $PII^{1/2}DD^{1/2}{}_L$: green).

5 Conclusions and Future Developments

The application of fractional-order schemes is a valuable option for improving the performance of position control of mechatronic devices, reducing the error with comparable maximum control effort. In particular, the potentiality of the $PII^{1/2}DD^{1/2}$ controller has been theoretically and experimentally demonstrated in [19], with application to a rotary axis.

In this paper, the influence of the digital filter order and sampling time for the calculation of the fractional-order derivatives and integrals is investigated. The results show that for the considered case study the sampling time is more critical than the filter order for keeping low the maximum control torque, and that the filter order should be selected doing a compromise, since its increase has a positive influence on the maximum torque and a negative effect on the error.

In the following of the work, this investigation will be deepened to obtain more general design indications, not limited to the specific case study.

References

1. Miller, K.S., Ross, B.: An Introduction to the Fractional Calculus and Fractional Differential Equations. John Wiley & Sons, New York (1993)
2. Podlubny, I.: Fractional-order systems and $PI^{\lambda}D^{\mu}$ controllers. IEEE Trans. Autom. Control **44**(1), 208–214 (1999)
3. Yeroglu, C., Tan, N.: Note on fractional-order proportional-integral-differential controller design. IET Control Theory Appl. **5**(17), 1978–1989 (2011)

4. Beschi, M., Padula, F., Visioli, A.: The generalised isodamping approach for robust fractional PID controllers design. Int. J. Control **90**(6), 1157–1164 (2017)
5. Kesarkar, A.A., Selvaganesan, N.: Tuning of optimal fractional-order PID controller using an artificial bee colony algorithm. Syst. Sci. Control Eng. **3**(1), 99–105 (2015)
6. Saidi, B., Amairi, M., Najar, S., Aoun, M.: Bode shaping-based design methods of a fractional order PID controller for uncertain systems. Nonlinear Dyn. **80**(4), 1817–1838 (2014). https://doi.org/10.1007/s11071-014-1698-1
7. Dimeas, I., Petras, I., Psychalinos, C.: New analog implementation technique for fractional-order controller: A DC motor control. AEU Int. J. Electron. Commun. **78**, 192–200 (2017)
8. Haji Haji, V., Monje, C.A.: Fractional-order PID control of a chopper-fed DC motor drive using a novel firefly algorithm with dynamic control mechanism. Soft. Comput. **22**(18), 6135–6146 (2017). https://doi.org/10.1007/s00500-017-2677-5
9. Bruzzone, L., Bozzini, G.: Application of the PDD$^{1/2}$ algorithm to position control of serial robots. In: Proceedings of the 28th IASTED International Conference Modelling, Identification and Control, MIC 2009, pp. 225–230. Innsbruck, Austria (2009)
10. Bruzzone, L., Bozzini, G.: PDD$^{1/2}$ control of purely inertial systems: Nondimensional analysis of the ramp response. In: Proceedings of the 31st IASTED International Conference Modelling, Identification, and Control, MIC 2011, pp. 308–315. Innsbruck, Austria (2011)
11. Bruzzone, L., Fanghella, P.: Fractional-order control of a micrometric linear axis. J. Control Sci. Eng. **2013**, 947428 (2013)
12. Corinaldi, D., Palpacelli, M., Carbonari, L., Bruzzone, L., Palmieri, G.: Experimental analysis of a fractional-order control applied to a second order linear system. In: Proceedings of the 10th IEEE/ASME International Conference on Mechatronic and Embedded Systems and Applications, MESA 2014, 6935535. Senigallia, Italy (2014)
13. Bruzzone, L., Fanghella, P.: Comparison of PDD$^{1/2}$ and PD$^{\mu}$ position controls of a second order linear system. In: Proceedings of the IASTED International Conference on Modelling, Identification and Control, MIC 2014, pp. 182–188. Innsbruck, Austria (2014)
14. Bruzzone, L., Fanghella, P., Baggetta, M.: Experimental assessment of fractional-order PDD$^{1/2}$ control of a brushless DC motor with inertial load. Actuators **9**(1), 13 (2020)
15. Bruzzone, L., Fanghella, P.: Fractional order control of the 3-CPU parallel kinematics Machine. In: Proceedings of the 32nd IASTED International Conference Modelling, Identification and Control, MIC 2013, pp. 286–292. Innsbruck, Austria (2013)
16. Bruzzone, L.E., Molfino, R.M., Zoppi, M.: An impedance-controlled parallel robot for high-speed assembly of white goods. Ind. Robot. **32**(3), 226–233 (2005)
17. Kizir, S., Elşavi, A.: Position-based fractional-order impedance control of a 2 DOF serial manipulator. Robotica **39**(9), 1560–1574 (2021)
18. Liu, X., Wang, S., Luo, Y.: Fractional-order impedance control design for robot manipulator. In: Proceedings of the ASME 2021 International Design Engineering Technical Conferences and Computers and Information in Engineering Conference, vol. 7, V007T07A028. Virtual (2021)
19. Bruzzone, L., Baggetta, M., Fanghella, P.: Fractional-order PII$^{1/2}$DD$^{1/2}$ Control: theoretical aspects and application to a mechatronic axis. Appl. Sci. **11**(8), 3631 (2021)
20. Das, S.: Functional Fractional Calculus. Springer-Verlag, Berlin/Heidelberg, Germany (2011)
21. Machado, J.T.: Fractional-order derivative approximations in discrete-time control systems. Syst. Anal. Model. Simul. **34**(4), 419–434 (1999)

Fractional Order Calculus-Inspired Kinematic Design in Adaptive Control

Bence Varga[1,2], Richárd Horváth[1,2], and József K. Tar[1,3,4(✉)]

[1] Doctoral School of Applied Informatics and Applied Mathematics, Óbuda University, H-1034 Budapest, Bécsi út 96/B, Budapest, Hungary
varga.bence@bgk.uni-obuda.hu, tar.jozsef@nik.uni-obuda.hu
[2] Bánki Donát Faculty of Mechanical and Safety Engineering, Óbuda University, Budapest, Hungary
horvath.richard@bgk.uni-obuda.hu
[3] John von Neumann Faculty of Informatics, Óbuda University, Budapest, Hungary
[4] Antal Bejczy Center of Intelligent Robotics, Óbuda University, Budapest, Hungary

Abstract. In control applications asymptotic convergence of the tracking error to zero can be formulated in various manners. The traditional PD or PID controllers use the error or its time-integral that normally produce huge overshoots and undershoots in the initial session of the controlled motion. It is well known that the use of special fractional order (FO) integrals and derivatives can improve this situation. The main point of the FO-based dynamic system models consists in their long term memory effects. This formalism mathematically can be very complicated because it is burdened by trying to maintain formal relationships between the various concepts of integrals and derivatives. Also, it often works with integrands that are singular at the upper boundary of integration. In digital controllers continuous signals are replaced by numerical sequences. In this paper the use of simple sequences characterized by "forgetting speed", "memory length", and "dominance of the latest value" qualitative properties is suggested. They reflect similar properties that are generally present in FO calculus, however, they are completely liberated from its mathematical burdens. The suggested method is illustrated via simulations in which a 3 degree of freedom robot arm is controlled by a Fixed Point Iteration-based adaptive controller. In this adaptive approach avoidance of the initial overshoots is very important for its realization by simple embedded systems.

Keywords: PID control · Fractional order calculus · Adaptive control · Lyapunov's 2nd method · Fixed point iteration · Banach space

1 Introduction

Model-based controllers frequently apply the classic "Proportional, Integral, Derivative (PID)" error feedback scheme (e.g., [3]). In the possession of precise

Supported by the Doctoral School of Applied Informatics and Applied Mathematics of Óbuda University.

A. Müller and M. Brandstötter (Eds.): RAAD 2022, MMS 120, pp. 218–225, 2022.
https://doi.org/10.1007/978-3-031-04870-8_26

dynamic model of the given robot these LTI-based considerations can be applied in the "Computed Torque Control (CTC)" [1]. If the available dynamic model is imprecise then modeling errors degrade the quality of the control. To improve precision, adaptive approaches using the PID feedback terms and Lyapunov's stability theorem were developed (e.g., [15]) that can guarantee monotonic decrease or asymptotic convergence to 0 of a quadratic error norm the individual components of which can behave in non-monotonic manner. In [17] an alternative adaptive approach was introduced that tried to control these individual components on the basis of Banach's "Fixed Point Theorem" [2]. By kinematically determining the desired actual 2nd time-derivatives of the generalized coordinates that can be immediately set by the control forces, as in the case of the CTC controllers, these derivatives were "adaptively deformed" before being introduced in the model of the CTC scheme. The deformation in a given control step was determined by the use the actually desired value, and the deformed and observed values of the previous control cycle. "Mathematically", finding the control signal was formulated as finding the fixed point of a contractive map $\mathcal{B} \mapsto \mathcal{B}$ in a Banach space so that during one digital control step only one step of the adaptive iteration could be done. Since the appropriate fixed point drifts much slower than the convergence speed of the fixed point iteration, by remaining in the close vicinity of the actual fixed point, the kinematically formulated tracking error damping can be well approximated. For transforming the problem into a fixed point task various "Fixed Point Transformation Functions" can be applied, in this paper that of [7] is used. In this manner the PD or PID-type kinematic designs can be precisely realized but they also suffer from the overshooting/undershooting issues that originates from using integer order calculus in their design. The kinematic tracking requirement is formulated as follows: let be $q^N(t) \in \mathbb{R}^n$ the *nominal trajectory to be tracked*, and let $q(t) \in \mathbb{R}^n$ denote the motion realized by the controlled system. Let $e(t) := q^N(t) - q(t)$ be the tracking error, and $\mathbb{R} \ni \Lambda > 0$ be a *positive constant*, and let (1)

$$
\begin{aligned}
&e_{int}(t) := \int_{t_0}^{t} e(\xi)\mathrm{d}\xi \left(\Lambda + \frac{\mathrm{d}}{\mathrm{d}t}\right)^3 e_{int}(t) \equiv 0 \text{ resulting in} \\
&\ddot{q}^{Des}(t) = \ddot{q}^N(t) + \Lambda^3 e_{int}(t) + 3\Lambda^2 e(t) + 3\Lambda \dot{e}(t),
\end{aligned}
\tag{1}
$$

be used for a robot as a 2nd order dynamical system, in which the controller starts it operation at time instant t_0. The PID feedback gains can be identified in (1). (By using the Lyapunov equation these gains can be set independently, too.) If precisely realized, this solution, following an overshoot and a subsequent undershoot, can drive the tracking error to 0 asymptotically. The second integral of the tracking error

$$
e_{int2}(t) := \int_{t_0}^{t} e_{int}(\xi)\mathrm{d}\xi \text{ with } \left(\Lambda + \frac{\mathrm{d}}{\mathrm{d}t}\right)^4 e_{int2}(t) \equiv 0 \tag{2}
$$

can also yield a control solution producing more overshoots and undershoots, etc. A possible way for introducing fractional order integrals is the generalization of the *"Riemann-Liouville n-fold Integral"* based on the observation that the sequence of multiple integrals can be expressed by a single formula in (3) [14]:

$$F_{n+1}(t) = \int_{t_0}^{t} F_n(\xi)\mathrm{d}\xi, \ \ F_n(t) = \frac{1}{(n-1)!}\int_{t_0}^{t} f(\xi)(t-\xi)^{n-1}\mathrm{d}\xi. \tag{3}$$

Let $\alpha > 0$, $f \in \mathcal{C}([a,b])$, and $x \in (a,b)$. The *Riemann-Liouville Fractional Integral and Derivative of order* α are defined as the generalization of the *Riemann-Liouville n-fold integral* for $f \in \mathcal{C}^1([a,b])$ as follows:

$$\begin{aligned} &{}_aI^{\alpha} f(t) \overset{def}{=} \frac{1}{\Gamma(\alpha)}\int_{a}^{t} f(\xi)(t-\xi)^{\alpha-1}\mathrm{d}\xi, \quad (s-1)! \equiv \Gamma(s), \\ &{}_aD^{\alpha} f(t) \overset{def}{=} \frac{1}{\Gamma(1-\alpha)}\frac{\mathrm{d}}{\mathrm{d}t}\int_{a}^{t} f(\xi)(t-\xi)^{-\alpha}\mathrm{d}\xi, \quad \Gamma(s) \overset{def}{=} \int_{0}^{\infty} \xi^{s-1}e^{-\xi}\mathrm{d}\xi \end{aligned} \tag{4}$$

in which Euler's Γ function generally can be computed for a complex variable $s \in \mathbb{C}$. Since the integrand in (4) is singular at $\xi = t$, the calculation of this derivative is not "easy". Historically various definitions of fractional order operations were developed, the most popular ones are Liouville's definition [13], the Grünwald-Letnikov form [12], and the Caputo model [6]. (Detailed historical overview of the concept can be found in [8,18].) For tackling the problem of the initial phase of the controlled motion in process control and robotics the use of successful combinations of various FO integral and feedback terms were suggested (e.g., [4,5,9,11]). To evade the differentiation of the integral with singular integrand in (4) in [16] the following idea was applied: in the PID feedback term the first derivative one was considered as the integral of the second derivative, and it was replaced by its modification in (5), in which the integral of the time-derivative eliminated this problem.

$$\ddot{q}^{Des} = \ddot{q}^{N}(t) + \Lambda_1^3 e_{int}(t) + 3\Lambda_1^2 e(t) + 3\Lambda_2 \int_{t_0}^{t} \left[{}_{t_0}D^{\alpha}\dot{e}(\xi)\right]\mathrm{d}\xi. \tag{5}$$

The integral in the last term was numerically approximated by allowing *finite memory length* $T = K\delta t$, and a grid of points $\{t - K\delta t, \ldots, t - \delta t, t\}$. It was assumed that $\dot{e}(t)$ did not vary considerably between the grid points. The feedback obtained was a sum of type

$$\sum_{\ell=0}^{\ell_{max}} \dot{e}(t_i - \ell\delta t)C_\ell\,,\ C_0 > C_1 > \ldots > C_{\ell_{max}} > 0, \tag{6}$$

in which the C_ℓ coefficients were yielded by the integrals in closed form. The *"speed of forgetting the past"* was determined by α, the *"memory length"* T became independent of the resolution of the time-grid δt, and the singularity of the integrand at $\xi = t$ in (4) manifested itself in the fact that $C_0 \gg C_1$. The present idea is the application of forgetting weighted sum over the grid points as feedback in the kinematic design that is liberated from any deeper connection/relationship with the mathematical restrictions prevailing in the realm of fractional order calculus.

2 Simple Weighted Sum as Kinematic Error Feedback

Consider the discrete time approximation of the *monotonic exponential decay* $\dot{q}(t) = -\Lambda q(t)$ as $q(t_{i+1}) = q(t_i)\,(1 - \delta t\Lambda)$ that is exempt of any memory effect. Add to it long memory effects by the modification as follows:

$$q(t_{i+1}) = q(t_i)\,(1 - \delta t\Lambda C_0) - \sum_{\ell=1}^{H-1} \delta t\Lambda C_\ell q(t_{i-\ell}) = a_0 q(t_i) - \sum_{\ell=1}^{H-1} a_\ell q(t_{i-\ell}) \quad (7)$$

by the use of the $\{C_\ell\}$ coefficients also having the properties as defined in Sect. 1. The *discrete memory length* evidently is $H-1$. In the above sequence $1 \approx a_0 < 1$, and for $\ell \in \{1, \dots, H-1\}$ $0 \approx a_\ell > 0$ for small δt and reasonable $\Lambda > 0$. The behavior of the so obtained sequence can be described in a matrix form, e.g., for a finite size with $a_{11} = a_0 = 1 - \delta t\Lambda C_0$, $a_{12} = -a_1 = -\delta t\Lambda C_1$, etc., as

$$\begin{bmatrix} q(t_{i+1}) \\ q(t_i) \\ q(t_{i-1}) \\ q(t_{i-2}) \end{bmatrix} = \left[\begin{array}{ccc|c} a_{11} & a_{12} & a_{13} & a_{14} \\ 1 & 0 & 0 & 0 \\ 0 & 1 & 0 & 0 \\ \hline 0 & 0 & 1 & 0 \end{array}\right] \begin{bmatrix} q(t_i) \\ q(t_{i-1}) \\ q(t_{i-2}) \\ q(t_{i-3}) \end{bmatrix}. \quad (8)$$

Its properties can be investigated by bringing it into the Jordan canonical form [10] via similarity transformation. In its Jordan blocks of the form $\lambda I + \Delta$ $\lambda \in \mathbb{C}$ is an eigenvalue of the matrix in (8), and Δ is a *nilpotent upper triangular matrix* of size $m \times m$ with the property of $\Delta^m = 0$. From this structure it immediately follows that

$$(\lambda I + \Delta)^n = \sum_{\ell=0}^{m-1} \frac{n!}{\ell!(n-\ell)!} \lambda^{n-\ell} \Delta^\ell \to 0 \text{ as } n \to \infty \text{ if } |\lambda| < 1\,. \quad (9)$$

The eigenvalues of this matrix are the roots of the characteristic polynomial. E.g., for a matrix of size 4×4 it is given in (10)

$$p_{(4)}(\lambda) = \left|\begin{array}{ccc|c} \lambda - a_{11} & -a_{12} & -a_{13} & -a_{14} \\ -1 & \lambda & 0 & 0 \\ 0 & -1 & \lambda & 0 \\ \hline 0 & 0 & -1 & \lambda \end{array}\right| = (\lambda - a_{11})\lambda^3 - a_{12}\lambda^2 - a_{13}\lambda - a_{14}. \quad (10)$$

By computing the determinant of the size 4×4 matrix according to its last column it is obtained that $p_{(4)}(\lambda) = \lambda p_{(3)}(\lambda) - a_{14}$ and this equation can be continued for higher matrix sizes as e.g., $p_{(5)}(\lambda) = \lambda p_{(4)}(\lambda) - a_{15}$, etc. For finding the roots of this polynomial it can be observed that in the $\delta t \to 0$ limit $a_{11} \to 1$, and for $k > 1$ $a_{1k} \to 0$. In this limit $p_{(5)}(\lambda) = (\lambda - 1)\lambda^4$ has the roots $\lambda_1 = 1$, and with greater multiplicity $\lambda_2 = 0$. For a small but non-zero δt this polynomial varies continuously and its roots remain in the vicinity of 1 and 0. From the point of view of convergence the modification of the root near 1 is important: its absolute value must be smaller than 1. For the estimation of its modification the form $\lambda_1 : 1 \to 1 + \mu\delta t$ can be considered in the first order of δt while the other terms are modified as $a_{11} : 1 \to 1 - \Lambda C_0 \delta t$, and $a_{12} : 0 \to -\Lambda C_1 \delta t$, etc. Consequently $(1-a_{11})$ varies as $0 :\to 1+\mu\delta t-(1-\Lambda C_0\delta t) = (\mu+\Lambda C_0)\delta t$ that has first order in δt. Therefore only the order 0 terms must be taken into account in λ^3, in λ^2, and in λ^1 as 1. For instance, for the equation $p_{(4)}(1 + \mu\delta t) = 0$ the approximation $\mu + \Lambda C_0 + \Lambda C_1 + \Lambda C_2 + \Lambda C_3 = 0$ is obtained leading to $\lambda_1 : 1 \to 1 - \delta t(\Lambda C_0 + \Lambda C_1 + \Lambda C_2 + \Lambda C_3) = 1 - \delta t \Lambda < 1$ that guarantees convergence for small δt and reasonable $\Lambda > 0$ values. In the sequel simulation examples are given for the control of a 3 DoF robot arm that is adaptively controlled by using a fixed point transformation made of abstract rotations.

3 Simulation Results and Conclusions

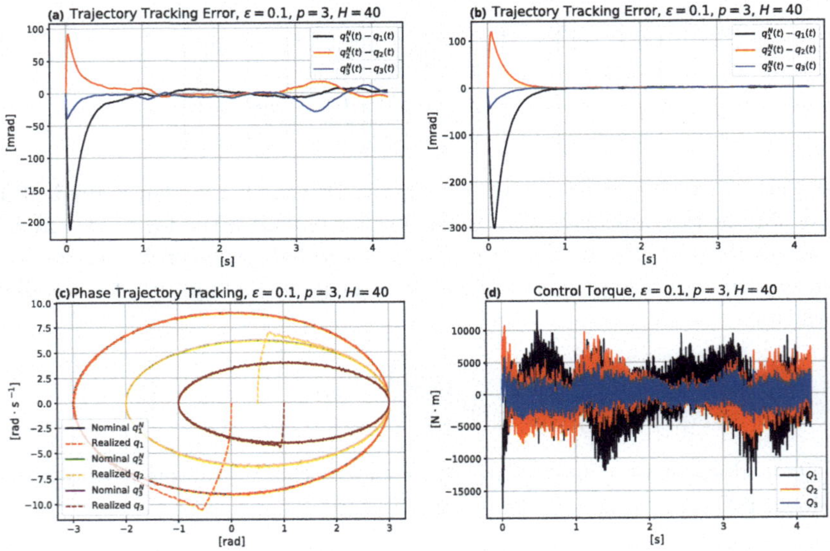

Fig. 1. The non-adaptive **(a)** and the adaptive **(b)** trajectory tracking errors for noisy joint coordinate measurements. The phase trajectory tracking **(c)** and control forces **(d)** in the adaptive and noisy case.

In the lack of enough room the dynamic model of the first 3 joints of a PUMA-type robot are not detailed here. In time instant t_i the controller computed the desired tracking error value for the next one as $e^{Des}(t_{i+1})$ resulting in

$$\begin{aligned} q^{Des}(t_{i+1}) &= q^N(t_{i+1}) - e^{Des}(t_{i+1}), \\ \ddot{q}(t_{i-1})^{Des} &\approx (q^{Des}(t_{i+1}) - 2q(t_i) + q(t_{i-1}))/\delta t^2, \end{aligned} \tag{11}$$

if the forward differences are used for differentiation and Euler integration. The *approximate dynamic model* of the controlled system yields the control force to be exerted in t_{i-1} as $Q(t_{i-1}) = F(q(t_{i-1}), \dot{q}(t_{i-1}), \ddot{q}^{Des}(t_{i-1}))$. If it is exerted on the system at time instant t_{i-1}, the realized 2nd time-derivative at that time will be determined by the exact model and it will determine the realized coordinate $q(t_{i+1})$ from the approximation $\ddot{q}(t_{i-1}) \approx (q(t_{i+1}) - 2q(t_i) + q(t_{i-1}))/\delta t^2$. The adaptively deformed second order derivative at t_{i-1} can be computed from the newest desired value $\ddot{q}^{Des}(t_{i-1})$, and from the observed and deformed previous values, i.e., $\ddot{q}(t_{i-2})$, and $\ddot{q}^{Def}(t_{i-2})$. In the calculations the time resolution was $\delta t = 10^{-3}\,\mathrm{s}$, the discrete memory length parameter was $H = 40$, $C_\ell \propto (\delta t(\varepsilon + \ell))^{-p}$ with $p = 3$ "forgetting speed", $\varepsilon = 0.1$ "dominance parameter for C_0", and $\Lambda = 6.0\,\mathrm{s}^{-1}$ "tracking speed parameter". In Fig. 1 results are given for an assumed Gaussian noise of zero mean and $\sigma = 10^{-4}\,\mathrm{rad}$ "noise amplitude" in measuring q_1, q_2, and q_3.

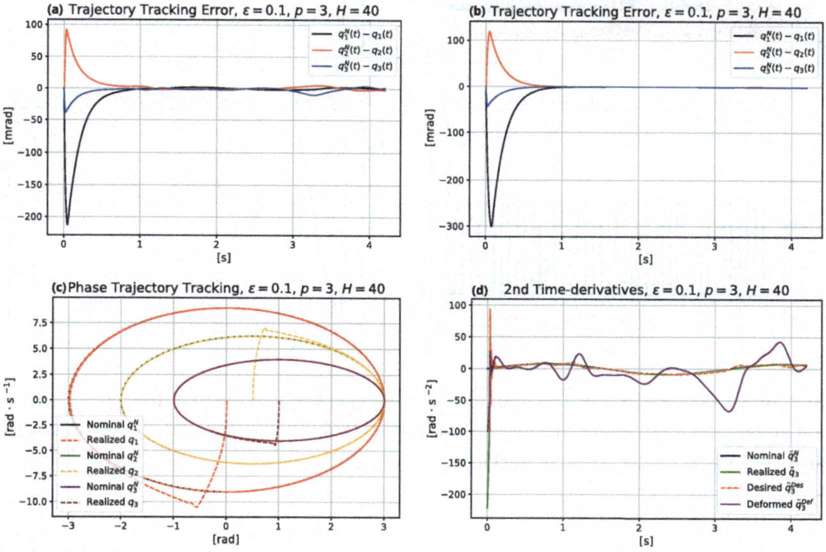

Fig. 2. Trajectory tracking error: **(a)**: non-adaptive, **(b)**: adaptive. Phase trajectory tracking of the adaptive controller **(c)**, and the appropriate $\ddot{q}_3$ values without measurement noises **(d)**.

Figure 2 displays the same results as that given in 1 without measurement noises in which graph **(c)** well reveals the operation of adaptivity. For instance the "deformed" $\ddot{q}_3^{Def}(t)$ signal is drastically different to the "desired" $\ddot{q}_3^{Des}(t)$. In this manner the controller guaranteed the good approximation $\ddot{q}_3^{Des}(t) \approx \ddot{q}_3(t)$, consequently only minimal PID error corrections were fed back that resulted in the approximation $\ddot{q}_3^{Des}(t) \approx \ddot{q}_3(t) \approx \ddot{q}^N(t)$.

In Fig. 3 the effect of the memory length on the computational time of the kinematic requirements and that of the adaptive rotation in the case of a simple `DELL inspiron 15R` laptop operating under `Manjaro Linux` are given when practically the operating system and the text editor (this latter in passive use) were running besides the Julia. Since it is a garbage collecting language, the figures yield rather qualitative information and reveal the dominance of the kinematic block's needs over that of the adaptive rotations. It can be seen well that a long horizon and slow forgetting speed makes the signal tracking more "clumsy", however, the convergence is guaranteed in this latter case, too.

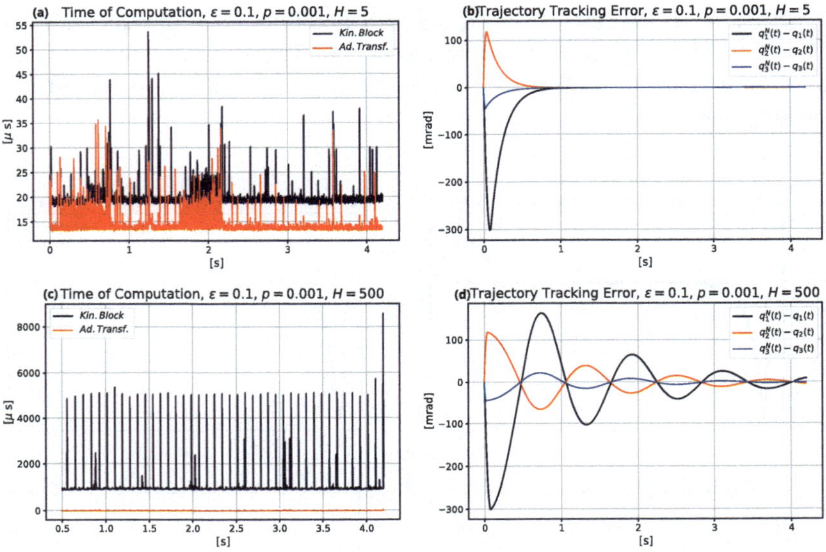

Fig. 3. Computational time **(a)** and trajectory tracking error **(b)** for $H = 5$. Computational time **(c)** and trajectory tracking error **(d)** for $H = 500$ of the adaptive control for noise-free signal and slow forgetting $p = 10^{-3}$.

It can be concluded that the suggested simple method is appropriate to making monotonic tracking error damping design in fixed iteration-based adaptive control under noisy environment. Though it was inspired by the concept of FO calculus, it is exempt of its various complicated restrictions. In the future it seems to be expedient to make experimental investigations for sequences of various qualitative and quantitative features.

References

1. Armstrong, B., Khatib, O., Burdick, J.: The explicit dynamic model and internal parameters of the PUMA 560 arm. In: Proceedings of IEEE Conference on Robotics and Automation 1986, pp. 510–518 (1986)
2. Banach, S.: Sur les opérations dans les ensembles abstraits et leur application aux équations intégrales (About the operations in the abstract sets and their application to integral equations). Fund. Math. **3**, 133–181 (1922)

3. Bennett, S.: Nicholas Minorsky and the automatic steering of ships. IEEE Control Syst. Mag. **4**(4), 10–15 (1984)
4. Bruzzone, L., Belotti, V., Fanghella, P.: Implementation of a fractional-order control for robotic applications. In: Ferraresi, C., Quaglia, G. (Eds.) Advances in Service and Industrial Robotics - Proceedings of the 26th International Conference on Robotics in Alpe-Adria-Danube Region (RAAD 2017), 21–23 June 2017, Torino, Italy, pp. 160–167 (2017)
5. Bruzzone, L., Fanghella, P.: Comparison of $PDD^{1/2}$ and PD^{μ} position controls of a second order linear system. In: Proceedings of the 33rd IASTED International Conference on Modelling, Identification and Control (MIC 2014), pp. 182–188 (2014)
6. Caputo, M.: Linear model of dissipation whose Q is almost frequency independent ii. Geophys. J. Roy. Astronom. Soc. **13**, 529–539 (1967)
7. Csanádi, B., Galambos, P., Tar, J., Györök, G., Serester, A.: A novel, abstract rotation-based fixed point transformation in adaptive control. In: Proceedings of the 2018 IEEE International Conference on Systems, Man, and Cybernetics (SMC 2018), 7–10 October 2018, Miyazaki, Japan, pp. 2577–2582 (2018)
8. de Oliveira, E.C., Tenreiro Machado, J.A.: A review of definitions for fractional derivatives and integral. Math. Probl. Eng. **2014**, 6 (2014). https://doi.org/10.1155/2014/238459. (Article ID 238459)
9. Deutschmann, B., Ott, C., Monje, C., Balaguer, C.: Robust motion control of a soft robotic system using fractional order control. In: Ferraresi, C., Quaglia, G. (Eds.) Advances in Service and Industrial Robotics - Proceedings of the 26th International Conference on Robotics in Alpe-Adria-Danube Region (RAAD 2017), 21-23 June 2017, Torino, Italy, pp. 142–149 (2017)
10. Dieudonné, J.: Oeuvres de Camille Jordan I-IV. Gauthier-Villars, Paris (1961–1964)
11. Dumlu, A., Erenturk, K.: Trajectory tracking control for a 3-DOF parallel manipulator using fractional-order $PI^{\lambda}D^{\mu}$ control. IEEE Trans. Indust. Electron. **61**(7), 3417–3426 (2014)
12. Grünwald, A.: Über "begrenzte" Derivationen und deren Anwendung (About "limited" derivations and their application). Zeitshrift für angewandte Mathematik und Physik **12**, 41–48 (1867)
13. Liouville, J.: Mémoire sur le calcul des différentielles a indices quelconcues (Dissertation on the calculus of differentials with any indices). J. Ecole Polytechn. **3**, 71–162 (1832)
14. Munkhammar, J.: Riemann-Liouville Fractional Derivatives and the Taylor-Riemann Series. Uppsala University, Sweden, Department of Mathematics (2004)
15. Slotine, J.J.E., Li, W.: Applied Nonlinear Control. Prentice Hall International Inc., Englewood Cliffs, New Jersey (1991)
16. Tar, J., Bitó, J., Kovács, L., Faitli, T.: Fractional order PID-type feedback in Fixed Point Transformation-based adaptive control of the FitzHugh-Nagumo neuron model with time-delay. In: 3rd IFAC Conference on Advances in Proportional-Integral-Derivative Control, 9-11 May 2018, Ghent, Belgium, pp. 906–911. IFAC (2018)
17. Tar, J., Bitó, J., Nádai, L., Tenreiro Machado, J.: Robust fixed point transformations in adaptive control using local basin of attraction. Acta Polytechn. Hungarica **6**(1), 21–37 (2009)
18. Tenreiro Machado, J., Kiryakova, V.: The chronicles of fractional calculus. Fract. Calc. Appl. Anal. **20**(2), 307–336 (2017). https://doi.org/10.1515/fca-2017-0017

Robot Design and Synthesis

Anatomy Categorization of a Serial Metamorphic Manipulator for Optimized Robust Controller Performance

Nikolaos Stravopodis(✉) and Vassilis C. Moulianitis

University of the Aegean, 84100 Ermoupolis, Syros, Greece
n.stravopodis@syros.aegean.gr, moulianitis@aegean.gr
https://www.syros.aegean.gr

Abstract. In this work the effect of system reconfiguration on the total dynamic model uncertainty of a serial metamorphic manipulator (SMM) is studied. An index related to the mass distribution characteristics of the anatomies of a SMM is defined, and given the bounded index score, groups of anatomies are identified. The ability of the proposed index to realize anatomies with lower overall system uncertainty is investigated and the tracking performance, of the emerging anatomy groups, under a centralized robust controller and a classical centralized joint control method is evaluated. The utilization of the proposed index for a rapid, off-line dynamic performance evaluation procedure and the determination of anatomies with adequate tracking performance under the same controller parameters for the specified tasks is discussed.

Keywords: Metamorphic manipulator · Dynamic model uncertainty

1 Introduction

Modular reconfigurable robots (MRR) overcome the operational limitations of fixed structure manipulators, by achieving a variety of system assemblies through the standardized connectivity interfaces of the modules used. SMM, as a class of MRR, are able to modify their own shape by altering the configuration of special connecting elements (passive modules) of their structure and by offline rearranging their structural elements, depending on the defined connectivity properties of the structure [1]. In this class of robots, the topology changes induced by the restructuring and reconfiguration process lead to a robotic system with enhanced properties [2,3].

Developing a generalized control scheme for MRR is a challenging task since reconfiguration leads to a great variation in system's dynamics, high parametric uncertainty etc. Centralized robust controllers have attracted many researchers through the years and numerous notable scientific studies have been presented. In case of MRR, special consideration is given in the automatic generation of the dynamic model from the module data. In [4,5] homogeneous structures are considered, in [6] a unified approach for modular robot description with arbitrary

A. Müller and M. Brandstötter (Eds.): RAAD 2022, MMS 120, pp. 229–238, 2022.
https://doi.org/10.1007/978-3-031-04870-8_27

module data is presented and in [7] the adaptive generation of the dynamic model and controller for a reconfigurable manipulator is addressed. In [8] an automatic centralized controller design for MRR is proposed. Heterogeneous modules are considered and information is distributed to the modules, which eliminates the requirement of a large database and successive updates on the central control unit. In [9] a robust controller for a MRR that can achieve multiple configurations and needs frequent conversion is presented. The controller scheme demands the uncertainty bounds for the dynamic model parameters of the MRR and can be easily integrated with a PID controller. Numerical analysis results of the proposed controller present a significant performance improvement for all configurations.

In this work, similar to [9], SMM structures that contain multiple anatomies are studied. The main motivation of this work is to define a procedure for the rapid performance evaluation of a SMM, under the user-defined uncertainty bounds of passive-module and load parameters. Parametric uncertainty bounds are introduced for the type of passive module used to structure the SMM and the effect of system reconfiguration on the model mismatch is studied for a given task. An index, applicable to modular serial robots is proposed, in order to effectively group the anatomies of a SMM based on their inertial characteristics. Since a SMM has the ability to perform system reconfiguration, without the need of user-interference for dis/reassembling the modular structure components, utilizing an anatomy-specific index that could quantify its dynamic performance capability, without executing off-line simulations, would further enhance the ability of reducing the overall time cost. It will be shown that, the extracted groups of anatomies, under the same robust controller gains and tuning parameters, present adequate performance with better position tracking capability than a simple PD controller. Consequently, this work provides tools for the robot designer in order to exploit the overall versatility of the SMM and determine groups of anatomies that can execute tasks with optimized performance.

2 Methodology

First, the uncertain parameters of the dynamic model of a SMM are presented and the uncertainty bounds are defined. The user-defined bounds are selected based on the CAD analysis simplifications of the passive modules (lack of electronic components, cables, uniform material properties for motor and the gearbox etc.). The total mass, the inertia tensor elements and the center of mass coordinates of the passive modules are defined as uncertain dynamic parameters. Additional parameter bounds for the uncertainty induced by the unknown load mass are considered. In this work two types of 3DoF SMM structures which contain the minimum (two) and the maximum (four) number of passive modules and a simple task is selected as case study. The Resolved Motion Rate Control (RMRC) trajectory generation algorithm is used to extract the joint angles/velocities at the desired waypoints. At each way-point the interval inertia matrix (IIM) is calculated. The elements of the IIM are the infimum and supremum values calculated by the analytic dynamic model for the user-defined uncertainty vector. The inertia uncertainty of the SMM is defined as the Frobenius norm of the inertia model mismatch computed for a given joint configuration. The total uncertainty of the inertia matrix along the extracted path is defined as the sum of the norms at each way-point of the task. The proposed Mass Balancing Score (MBS) index is calculated for the feasible anatomies of the SMM under study. This index value is anatomy dependent, unit independent, bounded and task-independent, which significantly reduces the computation cost, but introduces the risk of sub-optimal task execution optimization. The MBS index at each anatomy has the ability to indicate anatomies' uncertainty variation without exhaustive calculations at the given task is evaluated. The bounded lower and higher limits of the index are exploited to group the anatomies of the structure under study. In order to confirm the validity of the dynamic performance expectation for each anatomy group and the ability of the proposed index to provide a rapid performance evaluation of the SMM, the selected task is simulated using two different centralized motion controllers. The motion tracking performance of a centralized robust, based on interval arithmetic and inverse dynamics control (IA-ID) and a classical Proportional Derivative Feedforward (PDFF) controller is assessed for the highest/lowest scoring anatomies of the anatomy groups. The proposed method is simulated using the MathWorks MATLAB R2021a software.

3 Passive Modules' Parameter Uncertainty

The first 3 DoF SMM structure under study is shown in Fig. 1a. Both SMM structures present a heterogeneous modular structure, since two different types of modules are integrated. Active (A) modules are used for active operation and passive (P) modules are used for system reconfiguration. Since active modules are market-ready solution, no parameters related to them are considered uncertain for the dynamic model formulation. However, the second type of modules, are designed and custom-made as it is shown in Fig. 1b, inducing a lot of uncertainty. The module mass m_i, the coordinates of center of mass (CoM) $\underline{c_i} = \left[c_i^x \; c_i^y \; c_i^z\right]^T$ and the barycentric inertia tensor elements $\underline{I_i} = \left[I_i^{xx} \; I_i^{yy} \; I_i^{zz} \; I_i^{xy} \; I_i^{xz} \; I_i^{yz}\right]^T$ of the ith passive module, are considered as the uncertain parameters. The CAD provided data are assumed as nominal values for the uncertain module parameters. Without any loss of generality friction is not considered in this work. Finally the load attached to the end-effector is a box with known dimensions and unknown mass m_l. Hereafter, if a total of j passive modules are used for structuring the SMM, the real unknown dynamical parameters of the robot are considered to be collected in a vector as $\Delta = \left[m_1, \ldots m_j, \underline{c_1}, \ldots \underline{c_j}, \underline{I_1}, \ldots \underline{I_j}, m_l\right]^T$. In Table 1 the unknown dynamic parameters, along with the nominal values and the user-selected uncertainty bounds, are presented.

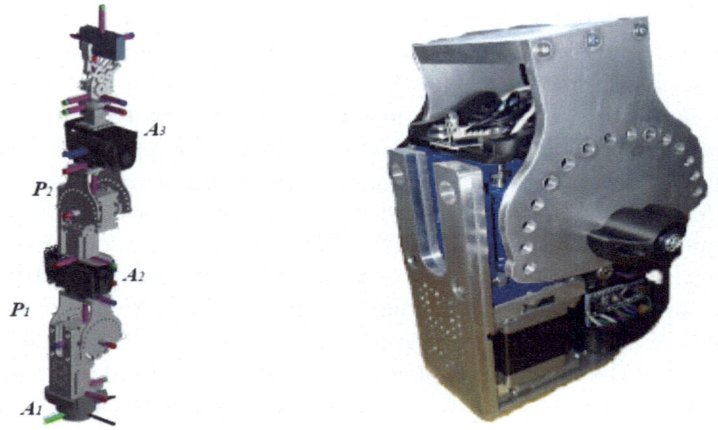

(a) SMM structure 1 (b) Pseudojoint prototype built

Fig. 1. Serial metamorphic structure and a passive module

Table 1. Uncertain dynamic parameters (m [kg], c_i [m]), I_i [kg $\cdot$ m^2].

Parameter (p_i)	m_i	c_i^x	c_i^y	c_i^z	m_l	–
Nominal (p_i^0)	2.65	0	−0.03	0.02	1	–
Uncertainty bound (u_i)	0.25	0.05	0.05	0.05	0.5	–
Parameter (p_i)	I_i^{xx}	I_i^{yy}	I_i^{zz}	I_i^{xy}	I_i^{xz}	I_i^{yz}
Nominal (p_i^0)	0.0186	0.0186	5e−04	3e−07	−3e−07	1e−07
Uncertainty bound (u_i)	1e−03	1e−03	1e−04	1e−05	1e−05	1e−05

4 Mass Balancing Scoring Index

Since a SMM contains a large amount of anatomies, an anatomy-specific measure is desired, that is posture independent, incorporates the structural modifications induced by anatomy metamorphosis and accounts for global attributes of the SMM. Here, a global intrinsic measure is proposed to quantify the mass balancing ability of each anatomy. For each anatomy a term $\mathbf{MBS_i}$ is calculated. This quantity is unit dependant, so it is divided by a structure-specific scaling factor to achieve unit independence. The scaling factor is constant for each SMM, since it stems from the product of the total mass of the SMM and the maximum sum of links' active length (distance between adjacent joint axes) that can be achieved by an anatomy of the SMM. The computed quantity is always positive but unbounded, so the exponential function is utilized to bound the final index value at $(0, 1]$. Therefore, when the final index $\mathbf{MBS_i^\star}$ is calculated higher values lead to "bad" scoring (closer to 0), while lower values lead closer to the ideal case of a mass balanced structure (closer to 1). A detailed view of the calculation process of the index is presented in Algorithm 1. In Fig. 2 the terms defined in Algorithm 1 are illustrated.

Algorithm 1. MBS index calculation algorithm for SMM

Given *structure string, assembly parameters, uncertainty bounds*
Implement *structure assembly*
Compute total mass M_a of SMM
for $i = 1 \colon n_{anat}$ **do**
    Implement *anatomy metamorphosis*
    Run collision detection algorithm for anatomy feasibility check
    if anatomy i is feasible **then**
        $n_{feas} \leftarrow n_{feas} + 1$
        Compute active ξ_{a_i} and passive ξ_{p_j} twists
        Compute total distance between active joint twists: $D_{a_i} \leftarrow \sum_{k=1}^{n_a - 1} D_k$
        for $c = n_a$: -1 : 1 **do**
            Compute aggregated mass: $m_{n_a,c} \leftarrow \sum_{i=c}^{n_a} m_i$
            Compute aggregated CoM: $M_{n_a,c}$
            Compute distance between: $M_{n_a,c}$ and next active twist : $dM_{n_a,c}$
            Compute the *mbs* term up to current link: $mbs_l \leftarrow M_{c,n_a} \cdot dM_{c,n_a}$
        end for
        Compute the aggregated term: $\mathbf{MBS_i} \leftarrow \sum_{k=1}^{n_a} mbs_l$
    else
        Discard anatomy i
    end if
end for
Extract the maximum active joint length D_a from all feasible anatomies
Compute the SMM scaling factor: $S \leftarrow M_a \cdot D_a$
for $i = 1 \colon n_{feas}$ **do**
    Compute the term: $r \leftarrow \mathbf{MBS_i}/S$
    Compute the final index value for anatomy i: $\mathbf{MBS_i^\star} \leftarrow exp(-r)$
end for

5 Task Simulations

The task considered in simulations is a linear path along the -z axis of a total distance of 20 cm and a constant end-effector velocity. The RMRC algorithm is utilized for generating the robot path of 10 intermediate waypoints, as presented in [3]. After introducing the interval vector $\Delta \in \mathbb{R}^{11}$ of the uncertain model parameters, and given the known uncertainty bounds the assumption that the uncertainty of the model parameters is bounded by the multidimensional interval $[\mathbf{\Delta}] : \Delta \in [\mathbf{\Delta}]$ is justified.

$$[\mathbf{\Delta}] := [\underline{\delta_i}, \overline{\delta_i}], \ \underline{\delta_i} = |p_i - u_i| \in \mathbb{R}, \ \overline{\delta_i} = |p_i + u_i| \in \mathbb{R} \tag{1}$$

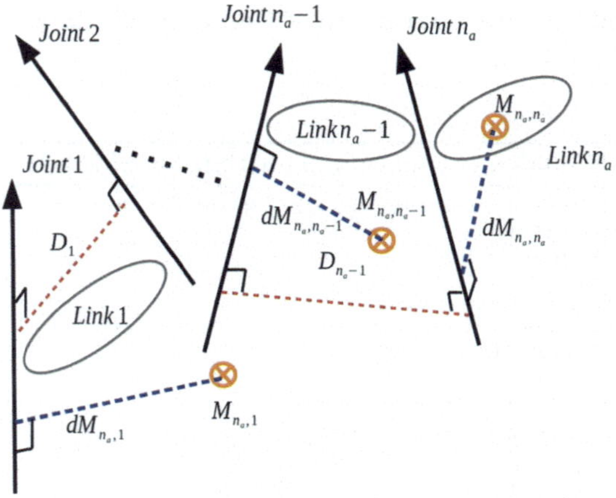

Fig. 2. Definition of MBS calculation terms.

The dynamic model of the system can be written as

$$\mathbf{M}(\mathbf{q}, \mathbf{q_p}, \mathbf{\Delta}) \cdot \ddot{\mathbf{q}} + \mathbf{C}(\mathbf{q}, \mathbf{q_p}, \dot{\mathbf{q}}, \mathbf{\Delta}) \cdot \dot{\mathbf{q}} + \mathbf{g}(\mathbf{q}, \mathbf{\Delta}) = \mathbf{u} + \mathbf{d} \tag{2}$$

where $\mathbf{q} \in \mathbb{R}^3$, $\mathbf{q_p} \in \mathbb{P}^j$ are the active and passive joint vectors respectively. $\mathbf{M} \in \mathbb{R}^{3\times 3}$ is the symmetric positive definite inertia matrix, $\mathbf{C} \cdot \dot{\mathbf{q}}$ is the vector containing the Coriolis and centrifugal terms, $\mathbf{g}$ is the gravity vector and $\mathbf{u}$, $\mathbf{d}$ are the actuator torque command and the disturbance vectors. In order to measure the model mismatch generated by the bounded uncertainty the interval inertia matrix is computed as the difference between the supremum inertia matrix calculated for the maximum $(\overline{\delta_i})$ and the infimum inertia matrix calculated for the minimum $(\underline{\delta_i})$ values of the uncertain parameters.

$$\widetilde{\mathbf{M}} = \mathbf{M_s} - \mathbf{M_i} \tag{3}$$

At each evaluated way-point the Frobenius norm of the interval inertia matrix

$$\|\widetilde{\mathbf{M}}\|_F = \sqrt{\sum_{i=1}^{3} \sigma_i^2}, \; \sigma_i : singular\ values\ of\ \widetilde{\mathbf{M}} \tag{4}$$

is calculated and the total uncertainty $\widehat{F}_{un}$ due to the model mismatch is extracted as the total sum of the computed norms along the task path. The inertia matrix of a SMM is analytically extracted given the structure assembly parameters and the current anatomy vector $\mathbf{q_p}$ [2]. The centralized IA-ID robust controller is derived in [10] and the PDFF controller is discussed in detail in [11].

5.1 Performance of the Proposed Index

First the exhaustive anatomy search comparison for both structures takes place. For the first (second) structure a step angle of 0.22 rad (0.45 rad) is considered

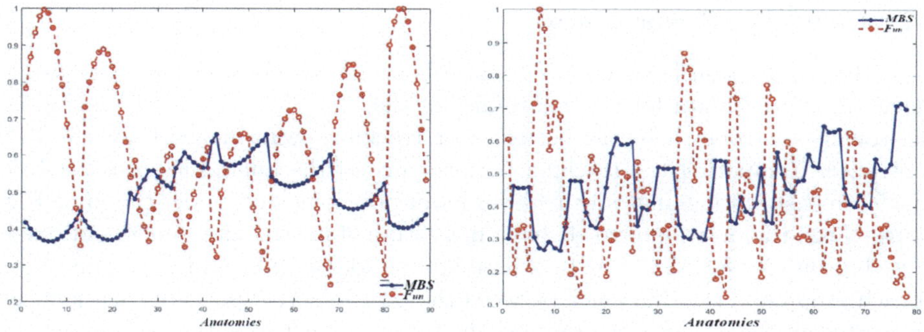

(a) SMM structure 1 (b) SMM structure 2

Fig. 3. MBS index and inertia matrix uncertainty

for anatomy metamorphosis and a total of 165 (2176) feasible anatomies arise and 89 (1031) of them are task-achievable. Next, the relation between the MBS index of each anatomy and the calculated uncertainty measure of the inertia matrix for task execution is studied. The results shown in Fig. 3 illustrate the effect of reconfiguration in the variation of the uncertainty of the dynamic model of the SMM for the given task. Under the same uncertainty bounds for module parameters, the total uncertainty of each anatomy highly differs. The proposed index can accurately track the variation of the total mass matrix uncertainty. Anatomies that achieve higher MBS tend to have a relative lower uncertainty norm. This index accuracy can be exploited in order to group the anatomies as it is presented in Fig. 4. From Fig. 4b it is evident that the second SMM which contains more passive modules, can achieve a greater number of best-scoring anatomies.

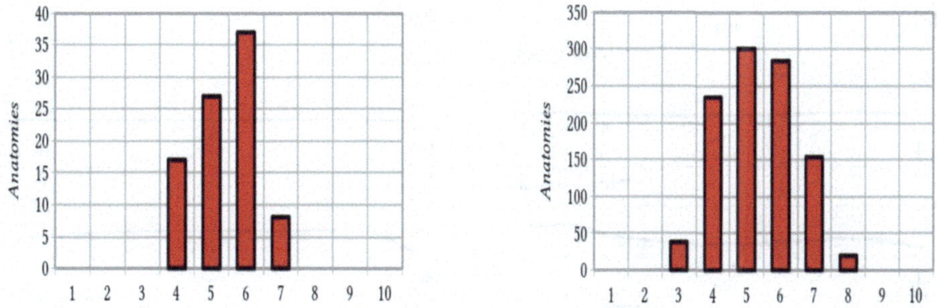

(a) SMM structure 1 (b) SMM structure 2

Fig. 4. Anatomy groups based on MBS index

5.2 Controller Performance

First the comparison between the IAID robust and the PDFF controller takes place. Simulation data for two anatomies of the first SMM are used. In Fig. 5a the robust controller achieves better error tracking performance than the PD controller. Comparing the tracking performance of the anatomies presented in Fig. 5 the best scoring anatomy achieves lower position error, under the same PD gains. Next, the position error behaviour is studied for the anatomies contained in each group for SMM 1. The task is simulated for the best and worst anatomy of each group and the difference between the position error norm of each joint is illustrated in Fig. 6. For both groups, the difference between the tracking error for the extreme anatomies of the group is below 0.1 rad, which is considered sufficient. Finally the actuator effort during task simulation is presented in Fig. 7,

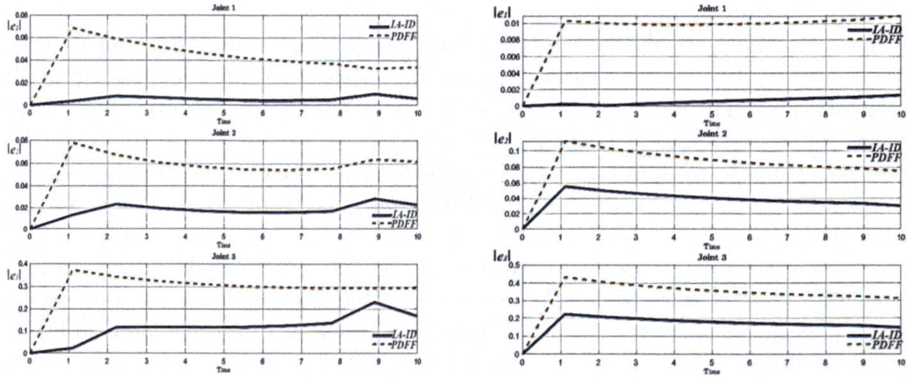

(a) Best MBS anatomy $[0, 1.5708]$ (b) Worst MBS anatomy $[1.5708, -0.2244]$

Fig. 5. Controller position error comparison

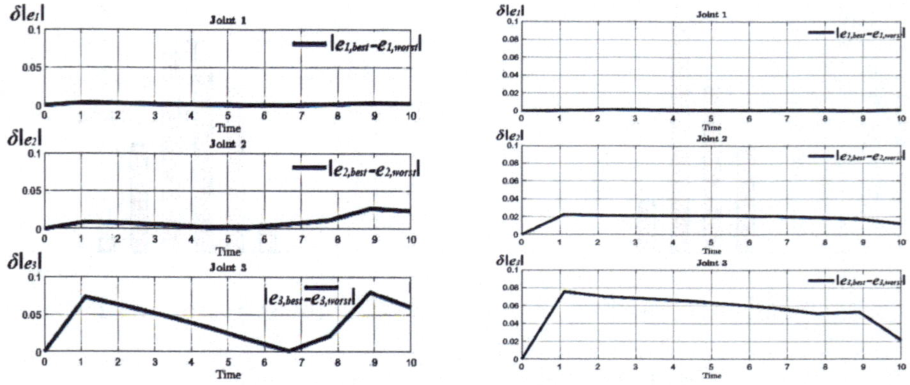

(a) Group 6 ($\mathbf{MBS} \in [0.5, 0.6)$) (b) Group 4 ($\mathbf{MBS} \in [0.3, 0.4)$)

Fig. 6. Position error difference for two of the defined anatomy groups of SMM 1

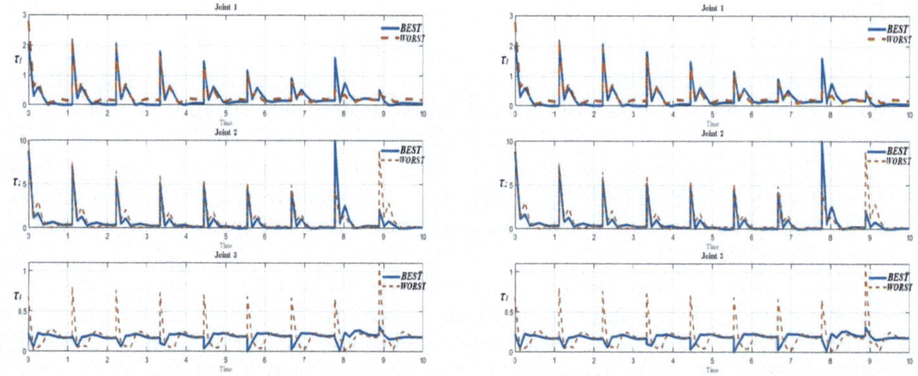

(a) Group 6 of SMM 1

(b) Group 5 of SMM 2

Fig. 7. Actuator effort (IA-ID controller) for different groups of SMM 1 & 2

for the extreme anatomies of a group. For both SMM structures, the anatomies of the group tested present similar torque effort for the first 2 joints. Although a maximum difference of approximately 0.8 Nm is observed for the last joint, this was expected since no pseudojoints are integrated in the last link and the index term calculated for this link remains constant for all anatomies of the SMM. Consequently the difference between the torque effort for the two anatomies only relates to the total rotational displacement.

6 Conclusion and Future Work

In this work, the effect of anatomy metamorphosis on the variation of the total dynamic model uncertainty of SMM structures was studied. The proposed index was found adequate for grouping the anatomies of the metamorphic structure with respect to their mass distribution characteristics. This anatomy categorization was evaluated by the error tracking performance of the grouped anatomies, extracted from simulation data for a simple task execution. Under the same controller gains the centralized robust controller achieved lower position error norms than the proportional derivative class controller. Moreover the joints' position error norm of extreme anatomies of the formed anatomy groups were compared and a relative small difference was observed. The proposed index can provide an adequate reference view for the expected performance of the anatomies of a SMM. Anatomy groups can be formed that lead to an optimized position tracking behaviour under a centralized robust controller. For future work, we consider the determination of optimal controller gains and tuning variables for each of the formed anatomy groups and the experimental verification of the extracted conclusions on the SMM prototype built.

Acknowledgment. This research has been financially supported by GSRT and HFRI (1184).

References

1. Valsamos, C., Moulianitis, V., Aspragathos, N.: Kinematic synthesis of structures for metamorphic serial manipulators. J. Mech. Robot. **6**(4), 041005 (2014)
2. Stravopodis, N.A., Katrantzis, L., Moulianitis, V.C., Valsamos, C., Aspragathos, N.A.: Evaluation of serial metamorphic manipulator structures considering inertia characteristics. In: Zeghloul, S., Laribi, M.A., Sandoval Arevalo, J.S. (eds.) RAAD 2020. MMS, vol. 84, pp. 574–587. Springer, Cham (2020). https://doi.org/10.1007/978-3-030-48989-2_61
3. Stravopodis, N.A., Moulianitis, V.C.: Rectilinear tasks optimization of a modular serial metamorphic manipulator. J. Mech. Robot. **13**(1), 011001 (2021)
4. Meister, E., Gutenkunst, A., Levi, P.: Dynamics and control of modular and self-reconfigurable robotic systems. Int. J. Adv. Intell. Syst. **6**(1), 66–78 (2013)
5. Wang, X., Zhang, M., Ge, W., Liu, J.: Dynamic modeling and configuration adaptive control for modular reconfigurable robot. Adv. Mech. Eng. **9**(10) (2017). https://doi.org/10.1177/1687814017726005
6. Nainer, C., Feder, M., Giusti, A.: Automatic generation of kinematics and dynamics model descriptions for modular reconfigurable robot manipulators. In: 2021 IEEE 17th ICASE, pp. 45–52. IEEE (2021)
7. Dogra, A., Padhee, S.S., Singla, E.: Towards dynamics and control of modular reconfigurable manipulators. In: Proceedings of the Advances in Robotics 2019, pp. 1–6 (2019)
8. Giusti, A., Althoff, M.: Automatic centralized controller design for modular and reconfigurable robot manipulators. In: 2015 IEEE/RSJ IROS, pp. 3268–3275. IEEE (2015)
9. Kirchoff, S., Melek, W.W.: A saturation-type robust controller for modular manipulators arms. Mechatronics **17**(4–5), 175–190 (2007)
10. Giusti, A., Liu, S.B., Althoff, M.: Interval-arithmetic-based robust control of fully actuated mechanical systems. IEEE Trans. Control Syst. Technol. 1–13 (2021)
11. Kelly, R., Davila, V.S., Perez, J.A.L.: Control of Robot Manipulators in Joint Space. Springer, London (2005). https://doi.org/10.1007/b135572

Design and Control of a Reclining Chair with Soft Pneumatic Cushions

Marco Bellusci[1], Carlo Ferraresi[1], and Giovanni Gerardo Muscolo[2(✉)]

[1] Department of Mechanical and Aerospace Engineering, Politecnico di Torino, Turin, Italy
[2] Department of Computer Science, University of Verona, Verona, Italy
giovannigerardo.muscolo@univr.it

Abstract. The object of this work is the control of a reclining chair with active cushions used to optimize the interaction of the human with the chair and to avoid the formation of sores on the skin. The cushion system, already presented in other works, consists of air-cell actuators distributed according to the risk that pressure ulcers on the skin can be generated in certain areas. In this work, four cushions are designed and controlled for four parts of the human body: 1) head; 2) back; 3) buttocks; 4) heels. Cushions are used to design a reclining chair that can be moved between two configuration limits: a) chair; b) bed. The four cushions can provide real-time pressure mapping with closed-loop control, which allows to identify critical points on the body surface where pressure ulcers could form. The control systems for the single air-cell and for all cushions are designed and simulated using the software MATLAB/Simulink, presenting very interesting results.

Keywords: Pressure injury · Automated seat cushion · Reclining chair · Bubble actuators · Soft actuators · Pneumatic actuators · Pressure ulcers

1 Introduction

Once called bedsores, Pressure Injuries (PIs) are localized damage to the skin and the underlying soft tissue. They are usually localized under bony prominences (like the tailbone and heels) [1]. The formation of PIs depends both on direct and indirect factors [2]. The most important cause of ulceration is a prolonged pressure between the bony prominence and the external surface that occluding the capillaries [3]. The formation of a PI is directly proportional to the pressure applied and the period for which it is applied: even a low pressure can generate a PI if applied for a long period [4]. Concomitant with pressure, shear is always present. Blood vessels are stretched and torn by this shearing, favoring the formation of PIs. Some other causes of PIs that must be considered are moisture, friction, the position of the patient, immobility, aging, diabetes, and neurological factors [3]. Pressure ulcers have been increasing in the last years due to an incremented average age of the population and this means that also the costs related to this problem are growing [5]. The National Pressure Injury Advisory Panel (NPIAP) classified the PIs in four major stages [1]. In particular, the average hospital treatment cost of stage 4 pressure ulcers is very high, around $129,124 for PIs acquired during one admission

A. Müller and M. Brandstötter (Eds.): RAAD 2022, MMS 120, pp. 239–246, 2022.
https://doi.org/10.1007/978-3-031-04870-8_28

at the hospital [6]. Different studies have shown how adopting a prevention protocol in long-term care is cost reducing with respect to treating PIs once they have developed [7, 8]. Different treatments for PIs are available in the literature. The first simple strategy is repositioning. Doctors suggest repositioning the patient every 2 h but there are no robust evidence on the best repositioning frequency and position [9]. Specific support surfaces like mattresses, beds, cushions, and overlays have been designed in order to reduce pressure magnitude and minimize shear. They can be classified as constant low-pressure (CLP) devices, which conform to body shape, or alternating pressure (AP) devices, where the pressure is mechanically varied [10]. CLP supports have been proven to be preferable with respect to standard foam hospital mattresses. Several alternatives have been studied like high-specification foam, bead-filled and water filled. However, no significant difference has been noticed between the different possible materials for a CLP support [10]. Using AP mattresses can delay 11 days the formation of a PI and have an 80% probability of being cost saving [11]. However, the AP algorithms consider only the body weight, neglecting other important parameters like body prominences, asymmetry, and bad posture [12]. Another drawback is the bottom out problem: the original shape and property of the air-filled compartments can be altered due to an excessive load or other external disturbances (aging, wear, weak maintenance) [13]. AP devices have a passive control mechanism where air-cells are repeatedly inflated or deflated regardless of the pressed regions and pressure level. To solve this problem, active support surfaces were introduced. These dynamic solutions are still characterized by air-filled compartments, but the air-cells are inflated and deflated based on the measured pressure above them [14, 15]. The reference pressure is compared with the measured one and the result is used to proportionally deflate or inflate the air-cells. It has been proven that this approach not only involves a redistribution of body weight but also better reperfusion and good comfort [16]. In [17], a first design of the air-cell actuators of one seat cushion has been presented and controlled. In this work, we optimized the control of the air-cell actuators and proposed a novel design for a reclined chair constituted of four cushions: Head, Back, Seat, Heel (see Fig. 1).

2 Design of the Automated Reclining Chair

Figure 1 shows the reclining chair with four cushions (head, back, seat, heel) connected through two joints and each of them can be controlled by an external input. The overall dimensions of the cushions are designed considering the sitting position of the standard man (age 30, weight 70 kg, height 1.72 m) [18]. The reclining chair has two rotary joints that can rotate of 90° allowing to assume different shapes between two limit configurations: a) chair, in which both the joints are at rest and the reclining chair has the same shape as shown in Fig. 1; b) bed, in which both joints are turned 90°, respect to Fig. 1, and the four cushions are aligned composing a sort of bed. The cushions composing the reclining chair have been designed considering the areas more at risk of developing the PIs when lying in a bed, in particular: back of head and ears, shoulder and back, elbows, lower back and buttocks, greater trochanter, heel [19]. An innovation introduced in the cushions of this work and presented in [17], respect to the proposed in literature [14–16], is that the air-cells are inflated or deflated in group making the system quicker to respond to the stimulus but also more accurate since these groups are correlated to the different areas, classified according to the risk of developing PIs. While sitting, the body weight mainly concentrates in the buttocks and the thigh regions, but it largely depends on the adopted sitting posture and on the type of chair [14].

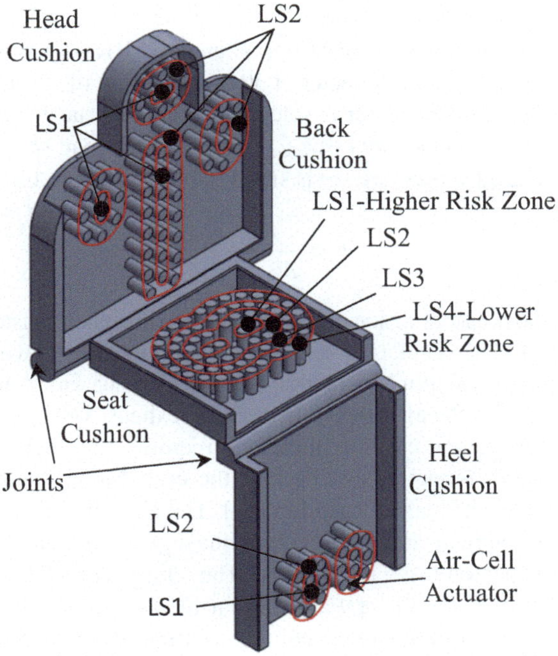

Fig. 1. 3D model of the reclining chair with four cushions: Head, Back, Seat, Heel.

Different studies using dynamic pressure mapping have proven that the highest-pressure values are below the Ischial Tuberosities (ITs) and the coccyx [16, 17]. Other studies have also confirmed that these regions are the more interested in shear and deformation effects [17, 18]. Starting from the buttock region more at risk, four Level Surfaces (LSs) have been designed, as shown in Fig. 1. The LSs are numbered starting from the internal one. The seat cushion, presented in [17], is composed of 66 air-cells distributed on the cushion in the following way: 4 cylinders inside the LS1; 14 cylinders inside the LS2; 22 cylinders inside the LS3; 26 cylinders inside the LS4 [17]. The pressure on the head, back and heel cushions goes from almost zero in the chair configuration to bigger values while the bed configuration is being reached. Two LSs have been introduced in the head, back and heel cushions. Each LS is constituted by a defined number of air-cells, which are activated and deactivated together. The two LSs are related to the higher and lower risk zones of pressure ulcers generation. The LS1 is delimited by the internal Level Curve (LC1) and the LS2 is delimited by the LC1 and the external Level Curve (LC2). In the head cushion, the air-cells of LS1 are two, while there are eight air-cells inside the LS2. In the back cushion, the air-cells of LS1 are eleven and they are disposed along the shoulder and the back, considering 500 mm the average distance between the arms. Thirty-two air-cells form the LS2. In the heel cushion, four air-cells form the LS1 and are disposed considering the average distance between the two legs equal to 150 mm. The LS2 instead contains sixteen air-cells.

3 Analytical Modelling of the Air-Cells

The air-cells have been optimized with a Finite Element (FE) analysis in [17]. They have a cylindrical shape with an outer diameter of 30 mm, a height of 65 mm and a thickness of 1.2 mm. Moreover, silicon rubber is selected as the material for the cylinders [20, 21]. If the change in the contact area ($A_{contact}$) between the load (F_{ext}) and the air-cell is neglectable, the interface pressure (p) is simply obtained by the following relation:

$$p = \frac{F_{ext}}{A_{contact}} \tag{1}$$

In our system, the pressure inside each level surface must be adjusted proportionally to a given command. The regulation is carried out by means of two two-port digital valves controlled in Pulse Width Modulation (PWM) (see Fig. 2a). A first valve (supply valve) connects the LS to the air supply, the second (exhaust valve) connects the LS to the ambiance. The entire seat cushion, therefore, requires only eight very inexpensive valves to be controlled, reducing significantly the cost. More details on the general control system of an air-cell can be found in [17]. The air-cells can be considered such as an elastic tank receiving an air flow $G(t)$ from the digital valves and an external force from the mass of the person using the cushion. The temperature T is assumed constant while the volume $V(t)$, the pressure $p(t)$, and the density ρ are free to change. Considering the transformation as polytrophic, the pneumatic continuity equation of the air-cell is

$$\frac{dp}{dt} = \frac{GnRT_i}{V}\left(\frac{p}{p_i}\right)^{\frac{1-n}{n}} - \frac{pn}{V}\frac{dV}{dt} \tag{2}$$

4 Control System Architecture

A closed-loop control system has been used to control the internal pressure of the air-cell. Therefore, the air-cell is the plant of our system while the two valves are the actuators (as shown in Fig. 2b). A manual switch allows to quickly change the pressure reference input from a step signal to a trapezoidal signal, or vice-versa. The tracking error e(t) enters the controller (red block), which generates the PWM commands to the digital valves. The computed airflows reach the plant (green block) composed of two blocks. In the air-cell pressure block the instantaneous pressure p(t) of the air-cell is calculated integrating Eq. 2). In the air-cell block, the volume V(t) of the air-cell, its derivative dV/dt, and the height h(t) are computed. In particular, the volume V(t) and the displacement z(t) were obtained through an interpolation of data measured with FE analysis [17].

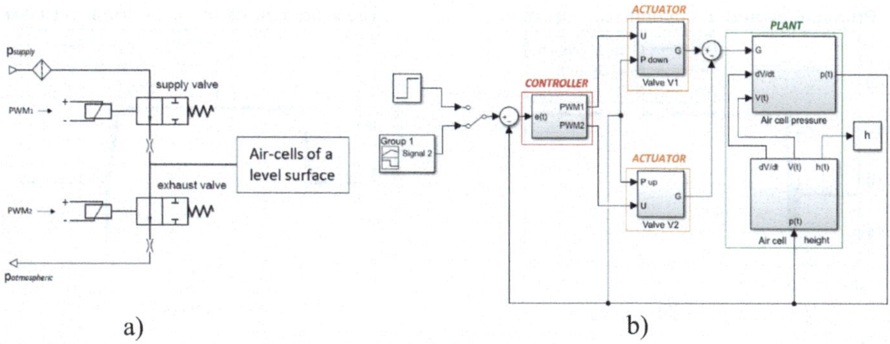

Fig. 2. Supply and exhaust valve connected to a level surface (a); Simulink feedback control system of the single air-cell (b).

Each cushion is studied considering a certain force distribution over it. In the seat cushion a 9N load on the LS1, 7N on the LS2, 5N on the LS3, and 3N on the LS4 are applied [17]. Each LS block is equal to the system of a single air-cell of Fig. 2b, but with different input data for volume, pressure and displacement, derived from the different FE analyses [17]. The total volume is set proportionally to the number of actuators inside the specific LS. Two different approaches have been studied. In the first one, the LSs are controlled in series; in the second one, the LSs are controlled in parallel. In the series configuration, the control of a LS starts only when the previous LS has reached the steady-state condition; in the parallel configuration the control of all the level surfaces starts simultaneously. This method is used to observe the inflation and the deflation of the four LSs in the seat cushion [17].

5 Results and Discussions

For the reference signal of the single air cell, two kinds of input have been used, a step signal and a trapezoidal signal. The results have been obtained implementing the control architecture of Fig. 2 and are shown in Fig. 3. The pressure p(t) tracks well the pressure set. From the step response, it can be noticed that the system is critically damped and has a rise time of 0.5 s. Figure 4 shows the simulation performed for the head, back and heel cushions. The simulations are performed using a supply pressure of 131.3 kPa for the LS1 and a supply pressure of 301.3 kPa for the LS2. Loads of 7N and 5N are applied on the air-cells of LS1 and LS2 respectively. The control system of the head cushion is faster than the other two because of the smaller number of air-cells to inflate/deflate but all the three cushions have a good ability in tracking the input. It is possible to note that in each cushion the control of LS1 is faster due to the smaller number of air-cells.

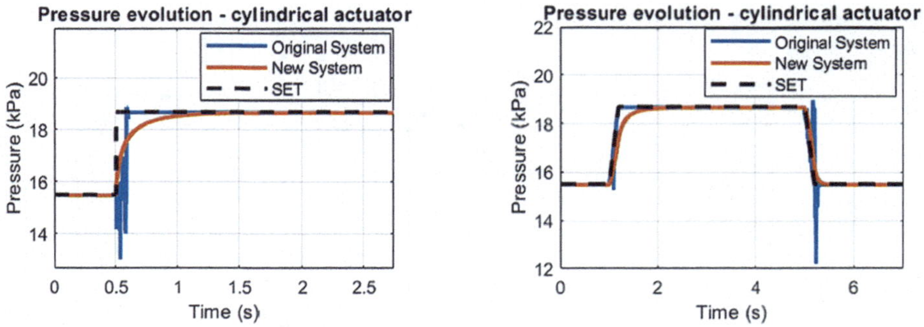

Fig. 3. System output with a step input (on the left) and a trapezoidal input (on the right) for the single air-cell.

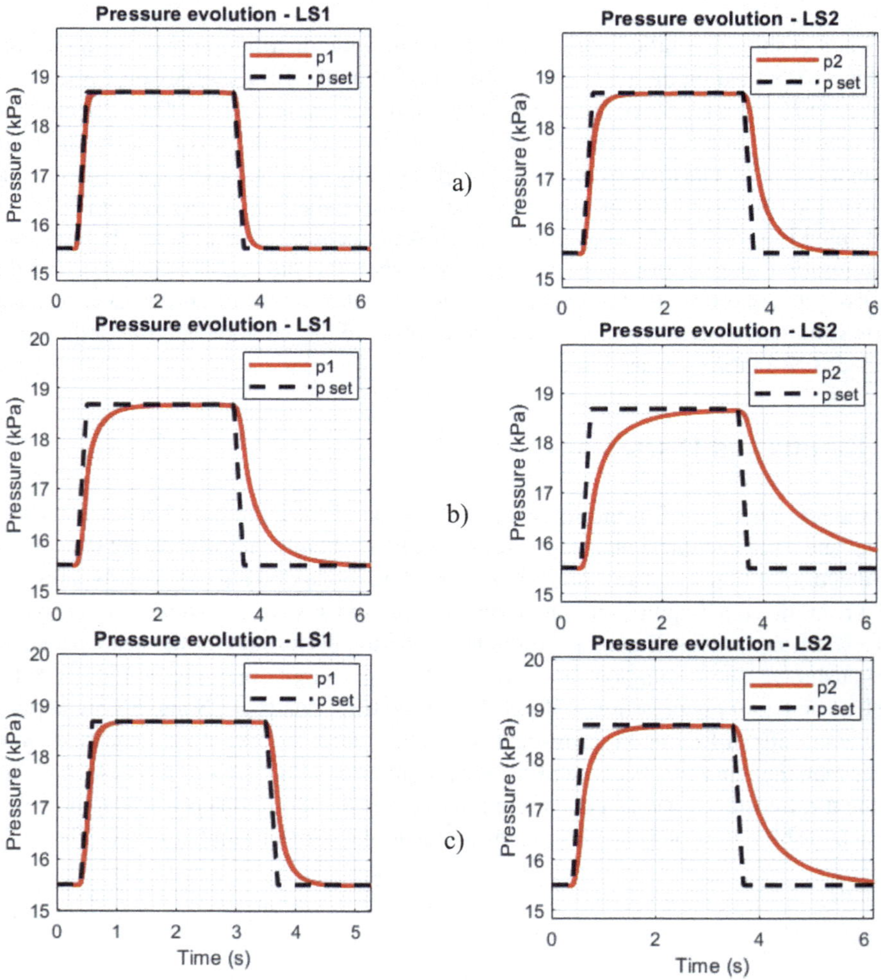

Fig. 4. Pressures evolution in head (a), back (b), and heel (c) cushions.

6 Conclusion

The objective of this work was the development of a reclining chair composed by automated cushions for pressure ulcer prevention. A previous work [17] was focused on improving the control system of a single air-cell that composes the contact surface of the seat cushion. In this work, the development of the control system for the four different cushions composing a complete reclining chair is presented. Head, back and heel cushions, together with the previously developed buttocks cushion, compose the reclining chair, which can be moved between two limits configuration: a) chair and b) bed. The three new cushions are designed such that, again, particular attention was given to the more at-risk area of the body (back of head and ears, shoulder and back, elbows, lower back and buttocks, greater trochanter, heel). Therefore, each cushion has its own number of air-cells and its own distribution of them. Only two-level surfaces were used since they have to respond to lower load levels with respect to the buttocks cushion. Even for these cushions, the results of simulations highlighted a good behavior of the developed control system, which is able to reach with good dynamic performance the optimal pressure with any external loads and initial pressure conditions.

References

1. Edsberg, L.E., Black, J.M., Goldberg, M., McNichol, L., Moore, L., Sieggreen, M.: Revised national pressure ulcer advisory panel pressure injury staging system: revised pressure injury staging system. J. Wound Ostomy Continen. Nurs. **43**(6), 585 (2016)
2. Bhattacharya, S., Mishra, R.K.: Pressure ulcers: current understanding and newer modalities of treatment. Indian J. Plastic Surg. **48**(01), 004–016 (2015)
3. Agrawal, K., Chauhan, N.: Pressure ulcers: back to the basics. Indian J. Plastic Surg. **45**(02), 244–254 (2012)
4. Bernabei, R., Manes-Gravina, E., Mammarella, F.: Epidemiologia delle piaghe da decubito. G Gerontol. **59**, 237–243 (2011)
5. Jaul, E.: Assessment and management of pressure ulcers in the elderly. Drugs Aging **27**(4), 311–325 (2010)
6. Brem, H., et al.: High cost of stage IV pressure ulcers. Am. J. Surg. **200**(4), 473–477 (2010)
7. Xakellis, J.G., Frantz, R.A., Lewis, A., Harvey, P.: Cost-effectiveness of an intensive pressure ulcer prevention protocol in long-term care. Adv. Wound Care J. Prevent. Heal. **11**(1), 22–29 (1998)
8. Pham, B., Stern, A., Padula, W.V., Mishra, M.K., Makic, M.B.F., Sullivan, P.W.: Improving the quality of pressure ulcer care with prevention: a cost-effectiveness analysis. Med. Care **50**(2), 188–190 (2012)
9. Gillespie, B.M., Chaboyer, W.P., McInnes, E., Kent, B., Whitty, J.A., Thalib, L.: Repositioning for pressure ulcer prevention in adults. Cochrane Database Syst. Rev. **22**, 108–109 (2014)
10. Mervis, J.S., Phillips, T.J.: Pressure ulcers: prevention and management. J. Am. Acad. Dermatol. **81**(4), 893–902 (2019)
11. Iglesias, C., et al.: Pressure relieving support surfaces (PRESSURE) trial: cost effectiveness analysis. BMJ **332**(7555), 1416 (2006)
12. Saegusa, M., Noguchi, H., Nakagami, G., Mori, T., Sanada, H.: Evaluation of comfort associated with the use of a robotic mattress with an interface pressure mapping system and automatic inner air-cell pressure adjustment function in healthy volunteers. J. Tissue Viabil. **27**(3), 146–152 (2018)

13. Hollington, J., Hillman, S.J., Torres-Sánchez, C., Boeckx, J., Crossan, N.: ISO 16840–2: 2007 load deflection and hysteresis measurements for a sample of wheelchair seating cushions. Med. Eng. Phys. **36**(4), 509–515 (2014)
14. Carrigan, W., Nuthi, P., Pande, C., Nothnagle, C.P., Wijesundara, M.B.: A Pressure modulating sensorized soft actuator array for pressure ulcer prevention. In ASME 2017 International Design Engineering Technical Conferences and Computers and Information in Engineering Conference. American Society of Mechanical Engineers Digital Collection (2017)
15. Wijesundara, M., Cooper, R.: Automated Seat Cushion for Pressure Ulcer Prevention Using Real-Time Mapping, Offloading, and Redistribution of Interface Pressure. University of Texas, Arlington Arlington United States (2016)
16. Carrigan, W., et al.: Design and operation verification of an automated pressure mapping and modulating seat cushion for pressure ulcer prevention. Med. Eng. Phys. **69**, 17–27 (2019)
17. Mannella, D., Bellusci, M., Graziani, F., Ferraresi, C., Muscolo, G.G.: Modelling, design and control of a new seat-cushion for pressure ulcers prevention. Proc. Inst. Mech. Eng. Part H J. Eng. Med. (2022). https://doi.org/10.1177/09544119211068908
18. Maiorino, A., Muscolo, G.G.: Biped robots with compliant joints for walking and running performance growing. Front. Mech. Eng. **6**, 11 (2020). https://doi.org/10.3389/fmech.2020.00011
19. Sonenblum, S.E., Sprigle, S.H., Cathcart, J.M., Winder, R.J.: 3D anatomy and deformation of the seated buttocks. J. Tissue Viabil. **24**(2), 51–61 (2015)
20. Muscolo, G.G., Fontana, M.: A novel linear pneumatic actuator with tunable-compliance constraint. Int. J. Mech. Control. **21**(02), 73–85 (2020). (ISSN: 1590–8844)
21. Bottero, S., Muscolo, G.G., Ferraresi, C.: A new soft RCC device with pneumatic regulation. Robotics **9**, 98 (2020). https://doi.org/10.3390/robotics9040098

Kinematic Modelling of a Bioinspired Two Sections Serial Continuum Robot (SCR)

Elie Gautreau[1(✉)], Juan Sandoval[1], Marc Arsicault[1], Xavier Bonnet[2], Saïd Zeghloul[1], and Med Amine Laribi[1]

[1] Department of GMSC, Pprime Institute CNRS, ENSMA, University of Poitiers, UPR 3346, Poitiers, France
{elie.gautreau,juan.sandoval,marc.arsicault,said.zeghloul, med.amine.laribi}@univ-poitiers.fr

[2] CEBC Center for Biological Studies of Chizé, CNRS, University of La Rochelle, La Rochelle, France
xavier.bonnet@cebc.cnrs.fr

Abstract. Designing and modelling bioinspired robots enables mimicking complex biomechanical process as snake swimming locomotion. Cable-driven compliant continuum robot made of separated modules was specifically designed as biology-push process to accurately reproduce snake's locomotion. This paper aims to present the design and modelling of a bioinspired two sections Serial Continuum Robot (SCR). This latter is a hybrid snake-arm composed of two sections assembled in series as a serial standard manipulator. Each section of the snake-arm is independently actuated and consists of 2-DoF. This polyvalent design combines the advantage of flexible continuum robot and serial manipulator. Swimming snake-like robot is presented as an example of biomimetic application, where the robot is able to significantly mimic snake locomotion with a homogeneous distribution of actuators.

Keywords: Continuum robots · Kinematic model · Piecewise Constant Curvature · Biomimetics

1 Introduction

Biomimicry is expending rapidly in various academic research areas. Bio-inspiration is handled in two main ways: (1) Top-down approach or technology pull process [1] to describe a bio-inspired design process initiated by a technical task or problem. (2) Bottom-up or solution driven approach or biology-push process depicts the emergence of technical applications initiated by biological hypotheses [2] as the study of swimming snake locomotion to develop snake like robots.

Snake-like robot field is widening over the past years. It requires the study of snake's anatomy and swimming snakes' analysis [3, 4]. The latter is frequently performed with video processing to extract qualitative and quantitative data from swimming behavior.

Extracted data from measurements are analyzed and transferred to design various snakebots according to its application. Multi-rigid module snake-like robots have the

A. Müller and M. Brandstötter (Eds.): RAAD 2022, MMS 120, pp. 247–255, 2022.
https://doi.org/10.1007/978-3-031-04870-8_29

capacity to evolve autonomously in complex environment. For instance the snake robot ACMR5 developed by Hirose [5] can be used for maintenance in nuclear environment. It appears that most of snake-like robots are made of a series of identical modules [6, 7, 8]. Multi-sections continuum snake robots are actively developed for medical applications such as minimally invasive surgery thanks to their low diameter and their capacity to evolve in constrained areas [9]. Designing a multi-section snake-like robot implies coupling between modelling and experimentations for validation with snakes' measurements. The modelling of continuum robots is frequently based on the hypothesis of Piecewise Constant Curvature (PCC). Further models as Variable Curvature (VC), Finite Element Models (FEM) and Rigid Link models are applied. According to the PCC assumption due to continuum robot design, motion control can be achieved throughout a large variety of formalisms for robot-independent mapping: arc geometry, DH parameters, integral representation and robot-specific mapping as robots shaped by tendons for instance [10]. Moreover, kinematic modelling is performed for both a two degrees of freedom single-section and redundant multi-sections continuum cable-driven robot [10].

In order to faithfully reproduce the snake's swimming locomotion, a bio-inspired multi-section continuum cable-driven robot has recently been developed [11]. Unlike classical designs using rigid sections, the proposed continuum robot adopts constant curvature sections enhancing to accurately reproduce the pre-studied snake shape. Preliminary validations of the new snake-arm are presented here, through a two-sections kinematic modelling and mapping.

The main contribution of this paper is to present the robots mapping and the kinematic modelling of a new bio-inspired two-sections Serial Continuum Robot (SRC) design. Section 2 presents the robot designing method through kinematic measurements of biological snakes as well as the proposed prototype. Section 3 depicts the kinematic modelling throughout the robot-independent and robot-specific mapping. Preliminary results from the coupling of modelling-experimentations are reported in Sect. 4. The last Sect. 5 summarizes the paper content and concludes with future work considerations.

2 Motivations

2.1 Snake Study

To bring snake swimming efficiency to light, we performed studies on anguilliform swimming of grass and whip snake. A video analysis algorithm was developed to quantify snake swimming motions [11]. A circle arc interpolation on snake's backbone was performed, proving that multiple circle arc section faithfully interpolates snake undulation. These circle arcs were the starting point for the mechanical design of the snake robot's sections. Thus, a constant curvature continuum design was chosen to reproduce the snake body curvature. Finally, the assembly of multiple sections will result on the snake-like robot.

2.2 Mechanical Design

Figure 1a shows one section of the Serial Continuum Robot. The flexible module of a section is designed through the RBR (Rigid Body Replacement) method [11, 12] where we reproduced the movement of snake vertebrae through compliant joints [13]. Each bioinspired section is driven by four antagonistic cables two by two (see Fig. 3a) and wounded by two servomotors Dynamixels® 2XL-430-W250T. Figure 1b shows the two-section SCR. Thus, the prototype is composed of 3 servomotors and two flexible and continuous sections. Each section is identical and can be assembled with the previous one to form a multi-section Serial Continuum Robot (SCR). Figure 1c depicts the SCR technical characteristics.

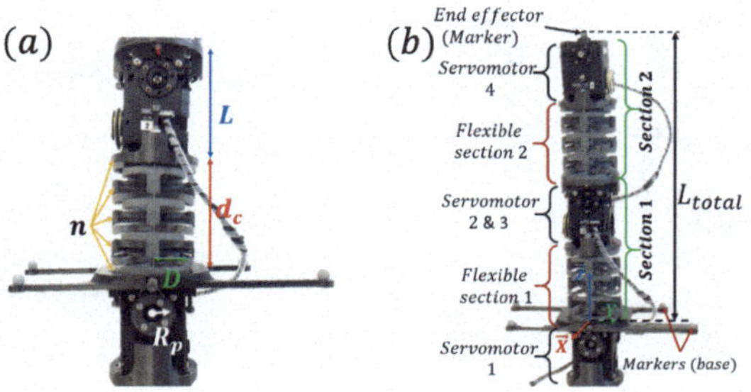

(c)

n	5
L	66 [mm]
D	23 [mm]
dc	60 [mm]
R_p	10 [mm]
L_{total}	252 [mm]

Fig. 1. (*a*) Single section Continuum Robot. (*b*) two-section Serial Continuum Robot. (*c*) Technical characteristics of the two-sections Serial Continuum Robot (SCR).

3 Modelling

In this section, we present robot modelling based on the assumption of Constant Curvature (CR depicts circle arcs). Coordinates are obtained from arc geometry formulation. The CC robot modelling is described by: (1) the *robot-independent mapping*, defined between configuration and cartesian space, (2) The *robot-specific mapping*, between configuration and actuator space, for a single section (see Fig. 2). Furthermore, we present the direct kinematic modelling of the two-section SCR.

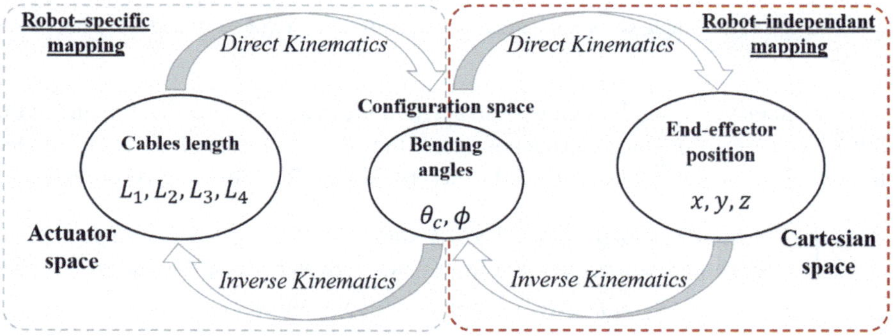

Fig. 2. Continuum robot's mapping.

3.1 Single Section Mapping: Direct and Inverse Kinematic Modelling

The kinematic scheme equivalency of a single section is given in Fig. 3b. Let us define three translations $\mathbf{T}(x)$, $\mathbf{T}(y)$, $\mathbf{T}(z)$ and three rotations $\mathbf{R}(\phi, z)$, $\mathbf{R}(\theta_c, y)$, $\mathbf{R}(-\phi, z)$. The geometric transformation matrix, ${}^{i-1}\mathbf{T}_i$, from section $i-1$ to section i is given by:

$$ {}^{i-1}\mathbf{T}_i = \mathbf{T}(x)\cdot\mathbf{T}(y)\cdot\mathbf{T}(z)\cdot\mathbf{R}(\phi, z)\cdot\mathbf{R}(\theta_c, y)\cdot\mathbf{R}(-\phi, z) \quad (1) $$

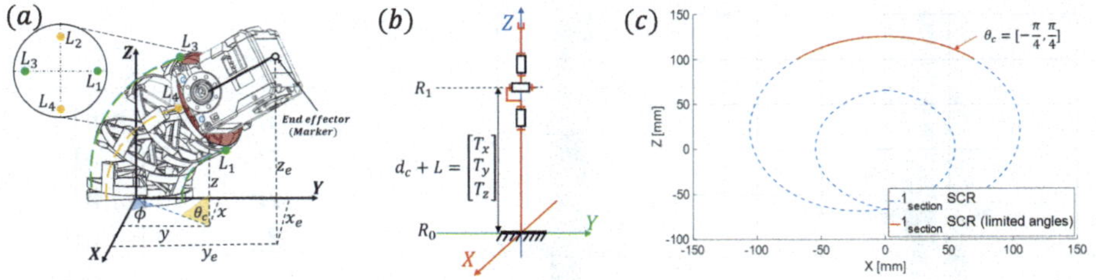

Fig. 3. (*a*) Single section modelling in configuration space ("θ_c", angles mechanically limited of the flexible module). (*b*) Single section kinematic scheme equivalency. (*c*) Single-section workspace (End effector position).

The Direct Kinematic Modelling (robots-independent mapping) of a single-section is geometrically identified (see Fig. 4) and its cartesian workspace is shown in Fig. 3c.

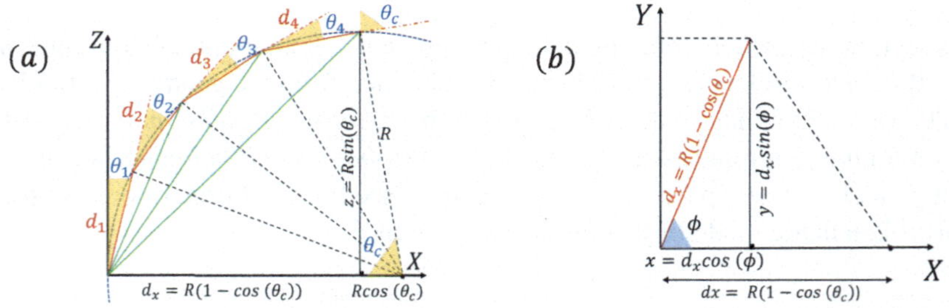

Fig. 4. (a) (b) Single section illustrations respectively in $\boldsymbol{XZ}(\phi = 0)$ and $\boldsymbol{XY}$ plan.

The geometry of a single section is depicted in the plane $\boldsymbol{XZ}$ (see Fig. 4a) and in the plane $\boldsymbol{XY}$ (see Fig. 4b). The cartesian position of the end-effector (see Fig. 3c) positioned at the top of the actuator 4 (see Fig. 3a) is described with the following coordinates:

$$ \begin{cases} x_e = L\cos(\phi)\sin(\theta_c) + x = L\cos(\phi)\sin(\theta_c) + R(1-\cos(\theta_c))\cos(\phi) \\ y_e = L\sin(\phi)\sin(\theta_c) + y = L\sin(\phi)\sin(\theta_c) + R(1-\cos(\theta_c))\sin(\phi) \\ z_e = L\cos(\theta_c) + z = L\cos(\theta_c) + R\sin(\theta_c) \end{cases} \quad (2) $$

with $R = \frac{d_c}{2nsin(\frac{\theta_c}{2n})}$ the radius of curvature of the flexible section center, d_c the length of the center of the flexible section, θ_c the bending angle of a flexible section, ϕ the angle giving the direction of curvature and L the servomotor length. We note that n the number of cable contact points (see Fig. 1a).

We solve the Inverse Kinematic Model (IKM) for a given end-effector position (robots-independent mapping):

$$\phi = \tan^{-1}\left(\frac{y}{x}\right) \tag{3}$$

$$\theta_c = 2\tan^{-1}\left(\frac{\sqrt{x^2 + y^2}}{z}\right) \tag{4}$$

The Direct Kinematic Model (robots-specific mapping) is applied to determine the angular parameters according to the cable lengths. We chose to control the continuous section using four antagonistic cables. The resulting direct model of these cables allows us to determine the angles ϕ et θ_c from the curvature radius $R_i = R - D \cdot \cos(\phi_i)$ and from the cable length $l_i = 2nR_i\sin(\frac{\theta_c}{2n})$, with i the cable number:

$$\phi = \tan^{-1}(\frac{l_4 - l_2}{l_3 - l_1}) \tag{5}$$

$$\theta_c = 2n\sin^{-1}\left(\frac{\sqrt{(l_3 - l_1)^2 + (l_4 - l_2)^2}}{4nD}\right) \tag{6}$$

From the previous equations, the IKM is applied to determine the cable lengths according to the angular parameters:

$$l_1 = -2n \cdot D \cdot \sin\left(\frac{\theta_c}{2n}\right)\cos(\phi) \tag{7}$$

$$l_2 = -2n \cdot D \cdot \sin\left(\frac{\theta_c}{2n}\right)\sin(\phi) \tag{8}$$

with D the distance between the point where the cable passes through and the center of the module. In view of the antagonistic aspect of the cables, we note the equivalence of the following lengths $l_4 = -l_2$ et $l_3 = -l_1$. The SCR (single and multi-sections) are equipped with that drive the cable winding. The movements of the continuous sections are achieved through the servomotor angles control. Thus, the conversion between cable lengths and sheave angles relative to the sheave radius R_p is given by $\gamma_i = \frac{L_i}{R_p}$.

3.2 Two-Section Direct Geometric Modelling

In this section, the direct geometric model is applied to two-sections SCR. The homogeneous transformation matrix of a single section is given by:

$$^{i-1}\mathbf{T}_i = \begin{bmatrix} \cos(\phi)^2(\cos(\theta_c)-1)+1 & \sin(\phi)\cos(\phi)(\cos(\theta_c)-1) & \cos(\phi)\sin(\theta_c) & x_e \\ \sin(\phi)\cos(\phi)(\cos(\theta_c)-1) & \cos(\phi)^2(1-\cos(\theta_c))+\cos(\theta_c) & \sin(\phi)\sin(\theta_c) & y_e \\ -\cos(\phi)\sin(\theta_c) & -\sin(\phi)\sin(\theta_c) & \cos(\theta_c) & z_e \\ 0 & 0 & 0 & 1 \end{bmatrix} \tag{9}$$

The robot-independent mapping of DKM for computing the effector positions in the Cartesian space is given by:

$$\begin{bmatrix} x_e \; y_e \; z_e \; 1 \end{bmatrix}^{\mathrm{T}} = {}^{0}\,\mathbf{T}_2 \cdot \mathbf{P} = {}^{0}\mathbf{T}_1 \cdot {}^{1}\mathbf{T}_2 \cdot \mathbf{P} \tag{10}$$

Note that $\mathbf{P} = [0001]^{\mathrm{T}}$ is the homogeneous coordinates of the manipulator's end effector. In our case, it coincides with the reflective marker (Fig. 1b). Let us note that this work is focused on position coordinates and the orientation coordinates are not considered.

4 Experimentations

In this section, the geometric model and the kinematic model are implemented in order to control the continuous robot through the angles defined in the configuration space. Further, a simulation has been developed under Matlab to visualize the robot's motion evolution. We carried out experiments reproducing two trajectories defined in a 3D space. The objective of this experimentation is to compare the simulated models with the experimental models in order to validate the mathematical developments.

4.1 Geometric Modelling

The DKM is achieved from the angles defined in the configurations space. The control process is implemented into 4 steps: (1) Angles limits of $\phi \; et \; \theta_c$ are defined in the spherical workspace. (2) The DKM is applied in order to scan the set of positions within the robot joint limits allowed by the robot dimensions. (3) The IKM of each section, computed according to a chosen trajectory, is defined in the cartesian workspace. (4) Cable lengths and sheave angles are determined by the IKM. At the beginning of each trial, the two-section SCR is set to the following initial configuration: $\phi = \theta_c = 0$. Thus R tends toward infinity (see Figs. 4a and 4b). Then, a circular trajectory defined outside the robot singularity along Z-axis is imposed on the structure. Figure 5a shows a comparison between the theoretical and experimental trajectories on the X, Y and $Z - axis$. One notes that the motion is close to theoretical trajectory despite an approximate initial configuration of the robot. This trajectory shows the ability of the Snake arm to perform a circular movement in a volume with a millimeter precision in **Y** and **Z**.

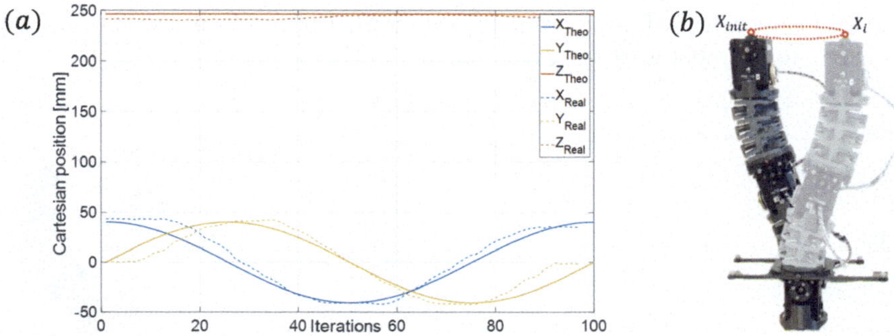

Fig. 5. (*a*) End-effector trajectory in Cartesian space for geometric modelling. (b) Achieved desired trajectory.

4.2 Kinematic Modelling

The kinematic Model is achieved through the infinitesimal motion method (Fig. 6a). The control process is implemented into 4 steps: (1) The two-section SCR is set in the initial configuration $\mathbf{q}_0 = \left[\theta_{10}\ \phi_{10}\ \theta_{20}\ \phi_{20}\right]$. This configuration allows to move the SCR away from a configuration close to the singularity. (2) An infinitesimal displacement is given along a direction in a Cartesian space $\mathbf{dX} = \left[d_{x_e}\ d_{y_e}\ d_{z_e}\right]$ in order to reach a final position $\mathbf{X_f}$. (3) The IKM allows to compute a small angle $\mathbf{dq}$ using the Moore-Penrose pseudo-inverse Jacobian method $\mathbf{J}^{+} = \mathbf{J}^{T} \cdot \left(\mathbf{J} \cdot \mathbf{J}^{T}\right)^{-1}$, with $\mathbf{J} = [\mathbf{J}_1\mathbf{J}_2]$. $\mathbf{J}_1$ and $\mathbf{J}_2$ the Jacobian matrix related to linear velocities of the end-effector due to joint velocities. (4) The cable lengths and pulleys angles are determined by the inverse geometric model.

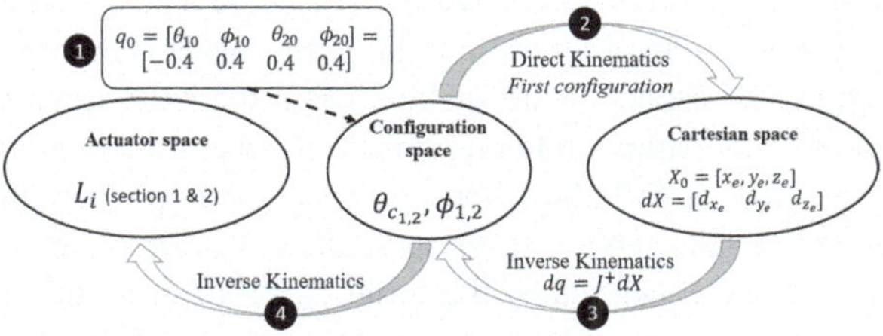

Fig. 6. Flow chart of the control design implemented in the kinematic modelling.

The module is set in the initial configuration $\mathbf{q}_0 = \left[-0.4\ 0.4\ 0.4\ 0.4\right]$. A linear trajectory is imposed with the following initial conditions: $\mathbf{dX} = \left[0\ 0.9\ 0\right]$ and $\mathbf{X_f} = \left[0\ 60\ 0\right]$ (mm). Theoretical and experimental trajectories are compared (see Fig. 7a). Trajectories along $Y-axis$ are similar while the final experimental position is slightly higher than the theoretical position. This issue is induced to both the small placement error of the robot in the initial configuration and the tension of the robot's cables. Note a deviation of the position along $X-axis$. Indeed, the initial configuration did not allow a unique movement according to $Y-axis$. An infinitesimal displacement along $X-axis$

was added to move the SCR away from its joint limits. This results in a shift in the plane XY inherent to the robot workspace.

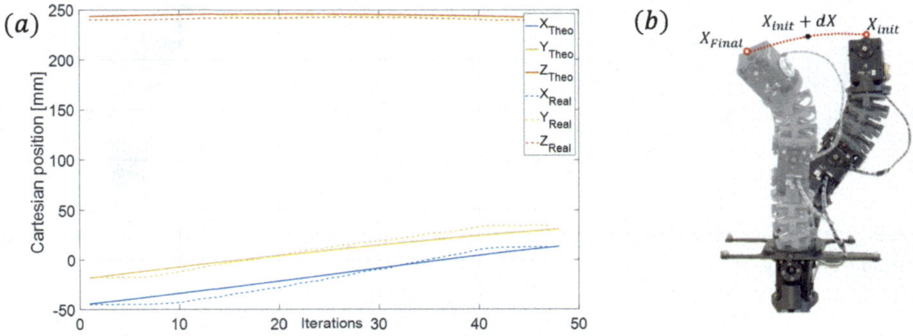

Fig. 7. (a) End-effector trajectory in Cartesian space for kinematic modelling. (b) Achieved desired trajectory.

5 Conclusion

In this study, a two-section Serial Continuum Robot (SCR) driven by servomotors was described and modelled according to geometric and kinematic models. A comparison between simulations and experiments was performed. This coupling has highlighted the ability of the proposed robot to perform the imposed trajectories despite small accuracy errors. This coupling also allows the validation of the geometric and kinematic models proposed in this work. However, it should be noted that a cable tension setting as well as an initial positioning fault disturbs the behaviour of the SCR. Thus, it is essential to ensure a flawless mechanical design in order to limit the position errors and backlashes in each section given by the serial aspect of the snake-arm. In the short term, this preliminary work opens the door to the study and comparison of several models of the snake-arm by adding further sections. It will allow both to pursue the implementation of the direct and inverse kinematic models and to explore several control strategies by exploiting the high degree of redundancy generated by the serial aspect of the continuous sections. In addition, a motion repeatability study will be performed to evaluate the accuracy of the SCR. In the long term, this study opens the way to the design of a snake drone in the framework of the ANR DRAGON2 project.

Acknowledgment. This research was funded by the French government by means of National Research Agency (ANR). This research is part of ANR DRAGON-2 project (ANR-20-CE02-0010).

References

1. Lindemann, U., Gramann, J.: Engineering design using biological principles. In: 8th International Design Conference, Dubrovnik, Croatia, pp. 355–360 (2004)

2. Farzaneh, H.H., Helms, M., Münzberg, C., Lindemann, U.: Technology-pull and biology-push approaches in bio-inspired design – comparing results from empirical studies on student teams (2016)
3. Munk, Y.: Kinematics of swimming garter snakes (Thamnophis Sirtalis). Comp. Biochem. Physiol. Mol. Integr. Physiol. **150**(2), 131–135 (2008)
4. Gillis, G.B.: Undulatory locomotion in elongate aquatic vertebrates: anguilliform swimming since Sir James Gray. Am. Zool. **36**(6), 656–665 (1996)
5. Hirose, S., Mori, M.: Biologically inspired snake-like robots. In: 2004 IEEE International Conference on Robotics and Biomimetics, pp. 1–7, Shenyang, China (2004)
6. Liljeback, P., Stavdahl, O., Pettersen, K.Y., Gravdahl, J.T.: Mamba - a waterproof snake robot with tactile sensing. In: 2014 IEEE/RSJ International Conference on Intelligent Robotics and Systems, pp. 294–301, Chicago, IL, USA (2014)
7. Porez, M., Boyer, F., Ijspeert, A.J.: Improved Lighthill fish swimming model for bio-inspired robots: modeling, computational aspects and experimental comparisons. Int. J. Rob. Res. **33**(10), 1322–1341 (2014)
8. Boyer, F., Chablat, D., Lemoine, P., Wenger, P.: The EeL-like robot. In: 33rd Mechanisms and Robotic Conference, Parts A and B, vol. 7, pp. 655–662, San Diego, USA (2009)
9. Orekhov, A.L., Abah, C., Simaan, N.: Snake-Like Robots for Minimally Invasive, Single Port, and Intraluminal Surgeries. arXiv:1906.04852, pp. 203–243 (2018)
10. Jones, R.B.: Design and kinematic modeling of constant curvature continuum robots: a review. Int. J. Robot. Res. **29**, 1661–1683 (2010)
11. Gautreau, E., Sandoval, J., Bonnet, X., Arsicault, M., Zeghloul, S., Laribi, M.A.: A new bio-inspired Hybrid Cable-Driven Robot (HCDR) to design more realistic snakebots. In: 2020 IEEE, International Conference on Robotic and Automation (ICRA) (2022). (in Press)
12. Gallego, J.A., Herder, J.: Synthesis methods in compliant mechanisms: an overview, pp. 193–214 (2010)
13. Trease, B.P., Moon, Y.-M., Kota, S.: Design of large-displacement compliant joints. J. Mech. Des. **127**(4), 788–798 (2004)

A Robot for Facade Cleaning Based on a Cartesian Configuration: Kinematic Analysis and Prototype Construction

G. M. García-Rodríguez(✉) and E. Castillo-Castañeda

Instituto Politécnico Nacional, CICATA-Unidad Querétaro, Santiago de Querétaro, México
ginnagarcia89@hotmail.com, ecastilloca@ipn.mx

Abstract. Despite advances in service robotics and automation, building facade cleaning continues to be performed manually by specialized technicians operating from complex scaffoldings. This form of cleaning, in addition to be inefficient, is expensive and dangerous for cleaning personnel who must work under the risk of falls from considerable height and exposed to toxic substances that are often required for cleaning. This work presents a robotic device with 5 degrees of freedom based on a Cartesian configuration and a pan-tilt type system; the latter allows directing a high-pressure jet of some specific substance to clean the facade. The robotic device is placed in front of the facade and an operator directs the cleaning from the ground in a teleoperated way. In a first stage, the cleaning substance can be water supplied by a high-pressure washer. The kinematic analysis of the robot is presented, as well as a prototype construction.

Keywords: Cartesian type robot · Facade cleaning · Service robot · High-pressure washer

1 Introduction

In recent decades, buildings have been constructed with new technologies and materials, which make them engineering masterpieces. Despite this technological advance, cleaning facades remains a traditional, manual, and dangerous practice. Correct and periodic cleaning can prevent structurally irreversible damage, since constant exposure to meteorological, environmental, and biological agents causes progressive wear and degradation of materials, which can even affect the health of the inhabitants of the buildings.

Currently, the cleaning of facades is carried out under different methods applied manually, through the propulsion of materials with high pressure jets [1]. For cleaning personnel to reach the entire surface of the facade, it is necessary to hook their bodies to ropes and robust scaffolds arranged at great heights, which is why it is classified as an unsafe, expensive [2] and time-consuming practice. Service robotics is an area of great opportunity that, since 2019, has been the one with the greatest development and the one that has generated the greatest increase in sales [3]. The cleaning tasks are part of the service robotics; however, there are very few commercial robots performing

A. Müller and M. Brandstötter (Eds.): RAAD 2022, MMS 120, pp. 256–263, 2022.
https://doi.org/10.1007/978-3-031-04870-8_30

cleaning tasks for building facades. Among the main developments, there are some examples such as: robot with guide rail integrated to the glass facades [4], robot with suction cup adhesion driven by pneumatic actuators [5], prototype with passive suction cup adhesion [6], and device that uses negative pressure between the robot and the wall, generally through vacuum pumps to adhere to the surface [7]. There are also robots with magnetic grip for oil tanks made of metal plates [8]. The rope or rail grab is another method of grabbing, where the robot is supported from the wall or ceiling through a rope or cable, allowing mobility and safe execution of tasks [9]. Robots with bio-inspired adhesion have also been proposed, with the ability to climb vertical walls even on irregular surfaces such as concrete [10]. However, the force of gravity and the irregularity of the surfaces (materials other than glass) of the facades, have become a challenge to achieve the high levels of adaptability and flexibility that climbing robots require in the execution of vertical tasks. Our proposal aims to optimize and improve this practice, through the development of a robot, made up of a light mobile "scaffolding type" structure, with wheels on its four support bases. In addition, tanks can be placed at the base of the structure, that contain the cleaning liquid, to provide support and stability to the structure.

2 Conceptual Design of the Proposal

The main contribution of this first stage is the complete definition of the mechanisms to generate translations and rotations of the robot, as well as the main mechanical elements to build a prototype. The mobility of our proposal is of the Cartesian type, which allows vertical and horizontal movements, Z and Y axes, respectively, driven by linear sliders, see Fig. 1.

Fig. 1. Full view of the conceptual design proposal.

Compared to other proposals for cleaning robots, where the robot climbs on the surface of the building, our concept consists of a robot that stands in front of the facade, without direct contact.

A high-pressure water jet gun is mounted on a pan-tilt type system as the end-effector of the robot, which is remotely manipulated with a joystick. The cleaning path will be carried out in a zig-zag pattern over the desired region automatically. This region will be covered manually by the operator who will be able to control speed and position of the device from a safe place.

3 Forward and Inverse Kinematics

This section discusses the movement of the robot with respect to a fixed reference frame located in the lower right corner of the scaffold-like structure. The reference frames for the device are indicated in Fig. 2.

a) **b)**

Fig. 2. Reference frames: a) Scaffold-like structure, b) pan-tilt and end-effector.

The kinematic analysis is simplified considering that the robot is composed of two main elements: a Cartesian configuration structure and a pan-tilt system. Figure 3 shows the diagram of the robot with 5 degrees of freedom: 3 translations and 2 rotations. The end-effector pose is defined by $P_x, P_y, P_z, \phi, \theta$.

The dimensions of the robot are defined by the constants a_1, a_2, a_3, a_4, a_5, a_6 and the joint variables by translational distances d_1, d_2, d_3 and the angles θ_1, θ_2. In this way, the corresponding homogeneous transformation matrices are:

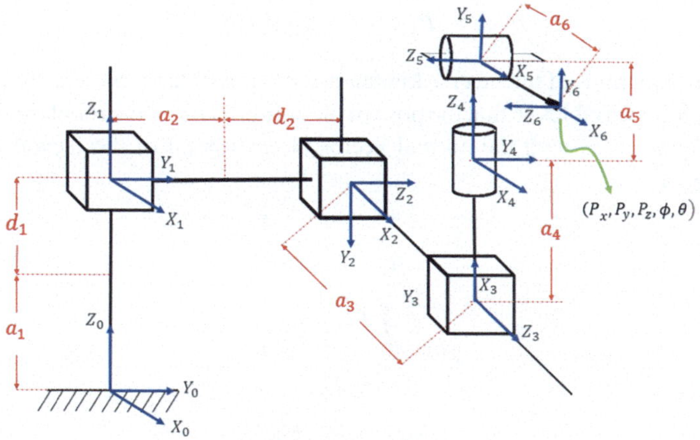

Fig. 3. Kinematic scheme of the 5-DOF robotic device.

$$^{0}T_3 = {}^{0}T_1\,{}^{1}T_2\,{}^{2}T_3 = \begin{bmatrix} 0 & 0 & 1 & a_3 + d_3 \\ 0 & -1 & 0 & a_2 + d_2 \\ 1 & 0 & 0 & a_1 + d_1 \\ 0 & 0 & 0 & 1 \end{bmatrix}$$

$$^{4}T_6 = {}^{3}T_4\,{}^{4}T_5\,{}^{5}T_6 = \begin{bmatrix} c\theta_1 c\theta_2 & -c\theta_1 s\theta_2 & s\theta_1 & a_6 c\theta_1 c\theta_2 \\ s\theta_1 c\theta_2 & -s\theta_1 s\theta_2 & -c\theta_1 & a_6 s\theta_1 c\theta_2 \\ s\theta_2 & c\theta_2 & 0 & a_4 + a_5 + a_6 s\theta_2 \\ 0 & 0 & 0 & 1 \end{bmatrix}$$

By multiplying the transformation matrices, the total transformation matrix is:

$$^{0}T_6 = \begin{bmatrix} c\theta_1 c\theta_2 & -c\theta_1 s\theta_2 & s\theta_1 & a_3 + d_3 + a_6 c\theta_1 c\theta_2 \\ s\theta_1 c\theta_2 & -s\theta_1 s\theta_2 & -c\theta_1 & a_2 + d_2 + a_6 s\theta_1 c\theta_2 \\ s\theta_2 & c\theta_2 & 0 & a_1 + d_1 + a_4 + a_5 + a_6 s\theta_2 \\ 0 & 0 & 0 & 1 \end{bmatrix}$$

Then, the inverse kinematics can be deduced:

$$\theta_1 = \theta$$

$$\theta_2 = -\phi$$

$$d_1 = P_z - a_1 - a_4 - a_5 - a_6 s\theta_2$$

$$d_2 = P_y - a_2 - a_6 s\theta_1 c\theta_2$$

$$d_3 = P_x - a_3 - a_6 c\theta_1 c\theta_2$$

Using the equations of the inverse kinematics it is possible to calculate the numerical values of the joint variables so that the pose of the end-effector of the robot (high-pressure washer gun) coincides with the normal vector of each point of the facade surface, as shown in Fig. 4.

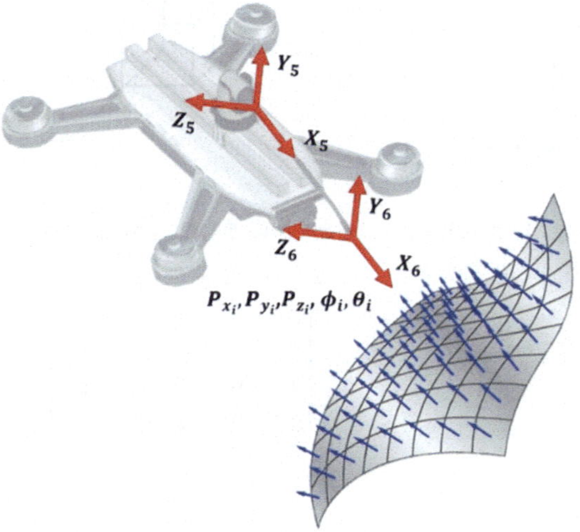

Fig. 4. End-effector pose as function of a normal vector of the facade surface.

The surface of the building facade will be digitally represented by $S \in \mathbb{R}^3 = \{x, y, z\}$, and then composed of a cloud of points $P_i \in S$. The set $P_i = \{x_i, y_i, z_i\}$ form a mesh with a spacing δ, which may or may not be uniform. For each element of the mesh, we have a vector N_i normal to the surface, see Fig. 4. This vector can be defined by three of the four vertices of the mesh, and calculated from the vector product:

$$N_i = \overrightarrow{P_i P_{i+1}} \times \overrightarrow{P_i P_{i+2}}$$

To define the design requirements, the following parameters were considered according to the needs that the robot must cover, such as: mobility to follow the required cleaning trajectory; adaptability to the surface considering the displacement ranges for reaching all points (including concavities) of the facade surface; horizontal and vertical maneuverability; ease of assembly; loads supported by the structure, including the repulsive force of the high pressure jet; cleaning speeds; and device position accuracy.

4 Prototype Construction

The device proposal considers that the cleaning activities will be carried out in buildings up to 15 m high and 20 m wide, with facade profile (concavities) of up to 40 cm. However, for the construction of a first laboratory prototype, the dimensions of the mobile structure were adjusted in a scalable and strategic way, considering the layout of the operational area of our laboratory facilities: 1.8 m high, 1.2 m wide, and remained 40 cm deep. Of course, the selected mechanical elements allow scaling to that required for cleaning buildings of the previously mentioned dimensions.

The mobile structure of the robot (Fig. 5a) is built in aluminum structural profile with a 40 × 40 mm T-slot; the profiles are connected through aluminum fasteners to build a mobile frame. At the bottom of the mobile frame, four 3″ rubber casters are attached at each end, fitted to a swivel plate with a brake. These wheels have thermoplastic properties, polypropylene rims, reinforced hubs, and a load capacity of up to 75 kg each. This mobile frame implements the Cartesian configuration (2 DOF).

Inside the mobile frame, a platform built with a standard 40 × 40 mm T-slot aluminum structural profile moves along the vertical Z-axis thanks to a multiple-pulley array.

The pan-tilt system is mounted on a square plexiglass plate (Fig. 5c), guided by a screw-nut linear displacement system that allows the movement of the robot in the X-axis, achieving precise adaptability to the different profiles of the facade.

The screw-nut system (Fig. 5b) is composed of: Two 400 mm linear motion rod shaft guides, an 8 mm lead screw, four SK8 shaft supports, four SCS8UU bearings, two KP08 flexible shaft couplings, and 2 flexible shaft couplings.

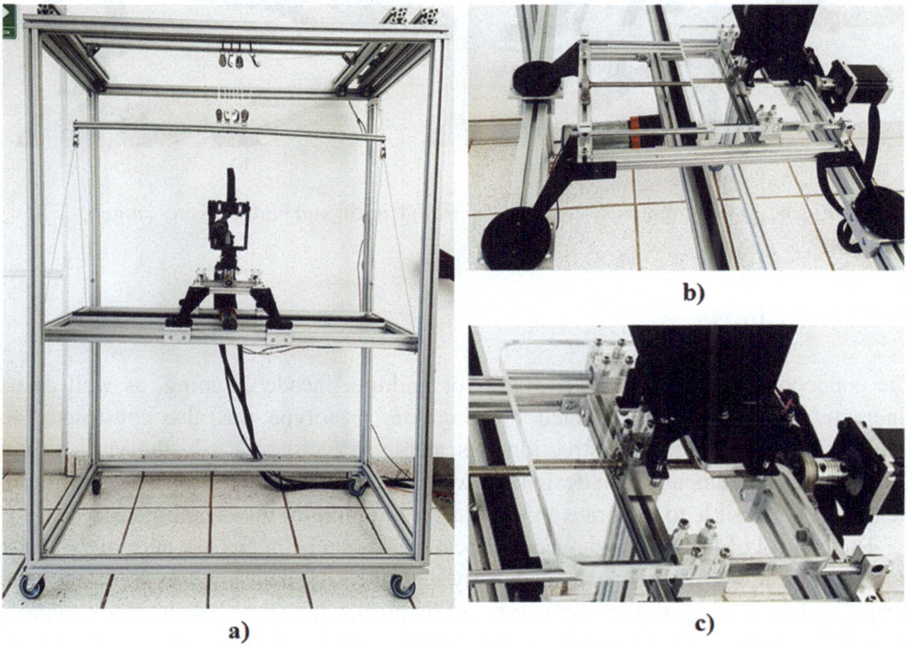

a) b) c)

Fig. 5. a) Mobile frame, b) Screw-nut linear system, c) Plexiglass plate.

The screw-nut system is attached to a rectangular structure built in structural aluminum profile with a 20 × 20 mm T-slot that moves on the Y-axis by means of a rack-and-pinion system (Fig. 6a) adjusted longitudinally in relation to the platform. The rack is made of low carbon steel, type TR with pressure angles of 20°, with a total length of 1120 mm and a width of 12 mm. The spur gear used in this system is cast iron which provides high strength and allows for a larger hole diameter. The rack-pinion system was chosen due to its mechanical simplicity, smooth transmission, load capacity and unlimited length.

The pan-tilt has two rotational movements, that allow the end effector to be oriented perpendicularly to any point on the facade surface, even if there are concavities in it. The pan-tilt system is coupled with a high-pressure jet gun from a commercial hydro washer (Fig. 6b), which will finally be able to reach any desired P_x, P_y, P_z position, thanks to the Cartesian configuration implemented in the robot, and to reach any desired ϕ, θ orientation to cover the entire operational area to be cleaned.

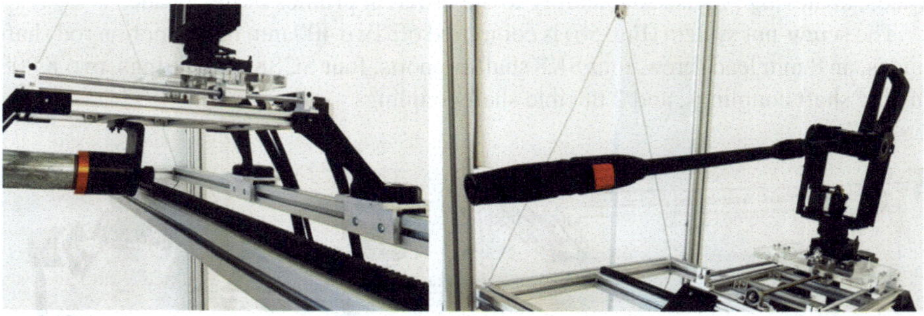

a) b)

Fig. 6. a) Rack-pinion system for Y-axis, b) Pan-tilt and high-pressure jet gun.

5 Conclusions

The conceptual design of 5 DOF robot for building facade cleaning, as well as its kinematic analysis, were presented. A laboratory prototype was also constructed to validate the concept; the construction is scalable to increase easily the vertical and horizontal displacements. The design allows to adapt to irregular profiles, common in facades. It is possible to integrate a video camera to monitor the cleaning task.

The control of the robotic device will be done remotely, in an easy and safe way for the operator. However, this work deals only with the kinematic model, the conceptual design, and the construction of a prototype; the control, where the hardware and actuators are immersed, will be addressed later, in a future work.

Based on the kinematic analysis and a detailed CAD design, our prototype is currently in under a patenting process. This proposal represents a technological innovation in the cleaning robot market that offers simplification of the task, minimization of risks as well as operational costs.

References

1. Caminiti, M.: Limpieza criogénica vs. métodos alternativos de limpieza (2019). https://blog-mx.coldjet.com/limpieza-criogenica-vs-otros-metodos
2. CronoShare (2020). https://www.cronoshare.com/cuanto-cuesta/limpieza-fachadas
3. IFR Press Conference (2020). https://ifr.org/DOWNLOADS/PRESS2018/PRESENTATION_WR_2020.PDF. Frankfurt
4. Zhang, H., Zhang, J., Zong, G.: Requirements of glass cleaning and development of climbing robot systems. In: Proceedings of the 2004 International Conference on Intelligent Mechatronics and Automation, pp. 101–106. Chengdu, China (2004)
5. Minmoon, S., Jaemyunghuh; Seunghoonlee, D., Soohan, Ch.: Vertical motion control of building façade maintenance robot with built-in guide rail (2015)
6. Kawasaki, S., Kikuchi, K.: Development of a small legged wall climbing robot with passive suction cups. In: proceedings of the 3rd international Conference on Design Engineering and Science—ICDES, pp. 112–116. Pilsen, Czech Republic (2014)
7. Wang, Y., Liu, S., Xu, D., Zhao, Y., Hao, S., Gao, X.: Development and application of wall-climbing robots. In: Proceedings of the 1999 IEEE International Conference on Robotics and Automation, vol. 2, pp. 1207–1212. Detroit, AL, USA, 10–15 (1999)
8. Xu, Z., Ma, P.: A wall-climbing robot for labelling scale of oil tank's volume. Robotica **20**, 209–212 (2002)
9. Zhang, H., Zhang, J., Liu, R., Zong, G.: Mechanical design and dynamics of an autonomous climbing robot for elliptic half-shell cleaning. Int. J. Adv. Robot. Syst. **4**, 437–446 (2007)
10. Funatsu, M., Kawasaki, Y., Kawasaki, S., Kikuchi, K.: Development of cm-scale wall climbing hexapod robot with claws. In: Proceedings of the 3rd International Conference on Design Engineering and Science—ICDES, pp. 101–106. Pilsen, Czech Republic (2014)

Study of a Three-Groove Kinematic Coupling for Precise Positioning in a Robotized Laser-Cutting Machine

Carlo De Benedictis(✉) and Carlo Ferraresi

Department of Mechanical and Aerospace Engineering, Politecnico di Torino, Turin, Italy
carlo.debenedictis@polito.it

Abstract. In this work, the positioning accuracy allowed by a three-groove kinematic coupling in a laser-cutting machine is studied. Kinematic couplings are designed to allow for improved accuracy and repeatability as compared to traditional elastic mediated coupling systems. A method for the estimation of positioning accuracy in a three-groove V shaped coupling is reported, and the results of the analysis carried out on a multi-purpose laser-cutting machine are presented. The results confirm that a positioning error lower than 0.02 mm is ensured by the coupling in any tested condition.

Keywords: Three-groove coupling · Kinematic coupling · Precise positioning · Machine design

1 Introduction

In robotic systems where automatic operations are carried out, it is often required to ensure a high-precision coupling between two parts that need to be positioned relatively. Nowadays, the designers have pushed the performance of the most common couplings used in manufacturing (elastic deformation methods) to the limit accuracy of about five microns.

The simplest coupling design consists of surface contact, which generally occurs with direct contact between large and relatively flat interface elements. Alignment pins have long been considered the simplest and least expensive method for aligning components [1]. If there is no clearance, that is, if there is a pressure mount between the pin and the hole, they fall into the category of couplings with elastic deformation [2]. The manufacture of precision couplings is expensive because the positioning of the centers and the sizing of the components must be subjected to tolerances that are more restrictive than the repeatability required by the coupling. Methods based on elastic deformation such as wedges, guides and grooves, dovetail joints, and pressure joints achieve forced couplings of an over-determined type, that is, with redundancy of constraints. In such condition, a good repeatability is achieved only at the end of a so-called wear-in phase. When a high load capacity is required, interface couplings can be designed with the appropriate contact area to achieve a rigid joint.

A. Müller and M. Brandstötter (Eds.): RAAD 2022, MMS 120, pp. 264–271, 2022.
https://doi.org/10.1007/978-3-031-04870-8_31

Kinematic couplings are static, structural, precision machined contacts used to precisely locate components relative to each other [2–5]. They are designed in such a way that the number of points of contact between two solid models corresponds to the number of degrees of freedom that are constrained. The coupling elements must be designed with great care and precision so that they can withstand high contact stresses at these points and maintain the integrity of the surface after repeated cycles [6, 7]. Repeatability can be very high, with errors lower than 1 micron.

In this work, a method aimed at determining the positioning accuracy guaranteed by a three-groove kinematic coupling is applied to a precision laser-cutting robotized machine. In particular, balancing is discussed to further improve the accuracy of the system for the several loading conditions allowed by the machine. In the following, a presentation of the main features of the machine is given. Then, the analysis of the kinematic coupling is carried out and the results are shown.

2 The Laser-Cutting Machine

The study was developed within a project developed by a company that is among the world leaders in the field of high-power laser-cutting and welding for industrial applications [8]. A CAD model of the machine is shown in Fig. 1.

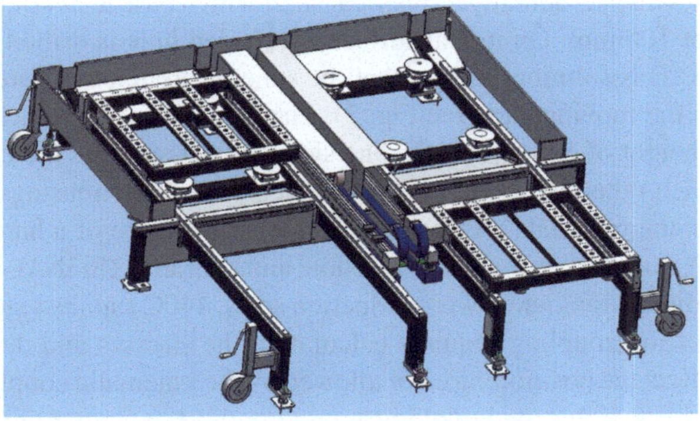

Fig. 1. The machine CAD model.

The machine consists of two independent modules located side by side. Each of these modules is mounted on a cart that translates along parallel routes between the workpiece loading station and the machining station, dragged by a gear motor mounted on a linear guide. At the four corners of the structure, height adjustable wheels are mounted to allow movement in the working area.

Once the working position is reached, the cart is decoupled from the guide and the vertical motion necessary to obtain the final positioning of the workpiece is allowed. The final positioning will be on the coupling elements only, so as to guarantee the required levels of precision (set as 0.02 mm by design specifications). The vertical movement is controlled by the inflation/deflation of two air springs for industrial applications interposed between the rails and the plane below (see Fig. 2). The elasticity of the air spring does not guarantee a straight vertical motion; therefore, side linear guides are required.

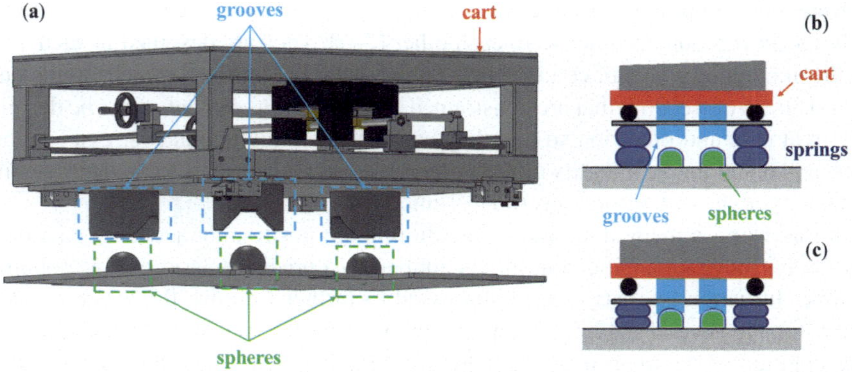

Fig. 2. Detail of the kinematic coupling: (**a**) CAD model; coupling in disengaged (**b**) and engaged (**c**) configurations.

3 Positioning Accuracy of a Three-Groove Kinematic Coupling

The final coupling takes place through a structure including three balls and three V-grooves [9, 10] (Fig. 2), mounted at the vertices of an equilateral triangle (see Fig. 3). The ball element, which is actually a spherical shell with a base at 3/4 of the total height, has a diameter of 100 mm. On its flat surface a threaded hole is drilled for connection to the fixed base. This solution implies that the spheres are fixed, whereas the grooves are integral with the moving cart (see Fig. 2). Each groove is a cylinder with diameter of 200 mm and height of 127.5 mm. On one side, contact surfaces are carved, inclined 90° relative to each other. Between the two surfaces, there are housings for two fixing screws. As far as the materials are concerned, the performance of a hardened stainless steel (440C SST), tungsten carbide (WC), Teflon, and a ceramic (Si3N4) were compared. To have smaller dimensions and lower production costs, 440C stainless steel was chosen.

The design of the coupling requires calculating the stresses and deflections at the contact, which affect the positioning error allowed by the kinematic coupling. Below, the most significant steps of this methodology are reported [9]. It starts from the knowledge of the following data: diameters of the spheres and radii of curvature of the grooves; the coordinates of the contact points of the spheres in the grooves; the orientations of the contact forces; the coordinates of the points of application of the preload forces; the preload forces on each ball/groove; the coordinates and components of an applied external load; the modulus of elasticity and Poisson coefficients of materials. To fulfill the theoretical stability requirement, for precise three-V kinematic couplings, stability and good overall stiffness will be ensured if the normals to the planes containing the contact force vectors for each ball bisect the internal angles between two adjacent sides of the coupling triangle. For instance, by considering the representation of Fig. 3, the normal to the π plane containing the contact force vectors for ball 2 must bisect the angle between the sides 12 and 23 (numbers refer to the spheres' nomenclature).

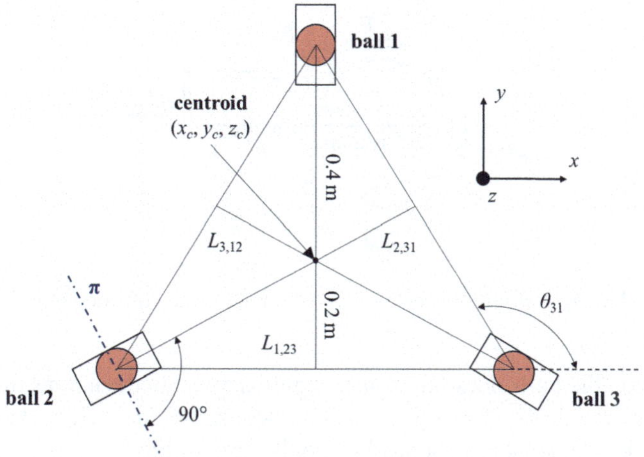

Fig. 3. Geometry of three-groove planar kinematic coupling.

By static analysis of the external forces F_e and moments M_e acting on the system, the calculation of the six contact forces F_{b_i} at the ball-groove interfaces can be derived (Eq. 1).

$$\sum \vec{F_e} = \vec{0}, \sum \vec{M_e} = \vec{0} \rightarrow F_{b_i}, \quad (i = 1 \ldots 6) \tag{1}$$

Once the forces are known, they can be used to determine the stresses and deformations at the points of contact using Hertz's theory. The equivalent modulus of elasticity E for each ball/groove is first determined, then the equivalent radius R of the system (Eq. 2) is calculated:

$$E = \frac{1}{\frac{\left(1-\nu_g^2\right)}{E_g} + \frac{\left(1-\nu_b^2\right)}{E_b}}, \quad R_e = \frac{1}{\frac{1}{R_{g\max}} + \frac{1}{R_{g\min}} + \frac{1}{R_{b\max}} + \frac{1}{R_{b\min}}} \tag{2}$$

where $R_{b_{max}}$ and $R_{b_{min}}$ are equal, while for the groove $R_{g_{max}}$ tends to infinity and $R_{g_{min}}$ is negative. The modulus of elasticity and Poisson coefficients are provided both for balls, E_b and ν_b, and grooves, E_g and ν_g. The factor $cos\theta$, necessary for the calculation of the contact region (which will be an ellipse, see Fig. 4) according to the Hertzian theory, is equal to $R_e/\left|R_{g_{\min}}\right|$. Having calculated the semi-major and semi-minor axes of the ellipse $R_{\max}$ and $R_{\min}$, the Hertzian contact stress q and deflection δ can be determined for each ball, as reported in Eq. 3:

$$q = \frac{3F}{2\pi R_{\max} R_{\min}}, \quad \delta = \lambda \left(\frac{2F^2}{3R_e E_e^2} \right)^{1/3} \tag{3}$$

where F is the overall external force and λ is one of the polynomial functions detailed for the application of Hertz theory.

Once the contact stresses and the corresponding deflections have been calculated, the positioning errors in the kinematic coupling can be determined. In the hypothesis that the

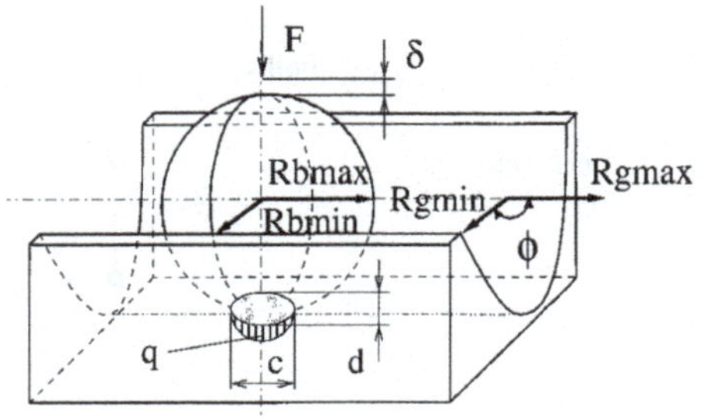

Fig. 4. Ball-groove interaction within the kinematic coupling.

distance between the balls does not change significantly, the deviations of the centroid of the coupling triangle $\delta_{\xi c}$ ($\xi = x, y, z$) can be calculated as the weighted average of the deflections of the balls that are obtained by application of Eq. 3.

Then, the calculation of the rotational errors is carried out: the estimation of the rotations ε_x and ε_y around the two corresponding axes requires to determine heights L and side orientation angles θ of the coupling triangle (Fig. 3) and leads to the following result (Eqs. 4–5):

$$\varepsilon_x = \frac{\delta_{z1}}{L_{1,23}} \cos\theta_{23} + \frac{\delta_{z2}}{L_{2,31}} \cos\theta_{31} + \frac{\delta_{z2}}{L_{3,12}} \cos\theta_{12} \tag{4}$$

$$\varepsilon_y = \frac{\delta_{z1}}{L_{1,23}} \sin\theta_{23} + \frac{\delta_{z2}}{L_{2,31}} \sin\theta_{31} + \frac{\delta_{z2}}{L_{3,12}} \sin\theta_{12} \tag{5}$$

The rotation error of the coupling around the z-axis (Eq. 6) is derived as the average of the rotations around the direction of z, with respect to the center of gravity of the coupling, calculated for each ball:

$$\varepsilon_z = \frac{\varepsilon_{z1} + \varepsilon_{z2} + \varepsilon_{z3}}{3} \tag{6}$$

Once the calculation of the rotation errors is completed, the translation errors δ_ξ ($\xi = x, y, z$) at any point around the coupling can be determined through the relation presented in Eq. 7:

$$\begin{bmatrix} \delta_x \\ \delta_y \\ \delta_z \\ 1 \end{bmatrix} = \begin{bmatrix} 1 & -\varepsilon_z & \varepsilon_y & \delta_{xc} \\ \varepsilon_z & 1 & -\varepsilon_x & \delta_{yc} \\ -\varepsilon_y & \varepsilon_x & 1 & \delta_{zc} \\ 0 & 0 & 0 & 1 \end{bmatrix} \cdot \begin{bmatrix} x - x_c \\ y - y_c \\ z - z_c \\ 1 \end{bmatrix} - \begin{bmatrix} x - x_c \\ y - y_c \\ z - z_c \\ 0 \end{bmatrix} \tag{7}$$

4 Analysis of the Coupling in the Laser-Cutting Machine

Payload of the cart is 450 kg. The mass of the cart, including the fixings and coupling elements, is equal to 300 kg. The inner part of the cart is used for the balancing of the external loads placed on the cart. The loading area is 1.1 m × 1.1 m. A 50 kg cast iron industrial weight was chosen as counterweight, 0.34 m × 0.2 m in plant view.

The first analysis concerned the study of deformations in the system, by considering pieces of different weight placed on the cart and located at its center of gravity. This analysis is necessary to assess that the required repeatability of 0.02 mm is achieved even in the case of maximum load. The results of the analysis are shown in Fig. 5. By considering a load of 4500 N, representative of the expected payload, the deformation at the center of gravity will be all-vertical, directed along the z-axis, and equal to about 0.006 mm. The stress ratio, i.e., the ratio between the contact pressure and the allowable Hertzian stress (equal to 1.5 times the ultimate tensile strength [11]), has been estimated at about 45%.

Subsequently, the point of application of the external load was changed. In this analysis, the center of gravity of the load was moved along the y axis, starting from ball 1 and approaching the other two balls (see Fig. 3 for reference). The results of the analysis are shown in Fig. 6. Ball 1 has a higher maximum stress ratio than the other two balls. The maximum value is reached when the load is applied directly on the ball (58%), whereas the contact stress is reduced at 34% when considering the midpoint between balls 2 and 3. As shown in Fig. 6, right, the trends related to balls 2 and 3 are overlapped, due to the applied load being equidistant to both spheres.

As far as the deformations are concerned, the overall deformation along the z-axis varies by the order of one thousandth of a millimeter, as shown in Fig. 6, left; however, it must be pointed out that the deformation changes locally on the three balls. By checking the individual balls, the maximum vertical deformation obtained at ball 1 was estimated equal to 0.0114 mm, therefore in any case lower than the required limit of 0.02 mm. The effect of rotation around x has also been verified. The rotation ε_x around the x-axis decreases as the external load approaches the center of gravity of the coupling. By using the maximum ε_x equal to 1.9e–5, obtained when the load was applied directly on ball 1, the overall displacement resulted equal to 0.0143 mm.

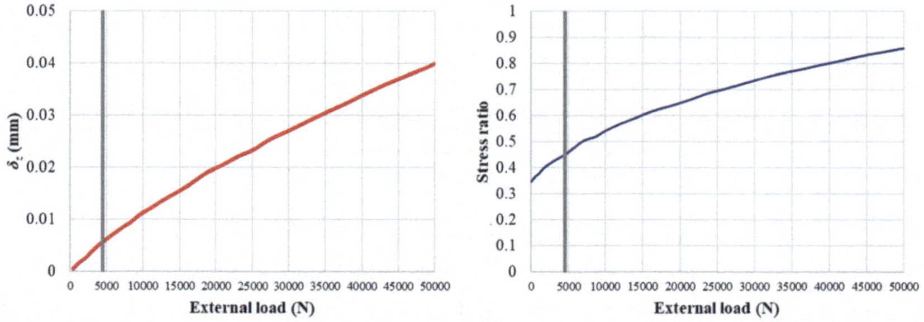

Fig. 5. Vertical deformation δ_z (left) and stress ratio (right) for different external load. Grey line refers to the expected load given by machine design specification.

Finally, the balancing of the system was evaluated, which can improve the precision of the coupling in the case of eccentric loads. Through the knowledge of the center of gravity of the cart and the positioning of the applied load, the field of action of the

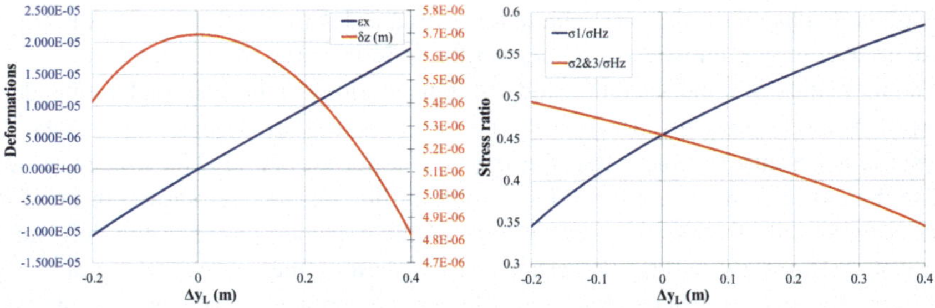

Fig. 6. Deformations (left) and stress ratios (right, for ball 1 and balls 2 & 3) for different positioning of the external load. Δy_L is the deviation from the centroid of the coupling, σHz is the allowable Hertzian stress.

counterweight was estimated by balancing the moments acting on the system, producing the result represented in Fig. 7. The extent of the blue area depends on the dimensions and weight of the counterweight selected. Balancing leads to uniformity of the deformations on the spheres, thus canceling the consequent rotation errors that would be amplified by moving away from the center of gravity of the coupling. Compared to the case without balancing, the deformations δ_z and ε_x were always lower, yielding a maximum accuracy error of 0.0138 mm.

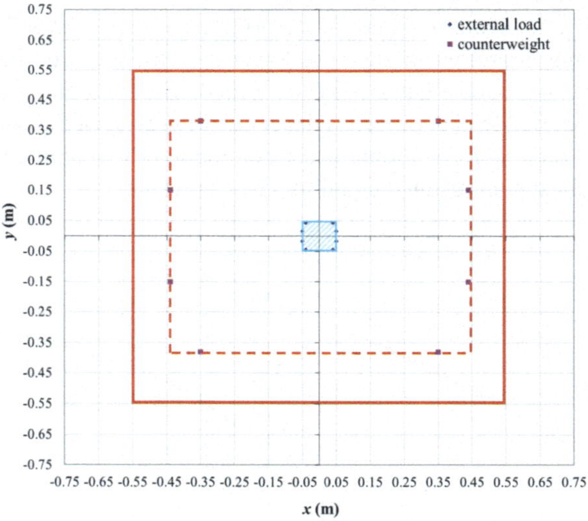

Fig. 7. Area of effect of the counterweight. Thick red line represents the cart, the red dashed line signals the area in which the counterweight must be confined to avoid obstacles. The area dotted in blue indicates where it is possible to locate the center of gravity of the load while perfect balance through the counterweight is still obtained.

5 Conclusion

The purpose of this paper was the verification of the precision guaranteed by a kinematic coupling with three grooves and spheres in a robotized laser-cutting machine, that requires a high level of accuracy.

A methodology used to estimate the positioning error was summarized, starting from the physical characteristics of the system and the application of Hertz's theory. The analysis was carried out considering typical loading conditions of the machine, that include unbalanced loads mounted on the cart in the working position. In each condition tested, the system has been shown to guarantee a positioning accuracy with errors lower than 0.02 mm at each point, as required by the application.

References

1. Vallance, R.R., Morgan, C., Slocum, A.H.: Precisely positioning pallets in multi-station assembly systems. Precis. Eng. **28**, 218–231 (2004)
2. Slocum, A.H.: Precision Machine Design. Prentice-Hall Inc., Englewood Cliffs (1992)
3. Hale, L.C., Slocum, A.H.: Optimal design techniques for kinematic coupling. Precis. Eng. **25**, 114–127 (2001)
4. Culpepper, M.L., Kartik, M.V., DiBiasio, C.: Design of integrated eccentric mechanisms and exact constraint fixtures for micron-level repeatability and accuracy. Precis. Eng. **29**, 65–80 (2005)
5. Slocum, A.H.: Kinematic couplings: a review of design principles and applications. Int. J. Mach. Tools Manuf. **50**, 310–327 (2010)
6. Slocum, A.H.: Kinematic couplings for precision fixturing—Part I: formulation of design parameters. Precis. Eng. **10**, 85–931 (1988)
7. Slocum, A.H., Muller, L., Braunstein, D.: Flexural mount kinematic couplings and method. US Patent No. US5678944A (2005)
8. Prima Industrie S.p.A. Homepage. https://www.primaindustrie.com/it. Accessed 31 Jan 2022
9. Slocum, A.H.: Design of three-groove kinematic couplings. Precis. Eng. **14**, 67–76 (1992)
10. Hale, L.C.: Testing the limiting coefficient of friction with an adjustable 3-V kinematic coupling. Precis. Eng. **64**, 200–209 (2020)
11. Roark, R.J., Young, W.C.: Formulas for Stress and Strain, 5th edn. McGraw-Hill Book Co., New York (1975)

Shape-Changing Manipulator Possibilities and the Effect of the Deformable Segment on the Size of the Working Area

Jakub Mlotek(✉), Zdenko Bobovský, Jiří Suder, Petr Oščádal, Michal Vocetka, and Václav Krys

Department of Robotics, Faculty of Mechanical Engineering, VSB - Technical University of Ostrava, 17. listopadu 2172/15, 708 00 Ostrava-Poruba, Czech Republic
jakub.mlotek@vsb.cz

Abstract. Research of customized serial manipulators seeks to robotize tasks in dense built environments, where arbitrary kinematic structures synethised for given trajectories can perform such tasks without collisions. This paper presents the effect of deformable manipulator links on manipulator working envelope and torque demand. Four manipulator structures were designed. One with unreformable segments, one with deformed segments, and two combined structures, with one deformed and one undeformed segment. The article discusses the effect of different structures on the size and shape of the manipulator working envelope. Furthermore, the flexibility of the system is mentioned in case of inserting a cubic obstacle into the manipulator's working envelope. Next, the influence of the shape of the manipulator when performing the path that all structures of the manipulator can perform is discussed. According to the necessary drive torques for the desired motion, it appears that the best manipulator is the manipulator with undeformed (straight) segments. For this reason, it is beneficial to deploy a deformable manipulator only in cases where the kinematic structure of the straight manipulator is not able to perform desired movement.

Keywords: Deformable segment · Manipulator kinematics · Adaptable robot · Robot working envelope

1 Introduction

Manipulators play an important role in today's manufacturing processes. Links between the joints are very often made of commonly available blanks, such as pipe, rod, cast, etc. All those profiles are commonly straight. The shapes of the profiles of the manipulator links define the manipulator working envelope. Manipulators with a rigid kinematic structure are not flexible to adapt for specific tasks that may occur, and this may change the requirements for the form of the working envelope. Specially manipulators on the mobile platform [1] must then adapt to conditions where the sampling point is mostly not fixed. There may be a situation where the mobile platform and the kinematic structure of the manipulator is unable to reach the pick point due to obstacles around the manipulation

A. Müller and M. Brandstötter (Eds.): RAAD 2022, MMS 120, pp. 272–280, 2022.
https://doi.org/10.1007/978-3-031-04870-8_32

point. One way is to change the manipulator configuration and change the manipulator working envelope by increasing the degree of freedom (DOF). The most used manipulators are those with 7 DOF. [2] Reshaping the manipulator's link also changes the manipulator's working envelope. The manipulator with the curved links can reach the positions, whereas the original straight-links manipulator cannot. Therefore, the possibility of curving the manipulator structure to produce an infinite number of degrees of freedom is considered [3, 4]. A deformable (can change the shape of its arm/link when needed) manipulator on a mobile platform would increase the flexibility of the system.

There exist several principles to achieve the curvature of the link segment. The first is a system which uses a friction between a lamella [5]. Vacuum is used to lock the lamellae. Another system consists of vacuum-controlled modules [6]. By scaling these modules, a modular manipulator can then be created. A solution with a low melting point material was used for the reconfigurable legs of a mobile robot [7]. Possible design of the deformable segment with low melting point material is given in [8].

This study presents the advantages of using a manipulator with deformed kinematic structure. Two shapes of deformed segments have been proposed using the principle of bending a segment following the involute in [8]. In this paper, four different manipulators structures are compared. A structure with undeformed segments and three structures with deformed segments. The segment is deformed to half and the maximum possible deformation.

2 Materials and Methods

The chapter defines the shapes of the deformed segments and manipulators kinematic structures. Furthermore, it presents how the manipulator working envelope was affected by the obstacle. Subsequently, how the given structures performed joint motion along a given path is described.

2.1 Segment and Manipulator Definition

The segments were deformed in 2D as assumed in [8], i.e. bending along the involute (Fig. 1). The radius of the base circle is given by the maximum bending radius r of the segment. In our case it is r = 50 mm. The maximum length of unwinding a line from a circle is equal to the length L of the straightened segment, in this case L = 106 mm. The diameter of segment D is 9.5 mm.

Three types of segments have been proposed and used to create the manipulator structures. Straight segment is in the position 0 (Fig. 2a), deformed segment is in position I (Fig. 2b) and second deformed segment is in position II (Fig. 2c).

Position I is the approximate midpoint of the maximum deformation of the segment. Position II is the maximum possible deformation of the segment. End-point E is designed in Table 1.

Manipulator structures with 3 degrees of freedom were designed (Fig. 3). The 3DOF were investigated because the first three joints of the angular 6DOF manipulators contribute to the resulting shape of the manipulator working envelope.

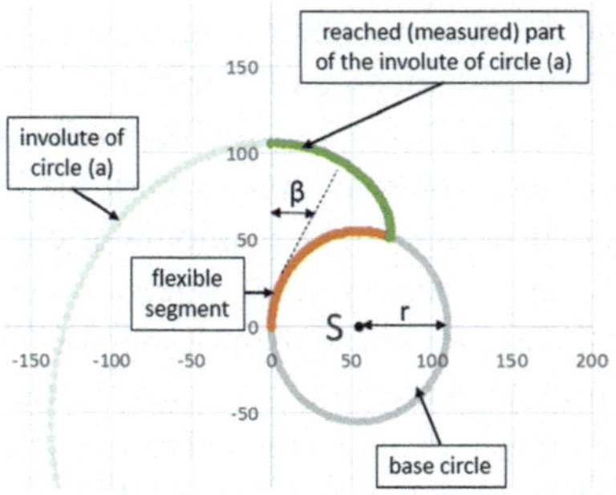

Fig. 1. Outer boundary curve of the working envelope for a deformable segment.

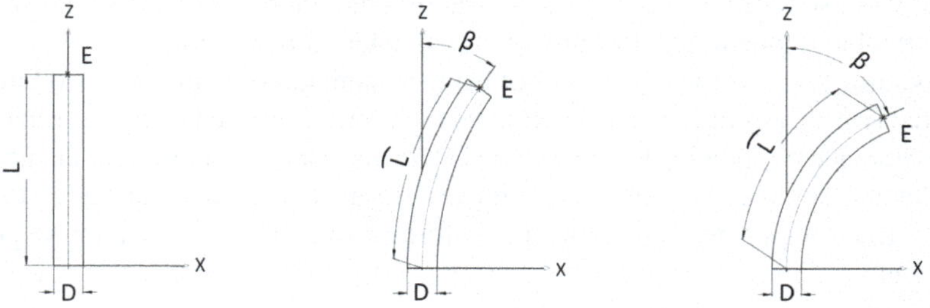

Fig. 2. (a) straight segment - position 0, (b) deformed segment - position I, (c) deformed segment - position II

Table 1. Point E coordinates

Position	X [mm]	Z [mm]	β [°]
0.	0	106	0
I.	31.99	99.27	35.72
II.	65.84	70.22	86.31

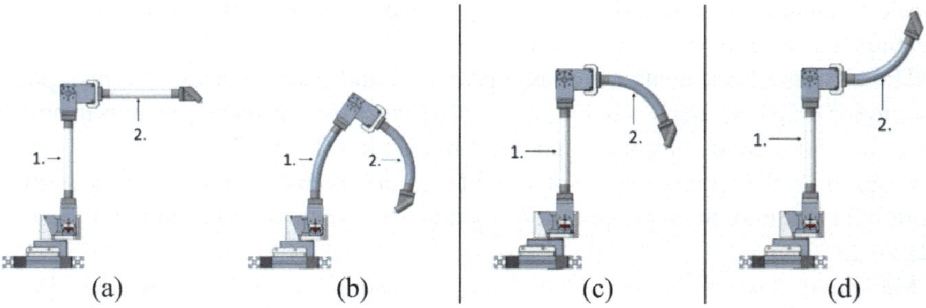

Fig. 3. Manipulator with segment (a) non-deformed (1., 2.) (b) deformed – 1. position I., 2. position II. (c, d) 1. non-deformed, 2. deformed – position II.

The first structure is composed of straight undeformed segments (Fig. 3a). The second structure (Fig. 3b) has the first two segments deformed into position I. and the third member into position II. The third structure (Fig. 3c) is then a combination of the first non-deformed segments and the third segment deformed also in to position II. The fourth structure (Fig. 3d) is the same as (Fig. 3c) with the difference that the third segment is deformed to the opposite side.

2.2 Changing the Working Envelope of the Manipulator When Blocked by an Obstacle and Performing a Shared Path

To determine how the manipulator working envelope change differs for each structure an obstacle was placed in the working envelope for individual structures. The obstacle is placed in the working envelope and configuration area of the manipulator, and it is a cube with a side size of 100 mm. The center of obstacle was placed into a grid in XZ plane (Fig. 4a) and a percentage remaining size of working enveloped was recorded. The points on the grid are 10 mm apart in the X and Z axes.

They were also simulating the passage of a circular path. The circle has a diameter of 125 mm. The path is oriented at two angles: 0° and 90° (Fig. 4b). The center of the path is placed in a 2D grid of 26 × 26 points in front of the manipulator (Fig. 4c), the grid is again in XZ plane, the circular paths are in XY and YZ planes. The points on the grid are 10 mm apart in the X, Z axes.

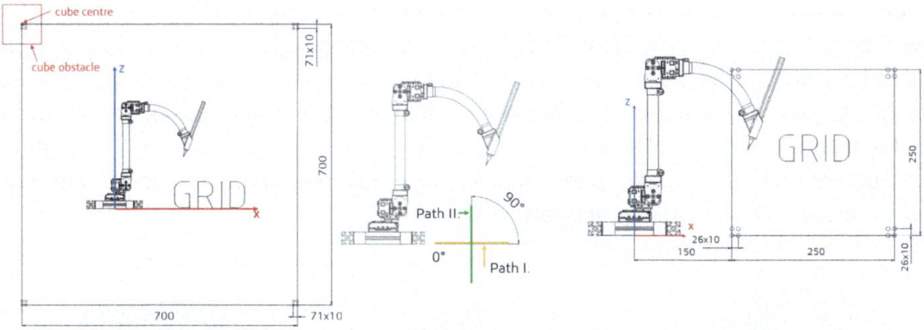

(a) (b) (c)

Fig. 4. (a) manipulator with the grid boundaries, the obstacle is represented by a red square that shifts position 71 times by 10 mm (b) projection of the paths in XZ plane (c) grid of path centers

3 Results

The percentage change of the working envelope when an obstacle is inserted into working envelope and configuration area of the manipulator is presented. The extension of the working envelope around the obstacle is shown if the segments could be deformed. Common paths that can perform all manipulators structures are found. Based on the measured torque, it was determined for which manipulators structure the motion is the least demanding.

3.1 Manipulator Working Envelope Results

The resulting working envelopes are displayed in (Fig. 5). With deformed segments (Fig. 5b), the reachable working envelopes near the base of the manipulator increases. The maximum range decreases. (Fig. 5c, d) represents a combination of these two properties. (Fig. 5e) shows sum of all working envelopes. Here, the presented advantage of using an on-demand curved manipulator can be seen. The overall working envelope has extended significantly to the base of the manipulator and below the base.

(a) (b) (c) (d) (e)

Fig. 5. Manipulator working envelope (a) non-deformed segments, (b) deformed segments, (c, d) combinations of non-deformed and deformed segments (e) union working envelope

3.2 Changing the Working Envelope of the Manipulator When Blocked by an Obstacle Result

In (Fig. 6a ÷ c) is shown the percentage remaining size of the working envelope when inserting an obstacle. Each position represents an obstacle location coordinate in the grid. The color shows what percentage of the original size of the envelope remains. The black area represents no possibility to calculate the working envelope due to the blockage of the mechanical structure by the obstacle. In (Fig. 6c) is shown the result for the structure C (Fig. 3c, d), the percentage change of the manipulator working envelope of the structure D is the same, only inverted.

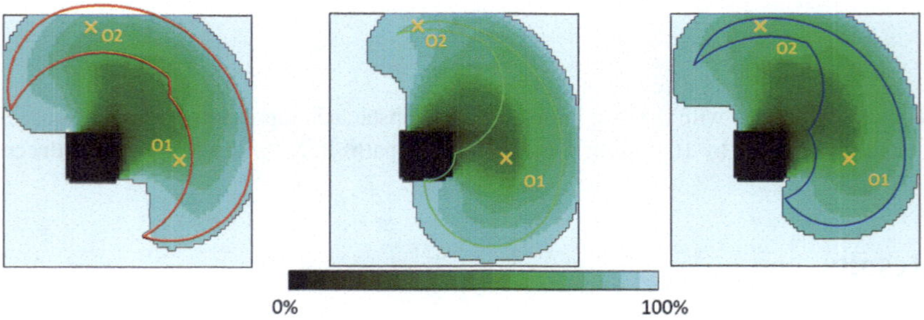

(a) (b) (c)

Fig. 6. Percentage of change in the size of the working envelopes, the colored lines represent the original envelopes; (a) non-deformed segment, (b) deformed segments, (c) combinations of non-deformed and deformed segments.

Based on the previous analysis (Fig. 6) we chose two points to demonstrate how the working envelope is affected when an obstacle is inserted in. The points are located at the coordinates [x, z] given in mm - O1 [250, 50] and O2 [0, 420]. The results are visualized in (Fig. 7). The deformed segments (Fig. 7b ÷ d) increase the reachable area near the obstacle. Only exception is the case of D structure (Fig. 7d) around point O1 where the opposite phenomenon occurs due to the shape of the robot. The intersection of all four working envelopes can be seen in (Fig. 7e).

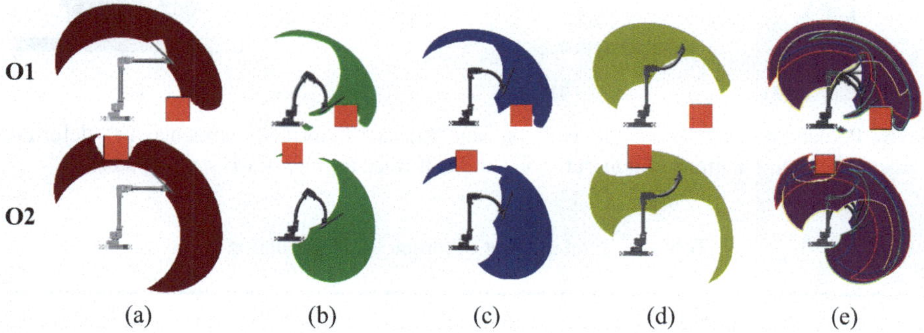

Fig. 7. Manipulator working envelope around a cubic obstacle (a) non-deformed segment, (b) deformed segments, (c, d) combinations of non-deformed and deformed segments (e) union working envelope

3.3 Performing a Shared Path Result

The passage through the circular path was simulated and recorded as a percentage of the passage success (Fig. 8a ÷ c), i.e., how much of the trajectory the manipulator reached. The simulations were performed using CoppeliaSim and the physics engine Newton Dynamics [9]. The passage was implemented only for the structure (Fig. 3a ÷ c). For the structure (Fig. 3d), this was not solved due to the minimal size of the working envelope in the path grid area. In (Fig. 8d), the passage of the paths was merged for all three structures and found common paths.

From the common paths (Fig. 8d), two paths were selected. The paths centers are located at the coordinates [x, z] given in mm - P1 [310, 210] for 0° and P2 [260, 250] for 90°. The common paths were re-simulated the course of torque for all three joints was recorded. This torque course has been integrated and displayed as total amount of the manipulator torque Table 2.

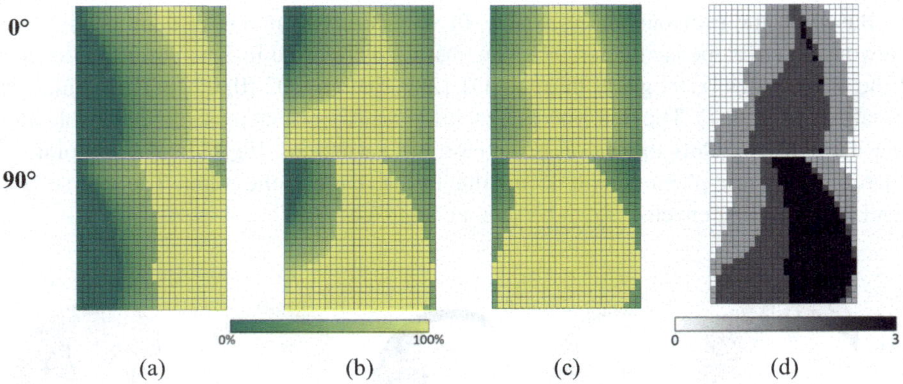

Fig. 8. Percentage success of the path passage (a) non-deformed segments, (b) deformed segments, (c) combinations of non-deformed and deformed segments, (d) shared paths

Table 2. Total amount of torque for manipulator

	J1 [Nm]		J2 [Nm]		J3 [Nm]		ΣJ [Nm]	
segments	0°	90°	0°	90°	0°	90°	0°	90°
straight	0.01	0.002	7.8	4.517	3.3	3.25	11.11	7.769
deformed	0.004	0.002	8.432	5.228	2.846	2.677	11.282	7.907
combination	0.004	0.002	9.765	6.598	2.87	2.709	12.639	9.309

4 Conclusions and Future Work

This study presents the advantage of the increase of the working envelope compared between the four individual manipulators which differ in the deformation of their links. The resulting working envelope is the sum of the individual working envelopes (Fig. 5e) and (Fig. 7e), which highlights the benefit of on-demand deformable manipulators that can change their kinematic structure between operations. To evaluate the difficulty of moving individual manipulators along a given path, simulations of the passage of a circular path were performed. The path relative to the manipulator was rotated by 0° and 90°. Using these simulations, common paths were determined for structures (Fig. 3a ÷ c). The torque was measured for these paths. If a structure can traverse through a given path, from the torque point of view the optimal manipulator is the one with straight segments as shown in Table 2. The advantage of deformable segments then lies in expanding the working envelope and the number of reachable paths.

The next step in solving the problem is to design and provide a deformable segment based on the data given in [8] and related methods. This includes experimental measurements of the cover for deformable core, FEM and real stress analysis in the segments. It is also necessary to investigate the effect of segment deformation on accuracy, repeatability, and maximum payload. The deformation of the core will affect the material parameters of the deformable core. Another future work includes an automatic shape-segment deformation. Now, there are several possible principles, for example to be

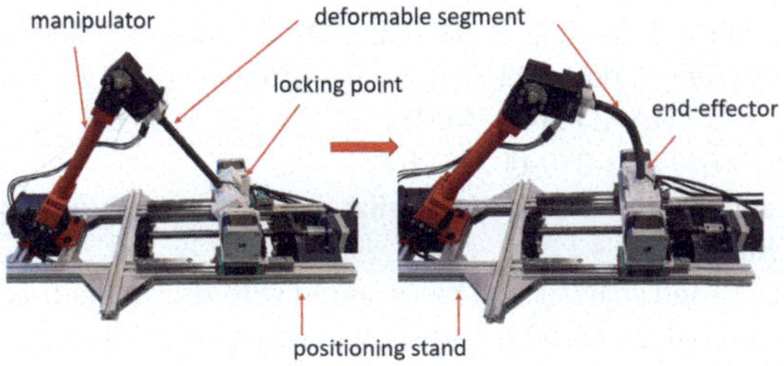

Fig. 9. Initial stand design. Left side image is manipulator before deformation. Right side image is after deformation.

able to deform the manipulator in a stand using the joints and the pre-defined movement of an end-effector locking point. The initial design of the stand can be seen in (Fig. 9). Another option is to use similar principles to hyper redundant robots. In [10] a methods of changing the shape of the control wires or a possible combination of control wires and springs were proposed.

Acknowledgements. This article has been elaborated under support of the project Research Centre of Advanced Mechatronic Systems, reg. no. CZ.02.1.01/0.0/0.0/16_019/0000867 in the frame of the Operational Program Research, Development and Education. This work has been also supported by specific research project SP2022/67and financed by the state budget of the Czech Republic.

References

1. Engemann, H., Du, S., Kallweit, S., Cönen, P., Dawar, H.: OMNIVIL—an autonomous mobile manipulator for flexible production. Sensors **20**, 7249 (2020). https://doi.org/10.3390/s20247249
2. Zheng, D., Wu, X., Pang, J.: Real-time whole-body obstacle avoidance for 7-DOF redundant manipulators. In: Proceedings of the 2021 33rd Chinese Control and Decision Conference (CCDC); Kunming, China, 22 May 2021, pp. 2626–2631. IEEE (2021)
3. Huczala, D., Kot, T., Pfurner, M., Heczko, D., Oščádal, P., Mostýn, V.: Initial estimation of kinematic structure of a robotic manipulator as an input for its synthesis. Appl. Sci. **11**, 3548 (2021). https://doi.org/10.3390/app11083548
4. Kot, T., Bobovský, Z., Brandstötter, M., Krys, V., Virgala, I., Novák, P.: Finding optimal manipulator arm shapes to avoid collisions in a static environment. Appl. Sci. **11**, 64 (2021). https://doi.org/10.3390/app11010064
5. Clark, A.B., Rojas, N.: Design and workspace characterisation of malleable robots. In: 2020 IEEE International Conference on Robotics and Automation (ICRA), pp. 9021–9027 (2020). https://doi.org/10.1109/ICRA40945.2020.9197439
6. Robertson, M.A., Paik, J.: New soft robots really suck: vacuum-powered systems empower diverse capabilities. Sci. Robot. **2**, eaan6357 (2017). https://doi.org/10.1126/scirobotics.aan6357
7. Nakai, H., Kuniyoshi, Y., Inaba, M., Inoue, H.: Metamorphic robot made of low melting point alloy. In: Proceedings of the IEEE/RSJ International Conference on Intelligent Robots and Systems, vol. 2, pp. 2025–2030, September 2002

8. Brandstötter, M., Gallina, P., Seriani, S., Hofbaur, M.: Task-dependent structural modifications on reconfigurable general serial manipulators. In: Aspragathos, N.A., Koustoumpardis, P.N., Moulianitis, V.C. (eds.) RAAD 2018. MMS, vol. 67, pp. 316–324. Springer, Cham (2019). https://doi.org/10.1007/978-3-030-00232-9_33
9. Dynamics Available online. https://coppeliarobotics.com/helpFiles/en/dynamicsModule.htm. Accessed 7 Mar 2022
10. Li, Z., Du, R.: Design and analysis of a bio-inspired wire-driven multi-section flexible robot. Int. J. Adv. Rob. Syst. **10**, 209 (2013). https://doi.org/10.5772/56025

Analysis of Impacts of the Design Parameters on the Working Frequency of the Compliant Devices

Jaroslav Hricko(✉) and Stefan Havlik

Institute of Informatics, Slovak Academy of Sciences, Banská Bystrica, Slovakia
{hricko,havlik}@savbb.sk

Abstract. This work aims to evaluate the dynamic properties and natural frequencies of one group of micro-robotic devices - flexible mechanisms. When designing such mechanisms the kinetostatic analysis and dynamic properties are calculated especially in the cases of the specific devices such as fast-moving positioning apparatus, motion/force amplifiers, deformable bodies of sensors, etc. The main task is to elaborate on how characteristic dimensions of flexural parts influence the natural frequency of proposed devices. Using the sensitivity analysis the dependence can be evaluated. It is shown that the quality of a device can be achieved by suitable form, geometry, and the dimension of flexural parts. Two examples are chosen to document the dependence of the mechanism's natural frequency: the Scott-Russell mechanism with elastically revolute joints and the flexural body of the force sensor. In the case of the first example is shown various forms of flexural joints and their dimensions have different influences on final frequency. For most frequently used forms of joint/notches the frequencies are calculated and the dependence on design dimensions is documented. For the force sensor, the sensitivities of natural frequency on dimensions of the flexural body are evaluated similarly.

Keywords: Compact compliant mechanisms · Dynamics of compliant devices · Sensitivity analysis · Micro-robotic devices · Sensors of mechanical quantities

1 Introduction

Micro-robotic devices that utilize for its motion and transformation of forces elastic deformation of its bodies (thru flexure hinges, and/or arms) belong to a group of compliant mechanisms [1]. Such devices provide many advantages, such as a movement without backlash, which enables to development of positioning devices with accuracy in the range of nanometers. There is a wide range of possible applications [1], like manipulation with small objects (with the greatness of cells), arranging/positioning of micro-mirrors [2, 3] and a wide group of their applications is in the area of the deformable bodies of sensors too (force, torque, accelerometers, gyroscopes, etc.). In the case of the manipulation tasks, such a device works in quasi-static modes, and analysis of its dynamics properties is not usually required.

A. Müller and M. Brandstötter (Eds.): RAAD 2022, MMS 120, pp. 281–289, 2022.
https://doi.org/10.1007/978-3-031-04870-8_33

On the other side, some tasks require higher working frequencies. For instance, micro-robotic positioning mechanisms, mainly various mechanical amplifiers of piezo-electric actuators [4, 5], where the working frequency is higher than *500* Hz. The deformable bodies of sensors (mainly sensors of mechanical quantities) use principles of compliant mechanical devices too. Their dynamical properties play a piece of essential information in the description of their performance. Except for accelerometers and gyroscopes that utilize torsions springs and achieve working frequencies in the range of kilohertz, the deformable sensor's bodies of the force/torque sensors are developed [6, 7, 15], where the working frequencies are around in a range of hundreds of hertz.

It is clear that dynamical analysis and natural frequencies play some task in the design of compliant micro-robotic devices. Unfortunately, most of the works deal with the basic calculation of dynamics and natural frequencies, but usually only on the solved device, without focusing on finding out which one parameter of mechanism has what kind of influence to own dynamics properties or natural frequency. This work aims to focus on solving the dynamics properties and natural frequencies of flexible mechanisms. The main task is to elaborate the sensitivity analysis where the influence of characteristic dimensions of the flexures on the natural frequency of proposed devices is studied.

2 Dynamics and Natural Frequency of Compliant Micro-robotic Devices

Calculation of the dynamics properties of compliant mechanisms utilize the Lagrange-based dynamic modeling approaches and among the most known methods can be included Pseudo-Rigid-Body (PRB) model described by Howell [1]. Such an approach is used for the kinetostatic description of the compliant mechanism and is based on the replacement of the flexure hinge by the torsion spring. Then the existing theory of rigid body mechanisms is applied. The equivalent stiffness K of the torsion spring that replaces the flexure hinge is expressed as [11]

$$K = \gamma K_{\theta} \frac{EI}{l} \tag{1}$$

where γ is the characteristic radius factor, K_{θ} is the stiffness component (corresponding with in-plane rotation), E is Young's modulus and I is the moment of inertia of the beam. Subsequently, the principle of the dynamic equivalence is utilized for the PRB dynamic model (PRBDM). Then the described equations of the kinetic energy T and potential energy U are substituted into the Lagrange equation

$$\frac{d}{dt}\left(\frac{\partial T}{\partial \dot{x}}\right) - \left(\frac{\partial T}{\partial x}\right) + \left(\frac{\partial U}{\partial x}\right) = Q \tag{2}$$

where Q is a generalized force acting to produce a change in coordinate x. According to such an approach can be achieved the lumped-parameter model (see (3)). Rem.: The described approach (PRBDM) is one of the possible descriptions and neglect influence of damping. To achieve particular matrices of lumped parameter model could be used different approaches too [8].

In general, the compliant device can be described as a mass-damping-spring system, where the system equation is in the general form

$$M\ddot{x} + B\dot{x} + Kx = 0 \tag{3}$$

where **M**, **B**, and **K** are mass, damping, and stiffness matrices respectively, and for damping and stiffness matrices non-diagonal elements represent the cross/transversal sensitivities. Compared with common kinematic structures, in the case of the compliant mechanisms, the values of the mass/damping/stiffness matrices should be estimated based on the physical variables like dimensions, material properties, the shape of flexures, etc. Rem.: the expression of the damping matrix **B** is in such devices problematic and some works neglected it, but for instance work [9] describes damping as a function of mass and stiffness, and work [10] consider Coriolis effects and artificial damping. A possible approach of the calculation of the stiffness matrix **K** is described above where PRBDM is used, or the stiffness matrix method [12] can be utilized too.

One of the important characteristics in the dynamic analysis is the natural frequency ω_0. It can be expressed, when (3) is rearranged in the form [13]

$$\lambda[M\{V\} + B\{V\} + K\{V\}]\{X\} = 0 \tag{4}$$

where $\{\mathbf{V}\} = \lambda\{\mathbf{X}\}$, what correspondent with velocity eigenvector and then (4) is

$$\lambda \begin{pmatrix} X \\ V \end{pmatrix} = \begin{pmatrix} 0 & I \\ -M^{-1}K & -M^{-1}B \end{pmatrix} \begin{pmatrix} X \\ V \end{pmatrix} \tag{5}$$

where $\boldsymbol{I}$ is an identity matrix, and component $-\mathbf{M}^{-1}\mathbf{K}$ express eigenvalue of which is expressed natural frequency

$$\omega_0 = \sqrt{M^{-1}K} \tag{6}$$

3 Sensitivity Analysis

To achieve optimal design of the compliant micro-robotic devices with the desired performance in dynamic modes it is necessary to perform multi-criteria optimization. To identify which one design parameter has the majority impact on dynamical properties and mainly on a natural frequency, the sensitivity analysis should be carried out.

Equation (6) show that only mass and stiffness have an impact on the value of natural frequency. In the case of the moving mass are limited possibilities. For instance, the advantages of additive manufacturing (3d printing) to minimize moving mass are shown in [7]. Another possible change of the moving mass is the utilization of the topology optimization of the rigid bodies, where the parameters (like rigidity) stay equal, but manufacturing of such structures can be complicated.

The influence of stiffness on the dynamical properties of the compliant device should be carefully analyzed. In such a case, the stiffness depends on design parameters. For instance, if only main building blocks are considered, i.e. flexure hinges then the design variables are t – the thickness (the thinnest place of flexure), l – length, w – width of the flexure hinge, and variables connected with notch type (e.g. r – radius of a notch for right-circular hinge, c – as a parameter of the elliptical/hyperbolic notch, etc.). It is expected that the Young modulus will be constant. In the dynamical analysis of the whole mechanism, dimensions of the rigid parts (that connect flexure hinges) should take into account too. Otherwise considering the assumption that the proposed device is optimized from the view of desired motions, then such dimensions will have minimum impact on eventual dynamics. It results that, the sensitivity analysis should be focused on the design parameters of the flexure hinges.

The mathematical description of flexure hinges with a notch, based on Castigliano's theorems is elaborated in [14]. It is shown that for one flexure hinge is expressed 6 × 6 compliance/stiffness matrix that includes in-/out-plane displacements. Considering that the main part of a movement is expressed by in-plane bending then the sensitivity analysis is focused on such component of stiffness matrix that is expressed as

$$K_{\theta zMz} = \frac{C_{yFy}}{C_{\theta zMz}C_{yFy} - C_{yMz}^2} \tag{7}$$

where C_{iy} are components of compliance matrix where index i correspondent with displacement/bending in particular axis and index j correspondent with a load that such movement causes. After substitution (7) to (6) can be expressed sensitivity function as

$$\boldsymbol{S}_{\omega 0} = \frac{\partial \omega_0(\boldsymbol{p})}{\partial \boldsymbol{p}} \tag{8}$$

where **p** is a vector of design parameters, i.e. for the case of the right circular hinge is **p** $= [m, w, r, t]^{\mathrm{T}}$. Considering that the natural frequency is not a time-dependent function, the design parameters should be changed in the allowable range.

The influence of design parameters on the natural frequency of the selected flexure hinges with a notch is studied as an example. Considering the complex mathematical description of the stiffness $K_{\theta zMz}$, the numerical values have been chosen as follows. Young modulus *210* GPa (steel), the width of the hinge $w = 10$ mm, length of the hinge $l = 2$ mm, the thickness of the flexure hinge $t = 0.2$ mm. As a moving mass is expected beam from steel with dimensions width $= w$, height $= l + t$, length $= l$, that is connected to a free end of the flexure hinge. The value of moving mass is $m = 0.3454$ g. In the analysis is expected the change of the design parameters in the range of 80% to 120% of their default value. The influence of the design parameters on the natural frequency of chosen flexure hinges show in Fig. 1. (Rem.: evaluation of results are discussed in Conclusion).

4 Sensitivity Analysis of the Scott-Russell Linkage

As an example of the sensitivity analysis of the micro-robotic device was chosen Scott-Russell (S-R) straight-line mechanism. Such a mechanism is often used in nano-manipulation applications [19] or by the design of micro-grippers [20]. The sensitivity analysis has been carried out by FEM, where the whole mechanism's body was from

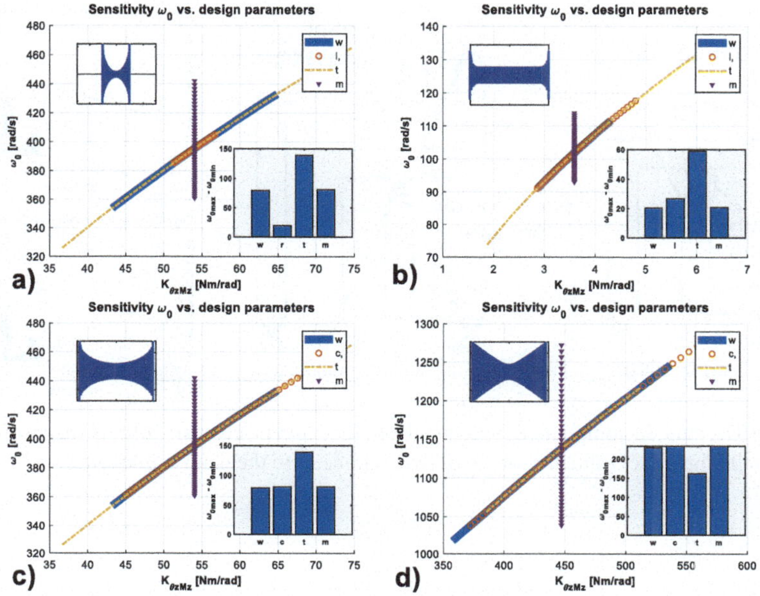

Fig. 1. The influence of the design parameters to the natural frequency for chosen flexure hinges. a) right-circular hinge ($r = l/2$), b) corner-filleted flexure hinge ($r = 0.1*l$), c) elliptical flexure hinge ($c = 0.0003$ m), d) hyperbolic hinge ($c = 0.0003$ m)

3D printed polyamide (nylon) with $E = 580$ MPa, with density $\rho = 1140$ kg/m^3. As is shown in Fig. 2, right-circular hinges were used with default dimensions $r = 1.25$ mm and $t = 1$ mm. The main dimensions of the mechanism are $30 \times 55 \times 10$ mm. To minimize the influence of the rigid parts on the mechanism's dynamic properties, there are their dimensions constant and they are connected with blocks that include flexure elements (see Fig. 2). The dimensions of such block are independent of the dimensions of the hinge. The added parallelogram to the Scott-Russell mechanism is one of the preferred solutions to minimize the rotation of the output point that moves on the straight line. Therefore, the sensitivity analysis was carried out on the compliant Scott-Russell mechanism with added parallelogram too.

5 Sensitivity Analysis of Sensor's Deformable Body

In our previous works [15, 16] we have developed a new wireless force sensor that utilizes change of the parameters of the electromagnetic field for identifying the size of the acting load. With this approach, it is possible to achieve very high resolution (up to 10^{-6} N and better) in the large range of force measurement. The principle has been verified on laboratory prototypes from nylon made by 3D printing technology. The sensor deformable-body for industrial application was produced from intrinsic Silicon (i-Si) with dimensions 3.2×1.5 mm (see Fig. 3). The width of the sensor body is a direct correspondent with the range of measured load (e.g. for $w = 0.4$ mm is range 25 N, but

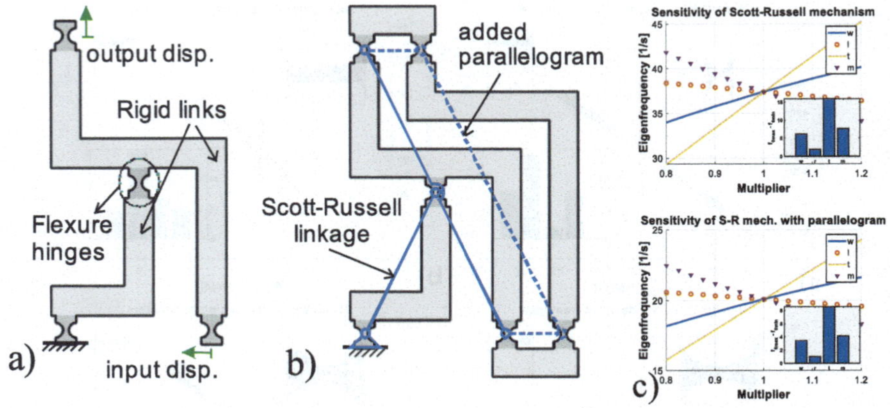

Fig. 2. a) The studied compliant Scott-Russell straight-line mechanism, b) S-R mechanism with added parallelogram, c) influence of the design parameter to the natural frequency.

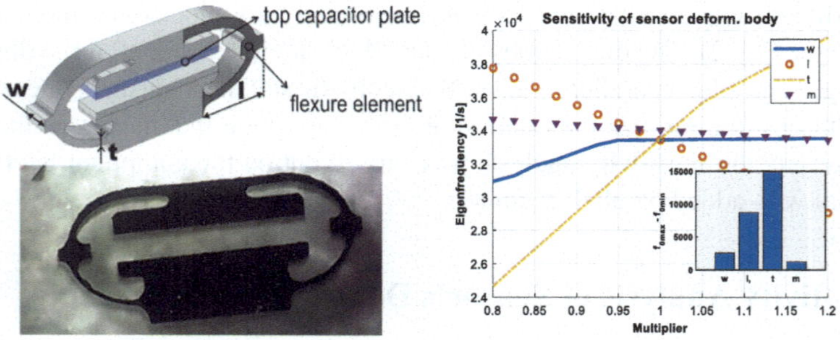

Fig. 3. The new force sensor's deformable body (left top - main design parameters, left bottom – prototype made from i-Si), right - the influence of the design parameters to the natural frequency for proposed sensor body

for $w = 1.5$ mm it is *120* N). The reactive-ion etching [17, 18] has been used for the production of the sensor body.

The sensitivity analysis of deformable-body has been executed by FEM analysis in Comsol Multiphysics. The influence of four characteristic design parameters on the eigenfrequency was studied. Similarly even, in this case, the range of parameters was $\pm 20\%$. In the case of the moving mass, the density of the top capacitor plate was changing, therefore, in the results (see Fig. 3); this dependence does not overlap with the other dependencies in the parameter's multiplier equal *1.0*.

From the results of sensitivity is clear, that the main influence on the eigenfrequency of the sensor body has the thickness and length of the flexure element. Simultaneously can be concluded that nonlinear behavior is present mainly in the case of the change of width of the sensor body, and for higher values of it is present minimum influence to the eigenfrequency.

6 Conclusion

This work deals with the analysis of the dynamical properties of the compliant micro-robotic devices. It is considered the basic description based on the pseudo-rigid body dynamic model and then application of the lumped parameter model for final the mass-spring-damper system. The main result achieved from the sensitivity analysis is the evaluation of characteristic dimensions of the flexure hinges on the natural frequency. From this result (Fig. 1) can be seen that the thickness of the compliant joint has the main influence on natural frequency, as expected. But the value of hinge thickness cannot change, there are technological constraints on its minimal value. Therefore, the other forms and geometry should be taken into account. Interesting results are achieved when comparing the characteristics or sensitivity of hinges with the circular and elliptical both-side notch. As it is shown the radius of a circular notch has only a small influence on natural frequency. Rem.: It is only the specific type of elliptical notch. In addition; the thickness in the case of the hyperbolic notch has the smallest influence on natural frequency compared to other notches.

Two examples were selected to document the influence of some design parameters on dynamical properties: the Scott-Russell mechanism with elastic joints and the elastic body of the force sensor. In the case of the Scott-Russell mechanism, the hypothesis that the design parameters of elastic joints have the main influence on the natural frequencies was confirmed. This conclusion is obvious by comparing sensitivity graphs in Fig. 1 and Fig. 2. The different type of flexible body for the force sensor deflects as the curved beams. The influence of design parameters is clear from Fig. 3, where again the thickness has the most important influence on own frequency. Further important conclusion is that another parameter, i.e. the width of the flexural sensor body has a small influence on frequency.

Acknowledgment. This work was supported by the national scientific grant agency VEGA under project No.: 2/0155/19 – "Processing sensory data via Artificial Intelligence methods" and by project APVV-20-0042 – "Microelectromechanical sensors with radio frequency data transmission" and by project APVV-14-0076.

References

1. Howell, L.L.: Compliant Mechanisms. Wiley-IEEE, p. 459 (2001). ISBN 047138478X
2. Stepanovsky, M.: A comparative review of MEMS-based optical cross-connects for all-optical networks from the past to the present day. IEEE Commun. Surv. Tutor. **21**(3), 2928–2946 (2019). https://doi.org/10.1109/COMST.2019.2895817
3. Furlong, C.: Analysis and optomechanical characterisation of a MEMS micromirror device. Strain **45**, 259–266 (2009). https://doi.org/10.1111/j.1475-1305.2008.00476.x
4. Chen, S., Ling, M., Zhang, X.: Design and experiment of a millimeter-range and high-frequency compliant mechanism with two output ports. Mechanism and Machine Theory **126**, 201–209 (2018). https://doi.org/10.1016/j.mechmachtheory.2018.04.003. ISSN 0094-114X
5. Ouyang, P.R., Zhang, W.J., Gupta, M.M.: Design of a new compliant mechanical amplifier. In: ASME. International Design Engineering Technical Conferences and Computers and Information in Engineering Conference, Volume 7: 29th Mechanisms and Robotics Conference, Parts A and B, pp. 15–24 (2005)
6. Jun, J., Jingmeng, L., Weihai, C., Wenjie, C.: Development of a novel uniaxial force sensor with two-stage force resolutions. Int. J. Appl. Electromagnet Mech. **42**, 27–39 (2013). https://doi.org/10.3233/JAE-121642
7. Hricko, J., Havlik, S.: Design and optimization of mechanics for two axes force/displacement E-M sensor. In: MEASUREMENT 2017: Proceedings of the 11th International Conference on Measurement, pp. 99–102 (2017). https://doi.org/10.23919/MEASUREMENT.2017.7983545. ISBN 978-80-972629-0-7
8. Ling, M., Howell, L.L., Cao, J., Chen, G.: Kinetostatic and dynamic modeling of flexure-based compliant mechanisms: a survey. ASME. Appl. Mech. Rev. **72**(3), 030802 (2020). https://doi.org/10.1115/1.4045679
9. Hricko, J.: Modelling compliant mechanisms - comparison of models in MATLAB/SimMechanics vs. FEM. In: RAAD 2012: 21th International Workshop on Robotics in Alpe-Adria-Danube Region. - Naples : Edizioni Scientifiche e Artistiche, pp. 57–62 (2012). ISBN 978-88-95430-45-4
10. Vedant, F., Allison, J.T.: Pseudo-rigid body dynamic modeling of compliant members for design. In: Proceedings of the ASME 2019 International Design Engineering Technical Conferences and Computers and Information in Engineering Conference, Vol. 2A: 45th Design Automation Conference. Anaheim, California, USA. August 18–21 (2019). https://doi.org/10.1115/DETC2019-97881
11. Yu, Y., Howell, L.L., Lusk, C., Yue, Y., He, M.: Dynamic modeling of compliant mechanisms based on the Pseudo-rigid-body model. ASME. J. Mech. Des. **127**(4), 760–765 (2005). https://doi.org/10.1115/1.1900750
12. Ottaviano, E., Carbone, G., Ceccarelli, M.: Workspace analysis and performances of a binary actuated parallel manipulator with flexural joints. J. Mech. Eng. Sci. **217**, 313–330 (2003)
13. Smith, T.S.: Flexures: Elements of Elastic Mechanisms., Gordon and Breach Science Publishers (2000). ISBN 90-5699-261-9
14. Lobontiu, N.: Compliant Mechanisms: Design of Flexure Hinges, CRC Press (2003). ISBN 0849313678
15. Hartansky, R., et al.: Novel method of contactless sensing of mechanical quantities. In: Measurement Science Review, vol. 20, no. 3, pp. 150–156 (2020). https://doi.org/10.2478/msr-2020-0018. ISSN 1335-8871
16. Hartansky, R., Mierka, M.: New method of one-component force measurement. In Techničeskije universitety: integracija s evropejskimi i mirovymi sistemami obrazovanija : materialy VIII Meždunarodnoj konferencii, Rossija, Iževsk, 23. - 24. 04. 2019. Iževsk : Izdateľstvo IžGTU imeni M.T.Kalašnikova (2019). ISBN 978-5-7526-0832-2

17. Škriniarová, J., et al.: Investigation of volume fraction of GaP nanowires by SEM characterization and spectroscopic ellipsometry. In: Optik: International Journal for Light and Electron Optics, vol. 234 (2021). https://doi.org/10.1016/j.ijleo.2021.166572. ISSN 0030-4026
18. Kostic, I., et al.: Experimental and theoretical study on chemically semi-amplified resist AR-P 6200. In: Journal of Physics: Conference Series, vol. 992 (2018). https://doi.org/10.1088/1742-6596/992/1/012057. ISSN 1742-6588
19. Tian, Y., Shirinzadeh, B., Zhang, D., Alici, G.: Development and dynamic modelling of a flexure-based Scott-Russell mechanism for nano-manipulation. Mech. Syst. Sig. Process. **23**(3), 957–978 (2009). https://doi.org/10.1016/j.ymssp.2008.06.007
20. Zhu, J., Hao, G.: Design and test of a compact compliant gripper using the Scott–Russell mechanism. Arch. Civil Mech. Eng. **20**(3), 1–12 (2020). https://doi.org/10.1007/s43452-020-00085-3

A Robotics Perspective on Architecture: Modelling and Control of Reconfigurable Buildings

Eftychios G. Christoforou[1(✉)], Loukas Georgiou[1], Marios C. Phocas[2], Loucas S. Louca[1], and Andreas Müller[3]

[1] Department of Mechanical and Manufacturing Engineering, University of Cyprus, Nicosia, Cyprus
e.christoforou@ucy.ac.cy
[2] Department of Architecture, University of Cyprus, Nicosia, Cyprus
[3] Institute of Robotics, Johannes Kepler University, Linz, Austria

Abstract. Reconfigurable architecture is a disruptive concept with unique opportunities for creativity in design towards more usable buildings, better energy efficiency and improved comfort for the occupants. From an engineering perspective, implementations involve mechanism design and control, while analytical approaches from robotics become relevant. A structural and control concept is presented, which allows stepwise reconfigurations for a building between an initial and a final position. Simulation and experimental results are presented to demonstrate the applicability of the approach. Dynamic analysis of the basic mechanism is also pursued and its role in providing design input is demonstrated.

Keywords: Reconfigurable architecture · Adaptive architecture · Reconfigurable mechanisms · Mechanism synthesis · Motion planning

1 Introduction

In response to constantly changing needs of modern societies many scientific and engineering fields often pursue new directions, while traditional concepts are reconsidered. In architecture the importance of reconfigurable buildings is often acknowledged and has recently attracted the interest of the scientific and engineering communities. Advantages include better space utilization, improved comfort for the occupants by adjusting lighting and ventilation conditions, and improved performance of a photovoltaic roof. Also, unique features including foldability and erectability become possible which can be useful for transportable structures, emergency shelters, storage rooms for seasonal products, etc. Developments in structural engineering, materials, robotics as well as other enabling technologies, constitute reconfigurable architecture a feasible option. Reconfigurable mechanisms provide a framework for such realizations and relevant methodologies from the field of robotics are readily applicable. Note that reconfigurable mechanisms find applications in various engineering fields including robotics

A. Müller and M. Brandstötter (Eds.): RAAD 2022, MMS 120, pp. 290–297, 2022.
https://doi.org/10.1007/978-3-031-04870-8_34

(e.g., manipulation, locomotion) and aerospace engineering. Relevant implementations may involve linkage systems, scissor-like elements and tensegrity structures [1]. Artistic techniques are often relevant including origami (the art of paper folding), from which inspiration and implementation ideas can be drawn. Reconfigurable buildings provide a basis for adaptive architecture, which refers to buildings capable of automatically adjusting their shape in response to external stimuli, including the sun motion or the wind conditions. In the latter case, the shape is accordingly adjusted towards reducing the aerodynamic loads on the structure.

A structural and control concept for reconfigurable architecture was presented in [2] and was further examined in [3–7]. A parallel arrangement of planar mechanisms with interconnections between them generate the skeleton structure of the building. A membrane material can be stretched over the structure to define the building envelope, as shown in Fig. 1. Alternatively, other elements can be installed on the structure including shades or photovoltaic panels. Note that the design of such solutions can be an important part of architectural courses [8]. Simultaneous reconfigurations of the planar mechanisms accordingly adjust the overall shape of the building. The control method requires a single motion actuator and brakes installed on each rotational joint. In general, it is desirable that a minimum number of actuators is involved and that these are stationary located at the ground, to avoid multiple moving drive aggregates. Two types of basic planar mechanisms were proposed and the corresponding methods were the "effective 4-bar" (E4B) and the "effective crank-slider" (ECS), respectively.

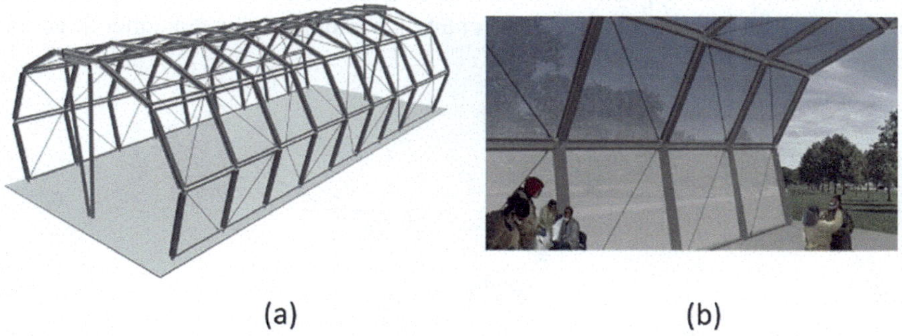

Fig. 1. Representation of the structural concept. (a) Skeleton structure of the building. (b) Rendering of the building based on the proposed concept.

In the present work, a different, structurally more robust basic planar mechanism is proposed called the "coupled effective crank-slider" (CECS) which is subject to a conceptually similar control approach, as explained in Sect. 2. In Sect. 3, a simulation study is presented to demonstrate the concepts along with the actual experimental implementation. The dynamic analysis of the mechanism is presented in Sect. 4, employing a bond graphs modelling approach to provide useful design insight for performance optimization. The last section presents the conclusions together with directions for further research.

2 Structural and Reconfiguration Concept

The proposed structural concept of the building, as depicted in Fig. 1, is composed of a number of planar, closed-chain mechanisms, which change the overall shape of the building as they are adjusted. The basic structural and kinematics element is a planar mechanism, the reconfiguration of which is the focus of the present work. This mechanism consists of one actuated sliding block and two chains of serially-connected links on either side, which are pinned to the ground, as shown in Fig. 2a. All rotational joints are installed with a brake whose operation is managed by the control system. Application of the brakes is used to define "effective links" (see Fig. 2). Note that upon releasing two brakes on either side of the chain the mechanism is reduced to a CECS as shown in Fig. 2b, which is an inherently 1-DOF mechanism. A control approach is proposed for the stepwise reconfiguration of the mechanism between an initial and a final configuration. The control concept is explained here based on the example of Fig. 2. Each intermediate reconfiguration step involves the selective application of the brakes in order to result in a CECS mechanism. Then, one joint angle is adjusted to its final value using the linear actuator. This joint remains locked during the subsequent steps. The procedure is repeated until the adjustment of all joints is completed. As part of motion planning attention is required for singular configurations to be avoided. In particular, singularities are related to "limit positions", which may occur on one side or simultaneously at both sides of the slider block. They correspond to fully-extended or fully-retracted positions, where the slider is locked without being able to move in either left or right direction. In that instance, any force applied to the slider side cannot turn the crank (link pinned to the ground) and when slightly disturbed, the crank is possible to move in either direction.

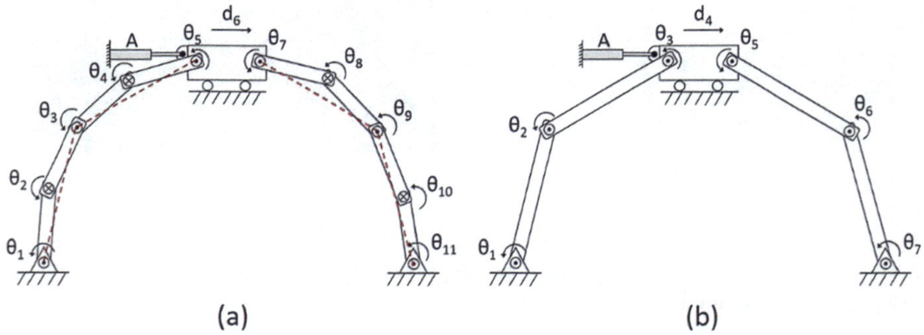

Fig. 2. (a) The basic planar mechanism and the control concept. (b) The corresponding coupled effective crank-slider mechanism. (⊗: locked joint, ⊙: unlocked joint, △: pivoted–to–the–ground joint, dashed lines represent effective links, A: linear actuator).

The number of degrees-of-freedom (DOF) for the coupled crank-slider (Fig. 2b) is calculated with the Grübler's formula: $M = 3(L\text{-}1)\text{-}2J_1\text{-}J_2$, which provides the DOF for a planar mechanism [9]. For this mechanism the number of links $L = 6$, the number of full joints (1-DOF) $J_1 = 7$, and the number of half joints (2-DOF) $J_2 = 0$. Therefore, the mechanism has only 1 DOF ($M = 1$).

3 Simulation and Experimental Study

The concept is demonstrated using the above mechanism, but with 3 links on either side of the slider. Both the initial and the final configurations are symmetric (this is not necessary), as shown in Fig. 3. The selected control sequence is represented by the scheduling table in Fig. 4 and includes three steps. During each step, one joint angle is completely adjusted to its final value. Note that the last step includes the simultaneous adjustment of 6 joints and the position of the slider itself to their target values. Motion planning includes the selection of appropriate sequences which are required to respect the motion limits at the joints, while avoiding singular configurations as well as collisions. Normally, many alternative feasible reconfiguration sequences exist and an optimization procedure can be applied in order to select the most appropriate one. Relevant optimization criterial may include energy consumption, actuation effort and reconfiguration times, which can be evaluated using the models developed in Sect. 4.

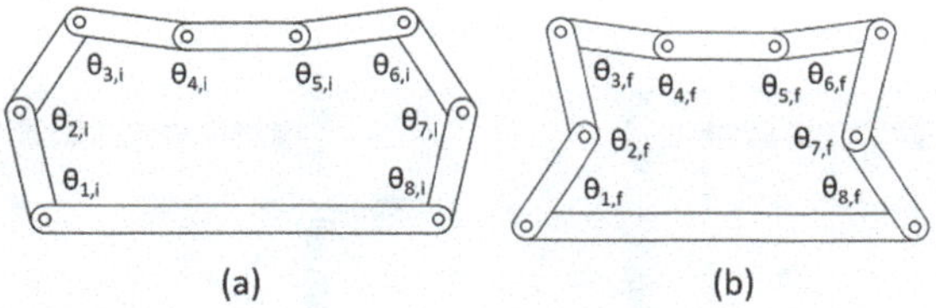

Fig. 3. (a) Initial configuration. (b) Final (target) configuration.

	θ_1	θ_2	θ_3	θ_4	d	θ_5	θ_6	θ_7	θ_8
Step 1	△	⊙	⊗	⊙		⊙	⊙	⊗	△
Step 2	△⊗	⊙	⊙	⊙		⊙	⊗	⊙	△
Step 3	△⊗	⊙	⊙	⊙		⊙	⊙	⊙	△⊗

Fig. 4. Scheduling table representing the selected control sequence. (⊗: locked joint, ⊙: unlocked joint, △: pivoted–to–the–ground joint, symbols in red colour represent the adjusted joints).

The reconfiguration and the intermediate positions of the mechanism are shown in Fig. 5a, while the corresponding time histories for the joint angles are shown in Fig. 5b. The line pairs depart from the same value and finally meet again when the configuration is completed, given that both the initial and final configuration have symmetrical shapes.

An experimental setup was built to verify the results of the simulation model. Actuation involves a linear motion ball screw mechanism and all joints are equipped with electromagnetic brakes. The motion of the actuator and the application of the brakes are computer-controlled. Optical encoders installed on each rotational joint are used for position feedback. The same reconfiguration sequence as before is also used for the experiment, and the intermediate reconfiguration steps are depicted in Fig. 6. The similarity with the simulation results in Fig. 5 is obvious.

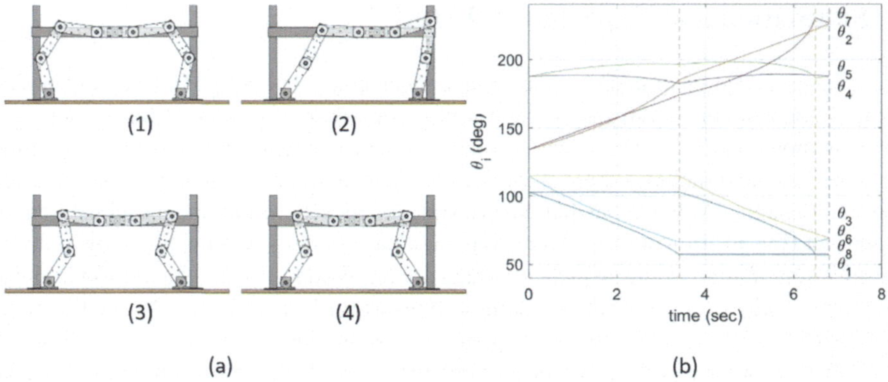

Fig. 5. Simulation results. (a) Stepwise reconfiguration of the planar mechanism between the initial and final configuration. (b) Time histories for the (internal) joint angles during the reconfiguration. The vertical zones separate the reconfiguration steps. Fixed velocity for the actuator was used (20 mm/s).

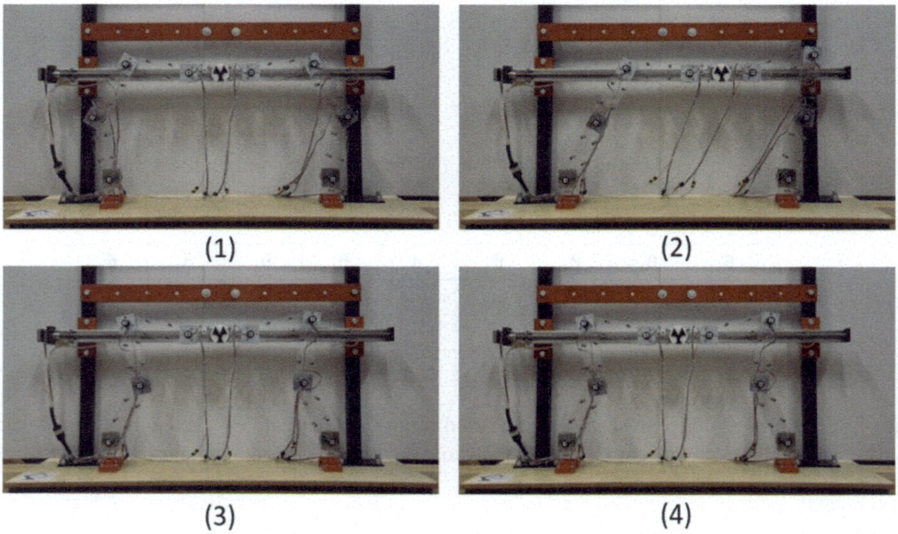

Fig. 6. Experimental results. Stepwise reconfiguration between the initial and final positions.

4 Dynamics Modelling

The bond graph methodology [10] is chosen for developing the dynamic system models in this work. This selection is based on the ability of the bond graph methodology to create a power topography of the system, which is useful in identifying the dominant dynamic elements of the system during the reconfiguration stage. Using this methodology model complexity can be easily addressed in order to produce a reduced model, which is useful for controller design purposes. More details about the model reduction can be found in

[11]. Also, 20SIM is chosen as the bond graph user interface and simulation software that allows the assembling and combining of bond graphs and block diagrams from modular element libraries [12]. 20SIM automatically converts the bond graph into an equation model and can perform the numerical integration with one of several integration schemes for either explicit or implicit systems of equations.

The dynamic model of the left half of the CECS mechanism in Fig. 2a during reconfiguration is developed, as shown on Fig. 7. This bond graph includes ideal elements along with multiports modelling the two links. The model consists of the two links where the first joint of Link 1 is pivoted to the ground. Next, Link 2 connects the other joint of Link 1 to the Slider. The three **R** elements represent the bearing friction of the brakes when they are not locked. The model of each link includes the kinematics, inertial effects in both rotation and translation, and gravity forces. This model is a multiport and its detailed bond graph model is shown in Fig. 8, where the **MTF** junction elements are used to model the kinematics and **I** elements the linear and rotational inertial effects. The slider is connected to the driving motor with a ball screw and the model includes its mass and friction with the ground. Finally, the ball screw is driven by the motor, and its model includes inertial and dissipative effects from both the electrical and mechanical parts. The dynamic model includes inertial and dissipative effects that are necessary for calculating the required motor torque and energy consumption.

Note that the bond graph model resembles the physical connections of the mechanism, and it is easy to build and understand. The model of each link is modular allowing the addition of more links if required for other analyses. More importantly, the power distribution is visualized in order to identify the dominant dynamic effects that can be used in making design modifications to improve performance of the mechanism.

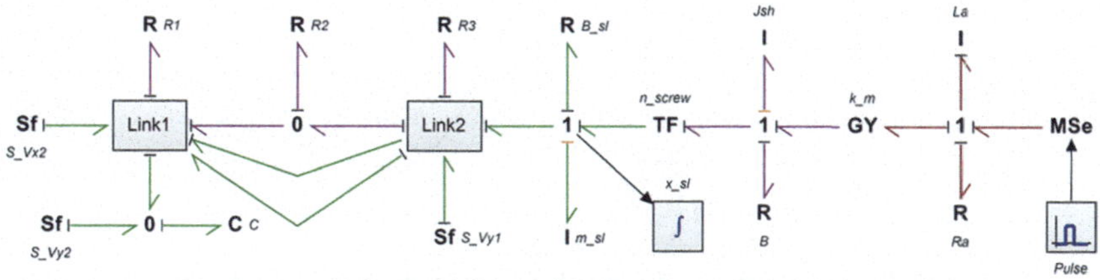

Fig. 7. Bond graph model of CECS mechanism with the driving motor/ball screw.

A simulation performed for a 0.5 s long pulse voltage input activated at 0.1 s and the time response for the two joint angles and slider position are shown in Fig. 9. The model doesn't include the locking action of the brakes, thus, the mechanism is moving due to its own weight until 0.1 s since no torque is provided by the motor. After the activation of the motor the links start decelerating, and the motion is reversed at about 0.25 s in order to move the links at their final configuration at about 0.75 s. After that the links continue to move since the brakes are unlocked and the weight of the links and brakes can cause the rotation of the links. The links have length 0.2 m and mass 0.218 kg, the brakes have mass 0.361 kg, the ball screw has ratio 2.5 mm/rev and the unlocked brakes have damping 0.1 Nms/rad.

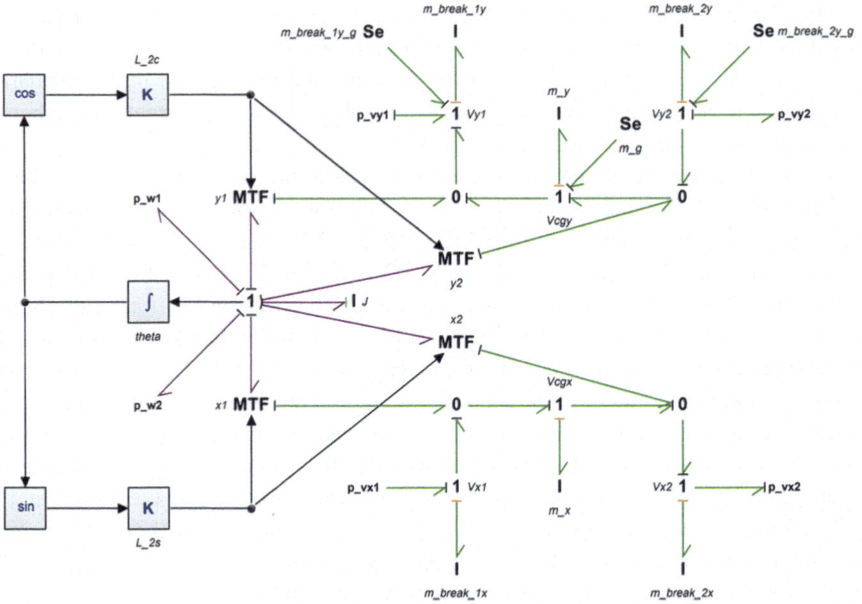

Fig. 8. Bond graph model of a link with 2 joints.

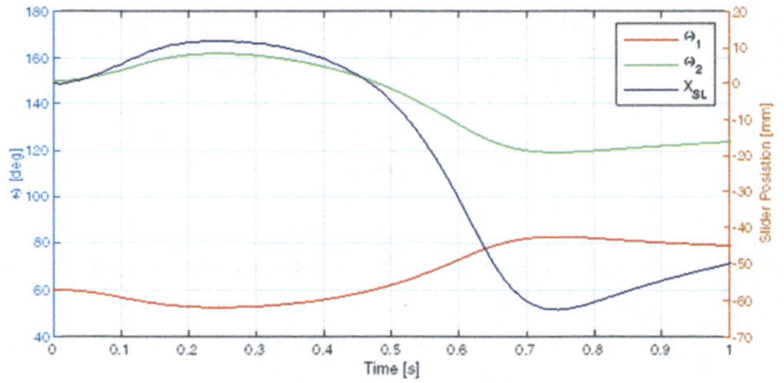

Fig. 9. Time response for a pulse voltage applied to the motor.

5 Conclusions

Exploitation of reconfigurability is expected to bring innovations to architectural design. A structural and control concept for responsive buildings has been presented. Implementation is based on linkages and is subject to a conceptually simple control approach. A simulation and experimental study demonstrated the applicability and the characteristics of the method. A dynamics model of the system has been derived using a bond graphs methodology. This model can provide useful design input by identifying critical elements which require particular attention with respect to inertial effects, power consumption and vibrations. The dynamics model may also provide a basis of an optimization procedure, which can become an integral part of motion planning for the systematic selection of the most favorable control sequences. This possibility will be further pursued as part of future work.

References

1. Dai, J.S., Zoppi, M., Kong, X.: Advances in Reconfigurable Mechanisms and Robots I. Proceedings for Second International Conference on Reconfigurable Mechanisms and Robots. Springer, London (2012). https://doi.org/10.1007/978-1-4471-4141-9
2. Christoforou, E.G., Müller, A., Phocas, M.C., Matheou, M., Arnos, S.: Design and control concept for reconfigurable architecture. J. Mech. Des. **137**(4), 042302 (2015)
3. Phocas, M.C., Christoforou, E.G., Matheou, M.: Design, motion planning and control of a reconfigurable hybrid structure. Eng. Struct. **101**(10), 376–385 (2015)
4. Matheou, M., Phocas, M.C., Christoforou, E.G., Müller, A.: On the kinetics of reconfigurable hybrid structures. Build. Eng. **17**, 32–42 (2018)
5. Christoforou, E.G., Phocas, M.C., Matheou, M., Müller, A.: Experimental implementation of the 'effective 4-bar method' on a reconfigurable articulated structure. Structures **20**, 157–165 (2019)
6. Phocas, M.C., Christoforou, E.G., Dimitriou, P.: Kinematics and control approach for deployable and reconfigurable rigid bar linkage structures. Eng. Struct. **208**, 110310 (2020)
7. Phocas, M.C., Christoforou, E.G., Theokli, C., Petrou, K.: Reconfigurable linkage structures and photovoltaics integration. Build. Eng. **43**, 103201 (2021)
8. Matheou, M., Couvelas, A., Phocas, M.C.: Transformable building envelope design in architectural education. Procedia Manuf. **44**, 116–123 (2020)
9. Norton, R.: Design of Machinery, 4th edn. McGraw-Hill, New York (2008)
10. Karnopp, D.C., Margolis, D.L., Rosenberg, R.C.: System Dynamics: Modeling and Simulation of Mechatronic Systems. Wiley, New York (2012)
11. Louca, L.S., Stein, J.L., Hulbert, G.M.: Energy-based model reduction methodology for automated modeling. J. Dyn. Syst. Meas. Control **132**(6), 061202 (2010)
12. Controllab Products BV, 20-sim Pro Users' Manual, Version 3.6. The University of Twente - Controllab Products BV. Enschede, The Netherlands (2005)

Static Modeling of an Inflatable Robotic Arm for Aerospace Applications

Mario Troise(✉), Matteo Gaidano, Pierpaolo Palmieri, Andrea Ruggeri, and Stefano Mauro

Department of Mechanical and Aerospace Engineering, Politecnico di Torino, C.so Duca degli Abruzzi 24, 10129 Turin, Italy

{mario.troise,matteo.gaidano,pierpaolo.palmieri,andrea.ruggeri, stefano.mauro}@polito.it

Abstract. Robots designed with soft material represent a growing trend in recent years and their use in space applications could minimize bulk and mass, reducing space mission costs. To achieve this goal the structure of soft robots must guarantee the possibility to be stored using deployment mechanisms. In this paper, the concept of an innovative inflatable robotic arm is proposed. The robot consists of two inflatable links and three actuated rigid joints. It can be stored in a relatively small package, if compared to its extended configuration, and can be inflated when required. After an overview of the robotic system, a radial deformation model is proposed to describe the link behavior, aiming for developing its control. The last section shows experimental tests performed on the link prototype to evaluate the static characterization with respect to the supplied pressure. Results suggest the validity of the adopted approach to model the inflatable link, putting the basis for the development of a suited control algorithm for the robotic system.

Keywords: Soft robotics · Inflatable structures · Characterization · Deformation model · Space robotics · Manipulators

1 Introduction

In industrial environments, traditional robot manipulators are essential to accomplish many tasks, while robotic technologies begin to be essential in space applications [1]. Industrial robots can generate high contact forces, achieving desired tasks with great accuracy and repeatability. However, these manipulators are relatively bulky and heavy that is a disadvantage if embarked on space vehicles. Typically space applications require specifically designed robots, with a special attention to mass and encumbrance in retracted position.

Robots designed with soft material are a growing trend in recent years and they could better fit the requirements in space exploration permitting to resolve aerospace issues. Soft robots can be suitable for on-board, as robotic assistants, and for off-board applications, as berthing activity or space debris collection. Their design can be gathered into two categories, articulated [2] and continuum robots [3]. Furthermore, they can

A. Müller and M. Brandstötter (Eds.): RAAD 2022, MMS 120, pp. 298–305, 2022.
https://doi.org/10.1007/978-3-031-04870-8_35

span along all the spectrum between totally soft and rigid structures [4]. The links of articulated manipulators can be inflatable, guaranteeing the possibility to be stored in a limited volume, and the joints can be rigid [5], soft [6] or with variable stiffness [7]. However, the use of soft joints significantly reduces the load capacity of the robot.

In the last years, the development of collaborative robot and collision avoidance algorithms [8–11] has increased the chances for a successful Human-Robot Collaboration (HRC) [12]. However, robots with intrinsic soft structure can improve safety in the event of an undesired collision, since the soft structure has the effect of buffering energy during a collision allowing a safer HRC [13].

In literature inflatable link has been described in several methods: in [14] the authors used pseudo-rigid body model to extract its bending stiffness, in [15] a vision sensor was placed inside the link to provide information about the structural deformation and, in [16] the stiffness and maximum load values of inflatable cylindrical tubes under different inflation pressures was obtained.

The design of soft manipulators influences the choice of the control strategy and dynamic parameters are crucial for the design an efficient controller, their identification is possible through appropriate algorithms [17] and considering the effects of temperature and mounting configuration [18]. However, in typically space applications the movements of the robots are very slow and to control them a quasi-static approach can be considered.

This paper presents a static model to account the deformation of a novel space robotic arm, consisting in inflatable links and electric actuation. This work distinguishes from the state of the art since it describes the bending stiffness of inflatable links considering their radial deformations due the air pressure level. Radial deformation model allows taking into account the variation of the moment of inertia as function of air pressure supply, improving the estimate of bending stiffness of the robot. Results suggest the validity of the deformation model and they will be used to improve the control algorithm showed in [19], and sensor and control strategy showed in [20].

2 Robotic System

The anthropomorphic soft arm consists in three revolute joints and two inflatable links. The links, made of a soft material, have rigid cups that guarantee to be connected with joints allowing them to assume a cylindrical shape when inflated. The structure of the extended configuration is shown in Fig. 1. The links are designed according to considerations discussed in previous work [21] about the influence of the internal pressure and the radius on the performances. The joint structural part hosts the motors and will be fabricated throughout additive manufacturing techniques, minimizing the weight and guaranteeing the resistance to loads at the same time. In order to save space on the stowing bay, soft manipulator can be deflated and stored in a relatively small package. A pneumatic line is responsible to control the inflation and deflation stages. Once reached the deployed configuration, the pneumatic supply is cut off and the robot is ready for the working phase. However, the using of inflatable links entail non-negligible deformations related to the robot payload. Defining the correlation between the force applied on the link and the generated deformation would allow to increase the performance of the robot control system.

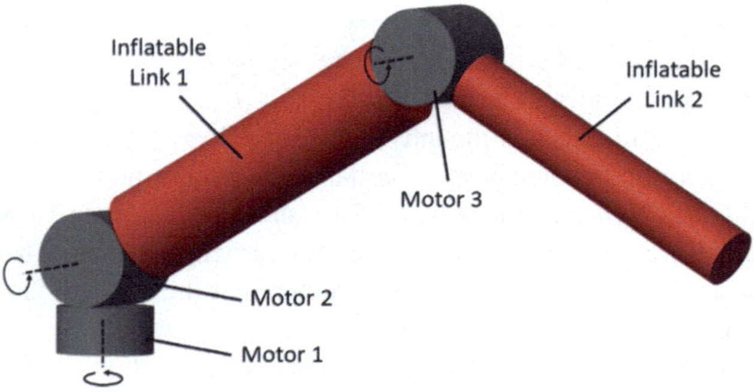

Fig. 1. Functional scheme of the anthropomorphic inflatable robot

3 Deformation Model

Since links are maintained stressed by the pressure they are modeled as a rigid beam until the moment acting on them reaches the wrinkling moment. In literature, several expressions are used to define the wrinkling moment [22–24]; in this work the Veldman's model [22] has been chosen. The contribution of the material and membranes has been considered combining the Wielsgosz [23] and Wood [24] models with an adaptation for orthotropic materials. The first term is the same as that proposed by Wielsgosz and represents the membrane contribution, while the second is an adaptation of the Wood's model. The expression of wrinkling moment M_{wr} are reported below:

$$M_{wr} = \frac{1}{2}\left(\frac{\pi}{2}\right)^2 pr^3 + \frac{1}{2}\frac{2\sqrt{2}}{9}E_x \pi rt^2 \sqrt{\frac{E_\theta}{E_x}}\sqrt{\frac{1}{1-\nu_{x\theta}\nu_{\theta x}} + 4\cdot\frac{p}{E_\theta}\left(\frac{r}{t}\right)^2} \quad (1)$$

where p is the internal link pressure, r is radius of the link, ν is the Poisson coefficient, t is the thickness of the link, E_x and E_θ are the longitudinal and torsional Young's modules. The deflection of an inflatable beam, below the wrinkling moment, can be determined using analogous expression to the case of a rigid beam:

$$f = Q\cdot\frac{L^3}{3\cdot E_x\cdot I} \quad M < M_{wr} \quad (2)$$

where Q is the applied load, I is the moment of inertia and L the length. The deflection would appear independent of the inflation pressure. Furthermore, the mechanical properties of each links are unknown because the value of longitudinal Young's module can be different for the robot links, it must be determined through experimental tests. To improve the accuracy of the inflatable link model the influence of internal pressure on the deflection is introduced. When the robot is inflated the internal pressure radially deforms the link leading to a variation of its moment of inertia. The expression of radial deformation produced by internal pressure is reported below:

$$u = \rho\left[(1-\nu) + \frac{1}{\rho^2}(1+\nu)\right]\cdot\frac{\beta^2}{1-\beta^2}\cdot\frac{r\cdot p_i}{E_x} \quad (3)$$

$$\begin{cases} R_e = r_e + u(r_e) \\ R_i = r_i + u(r_i) \end{cases} \tag{4}$$

$$\begin{cases} R_m = \frac{(R_e + R_i)}{2} \\ t' = R_e - R_i \end{cases} \tag{5}$$

where u is the radial deformation, E_x is the Young's modulus, r is the radius value in a generic surface point, r_e is nominal external radius of the link, r_i is nominal internal radius of the link, ρ is the ratio between r andr_e, β is the ratio between r_e and r_i, R_e and R_i are the deformed external and internal radius. However, using a thin material sheet allows assuming that thin does not change because the radial deformation will be the same for both internal and external radius. Solving the following equation, the value of Young's modulus of the link is calculated:

$$E_x = \frac{L^3}{3 \cdot I(p)} \cdot \frac{Q}{f} \tag{6}$$

4 Inflatable Link Prototype and Experimental Set-up

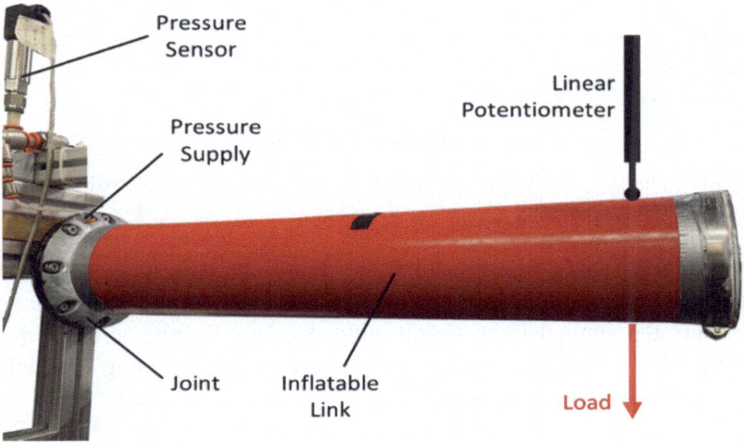

Fig. 2. Inflatable link prototype and experimental set-up of the static tests

A prototype of the inflatable link has been made by rolling and gluing a flexible PVC sheet, obtaining a cylindrical shape and good air pressure resistance. The link is connected to the test bench with screws. On the plug, a hole for the pressure supply is present. The pressure can been regulated with a pressure reducing valve. The link has length L = 600 mm, radius r = 55 mm and wall thickness t = 0.5 mm. During the tests, the bending load is applied using calibrated weights at a known position, d = 470 mm. The deflection has been measured using a resistive potentiometer at the application's point of the load. Then, a map with deflections and applied forces is built. The experimental set-up of the static test is shown in Fig. 2. The pressure is maintained constant during the test using a pressure regulator and measured using a pressure sensor.

5 Results

During the static tests, for each level of internal pressure, a load has been added till the collapse of the structure. The static characteristic of the link can be built using the data obtained from the tests. Results are shown in Fig. 3.

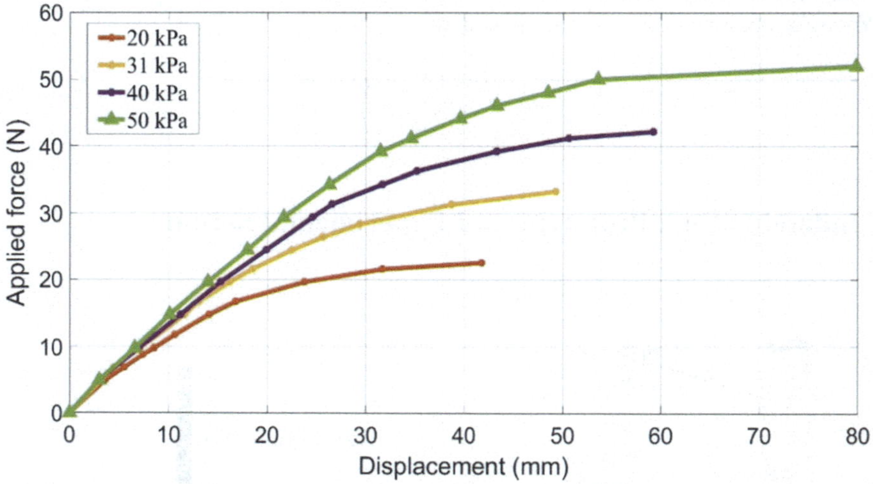

Fig. 3. Static characteristic of the link prototype depending on the internal pressure level

Figure 3 was built using average data obtained carrying out several tests for each level of pressure in which the application point of the load is the same, when the wrinkling moment occurs the applied force significantly does not change as the displacement increases. In Table 1 both the average value and variance of Young's modulus using radial deformation model are reported.

Table 1. Average value and variance of Young modulus with the radial deformation model

	20 kPa	31 kPa	40 kPa	50 kPa
$\overline{\overline{E_x}}$	164	182	177	182
$\sqrt{\sigma_E^2}$	12.0	17	21	14

where $\overline{\overline{E_x}}$ and $\sqrt{\sigma_{\bar{E}}^2}$ are the average value and variance of Young's modulus for each level of pressure. Table 1 has been obtained carrying out several tests for each level of pressure. For each of them, the radial deformations of the link, using Eq. 3, have been estimated and, they have been used to calculate the average value and variance of Young's modulus. The use of the radial deformation model allows changing the moment of inertia when the pressure changes, achieving the description of the slope of the experimental curves using the Veldman's model, as showed in Fig. 4 and Fig. 5.

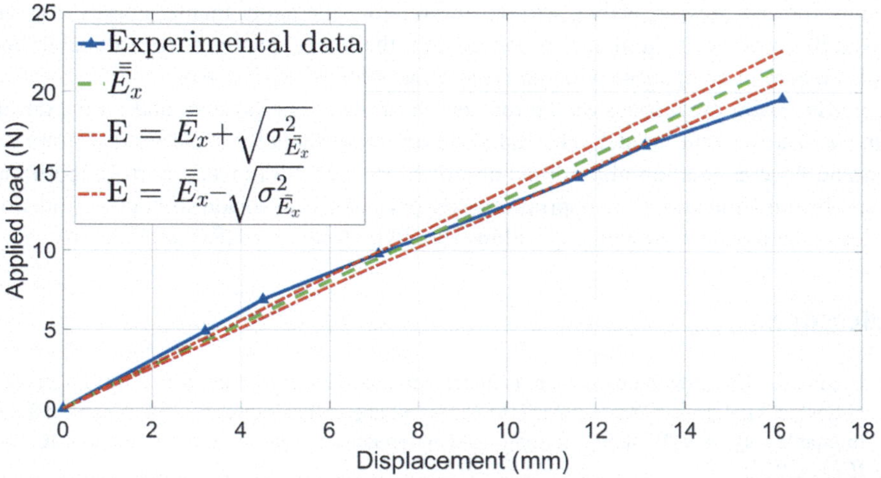

Fig. 4. Comparisons between Veldman's model with radial deformation and experimental static characteristics, $p = 31\,kPa$

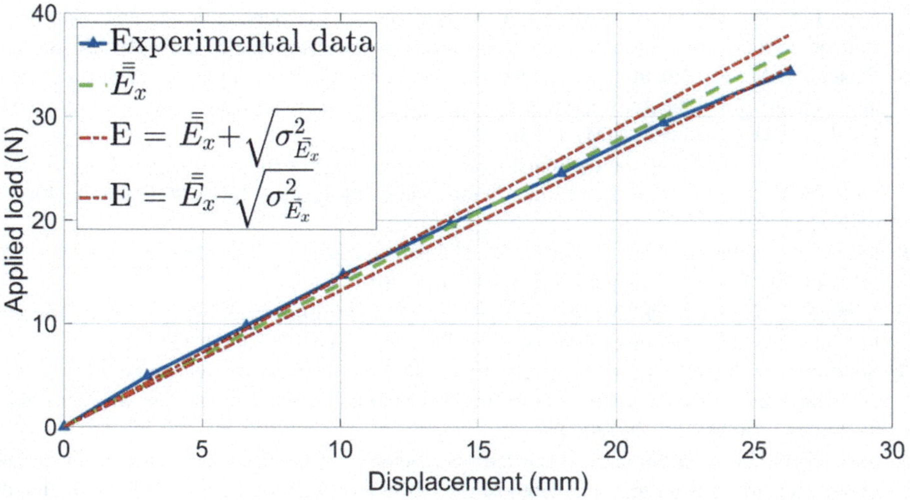

Fig. 5. Comparisons between Veldman's model with radial deformation and experimental static characteristics, $p = 50\,kPa$

6 Conclusion

This paper presented a model to account the deformation of a novel inflatable robot for space applications. The robot has two inflatable links and three revolute joints, it can be enclosed in a package and deployed when required. The model of the radial deformations of the links due to the air pressure was discussed. Radial deformation model allowed taking into account the variation of the moment of inertia as function of air pressure supply. The prototype of the inflatable link was built and statically characterized. Using a radial deformation model to describe inflatable link allows determining its Young's module for any pressure value, without discretizing the experimental curves. Results proved the validity of the model proposed and, the load capacity of the inflatable link has been proven compatible with the typical space loads requirements.

Further works will focus on the realization of the overall system and on the development of a dynamic model. The radial deformation model will be used to improve load and pose estimation algorithms, described in [19]. Moreover, control algorithms should be implemented using appropriate sensors and sensor fusion strategies. Collision avoidance and hand-following algorithms should be tested for HRC applications.

References

1. Bogue, R.: The growing use of robots by the aerospace industry. Ind. Robot. Int. J. (2018)
2. Gillespie, M.T., Best, C.M., Killpack, M.D.: Simultaneous position and stiffness control for an inflatable soft robot. In: IEEE International Conference on Robotics and Automation (ICRA), IEEE (2016)
3. Giannaccini, M.E., et al.: Novel Design of a Soft Lightweight Pneumatic Continuum Robot Arm with Decoupled Variable Stiffness and Positioning. Soft Robotics (2018)
4. Hughes, J., et al.: Soft manipulators and grippers: a review. Front. Robot. AI **3**, 69 (2016)
5. Kim, H.J., Kawamura, A., Nishioka, Y., Kawamura, S.: Mechanical design and control of inflatable robotic arms for high positioning accuracy. Adv. Robot. **32**(2), 89–104 (2018)
6. Palacio, J. M. A., Riwan, A., Mechbal, N., Monteiro, E., Voisembert, S.: A novel inflatable actuator for inflatable robotic arms. In: IEEE International Conference on Advanced Intelligent Mechatronics (AIM), IEEE (2017)
7. Mengacci, R., et al.: On the motion/stiffness decoupling property of articulated soft robots with application to model-free torque iterative learning control. Int. J. Robot. Res. **40**(1), 384–374 (2021)
8. Mauro, S., Scimmi, L.S., Pastorelli, S.: Collision avoidance algorithm for collaborative robotics. Int. J. Autom. Technol. **11**(3), 481–483 (2017)
9. Scimmi, L. S., Melchiorre, M. Mauro, S., Pastorelli, S.: Multiple collision avoidance between human limbs and robot links algorithm in collaborative tasks. In: ICINCO (2018)
10. Scimmi, L. S., Melchiorre, M. Mauro, S., Pastorelli, S.: Implementing a vision-based collision avoidance algorithm on a UR3 robot. In: 23rd International Conference on Mechatronics Technology (ICMT), IEEE (2019)
11. Melchiorre, M., Scimmi, L.S., Pastorelli, S., Mauro, S.: Collison avoidance using point cloud data fusion from multiple depth sensors: a practical approach. In: 23rd International Conference on Mechatronics Technology (ICMT). IEEE (2019)
12. Scimmi, L.S., Melchiorre, M., Troise, M., Mauro, S., Pastorelli, S.: A practical and effective layout for a safe human-robot collaborative assembly task. Appl. Sci. **11**(4), 1769 (2021)

13. Pang, G., et al.: Development of flexible robot skin for safe and natural human-robot collaboration. Micromachines **9**(11), 576 (2018)
14. Sanan, S., Moidel, J. B., Atkeson, C. G.: Robots with inflatable links. In: IEEE/RSJ International Conference on Intelligent Robots and Systems. IEEE (2009)
15. Oliveira, J., Ferreira, A., Reis, J.C.P.: Design and experiments on an inflatable link robot with a built-in vision sensor. Mechatronics **65**, 102035 (2020)
16. Okda, S., Akl, W., Elsabbagh, A.: Structural behaviour of inflatable PVC fabric cylindrical tubes. In: IOP Conference Series: Materials Science and Engineering, vol. 610, no. 1 (2019)
17. Raviola, A., De Martin, A., Guida, R., Pastorelli, S., Mauro, S., Sorli, M.: Identification of a UR5 collaborative robot dynamic parameters. In: Zeghloul, S., Laribi, M.A., Sandoval, J. (eds.) RAAD 2021. MMS, vol. 102, pp. 69–77. Springer, Cham (2021). https://doi.org/10.1007/978-3-030-75259-0_8
18. Raviola, A., De Martin, A., Guida, R., Pastorelli, S., Mauro, S., Sorli, M.: Effects of temperature and mounting configuration on the dynamic parameters identification of industrial robots. Robotics **10**(3), 83 (2021)
19. Troise, M., Gaidano, M., Palmieri, P., Mauro, S.: Preliminary analysis of a lightweight and deployable soft robot for space applications. Appl. Sci. **11**(6), 2558 (2021)
20. Palmieri, P., Gaidano, M., Ruggeri, A., Salamina, L., Troise, M., Mauro, S.: An Inflatable Robotic Assistant for Onboard Applications. In: 72nd IAC 2021, Dubai, UAE (2021)
21. Palmieri, P., Gaidano, M., Ruggeri, A., Salamina, L., Troise, M., Mauro, S.: A deployable and inflatable robotic arm concept for aerospace applications. In: 2021 IEEE 8th International Workshop on Metrology for AeroSpace (MetroAeroSpace), IEEE (2021)
22. Veldman, S.L., Bergsma, O.K., Beukers, A.: Bending of anisotropic inflated cylindrical beams. Thin-walled structures **43**(3), 461–475 (2005)
23. Wielgosz, C., Thomas, J.C.: Deflections of inflatable fabric panels at high pressure. Thin-Walled Structures **40**(6), 523–536 (2002)
24. Wood, J.: The flexure of a uniformly pressurized, circular, cylindrical shell. J. Appl. Mech. **25**(12), 453–458 (1958)

Static Balance of a Flexure-Based Four-Bar Mechanism: Less Torque with More Preload

Jan J. de Jong(✉) and Ronald G. K. M. Aarts

Department of Mechanics of Solids, Surfaces and Systems, University of Twente, Enschede, The Netherlands
j.j.dejong@utwente.nl

Abstract. Static balance eliminates the inherent actuation stiffness of flexure-based mechanisms. In this paper we investigate how, and under which conditions, preloading in the joints can approximately balance a flexure-based four-bar mechanism. We show that actuation reduction is possible when the preload leverages kinematic non-linearities of the system, resulting in a negative kinematic stiffness. The static balance conditions are formulated such that they can be solved using linear algebra and also permit kinematic insight. Application to a flexure-based Watt's linkage shows a reduction factor of 500 in actuator torque.

Keywords: Compliant mechanisms · Zero stiffness · Geometric non-linearities

1 Introduction

Flexure-based mechanisms—also known as compliant mechanisms—consist of thin strips of deforming material that allow motion in compliant directions while supporting forces in other stiff directions [6]. Since this allows for a backlash-free motion without friction, these type of hinges are used in precision applications where high repeatability is desired. However, the inherent stiffness of flexures requires a continuous actuation force to deflect the mechanism, necessitating larger motors with higher energy consumption and unwanted thermal loads.

To eliminate the actuation stiffness, a mechanism may be statically balanced. In that case it has no preference pose and is therefore in an indifferent equilibrium position [5]. Historically, static balance was used to counteract the pull of gravity, but it can also be used to counteract other conservative energy fields such as magnetic force and joint compliance. A mechanism is statically balanced when its potential energy is constant over deflection. Therefore, a potential energy storage, i.e. pretension, is needed with a synchronized energy release to balance the positive energy build-up in the flexures.

Static balance of compliant structures is often performed on joint level by adding different type of energy storages like, prestressed flexures [7], buckled beams [4] and linear springs [9]. Design methodologies on mechanism level are

A. Müller and M. Brandstötter (Eds.): RAAD 2022, MMS 120, pp. 306–313, 2022.
https://doi.org/10.1007/978-3-031-04870-8_36

typically based on optimization or visual inspection [8]. In [3] analytic rigid body approximation methods were used to statically balance compliant mechanisms with an additional zero-free-length spring.

In [1,2] redundantly actuated mechanisms with flexure joints were presented that show the potential of static balancing. Initially, actuator redundancy of a 2-DoF compliant mechanism was used to increase its effective workspace and support stiffness. Surprisingly, optimization showed and measurements confirmed that the actuation stiffness of this mechanism could be reduced with a factor of 1.9 by applying torsional preload in the flexure joints [1]. From the cited literature it is not clear how it is possible that more parallel connected flexure joints could lead to a reduction of actuation stiffness, nor is it clear how such mechanism could be designed.

In this paper we show how torsional preload in the flexure joints of a four-bar mechanism can lead to approximate static balance. We present a new formalism that allows for a closed form description of the balancing solution. Through simulation we demonstrate the procedure on a flexure-based Watt's linkage. This method eliminates the actuation stiffness in one desired pose, which is considered to be sufficient for most applications as compliant mechanisms typically are used in a rather limited working range. We rely on the pseudo-rigid body model (PRBM) [6] to be able to leverage kinematic insights from screw theory. Although the findings here are confined to a four-bar linkage, the procedure may be readily extended to other (multi-DOF) flexure-based structures as long as the PRBM assumptions are valid.

2 Methods

2.1 Notation

In this paper we will make use of screw theory notation to describe the kinematics and statics of the mechanism. Here, the generalized velocity of body i is denoted by the twist t_i which is the concatenation of the angular velocity ω_i of the body and the linear velocity v_i of the origin of the reference frame. When the body is in a kinematic chain, its twist may be described by the joint velocities $\dot{q}$ and screw vectors s_j of the joints lower in the chain

$$t_i = \begin{bmatrix} \omega_i \\ v_i \end{bmatrix} = \begin{bmatrix} s_1 \cdots s_i \; 0 \end{bmatrix} \dot{q}, \qquad s_i = \begin{bmatrix} n_z \\ o_i \times n_z \end{bmatrix}. \tag{1}$$

Here, the screw vectors are the twist of the connected bodies if only that particular joint rotates with unit velocity. In this paper we confine ourselves to planar revolute joints where n_z is the out-of-plane unit vector and o_i is the joint location.

The acceleration twist of a body under constant joint velocities $\ddot{q} = 0$ may be expressed similarly

$$\dot{t}_i = \begin{bmatrix} \dot{s}_1 \cdots \dot{s}_i \; 0 \end{bmatrix} \dot{q}, \qquad \dot{s}_i = \sum_{j=1}^{i-1} \mathrm{ad}(s_j) s_i \dot{q}_j, \tag{2}$$

where s_j are the screw vectors of the joints lower in the chain. The $\mathrm{ad}(\mathrm{s_j})$ denotes the adjoint twist representation of s_j

$$\mathrm{ad}(s_j) = \begin{bmatrix} [n_z\times] & 0 \\ [(o_j \times n_z)\times] & [n_z\times] \end{bmatrix}, \tag{3}$$

and $[a\times]$ is the skew symmetric matrix.

2.2 Kinematics of the Four-Bar Mechanism

The four-bar linkage under investigation (Fig. 1) has four revolute joints and one degree of freedom denoted by x. To study the static balance conditions we need the relation between the joint angles on velocity level, i.e. Jacobian J, and on acceleration level, i.e. Hessian H. These may be obtained from the velocity loop closure condition matrix F. This states that the fourth body is fixed and its twist is zero

$$t_4 = FJ\dot{x} = 0, \qquad F = \begin{bmatrix} s_1 \cdots s_4 \end{bmatrix}, \qquad \dot{q} = J\dot{x}. \tag{4}$$

The Jacobian follows from the choice of independent coordinates. Here we choose the angle of the first joint as independent coordinate $x = q_1$. The twist of the coupler t_2 passes through pole g, while the difference in twist between the crank and the follower $t_1 - t_3$ passes through pole h.

The Hessian H may obtained by equating the acceleration of the loop closure to zero with the condition that $\ddot{x} = 0$

$$\frac{\partial}{\partial x}(FJ) = \underbrace{\sum_{i=1}^{4} \frac{\partial}{\partial q_i}(F) J \frac{\partial q_i}{\partial x}}_{a_v} + FH = 0, \qquad H = \frac{\partial}{\partial x}(J), \tag{5}$$

Although this equation looks daunting it permits a geometrical interpretation. The a_v is the acceleration violation of the loop closure constraint (in this case, the acceleration of body 4) if the joint velocities were kept constant, i.e. if $\ddot{q} = 0$. The Hessian part FH then compensates this acceleration in order to satisfy the loop closure under acceleration. The a_v term may be found by observing that

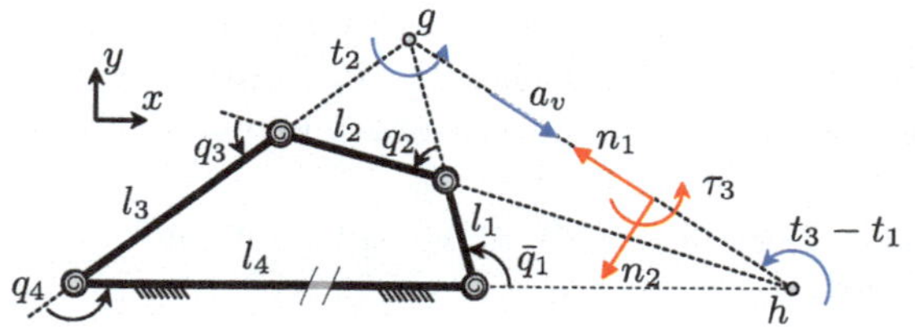

Fig. 1. The definition of a pseudo rigid model of a flexure-based four-bar linkage where all joints are modelled as pin joints with rotational springs.

only the location of joint 2 and 3 are changing, depending on the velocity of joints 1 and 4 respectively. By including the actual twists of the bodies i.e. $s_1\dot{q}_1 = t_1$, $s_2\dot{q}_2 = t_2 - t_1$, $s_3\dot{q}_3 = t_3 - t_2$, and $s_4\dot{q}_4 = t_3$ we find the violation term to be

$$a_v = \mathrm{ad}(s_1)s_2\dot{q}_1\dot{q}_2 + \mathrm{ad}(s_3)s_4\dot{q}_3\dot{q}_4 = \mathrm{ad}(t_2)(t_1 - t_3) = \alpha_v \begin{bmatrix} 0 \\ n_1 \end{bmatrix}. \tag{6}$$

This is a pure translational acceleration along vector n_1, which points from pole g to pole h (Fig. 1). When either g or h is at infinity this n_1 can still be found by inspecting the limit case. Yet, when both g and h are at infinity, i.e. the mechanism is a parallelogram, the Hessian H becomes zero and a_v is zero as the crank and the follower have the same velocity $t_1 - t_3 = 0$.

2.3 Static Balance Conditions

A mechanism is statically balanced when the potential energy V is constant for any motion. The potential energy of a mechanism with compliance in the joints is given by

$$V = \sum_{i=1}^{4} \frac{1}{2} k_i (e_i + q_i)^2 = \sum_{i=1}^{4} \frac{1}{2k_i} (p_i + k_i q_i)^2, \tag{7}$$

where q_i is the deflection of each joint, e_i its rest length/angle, and k_i its actuation stiffness. Later on we use the preload $p_i = k_i e_i$ in the joint as a design variable. The flexure joints are here assumed to be ideal joints with no pivot shift, constant actuation stiffness and no load dependencies. Notice also that we assume that gravity and other conservative forces do not play a role.

In this paper we are interested in a Taylor-like approximate balance where we set the first (resultant force/torque) and second derivatives (effective stiffness) of the potential energy with respect to the generalized coordinates x to zero. For first-order approximate static balance it is then required that the system is 1) in equilibrium and 2) that this equilibrium is locally indifferent i.e. that the stiffness is zero. Therefore, we impose two conditions on the potential energy. The "equilibrium condition" reads

$$\frac{\partial V}{\partial x} = \sum_{i=1}^{4} \frac{\partial V}{\partial q_i} \frac{\partial q_i}{\partial x} = \sum_{i=1}^{4} (p_i + k_i q_i) \frac{\partial q_i}{\partial x} = \sum_{i=1}^{4} p_i \frac{\partial q_i}{\partial x} = 0, \tag{8}$$

where we evaluated this condition around the origin ($x = 0$ and $q = 0$). In here the partial derivate $\frac{\partial q_i}{\partial x}$ are the elements of J, the joint Jacobian.

The "zero stiffness condition" is given by the second derivative of V

$$\frac{\partial^2 V}{\partial x^2} = \sum_{i=1}^{4} k_i \left(\frac{\partial q_i}{\partial x} \right)^2 + p_i \frac{\partial^2 q_i}{\partial x^2} = 0, \tag{9}$$

where $\frac{\partial^2 q_i}{\partial x^2}$ are the elements of H, the Hessian. Here it should be noticed that the mixed derivatives vanish since $\frac{\partial^2 V}{\partial q_i \partial q_j} = 0$ for $i \neq j$, while $\frac{\partial^2 V}{\partial q_i^2} = k_i$.

These balance conditions are linear in the parameters p_i and k_i, which may therefore be directly solved for using linear algebra, particularly pseudo-inverse and null-space operations. In matrix format these conditions read

$$J^\top p = 0, \qquad H^\top p + (J^{\circ 2})^\top k = 0, \tag{10}$$

where $p^\top = \begin{bmatrix} p_1 \cdots p_4 \end{bmatrix}^\top$ is the array of preloads in the joints, $k^\top = \begin{bmatrix} k_1 \cdots k_4 \end{bmatrix}^\top$ is an array with the joint stiffnesses, and $J^{\circ 2}$ denotes the element-wise square of the Jacobian.

It may be noticed that the positive elastic stiffness induced by k is to be balanced by the preload p. This negative stiffness leverages the non-linearities in the geometric transfer function as captured by the Hessian. This indicates that if this non-linearity is weak, a large preload is required for static balance. It should be emphasized that by selection of coordinates and reference frame, the origin $q = 0$ can always be selected as the evaluation point.

2.4 Static Balance Solution

Although the solution for p may be readily found through linear algebra, it may offer little insight in the meaning of the solution space of p. In this section we include the loop closure conditions F to retain some of this insight which might help to find the desired solution. The first-order loop closure condition denotes that a wrench w on the constrained body does not generate any work. Therefore without loss of generality we may therefore select the preload as

$$p = F^\top w, \tag{11}$$

where w is our new design parameter. It may be observed that w is the reaction wrench induced by p at the loop closure. So, this is the wrench required to keep the loop intact. Therefore, with Eq. 4 we eliminate the equilibrium condition of Eq. 10 and we obtain the following "zero-stiffness" condition

$$(FH)^\top w + (J^{\circ 2})^\top k = 0. \tag{12}$$

The FH term is the opposite of the acceleration violation term (Eq. 5 and Eq. 6) and the zero-stiffness condition finally reads

$$- a_v^\top w + (J^{\circ 2})^\top k = 0. \tag{13}$$

Clearly, if the closure reaction wrench w is selected to be orthogonal to a_v, the preload does not affect the stiffness of the system. However, if w is selected along a_v a geometric stiffness is induced and the effective stiffness of the four-bar is altered or even eliminated. We define the matrix W to span the reaction wrenches that are orthogonal to a_v. Finally, for first-order static balance we select w as

$$w = \frac{(J^{\circ 2})^\top k}{|a_v|^2} a_v + W\gamma, \qquad W = \begin{bmatrix} 0 & n_z \\ n_2 & 0 \end{bmatrix}, \qquad a_v^\top W = 0. \tag{14}$$

This solution approach shows that a parallelogram cannot be statically balanced through this approach as its a_v and Hessian are zero. Therefore, it does not possess sufficient non-linearity for the preload to act on. Other types of four-bar do permit such a static balance approach.

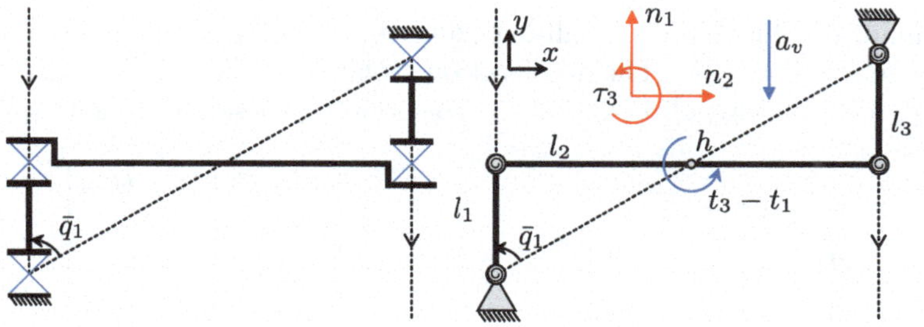

Fig. 2. The flexure-based Watt's linkage (left) consisting of cross-flexure pivots and its pseudo-rigid body model (right).

3 Case Study: Balancing the Watt's Linkage

To show the effectiveness of the static balance procedure we will balance a particular four-bar mechanism, the Watt's linkage (Fig. 2). This mechanism can be used as an approximate straight line guidance in precision applications. The joints consist of cross-flexures that have a maximum bending angle of 8°. The dimensions and design parameters can be found in Table 1. We first derive the PRBM balance conditions after which we simulate the PRBM as well as a fully non-linear flexible multibody system (FMBS) describing the mechanism for different preloads.

Due to the simple structure the Jacobian and the Hessian are

$$F = \begin{bmatrix} n_z & n_z & n_z & n_z \\ -l_1 & 0 & 0 & l_1 \\ -l_2/2 & l_2/2 & l_2/2 & l_2/2 \\ 0 & 0 & 0 & 0 \end{bmatrix}, \qquad J = \begin{bmatrix} -1 \\ -1 \\ -1 \\ -1 \end{bmatrix}, \qquad H = 2 \begin{bmatrix} 0 \\ -l_1/l_2 \\ -l_1/l_2 \\ 0 \end{bmatrix}. \tag{15}$$

Table 1. Parameters of the statically balanced flexure-based Watt's linkage.

Name	Symbol	Value	Unit
Crank and follower length	$l_1 = l_3$	100	mm
Coupler length	l_2	200	mm
Base length	l_4	282	mm
Flexure thickness	t_f	0.40	mm
Flexure length	l_f	7.07	mm
Flexure width	w_f	6.66	mm
Young's modulus	E	200	MPa
Joint stiffness	k_i	1.00	Nm/rad
Midpoint displacement	x	14.0	mm
PRMB preload	$p_2 = -p_3$	−2.00	Nm
FMBS preload	$p_2 = -p_3$	−2.15	Nm

As the quadratic Jacobian is an all-ones matrix, the original unbalanced stiffness of the system is $k = \sum k_i$. The acceleration violation twist is $a_v^\top = -2l_1 \begin{bmatrix} 0^\top & n_y^\top \end{bmatrix}$ such that reaction wrench w and the corresponding torsional preload p become

$$w = -\frac{k}{2l_1}\begin{bmatrix} 0 \\ n_y \end{bmatrix} + \begin{bmatrix} 0 & n_z \\ n_x & 0 \end{bmatrix}\gamma, \qquad p = \frac{kl_2}{4l_1}\begin{bmatrix} -1 \\ -1 \\ 1 \\ 1 \end{bmatrix} + \begin{bmatrix} \gamma_1 + l_1\gamma_2 \\ \gamma_1 \\ \gamma_1 \\ \gamma_1 - l_1\gamma_2 \end{bmatrix}. \tag{16}$$

Figure 3 shows PRMB and FMBS simulation results of a flexure-based Watt's linkage under no load condition and under PRBM static balance preload. Preload in base joints is set to zero $p_1 = p_4 = 0$. The mechanism is actuated by prescribing an x–displacement of the coupler's midpoint. It may be clearly seen that the maximum force required for deflection is reduced significantly from a maximum of 5.7 N to a maximum of 9.9 mN for the PRBM. This implies a reduction of a factor more than 500. When the preload computed for the PRBM is applied in the FMBS-model the balance quality is significantly less, i.e. a maximum force of 0.4 N and reduction factor of only 14. A potential explanations for this difference stems from the fact that PRBM does not take pivot-shift into account, leading to a significant different kinematic behaviour at this stroke. When the preload is slightly increased, the FMBS preload in Table 1, static balance is restored and the maximum force is reduced to 7.5 mN, a reduction factor of 760.

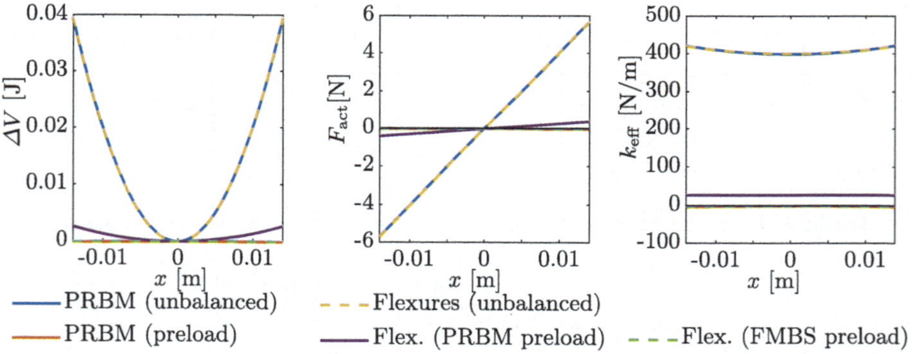

Fig. 3. The potential energy (left), actuator force (middle) and effective stiffness (right) of the flexure-based Watt's linkage with and without preload.

4 Conclusion

This paper untangles the paradoxical behaviour of static balance where higher preloads may lead to less driving torques. It is observed that if the preload is applied in a particular manner, the change in geometry results in a negative stiffness effect. To alter the effective stiffness, the preload-induced reaction forces at the loop closure should be directed along the acceleration violation twist vector, i.e. the acceleration of the loop closure constraint when joint-velocities are kept constant. Any preload that causes reaction force in another direction does not change the effective stiffness. The magnitude of these non-linear effects is inversely proportional to the required preload; highly linear systems require large preloads. In particular, the parallelogram cannot be statically balanced with just torsionally preloaded joints as it has a constant linear relation between the joint velocities. It is shown that the validity of the PRBM assumptions are critical in the effectiveness of the proposed approach. A more advanced FMBS analysis may be needed to account for e.g. pivot-shift. Numerical simulation of a compliant Watt's linkage show a potential actuation force reduction of more than 99.8%, indicating that static balance is obtained.

References

1. Berendsen, D., Sridhar, A., Aarts, R.: A compliant and redundantly actuated 2-DOF 3RRR PKM: less is more. In: Kövecses, J., Stépán, G., Zelei, A. (eds.) Proceedings 10th ECCOMAS Thematic Conference on Multibody Dynamics, pp. 246–256 (2021)
2. Cornelissen, R., Müller, A., Aarts, R.: A compliant and redundantly actuated 2-DOF 3RRR PKM: best of both worlds? In: Kecskeméthy, A., Geu Flores, F. (eds.) ECCOMAS 2019. CMAS, vol. 53, pp. 163–171. Springer, Cham (2020). https://doi.org/10.1007/978-3-030-23132-3_20
3. Deepak, S.R., Hansoge, A.N., Ananthasuresh, G.K.: Application of rigid-body-linkage static balancing techniques to reduce actuation effort in compliant mechanisms. J. Mech. Robot. **8**(2), 1–12 (2016). https://doi.org/10.1115/1.4031192
4. van Eijk, J., Dijksman, J.F.: Plate spring mechanism with constant negative stiffness. Mech. Mach. Theory **14**(1), 1–9 (1979). https://doi.org/10.1016/0094-114X(79)90036-3
5. Herder, J.L.: Design of spring force compensation systems. Mech. Mach. Theory **33**(1), 151–161 (1998). https://doi.org/10.1016/S0094-114X(97)00027-X
6. Howell, L.L., Midha, A.: A method for the design of compliant mechanisms with small-length flexural pivots. J. Mech. Des. Trans. ASME **116**(1), 280–290 (1994). https://doi.org/10.1115/1.2919359
7. Morsch, F.M., Herder, J.L.: Design of a generic zero stiffness compliant joint. In: International Design Engineering Technical Conferences and Computers and Information in Engineering Conference, vol. 2, pp. 427–435 (2011). https://doi.org/10.1115/detc2010-28351
8. Radaelli, G., Gallego, J.A., Herder, J.L.: An energy approach to static balancing of systems with torsion stiffness. J. Mech. Des. Trans. ASME **133**(9), 1–8 (2011). https://doi.org/10.1115/1.4004704
9. Yang, Q., Bi, S., Liu, T.: Static balancing of flexural pivots with two symmetrically arranged pre-compressing springs. In: IOP Conference Series: Materials Science and Engineering, vol. 892 (2020). https://doi.org/10.1088/1757-899X/892/1/012121

Assistive and Medical Robots

Multibody Analysis and Design Optimization of a Full-Scale Biped-Wheeled Exoskeleton

Giuseppe Nigido[1], Elvio Bonisoli[1], and Giovanni Gerardo Muscolo[2(✉)]

[1] Department of Mechanical and Aerospace Engineering, Politecnico di Torino, Turin, Italy
[2] Department of Computer Science, University of Verona, Verona, Italy
giovannigerardo.muscolo@univr.it

Abstract. This work presents the design optimization of the HANDSHAKE multibody model, a full-scale biped-wheeled exoskeleton, conceived in previous works. The ambition of the HANDSHAKE biped-wheeled exoskeleton is to give new locomotion perspectives at people with reduced mobility or people with complete paralysis of the lower body. In this paper, a critical analysis of the multibody simulation of the latest version of the model, performed using MATLAB® and Simulink®, allowed to design and optimize the new multibody model. A comparison of multibody simulation during the walking gait between the new and the latest version of the HANDSHAKE model underlined a reduction of motor torques and power supply in the new model. These results confirm the innovative solution proposed with the HANDSHAKE biped-wheeled exoskeleton.

Keywords: Biped-wheeled exoskeleton · Wearable machine · Wearable robot · Lower body exoskeleton

1 Introduction

Exoskeletons and wearable machines are conceived and designed for helping people to perform complex or dangerous tasks, or to increase human physical power [1–8]. During the last years of the '60s, some prototypes of modern exoskeletons have been realized. Professor Miomir K. Vukobratovic has been one of the pioneers to define models and control principles of the robotic biped locomotion [9]. Nowadays, many exoskeletons are developed for rehabilitation purposes and used in the medical field such as EksoNR™ [10], Indego® [11], HAL® [12], Honda Walking Assist Device® [13], and many others. Generally speaking, the lower-body exoskeleton is a robot composed of two legs connected each one to the waist using the hip joints. Each leg is constituted of three links (thigh, shank, and foot) connected by two joints (knee and ankle). Many of the lower body exoskeletons available in commerce force the user to assume a specific posture and crutches are used for helping users during walking. In many other exoskeletons developed in commerce, the user is obliged to perform the same trajectory as the robot, and in many cases, problems on human joints are created. In some cases, these problems can be caused by the reduction of the volume of muscles in people with reduced mobility, and many loads are transferred directly to the joints (i.e. hip, knees, ankles), instead of

A. Müller and M. Brandstötter (Eds.): RAAD 2022, MMS 120, pp. 317–324, 2022.
https://doi.org/10.1007/978-3-031-04870-8_37

muscles, with dangerous consequences. In the HANDSHAKE robot, this problem is avoided because the interfaces between the human body and the exoskeleton are limited to only the two feet and the trunk of the user. HANDSHAKE [14–17] is a completely new generation of biped-wheeled wearable machine composed of two wheeled supports for each foot and one support for the trunk, as shown in Fig. 1. Two versions of a full-scale HANDSHAKE multibody model [14, 16] and three versions of the reduced-scale prototypes [14, 17] have been conceived and presented in other works. HANDSHAKE is an evolution of the ROLLO robot, the first biped-wheeled robot with flexible legs [18, 19]. One of the most important innovations introduced with the HANDSHAKE is that the user can flex and move the legs, without perceiving directly the constraints induced by the exoskeleton.

In this paper, the design optimization of the latest full-scale HANDSHAKE model is performed reducing motor torques and power supply. This paper is structured as follows: Sect. 2 presents the old design version of the exoskeleton; Sect. 3 shows the new version with design optimization; Sect. 4 presents comparison and discussion between the old and new versions. The paper ends with conclusion.

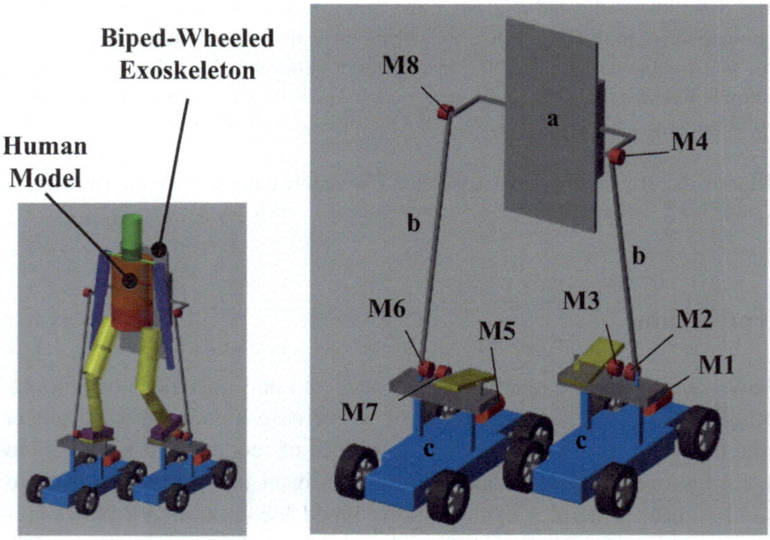

Fig. 1. Old version [16] of the HANDSHAKE multibody model in a full scale constituted of a trunk support (a), two rigid links (b), and two wheeled chassis (c).

2 HANDSHAKE Multibody Model in Full-Scale: Old Version

The HANDSHAKE biped-wheeled robot has been presented in other works [14–17]. The multibody dynamic models of the exoskeleton and the dummy on board have been presented in [16], analyzing torques and powers that the motors must provide in the full-scale model. The multibody model of the old version is shown in Fig. 1 and it is constituted of a trunk support (a), connected with the human trunk, and two rigid links (b) colligated with the two wheeled chassis (c) connected with the human feet. A total of eight electric motors are included in the design: two DC motors (M1, M5) are connected to the chassis to give the motion to the wheels; two servomotors (M3, M7) are connected to foot supports giving motion to the dummy's ankle joints. The other servomotors are at the extremities of the rigid links, respectively for the motion of the ankle joints of the exoskeleton (M2, M6) and for the motion of the trunk support (M4, M8). Figure 2 shows the two configurations used for the physical analysis of the model: standing position (left) and walking (right).

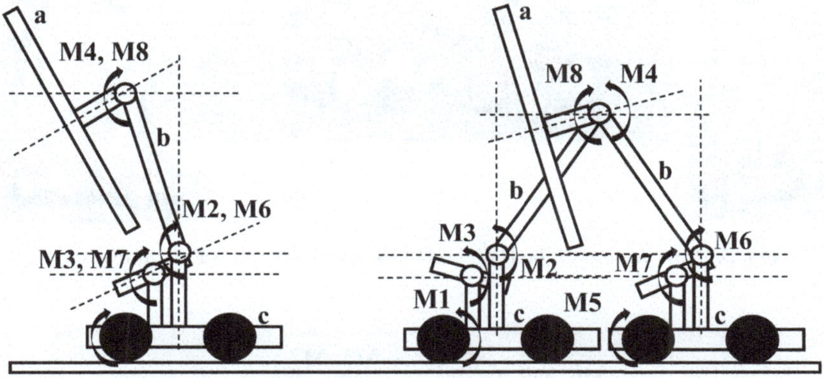

Fig. 2. Two general configurations for the HANDSHAKE model: standing position (left) and walking (right).

3 HANDSHAKE Multibody Model in Full-Scale: New Version

An analysis of the developed multibody model shown in Fig. 1 underlined high torques and powers of the electric motors [16]. The new design of the HANDSHAKE multibody model is shown in Fig. 3, and it is constituted of two chasses with three wheels each one, instead of four wheels like the old version. Figure 4 shows the new designed chassis. The rear wheel motor (M1, and M5) is a commercial product (XTI-HUB MOTORS 280-1367M) used typically in medical devices like electric wheelchairs. The front wheels are free to rotate. In order to reduce the height of the center of mass of the user and machine from the ground, and to give more stability to the multibody model and machine, the suspension system of the new model has a limited thickness, respect to the one in the old version and it is constituted of four cylinders (with pistons inside) for each chassis. The stiffness coefficient and the damping coefficient of each sliding system are respectively $k = 7000$ N/m, and $d = 150$ Ns/m. The foot support and the vertical link are attached by a plate to the suspensions. The servosystem used to move the foot support (M3, and M7) is a commercial product (HARMONIC DRIVE FHA-11C-50 MINI) linked directly to the foot support. Other two used commercial servomotors are Kollmorgen AKM 12E linked to a speed reducer Harmonic Drive CSD-20-2A with a ratio of 100 for the ankles of the exoskeleton (M2, and M6), and two Kollmorgen AKM 21E paired with two Harmonic Drive CSD-20-2A with 160 as ratio moving the hip joints (M4, and M8). The total mass of the new exoskeleton is reduced to around 51 kg respect to the old version of 78 kg.

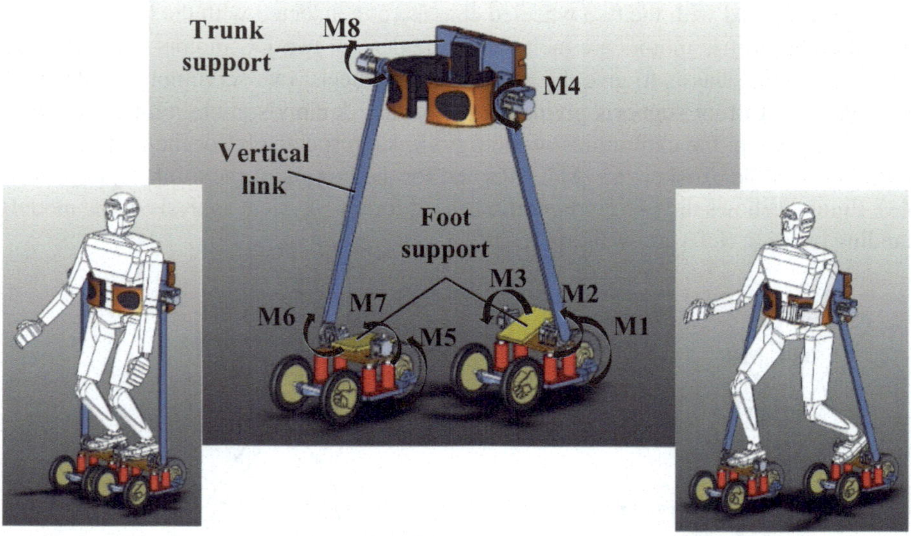

Fig. 3. New version of the HANDSHAKE multibody model in a full scale.

Fig. 4. New version of the chassis with three wheels.

4 Results and Discussion

The new version of the HANDSHAKE full-scale multibody model is obtained by an optimization of the old model: defining a correct walking gait for the user, reducing inertia of the model; giving smooth input in the walking gait trajectory (Fig. 5).

Figure 6 shows that the new model reduces differences between left and right robotic legs, allowing a repeatable and symmetric walking pattern for each leg. This symmetry can be noted by the graphs of torques of M4 and M8 (Fig. 6, New.torque), in which the same range of oscillation of torque from −20 Nm to 25 Nm is registered for the two motors. In the old model (Fig. 6, Old.torque), the torque requested by M4 has a range of oscillation from 10 Nm to 40 Nm, but M8 has a range from −15 Nm to 45 Nm. This asymmetry is not a good result and explains well why during simulation of the old model the robot was not able to follow a rectilinear trajectory. Figure 7 shows some small differences in the old model (Fig. 7, Old.torque) between the range of oscillation of torques in M2 and M6. These differences are reduced in the new model (Fig. 7, New.torque).

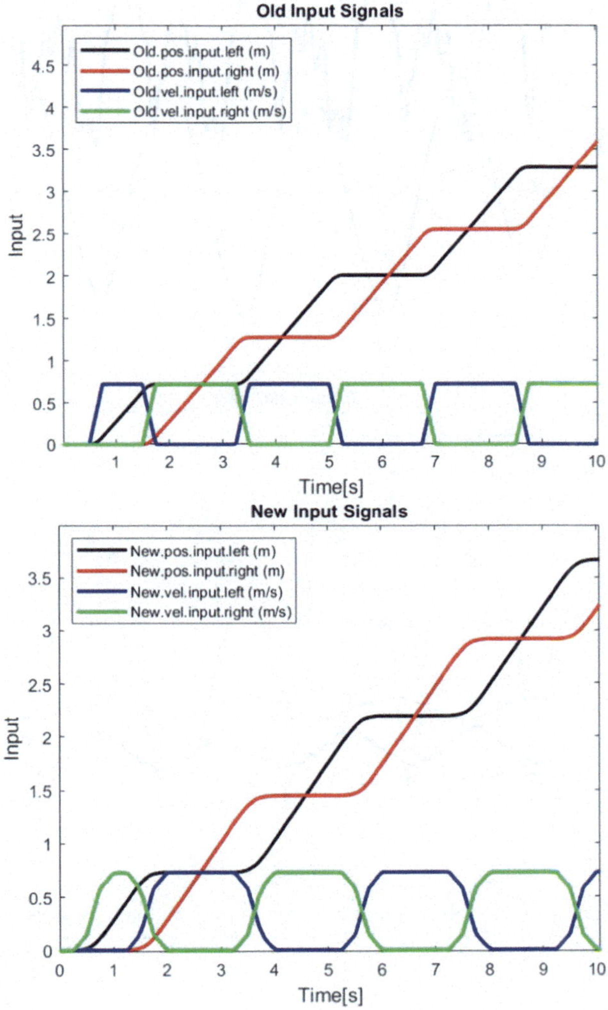

Fig. 5. Walking gait (input of M1 and M5) of the old and new HANDSHAKE multibody model.

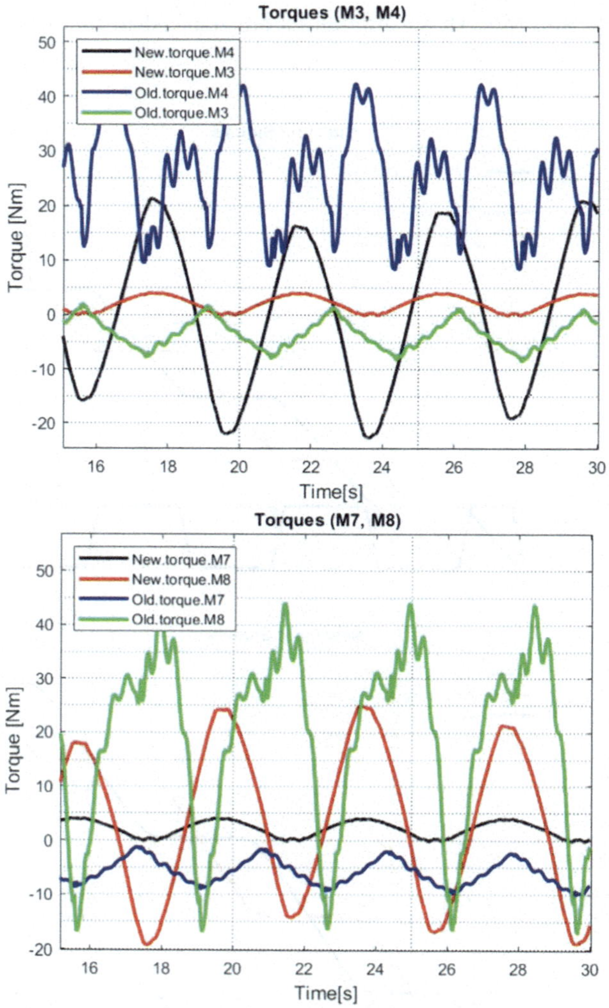

Fig. 6. Comparison of torque (Nm) output between the old and new models for M3, M4, M7, M8.

5 Conclusion

The new version of the full-scale HANDSHAKE model is proposed in this work and it is compared with the latest version, underlining optimization in reducing torque values. The new version of the HANDSHAKE multibody model shows a symmetric walking gait, overcoming limits underlined in the old version.

The objective to reduce the weight of the robot (from 78 kg to 51 kg), to give as input a smoother walking gait trajectory, and to perform a walking pattern of the user more in line with the standard values, permitted to obtain motor torques in the new version of the HANDSHAKE model lower than the latest version model.

These results must be confirmed by real experimentation. Future works are planned in this direction.

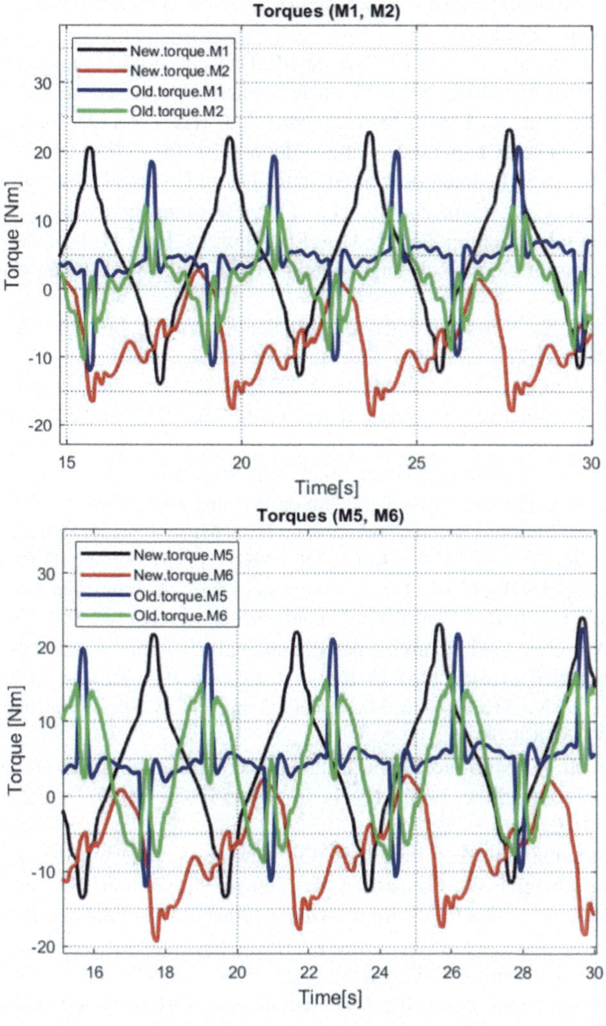

Fig. 7. Comparison of torque (Nm) output between the old and new models for M1, M2, M5, M6.

References

1. Herr, H.: Exoskeletons and orthoses: classification, design challenges and future directions. J. Neuroeng. Rehabil. **6**(1), 1–9 (2009)
2. Panero, E., Muscolo, G.G., Pastorelli, S., Gastaldi, L.: Influence of hinge positioning on human joint torque in industrial trunk exoskeleton. In: Uhl, T. (eds.) Advances in Mechanism and Machine Science. IFToMM WC 2019. Mechanisms and Machine Science, vol. 73, pp. 133–142. Springer, Cham (2019). https://doi.org/10.1007/978-3-030-20131-9_14

3. Mizen, N.J., Inventor, Calspan Corp., Assignee: Powered exoskeletal apparatus for amplifying human strength in response to normal body movements. United States patent US 3,449,769. 1969 June 17
4. Bogue, R.: Exoskeletons and robotic prosthetics: a review of recent developments. Ind. Robot Int. J. **36**(5), 421–427 (2009)
5. Magnetti Gisolo, S., Muscolo, G.G., Paterna, M., De Benedictis, C., Ferraresi, C.: Feasibility study of a passive pneumatic exoskeleton for upper limbs based on a mckibben artificial muscle. In: Zeghloul, S., Laribi, M.A., Sandoval, J. (eds.) RAAD 2021. MMS, vol. 102, pp. 208–217. Springer, Cham (2021). https://doi.org/10.1007/978-3-030-75259-0_23
6. Kelley, L.C.: Inventor. Pedomotor. United States Patent US 1,308,675. 1 July 1919
7. Chen, G., Chan, C.K., Guo, Z., Yu, H.: A review of lower extremity assistive robotic exoskeletons in rehabilitation therapy. Crit. Rev. Biomed. Eng. **41**(4–5) (2013)
8. Pons, J.L.: Rehabilitation exoskeletal robotics. IEEE Eng. Med. Biol. Mag. **29**(3), 57–63 (2010)
9. Vukobratovic, M.K.: When were active exoskeletons actually born? Int. J. Humanoid Rob. **4**(03), 459–486 (2007)
10. Ekso Bionics: Webpage visited on March 2022. https://eksobionics.com/eksonr/
11. Parker Huffin Corporation: Webpage visited on March 2022. https://www.indego.com/indego/us/en/home
12. Cyberdyne. Webpage visited on March 2022. https://www.cyberdyne.jp
13. Koseki, K., et al.: Gait training using the Honda Walking Assist Device® for individuals with transfemoral amputation: a report of two cases. J Back Musculoskelet Rehabil. **33**(2), 339–344 (2020). https://doi.org/10.3233/BMR-191726.PMID:31929139;PMCID:PMC7175943
14. Muscolo, G.G.: HANDSHAKE: HANDling system for human autonomous KEeping. Int. J. Humanoid Robot. **18**(01), 2150003 (2021). https://doi.org/10.1142/S0219843621500031
15. Zoccali, A., Muscolo, G.G.: Comfort perception analysis of human models interfacing with novel biped-wheeled-exoskeletons. In: Rauter, G., Cattin, P.C., Zam, A., Riener, R., Carbone, G., Pisla, D. (eds.) MESROB 2020. MMS, vol. 93, pp. 21–28. Springer, Cham (2021). https://doi.org/10.1007/978-3-030-58104-6_3
16. Trono, G.; Nicolì, A.; Muscolo, G.G.: Sustainable compliant physical interaction in a biped-wheeled wearable machine. Front. Mech. Eng. **6,** 581626 (2020). https://doi.org/10.3389/fmech.2020.581626
17. Loschi, A., Smerchinich, A., Muscolo, G.G.: New sustainable biped-wheeled exoskeleton prototypes. In: Quaglia, G., Gasparetto, A., Petuya, V., Carbone, G. (eds.) I4SDG 2021. MMS, vol. 108, pp. 158–165. Springer, Cham (2022). https://doi.org/10.1007/978-3-030-87383-7_17
18. Muscolo, G.G., Recchiuto, C.T.: Flexible structure and wheeled feet to simplify biped locomotion of humanoid robots. Int. J. Humanoid Rob. **14**(01), 1650030 (2017)
19. Lisitano, D., Bonisoli, E., Recchiuto, C.T., Muscolo, G.G.: Dynamic balance of the head in a flexible legged robot for efficient biped locomotion. MDPI Appl. Sci. **11**(7), 2945, 1–19 (2021). https://doi.org/10.3390/app11072945

Null-Space Compliance with Non-linear Behavior: Application to Spine Surgery Robotic Platform

Alizée Koszulinski[1(✉)], Juan Sandoval[1], Térence Essomba[2], Tanguy Vendeuvre[1,3], Saïd Zeghloul[1], and Med Amine Laribi[1]

[1] Department of GMSC, Pprime Institute CNRS, ENSMA, UPR 3346, University of Poitiers, Poitiers, France
alizee.koszulinski@univ-poitiers.fr
[2] Department of Mechanical Engineering, National Central University, No. 300, Zhongda Road, Zhongli, Taoyuan 32001, Taiwan
[3] Department of Orthopaedic Surgery and Traumatology, University Hospital, Poitiers, France

Abstract. In this paper, we develop a novel control strategy to constrain the internal motion of a robotic platform intended to assist surgeons in spine surgery. The targeted task consists in drilling into a specific region of vertebrae in order to position spinal implants and thus re-ensure the spine stability. The present platform includes a redundant robotic arm with 7 degrees of freedom (DoF) directly manipulated by the surgeon in a comanipulation scheme. A torque-based impedance control has been chosen to generate a compliant behavior on the whole body of the robot and thus increase the degree of safety of the robotic device. In this regard, a new control law was implemented to constrain the internal motion of the robot, i.e. the movements of its elbow, allowed by its degree of redundancy. Unlike the classical control approaches, the proposed control law implements a non-linear response of the stiffness in the null-space to better comply with the motion constraints. In order to validate the behavior of the robot using this new control law, experimental tests have been carried out and their results were compared with those obtained with a classic null-space compliance law defined in joint coordinates.

Keywords: Orthopedic surgical robot · Spine surgery · Safety · Null-space compliance · Non-linear behavior

1 Introduction

In the case of significant cervical spine deformities, surgery may be recommended to relive patients. The procedure consists of fusing at least two vertebrae by fixing a spinal implant composing of screws and rods. By guaranteeing a normal space between the vertebrae concerned, the previously compressed nerves and/or spinal cord are released and the spine is stabilized [1]. Drillings in specific areas of the vertebrae are then essential in order to position the screws. Due to the proximity to the spinal cord in particular, this procedure requires great precision.

A. Müller and M. Brandstötter (Eds.): RAAD 2022, MMS 120, pp. 325–332, 2022.
https://doi.org/10.1007/978-3-031-04870-8_38

In order to improve the surgeon's abilities and to assist orthopedic surgery, various robotic devices have been developed. The TSolution One developed by THINK Surgical is an autonomous robot capable of performing the bone cavities and joint surfaces preparation for both of hip and knee implant placement [2]. The Monogram from Monogram Orthopedics has also been designed to execute milling procedures autonomously. Compared to the TSolution One, the Monogram is combined with a tracking system that allows the robotic device to adapt its trajectory in real time according to the patient's recorded movements [3]. Comanipulation robots, directly manipulated by surgeons, also use this tracking technology and are currently the most widespread on the market of robotic devices to assist orthopedic surgery. First, the ROSA Spine [4] by Medtech, the Mazor X Stealth [5] from Medtronic and the ExcelsiusGPS [6] by Globus Medical were developed to be used essentially as a guide during drilling operations on the lumbar and thoracic spine. In fact, for each drilling, the surgeon slides the tool inside the robot end-effector allowing the surgeon to keep a complete sensation during the drill insertion in the bones. On the contrary, other cobots directly attach the surgical tool to the end-effector, such as the MAKO [7] produced by Stryker and the VELYS robot [8] designed by DePuySynthes from Johnson & Johnson. On this way, it is possible to retrieve the tool position and orientation in real time through the information provided by the robot controller, without needing an additional external tracking system.

In this context, we recently proposed a new robotic platform intended to assist surgeon during drilling tasks [9]. The last above-mentioned comanipulation strategy, attaching the surgical tool to the end-effector, was chosen for the development of this new device in order to take advantage of the surgeons' dexterity and experience combined with the precision of the robotic arm. Compared to existing devices for spine surgery, this platform offers an active motion of the robotic arm that allows to accompany and control the surgeon's gestures throughout the desired procedure. Vertebral motion tracking combined with virtual guidance thus allows the tool to remain aligned with the trajectory selected by the surgeon. In addition, a haptic depth motion limitation is provided in the robot control by modifying the stiffness of the robotic arm according to the tool advance along the drilling axis. A torque control approach, based on a compliance law, then allows to generate a compliant behavior both in the working space and in the null-space of the robotic device, a Franka Emika (FE) with 7 degrees of freedom (DoF).

This paper more particularly focuses on the behavior of the robot into the null-space in order to constrain its internal motion. This internal motion is allowed by the degree of redundancy of the FE and is represented by the motion of its elbow. By restricting the range of the elbow tilt, it is possible to avoid any risk of collision with other equipment or medical staff and thus increase the degree of safety of the platform in the operating room. Several works exist in which a linear compliance law is defined in order to respect a restricted null-space [10, 11]. The main contribution of this paper is to propose a non-linear law inspired by the viscoelastic behavior of the most compliant part of the human body, namely the abdomen, and thus generate better responsiveness of the robotic arm to external forces.

This paper is structured as follows: Sect. 2 describes the medical needs and presents the actual robotic platform. Section 3 presents the overall control strategy and details the proposed null-space compliance approach. Experimental results validating the behavior of the robotic platform implementing the proposed control approach are presented and discussed in Sect. 4. The last section concludes this paper.

2 Robotic Platform

2.1 Medical Needs/Gestures

First, the requirements associated with the surgical gesture to be performed were identified. Each vertebra must be drilled according to an optimal trajectory set by the surgeon and defined by an entry point, a direction, a diameter and a specific depth. In medical terms, the positioning of the drilling tool according to this optimal trajectory is called pedicle aiming [12] and is illustrated in Fig. 1(a).

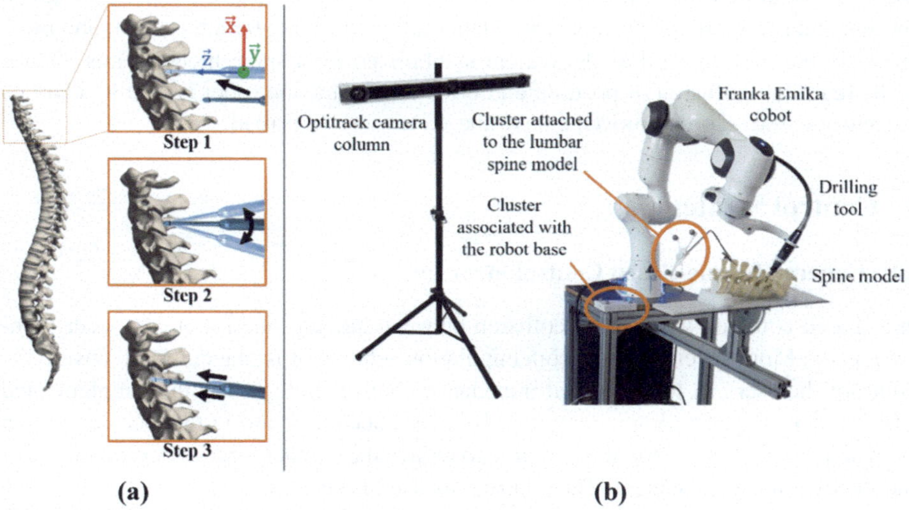

Fig. 1. (a) Analysis of the surgical procedure from the positioning of the tool according to the desired trajectory (steps 1 and 2) to the realization of the drilling (step 3). (b) Main elements of the robotic platform.

Some parameters of this gesture, such as the direction or the drilling depth, have been considered within the control mode of the robot in order to fully assist the surgeon and to respect the desired trajectory.

Therefore, the surgical procedure was divided into three phases and the DoF required for each one were determined. As shown in Fig. 1(a), the first step consists in moving the tool to the desired drilling entry point and thus mainly requires the three DoF in translation. Subsequently, the surgeon must adapt the orientation of the tool so that the screw is inserted perfectly into the pedicle of the vertebra to be drilled. During this phase, the rotation along the longitudinal axis of the tool can be neglected in terms of haptic feedback since the $\overrightarrow{z}$ axis corresponds to the rotation axis of the drill. However, this rotation remains interesting in order to be able to move away from any singularity at the level of the robotic arm when positioning of the tool. Finally, the last step of the procedure consists of the drilling, i.e. the insertion of the drill into the bone along the $\overrightarrow{z}$ axis of the tool, which must remain as stable as possible to fully respect the desired trajectory. In addition, the translation of the tool along its longitudinal axis is limited to a certain depth so that the drilling performed does not protrude from the vertebral body.

2.2 Description of the Robotic Platform

The proposed robotic platform is composed of a 7-DoF collaborative robot, more precisely a FE cobot, on which is fixed a drilling tool as shown in Fig. 1(b). The robot's movements are coordinated on the one hand by a control program implemented under the Robot Operating System (ROS) environment in Ubuntu 16.04 with PREEMPT-RT, and on the other hand by the data recorded with a motion capture system and streamed through a local network. This system is composed of an Optitrack V120: Trio camera column and two markers clusters. These latter allow to locate, both the robot, and more precisely the tool, as well as the patient symbolized by a spine model. Thus, thanks to the real time diffusion of position and orientation data, the robot is able to adapt its trajectory to the recorded movements of the vertebra to be drilled.

3 Control Strategy

3.1 General Torque-Based Control Strategy

In a shared common workspace, collision between humans and robot can occur unintentionally. More specifically, in comanipulation scheme, it is mandatory to ensure the safety of the users. A torque-based impedance control strategy was thus implemented [10]. In fact, by generating a compliant behavior, such as a spring-damper system, on the whole body of the robot, it is possible to reduce the contact forces between the robot and the external environment. The control law used is such as:

$$\boldsymbol{\tau}_{\mathrm{c}} = \boldsymbol{\tau}_{\mathrm{task}} + \boldsymbol{\tau}_{\mathrm{null}} + \boldsymbol{\tau}_{\mathrm{comp}} \tag{1}$$

where $\boldsymbol{\tau}_{\mathrm{c}}$ is the vector of control torques, $\boldsymbol{\tau}_{\mathrm{task}}$ is associated with the task, $\boldsymbol{\tau}_{\mathrm{null}}$ allows the exploitation of the robot's degree of redundancy and $\boldsymbol{\tau}_{\mathrm{comp}}$ is corresponding to the gravitational compensation.

In order to perform the surgical task presented in the previous section, the robot's trajectory is defined in particular as a function of the desired Cartesian trajectory $\mathbf{X}_{\mathrm{d}}$ as follows:

$$\boldsymbol{\tau}_{\mathrm{task}} = \mathbf{J}^{\mathrm{t}}.\mathbf{F}_{\mathrm{task}} = \mathbf{J}^{\mathrm{t}}.\left[\mathbf{K}_{\mathrm{px}}.(\mathbf{X}_{\mathrm{d}} - \mathbf{X}) - \mathbf{K}_{\mathrm{dx}}.\dot{\mathbf{X}}\right] \tag{2}$$

where $\mathbf{F}_{\mathrm{task}}$ is the vector of desired forces on 3D cartesian space associated to the impedance law. $\mathbf{K}_{\mathrm{px}}$ and $\mathbf{K}_{\mathrm{dx}}$ are diagonal matrices associated with the stiffness and damping values, respectively, allowing to adjust the compliant behavior of the robot. $\mathbf{J}$ is the Jacobian matrix, $\mathbf{X}$ and $\dot{\mathbf{X}}$ correspond respectively to the current cartesian positions and velocities of the robot.

3.2 Null-Space Control Strategy

The additional DoF of the FE cobot allows it to perform internal motion without changing the effector pose, i.e. in the case of cervical surgery the pose of the drilling tool. This degree of redundancy, represented by the tilt of the robot elbow, offers several advantages including joint limits avoidance or the possibility to move away from singularities. However, it is necessary to constrain this internal motion in order to prevent any collision with other equipment positioned near the robot or members of the medical team.

As reported in [10], the control law of a classical null-space compliance approach defined in joint coordinates is expressed as follows:

$$\tau_{\text{null}} = \begin{cases} \mathbf{N}(\mathbf{q}).\left[\mathbf{K}_{\mathrm{p}_{\aleph}}.\left(\mathbf{q}_{\psi_{min}} - \mathbf{q}\right) - \mathbf{K}_{\mathrm{d}_{\aleph}}.\dot{\mathbf{q}}\right] \text{ for } \psi \leq \psi_{min} \\ \mathbf{N}(\mathbf{q}).\left[\mathbf{K}_{\mathrm{p}_{\aleph}}.\left(\mathbf{q}_{\psi_{max}} - \mathbf{q}\right) - \mathbf{K}_{\mathrm{d}_{\aleph}}.\dot{\mathbf{q}}\right] \text{ for } \psi \geq \psi_{max} \end{cases} \tag{3}$$

where $\mathbf{N}(\mathbf{q}) = \mathbf{I} - \mathbf{J}^{\mathrm{T}}\mathbf{J}^{+\mathrm{T}}$ is the null-space projector, $\mathbf{K}_{\mathrm{p}_{\aleph}} = k_p.\mathbf{I}_{(7\times7)}$ with k_p the null-space stiffness value fixed at 15 N.m/rad for the needs of our application, $\mathbf{K}_{\mathrm{d}_{\aleph}} = k_d.\mathbf{I}_{(7\times7)}$ with $k_d = 2\sqrt{k_p}$ the damping value, $\mathbf{q}_{\psi_{min}}$ and $\mathbf{q}_{\psi_{max}}$ the joint coordinates associated with the desired minimum or maximum ψ elbow angle limit, $\mathbf{q}$ and $\dot{\mathbf{q}}$ the current joint positions and speeds of the robot. The null-space stiffness is thus fixed and the effort feedback evolution is in particular function of the difference between $\mathbf{q}_{\psi_{min}}$, or $\mathbf{q}_{\psi_{max}}$, and the current joint coordinates.

In addition, ψ can be defined by considering the angle between the vertical axis and the plan formed by the shoulder, elbow and wrist of the robot around the shoulder-wrist axis as shown in Fig. 2. Then, $\mathbf{q}_{\psi_{min}}$ and $\mathbf{q}_{\psi_{max}}$ are obtained by positioning the robot elbow according to the desired tilt and by recording the associated joint coordinates.

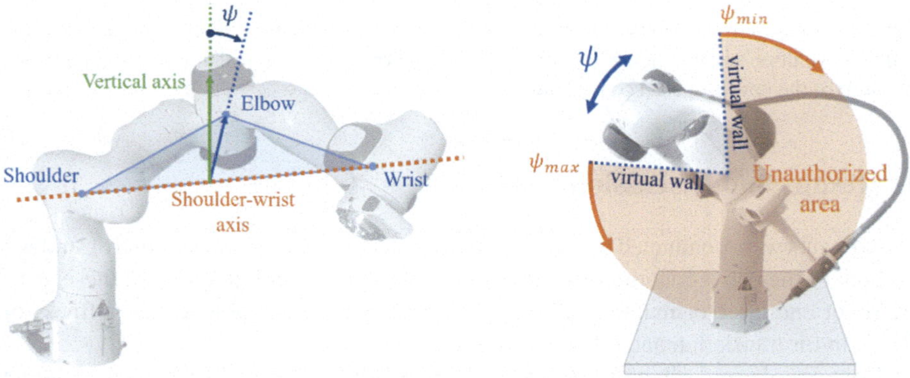

Fig. 2. Redundancy representation of the FE cobot. The desired virtual walls of the robot null-space are defined from the values of the limit angles ψ_{min} and ψ_{max} of the robot elbow.

In order to improve the robot performance in the null-space, a new compliance motion strategy has been implemented allowing a non-linear evolution of the stiffness as a function of the elbow angle. The new null-space law is such as:

$$\tau_{\text{null}} = \mathbf{N}(\mathbf{q}).\left[\mathbf{K}_{\mathrm{p}_{\aleph}}(\psi).\left(\mathbf{q}_{\psi_{\text{mean}}} - \mathbf{q}\right) - \mathbf{K}_{\mathrm{d}_{\aleph}}(\psi).\dot{\mathbf{q}}\right] \tag{4}$$

where $\mathbf{K}_{p_{\aleph}}(\psi) = k_p(\psi).\mathbf{I}_{(7\times7)}$ with $k_p(\psi)$ is the null-space stiffness as a function of the elbow angle noted ψ, $\mathbf{K}_{d_{\aleph}} = k_d.\mathbf{I}_{(7\times7)}$ with $k_d = 2\sqrt{k_p}$ the damping value and $\mathbf{q}_{\psi_{mean}}$ the joint coordinates associated with the mean elbow angle of the range $[\psi_{min}, \psi_{max}]$.

The stiffness $k_p(\psi)$ definition is inspired of the viscoelastic behavior of the most compliant part of the human body, i.e. the abdomen, studied in [13, 14] for the development of a safety prismatic joint. In order to adapt the mathematical response proposed in these works to our application, a tuning coefficient had to be applied so that the threshold null-space stiffness reached is equal to 15 N.m/rad, i.e. the stiffness value applied in the first approach defined in Eq. (3). A polynomial interpolation of the obtained curve was then carried out in order to obtain an equation suitable within the robot control program.

Figure 3 illustrates the main differences between the two null-space compliance laws studied in this paper. More precisely, the curves report the implantation of the stiffness according to the elbow tilt angle and the evolution of the joint coordinates range in relation to the current joint coordinates.

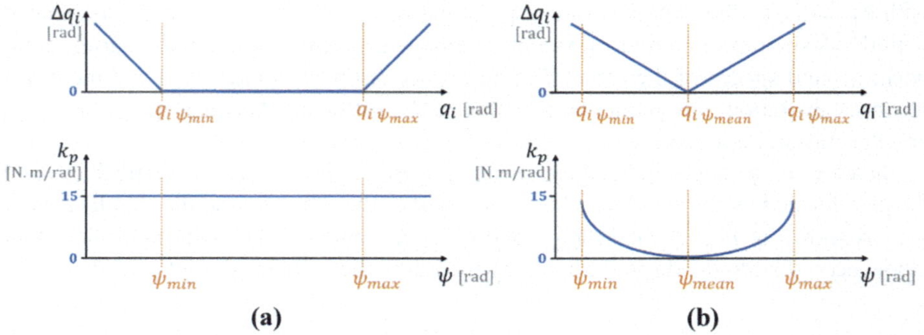

Fig. 3. Evolution of the joint coordinate range according to the current joint coordinate and the stiffness evolution as a function of the robot elbow angle ψ: (a) the classical null-space compliance law and (b) the new proposed control law.

4 Results and Discussion

Experiments were conducted to verify the performance of the null-space control strategy for both of the above-mentioned control laws. During the entire duration of these tests, the robot end-effector maintained a fixed position and orientation with a stiffness of 700 N/m for translation and 60 N/rad for rotation.

First, the desired elbow tilt limits were manually selected and the corresponding necessary ψ limit angles and joint coordinate values were calculated and recorded. The acquisition of the curves presented in this section began after replacing the robot elbow in the center of the selected range $[\psi_{min}, \psi_{max}]$ and activating one of the two null-space control strategies. Figures 4(a) and (b) show the operator action on the robot by the variation of the robot elbow angle from the first contact at about 4.8 s and 8.8 s respectively.

In both experiments, oscillations at the virtual walls have been voluntarily performed by the operator in order to highlight the limits imposed in the null-space.

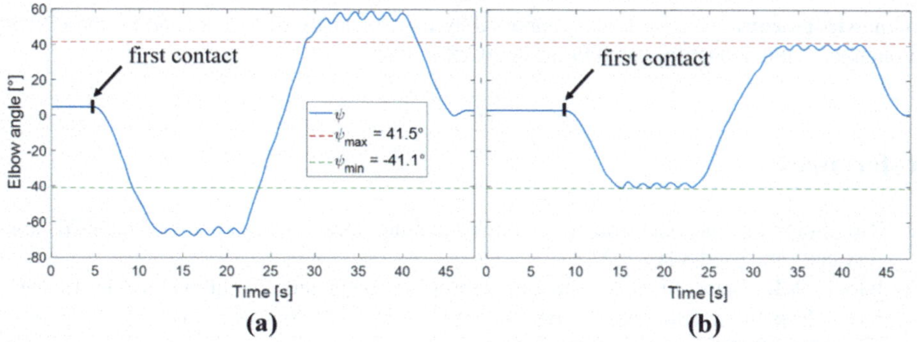

Fig. 4. Elbow angle evolution obtained with (a) the classical null-space compliant law which the applied stiffness is fixed and with (b) the new null-space law with a non-linear evolution of the stiffness.

The results show that due to the definition of the classical null-space law, the desired minimum and maximum limits are not respected. Moreover, as the stiffness is not defined as a function of the angle ψ, the possible exceeding of the elbow angular limits must necessarily be considered at the time of the manual selection of the virtual walls. The definition of these latter seems to be difficult to implement and then imprecise.

Conversely, the new proposed approach allows to constrain the internal motion of the robot between the desired angles by increasing the stiffness upstream at these limits. The configuration procedure of the robotic platform is thus simplified, easier to set up, and confers an improved safety with a strict limits preservation. In the operating room, this advantage is not negligible in view of the simplicity of the limits selection, which is therefore not a time-consuming step. It is possible to avoid, for instance, any contact with a member of the medical team or other equipment in the robot's environment, which must remain sterile and could interfere with the efficient running of the surgical procedure otherwise.

5 Conclusion

This paper proposes a novel null-space control strategy for a redundant robotic platform intended to assist surgeons in performing drilling procedures in orthopaedic surgery and more specifically in spine surgery. A new compliance law, based on a torque control, was thus implemented to limit the movements of the elbow of the robotic arm representing the zero space of the robot. Compared to existing laws, the response of the robot body to external forces is controlled by the evolution of the stiffness as a function of the angle of the robot elbow. A non-linear evolution of this stiffness is generated between two angular limits, defined by the surgeon, and customizable at each moment according to the changes that can be applied to the robot environment during a surgical operation. This new definition ensures that the robot's internal motion is perfectly constrained within the desired range, thus limiting the risk of collision with any other equipment or medical personnel present in the operating room.

Acknowledgments. This work was supported by the University of Poitiers and by the CNRS through the International Research Project RACeS.

References

1. Paris Spine Institute, Cervical spine. https://institutdurachis.com/pathologies-rachis/rachis-cervical/. Accessed 3 Jan 2022
2. Bargar, W.L., Netravali, N.A.: Total hip arthroplasty technique: TSolution One. In: Lonner, J. (ed.) Robotics in Knee and Hip Arthroplasty, pp. 219–224. Springer, Cham (2019). https://doi.org/10.1007/978-3-030-16593-2_21
3. Monogram Orthopedics, Presentation of the Monogram platform. https://www.monogramorthopedics.com/. Accessed 3 Jan 2022
4. Lonjon, N., Chan-Seng, E., Costalat, V., Bonnafoux, B., Vassal, M., Boetto, J.: Robot-assisted spine surgery: feasibility study through a prospective case-matched analysis. Eur. Spine J. **25**(3), 947–955 (2015). https://doi.org/10.1007/s00586-015-3758-8
5. Khan, A., et al.: Next-generation robotic spine surgery: first report on feasibility, safety, and learning curve. Oper. Neurosurg. **17**(2) (2019)
6. Elswick, C.M., et al.: Robotic-assisted spinal surgery: current generation instrumentation and new applications. Neurosurg. Clin. North Am. **31** (2020)
7. Hagag, B., et al.: RIO: robotic-arm interactive orthopedic system MAKOplasty: user Interactive haptic orthopedic robotics. In: Rosen, J., Hannaford, B., Satava, R. (eds.) Surgical Robotics. Springer, Boston (2011). https://doi.org/10.1007/978-1-4419-1126-1_10
8. Johnson & Johnson, VELYS Digital Surgery. https://www.jnjmedicaldevices.com/en-US/velys. Accessed 3 Jan 2022
9. Koszulinski, A., et al.: Comanipulation robotic platform for spine surgery with exteroceptive visual coupling: development and experimentation. ASME. J. Med. Dev. (2022)
10. Sandoval, J., et al.: Collaborative framework for robot-assisted minimally invasive surgery using a 7-DoF anthropomorphic robot. Robot. Auton. Syst. **106**, 95–106 (2018)
11. Xiong, G., Zhou, Y, Yao, J.: Null-space impedance control of 7-degree-of-freedom redundant manipulators based on the arm angles. Int. J. Adv. Robot. Syst. **17**(3), 1729881420925297 (2020)
12. Farooq Usmani, M. et al.: Management of Cranio-cervical injuries: C1-C2 posterior cervical fusion and decompression. Semin. Spine Surg. **32**(1) (2019)
13. Ayoubi, Y., et al.: A complete methodology to design a safety mechanism for prismatic joint implementation. In: 2016 IEEE/RSJ International Conference on Intelligent Robots and Systems (IROS), pp. 304–309 (2016)
14. Ayoubi, Y., et al.: Complete design methodology of biomimetic safety device for cobots' prismatic joints. Robot. Auton. Syst. **102**, 44–53 (2018)

Optimal Multi-robot Placement Based on Capability Map for Medical Applications

Amir Trabelsi[1,2](✉), Juan Sandoval[1], Abdelfattah Mlika[2], Samir Lahouar[3], Said Zeghloul[1], Jérôme Cau[4], and Med Amine Laribi[1]

[1] Department of GMSC, Pprime Institute CNRS, ENSMA, University of Poitiers, UPR 3346, Poitiers, France
amir.trabelsi@univ-poitiers.fr

[2] National Engineering School of Sousse, LMS, University of Sousse, BP 264 Sousse Erriadh, 4023 Sousse, Tunisia

[3] National Engineering School of Monastir, LGM, University of Monastir, 5019 Monastir, Tunisia

[4] GIE Group Sainte Marguerite, Vascular Laparoscopic Surgeon, Marseille, France

Abstract. Multi-robot systems have become important in various applications. These robotic platforms consisting of several robotic arms collaborating in a common workspace are a suitable tool to perform complex tasks, such as surgical applications. However, the placement of each robotic arm has an impact on the whole system kinematics and dynamic performance. In this paper, we present a novel technique to optimize the base location of the robot manipulators based on the capability map to develop a bi-cobot system dedicated to medical applications. This technique is implemented to simultaneously identify the base placement of the two collaborative robots while optimizing their kinematic criteria and the system compactness. The obtained results are evaluated using a cobot simulation platform for the execution of minimally invasive surgical tasks.

Keywords: Collaborative robots · Optimization · Robot placement · Capability map · Multi-robot systems

1 Introduction

Advanced medical applications are carried out nowadays using collaborative robots. In the most complex applications, robots assist doctors in the intricate procedures by providing them with accuracy, precision, and dexterity and helping them to perform their gestures in the best possible circumstances. Many robots have revolutionized the field of medicine by their flexibility and safety, such as the Kuka iiwa and Franka Emika robots used in minimally invasive surgery [1], teleoperation applications such as doppler echography [2], and robotized craniotomy [3].

Recently, several systems with different robot arms have been proposed. Complex tasks can be accomplished by the cooperation of the different arms working together [4]. In medical applications, a proper collaboration between multiple robots would ensure

A. Müller and M. Brandstötter (Eds.): RAAD 2022, MMS 120, pp. 333–342, 2022.
https://doi.org/10.1007/978-3-031-04870-8_39

that the medical operation is performed correctly. In our case, a multi-robot system consists of two Kuka IIWA redundant robots used to manipulate the surgical instruments in the context of minimally invasive surgery. Each robot arm is guided by the surgeon via a haptic interface since in most cases the two robots are required to perform the surgery. The ability of the robot arm to manipulate the objects depends on the position and orientation of the tools attached to the robot end-effector and is defined relative to its base coordinate system. To develop a multi-robot system for collaborative applications, it is necessary to have knowledge about the capabilities of individual arms. This information can be handled using a capability map [5], that connects the robot's end effector to its base. This map can be used to identify the base location of each robot constituting a multi-robot system.

Several strategies have focused on the study of the optimal robot placement in order to optimize its kinematic and dynamic behavior or establish the best path plan for pre-programmed tasks. In this article, we study the optimal placement of two redundant robots in a multi-robot platform context and with a common workspace for a medical application. To properly place our robots, we first have to understand the directions in which the objects may be approached and grasped by the robot arm. Information regarding the capabilities of individual arms is required to design multi-robot systems for cooperative tasks. Then, using the capacity map for each robot arm, we develop an optimization algorithm for simultaneously positioning the bases of multiple robots in a shared workspace.

This paper is organized as follows: Sect. 2 represents an optimization technique based on the robot's capability map for optimal multi-robot base placement. The robot capability map generation is detailed in Sect. 3. Section 4 carries out the formulation of the optimization problem using genetic algorithm techniques. The discussion of the obtained results is presented in Sect. 5. The conclusion and the perspectives of the current study are presented in the last section.

2 Problematic

Most of the techniques related in the literature for solving robot placement problems rely on algorithms that are limited to optimizing specific criteria, such as kinematic performances [6] and optimizing predefined trajectories [7]. However, these techniques do not deal with complex systems such as redundant robots or multi-robot systems. In this paper, we propose a new methodology for studying the optimal base placement of collaborative redundant robot platforms for minimally invasive surgery tasks. This approach is inspired by a recent online positioning technique used for humanoid robots to detect and manipulate objects automatically [8]. Our system consists of two 7 degrees of freedom (DoF) collaborative robots carrying surgical instruments and imitating the surgeon's gestures performed via haptic interfaces. This research aims to identify the 6 degrees of freedom base placement of each robot to optimize the robot kinematic behavior, based on the use of capability maps.

3 The Capability Map

The ability of a robotic arm to manipulate objects depends on its position and orientation with respect to the robot's base coordinate system. To develop multi-robot systems for programmed cooperative applications, it is necessary to understand the capabilities of individual arms. Knowing where to perform certain tasks helps to identify the best layout for the shared workspace between individual robotic arms. Therefore, a detailed representation of the working space is required for each arm inside a multi-robot system. This is called the capability map [9].

In order to generate a representation of the robot's capabilities in terms of accessibility, the 6D workspace (position and orientation of the robot's end-effector) is voxelized. The resulting voxel map is filled with pose information. Each voxel represents all possible spatial orientations of the end effector for which the tool center point (TCP) is reached within the voxel volume [8]. The robot capability map represents the reachable positions of the TCP and is presented in the robot base coordinate system. The TCP pose is described by a homogeneous matrix **T** defined as:

$$\mathbf{T} = \begin{bmatrix} \mathbf{R} & \mathbf{P} \\ 0 & 1 \end{bmatrix} \tag{1}$$

where **T** is the transformation matrix between the tool frame and the robot base frame and consists of the rotation matrix $\mathbf{R} \in So(3)$, and the position vector $\mathbf{P} \in R^3$.

To properly describe the robot's capacity map. First, a R^3space is discretized using voxelization. Second, the spatial orientations of TCP represent the discretization of So(3). Then, using an inverse kinematics process, accessible TCP position is validated to obtain the robot's generated capability map.

3.1 Discretization of $\mathbf{R}^3$

The physical workspace of the robot arm can be encapsulated in a box (V_{ws}) with a lateral length corresponding to the maximum reach of the manipulator l_{ws}, centered on the base of the robot arm as shown in Fig. 1a. The axes of the box are aligned with the robot arm base coordinate system. The V_{ws} volume is divided into a regular grid of cubes V_e aligned on the coordinate axes. The homogeneous partitioning is called voxelization of R^3 and the discrete elements of the volume are called voxels [5].

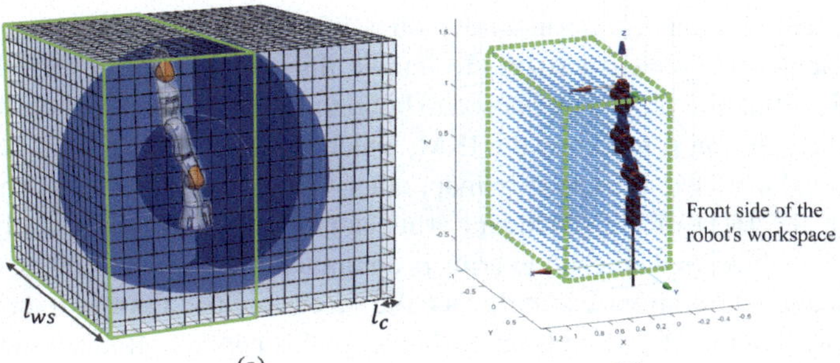

(a) (b)

Fig. 1. A homogeneous voxelization of the robot workspace

Each voxel contains information about the accessible orientations of the TCP relative to the occupied volume. Therefore, the Cartesian workspace of the robot is subdivided into N voxels.

$$N = V_{ws}/V_e \tag{2}$$

In our study, we constructed our multi-robot system using a Kuka iiwa robot with seven DoF. To properly position the poly-articulated system, which consists of two redundant collaborative robots, the capability map for each robotic arm has to be created. To simplify and reduce the map size, we focused only on the robot's front side of its workspace, as shown in Fig. 1b, and we picked $l_c = 0.1$ m as the elementary voxel V_e edge length. As a result, the capability map describing the robot contains 4130 voxels.

3.2 Discretization of So(3)

The capacity map's voxels serve as containers for the data. This information is represented by discretizing the orientation space So(3). As shown in Fig. 2, a sphere with a diameter equal to the voxel's edge length l_c is inserted into each voxel to produce uniformly distributed orientations. The center of the sphere is identical to the center of the voxel. A set P of uniformly distributed points is created using the spiral point technique on a sphere [10]. For each point of the set P on the sphere, a direction has been established as shown in Fig. 2b.

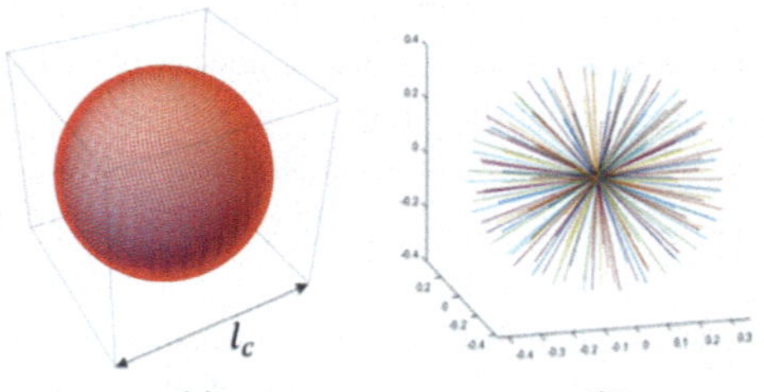

(a) (b)

Fig. 2. Discretization of the orientation space

3.3 Capability Map Generation

To create a capability map for a Kuka iiwa robot, it is necessary to determine the TCP's reachable orientations for each voxel. To handle this problem, an inverse kinematics model (IKM) validation procedure is used to build the capacity map. For each examined pose, if a valid solution is found by the IKM, which means that the pose is reachable, it is assigned a value of 1 in the capacity map, otherwise, a value of 0 is assigned.

However, the IKM of the redundant arms cannot be solved analytically. Usually, the IKM of 7-DoF redundant robotic arms is computed through an iterative model, but different initial configurations can influence the success of the algorithm and its failure does not imply that the TCP pose is unreachable. In this paper, optimization procedures are applied to identify one of the feasible configurations for the desired pose [11].

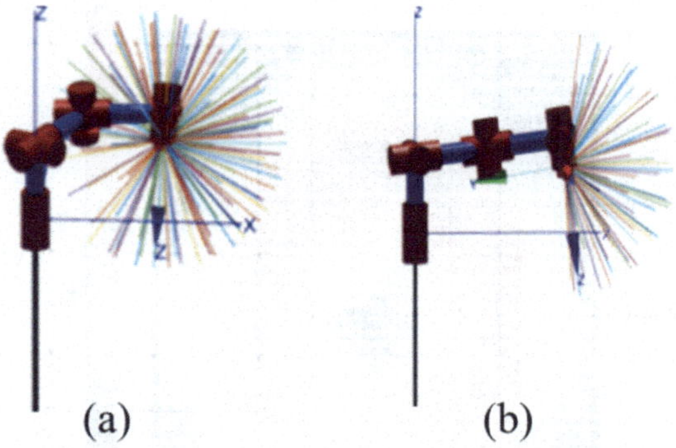

Fig. 3. Capability map: IKM validation and graphical representation

Figure 3 represents the data of two voxels in the robot's operating space after validation by the IKM. The voxel presented in the Fig. 3a shows the high accessibility at the TCP. On the other hand, the voxel on the Fig. 3b located at the boundary of the robot workspace shows limited orientation accessibility due to the kinematic limitations of a serial manipulator.

The capability map represents the grouping of all voxel data. It can be used to present the kinematic capabilities of the robot and visualize the accessibility of the workspace. All these data are stored in a database which then will be used to optimize the robot base placements. Figure 4 depicts the capacity map data in a graphical form. According to the reachability index, which is defined by the percentage of directions in the voxel (v) with an inverse geometric solution with respect to the full number of discretized orientations m, the colors of the spheres are determined as follows:

$$A(v) = 100\frac{\sum_{i=1}^{m} R(v,i)}{m}$$

where:

$$R(v,i) = \begin{cases} 1 & \text{Accessible direction} \\ 0 & \textit{otherwise} \end{cases} \tag{3}$$

4 Formulation and Optimization

The base placement of a serial manipulator robot is a significant challenge in many complex applications. The kinematic performance of the robot during the execution of a given operation is dependent on the robot's relative position to the task. The placement of the robot's base with respect to the global reference frame associated with the global workspace improves the robot's kinematic behavior for a given application.

The purpose of this paper is to propose a new algorithm based on the capability map that aims to optimize and identify the six DoF of each robot's base related to a global world frame, in order to design a multi-robot system for minimally invasive surgical applications. The recording of the surgeon's motion allows us to characterize

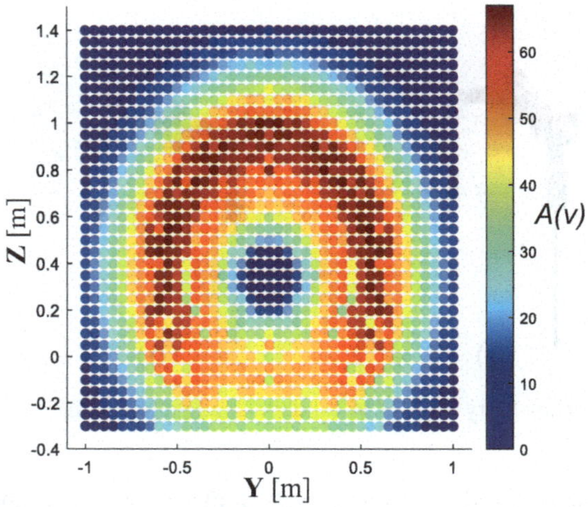

Fig. 4. Graphical representation of a capacity map section

the movement's degrees of freedom. According to these observations [12], we consider that each surgical instrument performs 4-DoF, three rotational movements, and one translation, within a cone defined by a remote center of motion (RCM) and an apex angle α as described in Fig. 5a. Each RCM is located at the incision point. The task workspace *Ws* is considered as a pair of cones swept by the tools tilted by angle β and located at a specific distance d based on the experimental data.

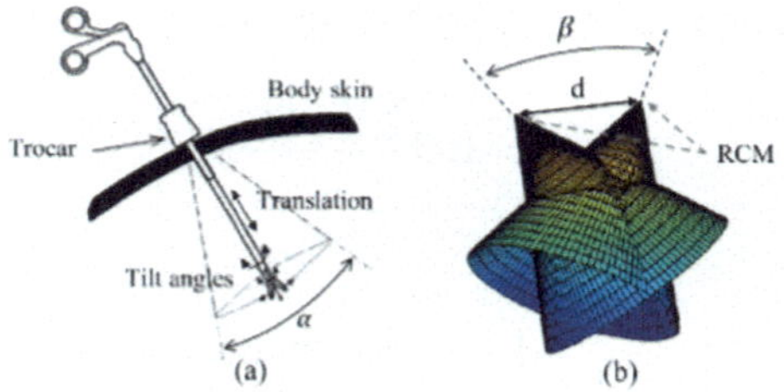

Fig. 5. Tools workspaces

4.1 Criteria Definition

The optimal robot's placement problem can be formulated using a variety of kinematic performance criteria. In this work, we focus on the two most suitable criteria for our medical application, which are the system's compacity and dexterity.

- *Compactness*
  This criterion aims at reducing the volume occupied by the multi-robot system. Thus, compactness (V) is described by the volume covering the whole system, i.e. by the parallelepiped volume containing the two robotic arms, and deduced as follows:

$$V = (X_{max} - X_{min}) \cdot (2Y_{max}) \cdot (Z_{max} - Z_{min}) \tag{4}$$

where the robot workspace is described in the world frame by the coordinates X, Y, and Z.

- *Manipulability*
  Yoshikawa's manipulability index represents the robot's TCP ability to change its position and orientation arbitrarily. Robot redundancy can be used to promote positions with better dexterity, leading to more efficient adaptation during task execution. The manipulability index (w) is computed as a function of the robot Jacobian matrix $\mathbf{J}(\mathbf{q}_r)$, where $\mathbf{q}_r$ is the instantaneous joint configuration of the manipulator.

$$w(\mathbf{q}_r) = \sqrt{\det \mathbf{J}(\mathbf{q}_r)\mathbf{J}(\mathbf{q}_r)^{\mathrm{T}}} \tag{5}$$

4.2 Formulation of the Optimization Problem

In order to identify the optimal placement and compute the homogeneous transformation matrix of the robots bases, the multicriteria optimization problem is described as follows:

$$\begin{cases} minimize(F_1, F_2)(X) \\ X = [x, y, z, Rx, Ry, Rz] \\ \mathbf{X}_{\min} < \mathbf{X} < \mathbf{X}_{\max} \end{cases} \tag{6}$$

The described mathematical model is solved using a Genetic Algorithm (GA) [13] method implemented under Matlab. This algorithm enables the identification of the position parameters required to optimize the placement of the two robots simultaneously to perform the predefined surgical application with the best kinematic performance. A multi-robot system symmetry is used to reduce the number of unknown parameters of the design vector (**X**). Thus, only the six DoF of one robot position has to be identified whereas the second one is deduced by symmetry. The goal is to maximize the system's dexterity while improving its compactness. Therefore, the two objective functions to be minimized are denoted by F_1 and F_2, respectively, and are defined as follows:

$$F_1 = -\frac{1}{gd}\sum_{i=1}^{g}\sum_{j=1}^{d}\left(w(\mathbf{q}_{l_i}) + w(\mathbf{q}_{r_i})\right) + C_1 + C_2$$

$$F_2 = V + C_1 + C_3 \tag{7}$$

The minimization is performed under the condition that the constraints are satisfied leading to the best solutions. The reachability of the tool's workspaces defined by the pair of RCMs is defined within these constraints.

$$C_1(\mathbf{X}, Ws) = \begin{cases} 0 & Ws \text{ is reachable} \\ K_1 & \text{otherwise} \end{cases} \tag{8}$$

To avoid the robot's workspace boundary and singular position, the Ws must be positioned in places with a high accessibility index. as indicated in Fig. 4, yellow and red areas where the index of accessibility is greater than 20%

$$C_2(\mathbf{X}, Ws) = \begin{cases} 0 & A(v_{Ws}) > 20\% \\ K_1 & \text{otherwise} \end{cases} \tag{9}$$

Parallel configuration when the robot's base z axes are collinear should be avoided since it interferes with other criteria and increases the risk of collisions:

$$C_3(\boldsymbol{X},\ Ws) = K_2\ log(\frac{1}{\|\mathbf{Z}_{Rr}\ \mathbf{Z}_{R1}\|}) \tag{10}$$

Where K_1 and K_2 are equal to 1^{e5} and 1^{e3}, and $\boldsymbol{Z}_R$ is the z axes of the robot base frame.

5 Results and Discussion

The GA method implemented under Matlab was used to solve the optimization problem. The results were obtained by minimizing the two objective functions F_1 and F_2. As result, a set of optimal non-dominated solutions that represent the best compromise between the two objectives is generated. Figure 6 depicts the obtained solutions in the form of a Pareto front.

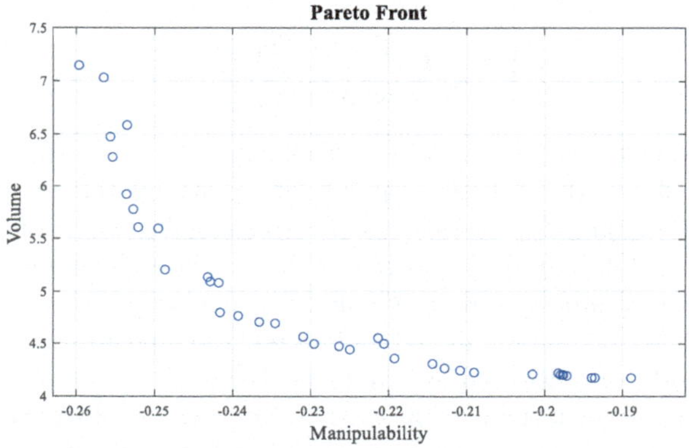

Fig. 6. The multi-objective optimization Pareto front results

The Pareto front presents multiple solutions, each one gives potential optimal robots bases placements. The selection of the optimal solution could be made by choosing a weight for each criterion, either by giving the same importance to the two criteria or by privileging one over the other [14].

In this paper, we propose to choose a solution on the Pareto front with equal weight for the two criteria and validate it using a cobot simulation platform (CSP) [15].The simulation platform presented in Fig. 7 is used to investigate the mechanical behaviors of two KUKA collaborative robots in a medical application dedicated to minimally invasive surgery.

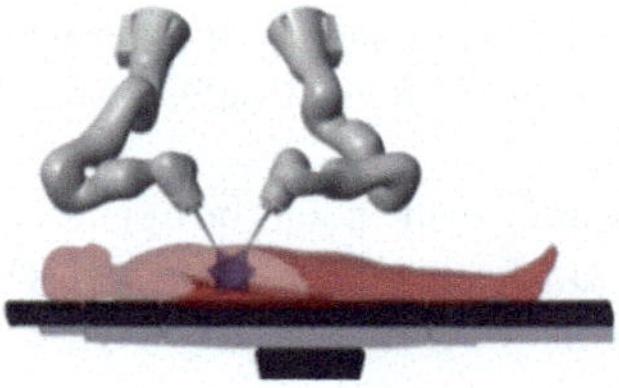

Fig. 7. Robot placement validation using the CSP platform

6 Conclusion

An identification method for determining the optimal base placement of two Kuka iiwa robots composing a bi-cobot platform for surgical applications is presented in this work. The optimization algorithm based on the robot capability map is used to improve the kinematic performance of the multi-robot system. To address this problem, we employed a multi-objective Genetic Algorithm based on two criteria: the bi-robot platform dexterity and system compactness.

The capability map of the Kuka iiwa robot is generated, representing the position and orientation information of the robot's TCP with respect to the base frame. This capability map is used in the optimization process and enhances the genetic algorithm to achieve optimal results. The obtained optimal base locations are evidenced via a Pareto front, letting the user choose an optimal solution trading-off between the two desired criteria. In future works, further criteria will be included in the robot placement problem, such as the robot's dynamic performances or the obstacle and collision avoidance.

References

1. Sandoval, J., Laribi, M.A., Zeghloul, S.: Autonomous robot-assistant camera holder for minimally invasive surgery. In: Kuo, C.-H., Lin, P.-C., Essomba, T., Chen, G.-C. (eds.) ISRM 2019. MMS, vol. 78, pp. 465–472. Springer, Cham (2020). https://doi.org/10.1007/978-3-030-30036-4_42
2. Sandoval, J., Laribi, M.A., Zeghloul, S., Arsicault, M.: On the design of a safe human-friendly teleoperated system for Doppler sonography. Robotics **8**, 11 (2019)
3. Zhan, Y. Duan, X.G. Craniotomy robot system based on human-machine parallel collaboration. In: IEEE International Conference Mechatronics Automation ICMA, August 2016, pp. 1119–1124 (2016)
4. Simorov, A., Otte, R.S., Kopietz, C.M., Oleynikov, D.: Review of surgical robotics user interface. Surg. Endosc. **26**(8), 2117–2125 (2012)
5. Zacharias, F., Borst, C., Wolf, S., Hirzinger, G.: The capability map: a tool to analyze robot arm workspaces. Int. J. Humanoid Robot. **10**(4) (2013)
6. Pamanes, J.: A criterion for the optimal placement of robotic manipulators. In: 6th IFAC/IFIP/IFORS/IMACS Symposium on Information Control Problems in Manufacturing Technology (1989)
7. Spensieri, J., Carlson, S., Bohlin, R., Kressin, J., Shi, J.: Optimal robot placement for tasks execution. Procedia CIRP **44**, 395–400 (2016)
8. Vahrenkamp, N., Asfour, T.: Robot placement based on reachability inversion. IEEE Int. Conf. Robot. Autom. **2**, 1970–1975 (2013)

9. Borst, C., Zacharias, F., Borst, C., Hirzinger, C.: Capturing robot workspace structure : representing robot capabilities. In: International Conference on Intelligent Robots and Systems, pp. 3229–3236, December 2016
10. Saff, E., Kuijlaars, A.: Distributing many points on a sphere. Math. Intell. **19**(1), 5–11 (1997)
11. Konietschke, R., Hirzinger, G.: Inverse kinematics with closed form solutions for highly redundant robotic systems. In: International Conference on Robotics and Automation, pp. 2945–2950 (2009)
12. Laribi, M.A., Arsicault, M., Riviere, T., Zeghloul, S.: Toward new minimally invasive surgical robotic system. In: IEEE International Conference on Industrial Technology, pp. 504–509 (2012)
13. Shorman, S.M., Pitchay, S.A.: Significance of parameters in genetic algorithm. ARPN J. Eng. Appl. Sci. **10**(2), 585–593 (2015)
14. Papadopoulos, C.E., Yeung, H.: Uncertainty estimation and Monte Carlo simulation method. Flow Meas. Instrum. **12**(4), 291–298 (2001)
15. Trabelsi, A., Sandoval, J., Ghiss, M., Laribi, M.A.: Development of a Franka Emika Cobot simulator platform (CSP) dedicated to medical applications. In: Zeghloul, S., Laribi, M.A., Sandoval, J. (eds.) RAAD 2021. MMS, vol. 102, pp. 95–103. Springer, Cham (2021). https://doi.org/10.1007/978-3-030-75259-0_11

EMG Driven Robotic-Aided Arm Rehabilitation

Daniel Bonilla[1(✉)], Julian D. Colorado[1], Med Amine Laribi[2], Juan Sandoval[2], and Catalina Alvarado-Rojas[1]

[1] Department of Electronics Engineering, Pontificia Universidad Javeriana Bogota, Cra. 7 No. 40-62, Bogota 110231, Colombia
{daniel-bonillab,coloradoj,catalina_alvarado}@javeriana.edu.co
[2] Department of GMSC, Pprime Institute CNRS, ENSMA, UPR 3346, University of Poitiers, Poitiers, France
{med.amine.laribi,juan.sandoval}@univ-poitiers.fr

Abstract. Robotic-assisted systems have been gaining significant traction in supporting rehabilitation tasks, enabling patients to manipulate objects by using a robotic arm controlled by the means of biological signals. In this regard, electromyography (EMG) signals are key for detecting the patient's intention of motion, that can be replicated by the robotic arm with higher accuracy and precision. In this paper, we present an integrated EMG-driven robotic system capable of performing manipulation tasks by understanding 3 hand gestures associated with certain pick and place commands. Comprehensive experimental tests were conducted to demonstrate that the proposed system can decode EMG-based commands with an accuracy >93%, undergoing precise robotic-assisted object manipulation.

Keywords: EMG signals · Robot-assisted systems · Machine learning

1 Introduction

Decoding the information of muscle activity during movement is of increasing interest for different applications, such as the control of robotic systems in rehabilitation. For instance, robotic arms and prostheses could assist people with high-precision and repetitive tasks. In this regard, the integration of wearable technologies to detect movement with robotic devices could have many applications in clinical research [1,2].

Alternatives for capturing the movement of the hand include displacement sensors, force sensors, inertial sensors, cameras, and electrodes. For instance, electromyography (EMG) is a technique to record the electrical activity of the muscle with electrodes. Because the EMG signals are generated by motor neurons from the nervous system, they reflect the intention of movement of the subject. Generally, the obtained EMG signals are strong and stable, which makes them useful to control dexterous robotic arms and prostheses [3]. In comparison with

A. Müller and M. Brandstötter (Eds.): RAAD 2022, MMS 120, pp. 343–350, 2022.
https://doi.org/10.1007/978-3-031-04870-8_40

other control sensors (e.g. force sensors), EMG has many advantages such as ease of use, comfort, promotion of muscular tone and intuitive use [4].

Different studies have implemented the detection of gestures from EMG signals, using signal processing and machine learning techniques. The most common features in the time domain used for gesture classification are: Mean Absolute Value (MAV), Willison Amplitude (WAMP), Waveform Length (WL), and Zero-Crossing (ZC) [5]. According to different studies, temporal features perform better than features in the frequency domain for gesture classification. Besides, the extraction of features in the time domain generally have lower computational cost [5].

Automatic classification methods have been implemented in previous studies for gesture recognition. These studies are highly variable in the number of subjects, channels, and movements. The performance of the classification reported varies between 89% for k-Nearest Neighbors (kNN) [6], 88%–97% for Artificial Neural Networks (ANN) [7], 99% for Linear Discriminant Analysis (LDA) [8], and 98% for Support Vector Machines (SVM) [9]. Whereas the results of classification of finger movements seem promising, most of the studies used multi-channel systems with more than 10 electrodes, were off-line or decoded broad hand instead of individual finger movements [10,11].

Although the results of gesture decoding from EMG signals are accurate, the integration of EMG signals to control robotic systems is very limited. Most of the mentioned studies used multi-channel wired EMG systems that require a complex setup. Recently, wearable and wireless sensors were proposed to increase the flexibility of the applications. For instance, a control based on EMG signals, using the Myo Armband (8 channels, Bluetooth) was developed to teleoperate a 5-DoF robotic arm. The authors reported an accuracy between 86% and 95% [12]. In this context, we propose a system to control in real time 3 movements of the robotic arm, with the Myo Armband.

This paper is organized as follows: Sect. 2 describes the experimental protocol proposed to identify the hand gestures based on the myoelectric signals and to control the robotic arm. Section 3 presents the experimental results validating the performance of the system. These results are discussed in Sect. 4.

2 Methodology

An experimental protocol was designed to control 3 movements of the robotic arm with myoelectric signals from the forearm. The overall architecture of the solution is shown in Fig. 1, and detailed in the following subsections.

2.1 Data Acquisition

Myoelectric signals were recorded with the Myo Armband (Thalmic Labs, Waterloo, Canada) from the dominant forearm of volunteers. This wearable sensor has 8 differential electrodes, with a sampling rate 200 Hz, digitized with 8-bits and

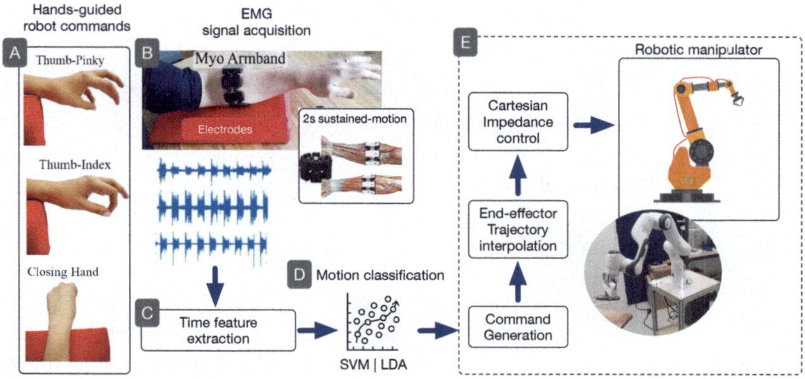

Fig. 1. Proposed system architecture.

connected through Bluetooth to the computer, as can be seen in Fig. 1B. The data was further processed with MATLAB 2021b (Mathworks, US).

The objective of the experimental protocol was to control the movement of the robotic arm in real time with EMG signals. Four volunteers executed 3 gestures of the hand: thumb-pinkie pinch, thumb-index pinch, and closing hand (Fig. 1A). The movements detected were interpreted as a specific movement of the robotic arm: a (move to home position), b (move to pick-zone and pick-up the object), and c (move to the place-zone and release the object), respectively. The robotic arm located in Poitiers (France) was teleoperated with commands from EMG signals of the volunteers in Bogotá (Colombia). The experiment was performed according to the Ethics and Research Committee of the School of Engineering at Pontificia Universidad Javeriana Bogotá.

The database contained 840 events from the 4 subjects, with 210 events per subject (70 repetitions × 3 gestures). Around 70% of events (n = 600, 150 per subject) were used for training the classification algorithms, whereas the remaining 30% of events (n = 240, 60 per subject) were used for testing the gesture classification. During training, the subjects were asked to perform 50 repetitions per gestures. The duration of each repetition was 2 s. Furthermore, after 10 repetitions of the same gesture, a rest window of 2 min was applied to reduce fatigue of the muscle. Between the 3 gestures, there was an additional rest of 5 min. The subjects reported no fatigue during the whole experiment.

2.2 EMG Signal Processing

The pre-processing of data was required to extract relevant features of the signal that discriminate between the 3 gestures (Fig. 1C). First, the raw signal was divided in 2-s windows. Then, 3 features were computed for each event, including the Mean Absolute Value (MAV), the Wavelength (WL) and the Variance (VAR). The equations of these features for signal x of length N are described in Table 1. A total of 24 features per event were obtained (3 features × 8 channels).

Table 1. Features extracted from the EMG signals.

Feature	Equation
Mean Absolute Value (MAV)	$\frac{1}{N}\sum_{i=1}^{N}\lvert x_i\rvert$
Wavelength (WL)	$\sum_{i=1}^{N-1}\lvert x_{i+1}-x_i\rvert$
Variance (VAR)	$\frac{1}{N-1}\sum_{i=1}^{N}x_i^2$

The obtained feature matrix was used to train Machine Learning models to classify the 3 gestures: thumb-pinkie pinch, thumb-index pinch, and closing hand (Fig. 1D). The training of the models was done for each subject separately. As described above, 70% (n = 600) of events were used for the model training. A cross-validation with 5 folds was implemented in order to reduce over-fitting of the model. We compared six classification models: Linear Discriminant Analysis (LDA); Support Vector Machines (SVM) with linear and gaussian kernels; k-nearest neighbors (kNN) with fine and random subspaces; and Subspace Discriminant (SD).

Once the trained models were calculated, the testing phase evaluated the ability of the system to control the movement of the robotic arm. Here, the subjects were asked to execute successive repetitions of the 3 gestures. Thus, 20 repetitions of each gesture were randomly alternated, for a total of 60 events per subject designated to this final stage, thus 240 in total.

2.3 Robotic Arm

A 7 degrees of freedom (DoF) Franka Emika collaborative robot was used for the experimental setup, with a pick-and-place process controlled by the hand gestures. As shown in Fig. 1E, the robot's controller receives the gesture identified by the machine learning mode, and generates a motion command accordingly. Figure 2 depicts the relation between the hand-gestures and the movements of the robotic arm.

An online trajectory interpolation library (i.e. Ruckig [13]) was used to calculate the desired trajectory of the end-effector $\mathbf{X}_d \in \Re^{6\times 1}$. Then, a cartesian impedance controller executed the desired trajectory with a compliant behavior, ensuring smooth robot movements, while guaranteeing a safe behavior in case of direct interaction with humans.

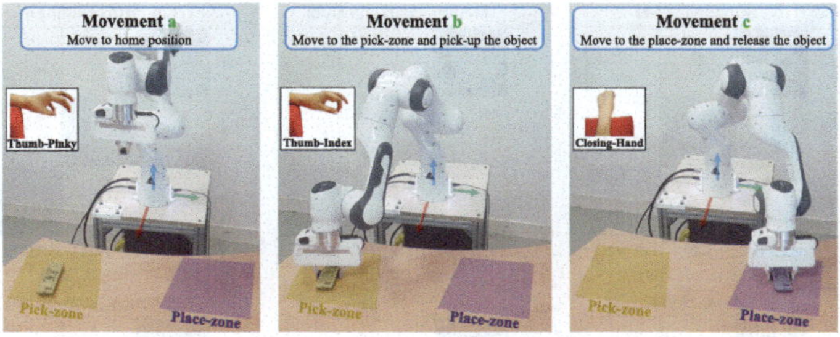

Fig. 2. Pick-and-place process. The 3 gestures of the hand (Thumb-Pinky, Thumb-Index, Closing hand) are traduced to 3 specific movements of the robotic arm (movement a, b and c), respectively.

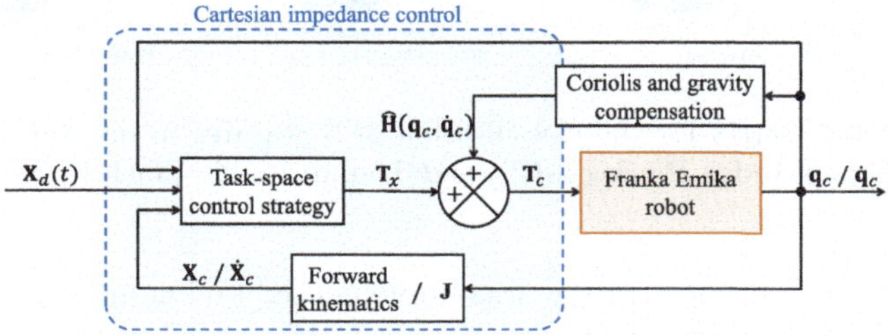

Fig. 3. Closed loop torque control approach. An impedance control approach is implemented for task-space.

The torque-control architecture for the proposed robotic system is detailed in Fig. 3. The torque control input $\mathbf{T}_c \in \Re^{7\times1}$ was defined as follows,

$$\mathbf{T}_c = \mathbf{T}_x + \hat{\mathbf{H}}(q_c, \dot{q}_c) \tag{1}$$

where $\hat{\mathbf{H}}(q_c, \dot{q}_c) \in \Re^{7\times1}$ is the torque estimation of Coriolis, Centrifugal and Gravity compensation effects. The torque control vector $\mathbf{T}_x \in \Re^{7\times1}$ implementing the compliant behavior yielded,

$$\mathbf{T}_x = \mathbf{J}^T \left[\mathbf{K}_p(\mathbf{X}_d - \mathbf{X}_c) - \mathbf{K}_d\dot{\mathbf{X}}_c\right] \tag{2}$$

where $\mathbf{J} \in \Re^{6\times7}$ is the Jacobian matrix and $\mathbf{X}_c, \dot{\mathbf{X}}_c \in \Re^{6\times1}$ are the current end-effector position and velocity, respectively. The level of compliance is regulated through the diagonal constant matrices $\mathbf{K}_p, \mathbf{K}_d \in \Re^{6\times6}$.

3 Results

3.1 Gesture Classification

In order to evaluate the performance of the system to classify 3 hand gestures, training and testing phases were implemented. Firstly, we trained different classification methods with 600 events, each with 24 features extracted from the EMG

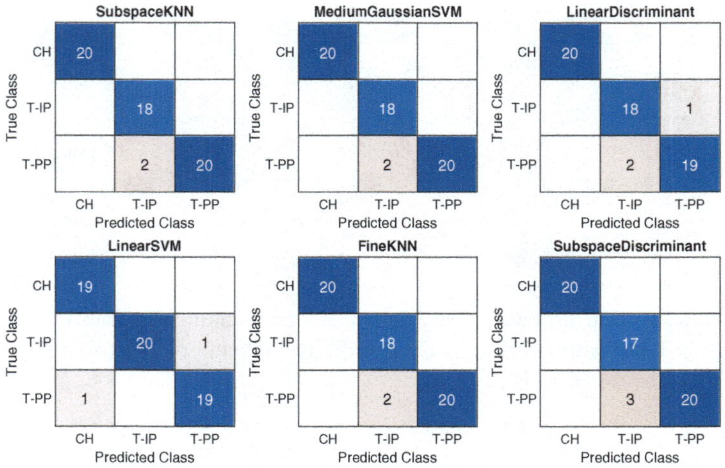

Fig. 4. Confusion Matrix for the classification of 3 gestures in one subjects: Closing Hand (CH), Thumb-Index Pinch (T-IP), and Thumb-Pinkie Pinch (T-PP)

signals. A cross-validation strategy was implemented giving an average accuracy of 100% for the 6 classifiers. Secondly, the optimized models were used to predict the gestures with new online recordings from Myo Armband. A sequence of 60 movements was executed by each subject, containing 20 movements per gesture. The predictions were compared to the labels of the real movement, obtaining the confusion matrices shown in Fig. 4. Table 2 reports the methods and the accuracy in the training and validation phases. The accuracy of testing phase was between 93.3% and 96.3%.

Table 2. Classification methods in training. Average for 4 subjects

Method	Description	Accuracy in validation (avg)	Accuracy in testing (avg)
LDA	–	100%	96.2%
SVM	Linear kernel	100%	95.0%
SVM	Gaussian kernel	100%	93.3%
kNN	Fine	100%	94.2%
kNN	Random subspaces	100%	96.2%
SD	–	100%	95.4%

3.2 Robotic System Execution

In order to guarantee a smooth execution of the robot movements, the maximum Cartesian velocity and acceleration were limited 0.1 m/s and 10 m^2/s, respectively. Moreover, the Cartesian motion was constrained to only translational

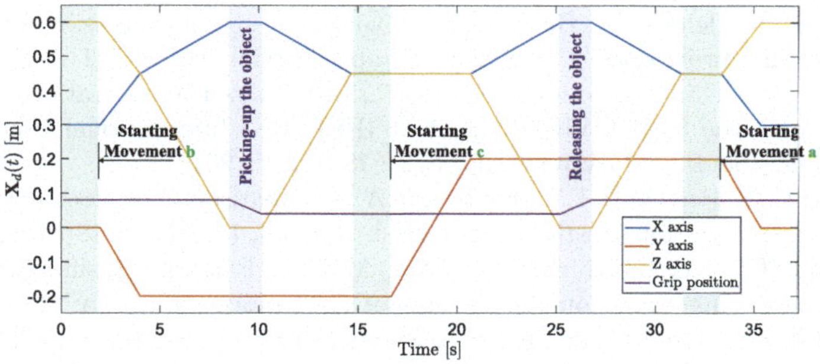

Fig. 5. Example of the desired robot's trajectory for a sequence of movements. Robot's motion starts and ends at home position. A resting time of 2 s (green region) is imposed before starting a new movement.

motion. Figure 5 depicts an exemplary sequence of robot's movements. Once a movement is achieved, a resting time of 2 s is imposed before executing the next received command.

4 Conclusions

In this study, we integrated a wearable myoelectric sensor and machine learning algorithms to manipulate a robotic arm. The proposed system is a first approach in the design of assisted-robotic prostheses. The flexibility of the wearable sensor and the advantages of the EMG signals (robustness, intention of movement) make this approach suitable for robotic-assisted rehabilitation contexts.

The EMG signals in the forearm are suitable to classify gestures of the hand such as fingers pinch or closing hand. Although the database size was limited (840 events, 210 per subject and 280 per class) we have obtained a performance above 93.3%, which suggest the feasibility of automatic online gesture classification. An extended database with more subjects and repetitions could be done in the future.

A compliant control approach was implemented in the robot to smoothly execute the end-effector movements, according to the hand gestures. This approach guarantees a safe robot's behavior in the case where the robot shares a common workspace with humans.

Acknowledgement. This work was partly funded by the Pontificia Universidad Javeriana in Bogota Colombia, under the project *Sistema de rehabilitación inteligente y progresivo para terapias de rehabilitación articular de la mano por medio de un exoesqueleto robótico* (ID 20128), and by the Pprime Institute.

References

1. Ferguson, S., Dunlop, G.R.: Grasp recognition from myoelectric signals. In: Proceedings of the Australasian Conference on Robotics and Automation, vol. 1. Auckland, New Zealand (2002)

2. Parker, P., Englehart, K., Hudgins, B.: Myoelectric signal processing for control of powered limb prostheses. J. Electromyograp. Kinesiol. **16**(6), 541–548 (2006)
3. Vogel, J., Castellini, C., van der Smagt, P.: EMG-based teleoperation and manipulation with the DLR LWR-III. In: 2011 IEEE/RSJ International Conference on Intelligent Robots and Systems, pp. 672–678. IEEE (2011)
4. Corbett, E.A., Perreault, E.J., Kuiken, T.A.: Comparison of electromyography and force as interfaces for prosthetic control. J. Rehabil. Res. Dev. **48**(6), 629 (2011)
5. Robinson, C.P., Li, B., Meng, Q., Pain, M.T.G.: Pattern classification of hand movements using time domain features of electromyography. In: Proceedings of the 4th International Conference on Movement Computing, p. 27. ACM (2017)
6. Cipriani, C., et al.: Online myoelectric control of a dexterous hand prosthesis by transradial amputees. IEEE Trans. Neural Syst. Rehabil. Eng. **19**(3), 260–270 (2011)
7. Ariyanto, M., et al.: Finger movement pattern recognition method using artificial neural network based on electromyography (EMG) sensor. In: 2015 International Conference on Automation, Cognitive Science, Optics, Micro Electro-Mechanical System, and Information Technology (ICACOMIT), pp. 12–17. IEEE (2015)
8. Adewuyi, A.A., Hargrove, L.J., Kuiken, T.A.: An analysis of intrinsic and extrinsic hand muscle EMG for improved pattern recognition control. IEEE Trans. Neural Syst. Rehabil. Eng. **24**(4), 485–494 (2016)
9. Wahid, F.Md., Tafreshi, R., Al-Sowaidi, M., Langari, R.: Subject-independent hand gesture recognition using normalization and machine learning algorithms. J. Comput. Sci. **27**, 69–76 (2018)
10. Al-Timemy, A.H., Bugmann, G., Escudero, J., Outram, N.: Classification of finger movements for the dexterous hand prosthesis control with surface electromyography. IEEE J. Biomed. Health Inform. **17**(3), 608–618 (2013)
11. Naik, G.R., Al-Timemy, A.H., Nguyen, H.T.: Transradial amputee gesture classification using an optimal number of SEMG sensors: an approach using ICA clustering. IEEE Trans. Neural Syst. Rehabil. Eng. **24**(8), 837–846 (2016)
12. Hassan, H.F., Abou-Loukh, S.J., Ibraheem, I.K.: Teleoperated robotic arm movement using electromyography signal with wearable MYO armband. J. King Saud Univ.-Eng. Sci. **32**(6), 378–387 (2020)
13. Berscheid, L., Kröger, T.: Jerk-limited real-time trajectory generation with arbitrary target states. Robotics: Science and Systems XVII (2021)

Evaluation on Implementing an Active Braking System in Wheelchair Rear-Mounted Power-Assisted Device

Valerio Cornagliotto(✉), Flaminia Perino, Laura Gastaldi, and Stefano Pastorelli

Department of Mechanical and Aerospace Engineering, Politecnico di Torino, Turin, Italy
valerio.cornagliotto@polito.it

Abstract. Power-Assisted Devices (PADs) for wheelchairs are becoming popular tools to enhance propulsion capabilities and to assist wheelchair users to perform daily activities. PADs include Pushrim Activated Power-Assisted Wheelchairs, joy-stick-driven wheels, front-end attachments, and rear-end attachments. Considering the latest, they are not equipped with any active braking system. This could affect the handling of the wheelchair and introduce safety concerns.

The paper aims to assess the performance of a rear add-on during driving and braking conditions, and to investigate the implementation and effectiveness of a servo braking system. A dynamic multibody model of a wheelchair has been developed and the dynamic of the system has been analyzed. To enhance the braking effectiveness, an additional preload torque between the wheelchair and the device has been modelled. Simulations have been performed for different braking torques. The results show that the introduction of a mounting preload positively affects the braking effectiveness, and it assists the user to perform part of the braking action.

Keywords: Add-on · PAPAW · Braking system · Power-assisted devices

1 Introduction

Many manual wheelchair (MWC) users might have to struggle to move and to accomplish everyday activities. The difficulty could be due to reduced physical ability, upper body weakness, pain, injuries, or fatigue due to wheelchair propulsion for a prolonged time. Facilitating wheelchair users' mobility is in line with the aim of the third sustainable development goal (SDG3) - healthy lives and promotion of well being. Active assistive devices to reduce wheelchair users' effort have been developed [1]. Previous studies demonstrated that power-assisted devices (PADs) positively influenced the propulsion capabilities, reducing the biomechanical and physiological effort associated with manual wheelchair self-propulsion [2–4] and improved mobility [5]. Currently, different kinds of PADs are commercially available. PADs include Pushrim Activated Power-assisted Wheelchairs (PAPAWs), joystick-driven wheels, front-end attachments, and rear-end attachments. PAPAWs consist of motorized wheels which provide a propelling torque

A. Müller and M. Brandstötter (Eds.): RAAD 2022, MMS 120, pp. 351–358, 2022.
https://doi.org/10.1007/978-3-031-04870-8_41

proportional to the torque exerted on pushrims by the user (e.g., Servo®, Twion®, e-motion®). Joystick controlled wheels (e.g., e-fix®, Solo®), which might be mounted on a personal wheelchair, allow the user to control the MWC motion through a joystick mounted on the armrest. Front-end add-on devices are attached to the front of a MWC lifting the caster wheels off the ground thus converting the MWC into a powered tricycle (e.g., Batec®, etc.). At last, the rear-mounted devices consist of an additional powered wheel, attached to the MWC wheel axis, that provides longitudinal thrust. Rear-mounted devices are generally controlled through a wristband or knobs (e.g., SmartDrive MX®, SmoovONE®, YOMPER®).

In previous studies, the characteristics and the users' perception of some PADs have been investigated [6–12]. The mounting procedure of the front-end devices turned out to be easier than the others PADs, on the other hand, the PAPAW or rear-mounted devices footprint is smaller [8]. Moreover, rear add-on devices are lighter than the other solutions taken into account [6]. Some differences also emerged concerning the operation and control of the devices. The PAPAWs usually detect the user intention through the evaluation of the torque exerted on the pushrim. Then, the device provides the additional torque proportional to the one exerted by the user at each actuated wheel. Thus, the user perceives a lower effort to thrust the wheelchair. The user has to repeat the push on the handrims to continue the movement. With front-end attachments, the input is given by a handlebar, similarly to a scooter or tricycle. Rear-mounted solutions are controlled in different ways, depending on the model and the selected driving mode. Some devices adopt a wristband or a smartwatch as input interface, others detect the user's intention to activate the device monitoring the add-on motor motion. In all the rear-mounted solutions a velocity control is implemented, hence the devices sustain the propulsion at a given speed until the user provides an input to change the speed or arrest the device. As regards the braking function, each PAD model requires a different user's input. In PAPAWs and front-mounted solutions, the device itself provides an active action (mechanical or electrical), and the user only needs to give the braking command as input to the device. In PAPAWs, since the torque supplied by the device is proportional to the torque provided by the user, when the hand exerts resistance torque, the device applies a braking torque. In the front-end attachments, the braking input is given directly through the handlebar brake which controls a mechanical braking system. In rear add-ons there is no active assisted braking. The user who intends to stop the wheelchair has to switch off the device and exert theirself the braking force directly on the pushrims. Unlike PAPAWs, the rear-mounted device is not controlled by sensors on the handrims; therefore, the user is requested to switch off the device by using a button or performing different gesture with the wristband. Indeed, some devices detect the user intention by monitoring the current absorbed by the motor. Since the motor doesn't exert an active braking torque, the user has to exert theirself the whole amount of force required to stop the wheelchair. The physical effort required to handle the wheelchair equipped with rear-mounted devices has been investigated in [8] and [9]. It turned out that users suffered from hands pain due to the required force that has to be exerted on the handrims to halt the wheelchair. The braking action has also been revealed to be an issue in terms of intuitiveness and multi-task procedure required to stop the wheelchair. Moreover, the lack of an active brake reduces the effectiveness of wheelchair handling during the downhill. Indeed, unlike the other solutions, rear add-on devices don't provide any braking assistance during the downhill. This aspect has been highlighted as a safety concern by the users who tested the devices [9].

The introduction of an active braking system in the rear add-on devices could make them safer and would increase wheelchair handling. For this reason, the following work aims to investigate the possibility of implementing servo braking in rear-mounted devices and analyse the effectiveness of the braking force.

2 Wheelchair and Add-On Models

A dynamic multibody model of a wheelchair was developed and implemented in Simulink Simscape multibody environment, as in Fig. 1a. A standard reference wheelchair (14 kg and 24 inch. wheels diameter) and a multibody dummy 50th percentile Italian male were considered. The contact between tyres and ground was modelled as a visco-elastic contact model. The stiffness and damping coefficients of the tyres were defined [13]. The friction between the ground and the wheels was modelled as Coulomb and rolling friction [14]. The attached wheel was assumed as an omni wheel, hence the tangential friction force was modelled only in the rolling direction (i.e., orthogonal to the wheel axis). In the wheel axis direction, both friction force and rolling resistance were considered negligible. Table 1 shows the coefficients used for the wheelchair tyres and the add-on tyre contact models. The add-on influences the dynamics of the system pushing the wheelchair in the longitudinal direction. A parametric simplified model of an add-on device has been implemented.

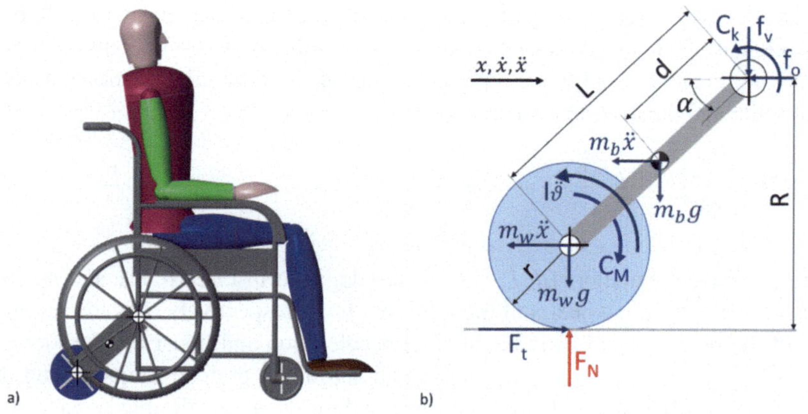

Fig. 1. (a) Multibody model of wheelchair, dummy, and add-on; (b) PAD free body diagram

The device model has been parametrized as a function of its geometrical parameters (d, L, r, R, α); Fig. 1b shows the PAD free body diagram. In the multibody model, the add-on has been attached at the wheels' axis through a joint as depicted in Fig. 1a.

Table 1. Friction, stiffness, and damping coefficient of the contact model

Tyre contact and friction model		
Stiffness	6.8 10^4	N/m
Damping	1.0 10^4	Ns/m
μ_s (static friction)	0.65	
μ_d (dynamic friction)	0.5	
f (rolling friction coeff.)	0.0099	

3 Tests and Simulations

The rear-mounted wheel provides thrust to the wheelchair. In order to investigate the device's performance, various simulations have been carried out. Firstly, the driving mode has been considered and the dynamic of the system has been investigated as a function of the add-on's geometric parameter. Then, the braking mode has been studied and the forces that act on the system have been analysed. Considering the results obtained in the two different cases, a deeper evaluation of the effect of the braking torque and a possible design solution have been examined.

3.1 Test 1: Driving Torque

To evaluate the maximum torque transmissible to the ground without slipping, the trend of the normal force (F_N) exerted by the device to the ground, as the driving torque (C_m) varies has been observed. The simulation has been performed with the devices hooked to the wheelchair's axle in steady-state conditions. Results for add-on geometrical parameters $L = 0.3$ m and $r = 0.1$ m are reported. Equation (1) shows the relation between the normal force and the driving torque:

$$F_N = \left(m_w + m_b \cdot \frac{d}{L}\right) \cdot g - \left(\frac{I \cdot \ddot{\theta}}{r} + \left(m_w + m_b \cdot \frac{d}{L}\right) \cdot \ddot{x}\right) \cdot tg(\alpha) + C_m \cdot \left(\frac{tg(\alpha)}{r} + \frac{1}{L \cdot \cos(\alpha)}\right) + \frac{C_k}{L \cdot \cos(\alpha)} \quad (1)$$

where geometric parameters (d, L, r, R, α) are depicted in Fig. 1b, m_w and m_b are the masses of the add-on's wheel and the add-on's body respectively. I is the moment of inertia of the wheel. $\ddot{\theta}$ and $\ddot{x}$ are the angular acceleration and linear acceleration of the add-on's wheel respectively. C_k is an eventual additional preload torque acting on the device. The contributions of the inertial force and torque are negligible because the test has been performed in steady-state conditions. No additional torque has been considered, hence, Eq. (1) could be rearranged as a function of two coefficients, A and K:

$$F_N = A + C_m \cdot K \quad (2)$$

Normal force (red line) and driving torque (blue line) tendency during a startup simulation are shown in Fig. 2a. According to Eq. (2), the result shows a linear trend of the normal force as the driving torque varies, as represented in Fig. 2b. The initial offset of the plots is represented by A. The normal force linearly increases with the increase of the driving torque. The slope K of the curve is only influenced by the add-on geometrical parameters. With the geometric parameters chosen for the driving test, K is 13.4 m^{-1}. In Fig. 2b the black dotted line represents the slipping limit computed according to the static friction coefficient and the wheel's radius, as in Eq. (3):

$$F_{N_{lim}} = C_m/(\mu_s \cdot r) \tag{3}$$

The angular coefficient of the slipping limit line is 15.4 m^{-1}. Hence, the two lines are incident, and the crossing point is given as follow:

$$C_{m_{lim}} = \left(\frac{\left(m_w + m_b \cdot \frac{d}{L}\right)}{\frac{1}{(\mu_s \cdot r)} - \frac{tg(\alpha)}{r} + \frac{1}{L \cdot \cos(\alpha)}} \right) \tag{4}$$

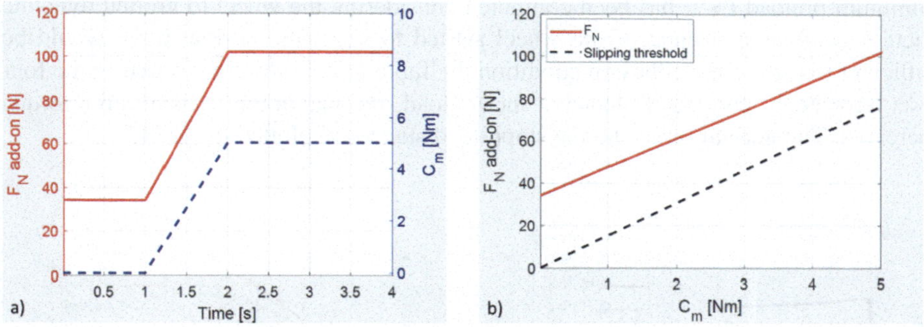

Fig. 2. (a) Normal force F_N and the driving torque C_m; (b) Relation between F_N and C_m.

3.2 Test 2: Braking Torque

In order to simulate a braking effect on the system, the direction of the torque delivered by the motor was inverted. Considering a negative C_m exerted by the wheel, the relation between F_N and C_m expressed in Eq. (1) remains unchanged. For the sake of clarity from now on the braking torque is called C_b and corresponds to $-C_m$.

Hence, when the braking torque increases, the normal force decreases with a slope that changes when the slipping threshold is achieved, as shown in Fig. 3a and 3b, without any additional torque C_k. When the wheel starts to slip, inertial effects of the add-on wheel change the normal force trend.

As it can be deduced by the results of the previous section, the normal force during the braking action is not sufficient to generate an effective braking force. Therefore, an additional preload torque C_k was applied between the wheelchair and the frame of the device. Four tests have been performed increasing the maximum braking torque by 2 Nm steps, setting the initial torque equal to 1 Nm and the initial speed equal to 8 km/h for each test.

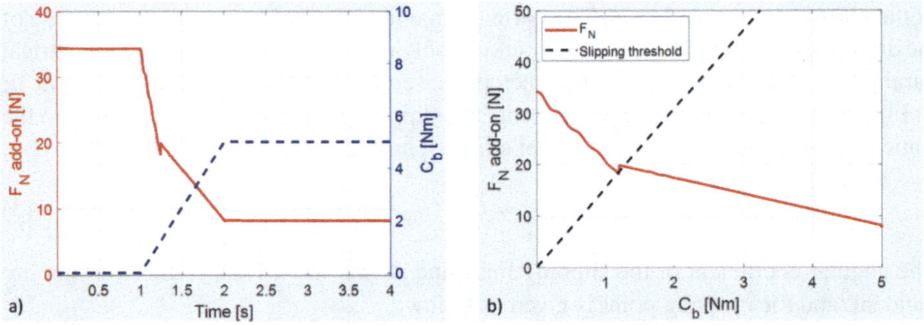

Fig. 3. (a) Normal force F_N and the braking torque C_b; (b) Relation between F_N and C_b.

During the tests, to simulate an instantaneous brake, the torque was increased from 0 to the set value in 0.1 s. To compare the results, the distance x_f and time t_f needed to stop the wheelchair have been measured. Aiming to ensure pure rolling condition, the minimum preload $C_{k_{min}}$ has been computed considering the wheel to ground dynamic friction coefficient. Hence, if the wheel started to slide, the normal force would be sufficient to restore the adhesion condition. In Table 2, the values measured in the four tests have been reported. Enhancing the preload, the maximum transmissible torque increases. The normal force and the stopping distance are plotted in Fig. 4.

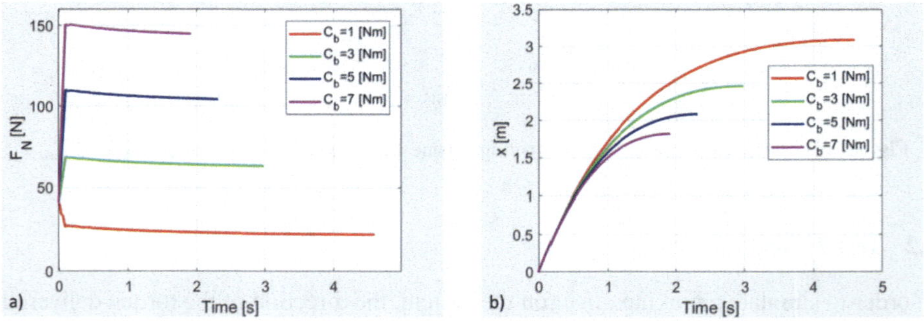

Fig. 4. Normal force (a) and stopping distances (b) for different braking torques.

As shown in Fig. 4, in the test T4 t_f and x_f are 1.9 s and 1.8 m respectively. Whereas, in the test T1 conducted without preload and with a braking torque equal to 1 Nm, t_f and x_f are almost twice. Hence results show that the introduction of a mounting preload would enhance the braking effectiveness.

Table 2. Stopping time and distances evaluated introducing the preload

Test	C_b [Nm]	$C_{k_{min}}$ [Nm]	t_f [s]	x_f [m]
T1	1	0	4.7	3.1
T2	3	15	3	2.5
T3	5	30	2.3	2.1
T4	7	44	1.9	1.8

4 Discussion and Conclusions

In this paper, the dynamic of a rear add-on device in driving and braking conditions has been analysed. The simulation results suggest that the normal force has a linear dependency on the driving torque represented by the K factor, whose value depends only on the geometric configuration. Increasing the radius or the length of the rear-mounted device, the K factor decrease. Generally, the maximum transmissible torque during the pushing mode is sufficient to ensure an effective thrust. Then, the effectiveness of the braking action has been investigated. Results show that the normal force exceeds the slipping threshold. The normal force decreases as the braking torque act due to the negative contribution of the torque and inertia.

Considering the results obtained in the two different dynamic configurations, a preload torque C_k has been added to guarantee effective braking. Four tests have been performed introducing the maximum admissible braking torque accordingly to the slipping threshold. The presence of the preload positively affects the normal force ensuring pure rolling also during the braking phase. However, the constant presence of a preload would inevitably increase the tire wear, causing a shortening of the useful life of the device's wheel. For this reason, would be suggested to introduce C_k only during the braking phase. The effectiveness of a braking force is quite evident by looking at the braking distances and times reported in Table 2. Moreover, the servo braking system would reduce the physical effort of the user, as the device assists the user to perform part of the braking action.

Acknowledgement. This research was partially conducted within the project "Advanced Light Body Assistants - Sistema avanzato leggero per l'assistenza a persone diversamente abili"- P.O.R. FESR 2014/2020 - Azione I.lb.2.2 Bando Pi.Te.F.

References

1. Cavallone, P., Botta, A., Bonisoli, E., Quaglia, G.: Preliminary experimental test of a cable-driven wheelchair in different configurations. In: Quaglia, G., Gasparetto, A., Petuya, V., Carbone, G. (eds.) I4SDG 2021. MMS, vol. 108, pp. 166–173. Springer, Cham (2022). https://doi.org/10.1007/978-3-030-87383-7_18
2. Algood, S.D., Cooper, R.A., Fitzgerald, S.G., Cooper, R., Boninger, M.L.: Effect of a pushrim-activated power-assist wheelchair on the functional capabilities of persons with tetraplegia. Arch. Phys. Med. Rehabil. **86**(3), 380–386 (2005). https://doi.org/10.1016/j.apmr.2004.05.017

3. Cooper, R.A., et al.: Evaluation of a pushrim-activated, power-assisted wheelchair. Arch. Phys. Med. Rehabil. **82**(5), 702–708 (2001). https://doi.org/10.1053/apmr.2001.20836
4. Algood, S.D., Cooper, R.A., Fitzgerald, S.G., Cooper, R., Boninger, M.L.: Impact of a pushrim-activated power-assisted wheelchair on the metabolic demands, stroke frequency, and range of motion among subjects with tetraplegia. Arch. Phys. Med. Rehabil. **85**(11), 1865–1871 (2004). https://doi.org/10.1016/j.apmr.2004.04.043
5. Levy, C.E., Buman, M.P., Chow, J.W., Tillman, M.D., Fournier, K.A., Giacobbi, P., Jr.: Use of power assist-wheels results in increased distance traveled compared to conventional manual wheeling. Phys. Med. Rehabil. Serv. **89**(8), 625–634 (2010). https://doi.org/10.1097/PHM.0b013e3181e72286
6. Khalili, M., Eugenio, A., Wood, A., Van der Loos, M., Ben Mortenson, W., Borisoff, J.: Perceptions of power-assist devices: interviews with manual wheelchair users. Disabil. Rehabil. Assist. Technol. 1–11 (2021), https://doi.org/10.1080/17483107.2021.1906963
7. Khalili, M., Kryt, G., Ben Mortenson, W., Van der Loos, H.F.M., Borisoff, J.: Comparison of manual wheelchair and pushrim-activated power-assisted wheelchair propulsion characteristics during common over-ground maneuvers. Sensors, **21**(21) (2021), https://doi.org/10.3390/S21217008
8. Flockhart, E.W., Miller, W.C., Campbell, J.A., Mattie, J.A., Borisoff, J.F.: Evaluation of two power assist systems for manual wheelchairs for usability, performance and mobility: a pilot study. Disabil. Rehabil. Assist. Technol. 1–11 (2021). https://doi.org/10.1080/17483107.2021.2001063
9. Sawatzky, B., Ben Mortenson, W., Wong, S.: Learning to use a rear-mounted power assist for manual wheelchairs. Disabil. Rehabil. Assist. Technol. **13**(8), 772–776 (2017). https://doi.org/10.1080/17483107.2017.1375562
10. Kloosterman, M.G.M., Buurke, J.H., Schaake, L., Van Der Woude, L.H.V., Rietman, J.S.: Exploration of shoulder load during hand-rim wheelchair start-up with and without power-assisted propulsion in experienced wheelchair users. Clin. Biomech. **34**, 1–6 (2016). https://doi.org/10.1016/j.clinbiomech.2016.02.016
11. Cooper, R.A., et al.: Performance assessment of a pushrim-activated power-assisted wheelchair control system. IEEE Trans. Control Syst. Technol. **10**(1), 121–126 (2002). https://doi.org/10.1109/87.974345
12. Best, K.L., Kirby, R.L., Smith, C., Macleod, D.A.: Comparison between performance with a pushrim-activated power-assisted wheelchair and a manual wheelchair on the Wheelchair Skills Test. Disabil. Rehabil. **28**(4), 213–220 (2009). https://doi.org/10.1080/09638280500158448
13. Gordon, J., Kaualarich, J.J., Thacker, J.G.: Tests of two new polyurethane foam wheelchair tires. J. Rehabil. Res. Dev. **26**(1), 33–46 (1989)
14. Hoffman, M.D., Millet, G.Y., Hoch, A.Z., Candau, R.B.: Assessment of wheelchair drag resistance using a coasting deceleration technique. Am. J. Phys. Med. Rehabil. **82**(11), 880–889 (2003). https://doi.org/10.1097/01.PHM.0000091980.91666.58

Research and Control of Wearable Robot for Wrist Rehabilitation

Denis Chikurtev(✉) and Petko Stoev

Institute of Information and Communication Technologies, Bulgarian Academy of Science, Sofia, Bulgaria

denis.chikurtev@iict.bas.bg

Abstract. The article presents a rehabilitation robot for wrist movement. The robot is designed to move the wrist in all directions - flexion-extension and radial-lunar. The mechanical construction of the robot, its hardware, software and control systems are presented. The kinematic equations for control of the final unit are described. Based on the study of the kinematic movements of the wrist, a control method is created. After simulation experiments, the obtained result of the robot's work is presented, where it can be seen that the robot successfully performs movements close to those of the natural wrist.

Keywords: Rehabilitation robot · Wearable robot · Control · Tele-operation

1 Introduction

Rehabilitation is a procedure that involves a series of repetitive movements. Therefore, robots and robotic machines are widely used in these fields [1]. The purpose of rehabilitation is to fully or partially restore the mobility of the joints of the human limb - shoulder, elbow, wrist, knee, ankle and others. Rehabilitation procedures have in the past been performed manually by health professionals, therapists and specialists. Then mechanical devices enter, which facilitate the work of specialists and contribute to improving the results of rehabilitation. When it is noticed that the procedure is significantly shortened, we move on to research and development of mechatronic robotic devices. The idea is for these devices to further improve procedures and allow more patients to have timely therapy [2, 3].

As we said, rehabilitation is applied to different limbs. In most cases, one procedure involves moving 1 or 2 joints. This means that rehabilitation devices do not need to be complex and be able to move many limbs at the same time, but to be reliable and to be able to reproduce the natural movements of each joint as accurately as possible [9]. This condition, requires a detailed study of the movements, connections, kinematics of the joint for which the rehabilitation robot will be designed.

In this study, we focus on rehabilitation processes in restoring wrist movement. The most common causes of wrist problems are due to injuries - fractures, immobilization, sprains and as a result of stroke - lack of sensitivity and mobility [8]. Whatever the reason, the rehabilitation procedure involves repetitive movements to move the wrist. In order

A. Müller and M. Brandstötter (Eds.): RAAD 2022, MMS 120, pp. 359–366, 2022.
https://doi.org/10.1007/978-3-031-04870-8_42

to apply this procedure, an appropriate mechanism must first be modeled to reproduce the movements according to the anatomy of the wrist joints. Then drive mechanisms and devices must be applied to perform the movements. In addition, in order to achieve good efficiency and, most importantly, safety, a sensor system and a control system must be developed.

We can distinguish 2 types of robots for moving hands. The first type are exoskeleton robots [4, 5]. The second type are "platform" robots [6]. In the first type, the person puts on the robot and he begins to move the hand through natural movements. This type of robot is currently more popular and researched because it allows separate movement of each joint [7]. In addition, due to the significant development of technology in hardware, they can be controlled by small and compact microcontrollers and driven by various systems - pneumatics or electric motors. The second type uses a manipulator or mechatronic device-platform, which moves the patient's arm. These systems are usually large in size or are built from a single manipulator with 5–7 degrees of freedom. Due to their size, they are applicable only in hospitals and rehabilitation centers, while the first type of robots can be portable and a person can perform procedures at home or elsewhere.

In this article we describe the processes of design and research of a wearable rehabilitation robot type exoskeleton. The robot is designed to be used for wrist rehabilitation, powered by electric motors, equipped with various sensors and designed to meet specific wrist movements. It is envisaged that the robot will be able to connect to the Internet and become a smart device, part of the Internet of Things. The robot has the ability to perform all three methods of rehabilitation: assisted, resistive and passive. At the moment, passive and assisted rehabilitation have been realized. The resistive method is intended for future work and research.

2 Mechanical Construction of the Robot

This rehabilitation robot is a improved version of a previously presented robot, with a degree of freedom and performing only flexion and extension in the wrist joints [10]. The upgrade in the current device provides two degrees of freedom, thus aiming to achieve complete rehabilitation of the wrist joints. Circular motions are performed by interpolation from both axes.

The realized assembly aims to be universal and to be able to carry out rehabilitation on both hands without the need to add additional units or other changes in the structure. The device is designed to be ergonomic to ensure simplicity and ease of use. The product is intended to be realized as a prototype by means of 3D printing.

The developed robot is designed according to the anatomical and biomechanical requirements of the wrist [11]. The dimensions are in accordance with the anthropometric data as the weight is transferred to the forearm to protect the wrist. In Fig. 1 is illustrated the mechanical structure of the robot. The main components are the mounting plate – 1; movable unit – 2: contacting with the hand, by means of which the movement is carried out; force measuring unit – 3. The mounting plate is fastened to the movable unit by means of the holders – 4. The force measuring unit is attached with a Velcro turnstile to the mobile unit. In the same way, the mounting plate is attached to the forearm.

The servomotors, the holders and the two spools are mounted to the mounting plate. The worm gear is fixed by the holders and provides the wrist flexion/extension movement.

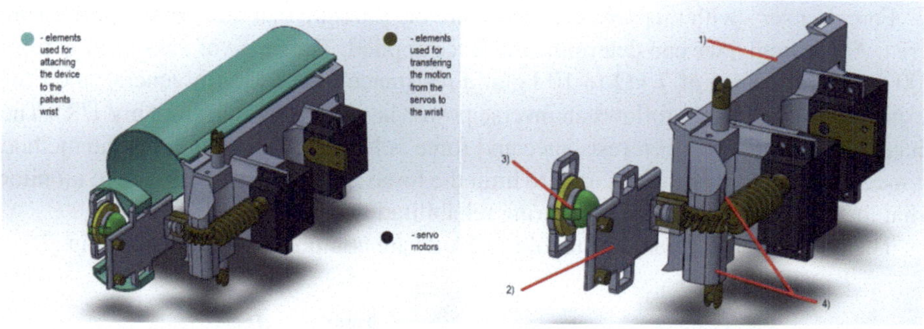

Fig. 1. Mechanical structure of the robot.

Thanks to the 3D print technology, a socket has been designed for the gear, in which the movable unit is placed and mounted. Placed in the socket, the movable unit has the ability to rotate around the axis, which is used to fasten it. This socket provides the wrist radial/ulnar movement.

3 Hardware and Software Systems of the Robot

3.1 Hardware System of the Robot

The robot is controlled by a NodeMCU-32S microcontroller. It is built with a wireless module AI Thinker ESP32S, working with microcontroller ESP32. At this stage of robot development, we use this microcontroller due to the presence of a Wi-fi module that allows it to connect to a router. In this way the remote control of the robot is realized by means of web technologies.

The robot is driven by two servomotors, one for each mobile joint. The selected motors are continuously rotating and in practice function as DC motors with gearboxes. These motors are suitable for application purposes due to their characteristics: low supply voltage, low power consumption and good power/torque.

The servomotors are powered by the Pololu Maestro driver. This is a specialized driver for controlling up to 6 servomotors. It supports three control methods: USB for direct connection to a computer, TTL serial for use with embedded systems, and internal scripting for self-contained, host controller-free applications.

Position sensor - we use rotary potentiometers to determine the position of the robot's joints. High-sensitivity potentiometers are selected, which give detailed information about the position of each joint. In order for the robot to function properly, a calibration procedure must be performed at initial start-up – each joint move to its maximum positions and then goes in the center. This procedure is done before the robot is placed on the arm. For safety reasons, without this procedure, the robot will not start any rehabilitation actions.

In addition, we mounted six DoF accelerometer and gyroscope. This sensor will give direct position and orientation of the wrist. It communicates with the controller using I2C interface.

Force sensor - with this sensor we measure the pressure that the robot applies to the wrist. Accordingly, we can determine the force applied. The range of the sensor is from 100 g at a resistance of 7 kΩ to 10 kg at a resistance of 0.4 kΩ. In general, the FSR response approximately follows an inverse power-law characteristic (roughly 1/R). The change in the ratio between resistance and force is linear, which simplifies the method of measurement. This sensor is used to limit the force applied to the wrist and to monitor what load the patient has reached during rehabilitation.

The connection diagram of the robot's hardware components is shown in Fig. 2.

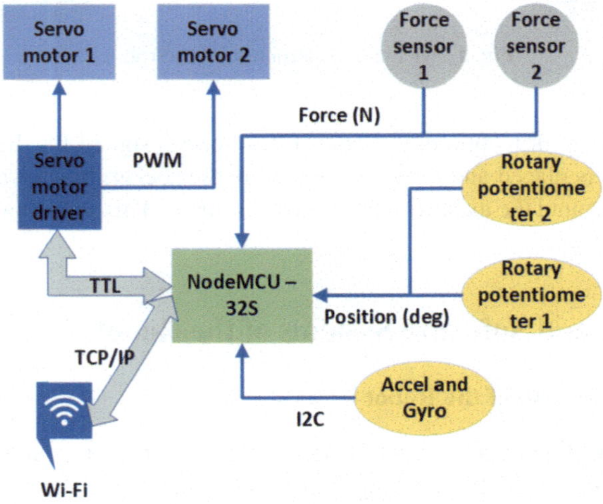

Fig. 2. Structure and connections of the robot hardware system.

3.2 Software System of the Robot

The robot software is implemented and executed entirely on the microcontroller. There are two main software blocks: one for motor control and sensor reading and other for a web interface and remote access. The control of the motors is performed by a specialized function. In this function the kinematic model of the robot is set, so that it describes circular movements according to set restrictions. The algorithm that is executed is shown in Fig. 3.

The principle of operation of the algorithm is as follows: the number of repetitions, the power limit and the limit of the maximum angles for each joint are set. The procedure starts and the motors start moving the joints. Any time the angle limit is reached, means that one repetition is done. If the maximum force is reached, the movement is stopped immediately, the motors return to zero position and the voltage is released so that they do not put pressure on the wrist.

The web interface of the robot is implemented in standard web programming languages: HTML and JavaScript, as well as web libraries such as Ajax. An asynchronous website has been developed, which allows the information to be updated without the need for a refresh. At any time, we can send control signals and thus start, set up and terminate procedures (Fig. 4).

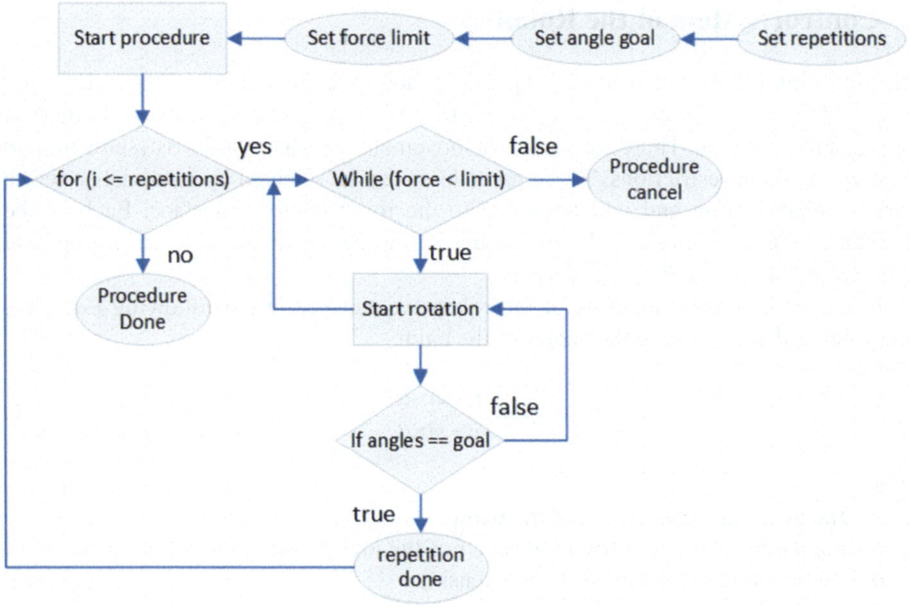

Fig. 3. Block diagram of the software algorithm.

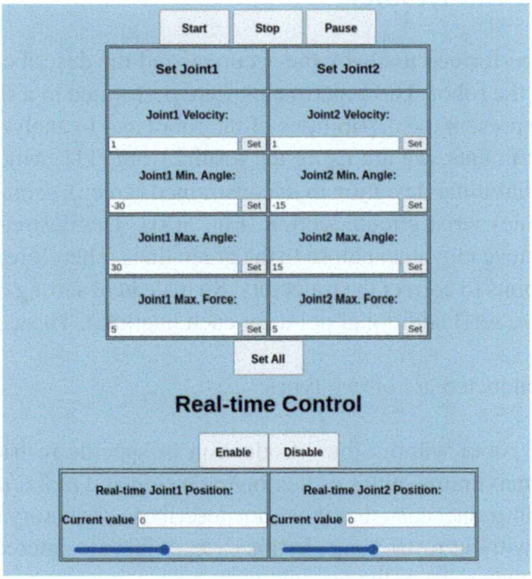

Fig. 4. Web graphical interface for distant robot control.

4 Control System of the Robot

The algorithm is based on a study [11], which shows the movements of the wrist under different loads and restrictions. Based on the results of this study, we developed a method for gradual movement. There are 3 levels of movement: with a limited wrist, semi-limited wrist and without restrictions. These procedures can be selected by a specialist before starting rehabilitation and will be added to the robot control interface. Each of the procedures has limitations in the maximum unfolding angles so as to ensure optimal progress and minimize the risk of injury.

From the kinematic analysis of the robot, we can derive the following Eq. 1 for determining the position of the center of the palm:

$$p = \begin{matrix} r * \cos\alpha \\ r * \sin\beta \end{matrix} \tag{1}$$

where

p – the position of the center of the palm

r – the length of the unit from the center of the joint to the center of the palm

α – angle along the z-axis, flexion/extension

β – angle along the y-axis, radial/ulnar

The control system is implemented through a PID controller with position feedback. The maximum power limit is an additional component, set outside the regulator. This is done to prevent overwhelm of the patient.

5 Experiments and Results

Experiments were performed to check the accuracy and the described trajectory of the kinematic model of the robot. The experiments were performed in a simulation environment. The aim is to present the capabilities of the robot and to analyze its applicability. As part of the experiments, we are using the results from [11], which represents flexion/extension vs radial/ulnar deviation in unconstrained (green), semi-constrained (red), and constrained (blue) wrist circumduction (Fig. 5(a)). The desired trajectory in our experiment is to achieve movement close to the green lines. Therefore, for each quadrant we include middle pints to correct the trajectory. So instead of setting only minimum and maximum angles, we set 3 additional point for each quadrant. These point are gathered from the Fig. 5(a).

The experiments conducted are of two types:

- movement of the robot without the introduction of specific restrictions, only those according to the maximum values of flexion/extension and radial/ulnar deviation;
- with restrictions introduced, so that the robot describes a trajectory similar to that of a wrist movement without restrictions. In this case, points are entered in each quadrant according to the change from the original trajectory.

The results of the first experiment shows that the robot is able to represent a simple trajectory in the set limit: from -70° to 70° for flexion/extension and -18° to 36° for

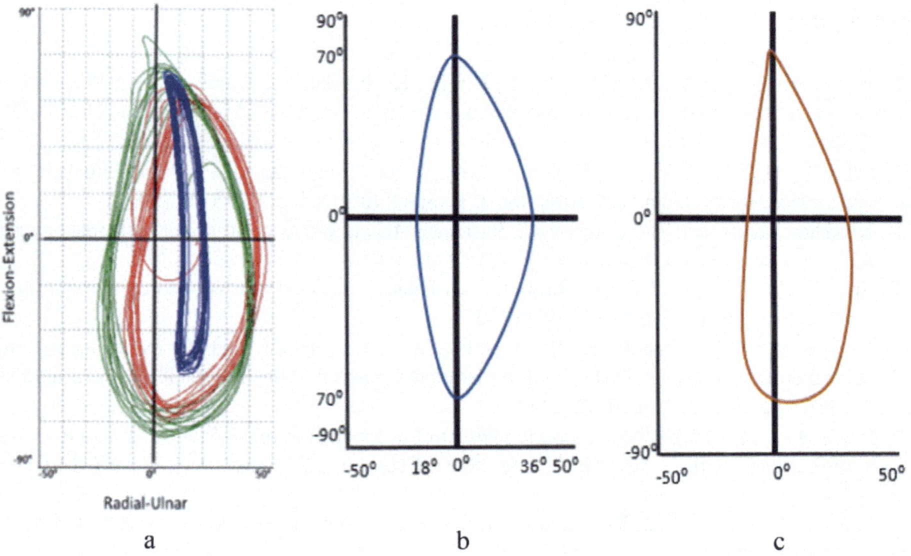

Fig. 5. Trajectory performed by the wrist and the rehabilitation robot. Sub-figure a) shows the trajectory of the wrist [11]. Sub-figure b) shows the trajectory of the robot at the entered limits. Sub-figure c) shows the trajectory of the robot after additional restriction points have been introduced.

radial/ulnar. This trajectory is safe and suitable for rehabilitation. But to achieve precise and more effective rehabilitation, this is not enough.

The results of the second experiment are close to the trajectory from the real wrist movement. The maximum angles are also in the limits, but this time the trajectories in the different quadrants are similar to the green ones in Fig. 5(a). This method provides better representation of the natural unconstrained wrist movement. The maximum positions of the robot ill not exceed the wrist limitation, so the robot will provide safe and optimal rehabilitation.

6 Conclusion

The article presented a robot for wrist rehabilitation. After simulation experiments, it was proved that the robot is structurally functional and can successfully meet the needs of rehabilitation procedures. Control systems ensure precise movements with good accuracy.

For future work, it is planned to develop functionalities that will recreate the other two loads on the wrist: semi-constrained and constrained. In addition, the implementation of the robot and testing on patients is planned.

Acknowledgement. The work is support by the Grant No BG05M2OP001-1.002-0011 Center of Competence "MIRACle (Mechatronics, Innovation, Robotics, Automation, Clean technologies)", financed by the Science and Education for Smart Growth Operational Program (2014–2020) and co-financed by the European Union through the European structural and Investment funds.

References

1. Yamamoto, I., Matsui, M., Higashi, T., Iso, N., Hachisuka, K., Hachisuka, A.: Wrist rehabilitation robot system and its effectiveness for patients. Sens. Mater. **30**(8), 1825–1830 (2018)
2. Chay, K.-H., Lee, J.-V., Chuah, Y.-D., Chong, Y.-Z.: Upper extremity robotics exoskeleton: application, structure and actuation. Int. J. Biomed. Eng. Sci. **1**(1), 35–45 (2014)
3. Khokhar, Z.O., Xiao, Z.G., Menon, C.: Surface EMG pattern recognition for real-time control of a wrist exoskeleton. Biomed. Eng. **9**(1), 1–17 (2010)
4. Shalal, N.S., Aboud. W.S.: Smart robotic exoskeleton: a 3-DOF for wrist-forearm rehabilitation. J. Robot. Control **2**(6) 476–483 (2021)
5. Gopura, R.A.R.C., Kiguchi, K., Bandara. D.S.V.: A brief review on upper extremity robotic exoskeleton systems. In: 2011 6th international Conference on Industrial and Information Systems, pp. 346–351. IEEE (2011)
6. Fraile, J.C., et al.: E2Rebot: a robotic platform for upper limb rehabilitation in patients with neuromotor disability. Adv. Mech. Eng. 8(8) (2016). https://doi.org/10.1177/1687814016659050
7. Gull, M.A., Bai, S., Bak, T.: A review on design of upper limb exoskeletons. Robotics **9**(1), 16 (2020)
8. Aggogeri, F., Mikolajczyk, T., O'Kane, J.: Robotics for rehabilitation of hand movement in stroke survivors. Adv. Mech. Eng. **11**(4), 1687814019841921 (2019)
9. Kwakkel, G., Kollen, B.J., Krebs, H.I.: Effects of robot-assisted therapy on upper limb recovery after stroke: a systematic review. Neurorehabil. Neural Repair **22**(2), 111–121 (2008)
10. Chikurtev, D., Stoev, P., Andonov, I., Chikurteva, A.: Research and design of a wearable robot for wrist rehabilitation. In: 2021 XXX International Scientific Conference Electronics (ET), pp. 1–5. IEEE (2021)
11. Rainbow, M.J., Wolff, A.L., Crisco, J.J., Wolfe, S.W.: Functional kinematics of the wrist. J. Hand Surg. (European Volume) **41**(1), 7–21 (2016)

Geometric Modeling of a New Modular Spherical Robotic System for Single Incision Laparoscopic Surgery

Calin Vaida, Iosif Birlescu, Alexandru Pusca, Bogdan Gherman, Paul Tucan, Tiberiu Alexandru Antal, and Doina Pisla(✉)

CESTER, Technical University of Cluj-Napoca, Cluj-Napoca, Romania
Doina.Pisla@mep.utcluj.ro

Abstract. The paper presents the geometric modeling of an innovative modular robotic system designed for SILS. The proposed robotic system uses a Cartesian positioning robot which aligns simultaneously all the surgical instruments with the access port. Each surgical instrument is positioned, kinematically constrained with respect to de access point by a spherical robot module. The geometric model is described by fully parametric equations defining the relations between the robotic modules, the patient, and the surgical instruments.

Keywords: Geometric modeling · Modular spherical robot · Robotic assisted SILS

1 Introduction

Single incision laparoscopic surgery (SILS) is a technique derived from minimally invasive surgery (MIS) that uses a single access port, i.e., all the surgical instruments are inserted within the operating field using a single (specialized) trocar (in contrast with classical MIS where each instrument is inserted through its own trocar). SILS was first described in the gynecological literature in 1969, where Wheesess reported the first 4.000 cases of tubal ligation [1]. The specific advantages of SILS (due to the single incision) are less tissue scarring, shorter patient recovery time, and reduced bleeding [2]. However, since all the instruments are inserted through a single port, SILS has a series of drawbacks: the need for a high degree of training for the surgeon before entering the operation room, reduced instruments workspace compared to standard MIS, lack of ergonomics due to the surgeon's hand positioning outside the operating field and the possible crossing effect of the surgical instruments [3, 4].

All the limitations of SILS can be minimized by using Master-Slave robotic systems, where the surgeon manipulates the surgical instruments (mounted on the Slave robotic system) through a Master control console [5]. Robotic systems have better accuracy than humans, eliminating hand tremors, enabling scalable motions, and increasing procedure ergonomics and patient safety [6].

The first robotic-assisted SILS was successfully performed with the da Vinci system in 2011, demonstrating its feasibility and highlighting a considerable potential [7].

A. Müller and M. Brandstötter (Eds.): RAAD 2022, MMS 120, pp. 367–374, 2022.
https://doi.org/10.1007/978-3-031-04870-8_43

In 2018 the da Vinci SP (single port) system, designed specifically for SILS, received FDA approval [7, 8]. Other SILS robotic systems were also presented in the scientific literature. Senhance [8] is a multi-arm robotic system able to perform SILS, which guides the surgical instruments independently. In [3], the authors presented an approach based on a hybrid robotic system, with a serial robot for positioning, having 7 Degrees of Freedom (DOF), and an orientation platform (with 3 DOF) for Remote Center of Motion (RCM) [9] manipulation. In [10], the authors presented a new SILS robotic system with 5 DOF, with the kinematic, dynamic, and stiffness analysis. In [11], the authors proposed a SILS robotic system with a novel elbow mechanism for optimum force transmission.

The paper presents the geometric modeling of a novel modular robotic system for SILS, consisting of a Cartesian robot that positions (simultaneously) three independent spherical robot modules, which in turn guides the surgical instruments within the operating field. The use of a spherical structure in minimally invasive medical procedures has been explored before, [12, 13] due to the natural achievement of the RCM in the center of the sphere. The design proposed in the paper can reduce the size of the robotic system and its footprint in the operating room while the use of three spherical modules positioned on a single platform ensures: (1) architecturally constraints RCMs for the three instruments in direct relation to the SILS trocar (which increases patient safety [14]) and (2) constant gripping force in the operational workspace (the tissue manipulation is achieved by the combined motion of the active instrument supported by the spherical module [3]). The motivation is to determine a complete parameterized mathematical model for further design stages, e.g., defining the operation workspace, avoiding modules collision, kinematics, and singularity analysis.

The paper is structured as follows: Sect. 2 introduces a new modular robotic system for SILS able to manipulate independently three surgical instruments (one endoscope and two active instruments). Section 3 presents the geometric modeling of the proposed robotic system. Section 4 presents the conclusions and further work.

2 An Innovative Modular Spherical Robot for SILS

The SILS task is illustrated in Fig. 1, whereas the CAD model of the robotic system is shown in Fig. 2 [15]. In STEP – 1, the Cartesian robot guides (simultaneously) all three spherical modules (using the active joints q_1, q_2, q_3) such that the SILS instruments are located at their entry points. In STEP – 2, the SILS instruments are inserted within the operating field (through the trocar). All active joints of the three spherical modules are used, but the Cartesian robot must remain fixed. In STEP – 3, the surgical task is performed. All the active joints of the three spherical modules are manipulated around the RCM points, together with the active motions of the surgical instruments' tips q_{instr_l}, q_{instr_r} (each with 4 DOF [16]). The mathematical model hereafter considers the instruments rigid (without active bends) to determine the constraint equations for the modular robotic system. The mathematical models for the 4 DOF active instruments can be added into the geometric models presented in this paper for the description of the new modular robotic system with active instruments mounted.

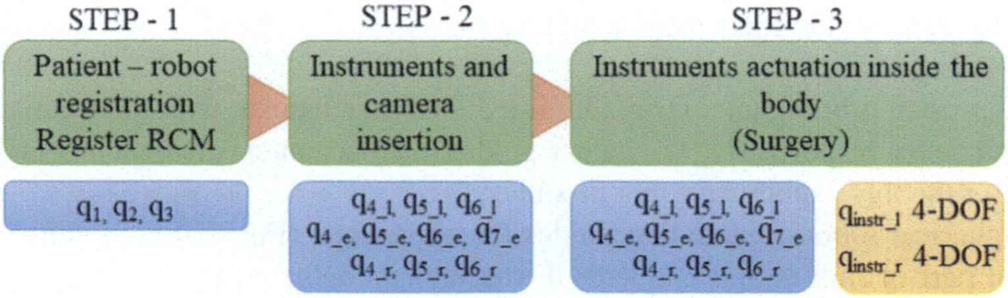

Fig. 1. The tasks of robotic-assisted SILS procedure

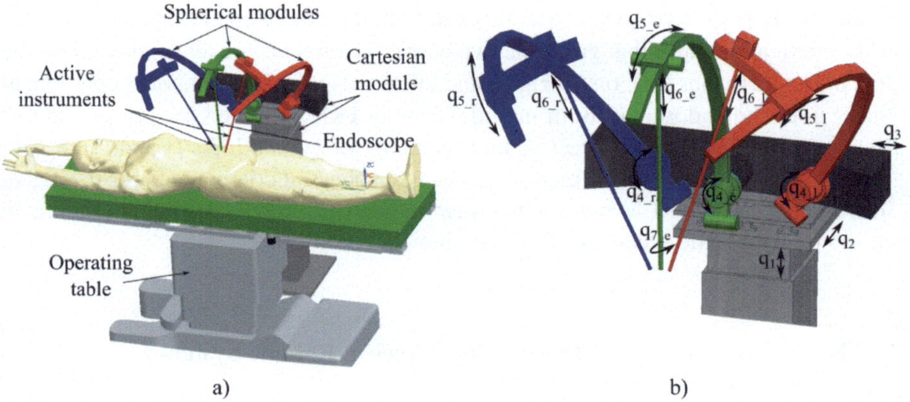

Fig. 2. CAD model of the proposed robotic system for SILS

3 The Geometric Model of the New Spherical Robot for SILS

The kinematic scheme of the proposed SILS robotic system is shown in Fig. 3. The geometric parameters are defined. The subscripts l, r and e are used to denote the left, right, and endoscope spherical modules, respectively. A fixed coordinate system $OXYZ$ is defined at the basis of the gantry robot and three mobile coordinate systems are defined at the tip of each surgical instrument. R_l, R_r, R_e represent the independent radii of each spherical arm; R_{e_disp} represent an OY displacement applied on the central arm towards the RCM as the endoscopic camera is shorter compared to the active instruments; d_{inst} represents the distance (on the OX) axis between the RCMs of the instruments with respect to the central one; α represents the rotation angle around OZ between the central arm and the side ones; β represents an offset for the active instruments mounting which allows the instruments to work on close to parallel trajectories without collisions on the spherical arms; the lengths of the instruments are not required to solve the mathematical model, being relevant only for the definition of the robot size; m_{ins}, m_e represent the minimum insertion length of the instruments and endoscopic camera; RCM_l, RCM_r and RCM_e represent the RCM point for the two active instruments and the endoscope, respectively; $link_l$, $link_r$ and $link_e$ represent the circular mechanical links for q_{5_l}, q_{5_r} and q_{5_e}, which are actuated on arcs with the centers in RCM_l, RCM_r and RCM_e, respectively.

The modules of the SILS robotic system are:

- The Cartesian robot (gantry) (Fig. 3a) is a 3 DOF mechanism with three active joints q_1, q_2, and q_3, respectively. The gantry guides the mechanical element m which is the basis for the three spherical robot modules;
- The spherical robot module that guides the endoscope (Fig. 3b) is mounted on the central part of link m and is a 4 DOF mechanism actuated by q_{4e} (active rotation motion), q_{5e} (active motion along the curved link l_1), q_{6e} (active linear motion) and q_{7e} (active rotational motion) added for the camera orientation around its longitudinal axis. The combined motion of the active joints achieves the RCM motion with respect to the RCM_e point (which is at a distance R from the link m on OY direction);
- The spherical modules that guide the two active instruments are also mounted on the link m (on a circular arc), on the right- and left-hand side of the spherical component that guides the endoscope. Both modules have 3 DOF, actuated by q_{4r}, q_{4l} (active rotational motions), q_{5r}, q_{5l} (active motion along the curved links l_2 and l_3, respectively), q_{6r}, q_{6l} (active linear motions). The two spherical components for the active instruments achieve RCM motion with respect to RCM_r and RCM_l points. Figure 3c shows the kinematic scheme of the left spherical module.

3.1 The Inverse Geometric Model of the Spherical Robotic System

The input parameters for the inverse geometrical model are the coordinates of the central RCM (RCM_e – for the endoscopic camera) and the coordinates of the tips of the three SILS instruments (i.e., the origins of the mobile frames O_e, O_l, O_r), whereas the outputs are the values of the active joints of the three spherical robot modules and the Cartesian robot. To achieve the geometrical model of the robotic system the instruments are considered rigid (even though they will include 4 DOF [3]) (Fig. 1). The supplementary motions of the active instruments include a bending area near the distal tip which will improve the reach of the instrument but do not affect the mathematical model of the modular robotic system. The model has the following input data:

$$\begin{aligned} RCM_e &= [X_{RCM}\ Y_{RCM}\ Z_{RCM}]^T, O_l = [X_l\ Y_l\ Z_l]^T, \\ O_e &= [X_e\ Y_e\ Z_e\ \varphi_e]^T, O_r = [X_r\ Y_r\ Z_r]^T \end{aligned} \tag{1}$$

The location of the RCMs for the active instruments is computed:

$$RCM_l = [X_{RCM} - d_{inst}\ Y_{RCM}\ Z_{RCM}]^T, RCM_r = [X_{RCM} + d_{inst}\ Y_{RCM}\ Z_{RCM}]^T \tag{2}$$

The input coordinates of the active instruments are transformed relative to the spherical arms which guide them, with respect to the rotation angle α:

$$\begin{aligned} X_{l\_rot} &= (X_l + d_{inst} - X_{RCM}) \cdot \cos(\alpha) + (Y_{RCM} - Y_l) \cdot \sin(\alpha) \\ Y_{l\_rot} &= (X_l + d_{inst} - X_{RCM}) \cdot \sin(\alpha) + (Y_l - Y_{RCM}) \cdot \cos(\alpha) + Y_{RCM} \\ Z_{l\_rot} &= Z_l \end{aligned} \tag{3}$$

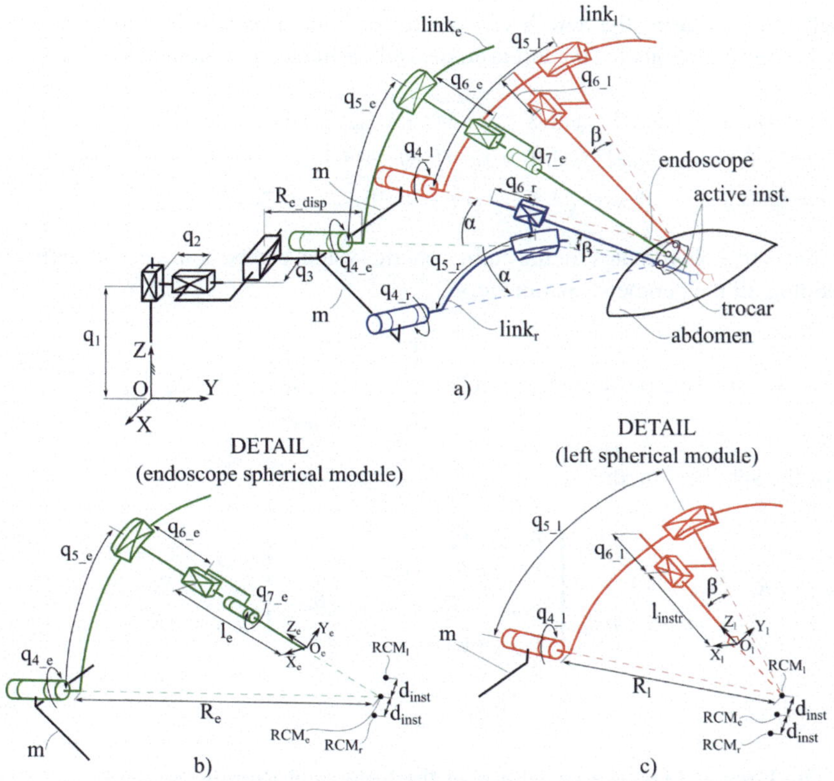

Fig. 3. The kinematic scheme of the SILS modular robotic system: a) the entire robotic system; b) the endoscope spherical module (detail); c) the left spherical module (detail).

For the right spherical arm, the equations are:

$$\begin{aligned}
X_{r_rot} &= (X_r - d_{inst} - X_{RCM}) \cdot \cos(-\alpha) + (Y_{RCM} - Y_r) \cdot \sin(-\alpha) \\
Y_{r_rot} &= (X_r - d_{inst} - X_{RCM}) \cdot \sin(-\alpha) + (Y_r - Y_{RCM}) \cdot \cos(-\alpha) + Y_{RCM} \\
Z_{r_rot} &= Z_r
\end{aligned} \tag{4}$$

The orientation angles of each individual instrument can be computed taking into consideration that the position of the instrument tip can be either above the RCM (outside the body) or below (inside the body).

$$\begin{aligned}
\theta_e &= \mathrm{sgn}(Z_{RCM} - Z_e) \cdot \mathrm{sgn}(X_{RCM} - X_e) \cdot \mathrm{atan2}(|X_{RCM} - X_e|, |Z_{RCM} - Z_e|) \\
\psi_e &= \mathrm{sgn}(Z_{RCM} - Z_e) \cdot \mathrm{sgn}(Y_{RCM} - Y_e) \cdot \mathrm{atan2}\left(|Y_{RCM} - Y_e|, \frac{|Z_{RCM} - Z_e|}{\cos(\theta_e)}\right)
\end{aligned} \tag{5}$$

and

$$ins_e = \mathrm{sgn}(Z_{RCM} - Z_e) \cdot \sqrt{(X_{RCM} - X_e)^2 + (Y_{RCM} - Y_e)^2 + (Z_{RCM} - Z_e)^2}$$

For the active instruments the orientation angles are computed using Eqs. (3) and (4) relative to the RCM of each instrument which represents the fixed rotation center of each spherical arm (the new X coordinate for both arms will be zero). Equation 6 illustrates the equations for the left arm, the right arm being computed similarly:

$$\begin{aligned} &\theta_l = \operatorname{sgn}(Z_{RCM\_l} - Z_{l\_rot}) \cdot \operatorname{sgn}(-X_{l\_rot}) \cdot \operatorname{atan2}(|X_{l\_rot}|, |Z_{RCM\_l} - Z_{l\_rot}|) \\ &\psi_l = \operatorname{sgn}(Z_{RCM\_l} - Z_{l\_rot}) \cdot \operatorname{sgn}(Y_{RCM\_l} - Y_{l\_rot}) \cdot \operatorname{atan2}\left(|Y_{RCM\_l} - Y_{l\_rot}|, \tfrac{|Z_{RCM} - Z_{l\_rot}|}{\cos(\theta_l)}\right) \\ &ins_l = \operatorname{sgn}(Z_{RCM\_l} - Z_{l\_rot}) \cdot \sqrt{X_{l\_rot}^2 + (Y_{RCM\_l} - Y_{l\_rot})^2 + (Z_{RCM\_l} - Z_{l\_rot})^2} \end{aligned} \tag{6}$$

Thus, the active coordinates for the entire robotic system can be computed straight away, embedding all the geometric parameters:

$$\text{For the positioning module}: \begin{cases} q_1 = Z_{RCM} \\ q_2 = Y_{RCM} - R_e - R_{e\_disp} \\ q_3 = X_{RCM} \end{cases} \tag{7}$$

For the spherical modules:

$$\begin{cases} q_{4\_l} = \theta_l + \beta \\ q_{5\_l} = R_l \cdot \left(\psi_l + \frac{\pi}{2}\right) \\ q_{6\_l} = (R_l - m_{ins}) + ins_l \end{cases} \begin{cases} q_{4\_e} = \theta_e \\ q_{5\_e} = R_e \cdot \left(\psi_e + \frac{\pi}{2}\right) \\ q_{6\_e} = (R_e - m_e) + ins_e \\ q_{7\_e} = \varphi_e \end{cases} \begin{cases} q_{4\_r} = \theta_r - \beta \\ q_{5\_r} = R_r \cdot \left(\psi_r + \frac{\pi}{2}\right) \\ q_{6\_r} = (R_r - m_{ins}) + ins_r \end{cases} \tag{8}$$

3.2 The Forward Geometric Model of the Spherical Robotic System

For the forward geometrical model, the inputs are the coordinates of the active joints of the robotic system (the Cartesian robot together with the spherical robot modules), and the outputs are the coordinates of the three rigid surgical instruments defined by the origins of the three mobile coordinate frames (O_e, O_l, O_r). The coordinates of the central RCM point (RCM_e) are computed using the first three active joint coordinates, whereas the coordinates of RCM_r and RCM_l are computed using the coordinates of RCM_e and the displacement d_{inst} between the RCM points:

$$\begin{cases} X_{RCM} = q_2 \\ Y_{RCM} = q_3 + R_e + R_{e\_disp} \\ Z_{RCM} = q_1 \end{cases} \begin{cases} X_{RCM\_l} = X_{RCM} - d_{inst} \\ Y_{RCM\_l} = Y_{RCM} \\ Z_{RCM\_l} = Z_{RCM} \end{cases} \begin{cases} X_{RCM\_r} = X_{RCM} + d_{inst} \\ Y_{RCM\_r} = Y_{RCM} \\ Z_{RCM\_r} = Z_{RCM} \end{cases} \tag{9}$$

Based on the modularity of the robotic system and considering all the geometric parameters defined above, the coordinates of each instrument can be computed with respect to the active joint values of each spherical module:

$$\begin{cases} X_e = X_{RCM} - (q_{6\_e} - R_e + m_e) \cdot \cos\left(\frac{q_{5\_e}}{R_e} - \frac{\pi}{2}\right) \cdot \sin(q_{4\_e}) \\ Y_e = Y_{RCM} - (q_{6\_e} - R_e + m_e) \cdot \sin\left(\frac{q_{5\_e}}{R_e} - \frac{\pi}{2}\right) \\ Z_e = Z_{RCM} - (q_{6\_e} - R_e + m_e) \cdot \cos\left(\frac{q_{5\_e}}{R_e} - \frac{\pi}{2}\right) \cdot \cos(q_{4\_e}) \\ \varphi_e = q_{7\_e} \end{cases} \tag{10}$$

$$\begin{cases} X_{l\_rot} = \left(q_{6\_l} - R_l + m_{ins}\right) \cdot \cos\left(\frac{q_{5\_l}}{R_l} - \frac{\pi}{2}\right) \cdot \sin\left(q_{4\_l} - \beta\right) \\ Y_{l\_rot} = Y_{RCM} + \left(q_{6\_l} - R_l + m_{ins}\right) \cdot \sin\left(\frac{q_{5\_l}}{R_l} - \frac{\pi}{2}\right) \\ Z_{l\_rot} = Z_{RCM} + \left(q_{6\_l} - R_l + m_{ins}\right) \cdot \cos\left(\frac{q_{5\_l}}{R_l} - \frac{\pi}{2}\right) \cdot \cos\left(q_{4\_l} - \beta\right) \end{cases} \quad (11)$$

$$\begin{cases} X_{r\_rot} = \left(q_{6\_r} - R_r + m_{ins}\right) \cdot \cos\left(\frac{q_{5\_r}}{R_r} - \frac{\pi}{2}\right) \cdot \sin\left(q_{4\_r} + \beta\right) \\ Y_{r\_rot} = Y_{RCM} + \left(q_{6\_r} - R_r + m_{ins}\right) \cdot \sin\left(\frac{q_{5\_r}}{R_r} - \frac{\pi}{2}\right) \\ Z_{r\_rot} = Z_{RCM} + \left(q_{6\_r} - R_r + m_{ins}\right) \cdot \cos\left(\frac{q_{5\_r}}{R_r} - \frac{\pi}{2}\right) \cdot \cos\left(q_{4\_r} + \beta\right) \end{cases} \quad (12)$$

Equations (11) and (12) represent the instrument tip coordinates relative to the spherical arms that guide them and using Eqs. (3) and (4), the coordinates relative to the fixed coordinate system can be finally computed.

4 Conclusions

The paper presents a fully parameterized geometric model of a modular robotic system designed for SILS which uses spherical modules for the independent positioning of the instruments inside the body. The fully parametric analytic solution for the geometric model has distinct advantages. On the one hand, it describes the sequential operation of the modular robotic system, and the models can be used for future studies (e.g., dimensional synthesis, operational workspace definition, kinematics for velocity and acceleration, singularity analysis). On the other hand, the analytics solution provides advantages for the real-time robot control, where both the inverse and forward geometric will be used (the position of the active instruments must always be known).

Acknowledgments. This work was supported by a grant of the Ministry of Research, Innovation and Digitization, CNCS/CCCDI – UEFISCDI, project number PCE 171/2021 - Challenge within PNCDI III, and project POCU/380/6/13/123927–ANTREDOC, "Entrepreneurial competencies and excellence research in doctoral and postdoctoral studies programs", a project co-funded by the European Social Fund through the Human Capital Operational.

References

1. Saidy, M.N., Tessier, M., Tessier, D.: Single-incision laparoscopic surgery-hype or reality: a historical control study. Perm. J. **16**(1), 47–50 (2012)
2. Leal Ghezzi, T., Campos Corleta, O.: 30 years of robotic surgery. World J. Surg. **40**(10), 2550–2557 (2016). https://doi.org/10.1007/s00268-016-3543-9
3. Pisla, D., et al.: New approach to hybrid robotic system application in single incision laparoscopic surgery. Acta Tehnica Napocensis **64**(3), 369–378 (2021)

4. Rao, P.P., Rao, P.P., Bhagwat, S.: Single-incision laparoscopic surgery - current status and controversies. J. Minim. Access Surg. **7**(1), 6–17 (2011)
5. Harky, A., et al.: The future of open heart surgery in the era of robotic and minimal surgical interventions. Heart Lung Circ. **29**(1), 49–61 (2020)
6. Greaves, N., Nicholson, J.: Single incision laparoscopic surgery in general surgery: a review. Ann. R. Coll. Surg. Engl. **93**(6), 437–440 (2011)
7. Barrera, K., Wang, D., Sugiyama, G.: Robotic assisted single site surgery: a decade of innovation. Ann. Laparosc. Endosc. Surg. **5**, 4 (2020)
8. Fahmy, O., et al.: Single-port versus multiple-port robot-assisted radical prostatectomy: a systematic review and meta-analysis. J Clin Med. **10**(24), 5723 (2021)
9. Pisla, D., et al.: PARASURG hybrid parallel robot for minimally invasive surgery. Chirurgia **106**(5), 619–625 (2011)
10. Tian, H., et al.: Design, kinematic, dynamic and stiffness analysis of a 5-DOF single-incision laparoscopic surgery robot. Int. J. Robot. Mechatron. **7**(1), 1–16 (2020)
11. Hwang, M., et al.: A single port surgical robot system with novel elbow joint mechanism for high force transmission. Int. J. Med. Robot. Comput. Assist. Surg. **13**(4), e1808 (2017)
12. Ramrath, L., Hofmann, U.G., Schweikard, A.: Spherical assistant for stereotactic surgery. In: International Conference on Intelligent Robots and Systems, pp. 859–864 (2007)
13. Yip, H.M., et al.: A new robotic uterine positioner for laparoscopic hysterectomy with passive safety mechanisms: design and experiments. In: Proceedings of the International Conference on Intelligent Robots and Systems, pp. 3188–3194 (2015)
14. Aksungur, S.: Remote center of motion (RCM) mechanisms for surgical operations. Int. J. Appl. Math. Electron. Comput. **3**(2), 119 (2015)
15. Vaida, C., Pisla, D., Birlescu, I., Gherman, B., Tucan, P., Plitea, N.: Family of modular robotic systems for SILS with kinematic constraint of the entrance point, OSIM A00734/03.12.2021
16. Vaida, C., Andras, I., Birlescu I., Crisan, N., Plitea, N., Pisla, D., Preliminary control design of a single-incision laparoscopic surgery robotic system. In: 25th International Conference on System Theory, Control and Computing (ICSTCC), pp. 384–289. IEEE (2021). https://doi.org/10.1109/ICSTCC52150.2021.9607128

Robotic Innovations in Support of the Healthcare Workers Against COVID-19 - DIH-HERO Perspective

Kosta Jovanovic[1](✉), Andrea Schwier[2], Eloise Matheson[3], Michele Xiloyannis[4], Esther Rodijk-Rozeboom[5], Nadine Hochhausen[6], Brecht Vermeulen[7], Birgit Graf[8], Peter Wolf[4], Zbigniew Nawrat[9], Jordi Escuder[10], Mare Mechelinck[6], Birgitte Sørensen[11], Paola Roberta Boscolo[12], Michael Obach[13], Selene Tognarelli[14], Milica Jankovic[1], Christophe Leroux[15], Giancarlo Ferrigno[16], Françoise Siepel[17], and Stefano Stramigioli[17]

[1] School of Electrical Engineering, University of Belgrade, Belgrade, Serbia
{kostaj,piperski}@etf.bg.ac.rs

[2] Institute of Robotics and Mechatronics German Aerospace Center (DLR), Oberpfaffenhofen, Cologne, Germany
andrea.schwier@dlr.de

[3] Imperial College London, London, UK
e.matheson@imperial.ac.uk

[4] Institute of Robotics and Intelligent Systems, ETH Zurich, Zürich, Switzerland
{michele.xiloyannis,peter.wolf}@hest.ethz.ch

[5] TechMed Centre, University of Twente, Enschede, The Netherlands
e.d.rodijk-rozeboom@utwente.nl

[6] Faculty of Medicine, RWTH Aachen University, Aachen, Germany
{nhochhausen,mmechelinck}@ukaachen.de

[7] imec - IDLab, Ghent University, Ghent, Belgium
brecht.vermeulen@imec.be

[8] Department of Robot and Assistive Systems, Fraunhofer IPA, Stuttgart, Germany
birgit.graf@ipa.fraunhofer.de

[9] Foundation of Cardiac Surgery Development, Zabrze, Poland
nawrat@frk.pl

[10] Fundació Eurecat, Barcelona, Spain
jordi.escuder@eurecat.org

[11] Danish Technological Institute, Odense, Denmark
bso@teknologisk.dk

[12] Icub Tech - Istituto Italiano di Tecnologia, Genova, Italy
paola.boscolo@iit.it

[13] Basque Research and Technology Alliance (BRTA), Health Division, TECNALIA, Mendaro, Spain
michael.obach@tecnalia.comxc

[14] Scuola Superiore Sant'Anna, Pisa, Italy
s.tognarelli@santannapisa.it

[15] CEA Paris-Saclay Robotics Unit, New Delhi, India
Christophe.leroux@cea.fr

[16] DEIB, Politecnico Di Milano, Milano, Italy
giancarlo.ferrigno@polimi.it

[17] Robotics and Mechatronics Group, University of Twente, Enschede, The Netherlands
{f.j.siepel,S.Stramigioli}@utwente.nl

A. Müller and M. Brandstötter (Eds.): RAAD 2022, MMS 120, pp. 375–382, 2022.
https://doi.org/10.1007/978-3-031-04870-8_44

Abstract. Coronavirus disease 2019 (COVID-19) has affected the conditions of work in healthcare institutions and the quality of patient care around the world. Emerging healthcare robotic technology may facilitate and improve the overall quality of life, as well as the diagnostics, rehabilitation, and intervention services. This paper reports the lessons learned from actions carried out in the implementation of the DIH-HERO project call across Europe for robotic solutions that support healthcare activities. Conclusion remarks are as follows: *i*) technology pull and the urgent need together accelerate the innovation development and deployment, *ii*) it takes time to establish safety and legal regulations for the deployment of human-machine devices, so the ethical, safety, and reliability aspects of robotic application need to be carefully considered, *iii*) technology adoption depends on the trust of the users to the technology, *iv*) existing robotic technology for prevention, diagnosis, hospital admission, rehabilitation, and intervention is mature enough for the adaptation and deployment, *v*) unskilled general population should be trained for the usage of robotic technology.

Keywords: COVID-19 · Diagnostic robotics · Interventional robotics · Rehabilitation robotics · Robotics supporting healthcare professionals · Robotics supporting patients

1 Introduction

Coronavirus disease 2019 (COVID-19) pandemic has strong negative effects on the quality of patient care and the organizational aspects of work in healthcare institutions. Non-covid healthcare institutions reduced and slowed down the work capacity because of COVID-19 epidemiological restrictions related to the decreasing of the physical interactions of people. The diagnostic services [1, 2] and continuum care for patients with chronic diseases were disrupted and postponed [3], as well as rehabilitation healthcare services [4] and surgical interventions [5]. New measures such as lockdowns, physical distancing, self-isolation, and quarantine have led to an increase in various psychological consequences such as increased anxiety, depression, insomnia, harmful alcohol, drug use, self-harm, or suicidal behavior, etc. [6].

especially in nursing homes where elderly people stay isolated from their families for months. Healthcare workers are particularly exposed to mental challenges as well as to physical efforts, and these facts indicate the need for implementation strategies to reduce their morbidity and mortality (e.g. reducing direct contact with patients, shorter shift lengths, mental health support, physical support).

The pandemic situation caused the urgent transfer to non-contact, digital communication in all spheres where it was possible (human health, healthcare, economy, industries, agriculture, etc.). Due to COVID-19, a new environment is created, in which we need to learn about necessity and challenges in human-human collaboration and interaction, but also in human-machine interaction and collaboration since we are faced with a completely new perspective in the communication. Emerging technologies in robotics can provide the implementation of numerous solutions to support the fight against the

COVID-19 pandemic. Robots can help in organizing the living environment, cleaning, keeping production going in the pandemic scenario, etc. During the COVID-19 pandemic, healthcare robots particularly may play an important role in reducing the spread of COVID-19 infection and providing quality patient care [9]. The current pandemic situation requires the deployment of practical, innovative robotic solutions in the different areas: diagnostic robotics, interventional robotics, rehabilitation robotics, robotics supporting patients, and robotics supporting healthcare professionals. Due to the urgent situation, it is expected that the expansion and adoption of new innovative helpful robotic technologies will encounter fewer barriers than under normal conditions.

This paper provides an overview of the lessons learned from concrete actions that the European network of Digital Innovation Hubs in Healthcare Robotics (DIH-HERO) [10] took across Europe in supporting healthcare workers through robotic innovations and overcoming all the challenges COVID-19 imposed on the community in the previous period.

2 DIH-HERO Open Call – Emerging Scenarios as Key Drivers for Innovations

In April 2020, DIH-HERO announced a call for Financial Support for Third Parties (FSTP) to support small and medium companies in their ideas and activities towards fighting COVID-19. The total call budget was €1 million for up to ten projects, which means 100.000€ per project.

The call was open for only one week. Typically, these types of calls for companies are open for 2–3 months. The call was launched in the second month of the pandemic, during the lockdown, so companies could not prepare a lot in fighting pandemics before the call. The requested duration of granted projects was only 2–3 months. At the end of this short period, companies should have a working prototype validated in clinical settings. Therefore, only those companies that had already had some working solution could fulfill the requirements and on top of that, they should have provided a letter of support from a clinic that was ready to adopt the developed solution at least for testing in clinical environments. Through the DIH-HERO network and with the support of the European Commission, 146 companies from 30 different European countries were mobilized to submit project proposals in such a short time frame.

Consequently, the first lesson learned is that ***the key drivers for innovations are technology pull and the urgent need together***. Throughout history, significant scientific discoveries have occurred during times of global challenges such as wars or pandemics. Although systematic and numerous actions have been made to support innovations and uptake of technology on nationals and EU levels, the technology has been rarely fully prepared to appropriately respond to emergent situations. In conclusion, dedicated measures do not impact the technology uptake and innovations as an emerging scenario does. This may be because of global challenges such as the COVID-19 pandemic affects all aspects of life, not just as a working task, or because technology adopters (such as medical doctors) reduce technology adoption barriers themselves in hopes of resolving problems urgently. Therefore, emerging scenarios often achieve improved communication between technology developers and adopters, and thus match technical innovations with the direct user needs.

3 Safety and Legal Regulation Affect the Robotic Innovation Deployment

Within healthcare robotics, DIH-HERO distinguished five major application domains: diagnostic robotics, interventional robotics, rehabilitation robotics, robotics supporting patients, and robotics supporting healthcare professionals, Fig. 1.

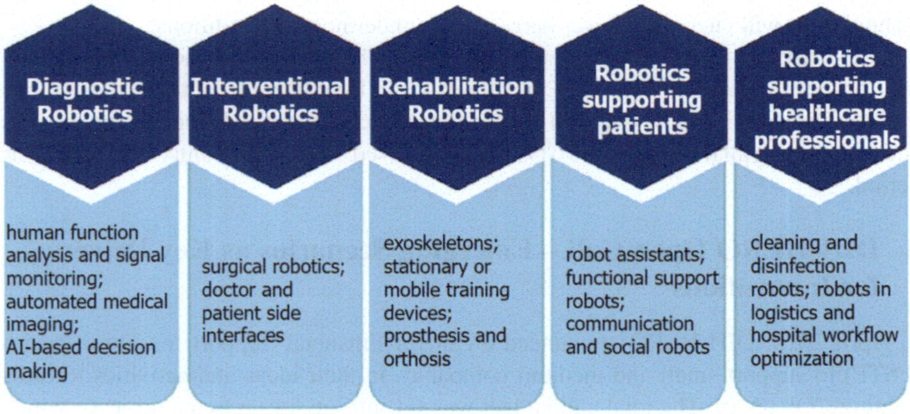

Fig. 1. Five major application areas of healthcare robots

Diagnostic robots are used for human signals monitoring by different sensor systems including medical imagining or micro-nano robots for endoscopy, but also functional analysis and intelligent decision making in diagnostics [11]. Interventional robotics stands mostly for surgical robotics applications, including robotics setups on the patient side (where a robot penetrates a human body) as well as intervention interfaces on a medical doctor's side (e.g. haptics) [12]. The domain of rehabilitation is also considered quite broad - robotics applied to restore health through training and therapy after or during illness (such as stroke or other injuries) or a range of mobile assistive robotic technologies (e.g. exoskeletons, prosthesis, orthosis) [13]. The fourth domain of robots supporting patients directly is considered for robotics solutions that provide direct support and assistance to patients (e.g. bringing the supplies, assisting in motion) but also robots that are used for communication and social interaction [14]. Finally, the robotics solutions which provide support to healthcare professionals are robots that replace healthcare professionals in some tasks such as cleaning or disinfection, assist in hospitals or home logistics including patient transportation [15].

In the scope of the presented DIH-HERO COVID-19 call for projects, project applications that targeted urgent needs at the start of the pandemic were expected such as disease prevention, COVID-19 diagnostics, as well as a few in patient care.

However, the distribution between application domains is presented in Fig. 2. Most of the project applications targeted the domain of robots directly supporting healthcare professionals (64 of 146), and then robots supporting patients (27 of 146), and automated diagnostics (29 of 146).

A

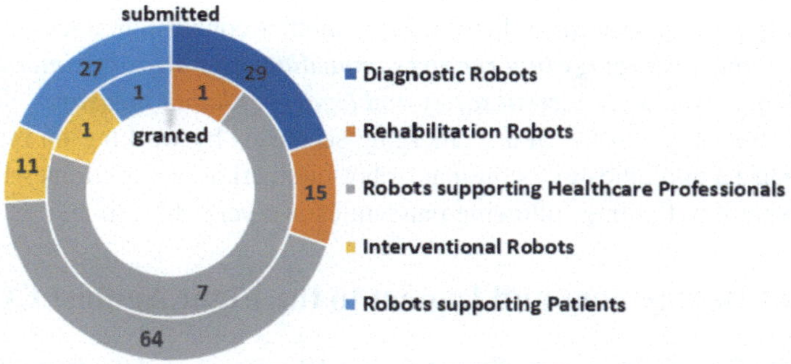

B

	COMPANY	PROJECT TITTLE
7	ACCREA Engineering (PL)	DISInfecting RoboT (DisiRT)
	Akara Robotics (UK)	Disinfectant RObots to Protect Against COVID (DROPAC)
	Jonker-Makis Robotics (NL)	Autonomous mobile disinfection robot SAM-Air
	KELO Robotics (DE)	Autonomous robots for disinfection (ARODIS)
	MetraLabs (DE)	Portable and agile autonomous disinfection robot (STERY)
	PAL Robotics (ES)	Fast Deployment of AVs in Hospitals
	Rubedo Sistemos (LT)	Unmanned Disinfection Solution (UDS)
1	Voxdale (BE)	Robot for Intradermal Drug Delivery (ROB-ID)
1	F&P Robotics (CH)	Autonomous Sanitiser and Assistant Lio
1	Hocoma (CH)	Sensor-based arm and hand functional tele-rehabilitation with Armeo®Senso (Tele-AX)

Fig. 2. A) The distribution of submitted and granted project proposals for DIH-HERO COVID-19 call, B) List of granted projects per application domains

The distribution of submitted and granted projects per application domains demonstrates the priorities of supporting frontline healthcare workers and intention of applicants to stop spreading the disease according to the number of applications in the robotics disinfection. After the submission process had been completed, each proposal was evaluated by the evaluation committee, at least 3 evaluators evaluated each proposal, its potential impact, maturity of the proposed solution, feasibility in the very limited time frame of 2–3 months, and clinical relevance. The evaluation committee also involved clinicians. The result of the evaluation was that only one project per domains intervention, rehabilitation, and support to patients was granted, whereas 7 projects were granted in the domains of robots supporting healthcare professionals. In the area of intervention, the granted project was about intradermal drug delivery, in the area of rehabilitation it was a telerehabilitation device, and in the area of support to patients, it was a personal robot assistant. In the domain of robotics supporting healthcare professionals, there were 6 projects in robotics disinfection and one in logistics. Surprisingly, no diagnostic robotics projects were funded due to different reasons such as difficulties in handling patient personal data and lack of trust of medical doctors for robotized diagnostic solutions.

According to the presented distributions, another conclusion, a lesson learned, was reached: ***robotic technology that requires immediate human-robot interaction cannot be quickly deployed with current safety and legal regulation measures***. So, the society must be better prepared for future emerging scenarios by working on regulations and off-the-shelf technologies in the human-robot interaction preparations which can have a strong benefit in fighting following pandemics in a very short time.

4 Robot Deployment and Impact to the Fight Against COVID-19

Robotic innovations developed within COVID-19 call included the first fully automated standardized intradermal vaccination robot, an upper-limb telerehabilitation setup with data-driven therapy protocols, and an option of semi-supervised rehabilitation protocols, multisensory SLAM robot systems for hospital logistics, a multimodal disinfection robot (spraying, UV lights, ozone), advanced visions systems for human and critical object detections for advanced safety and efficiency in robotics disinfection, etc. A complete overview of the developed technologies is available in the authors previous work [16]. However, the lessons learned in the deployment of robotics innovation in the pandemic scenario resulted in several lessons learned as a foundation for the uptake of advanced technologies in emerging scenarios yet to come.

Technology adoption depends predominantly on the trust of the users to the technology. This becomes evident when facing technology provider priorities in the disinfection robots' innovations (path planning, minimal disinfection time, object detection and recognition, area coverage) with technology adopter key requests (perfect disinfection outcome proven by indicators, reliability of the disinfection process).

Existing robotic technologies are mature enough to be deployed after the period of adaptation and contribute to all phases of the disease treatment – from prevention, diagnosis, hospital admission, intervention, to rehabilitation. The prospective use of robots at different stages of disease treatment is depicted in Fig. 3.

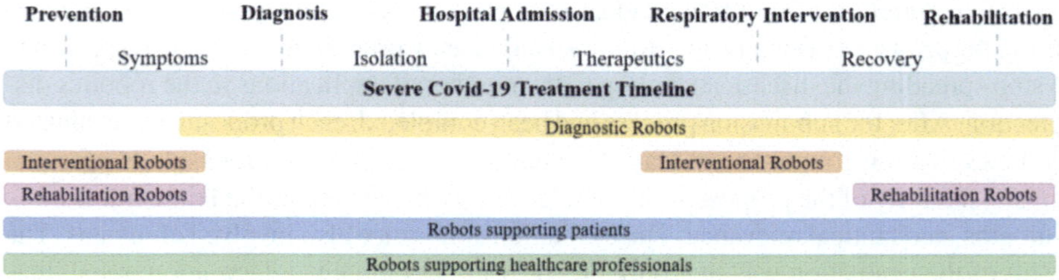

Fig. 3. Prospective use of robots at different stages of disease treatment.

The unskilled general population should be trained for the usage of robotic technology. Although the robotics technology demonstrated maturity for quick deployment in clinical settings and treatment of public spaces, the DIH-HERO COVID-19 experiments highlighted the importance of easy and intuitive commands of advanced technologies and possibly advancements in natural human-machine interaction, for the widespread utilization and general trust in robotic technologies.

5 Conclusion

In this paper, an overview of DIH-HERO perspectives about announced COVID-19 project call for companies is presented. Several conclusions were noticed regarding the realized experience in the call execution. The first remark was regarding the scenario when innovations happen. It was concluded that an emerging scenario, like an actual pandemic, in symbiosis with technological readiness inspires innovation development. The second remark concerns slow safety and legal regulations for the implementation of robotic technology that requires immediate interaction with humans. So, ethical, safety, and reliability concerns need to be carefully considered and exclusive policies for use of the technologies in emerging scenarios should be made. The existing robotic technologies are already mature enough to quickly adapt and contribute to all stages of disease treatment: prevention, diagnosis, hospital admission, intervention, rehabilitation. However, technology adoption depends on the trust of the users much more than the confidence of the technology providers. Finally, robot deployment in everyday living environments and operations with unskilled general population, impose challenges that need to be resolved.

References

1. De Vincentiis, L., Carr, R.A., Mariani, M.P., Ferrara, G.: Cancer diagnostic rates during the 2020 'lockdown', due to COVID-19 pandemic, compared with the 2018–2019: an audit study from cellular pathology. J. Clin. Pathol. **74**(3), 187–189 (2021)
2. Hamilton, W.: Cancer diagnostic delay in the COVID-19 era: what happens next? Lancet Oncol. **21**(8), 1000–1002 (2020)
3. Chudasama, Y.V., et al.: Impact of COVID-19 on routine care for chronic diseases: a global survey of views from healthcare professionals. Diabetes Metab. Syndr. **14**(5), 965–967 (2020)
4. Bettger, J.P., et al.: COVID-19: maintaining essential rehabilitation services across the care continuum. BMJ Global Health. **5**(5), e002670 (2020)
5. Moletta, L., et al.: International guidelines and recommendations for surgery during Covid-19 pandemic: a systematic review. Int. J. Surg. **79**, 180–188 (2020)
6. Kumar, A., Nayar, K.: R: COVID 19 and its mental health consequences. J. Ment. Health **30**(1), 1–2 (2021)
7. Shaukat, N., Ali, D.M., Razzak, J.: Physical and mental health impacts of COVID-19 on healthcare workers: a scoping review. Int. J. Emerg. Med. **13**(1), 1–8 (2020). https://doi.org/10.1186/s12245-020-00299-5
8. Javaid, M., Haleem, A., Vaish, A., Vaishya, R., Iyengar, K.P.: Robotics applications in COVID-19: a review. J. Indust. Integr. Manage. **5**(04), 441–451 (2020)
9. Di Lallo, A., Murphy, R., Krieger, A., Zhu, J., Taylor, R.H., Su, H.: Medical robots for infectious diseases: lessons and challenges from the COVID-19 pandemic. IEEE Robot. Autom. Mag. **28**(1), 18–27 (2021)
10. DIH-HERO project web site. https://dih-hero.eu/
11. Moglia, A., Menciassi, A., Schurr, M.O., Dario, P.: Wireless capsule endoscopy: from diagnostic devices to multipurpose robotic systems. Biomed. Microdevice **9**, 235–243 (2006)
12. Pieters, B., Armijo, P., Krause, C., Choudhury, S., Oleynikov, D.: Review of emerging surgical robotic technology. Surg. Endosc. **32**(1), 1636–1655 (2018)
13. Gassert, R., Dietz, V.: Rehabilitation robots for the treatment of sensorimotor deficits: a neurophysiological perspective. J. Neuroeng. Rehabil. **15**(46), 1–15 (2018)

14. Hung, L., et al.: The benefits of and barriers to using a social robot PARO in care settings: a scoping review. BMC Geriatr. **19**(232), 1–10 (2019)
15. Pino, M., Boulay, M., Jounen, F., Rigaud, A.S.: Are we ready for robots that care for us? - Attitudes and opinions of older adults toward socially assistive robots. Front. Aging Neurosci. **7**(141), 1–15 (2015)
16. Jovanovic, K., et al.: Digital innovation hubs in health-care robotics fighting COVID-19: novel support for patients and health-care workers across Europe. IEEE Robot. Autom. Mag. **28**(1), 40–47 (2021)

On the Kinematics and Dimensional Optimization of a Robotic System for Single Incision Laparoscopic Surgery

Bogdan Gherman[1], Paul Tucan[1], Calin Vaida[1], Nicolae Crisan[1,2], Gabriela Rus[1], Iosif Birlescu[1], and Doina Pisla[1](✉)

[1] Research Center for Robots Simulation and Testing, Technical University of Cluj-Napoca, Cluj-Napoca, Romania
doina.pisla@mep.utcluj.ro
[2] "Iuliu Hatieganu" University of Medicine and Pharmacy, Cluj-Napoca, Romania

Abstract. Single Incision Laparoscopic Surgery has become a self-standing approach within the minimally invasive surgery techniques due to its ability of reduced trauma and the almost invisible marks on the patient's body when achieved using as insertion port the umbilicus. The very tight available space to perform the surgery imposes a significant challenge for the developers of tools and devices to serve this type of intervention. The paper addresses the development of a robotic system for SILS, based on the master-slave architecture. The slave robotic system architecture is defined to fit the requirements of the targeted surgical approach. The kinematics is analyzed and prepared for implementation in the control considering that the master system is an Omega 7 haptic device. Based on a proposed algorithm, an optimization process of the main geometric elements of the robotic system is provided, considering the required workspace and the collision free mechanical configurations of the robot. Genetic algorithms are used to yield the optimal values for the selected parameters, which will be further considered in the design of the robotic system.

Keywords: Kinematics · Dimensional optimization · Robotic surgery · Teleoperation

1 Introduction

Single incision laparoscopic surgery (SILS) is a minimally invasive procedure, which seems to be a better option compared to the conventional laparoscopic surgery where a multiport approach is used, leading to a faster recovery, shorter hospital stays, less postoperative pain, and a better cosmetic outcome [1]. SILS assumes the use of a single trocar (the SILS port device) through the patient's navel, so the scar is barely noticeable. SILS is frequently used in surgical procedures such as colectomies, nephrectomy, appendectomy, etc. The most obvious challenge within this approach is the development and use of specific devices that must have a flexible, ergonomic, and articulating structure, which allows a convenient utilization without minimizing the workspace and

A. Müller and M. Brandstötter (Eds.): RAAD 2022, MMS 120, pp. 383–391, 2022.
https://doi.org/10.1007/978-3-031-04870-8_45

limits collisions between instruments [2]. In [3–6] several robotic systems for SILS have been proposed. However, their practical utility has not been proven within the medical environment (clinical trials). One of the renowned commercial systems is the da Vinci SP [7], but its reduced workspace and prohibitive price reduces its popularity for SILS procedures. Within the SILS robotic systems it is critical to generate the remote Center of Motion (RCM) to ensure the best circumstances for manipulating robotic instruments. Many RCM manipulator concepts have been presented [8, 9], the most well-known of which being the spherical and parallelogram mechanisms [10], which provides increased stiffness and good kinematic performance. The goal of this paper is to develop a robotic system for SILS and provide a dimensional optimization algorithm that leads to suitable values for the geometrical parameters of the proposed mechanical structure. The paper is organized as follows: Sect. 2 presents the proposed solution for SILS using parallelogram mechanisms; Sect. 3 presents the kinematic analysis and the control solution; Sect. 4 shows the optimization algorithm for a suitable design of the robotic system; several conclusions are presented in Sect. 5.

2 A Robotic System for SILS

The proposed solution implements a master-slave architecture, where the slave robotic system, based on the requirements defined in [11–13], should be able to guide an endoscopic camera (laparoscope) and two active instruments designed for SILS. The master is the Omega.7, a 7 Degrees of Freedom (DoF) haptic device by Force Dimension [14]. The robotic system is presented in Fig. 1, [15], and it consists of a positioning platform with three DoFs, three parallelogram mechanisms disposed in such a way that the RCM of each guiding mechanism fits the standard SILS port, the laparoscope and two active instruments. The 3 DoFs (actuated by q_1, q_2, q_3) positioning mechanism (in red color, Figs. 2a and 2b) is used to map the LP, IP1 and IP2 mechanism RCM to their respective trocars in the SILS port, after which its active joints are locked. The first parallelogram mechanism (LP) is designed to guide an endoscopic camera (the laparoscope), while the

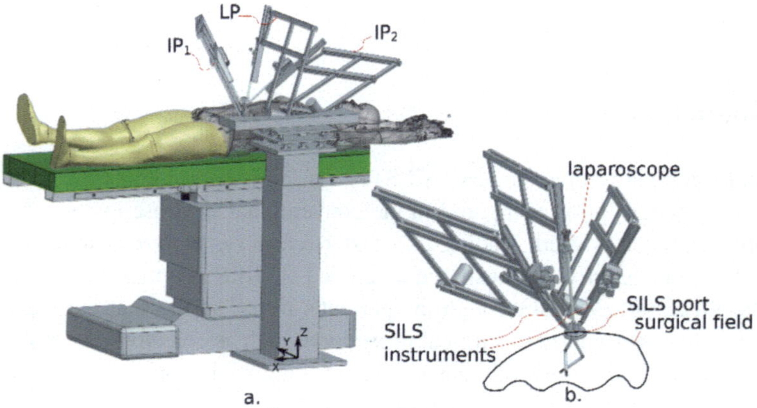

Fig. 1. The proposed robotic system architecture for SILS: a. the entire system; b. detail with the laparoscope and the two active instruments for SILS.

other two identical parallelogram mechanisms are designed to guide active laparoscopic instruments (IP).

3 Kinematic Analysis

The kinematic schemes of the LP and IP are presented in Fig. 2. Besides the 3 DoFs cartesian positioning mechanism, the active joints of the LP (in blue color) are $q_4{}'$ and $q_5{}'$ and finally $q_6{}'$, used to insert the laparoscope through the RCM_{L} (in green color). A similar approach is used to actuate the IP mechanism (q_4, q_5, q_6). Each of the active instruments have 3 active joints (q_7, q_8, q_9) used for the orientation of the distal head of the instrument and the active function (not symbolized in Fig. 2).

3.1 Forward Kinematics

The input data into the forward kinematics consists in the active joints' coordinates ($q_1, q_2, q_3, q_4, q_5, q_6, q_7, q_8, q_9$) and the geometric parameters (Figs. 2a and 2b), the output being the end-effector (active instrument or laparoscope) pose, i.e. the orientation angles ψ, θ, φ, considering the following convention for the rotation matrix: $R_x(\psi), R_y(\theta), R_z(\varphi)$ and the coordinates of the laparoscope tip. The variables required to compute the direct kinematics are summarized in Table 1 using the Denavit-Hartenberg convention [16]. The rotation angle α_3 applies only for the IP, as do all the joints from 7 onwards (with * in purple color). The direct kinematics is computed as in Eq. (1) for the LP Eq. (2) for the IP.

$$T_0^L = T_0^1 T_1^2 T_2^3 T_3^4 T_4^5 T_5^6 T_6^{6'} T_6^{6''} \tag{1}$$

$$T_0^I = T_0^1 T_1^2 T_2^3 T_3^4 T_4^5 T_5^6 T_6^{6'} T_6^{6''} T_7^8 T_8^{8'} T_8^9 T_9^{10} T_{10}^{10'} \tag{2}$$

where $T_i^{i'}$ represent a transformation matrix using only the geometric parameters of the robot. The laparoscope and instrument orientation (excluding the distal head) are given by Eqs. (3), (4) and (5), while the position of the laparoscope tip is easily obtained from Eq. (1). The values of the ψ and θ angles during SILS procedure will be less than 90°.

$$\theta = \mathrm{atan2}\left(S_\theta, \sqrt{1 - S_\theta{}^2}\right), k_j = \mathrm{atan2}\left(\sqrt{1 - C_{k_j}{}^2}, C_{k_j}\right), j = 1, 2 \tag{3}$$

$$S_\theta = c\alpha_6(s\alpha_3 sq_5 + c\alpha_3 cq_5 sq_4) + s\alpha_6(s\alpha_3 cq_5 - c\alpha_3 sq_4 sq_5), \tag{4}$$

$$k_1 = \varphi, k_2 = \psi, C_\varphi = \frac{c\alpha_3 cq_4}{\sqrt{1 - S_\theta{}^2}}, C_\psi = \frac{cq_4(c\alpha_6 cq_5 - s\alpha_6 sq_5)}{\sqrt{1 - S_\theta{}^2}}, \tag{5}$$

and $sq_i = \sin(q_i)$ and $cq_i = \cos(q_i)$.

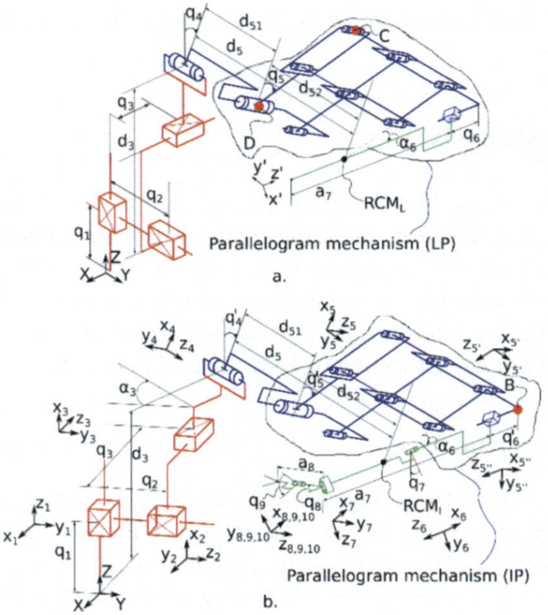

Fig. 2. The robotic system joint configuration: a. the laparoscope guiding arm; b. the active instrument guiding arm.

Table 1. Kinematics parameters of the robotic system (D-H).

Targeted mechanism	Joint i	a_{i-1} (mm)	α_{i-1} (°)	d_i (mm)	q_i (°/mm)
LP and IP mechanisms	1	0	−90	0	q_1
	2	0	90	0	q_2
	3	0	−90	d_3	q_3
	4	0	$90 + \alpha_3$*	0	q_4
	5	0	90	d_5	q_5
	6	0	90	0	q_6
IP mechanisms only	**7***	**0***	α_6*	**0***	q_7*
	8*	a_7*	**90***	**0***	q_8*
	9*	**0***	**90***	**0***	q_9*
	10*	a_9*	**0***	**0***	**0***

3.2 Inverse Kinematics

The input data into the inverse kinematics consists in the required pose of the distal head (element a_8) and the main rod (element a_7 or the laparoscope), namely the position in cartesian coordinates of the laparoscope tip or of the q_9 (q_8) revolute joint rotation center and their orientation (the angles ψ, θ, φ), the output being the active joints' coordinates.

Since the robotic system has a total of 9 DoFs, it is impossible to obtain a valid single solution for the inverse kinematics, which is why a modular approach is required. For a successful operation of the robotic system, a short protocol consisting of 5 steps has been defined:

1. Place the SILS port device in the vicinity of the patient's lesion.
2. Initialize the system (homing procedure).
3. Drive the two IP towards a safe position to avoid collision with the LP when finding the RCM. This step should be performed automatically.
4. With the laparoscope attached to the corresponding LP, drive the system to the RCM point using only the q_1, q_2, q_3 translation joints, while all other joints are locked into their current position. This leads towards an easy inverse kinematics:

$$T_{RCM} = T_0^1 T_1^2 T_2^3 T_6^{6'} \tag{6}$$

which provides the analytic solutions for the q_1, q_2, q_3 active joints.
5. Once the LP is in place (the RCM_L is set), since the standard SILS port device is used, the RCM_I of the two IP mechanisms is also determined.

From this moment onwards, the system is controlled visually, using the streaming video from the laparoscope. The 3 DoFs of the Omega.7 Delta mechanism is used to position the LP and the two IP (the two DoFs of the parallelogram mechanism and the insertion, namely the q_4, q_5, q_6 active joints), while the remaining Roll-Pitch-Yaw wrist DoFs are used to guide the distal head of the active instruments (q_7, q_8, q_9 and the active function: gripping, cutting, which will require an additional DoF). The active joints coordinates q_4, q_5, q_6 with a determined orientation of the laparoscope (or main rod of the active instrument), namely the angles ψ, θ, φ are:

$$\begin{cases} q_4 = \text{atan2}\left(-S_{q_4}, \sqrt{1 - {S_{q_4}}^2}\right) \\ q_5 = \text{atan2}\left(\sqrt{1 - {C_{q_5}}^2}, C_{q_5}\right) - \alpha_6 \text{ ,} \\ q_6 = \frac{(Z_E - d_3 - q_1)}{(cq_4 c(q_5 + \alpha_6))} \end{cases} \tag{7}$$

$$S_{q_4} = s\varphi s\psi - c\varphi c\psi s\theta, C_{q_5} = (c\varphi s\psi + s\varphi c\psi s\theta)/cq_4. \tag{8}$$

4 Dimensional Optimization of the IP Mechanism

Since the position mechanism of the LP and IP is used only to reach the desired RCM, after which its motors are locked, its optimal dimensioning is mostly related to the dimension of the surgical bed, namely reaching at full extent the opposite bed side, being also able to sustain the weight of the LP and the two IP. Because of these reasons, in this stage of development, the positioning mechanism dimensional optimization is not relevant, the focus being placed on the two IP and the LP, as well as on the active instruments. Since the SILS is performed within very tight-positioned orifices, there is a high probability of collision between the IP and LP mechanisms. Collisions may arise outside the patient's body as well as inside the abdominal cavity between the two active instruments. The latter case, although not desirable is far less dangerous since the surgeon can see (using the laparoscope) the two robotic arms and take measures so that the instruments don't collide. It is not the case with the IP and LP, which are not supervised during the surgery procedure. The optimization process is organized as follows: to cover as much as possible an imposed workspace and avoid external collisions. For the moment, the RCM position calculated considering that the SILS port is placed in the patient's umbilicus. Figure 3a presents the algorithm which leads to valid and collision free configurations of the robot (including both IP mechanisms). The workspace is discretized into n points based on an interval, and the inverse kinematics is used to compute the configuration for both IP mechanisms. To simplify the kinematics of the IP + active instruments, the following hypotheses have been used:

- The rotation of the distal head (joint q_9) has been disregarded within the inverse kinematics.
- Each discrete point of the robot workspace represents a sphere having the radius a_7 and the center the coordinates of point $i(x_i, y_i, z_i)$, with the laparoscope oriented by the $\overline{i, RCM}$ line, with the laparoscope tip height: $Z_{\text{laparoscope\_tip}} > Z_i + 100$. The active instruments joint q_8 will have the same Z_i coordinate, while the X_{IP_q8} and Y_{IP_q8} are calculated using Eq. (9). This assumption considers that the i discrete point can be reached by both IP mechanisms during the surgical intervention using q_7 and q_8 active joints. Figure 3b presents graphically this assumption.

$$\begin{cases} X_{IP\_q8} = X_i \pm a_7 \cdot \cos\beta \\ Y_{IP\_q8} = Y_i \pm a_7 \cdot \sin\beta \end{cases}, \ \beta = \operatorname{atan2}(X_i, Y_i) \tag{9}$$

The kinematics of the IP and active instrument is calculated using Eq. (10), where the orientation of the active instrument is determined using Eq. (11):

$$T_i^{RCM} = T_4^5 T_5^6 T_6^{6'} T_6^{6''} T_7^8 \tag{10}$$

$$\begin{cases} \psi = \operatorname{atan2}(Y_i - Y_{\text{RCM}}, X_i - X_{\text{RCM}}) \\ \theta = \operatorname{atan2}\left(\sqrt{(Y_i - Y_{\text{RCM}})^2 + (X_i - X_{\text{RCM}})^2}, Z_i - Z_{\text{RCM}}\right) \end{cases} \tag{11}$$

The collisions between the IP and the LP mechanisms are avoided by imposing a minimum distance between the point B belonging to the IP (Fig. 2a) and the plane defined by the points RCM_{L}, C, D belonging to the LP mechanism (Fig. 2b), namely nd_{min}. $\overline{CD}$ has the length of the laparoscope (430 mm), element d_5 is broken into $d_{51} = 150$ mm and d_{52}, so the coordinates of points C and B can be determined using the direct kinematics, which finally leads to the following:

$$\begin{cases} \hat{u} = \frac{(RCM_{\text{L}} - E) \times (C - E)}{|(RCM_{\text{L}} - E) \times (C - E)|} \\ nd_i = \hat{u} \cdot (B - RCM_{\text{L}}) \end{cases} \tag{12}$$

The imposed condition is: $min(nd_i) > nd_{min}$. The motion ranges for the robot parameters are presented in Table 2. The optimization function is defined as follows:

$$f(1) \equiv N_v = \frac{n}{n_v}; f(2) \equiv L = d_5 + a_7 + a_8; f(3) = \alpha_3 \quad (13)$$

which means that the total number of valid points should be as large as possible and the robot geometric parameters (including α_3) should be as low as possible.

Genetic Algorithm within MATLAB has been used as the global method for the optimization of the robotic system. In this case, the general boundary conditions have been set between 20–100, 0.4–0.9, and 0.001–0.1 for the population size, the crossover operator, and the mutation operator. After multiple running tests, the final values for these parameters have been found as: 20, 0.6 and 0.03. The obtained values for the geometric parameters of the robot are listed in Table 3.

Table 2. The range of the robot variables.

Parameter	$\alpha_3(°)$		$q_4(°)$		$q_5(°)$		$d_{52}(mm)$		$q_6(mm)$		$a_7(mm)$		$a_8(mm)$	
	min	max	min	max	min	max	min	max	min	max	min	max	min	max
Value	10	45	−90	90	−60	90	300	600	300	500	300	500	50	100

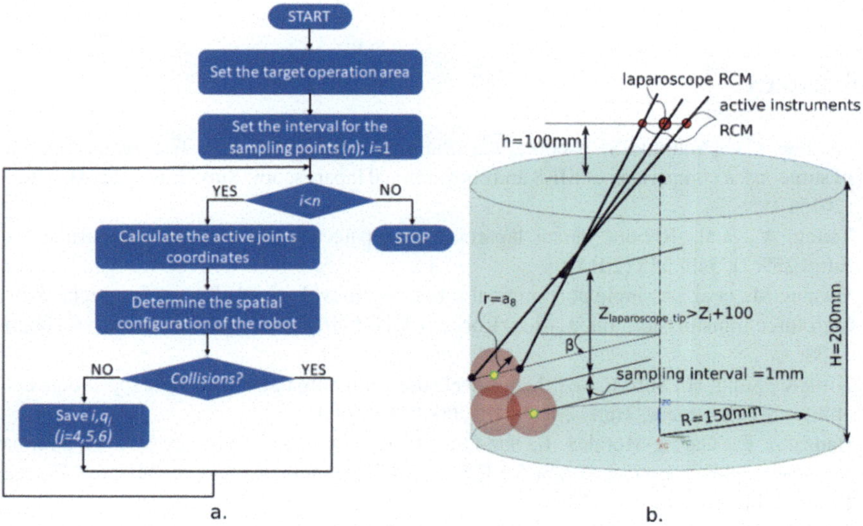

Fig. 3. a. The algorithm used to determine the collision free configuration of the robot. b. The imposed workspace used for the dimensional optimization of the robot [17].

Table 3. The results of the optimization process.

Parameter	α_3 (°)	d_{52} (mm)	a_7 (mm)	a_8 (mm)
LP value	0	440	430	-
IP value	25	425	380	90

5 Conclusions

Within the context of the accelerated development of robotic systems for minimally invasive surgery, the paper presents a robotic system for SILS based on a parallelogram architecture with three arms: one for guiding the laparoscope and the other two for the active surgical instruments. The inverse and direct kinematics have been analyzed, yielding analytic solutions, and considering that the teleoperation is performed using the Omega.7 haptic device. A dimensional optimization algorithm fitted for the specific SILS approach has been proposed, which yielded numerical values for the most important elements of the robotic system, as a starting point in the future development of the experimental model.

Acknowledgement. This work was supported by a grant of the Ministry of Research, Innovation and Digitization, CNCS/CCCDI – UEFISCDI, project number PCE171/2021 – Challenge within PNCDI III.

References

1. Alleblas, C.C., Velthuis, S., Nieboer, T.E., Sietses, C., Stegeman, D.F.: The physical workload of surgeons: a comparison of SILS and conventional laparoscopy. Surg. Innov. **22**(4), 376–381 (2015)
2. Tsung, A., et al.: Robotic versus laparoscopic hepatectomy: a matched comparison. Ann. Surg. **259**(3), 549–555 (2014)
3. Hwang, M., et al.: A single port surgical robot system with novel elbow joint mechanism for high force transmission. Int. J. Med. Robot. **13**(4), 1–10 (2017). https://doi.org/10.1002/rcs.1808
4. Kim, K., Song, H., Suh, J., Lee, J.: A novel surgical manipulator with workspace-conversion ability for telesurgery. Trans. Mech. **18**, 200–211 (2013)
5. Bauzano, E., Garcia-Morales, I., Saz-Orozco, P., Fraile, J.C., Muñoz, V.F.: A minimally invasive surgery robotic assistant for HALS–SILS techniques. Comput. Methods Programs Biomed. **112**(2), 272–283 (2013)
6. Choi, H., Kim, H., Lim, Y., Kwak, H., Jang, J., Won, J.: Conically shaped remote center-of-motion mechanism for single-incision surgery. Int. Conf. Intell. Robots. Syst. **2013**, 3604–3609 (2013)
7. Lefor, A.K.: Robotic and laparoscopic surgery of the pancreas: an historical review. BMC Biomed. Eng. **1**, 2 (2019). https://doi.org/10.1186/s42490-019-0001-4
8. Pisla, D., et al.: PARAMIS parallel robot for laparoscopic surgery. Chirurgia (Bucur) **105**(5), 677–83 (2010)

9. Vaida, C., et al.: Structural analysis and synthesis of parallel robots for brachytherapy. New. Trends. Med. Serv. Robots. **16**, 191–204 (2014)
10. Kong, K., Li, J., Zhang, H., Li, J., Wang, S.: Kinematic design of a generalized double parallelogram based remote center-of-motion mechanism for minimally invasive surgical robot. J. Med. Dev. **10**(4), 0410–06 (2016)
11. Pisla, D., et al.: New approach to hybrid robotic system application in single incision laparoscopic surgery. Acta. Tech. Nap. **64**(3), 369–378 (2021)
12. Pisla, D., et al.: A novel control architecture for robotic-assisted single incision laparoscopic surgery. Robot. J. Techn. Sci. **66**(2), 141–162 (2021)
13. Vaida, C., Andras, I., Birlescu, I., Crisan, N., Plitea, N., Pisla, D.: Preliminary control design of a single-incision laparoscopic surgery robotic system. In: 2021 25th International Conference on System Theory, Control and Computing (ICSTCC), pp. 384–389 (2021)
14. Force Dimension Gmbh. https://www.forcedimension.com. Accessed Feb 2022
15. Vaida, C., at al.: Family of modular robots for single incision laparoscopic surgery with kinematic remote center of motion. Patent Pending OSIM A00734/03.12.2021
16. Faria, C., Vilaça, J., Monteiro, S., Erlhagen, W., Bicho, E.: Automatic Denavit-Hartenberg parameter identification for serial manipulators. In: IECON 2019 - 45th Annual Conference of the IEEE Industrial Electronics Society, 2019, pp. 610–617 (2019)
17. Lee, H., Cheon, B., Hwang, M., Kang, D., Kwon, D.S.: A master manipulator with a remote-center-of-motion kinematic structure for a minimally invasive robotic surgical system. Int. J. Med. Robot. **14**(1), 1–10 (2018). https://doi.org/10.1002/rcs.1865

Inverse Dynamic Modeling of a Parallel Elbow Rehabilitation Robot for Spasticity Treatment

Alexandru Banica[1], Bogdan Gherman[1(✉)], Nicoleta Tohanean[2], Tiberiu Antal[1], Adrian Pisla[1], Cristian Abrudan[2], Giuseppe Carbone[3], and Doina Pisla[1]

[1] CESTER Technical University of Cluj-Napoca, Cluj-Napoca, Romania
Bogdan.GHERMAN@mep.utcluj.ro
[2] "Iuliu Hatieganu" University of Medicine and Pharmacy, Cluj-Napoca, Romania
[3] DIMEG, University of Calabria, Rende, Italy

Abstract. Given the increasing number of patients suffering from stroke related upper limb spasticity, there is a strong trend in the development of rehabilitation robotic systems. The paper focuses on the development of the inverse dynamic model for the ParReEx-elbow parallel robot, designed for elbow and forearm spasticity treatment. The model is based on the virtual work concept and the lumped masses paradigm. The generated model allows for a detailed examination of the ParReEx-elbow robot's dynamic capabilities and the generation of an accurate and efficient robotic-assisted protocol for spasticity treatment.

Keywords: Rehabilitation robot · Inverse dynamic model · Spasticity · Simulation

1 Introduction

In the medical field, the problem of rehabilitating patients which suffer from spastic hemiparesis presents itself as a huge challenge [1]. Spasticity is described as a velocity-dependent increase in tonic stretch reflex arising from stretch reflex over-excitation caused by upper motor neuron injuries [2]. It is common in people with post-stroke hemiplegia. Because of the growing number of patients in need of post-stroke recovery, medical robots are perceived as a viable solution, as they will facilitate faster and more efficient recovery protocols.

The method in which robots are perceived to work best, regarding spasticity treatment, is treating it by means of physical recuperation. According to [1], there exist three main characteristics that require improvement, which will generalize the process of assessing spasticity: the stages of the procedure, intensity and provided metrics.

Concerning physical rehabilitation using medical robots, patient-robot synergy is crucial [3]. The implementation of the dynamic model in the control system is required to better understand of how robotic rehabilitation robotic systems can provide a viable solution for the problem that is brought by spasticity. Several techniques have been used in developing comprehensible inverse dynamic models. The Lagrange-Euler formalism, which has been applied on parallel robots in [4], proving that this algorithm is useful for

A. Müller and M. Brandstötter (Eds.): RAAD 2022, MMS 120, pp. 392–400, 2022.
https://doi.org/10.1007/978-3-031-04870-8_46

closed-loop chains. Another favored technique is represented by the idea of virtual work [5], as well as the Newton-Euler technique [6]. The chosen technique for this paper is the virtual work method, taking advantage of its simplicity, leading to shorter computing time [7]. The goal of this paper is to develop the ParReEx-elbow control system capable of achieving an accurate torque monitoring, towards an increased accuracy and efficiency in the treatment of upper limb spasticity.

The layout of the paper contains, in Sect. 2, a description of the ParReEx-elbow parallel robotic system. Section 3 describes the inverse dynamic model, obtained applying the lumped masses technique. Finally, Sect. 4 focuses on the experimental results obtained in a laboratory environment, where a protocol for upper limb spasticity treatment was developed.

2 The ParReEx-Elbow Parallel Robot for Upper Limb Rehabilitation

The ParReEx-elbow was developed as a parallel robot [8, 9], with the purpose of performing elbow rehabilitation [10–12]. More precisely, it performs the flexion/extension of the forearm and the pronation/supination of the elbow complex. Figure 1 describes the kinematic scheme of the parallel robot, where: q_1 – flexion/extension active joint, α_1 - flexion/extension motion angle, q_2 – pronation/supination active joint, β_1 - pronation/supination motion angle, Rc - cardan coupling which helps achieve the pronation supination motion. X'Y'Z' represents the coordinates of the end effector.

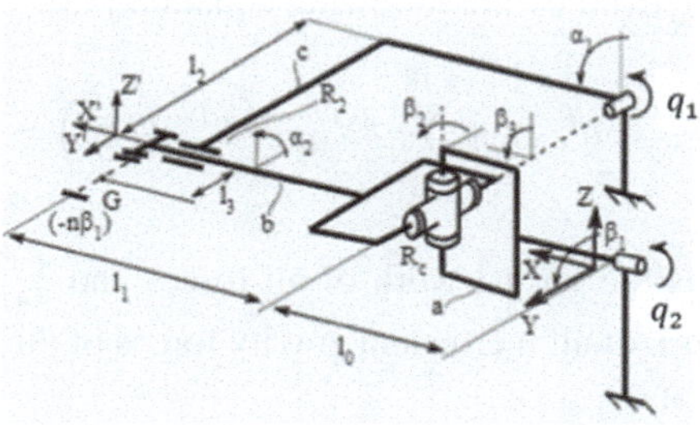

Fig. 1. ParReEx-elbow kinematic scheme

The main components of the ParReEx-elbow are presented in Fig. 2a and 2b and it is comprised of the following: 1.Flexion/extension motor; 2. Bevel gear mechanism for flexion/extension motion; 3. Flexion extension flange; 4. Flexion/extension connection rod; 5. Pronation/supination motor; 6. Cardan coupling; 7. Forearm support; 8. Wrist support; 9. Bevel gear mechanism for pronation/supination motion; 10. Hand anchor.

The experimental model has been designed and manufactured with a tendency towards a more lightweight structure, using machining and additive manufacturing processes (3D Printed). The control system has been designed and implemented in B&R Automation GMBH's Automation Studio [13]. The hardware control architecture connection includes: the PC/tablet connects to the PLC through VNC Viewer, which is connected to the driver, which controls the two servomotors that actuate the mechanical structure. Proximity sensors are employed for initialization as well as stroke limitations [14].

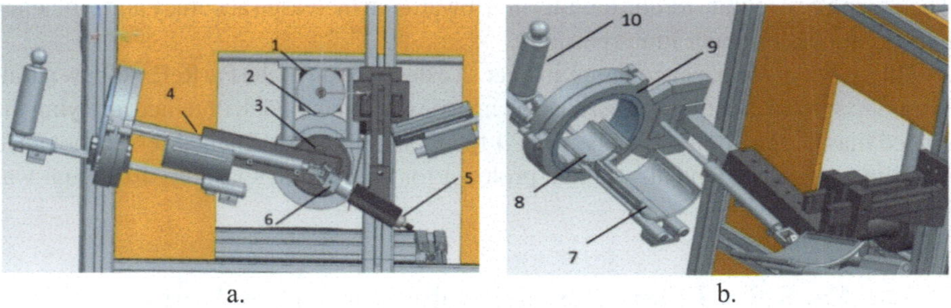

Fig. 2. Design of the ParReEx-elbow robot

3 Inverse Dynamic Model of the ParReEx-Elbow Robotic System

To define the inverse dynamic model, or more precisely, to determine the forces that act upon the active joints, several items of input are needed, such as the flexion, extension, pronation and supination motion ranges and motion parameters, which are characterized by speed and acceleration, all of which are defined in the robot's kinematics [15, 16]. It is also noted that friction forces will be ignored. The principle of virtual work was chosen, which is characterized by the following equation [13]:

$$\delta W = \delta q^T \cdot \tau + \sum_{i=1}^{18} \delta X_{Mi}^T \cdot \left(T_i^{T_r} + T_i^g\right) = 0 \tag{1}$$

where $\delta q^T \cdot \tau$ represents the virtual work of all forces and $\sum_{i=1}^{18} \delta X_{Mi}^T \cdot \left(T_i^{T_r} + T_i^g\right)$ represents the virtual work of all inertia and gravity forces of the robotic system. Inertia matrices are described as:

$$T_i^{T_r} = \begin{bmatrix} m_i^* & 0 & 0 \\ 0 & m_i^* & 0 \\ 0 & 0 & m_i^* \end{bmatrix} \begin{bmatrix} \ddot{X}_i \\ \ddot{Y}_i \\ \ddot{Z}_i \end{bmatrix} \tag{2}$$

The matrix of gravity forces is:

$$T_i^g = \begin{bmatrix} 0 \\ 0 \quad 0 \\ -m_i^* g_i \end{bmatrix} \tag{3}$$

$$T_i = T_i^{T_r} + T_i^g \tag{4}$$

To incorporate all the masses of the components in the inverse dynamic equations, the lumped masses principle was used, where the masses are concentrated into several different points (m_i^*), each body's mass being divided into 3 distinct points, one placed at the center of the body, while the other 2 placed at each end of the rigid. Figure 3 describes the visual representation of the lumped masses principle. The values of the new masses are (considering m being the mass of said body).

$$\frac{1}{6}m = m_1^*;\ \frac{2}{3}m = m_2^*;\ \frac{1}{6} = m_3^* \tag{5}$$

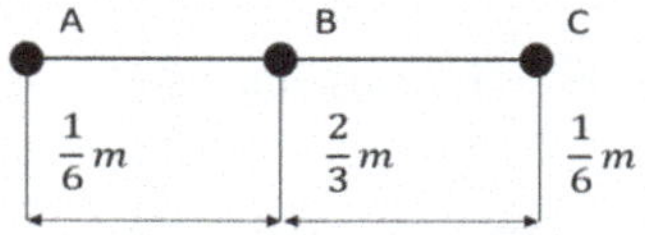

Fig. 3. Principle of lumped masses visual representation

The geometrical parameters which are needed for calculating the model are as follows: $D_i, i = 1..8$ - diameters of active components, $l_i, i = 1..54$- items of length that measure the distance of each equivalent mass to the CSYS(placed on the position of the flexion/extension motor), on each axis (Ox, Oy and Oz), m_1 - flexion/extension motor mass, m_2 flexion/extension upper bevel gear mass, m_3 - flexion/extension lower bevel gear mass, m_4 - flexion/extension flange mass, m_5 - flexion/extension rod mass, m_6 - pronation/supination motor mass, m_7 - cardan coupling mass, m_8 - forearm support mass, m_9 - wrist support mass, m_{10} - pronation/supination bevel gear system mass, m_{11} - hand anchor mass, m_{12} - pronation/supination rod, I_{mot1} - equivalent moment of inertia for flexion/extension motor, I_{mot2} - equivalent moment of inertia for pronation/supination motor, I_{r1} - moment of inertia for flexion/extension reductor, I_{r2} - moment of inertia for pronation/supination reductor, n_1 - transmission ratio for flexion/extension motor, n_2 - transmission ratio for pronation/supination motor, i_Δ - gyration Radius.

By using the lumped mass model theory, 18 equivalent masses have been chosen. Figures 4a and 4b reveal their distribution.

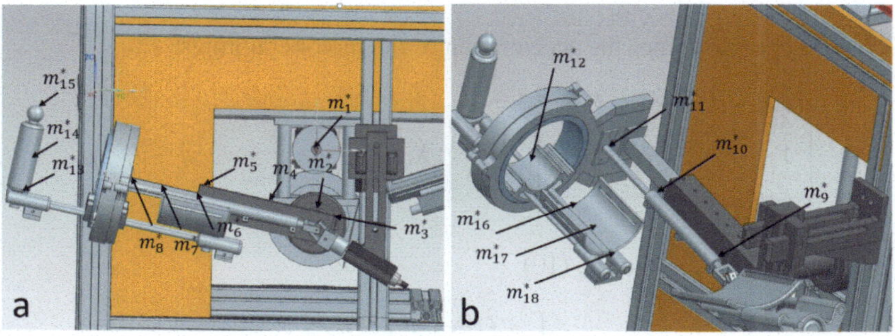

Fig. 4. Distribution of equivalent masses on the ParReEx-elbow parallel robot

The equivalent masses are:

$$\begin{cases} m_1^* = m_1; m_2^* = m_2; m_3^* = \frac{1}{6}m_5; m_4^* = \frac{2}{3}m_5; m_5^* = \frac{1}{6}m_5; \\ m_6^* = \frac{1}{6}m_{11}; m_7^* = \frac{2}{3}m_{11}; m_8^* = \frac{1}{6}m_{11}; m_9^* = \frac{1}{6}m_{12}; \\ m_{10}^* = \frac{2}{3}m_{12}; m_{11}^* = \frac{1}{6}m_{12}; m_{12}^* = m_{10}; m_{13}^* = \frac{1}{6}m_8; \\ m_{14}^* = \frac{2}{3}m_8; m_{15}^* = \frac{1}{6}m_8; m_{16}^* = \frac{1}{6}m_4; m_{17}^* = \frac{2}{3}m_4; m_{18}^* = \frac{1}{6}m_4 \end{cases} \tag{6}$$

The coordinates of each equivalent mass are as follows, considering the flexion extension motor is q_1 and the pronation/supination motor is q_2: $X_1 = 0; Y_1 = 0; Z_1 = 0;$

$$X_2 = -l_6 \sin q_2; Y_2 = l_6 \cos q_2 \sin q_1; Z_2 = -l_6 \cos q_1 \cos q_2;$$

$$X_3 = l_7 \cos q_2 + l_9 \sin q_2; Y_3 = l_7 \sin q_1 \sin q_2 - l_9 \cos q_2 \sin q_1; \tag{7}$$

$$Z_3 = l_9 \cos q_1 \cos q_2 - l_7 \cos q_1 \cos q_2$$

For $X_i, Y_i, Z_i, i = 4..18$, the position equations will be as such (also considering that $l_j, j = 10, 13, 16 \ldots 52$): $X_i = l_j \cos q_2 - l_{I+2} \sin q_2;$

$$Y_i = l_{j+1} \cos q_1 + l_{J+2} \cos q_2 \sin q_1 + l_j \sin q_1 \sin q_2 \tag{8}$$

$$Z_i = l_{J+1} \sin q_1 - l_{J+2} \cos q_1 \cos q_2 - l_j \cos q_1 \sin q_2$$

The following notations are used to define the equations in a compact form.: $q = [q_1, q_2]^T$-the position vector of the motor; $\delta_q = [\delta_{q1}, \delta_{q2}]^T$- virtual displacement of the motor position; $\dot{q} = [\dot{q}_1, \dot{q}_2]^T$ – velocity vector; $\ddot{q} = [\ddot{q}_1, \ddot{q}_2]$ – acceleration vector; $X_{Mi} = [X_i, Y_i, Z_i]^T, i = 1..18$ – vector of the concentrated mass points in M_i; $X_i, Y_i, Z_i, i = 1..18$ – coordinates of the concentrated mass pointsm_i^*; $X_i = [X_{Mi}, Y_{Mi}, Z_{Mi}]^T$- M_i position vector; $\delta X_i = [\delta X_{Mi}, \delta Y_{Mi}, \delta Z_{Mi}]^T$- virtual displacement vector of the M_i coordinates; $\dot{X}_i = [\dot{X}_{Mi}, \dot{Y}_{Mi}, \dot{Z}_{Mi}]^T$-velocity vector of the M_i point.; $\ddot{X}_i = [\ddot{X}_{Mi}, \ddot{Y}_{Mi}, \ddot{Z}_{Mi}]^T$-acceleration vector of the M_i point; $\tau = [\tau_1, \tau_2]^T = [F_1, F_2]^T$-motor forces vector.

Equation (9) determines the acceleration of the motor couplings:

$$\begin{bmatrix} \dot{X}_{Mi} \\ \dot{Y}_{Mi} \\ \dot{Z}_{Mi} \end{bmatrix} = \begin{bmatrix} \frac{\delta X_{Mi}}{\delta q_1} & \frac{\delta X_{Mi}}{\delta q_2} \\ \frac{\delta Y_{Mi}}{\delta q_1} & \frac{\delta Y_{Mi}}{\delta q_2} \\ \frac{\delta Z_{Mi}}{\delta q_1} & \frac{\delta Z_{Mi}}{\delta q_2} \end{bmatrix} \cdot \begin{bmatrix} \dot{q}_1 \\ \dot{q}_2 \end{bmatrix}; \dot{X}_i = J_i \cdot \dot{q}; \ddot{X}_i = J_i \cdot \ddot{q} + \dot{J}_i \cdot \dot{q} \tag{9}$$

We define the matrix $\dot{J}_i$ as in Eq. (10):

$$\dot{J}_i = \begin{bmatrix} \frac{d}{dt}\left(\frac{\delta X_g}{\delta q_1}\right) & \frac{d}{dt}\left(\frac{\delta X_g}{\delta q_2}\right) \\ \frac{d}{dt}\left(\frac{\delta Y_g}{\delta q_1}\right) & \frac{d}{dt}\left(\frac{\delta Y_g}{\delta q_2}\right) \\ \frac{d}{dt}\left(\frac{\delta Z_g}{\delta q_1}\right) & \frac{d}{dt}\left(\frac{\delta Z_g}{\delta q_2}\right) \end{bmatrix} \tag{10}$$

The elements of the $\dot{J}_i$ matrix are (for i = 1..18):

$$\begin{aligned}
\frac{d}{dt}\left(\frac{\delta X_{Mi}}{\delta q_1}\right) &= \frac{\partial^2 X_{Mi}}{\partial q_1^2}\dot{q}_1 + \frac{\partial^2 X_{Mi}}{\partial q_1 \cdot \partial q_2}\dot{q}_2;\ \frac{d}{dt}\left(\frac{\delta X_{Mi}}{\delta q_2}\right) = \frac{\partial^2 X_{Mi}}{\partial q_1 \cdot \partial q_2}\dot{q}_1 + \frac{\partial^2 X_{Mi}}{\partial q_2^2}\dot{q}_2 \\
\frac{d}{dt}\left(\frac{\delta Y_{Mi}}{\delta q_1}\right) &= \frac{\partial^2 Y_{Mi}}{\partial q_1^2}\dot{q}_1 + \frac{\partial^2 Y_{Mi}}{\partial q_1 \cdot \partial q_2}\dot{q}_2;\ \frac{d}{dt}\left(\frac{\delta Y_{Mi}}{\delta q_2}\right) = \frac{\partial^2 Y_{Mi}}{\partial q_1 \cdot \partial q_2}\dot{q}_1 + \frac{\partial^2 Y_{Mi}}{\partial q_2^2}\dot{q}_2 \\
\frac{d}{dt}\left(\frac{\delta Z_{Mi}}{\delta q_1}\right) &= \frac{\partial^2 Z_{Mi}}{\partial q_1^2}\dot{q}_1 + \frac{\partial^2 Z_{Mi}}{\partial q_1 \cdot \partial q_2}\dot{q}_2;\ \frac{d}{dt}\left(\frac{\delta Z_{Mi}}{\delta q_2}\right) = \frac{\partial^2 Z_{Mi}}{\partial q_1 \cdot \partial q_2}\dot{q}_1 + \frac{\partial^2 Z_{Mi}}{\partial q_2^2}\dot{q}_2
\end{aligned} \tag{11}$$

Based on the previous equation, we can rewrite the acceleration matrix as:

$$\begin{bmatrix} \ddot{X}_{Mi} \\ \ddot{Y}_{Mi} \\ \ddot{Z}_{Mi} \end{bmatrix} = \begin{bmatrix} \frac{\partial X_{Mi}}{\partial q_1} & \frac{\partial X_{Mi}}{\partial q_2} \\ \frac{\partial Y_{Mi}}{\partial q_1} & \frac{\partial Y_{Mi}}{\partial q_2} \\ \frac{\partial Z_{Mi}}{\partial q_1} & \frac{\partial Z_{Mi}}{\partial q_2} \end{bmatrix} \cdot \begin{bmatrix} \ddot{q}_1 \\ \ddot{q}_2 \end{bmatrix} + \begin{bmatrix} \frac{d}{dt}\left(\frac{\delta X_{Mi}}{\delta q_1}\right) & \frac{d}{dt}\left(\frac{\delta X_{Mi}}{\delta q_2}\right) \\ \frac{d}{dt}\left(\frac{\delta Y_{Mi}}{\delta q_1}\right) & \frac{d}{dt}\left(\frac{\delta Y_{Mi}}{\delta q_2}\right) \\ \frac{d}{dt}\left(\frac{\delta Z_{Mi}}{\delta q_1}\right) & \frac{d}{dt}\left(\frac{\delta Z_{Mi}}{\delta q_2}\right) \end{bmatrix} \cdot \begin{bmatrix} \dot{q}_1 \\ \dot{q}_2 \end{bmatrix} \tag{12}$$

Subsequently, the first equation can be written as:

$$W = \delta q^T \cdot \tau + \sum_{i=0}^{18} \delta q^T \cdot J_i^T \cdot \left(T_i^{Tr} + T_i^g\right) = 0 \tag{13}$$

The forces and inertia vectors are finally computed:

$$\tau = -\sum_{i=0}^{54} J_i^T \cdot \left(T_i^{Tr} + T_i^g\right) \tag{14}$$

4 Implementation of Inverse Dynamic Model and Experimental Results

A protocol for spasticity treatment has been established, using the ParReEx-elbow robotic system, undergoing testing in a laboratory environment, using healthy patients, and simulating average levels of spasticity in the upper limb:

1. The protocol begins with measurements of the affected limb with the dynamometer. The necessary torque will be determined at the level of the patient's elbow (T_M). The resulting torque will then be added to the necessary torque calculated by the inverse dynamic model (T_D), yielding T_{Max}. An initial trajectory (extension angle, 0° in our experiment) is imposed.
2. The patient's limb will be mounted on the ParReEx-elbow robot and the rehabilitation process will begin from an initial position (45° degrees in our experiment)
3. The robot will then start the rehabilitation motion, using the planned trajectory. The estimated torque is read (Read_actual_torque() from B&R Automation Studio).

   3a. If the torque will stay beneath the maximum accepted torque value (T_{Max}), the robot will perform 2 repetitions and the patient will be dismounted afterwards.

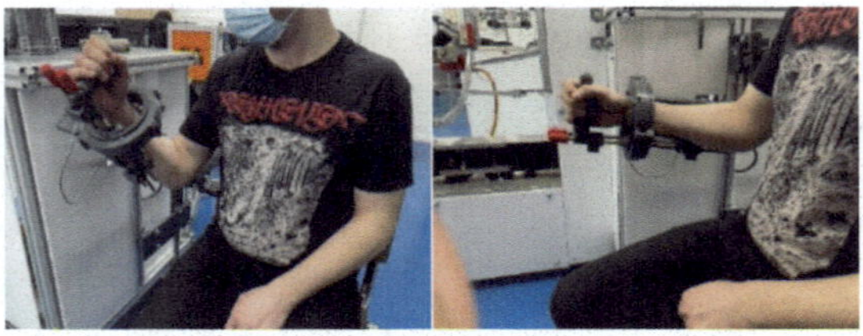

Fig. 5. ParReEx-elbow testing with a healthy subject.

3b. If the torque exceeds the maximum value, the robot will stop and revert to the initial position and will begin the next repetition (up to maximum 2 times) (Fig. 5).

The tests compared the torque needed for a full flexion/extension exercise for the upper limb. The simulated exercises utilized the imposed torque determined by a dynamometer. It is assumed that the healthy subject can move his arm freely within the maximum range of motion of the robot (45° for elbow flexion). Afterwards, an upper limb that simulated ordinary upper limb spasticity used the ParReEx-elbow robot and the necessary torque for a full flexion extension exercise was estimated (T_{Est}).

The graph described by Fig. 6 demonstrates that the maximum torque admitted for spasticity treatment cannot exceed $T_{Max} = 6Nm$. The velocity of the motion has been set to $\dot{q}_1 = 10°/s$.

Figure 7 describes the torque control architecture of the ParReEx-elbow robotic system. As the inverse dynamic model requires the input values of the position, velocity, and acceleration of the active joints, it will calculate the required torque for the rehabilitation motion. Adding the torque from the patient dynamics to the previous value, we will obtain the maximum torque value for each patient which will vary depending on the level of spasticity that afflicts the upper limb. T_{est} represents the value estimated by the Read_Actual_Torque() B&R Automation function, which will be used for monitoring the spasticity treatment exercised (it will be compared to the maximum admitted torque $T_{Max} = 6\,\text{Nm}$).

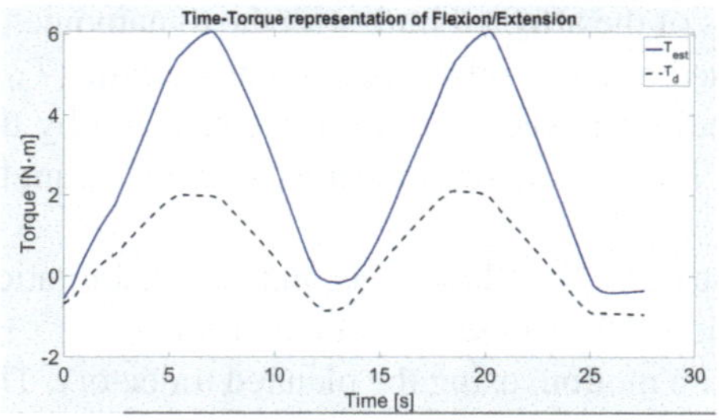

Fig. 6. T_D and T_{Est} comparison graph

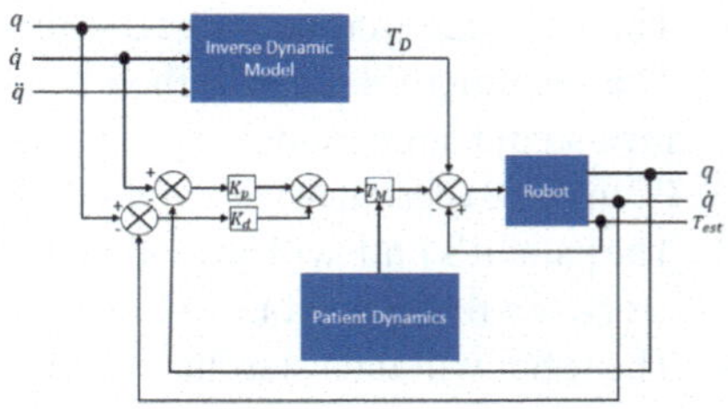

Fig. 7. Torque control architecture

5 Conclusion

The purpose of this work is to establish a comprehensive protocol for the treatment upper limb spasticity by using the ParReEx-elbow robotic system. To accomplish this purpose, the inverse dynamics of the ParReEx-elbow have been developed by combining the virtual work principle with the lumped masses model, which has proved to result in quick computations. A control strategy has been proposed to accurately monitor the required torque for the spasticity treatment exercises. The model was validated using the predicted torque values from Automation Studio for the two actuators.

Acknowledgement. This research was funded by a grant from the Romanian Ministry of Research and Innovation, CCCDI—UEFISCDI, project number PN-III-P2-2.1-PED-2019-3022/546PED/2020 (NeuroAssist) within PNCDI III, and by the project POCU/380/6/13/123927-ANTREDOC, "Entrepreneurial competencies and excellence research in doctoral and postdoctoral studies programs", co-funded from the European Social Fund through the Human Capital Operational Program 2014–2020.

References

1. Torre, R., Oña, E.D., Balaguer, C., Jardon, A.: Robot-aided systems for improving the assessment of upper limb spasticity: a systematic review. Sensors **20**(18), 5251 (2020)
2. Sheean, G.: The pathophysiology of spasticity. Eur. J. Neurol. **9**, 3–9 (2002)
3. Carbone, G., Gherman, B., Ulinici, I., Vaida, C., Pisla, D.: Design issues for an inherently safe robotic rehabilitation device. Mech. Mach. Sci. **49**, 1025–1032 (2018)
4. Abdelatiff, H., Heimann, B.: Heimann, B: Computational efficient inverse dynamics of 6 DOF fully parallel manipulators by using the Lagrangian formalism. J. Mech. Mach. Theory **44**(1), 192–207 (2009)
5. Codourey, A., Burdet, E.: A body oriented method for finding a linear form of the dynamic equations of fully parallel robot. In: Conference on Robotics and Automation, pp.1612–1619, IEEE Albuquerque, New Mexico, U.S. (1997)
6. Dasgupta, B., Mruthyunjaya, T.S.: Closed-form dynamic equations of the general Stewart platform through the Newton-Euler approach. J. Mech. Mach. Theory **33**(7), 993–1012 (1998)
7. Cheng, H.,Yiu, Y.K., Li, Z.: Dynamic analysis of a modified DELTA parallel robot for cardiopulmonary resuscitation. In: Proceedings of the International conference on Intelligent Robots and Systems, pp. 233–288 (2005)
8. Pisla, D. Plitea, N., Vidrean, A., Prodan, B., Gherman, B: Kinematics and design of two variants of a reconfigurable parallel robot. In: ASME/IFToMM International Conference on Reconfigurable Mechanisms and Robots, pp. 624–631. IEEE, London, UK (2009)
9. Pisla, D., Husty, M., Birlescu, I., Tucan, P., Vaida, C.: An algebraic parameterization approach for parallel robots analysis. Mech. Mach. Theory. Pergamon **140**, 245–257 (2019)
10. Tucan, P., et al.: Fuzzy logic-based risk assessment of a parallel robot for elbow and wrist rehabilitation. Int. J. Environ. Res. Public Health **17**(2), 654 (2020)
11. Gherman, B., et al.: Parallel robotic system for the medical rehabilitation of the upper limb. Patent OSIM: RO132234 -B1/30.03.2020
12. Tucan, P., Vaida, C., Plitea, N., Pisla, A., Carbone, G., Pisla, D.: Risk-based assessment engineering of a parallel robot used in post-stroke upper limb rehabilitation. Multidiscipl. Digital Publ. Inst. **11**(10), 2893 (2019)

13. B&R Automation. https://www.br-automation.com/en/. June 2021
14. Birlescu, I., et al.: On the singularities of a parallel robotic system used for elbow and wrist rehabilitation. In: Lenarcic, J., Parenti-Castelli, V. (eds.) ARK 2018. SPAR, vol. 8, pp. 203–211. Springer, Cham (2019). https://doi.org/10.1007/978-3-319-93188-3_24
15. Plitea, N., et al.: Kinematic analysis of an exoskeleton-based robot for elbow and wrist rehabilitation. In: Carvalho, J.C.M., Martins, D., Simoni, R., Simas, H. (eds.) Multibody Mechatronic Systems. MMS, vol. 54, pp. 424–433. Springer, Cham (2018). https://doi.org/10.1007/978-3-319-67567-1_40
16. Gherman, B., Pisla, D., Vaida, C., Plitea, N.: Development of inverse dynamic model for a surgical hybrid parallel robot with equivalent lumped masses. Robot. Comput. Int. Man. **2**(3), 402–415 (2012)

Autonomous Mobile Manipulation

Flexible Vision-Based Auto-docking Control System for Unmanned Ground Vehicles Equipped with Differential Chassis

Andrey Vukolov(✉), George Kourousias, and Roberto Pugliese

Elettra Sincrotrone Trieste S.C.p.A., Strada Statale 14 - 163.5km, AREA Science Park, 34149 Basovizza, Trieste, Italy
andrey.vukolov@elettra.eu
https://elettra.eu

Abstract. This paper proposes configurable, parametrically flexible, vision-based automatic docking control system for unmanned surface vehicles (mobile robots) equipped with a differential chassis. The proposed system uses only fiducial localization source to find and attach the robot to docking station. It exploits peculiarities of multiple tags setups to obtain final push-in docking motion straight with high geometrical precision. The trajectories with inverse curvatures are implemented to decrease the number of docking attempts with respect to the actual orientation of the charging slot.

Keywords: Mobile robot · Automatic docking · Charging station · Indoor UGV · Differential chassis

1 Introduction

Docking to the charging station is a very well-known problem in development of mobile robots. The algorithms used to solve this problem should implement a compromise solution between geometrical precision, speed and repeatability of the process [1]. Efficiency in terms of energy consumption is also an important factor of the development for such a system because the docking process is run when the robot/UGV's batteries are in low charge state. This requires very high (usually over 95%) probability of successful docking to prevent the robot from being shut off unexpectedly by the battery safety control system. According to these requirements, the auto-docking control system should be able to precisely localize the robot in the relative position from the charging station, and then to guide the chassis in real time during the docking operation. In such a case the general-purpose positioning systems may not provide the desired robustness and repeatability for docking, especially in case of absence of object snapping technologies. There are numerous ways to solve the problems described here and most of them introduce the different positioning system to localize the robot

A. Müller and M. Brandstötter (Eds.): RAAD 2022, MMS 120, pp. 403–411, 2022.
https://doi.org/10.1007/978-3-031-04870-8_47

in relative position to the charger. There are systems utilizing infrared beacons [14,15], ultrasonic transceivers [7,8], RFID markers and radio frequency beacons [5,16], depth scans [11]. It is also possible to implement a combined sensory system but with additional requirement of sophisticated data filtering to fuse measurements [20].

Previous research demonstrated [1–3] that the vision-based methods provide more robustness than the others. But they have their own cost in computational resources, especially in case of high resolution image processing. The vision-based methods could be divided contingently into the two large groups: non-deterministic methods utilizing AI solutions and neural network models [10] and deterministic marker-driven methods based on fiducial recognition. For automatic docking the fiducial control systems have more perspectives to use because the general purpose mobile robot during the docking operates in semi-determined environment.

High repeatability, robustness and success probability required for automatic docking can be obtained in the marker-driven systems using different approaches. The universal systems with high amount of available computational power may use high resolution 3D reconstruction and stereo vision with multiple cameras [17]. Less sophisticated systems dedicated for general purpose and industrial mobile robotics mostly use tag-driven computer vision software such as ArUco and AprilTag [6]. These systems can improve repeatability and robustness of the control using multiple tags [19]. Here the peculiarities of the given chassis and position of the charging slot on the robot's case creates the group of special cases for implementation of the automatic docking. For example the charging system equipped with flat vertical contact jaw requires high repeatability of the docking trajectory, at least on the last stage. The control system proposed in this paper implements the inverse curvature approach to automatically generate trajectory for the successful docking with the only requirement of tags the robot's camera is able to recognize.

2 Target Setting

The proposed docking control system is implemented over the lightweight (less than 60 kg) indoor UGV equipped with single frontal camera. Actually it could be adopted with ease on any ROS 1 compatible UGV with differential chassis. The charging station has a guiding rail, flat vertical magnetic contact jaw and the space for tags above the case of the charger. The Microsoft LifeCam HD-3000 camera (1280×720, 10 mm focal length, 45° aperture) is installed in the frontal wall of the robot's case. The charger is equipped with two predefined tags which are IDs 51 and 52 from AprilTag `t36h11` series. The main problem of this vendor-provided setup is that the predefined tags are partially out of the camera's vision area when the robot is docked (Fig. 1, *a*). The onboard computer of the robot is Intel NUC 7i7BNH equipped with Mint Linux 20.02 and ROS 1[1]. ROS has an open source library [9,18] for recognition of AprilTag markers.

[1] http://ros.org/.

Thus the target setting for this paper is development of ROS 1-based fiducial automatic docking control system for setup with AprilTag markers partially out of sight and vertical flat contact jaw located in the center of the robot's case front wall. The target repeatability is selected more than 90% with 95% probability of success for both docking and undocking operations. The additional tasks were also set to provide configurability and parametric flexibility of the system.

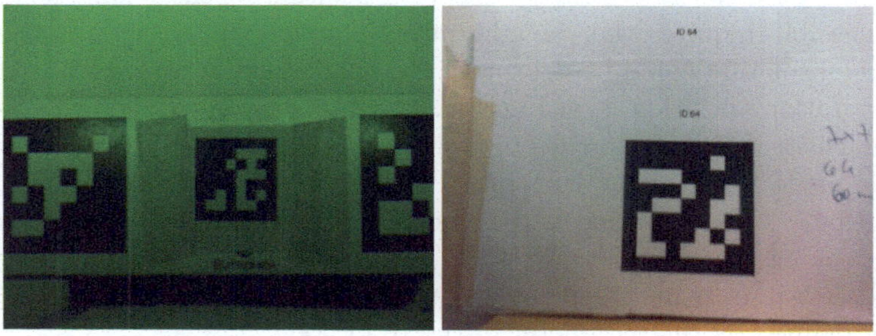

a *b*

Fig. 1. Camera vision area setup with AprilTag (*a*) and ArUco (*b*) markers

3 Docking Strategy and Algorithm

Usage of multiple tags to improve geometrical precision of the docking is widely spread practice [6,19]. The proposed control system uses the set containing 3 AprilTag markers with known IDs [9].

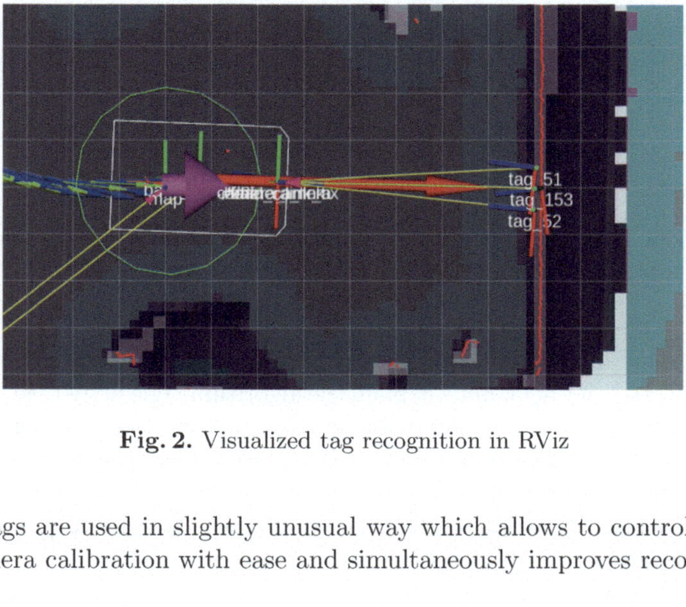

Fig. 2. Visualized tag recognition in RViz

The tags are used in slightly unusual way which allows to control the quality of camera calibration with ease and simultaneously improves recognition on closer distances. At first, the central tag is added to the setup to avoid the cases when one of the tags occasionally went away from the frame. As it is shown on Fig. 1 the central tag is printed smaller than the left and right ones. The actual distances between tags and camera are measured by AprilTag library and they should be observable the same when visualized in RViz (Fig. 2) to ensure the camera is well calibrated. This also allows the user to configure AprilTag library with right sizes of the tags.

The proposed docking strategy respects position of the connection jaw on robot's case. It requires the straight push-in action in the final stage of the docking. This action could be driven by mechanical guide placed in front of the charging station (position 2 on Fig. 3[2]). But the self-adjustment wheels of the UGV can easily get stuck on the guide (as it was observable in the experiments) when the angle between the lengthwise axis of the robot and charging station is more than 15° (the value is determined empirically during the tests with manual docking controlled by joystick). Then the decision was made to perform the final push-in only in straight lengthwise direction with configurable angular precision defined by the parameter in the ROS node.

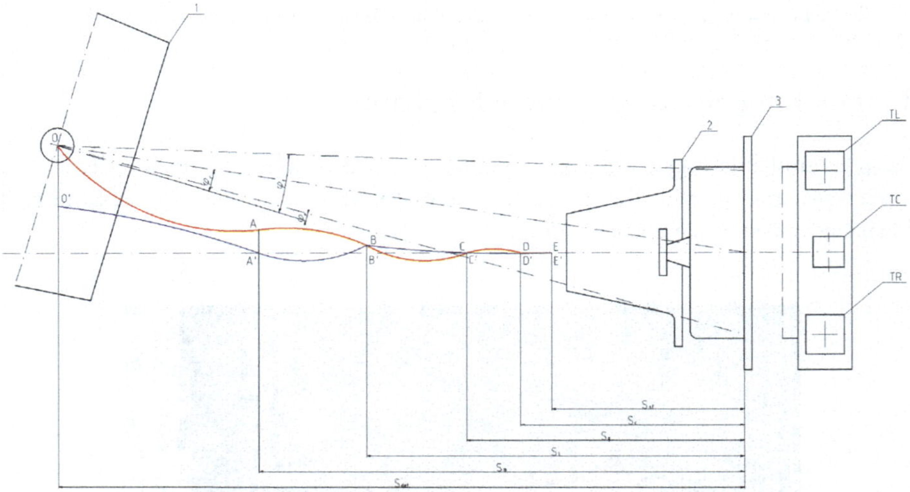

Fig. 3. Docking strategy. 1 – robot, 2 – guide, 3 – charging station, red line—forward movement trajectory, blue line—backward movement trajectory

The strategy divides the docking movements of the robot into 5 stages with lengths indicated on Fig. 3. Let's explain the algorithm. The first stage, *detection-driven adjustment*, starts at the distance S_{det}. In this stage the robot adjusts itself to drive into the position suitable to decrease further the angle ϕ_C between

[2] All distances are shown snapped to the robot's main coordinate system, usually called `base_link`.

robot and charger's lengthwise axis. Also the assumption is made that the central tag TC is placed to indicate position of this axis for the charger. The robot in this stage keeps the curvature radius of the trajectory directed side away from the charger (as the red line on Fig. 3 shows). Both left TL and right TR tags should be presented. The robot then decreases the angular distance to the closest side (*Left* or *Right*) tag driving forward with constant speed and keeping the opposite tag in sight. In the case illustrated on Fig. 3 the left tag is closest, so the robot drives towards it keeping the right tag in sight and the curvature radius of trajectory is directed towards the left tag too. The robot proceeds as well by the line OA until the second stage *landing maneuver* will start at the distance S_m. S_m is internally calculated distance and it actually should be defined as heuristic function that depends from the position of robot's default coordinate center within the case projected to the ground plane (O point on Fig. 3). In the landing maneuver the robot inverts the direction of the trajectory's curvature radius driving towards the distant tag (right tag TR and line AB on Fig. 3) keeping signs of the angles ϕ_l and ϕ_r different. This is needed to reduce the distance between the geometrical center of the robot and charger's lengthwise axis without loss of detection for one of the tags. When the distance S_l reached, the robot measures the *landing angle* $\phi_{landing}$ from ϕ_l and ϕ_r using approximation:

$$\phi_{landing} = \arctan\left(\frac{|l_{O-TL} - l_{O-TR}|}{\phi_r l_{O-TR} + \phi_l l_{O-TL}}\right) \quad (1)$$

After this measurement the decision is made about the success of the current docking attempt. If $\phi_{landing}$ is more than the preconfigured precision parameter (taken from the ROS node parameters list), the attempt is considered insufficient and the robot starts to move backwards in *reverse* mode which will be described below. Until the decision is not made, the recognition of the central tag is considered optional.

In case of sufficient attempt, the robot then enters the third, *landing* stage in which the guidance involves the central tag TC. The central tag should now be visible to the camera, if not, the algorithm fails. Moving towards the charger by the line BC the robot strictly keeps the signs of ϕ_l and ϕ_r angles different stopping the forward motion to adjust if these angles have the same signs. In addition, the robot on this stage constantly oscillates around the line O–TC similar to that the vacuum cleaning robots do [13] to keep the angle ϕ_c non-zero. The robot proceeds until entering the fourth, *guidance* stage at the distance S_g moving by the line CD. On this distance the detection for left and right tags is switched off and the robot starts to rely only to the central tag TC. This prevents the cases when AprilTag library loses the detection while the tags' images are too big and/or out of the frame. Also here the mechanical guide 2 is assumed to start working. The robot proceeds tracking continuously the central tag until the *connection* distance S_c is reached. Then the robot enters the last, *connection* stage and the final push-in starts straight ahead until the robot will connect to the charger or reach the safety distance S_{sf} where the motors will be forced to stop. In normal operation, the point E is not actually reachable.

3.1 Reverse Motion and Undocking

The reverse motion algorithm is designed to decrease the distance between the robot's position on the distance S_{det} and charger's lengthwise axis. The trajectory is traced back on Fig. 3 with the blue line $B'O'$. The robot starts to move backwards keeping the maximal distance for *most distant* tag (the right tag TR on Fig. 3) but preserving both left and right tags in the consequent half of the view field according to angles ϕ_l and ϕ_r. When the distance S_{det} is reached, the robot then aligns itself to the *closest* tag striving to hold it in the center of the view field as precise as possible. After the robot reaches the distance S_{det}, the reverse motion is turned off and the next attempt of docking starts then. In the similar way the **undocking** algorithm is implemented but before activating the described algorithm of reverse motion, the robot reaches distance S_l moving backwards and using only central tag TC to align itself by the line $E'B'$.

4 Implementation and Experiments

The described algorithms were implemented in C++ using latest available version of ROS 1 framework and existing AprilTag compatibility layer [9,12,18]. The system is designed keeping in mind the high efficiency requirements. It utilizes peculiarities of ROS internal API to load and unload AprilTag libraries on demand to prevent high CPU load. The compiled code is also optimized to exploit the possibilities granted by Intel Core i7 7567 SkyLake micro architecture. The system is contained within the daemonized ROS node exploiting feedback API provided by `actionlib` library. The ROS node issues the commands in ROS standard way using publishing on `/cmd_vel` topic on constant configurable frequency. All distances except S_m (Fig. 3) are also configurable. The motion velocities recorded into the `/cmd_vel` message structure are constant and configurable too because they allow the user to apply the actual motor characteristic to the docking process. The system inherits AprilTag compatibility layer configuration file and loads it automatically configuring IDs for TL, TR and TC tags independently. Also the rotation was added to initially find the presence and relative position of the tags around the robot. The system was installed on the robot's internal computer as the single ROS package with standard script for `roslaunch` the internal ROS launch server. ROS environment collects the parameters from the internal parameter server to determine the distances and precision factor described above.

The comparison experiments were then conducted to obtain overall characteristics of the system, especially probability of successful docking. The comparison was between the proposed system and Rover Robotics's Python-written vision-based auto docking system[3] utilizing ArUco tags [4], also compatible with existing UGV setup without changing. Every experiment included 3 blocks with 10 attempts each from random initial position with only two constraints: the distance between the robot and the charger $S_{det} = 1.5\,\text{m}$ and at least one of the

[3] http://wiki.ros.org/rr_auto_dock.

tags within view field of the camera. During the experiment the docking time was measured and only the cases where the robot actually started the charging process were considered successful. When Rover Robotics system was tested, the AprilTag markers were replaced with single ArUco tag in the same position with *TC* (Fig. 3) sized 60 mm on side (Fig. 1, *b*). The `roslaunch` XML scripts were created to run the tested systems independently on the same machine. Results of the experiments are presented in Table 1.

Here it should be mentioned that most of the Rover Robotics' system faults were conditioned by large angle between robot's and charger's lengthwise axis. In most of such cases the Rover Robotics' system has not obtained the electric contact between the charger and robot's contact jaw and then it dropped the attempt without prompt. Also there were the cases when ArUco marker of the given size was unrecognized by the system on close distances when the smaller marker was not recognized on the suitable S_{det}.

Table 1. Comparison experiment results

Docking system	Successful attempts	Total attempts	Average time [mm:ss]	Minimal time [mm:ss]
Rover Robotics	12	30	00:45	00:12
Proposed system	29	30	01:25	00:30

5 Conclusion

In comparison with Rover Robotics' vision-based system the proposed system presented itself as reliable solution with the probability of success ~96%. The inverse curvature approach and usage of *landing angle* (1) actually eliminated the main reason of fault in cost of increasing docking time. Also usage of the markers with different sizes eliminated the problem with bad recognition on close distances. The system shown average docking time about minute and a half which can be considered acceptable for machine cruising on speed ~0.6 m/s in industrial environment with presence of people. ROS package structure and API provides compatibility layer for conjoined usage with the other control software. The system is designed to yield the computational resources to more general control tasks loading and unloading AprilTag libraries on demand.

Acknowledgements. The authors acknowledge partial support by VisionQub.IT S.R.L. that provided a case study and discussion on computer vision, robotics and logistics.

References

1. Abdulov, A., Abramenkov, A.: Research of visual positioning system for mobile robots remote automatic control. In: 2020 International Conference on Industrial Engineering, Applications and Manufacturing (ICIEAM), pp. 1–6 (2020). https://doi.org/10.1109/ICIEAM48468.2020.9112035
2. Aqel, M.O., Marhaban, M.H., Saripan, M.I., Ismail, N.B.: Review of visual odometry: types, approaches, challenges, and applications. SpringerPlus **5**(1), 1–26 (2016). https://doi.org/10.1186/s40064-016-3573-7
3. Badue, C., et al.: Self-driving cars: a survey. Expert Syst. Appl. **165**, 113816 (2021). https://doi.org/10.1016/j.eswa.2020.113816. https://www.sciencedirect.com/science/article/pii/S095741742030628X
4. Garrido-Jurado, S., Muñoz-Salinas, R., Madrid-Cuevas, F., Marín-Jiménez, M.: Automatic generation and detection of highly reliable fiducial markers under occlusion. Pattern Recogn. **47**(6), 2280–2292 (2014). https://doi.org/10.1016/j.patcog.2014.01.005. https://www.sciencedirect.com/science/article/pii/S0031320314000235
5. Kim, M., Kim, H.W., Chong, N.Y.: Automated robot docking using direction sensing RFID. In: Proceedings 2007 IEEE International Conference on Robotics and Automation, pp. 4588–4593 (2007). https://doi.org/10.1109/ROBOT.2007.364186
6. Krejsa, J., Vechet, S.: Evaluation of visual markers detection used for autonomous mobile robot docking navigation. In: Szewczyk, R., Krejsa, J., Nowicki, M., Ostaszewska-Liżewska, A. (eds.) MECHATRONICS 2019. AISC, vol. 1044, pp. 229–236. Springer, Cham (2020). https://doi.org/10.1007/978-3-030-29993-4_29
7. Li, D., Fu, H., Wang, W.: Ultrasonic based autonomous docking on plane for mobile robot. In: 2008 IEEE International Conference on Automation and Logistics, pp. 1396–1401 (2008). https://doi.org/10.1109/ICAL.2008.4636372
8. Luo, R.C., Huang, C.H., Huang, C.Y.: Search and track power charge docking station based on sound source for autonomous mobile robot applications. In: 2010 IEEE/RSJ International Conference on Intelligent Robots and Systems, pp. 1347–1352 (2010). https://doi.org/10.1109/IROS.2010.5649993
9. Malyuta, D.: Guidance, navigation, control and mission logic for quadrotor full-cycle autonomy. Master thesis, Jet Propulsion Laboratory, 4800 Oak Grove Drive, Pasadena, CA 91109, USA, December 2017
10. Muse, D., Weber, C., Wermter, S.: Robot docking based on omnidirectional vision and reinforcement learning. In: Bramer, M., Coenen, F., Allen, T. (eds.) Research and Development in Intelligent Systems XXII, pp. 23–36. Springer, London (2006). https://doi.org/10.1007/978-1-84628-226-3_3
11. Poomarin, W., Chancharoen, R., Sangveraphunsiri, V.: Automatic docking with obstacle avoidance of a differential wheel mobile robot. Int. J. Mech. Eng. Robot. Res. **5**(1), 11 (2016)
12. Quigley, M., et al.: ROS: an open-source robot operating system. In: ICRA Workshop on Open Source Software, Kobe, Japan, vol. 3, p. 5 (2009)
13. Quilez, R., Zeeman, A., Mitton, N., Vandaele, J.: Docking autonomous robots in passive docks with infrared sensors and QR codes. In: International Conference on Testbeds and Research Infrastructures for the Development of Networks and Communities (TridentCOM), Vancouver, Canada (2015). https://hal.inria.fr/hal-01147332
14. Raj, K.V., Patil, K., Kariappa, D.K., Jakati, A.M.: A beacon-based docking system for an autonomous mobile robot. In: National Conference on Mechanisms and Machines (NaCoMM07), India, pp. 1–7. Citeseer (2007)

15. Rao, M.S., Shivakumar, M.: IR based auto-recharging system for autonomous mobile robot. J. Robot. Control (JRC), **2**(4), 244–251 (2021). https://www.koreascience.or.kr/article/JAKO200801440604077.page
16. Vallicrosa, G., Ridao, P., Ribas, D., Palomer, A.: Active range-only beacon localization for auv homing. In: 2014 IEEE/RSJ International Conference on Intelligent Robots and Systems, pp. 2286–2291. IEEE (2014)
17. Volden, Ø., Stahl, A., Fossen, T.I.: Vision-based positioning system for auto-docking of unmanned surface vehicles (USVs). Int. J. Intel. Robot. Appl. 1–18 (2021)
18. Wang, J., Olson, E.: AprilTag 2: efficient and robust fiducial detection. In: 2016 IEEE/RSJ International Conference on Intelligent Robots and Systems (IROS), pp. 4193–4198. IEEE (2016). https://doi.org/10.1109/IROS.2016.7759617
19. Xu, Z., Haroutunian, M., Murphy, A.J., Neasham, J., Norman, R.: An underwater visual navigation method based on multiple ArUco markers. J. Mar. Sci. Eng. **9**(12) (2021). https://doi.org/10.3390/jmse9121432. https://www.mdpi.com/2077-1312/9/12/1432
20. Zhang, J., Cai, L., Chu, Y., Zhou, Q.: A sectional auto-docking charging control method for the mobile robot. In: 2019 IEEE International Conference on Mechatronics and Automation (ICMA), pp. 330–335 (2019). https://doi.org/10.1109/ICMA.2019.8816514

Autonomous Driving System for Reversing an Articulated Rover for Precision Agriculture

Andrea Botta(✉), Eleonora Moreno, Lorenzo Baglieri, Giovanni Colucci, Luigi Tagliavini, and Giuseppe Quaglia

Politecnico di Torino, 10129 Turin, Italy
{andrea.botta,giovanni_colucci,luigi.tagliavini, giuseppe.quaglia}@polito.it,
{eleonora.moreno,lorenzo.baglieri}@studenti.polito.it

Abstract. This paper focuses on a unique articulated robot architecture for precision agriculture to develop an autonomous driving system for its reverse motion. This driving system employs a path planner based on Dubins path to generate an optimal trajectory which has to be tracked by a path following routine based on the pure pursuit algorithm while a controller ensures the stability of the vehicle. This control system has been successfully tested and assessed using kinematic and dynamic models of the rover.

Keywords: Autonomous reversing · Articulated vehicle · Path following · Agriculture robot

1 Introduction

Within the growing trend of autonomous driving, a large portion of the spotlight has been focused on articulated vehicles. This is mostly related to their significance in logistics, which is considered a key part of the economy [10]. There are two primary reasons for choosing articulated vehicles over non-articulated ones: 1) articulated vehicles can perform significantly sharper turns than rigid vehicles of similar size [11], and 2) they improve the wheels contact with the ground on uneven terrain [9]. However, articulated vehicles suffer from the jackknife effect, which causes the vehicle to fold into a position from which it cannot recover. While in reverse motion, the pushed module of the articulated vehicle (i.e., a trailer) naturally tends to jackknife since it is in an unstable equilibrium while the folded condition is the stable equilibrium of the vehicle [14].

This represents a safety hazard also in the agricultural field where several autonomous articulated robots could be employed in the precision agriculture framework. In particular, it is of great interest the autonomous navigation in orchards or vineyards, environments that feature tight spaces that could benefit from the improved manoeuvrability of articulated vehicles moving also in reverse.

A. Müller and M. Brandstötter (Eds.): RAAD 2022, MMS 120, pp. 412–419, 2022.
https://doi.org/10.1007/978-3-031-04870-8_48

Several studies have discussed various implementations of autonomous navigation in reverse for articulated vehicles [13]. Moreover, recent studies have proposed some alternatives to be employed in agriculture on conventional articulated rover architectures [1].

This work focuses on developing an autonomous driving system for reversing a specific articulated rover that consists of a path planner, a trajectory tracker, and a hitch angle controller to avoid the jackknife effect. The proposed solution was tested and evaluated through kinematic and dynamic models of the robot.

2 Articulated Rover Modelling

Agri.Q is a mobile robot developed for precision agriculture applications [2,7,8,12,15]. The prototype is shown in Fig. 1. The robot is made up of two modules that are linked together by several joints, resulting in an articulated vehicle. Each module is outfitted with two drive units, each with two wheels. As a result, each module is an independent skid-steered part. Yet, a freewheel mechanism constraint the rear module to contribute only to the forward motion. Finally, two PV panels that can be positioned using two joints are put on the robot to serve as a landing platform for drones as well as a method to extend the robot autonomy in a sustainable manner.

2.1 Kinematic Model

Fig. 1. Agri.Q in a vineyard

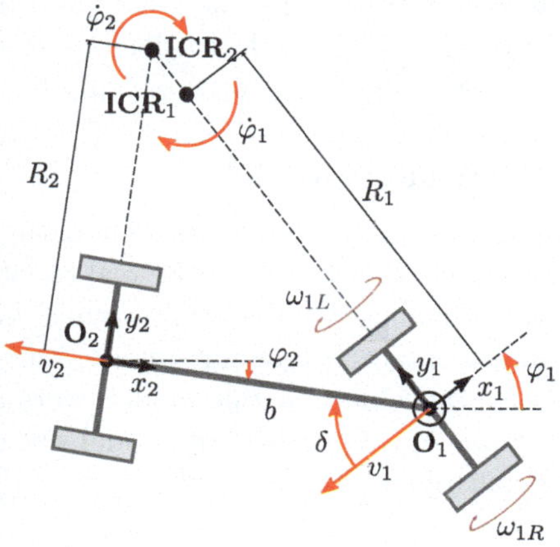

Fig. 2. Agri.Q simplyfied kinematic model

The kinematic model is based on the following hypothesis:

- the model considers only planar motion on a flat surface, out-of-plane motions are neglected;

- the locomotion units are simplified considering a single equivalent wheel per side with the axis passing through the centre of mass of the module.

Figure 2 represents the kinematic model, where (x_n, y_n, φ_n) is pose of the nth module ($n = 1$ for the front module, $n = 2$ for the back), δ is the relative yaw rotation between the modules, $i = 0.845\,\mathrm{m}$ is the wheel track of the module, $b = 1.3\,\mathrm{m}$ is the distances between the central joint and the rear module. v_n and $\dot{\varphi}_n$ are the longitudinal and the angular speeds respectively of the nth module in its reference frame. A characteristic of Agri.q is that the reference point of the front module $\mathbf{O}_1$ coincides with the relative yaw join; this feature makes reversing behaviour particularly unstable.

The kinematic model[4] of is defined as

$$\dot{\mathbf{q}} = \begin{bmatrix} \dot{x}_1 \\ \dot{y}_1 \\ \dot{\varphi}_1 \\ \dot{\delta} \end{bmatrix} = \begin{bmatrix} \cos\varphi_1 & 0 \\ \sin\varphi_1 & 0 \\ 0 & 1 \\ -\frac{1}{b}\sin\delta & 1 \end{bmatrix} \begin{bmatrix} v_1 \\ \dot{\varphi}_1 \end{bmatrix} \tag{1}$$

then, considering that each module behaves as a skid steering robot, it is possible to introduce the two following relations

$$v_1 = \frac{r}{2}\left(\omega_{1R} + \omega_{1L}\right) \quad \text{and} \quad \dot{\varphi}_1 = \frac{r}{i}\left(\omega_{1R} - \omega_{1L}\right) \tag{2}$$

By solving the equation

$$\dot{\delta} = \dot{\varphi}_1 - \frac{v_1}{b}\sin\delta_{eq} = 0 \tag{3}$$

it is possible to define the equilibrium hitch angle δ_{eq}. The equation has two solutions, representing the stable and unstable equilibrium points. While going in reverse it is possible to demonstrate that the equilibrium hitch angle of a given trajectory is an unstable equilibrium point of the system.

2.2 Dynamic Model

The hypothesis behind the kinematic model does not take into account the different characteristics of the real robot. The most important aspects, neglected in the kinematic model are: 1) wheel longitudinal and lateral slippage, 2) tyre-ground contact behaviour, and 3) the capabilities of the robot actuators. For this reason some dynamic models were developed by the authors in [3,6]. In this study, an Agri.Q multibody model was done with ADAMS to be able to improve the control simulation results.

3 Autonomous Reverse Control System

Figure 3 illustrates a block diagram of the proposed autonomous reverse control system with its main functional blocks. It starts with an off-line path planner that generates a suitable trajectory to minimise the jackknife occurrence. The generated path is then followed employing a pure pursuit algorithm that drives the robot motors together with a PD hitch angle controller to avoid jackknifing the robot. These main blocks are described in the following sections.

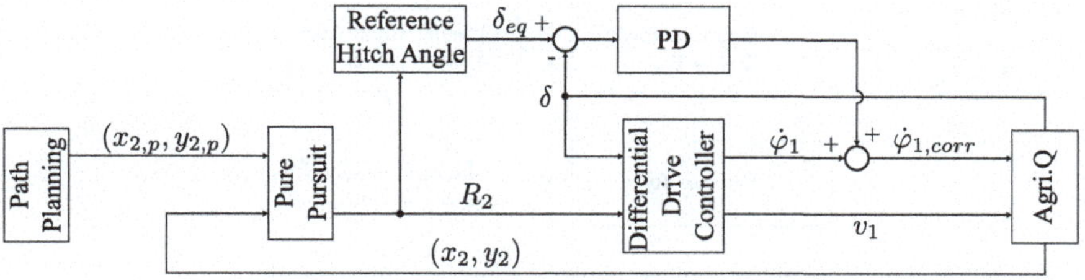

Fig. 3. Autonomous reverse control system block diagram

3.1 Path Planning

An off-line path planner based on Dubins path (i.e., a path composed only of straight segments and circular arcs) has been adopted to generate a suitable trajectory. As the typical Dubins algorithm, it is possible to achieve the minimum number of elemental paths without any obstacle, but the same approach can be adapted to an environment with obstacles. The first case is reported here. Since the path has to be done in reverse, the planning is done with respect to the rear module (i.e., the trailer), but through the kinematic model, it is possible to relate the required quantities to the front module as well.

Figure 4 depicts the working principle of the proposed planner. The algorithm receives as input the initial pose of the robot $\mathbf{P}_O$, the final pose $\mathbf{P}_R$ and the radius of the circular arcs is set to the minimum value possible, $R_{2,min}$. A typical Dubins path between the poses is then generated. Later, the first path element (always an arc) is simulated kinematically to predict the hitch angle δ_I at the end of the first path. If the magnitude of δ is above the threshold value δ_{max}, jackknifing may occur, hence the curvature radius is increased and the whole path is generated again. When the first path is verified, the second (it can be an arc or a segment) and the third (an arc) ones are checked similarly. However, if one of the following paths violates the hitch angle threshold, the previous verified path is kept and used as a starting point. This approach improves the path computation time and avoids increasing δ when not required. The planning ends when the final pose is reached.

The same planner can be adopted with some variations also when there are some obstacles. The two main differences are that a probabilistic roadmap planner is employed to estimate a simple path between the two robot poses and that the previously introduced planner is used considering that more than three elemental Dubins paths may be required and that such paths may differ from the typical Dubins optimal combinations.

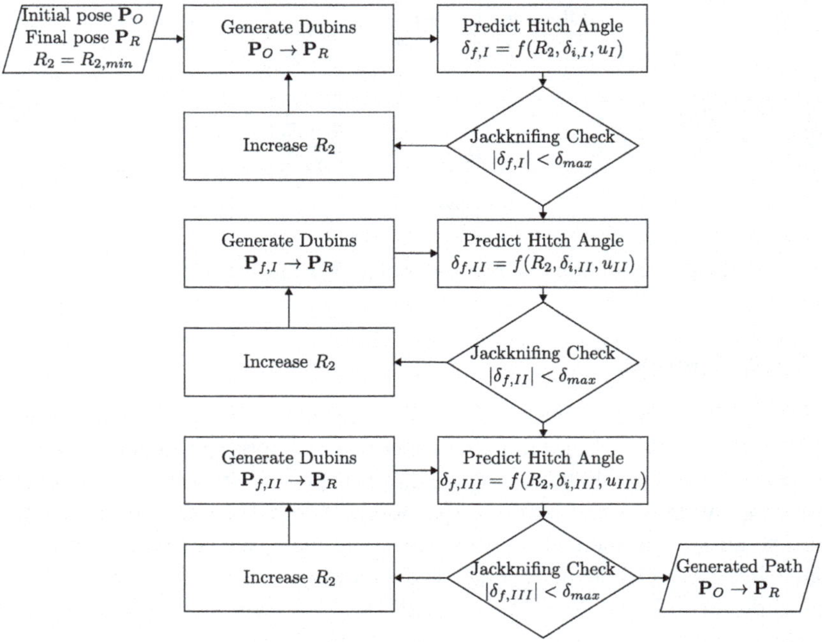

Fig. 4. Path planning flowchart

3.2 Path Tracking

Pure pursuit [5] was chosen among various existing tracking algorithms because it is simple and reliable. To bring the vehicle back and keep it on its trajectory (Fig. 5), the algorithm geometrically evaluates the circular arc that enables the vehicle to move from its current position towards a target point $\mathbf{G}$, which is located at a look-ahead distance $\mathbf{L}$ from the current projection of the vehicle on the path. The resulting curvature is then used to calculate the linear and angular velocity commands of the front module.

The planned path can be defined as a succession of waypoints ($\mathbf{P}_O$, $\mathbf{P}_1$, ..., $\mathbf{P}_R$) connected by segments $\mathbf{S}_n = \overrightarrow{\mathbf{P}_n\mathbf{P}_{n+1}}$. The algorithm locates the closest segment to the robot, but to speed up the process the heuristic assumption of checking only a part of the whole path is done. Then the objective point $\mathbf{G}$ is defined and the circular tracking trajectory between $\mathbf{O}_2$ and $\mathbf{G}$ is computed.

Differently from the traditional implementation of pure pursuit, the required robot longitudinal speed v_1 is not constant, instead it is defined to be a function of the tracking path curvature and the maximum yaw rate. By doing so, the pursuit algorithm adapts v_1 to the path, automatically slowing down before tight corners and speeding up on straight segments.

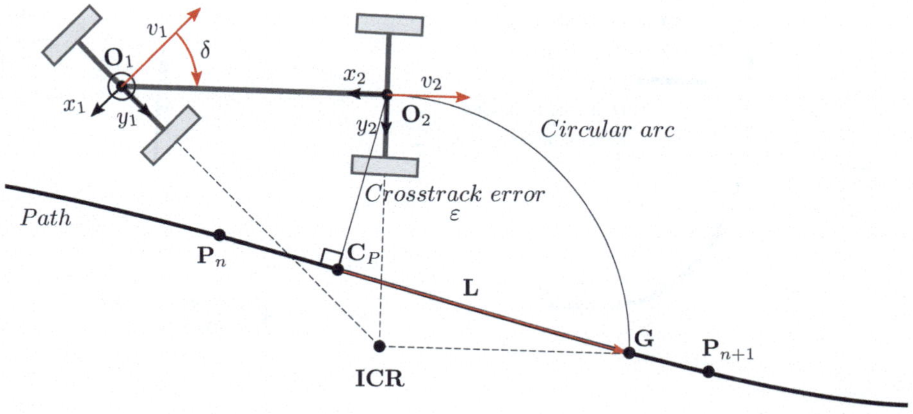

Fig. 5. Pure pursuit working principle

3.3 Hitch Angle Control

To avoid instability due to jackknifing, a PD controller has been implemented in parallel with the pure pursuit algorithm. The controller takes as input a reference angle δ_{eq}, that is computed from the path curvature R_2, and the measured hitch-angle δ. The controller could act both on the front module yaw rate $\dot{\varphi}_1$ and its longitudinal velocity v_1, however, a sensitivity analysis demonstrated that for Agri.Q and articulated vehicles where the hitch point is on the axle of the first module acting only on $\dot{\varphi}_1$ is more effective.

It is important to note that both the PD controller and the path tracking drive the front module yaw rate $\dot{\varphi}_1$, hence their contributions are added together to define the reference yaw rate of Agri.Q. This means that on some occasions the PD controller could slightly worsen the tracking of the planned path if it is required to compensate some deviation of δ from the reference value δ_{eq}.

4 Results

Figure 6 shows the results of the autonomous reverse control system applied to a simple manoeuvre. Kinematic and dynamic behaviours showed very close results. In particular, it observable a slightly larger tracking error in the dynamic model compared to the kinematic one (the peak errors were 43 cm and 41 cm respectively). This result comes with no surprise since the vehicle architecture is known to have an understeering behaviour, an effect that becomes clearer with tighter turns. Figure 6 also depicts the evolution in time of v_1 and $\dot{\varphi}_1$. Among the two, the yaw rate showed the result of a quite aggressive controller trying to stabilise δ while tracking the path during the turn (between 10 s and 28 s).

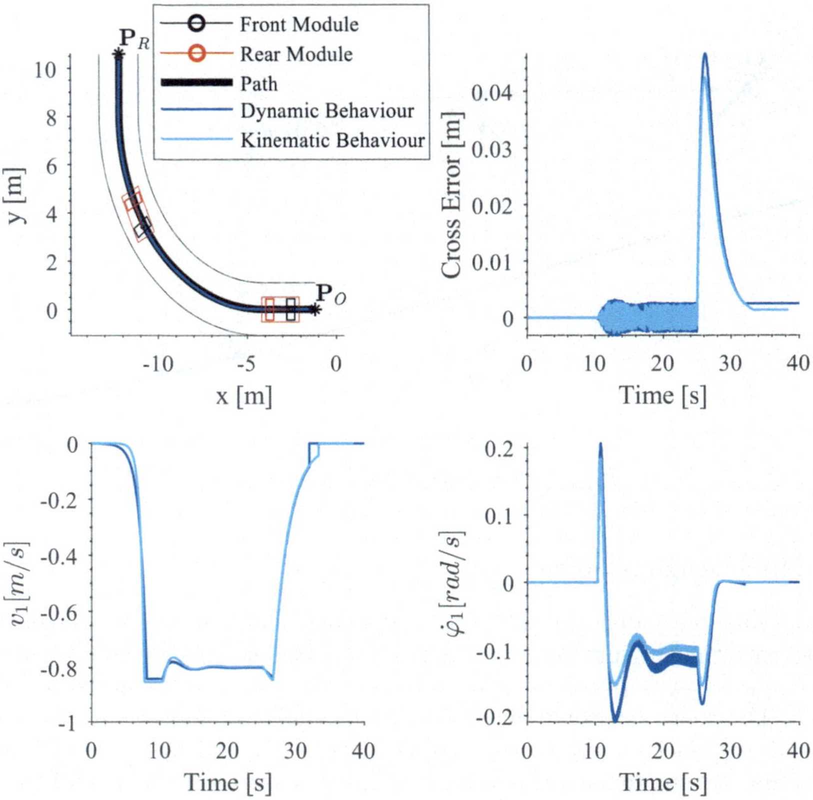

Fig. 6. Kinematic and dynamic results

5 Conclusions

In this work, an algorithm has been developed to plan, track, and control the reverse motion of the articulated mobile robot Agri.Q, with the particular goal of avoiding the occurrence of jackknifing (i.e., the folding of the articulated vehicle on itself). The autonomous driving system proved to be a robust algorithm for open field scenarios, but is easily adaptable to an environment with obstacles, through kinematic and dynamic simulation of the robot.

References

1. Bertolani, D.N., Bacheti, V.P., Sarcinelli-Filho, M.: Controlling a unicycle mobile platform pushing 1 and 2 passive trailers. J. Intell. Robot. Syst. **103**(3), 43 (2021). https://doi.org/10.1007/s10846-021-01478-z
2. Botta, A., Cavallone, P.: Robotics applied to precision agriculture: the sustainable Agri.Q rover case study. In: Quaglia, G., Gasparetto, A., Petuya, V., Carbone, G. (eds.) I4SDG 2021. MMS, vol. 108, pp. 41–50. Springer, Cham (2022). https://doi.org/10.1007/978-3-030-87383-7_5

3. Botta, A., Cavallone, P., Carbonari, L., Tagliavini, L., Quaglia, G.: Modelling and experimental validation of articulated mobile robots with hybrid locomotion system. In: Niola, V., Gasparetto, A. (eds.) IFToMM ITALY 2020. MMS, vol. 91, pp. 758–767. Springer, Cham (2021). https://doi.org/10.1007/978-3-030-55807-9_84
4. Botta, A., Cavallone, P., Tagliavini, L., Carbonari, L., Visconte, C., Quaglia, G.: An estimator for the kinematic behaviour of a mobile robot subject to large lateral slip. Appl. Sci. **11**(4), 1594 (2021). (Multidisciplinary Digital Publishing Institute)
5. Campbell, S.F.: Steering control of an autonomous ground vehicle with application to the DARPA urban challenge. Ph.D. thesis, Massachusetts Institute of Technology (2007)
6. Carbonari, L., Botta, A., Cavallone, P., Tagliavini, L., Quaglia, G.: Data-driven analysis of locomotion for a class of articulated mobile robots. J. Mech. Robot. **13**(5), 050905 (2021). (American Society of Mechanical Engineers)
7. Cavallone, P., Botta, A., Carbonari, L., Visconte, C., Quaglia, G.: The Agri.Q mobile robot: preliminary experimental tests. In: Niola, V., Gasparetto, A. (eds.) IFToMM ITALY 2020. MMS, vol. 91, pp. 524–532. Springer, Cham (2021). https://doi.org/10.1007/978-3-030-55807-9_59
8. Cavallone, P., Visconte, C., Carbonari, L., Botta, A., Quaglia, G.: Design of the mobile robot Agri.q. In: Venture, G., Solis, J., Takeda, Y., Konno, A. (eds.) ROMANSY 2020. CICMS, vol. 601, pp. 288–296. Springer, Cham (2021). https://doi.org/10.1007/978-3-030-58380-4_35
9. González-Cantos, A., Ollero, A.: Backing-up maneuvers of autonomous tractor-trailer vehicles using the qualitative theory of nonlinear dynamical systems. Int. J. Robot. Res. **28**(1), 49–65 (2009)
10. Hall, R.: Handbook of Transportation Science, vol. 23. Springer Science & Business Media, Cham (2012). https://doi.org/10.1007/978-1-4615-5203-1
11. Li, B., Zhang, Y., Acarma, T., Kong, Q., Zhang, Y.: Trajectory planning for a tractor with multiple trailers in extremely narrow environments: a unified approach. In: 2019 International Conference on Robotics and Automation (ICRA), pp. 8557–8562. IEEE (2019)
12. Quaglia, G., Visconte, C., Carbonari, L., Botta, A., Cavallone, P.: Agri.q: a sustainable rover for precision agriculture. In: Visa, I., Duta, A. (eds.) Solar Energy Conversion in Communities. SPE, pp. 81–91. Springer, Cham (2020). https://doi.org/10.1007/978-3-030-55757-7_6
13. Rahimi, A., He, Y.: A review of essential technologies for autonomous and semi-autonomous articulated heavy vehicles. In: Proceedings of the Canadian Society for Mechanical Engineering International Congress, pp. 21–24 (2020)
14. Roh, J., Chung, W.: Reversing control of a car with a trailer using the driver assistance system. Int. J. Adv. Robot. Syst. **8**(2), 23 (2011)
15. Visconte, C., Cavallone, P., Carbonari, L., Botta, A., Quaglia, G.: Design of a Mechanism with Embedded Suspension to Reconfigure the Agri_q Locomotion Layout. Robotics **10**(1), 15 (2021). (Multidisciplinary Digital Publishing Institute)

High Accuracy Data-Based Trajectory Tracking of an Omnidirectional Mobile Robot

Hannes Eschmann(✉), Henrik Ebel, and Peter Eberhard

Institute of Engineering and Computational Mechanics (ITM), University of Stuttgart, Stuttgart, Germany
{hannes.eschmann,henrik.ebel,peter.eberhard}@itm.uni-stuttgart.de

Abstract. Mobile robots play a key role in a number of different automation tasks. Robots capable of omnidirectional motion allow for a particularly very flexible operation. They are able to move in every direction, regardless of their orientation. A common task for industrial mobile robots is trajectory tracking. These trajectories often are periodic, e. g., if the same path is followed repeatedly through a large warehouse or at an assembly line. In the contribution, it is shown how data-based techniques can be used to tailor a motion model to a specific re-occurring reference, resulting in highly accurate tracking. The results of this work are finally validated in real-world hardware experiments. This novel contribution builds upon preliminary work but, for the first time, includes an adaptive component for high accuracy tracking of periodic references as well as a general approach to generating periodic input and state references based on a dynamic model of a robot.

Keywords: Mobile robots · Adaptive control · Trajectory tracking

1 Introduction

Mobile robots have long since become an integral part of many industrial applications due to their automation potential. These mobile robots often end up driving the same exact path repeatedly, e. g., in a parcel center. Moreover, mobile robots may serve as measurement platforms, e. g., to automate repeating measurement campaigns of wireless signals in development and product evaluation, where extremely high accuracy, with deviations smaller than the signal wavelength, and high reproducibility are paramount. Thus, it is shown how the information gained from these periodic paths can be leveraged to improve the tracking

This research is funded by the Deutsche Forschungsgemeinschaft (DFG, German Research Foundation) under Germany's Excellence Strategy – EXC 2075 – 390740016, project PN4-4 "Theoretical Guarantees for Predictive Control in Adaptive Multi-Agent Scenarios" and project EB195/32-1, 433183605 "Research on Multibody Dynamics and Control for Collaborative Elastic Object Transportation by a Heterogeneous Swarm with Aerial and Land-Based Mobile Robots".

A. Müller and M. Brandstötter (Eds.): RAAD 2022, MMS 120, pp. 420–427, 2022.
https://doi.org/10.1007/978-3-031-04870-8_49

performance of the mobile robots using an optimization and data-based approach. While the approach is conceptually similar to the work of [1], trajectory tracking instead of path following is considered and the model in this work is tailored to a specific reference trajectory, potentially reducing the tracking error further without significantly increasing the computational burden. In Sect. 2, the robotic hardware is shortly presented. In the following Sect. 3, the model setup is explained in detail while in Sect. 4 the proposed reference generator and the corresponding model-based controller are introduced. Finally, we show in Sect. 5 the effectiveness of the approach in real-world trajectory tracking experiments of an omnidirectional mobile robot. To our best knowledge, the approach taken is conceptually novel and unique in the fidelity it achieves.

2 The Omnidirectional Mobile Robot

Omnidirectional motion can be achieved using a number of different ways. In this work, a special type of wheel is used, called the Mecanum wheel, where several rollers are fixed to the circumference of the wheel at an angle. For omnidirectional mobile robots using Mecanum wheels, several different wheel arrangements are possible, the most popular ones being either the rectangular or the radial wheel arrangement. The custom build robot used throughout the hardware experiments in Sect. 5 has the radial arrangement using 4 equally spaced wheels. The robot takes the desired translational velocities v^{x} and v^{y} in the robot frame of reference and the angular velocity ω about the vertical axis as inputs $\boldsymbol{u}$. To compute the four resulting desired wheel speeds Ω_i for wheel i, the relationship

$$\boldsymbol{\Omega} := \begin{bmatrix} \Omega_1 \\ \Omega_2 \\ \Omega_3 \\ \Omega_4 \end{bmatrix} = \frac{1}{R} \begin{bmatrix} -1 & -1 & -\ell \\ -1 & 1 & -\ell \\ 1 & 1 & -\ell \\ 1 & -1 & -\ell \end{bmatrix} \begin{bmatrix} v^{\mathrm{x}} \\ v^{\mathrm{y}} \\ \omega \end{bmatrix} =: \boldsymbol{D}\boldsymbol{u} \tag{1}$$

is used, where $R = 3\,\mathrm{cm}$ is the radius of a wheel and $\ell = 11.5\,\mathrm{cm}$ is the distance between the robot's and the wheel's center in the motion plane. Given an input $\boldsymbol{u}$, over time, relationship (1) can be used to compute the corresponding set points for each wheel's individual proportional–integral–derivative (PID) controller, running at 100 Hz. The input $\boldsymbol{u}$ is updated at a rate of 5 Hz. The pose of the robot, i. e., its position in the movement plane in the inertial frame of reference and its orientation is measured using an external tracking system. For more information as well as pictures of the specific robotic hardware and its special wheels we refer to [2].

3 Data-Based Model of the Mobile Robot

A kinematic model of the planar omnidirectional mobile robot is introduced. This model is commonly used for robotics applications and will act as our nominal baseline system model. The position at time step $t \in \mathbb{N}_0$ in the inertial frame of reference can be described by the kinematic discrete-time relationship

$$\boldsymbol{x}_{t+1} = \boldsymbol{x}_t + T\boldsymbol{S}_{\theta_t}\boldsymbol{u}_t =: \boldsymbol{f}_{\mathrm{n}}(\boldsymbol{x}_t, \boldsymbol{u}_t), \tag{2}$$

where $\boldsymbol{x}_t := [x_t \; y_t \; \theta_t]^{\mathsf{T}}$ is the pose of the mobile robot at time t, i. e., its position $[x_t \; y_t]^{\mathsf{T}}$ in the inertial frame of reference and its orientation θ_t. The rotation matrix $\boldsymbol{S}_{\theta_t}$ describes a positive rotation about the robot's vertical axis by the angle θ_t. We use the shorthand notation $\boldsymbol{v}_t := \boldsymbol{v}(t)$ unless we want to highlight the explicit time dependency. The sampling time is chosen to $T = 0.2\,\mathrm{s}$. The simple kinematic model (2) is obviously only justified for reasonable input combinations. Therefore, the wheel speeds and their rates of change are constrained

$$-\Omega_{\max} \leq \|\boldsymbol{\Omega}_t\|_\infty \overset{(1)}{=} \|\boldsymbol{D}\boldsymbol{u}_t\|_\infty \leq \Omega_{\max}, \tag{3a}$$

$$-\Delta\Omega_{\max} \leq \|\boldsymbol{\Omega}_t - \boldsymbol{\Omega}_{t-1}\|_\infty \overset{(1)}{=} \|\boldsymbol{D}(\boldsymbol{u}_t - \boldsymbol{u}_{t-1})\|_\infty \leq \Delta\Omega_{\max}, \tag{3b}$$

$$-\omega_{\max} \leq |\omega_t| \leq \omega_{\max}, \tag{3c}$$

where $\Omega_{\max} = 20\,{}^{1}/_{\mathrm{s}}$ and $\Delta\Omega_{\max} = 3.3\,{}^{1}/_{\mathrm{s}}$ are the maximum wheel speed and its maximal change between two time instances, respectively. Additionally, the maximum magnitude of the angular velocity ω_t of the robot is limited to $\omega_{\max} = 0.2\,{}^{1}/_{\mathrm{s}}$. The maximum norm is indicated by $\| \bullet \|_\infty$. As the mobile robot shows interesting nonlinear behavior in hardware experiments, which is not captured by (2), an augmented model is defined where systematic mismatches between the actual system dynamics and the nominal model in the robot frame of reference are captured by a data-based nonlinear function $\boldsymbol{g}$. In the following, the unknown nonlinear function $\boldsymbol{g}(\boldsymbol{u}_t, \boldsymbol{u}_{t-1})$ is assumed to depend on the current and previous inputs. Experience with the robotic hardware motivates this choice. The data-based model of the omnidirectional robot is defined as

$$\boldsymbol{x}_{t+1} = \boldsymbol{f}_{\mathrm{n}}(\boldsymbol{x}_t, \boldsymbol{u}_t) + T\boldsymbol{S}_{\theta_t}\boldsymbol{g}(\boldsymbol{u}_t, \boldsymbol{u}_{t-1}) =: \boldsymbol{f}_g(\boldsymbol{x}_t, \boldsymbol{u}_t, \boldsymbol{u}_{t-1}). \tag{4}$$

The procedure for generating the training data is shortly explained in the following; more detail is available in [2,3]. The hit-and-run sampler from [4] is used to generate uniformly distributed random points from the six-dimensional convex polytope for $\boldsymbol{u}_k$ and $\boldsymbol{u}_{k-1}$ resulting from (3). The samples for the input combinations are then concatenated in a time series and applied to the robot. Collecting input-output data in this way ensures good generalization to all possible input combinations satisfying (3) while sufficiently covering the corresponding input space. Alternatively, if available, data collected during normal operation of the robot can be used. While in this case Gaussian process (GP) regression, see [5], is used due to its data efficiency, several other regression techniques could be used, too. The model (4) is able to capture the nature of the robot's motion for general trajectories. In Sect. 4 we will consider trajectory tracking for the mobile robot. The robot needs to follow a specific reference trajectory $\boldsymbol{x}_t^{\mathrm{r}}$ for the state and $\boldsymbol{u}_t^{\mathrm{r}}$ for the input. In Sect. 4.1 it is shown, how one can obtain these references for the periodic case. In the following, an additional function $\boldsymbol{h}$ is introduced to compensate mismatches that are specific to the current trajectory, e. g., effects of uneven ground or mismatches that only develop over larger time intervals and can, therefore, not be captured by $\boldsymbol{g}$. It is assumed that due to the choice of $\boldsymbol{g}$ the robot will be able to generally follow its desired trajectory and the trajectory-specific mismatches are expected to be small. Therefore, it is assumed that $\boldsymbol{x}_t \approx \boldsymbol{x}_t^{\mathrm{r}}$ and $\boldsymbol{u}_t \approx \boldsymbol{u}_t^{\mathrm{r}}$. This results in $\bar{\boldsymbol{h}}(\boldsymbol{x}_t, \boldsymbol{u}_t) \approx \tilde{\boldsymbol{h}}(\boldsymbol{x}_t^{\mathrm{r}}, \boldsymbol{u}_t^{\mathrm{r}}) \approx \boldsymbol{h}(\boldsymbol{u}_t^{\mathrm{r}}, t)$. The state dependency is now taken into account via the explicit time dependency of $\boldsymbol{h}$. The augmented dynamics taking the trajectory-specific mismatches into account, is defined as

$$\boldsymbol{x}_{t+1} = \boldsymbol{f}_g(\boldsymbol{x}_t, \boldsymbol{u}_t, \boldsymbol{u}_{t-1}) + T\boldsymbol{S}_{\theta_t}\boldsymbol{h}(\boldsymbol{u}_t^{\mathrm{r}}, t) =: \boldsymbol{f}_{gh}(\boldsymbol{x}_t, \boldsymbol{u}_t, \boldsymbol{u}_{t-1}, \boldsymbol{u}_t^{\mathrm{r}}, t). \tag{5}$$

The function $\boldsymbol{h}$ is trained using the corresponding model mismatch $\boldsymbol{e}_h(t)$, i. e.,

$$\boldsymbol{e}_h(t) := \tfrac{1}{T}\boldsymbol{S}_{\theta_t}^{\mathsf{T}} \left(\boldsymbol{x}_{t+1} - \boldsymbol{f}_g(\boldsymbol{x}_t, \boldsymbol{u}_t, \boldsymbol{u}_{t-1})\right). \tag{6}$$

Note that $\boldsymbol{e}_h(t)$ is the error to the baseline model $\boldsymbol{f}_g$ instead of the deviation to the state reference and is, again, computed in the robot frame of reference. For the training data, the robot drives several laps on the desired periodic trajectory. The input and state references are subsequently adapted for the new model and $\boldsymbol{h}$ can potentially be retrained using the data observed with the new as well as the previous input references, resulting in an iterative procedure. If the incremental changes to $\boldsymbol{u}^{\mathrm{r}}$ are negligible, however, one can simplify and assume a pure time dependence of $\boldsymbol{h}(t)$. Later, an optimization-based approach is used to compute the control input, see Sect. 4.2, and, interestingly, $\boldsymbol{h}$ can even compensate for delays introduced by the solver.

4 Trajectory Tracking of Periodic References

For trajectory tracking, the robot is generally required to follow a desired predefined trajectory $\boldsymbol{x}_t^{\mathrm{des}}$ in space and time. In contrast, for path following [1], the desired position is not explicitly dependent on the time t. For the nominal dynamics, the reference can be computed analytically via

$$\boldsymbol{x}_t^{\mathrm{r}} = \boldsymbol{x}_t^{\mathrm{des}} \text{ and } \boldsymbol{u}_t^{\mathrm{r}} = \tfrac{1}{T}\boldsymbol{S}_{\theta_t^{\mathrm{des}}}^{\mathsf{T}} \left(\boldsymbol{x}_{t+1}^{\mathrm{des}} - \boldsymbol{x}_t^{\mathrm{des}}\right), \tag{7}$$

where $\theta_t^{\mathrm{des}} := \left[\boldsymbol{x}_t^{\mathrm{des}}\right]_3$ is the third component of the desired trajectory indicating the desired orientation of the robot. When computing the nominal reference solution that way, $\boldsymbol{x}_t^{\mathrm{des}}$ should be chosen in a way that the nominal reference input does not exceed the constraints (3) on the control inputs. In this work, periodic reference trajectories are considered. A periodic reference of period N time steps needs to satisfy the following properties

$$\boldsymbol{x}_t^{\mathrm{r}} = \boldsymbol{x}_{t+N}^{\mathrm{r}}, \tag{8a}$$
$$\boldsymbol{u}_t^{\mathrm{r}} = \boldsymbol{u}_{t+N}^{\mathrm{r}} \quad \forall t \in \mathbb{N}_0, \tag{8b}$$

i. e., the final state and input should loop back to the initial state and input reference. Note that the methods presented are also applicable to more general trajectories as long as (8) holds for at least some sufficiently long time intervals of the overall trajectory.

4.1 Generating the Reference Solution

Due to the nonlinear dynamics (4) or (5), especially in the presence of input constraints (3) it may not be possible to perfectly follow the desired trajectory $\boldsymbol{x}_t^{\mathrm{des}}$. In this subsection, a method is proposed to compute reference trajectories $\boldsymbol{u}_t^{\mathrm{r}}$, $\boldsymbol{x}_t^{\mathrm{r}}$ for input and state, respectively, following a desired trajectory $\boldsymbol{x}_t^{\mathrm{des}}$ as close as possible in case the reference cannot be computed in closed form like in (7). Without loss of generality, $\boldsymbol{x}_0^{\mathrm{r}} := \boldsymbol{x}_0^{\mathrm{des}}$ is assumed. We will use the notation $\boldsymbol{v}_{t+i|t}$ to indicate the model-based prediction at time t of the quantity $\boldsymbol{v}$ for time $t+i$. The reference is computed in a receding horizon fashion, i. e., for each time step the trajectory is optimized over a horizon $H_{\mathrm{ff}} < N$, which, while possibly being suboptimal is computationally more efficient than a single large optimization problem optimizing the entire reference at once. Another benefit is that the total computation time scales only linearly with the period of the reference N. For each $t = 0 < t < N - H_{\mathrm{ff}}$ the reference can be computed via

$$\min_{\boldsymbol{x}_{\bullet|t}^{\mathrm{r}}, \boldsymbol{u}_{\bullet|t}^{\mathrm{r}}} \quad \sum_{k=t}^{t+H_{\mathrm{ff}}-1} \|\boldsymbol{x}_k^{\mathrm{des}} - \boldsymbol{x}_{k|t}^{\mathrm{r}}\|_{\boldsymbol{Q}_{\mathrm{ff}}}^2 + \|\boldsymbol{u}_{k|t}^{\mathrm{r}} - \boldsymbol{u}_{k-1|t}^{\mathrm{r}}\|_{\boldsymbol{R}_{\mathrm{ff}}}^2 \tag{9a}$$

$$\text{subject to} \quad \boldsymbol{u}_{t-1|t}^{\mathrm{r}} = \boldsymbol{u}_{t-1}^{\mathrm{r}}, \tag{9b}$$

$$\boldsymbol{x}_{t|t}^{\mathrm{r}} = \boldsymbol{x}_t^{\mathrm{r}}, \tag{9c}$$

$$\boldsymbol{x}_{k+1|t}^{\mathrm{r}} = \boldsymbol{f}_{g/gh}(\boldsymbol{x}_{k|t}^{\mathrm{r}}, \boldsymbol{u}_{k|t}^{\mathrm{r}}, \bullet), \tag{9d}$$

$$-k_\varepsilon \Omega_{\max} \leq \|\boldsymbol{D}\boldsymbol{u}_{k|t}^{\mathrm{r}}\|_\infty \leq k_\varepsilon \Omega_{\max}, \tag{9e}$$

$$-k_\varepsilon \Delta\Omega_{\max} \leq \|\boldsymbol{D}(\boldsymbol{u}_{k|t}^{\mathrm{r}} - \boldsymbol{u}_{k-1|t}^{\mathrm{r}})\|_\infty \leq k_\varepsilon \Delta\Omega_{\max}, \tag{9f}$$

$$-k_\varepsilon \omega_{\max} \leq |[\boldsymbol{u}_{k|t}^{\mathrm{r}}]_3| \leq k_\varepsilon \omega_{\max} \tag{9g}$$

with $\boldsymbol{v}^\mathsf{T}\boldsymbol{V}\boldsymbol{v} =: \|\boldsymbol{v}\|_{\boldsymbol{V}}^2$. Here, $\boldsymbol{Q}_{\mathrm{ff}}$ and $\boldsymbol{R}_{\mathrm{ff}}$ are positive definite weighting matrices and $\boldsymbol{f}_{g/gh}(\boldsymbol{x}_{k|t}^{\mathrm{r}}, \boldsymbol{u}_{k|t}^{\mathrm{r}}, \bullet)$ indicates either (4) or (5) where the state and input are replaced with their respective reference value. The constant $0 < k_\varepsilon < 1$ ensures that a controller can still compensate for possible mismatches in closed-loop operation. We penalize the input change in (9a) in order to encourage smooth input references. For $t = 0 < t < N - H_{\mathrm{ff}}$, optimization problem (9) is solved and the state and input reference are then set to $\boldsymbol{x}_{t+1}^{\mathrm{r}} := \boldsymbol{x}_{t+1|t}^{\mathrm{r}}$ and $\boldsymbol{u}_t^{\mathrm{r}} := \boldsymbol{u}_{t|t}^{\mathrm{r}}$. However, for the initial time step $t = 0$ and the tail end of the trajectory's period, some modifications of (9) are necessary, see Table 1. For $t = 0$, the input $\boldsymbol{u}_{-1}^{\mathrm{r}} = \boldsymbol{u}_{N-1}^{\mathrm{r}}$ is left as an optimization variable. For $t = N - H_{\mathrm{ff}}$, the additional constraints $\boldsymbol{u}_{N-1|N-H_{\mathrm{ff}}}^{\mathrm{r}} = \boldsymbol{u}_{N-1}^{\mathrm{r}}$ and $\boldsymbol{x}_{N|N-H_{\mathrm{ff}}}^{\mathrm{r}} = \boldsymbol{x}_0^{\mathrm{r}}$ are added to the optimization problem to ensure that (8) holds. The final reference states and inputs for time steps $N - H_{\mathrm{ff}} \leq t < N - 1$ are the corresponding open-loop solution from time step $N - H_{\mathrm{ff}}$, i. e., $\boldsymbol{x}_{t+1}^{\mathrm{r}} = \boldsymbol{x}_{t+1|N-H_{\mathrm{ff}}}^{\mathrm{r}}$ and $\boldsymbol{u}_t^{\mathrm{r}} = \boldsymbol{u}_{t|N-H_{\mathrm{ff}}}^{\mathrm{r}}$, respectively. Alternatively, instead of setting the tail end of the reference to the open-loop solution of time $N - H_{\mathrm{ff}}$, the prediction horizon H_{ff} could be reduced as time progresses. The constants are set to $\boldsymbol{Q}_{\mathrm{ff}} = 100 \cdot \mathrm{diag}(10, 10, 5)$, $\boldsymbol{R}_{\mathrm{ff}} = \mathrm{diag}(1, 1, 50)$, $H_{\mathrm{ff}} = 10$, and $k_\varepsilon = 0.95$.

Table 1. Different modifications to the optimization problem (9) to obtain a periodic state and input reference.

Time step t	Solution $\boldsymbol{x}^{\mathrm{r}}$, $\boldsymbol{u}^{\mathrm{r}}$	Additions to (9)
$t = 0$	$\boldsymbol{x}_0^{\mathrm{r}} := \boldsymbol{x}_0^{\mathrm{des}}$, $\boldsymbol{u}_0^{\mathrm{r}} := \boldsymbol{u}_{0\|0}^{\mathrm{r}}$, $\boldsymbol{x}_1^{\mathrm{r}} := \boldsymbol{x}_{1\|0}^{\mathrm{r}}$, $\boldsymbol{u}_{N-1}^{\mathrm{r}} := \boldsymbol{u}_{-1\|0}^{\mathrm{r}}$	Opt. variable $\boldsymbol{u}_{-1}^{\mathrm{r}} = \boldsymbol{u}_{N-1}^{\mathrm{r}}$, constraint $\boldsymbol{x}_{0\|0}^{\mathrm{r}} = \boldsymbol{x}_0^{\mathrm{des}}$
$0 < t < N - H_{\mathrm{ff}}$	$\boldsymbol{x}_{t+1}^{\mathrm{r}} := \boldsymbol{x}_{t+1\|t}^{\mathrm{r}}$, $\boldsymbol{u}_t^{\mathrm{r}} := \boldsymbol{u}_{t\|t}^{\mathrm{r}}$	-
$N - H_{\mathrm{ff}} \leq t < N - 1$	$\boldsymbol{x}_{t+1}^{\mathrm{r}} = \boldsymbol{x}_{t+1\|N-H_{\mathrm{ff}}}^{\mathrm{r}}$, $\boldsymbol{u}_t^{\mathrm{r}} = \boldsymbol{u}_{t\|N-H_{\mathrm{ff}}}^{\mathrm{r}}$	constraints $\boldsymbol{u}_{N-1\|N-H_{\mathrm{ff}}}^{\mathrm{r}} = \boldsymbol{u}_{N-1}^{\mathrm{r}}$, $\boldsymbol{x}_{N\|N-H_{\mathrm{ff}}}^{\mathrm{r}} = \boldsymbol{x}_0^{\mathrm{r}}$

4.2 Model Predictive Trajectory Tracking Controller

In order to follow the desired trajectory, a model predictive controller (MPC) is employed. The optimal control problem can be formulated as

$$\min_{\boldsymbol{x}_{\bullet|t},\, \boldsymbol{u}_{\bullet|t}} \quad \sum_{k=t}^{t+H-1} \|\boldsymbol{x}_k^{\mathrm{r}} - \boldsymbol{x}_{k|t}\|_{\boldsymbol{Q}}^2 + \|\boldsymbol{u}_k^{\mathrm{r}} - \boldsymbol{u}_{k|t}\|_{\boldsymbol{R}}^2 \tag{10a}$$

$$\text{subject to} \quad \boldsymbol{u}_{t-1|t} = \boldsymbol{u}_{t-1}, \tag{10b}$$

$$\boldsymbol{x}_{t|t} = \boldsymbol{x}_t, \tag{10c}$$

$$\boldsymbol{x}_{k+1|t} = \boldsymbol{f}_{\mathrm{n}/g/gh}(\boldsymbol{x}_{k|t}, \boldsymbol{u}_{k|t}, \bullet), \tag{10d}$$

$$-\Omega_{\max} \leq \|\boldsymbol{D}\boldsymbol{u}_{k|t}\|_\infty \leq \Omega_{\max}, \tag{10e}$$

$$-\Delta\Omega_{\max} \leq \|\boldsymbol{D}(\boldsymbol{u}_{k|t} - \boldsymbol{u}_{k-1|t})\|_\infty \leq \Delta\Omega_{\max}, \tag{10f}$$

$$-\omega_{\max} \leq |[\boldsymbol{u}_{k|t}]_3| \leq \omega_{\max} \tag{10g}$$

with positive definite weighting matrices $\boldsymbol{Q} = \boldsymbol{I}$ and $\boldsymbol{R} = \frac{1}{10}\boldsymbol{I}$, the current state measurement $\boldsymbol{x}_t$ and the prediction horizon $H = 5$. After (10) is solved, the current input is set to the solution $\boldsymbol{u}_t := \boldsymbol{u}_{t|t}$ for each time step. Note that, as the trajectory-specific mismatch $\boldsymbol{h}$ is not state and input dependent, the computational effort is similar to solving (10) for $\boldsymbol{f}_g$.

5 Experimental Results

The optimization problems (9) and (10) are set up in Matlab using CasADi [6] and are solved using the interior point optimizer IPOPT [7] and a sequential quadratic programming (SQP) approach using qpOASES [8], respectively. For the experiments, after an initial lap our custom robot drives five consecutive laps of a lemniscate trajectory with $N = 124$ and $\theta_t^{\mathrm{des}} \equiv 45°$, compare Fig. 1. The five laps that are driven using $\boldsymbol{g}$ are used to train the trajectory-specific mismatch $\boldsymbol{h}_0(\boldsymbol{u}_t^{\mathrm{r}}, t)$. This model is used for the generation of the adapted reference $\boldsymbol{u}_{h,0}^{\mathrm{r}}$. This procedure is repeated three more times until the input reference converged, resulting in $\boldsymbol{h}_3(\boldsymbol{u}_t^{\mathrm{r}}, t)$ and $\boldsymbol{u}_{h,3}^{\mathrm{r}}$. While, at a first glance, due to the size of the trajectory, all robot models seem to perform equally well, when the control errors are inspected in greater detail in Fig. 2 the differences become more evident. In the tracking error histogram in Fig. 2(b) the nominal model $\boldsymbol{f}_{\mathrm{n}}$ shows a non-vanishing control error due to the unmodeled mismatches, resulting in a root mean square error (RMSE) of 12.78 mm or 1.01° on position and orientation level, respectively. The general GP-based model $\boldsymbol{f}_{g}$ reduces the RMSE significantly to 4.34 mm or 0.73°. Still, the model additionally tailored to the specific trajectory $\boldsymbol{f}_{gh}$ manages to reduce the tracking RMSE further to 2.37 mm and 0.58°. This becomes evident in the histogram Fig. 2(b). The error on position level is reliably reduced for most time steps, compare Fig. 2(a) (top). The error for the orientation only improves slightly between $\boldsymbol{f}_{g}$ and $\boldsymbol{f}_{gh}$. It has to be noted that assuming a pure time-dependency $\boldsymbol{h}(t)$ showed similar results as $\boldsymbol{h}(\boldsymbol{u}_t^{\mathrm{r}}, t)$ since the input reference did not change massively to the one obtained with $\boldsymbol{f}_{g}$.

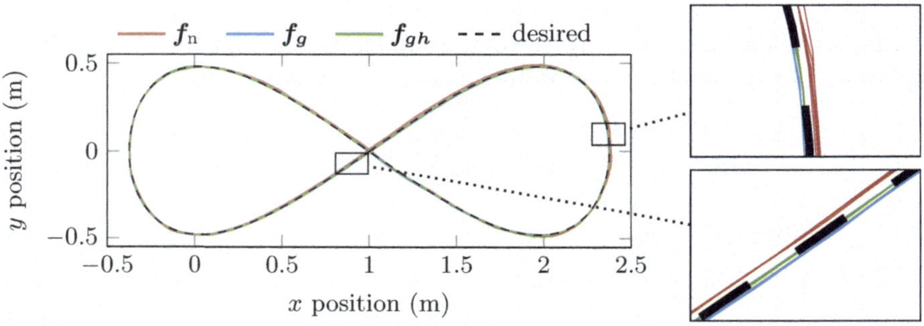

Fig. 1. For the hardware experiments, the robot drives five consecutive laps for each model on the lemniscate trajectory. The orientation of the robot is not depicted.

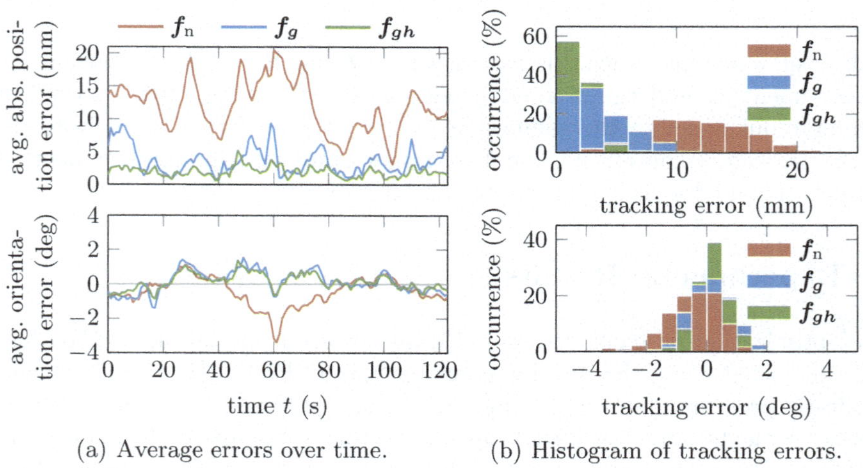

Fig. 2. Error for the deviations from the desired trajectory $\boldsymbol{x}_t^{\mathrm{des}} - \boldsymbol{x}_t$ on position level (top) and orientation level (bottom) using the three different robot models.

6 Conclusion and Outlook

In this paper, a more general, robot-specific data-based model and a trajectory-specific data-based correction term were combined to achieve high-accuracy tracking of periodic references. This allows for great flexibility without sacrificing control accuracy. While for the regression of both models Gaussian processes were used, the trajectory-specific term especially could be trained using any regression technique, as it is assumed to be explicitly time and reference specific. This results in significant reduction of the computational time of the employed model predictive controller. Moreover, an optimization-based approach was proposed to efficiently compute periodic input and state references taking the model dynamics into account. Future work may extend the theory to high-accuracy path following control or to the distributed case, e. g., for decentralized high-accuracy formation control [9].

References

1. Hewing, L., Liniger, A., Zeilinger, M.N.: Cautious NMPC with Gaussian process dynamics for autonomous miniature race cars. In: 2018 European Control Conference (ECC), pp. 1341–1348 (2018). https://doi.org/10.23919/ECC.2018.8550162
2. Eschmann, H., Ebel, H., Eberhard, P.: Trajectory tracking of an omnidirectional mobile robot using Gaussian process regression. Automatisierungstechnik **69**(8), 656–666 (2021). https://doi.org/10.1515/auto-2021-0019
3. Eschmann, H., Ebel, H., Eberhard, P.: Data-based model of an omnidirectional mobile robot using Gaussian processes. In: IFAC Symposium on System Identification (SYSID) - Learning Models For Decision and Control, pp. 1–6. Padova, Italy (2021). https://doi.org/10.1016/j.ifacol.2021.08.327
4. Kroese, D.P., Taimre, T., Botev, Z.I.: Handbook of Monte Carlo methods. John Wiley & Sons, Hoboken (2013)
5. Rasmussen, C.E., Williams, C.K.I.: Gaussian Processes for Machine Learning (Adaptive Computation and Machine Learning). The MIT Press, Cambridge (2005)
6. Andersson, J.A.E., Gillis, J., Horn, G., Rawlings, J.B., Diehl, M.: CasADi: a software framework for nonlinear optimization and optimal control. Math. Programm. Comput. **11**(1), 1–36 (2019). https://doi.org/10.1007/s12532-018-0139-4
7. Wächter, A., Biegler, L.T.: On the implementation of an interior-point filter line-search algorithm for large-scale nonlinear programming. Math. Program. **106**(1), 25–57 (2006). https://doi.org/10.1007/s10107-004-0559-y
8. Ferreau, H., Kirches, C., Potschka, A., Bock, H., Diehl, M.: qpOASES: a parametric active-set algorithm for quadratic programming. Math. Programm. Comput. **6**(4), 327–363 (2014). https://doi.org/10.1007/s12532-014-0071-1
9. Ebel, H., Eberhard, P.: A comparative look at two formation control approaches based on optimization and algebraic graph theory. Robot. Auton. Syst. **136**, 103686 (2021). https://doi.org/10.1016/j.robot.2020.103686

Recent Trends in Mobile Robotics for 3D Mapping in Agriculture

Diego Tiozzo Fasiolo[1,2], Lorenzo Scalera[1], Eleonora Maset[1], and Alessandro Gasparetto[1(✉)]

[1] University of Udine, Udine, Italy
{lorenzo.scalera,eleonora.maset,alessandro.gasparetto}@uniud.it
[2] University of Naples Federico II, Naples, Italy
diego.tiozzofasiolo@unina.it

Abstract. This article presents trends and future developments in mobile robotics for 3D mapping in agriculture. Recent examples of robotic platforms and sensors are first presented to highlight the technologies adopted for autonomous surveying in the agricultural field. Then, localization and mapping approaches are discussed, as well as path planning algorithms for the navigation of mobile robots in orchards and crops. Finally, insights into applications of artificial intelligence to robotic mapping are given to evaluate the potentiality of neural networks in this field. The results of the survey indicate research directions and suggest future applications of mobile robotics as an efficient tool for smart agriculture.

Keywords: Mobile robotics · 3D mapping · Precision agriculture · SLAM · Path planning · Artificial intelligence

1 Introduction

In recent years, mobile robotics has been applied in several different fields to autonomously acquire 3D information for surveying and monitoring. Notable applications include the mapping of hazardous and cluttered areas [13], indoor and outdoor buildings and structures [1,18], as well as archaeological sites [14]. Agriculture is a challenging field in which robotic systems are adopted for various tasks, as for instance, planting, sowing, weeding, harvesting and phenotyping [21]. Their use is expected to boost precision agriculture (PA), with significant advantages in terms of environmental impact of the crop production cycle, as they enhance speed, coverage, repeatability, and cost effectiveness [12].

In this context, mobile robotic platforms (often referred to as unmanned ground vehicles, UGVs) equipped with sensors and computational capabilities are increasingly used for generating 3D maps that provide information about health, number of plants, inflorescence, presence of parasites or diseases, water content as well as chemical and morphological characteristics [4]. More in detail, the near infrared (NIR) radiation reflected by vegetation can be measured by means of multi-spectral cameras, and indexes that provide information to assess

A. Müller and M. Brandstötter (Eds.): RAAD 2022, MMS 120, pp. 428–435, 2022.
https://doi.org/10.1007/978-3-031-04870-8_50

the crop health, such as the Normalized Difference Vegetation Index (NDVI) [11,29], can be computed. Moreover, from a 3D map it is possible to measure volume of plants and canopy height, which are necessary to monitor crop growth, reduce pesticide losses, and make agriculture more efficient and sustainable [24].

To navigate and monitor the environment, robots need to perform online localization and mapping. Positioning based on Global Navigation Satellite System (GNSS) is the most used method for localization in outdoor scenarios and, in particular, in agriculture [2]. However, when satellite signals are unavailable or inaccurate, the simultaneous localization and mapping (SLAM) approach is a promising solutions for localization reliability and availability.

Furthermore, autonomous and safe navigation is one of the hardest challenges in the agricultural context, since agrarian environments are intricate and unstructured [25]. For this reason, path planning strategies commonly used for indoor environments may not fit the agricultural requirements, leading to the necessity of implementing ad-hoc approaches to plan the route of a robot avoiding obstacles and optimizing the required task.

Artificial intelligence (AI) technologies, especially machine learning, are expected to play a relevant role in 3D mapping, and will be essential enablers for agricultural robots [9]. In the mapping context, deep learning methods are investigated for the interpretation of sensor data, recognition, classification and segmentation, especially coupled with vision systems, including depth perception and scanning sensors, such as Light Detection And Ranging (LiDAR).

The main contribution of this paper is a discussion on recent trends and future developments of mobile robotics for 3D mapping in agriculture. Actual examples of robotic platforms are first presented to highlight the technologies adopted for autonomous surveying in agricultural scenarios, with particular focus on the characteristics commonly found in recent agricultural applications (Sect. 2). Then, SLAM approaches are surveyed and compared in Sect. 3. Moreover, insights into path planning strategies (Sect. 4) and applications of AI (Sect. 5) are illustrated. Finally, the conclusions are given in Sect. 6.

2 Robotic Platforms and Sensors

In the following, recent examples of robotic platforms and sensors are described to outline the state of the art and the research trends for agricultural purposes. UGVs are developed with different kinematics to adapt to the navigation in agricultural harsh terrains: differential wheeled robots [11] (Fig. 1(a)), high-grip tracked vehicles [24] (Fig. 1(b), 1(c)), and solutions with multiple steering wheels [6] (Fig. 1(d)). Common sensor suites for UGVs in agriculture include 2D and 3D LiDAR, as well as stereo, time-of-flight, and monocular cameras as exteroceptive sensors to acquire data from the surrounding environment. Furthermore, Real-Time Kinematics positioning systems (RTK-GNSS), often coupled with Inertial Measurement Units (IMUs) are adopted for localization purposes.

A recent example is given by the remotely-controlled compact platform for in-row under-canopy crop monitoring developed in [17] (Fig. 1(e)). The novelty

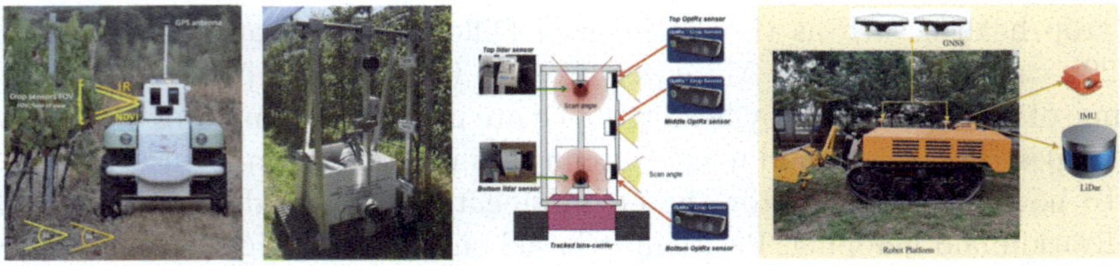

(a) Vinescout [11]. (b) ByeLab [24]. (c) Tracked robot [30].

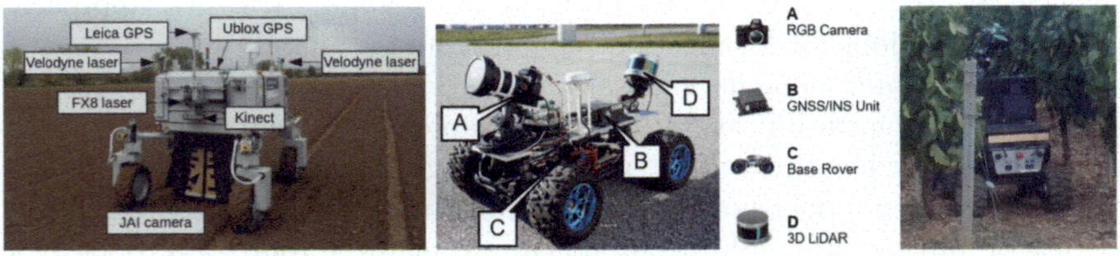

(d) Bonirob [6]. (e) Undercanopy UGV [17]. (f) Summit [8].

Fig. 1. Examples of robotic systems for mapping in agriculture.

of that system is the capability of capturing images on both sides of the UGV, by means of one front camera equipped with fisheye lens only.

Another interesting UGV is Bonirob [6] (Fig. 1(d)), which features a complete sensor suite, including a multi-spectral camera. It is used to collect an open-source large-scale agricultural data set. Nevertheless, only a paucity of robotic systems manage to superimpose radiometric information to the point cloud acquired with laser scanners. Byelab [29] (Fig. 1(b)), and the robot developed by Clamens et al. [8] (Fig. 1(f)) are among them.

AgRob V16 [25] is developed for vineyards monitoring and supports two interchangeable sensor suites: the former includes a monocular camera with special light filters to extract NDVI images; the latter configuration comprises a 3D LiDAR, a robotic arm manipulator, and a thermal camera. However, thermal data are not merged with the point cloud acquired by the LiDAR. To the best of our knowledge, no autonomous robotic system are able to directly superimpose thermal information to an agricultural 3D map.

BrambleBee [31] is similar to AgRobV16, and it is capable of pollinating with its end-effector: an endoscope camera is used for additional guidance of the robotic arm. Sensors carried by a robotic arm provide a flexible solution to the occlusion problem that is common in image-based phenotyping [4], as they manage to position and orient cameras at the best viewpoint. No LiDAR and multi-spectral cameras are found in literature in a eye-in-hand configuration yet. Most UGVs rely on storage memories to manage data, requiring a frequent operator support. Internet of Robotic Things will be a solution to this problem. Moreover, storing data on clouds will make them easily available from remote.

Complete sensor suites ensure for a flexible system, but increase the energy consumption. To solve this problem, multiple and modular sensor suites could be developed. Moreover, for large-scale operations renewable and sustainable energy sources, like solar panels and hydrogen fuel cells should be integrated. Swarm robotics could be another solution to extend the range of action, since the farmland can be divided into sub-maps covered by different robots. Finally, attention should be given to the durability and availability of these systems and sensors as they operate in harsh outside environments.

3 Localization and Mapping

In this section, an overview on recent localization and mapping strategies that overcome pose estimation issues in rough agricultural terrains is given. Classic localization methods (based on wheel encoders, IMUs and GNSS data) do not often provide adequate accuracy and SLAM algorithms are therefore needed. SLAM problem refers to the estimation of the state of the robot concurrently with the construction of a model of the environment. With this method, data deriving from the previously mentioned sensors are integrated with LiDAR (known as LiDAR SLAM) or cameras (i.e., Visual SLAM).

The strategy of Visual SLAM is to sequentially estimate the camera poses by tracking feature points in the image sequence. In this context, it is worth citing RTAB-Map [15], an open-source software that provides feature-based Visual SLAM with loop closure (i.e., the recognition of a previously mapped region, with the aim of reducing trajectory drift and increasing map accuracy). RTAB-Map is frequently used in the agricultural research field [19,20,23]; nevertheless, conflicting results regarding its performance are found in literature and additional tests in agricultural environments are needed. In [22], RTAB-Map is improved aligning LiDAR scans with a 2D map and integrating wheel odometry.

The main solution for LiDAR SLAM is the scan-matching approach, usually based on the Iterative Closest Point (ICP) algorithm for the alignment of pairs of point clouds, followed by graph optimization [30]. To obtain higher accuracy and lower computational requirements, in [3] an online clustering algorithm is integrated with the previous VineSLAM method [25] to reduce data size and speed up the refinement step based on ICP.

Fusing Visual and 3D LiDAR SLAM is a promising approach, but it is not fully implemented in agricultural UGVs yet. To this regard, a benchmark of the up-to-date SLAM algorithms in agricultural field is needed, together with a standard for the ground truth. Besides, the environment dictates the conditions on which sensor is the most advantageous, and, therefore, the robot should be able to choose the best data source according to the external conditions. The application of LiDAR and RGB cameras is a common practice and many algorithms are available (also open-source) that use data acquired by these sensors for navigation and mapping. On the contrary, multi-spectral and hyper-spectral cameras, fundamental in agriculture, are not extensively considered in the field of computer vision and robotics, and there is a lack of methods to adequately merge the data acquired with these sensors with most common ones.

4 Path Planning

Planning the path of a robot consists of finding a sequence of translations and rotations from a starting point to a goal one, while avoiding obstacles in the environment [26]. Coverage path planning is the task of determining a path that passes through all points of an area, and it is advanced in this field, given its need for operations such as harvesting, pruning and spraying. A common solution is to use cell decomposition, or graph Voronoi diagram [31], and row following (exploiting particle filters or Kalman filters [5]), with zig-zag patterns.

Regarding point-to-point path planning in static environment, several algorithms have been tested: artificial potential field, sampling based methods, genetic algorithms, and methods that fall into the Traveling Salesman Problem. In [16], for example, the Ant Colony algorithm, based on the behavior of ants searching for food, is used to reduce total measurement time, improve efficiency and save energy of an UGV operating in a potato field. Further discrete optimization algorithms (e.g., A* [25], Dijkstra, RRT* [7]) have been modified to rely on robot pose and kinematic constraints or to provide planning for automatic recharging system and a solution to avoid repetitive paths, minimizing the soil compaction effects. In [7], RRT* is used in combination with the Reeds-Shepp algorithm, considering non-holonomic constraints of the robot, to deal with narrow pathways of a greenhouse.

There is a lack of online solutions for path planning in dynamic environments. A popular tool is ROS Navigation Stack [31]: using Dijkstra for high level path planning and the dynamic window approach for obstacle avoidance. However, further tests are needed to provide a guideline on the best parameters for this navigation framework in each agricultural environment.

All the above methods require prior information: manual measures [31], satellite images [25], or 2D and 3D maps. To acquire these information, collaboration with unmanned aerial vehicle would be useful, as a substitute to the above sources to provide an updated view of the current conditions. Moreover, up-to-date exploration algorithms not relying on prior data should be exploited [28].

5 Application of Artificial Intelligence

Technologies from areas including data science and AI can be used alongside autonomous systems to automatically fuse and interpret collected data. Nowadays, classification and segmentation algorithms are increasingly applied to add semantic information to the map. In [23], a supervised learning method is proposed to perform terrain characterization, using color and geometric feature and measuring wheel slip, rolling resistance and vibration response experienced by the vehicle. Among the various alternatives provided by convolutional neural networks (CNNs), two can be highlighted: object detection models and image segmentation models. The former can build a bounding box corresponding to each investigated class in the image, whereas the latter create a pixel-wise mask for each object. With the pixel-wise segmentation technique it is possible to obtain a far more granular understanding of objects, with respect to the bounding boxes technique, with the disadvantage that annotating each class requires a greater effort. In [10], You Only Look Once (YOLO), the preeminent solution for object detection in PA, is used to build a hybrid topological map with semantic classification and key location estimation. Instead, segmentation is used in [20] to obtain a 3D map with semantic labels: this probabilistic information is used by the UGV to recognize leaves as traversable regions inside a greenhouse.

In real-world scenarios, data collection for the network training can be challenging and time-expensive. Moreover, the system must be robust to the variability of the field to avoid re-training. To overcome this problem, Generative Adversarial Networks (GANs) can be used to generate photorealistic agricultural images, based on semantic information [32]. The learning process can also be performed by transferring knowledge from a given task that is already learned, and the training can involve only a subset of layers of the CNN. This method, called transfer learning, is used in [25] to detect vine trunks and to develop a feature-based localization framework to navigate in the vineyard. YOLO afterward is used in [19] to perform person detection and following.

Future work will address the use of bounding boxes to improve the quality of point clouds and to remove workers and agricultural machines from it. Deep learning algorithms for timely classification and recognition of diseases are also required to increase the crop quality and yield. This could be achieved by preparing plant pathology data sets with diseases in the early stages. AI can also improve exteroceptive capabilities of the system. For example, Shu et al. [27] presented, to the best of our knowledge, the first evaluation results for monocular SLAM. Their work explores unsupervised depth estimation in a soy-bean field, by simulating RGB-D SLAM. Not relying on LiDAR, this method is less expensive and power consuming. Finally, as UGVs can accumulate a large quantity of data, cloud computing will enhance the data processing efficacy, offer higher data security and scalability, and minimize costs.

6 Conclusion

In this article, trends and future developments in mobile robotics for 3D mapping in agriculture have been surveyed. Recent examples of UGVs have been reported emphasizing the importance of developing platforms capable of moving in rough terrains, and sensors to provide information about volume, height and health of plants, among others. Then, SLAM approaches and path planning algorithms have been discussed, pointing out that more effort will be necessary to deal with unknown and dynamic harsh environment, during mapping and phenotyping tasks. Finally, insights into applications of AI to robotic mapping have been given to evaluate the potentiality of neural networks. The three topics covered in this paper must be integrated in a tightly coupled manner in the robotic system, to make it autonomous for mapping, and to compensate for the weaknesses of the standalone solutions. In the coming years there will be great progress in autonomous systems for PA thanks to multidisciplinary activities involving agronomy, robotics, and AI. Future works will address the creation of a taxonomy of hardware solutions and software capabilities of robotics systems in PA.

References

1. Adán, A., Quintana, B., Prieto, S.A.: Autonomous mobile scanning systems for the digitization of buildings: a review. Remote Sens. **11**(3), 306 (2019)
2. Aguiar, A.S., dos Santos, F.N., Cunha, J.B., Sobreira, H., Sousa, A.J.: Localization and mapping for robots in agriculture and forestry: a survey. Robotics **9**(4), 97 (2020)
3. Aguiar, A.S., dos Santos, F.N., Sobreira, H., Cunha, J.B., Sousa, A.J.: Particle filter refinement based on clustering procedures for high-dimensional localization and mapping systems. Robot. Autonom. Syst. **137**, 103725 (2021)
4. Atefi, A., Ge, Y., Pitla, S., Schnable, J.: Robotic technologies for high-throughput plant phenotyping: contemporary reviews and future perspectives. Front. Plant Sci. **12** (2021)
5. Blok, P.M., van Boheemen, K., van Evert, F.K., IJsselmuiden, J., Kim, G.H.: Robot navigation in orchards with localization based on particle filter and Kalman filter. Comput. Electr. Agric. **157**, 261–269 (2019)
6. Chebrolu, N., Lottes, P., Schaefer, A., Winterhalter, W., Burgard, W., Stachniss, C.: Agricultural robot dataset for plant classification, localization and mapping on sugar beet fields. Int. J. Rob. Res. **36**(10), 1045–1052 (2017)
7. Choudhary, A., Kobayashi, Y., Arjonilla, F.J., Nagasaka, S., Koike, M.: Evaluation of mapping and path planning for non-holonomic mobile robot navigation in narrow pathway for agricultural application. In: International Symposium on System Integration, pp. 17–22 (2021)
8. Clamens, T., Alexakis, G., Duverne, R., Seulin, R., Fauvet, E., Fofi, D.: Real-time multispectral image processing and registration on 3d point cloud for vineyard analysis. In: International Conference on Computer Vision Theory and Applications (2021)
9. Duckett, T., et al.: Agricultural robotics: the future of robotic agriculture. UK-RAS White papers (2018)
10. Emmi, L., Le Flécher, E., Cadenat, V., Devy, M.: A hybrid representation of the environment to improve autonomous navigation of mobile robots in agriculture. Precis. Agric. **22**(2), 524–549 (2021)
11. Fernández-Novales, J., et al.: Monitoring and mapping vineyard water status using non-invasive technologies by a ground robot. Remote Sens. **13**(14), 2830 (2021)
12. Fountas, S., Mylonas, N., Malounas, I., Rodias, E., Hellmann Santos, C., Pekkeriet, E.: Agricultural robotics for field operations. Sensors **20**(9), 2672 (2020)
13. Kim, P., Park, J., Cho, Y.K., Kang, J.: UAV-assisted autonomous mobile robot navigation for as-is 3D data collection and registration in cluttered environments. Autom. Constr. **106**, 102918 (2019)
14. Kurazume, R., Oshima, S., Nagakura, S., Jeong, Y., Iwashita, Y.: Automatic large-scale three dimensional modeling using cooperative multiple robots. Comput. Vis. Image Underst. **157**, 25–42 (2017)
15. Labbé, M., Michaud, F.: RTAB-MAP as an open-source lidar and visual simultaneous localization and mapping library for large-scale and long-term online operation. J. Field Robot. **36**(2), 416–446 (2019)

16. Mai, T., Shao, S., Yun, Z.: The path planning of agricultural AGV in potato ridge cultivation. Ann. Adv. Agric. Sci. **3**(2) (2019)
17. Manish, R., Lin, Y.C., Ravi, R., Hasheminasab, S.M., Zhou, T., Habib, A.: Development of a miniaturized mobile mapping system for in-row, under-canopy phenotyping. Remote Sens. **13**(2), 276 (2021)
18. Maset, E., Scalera, L., Beinat, A., Cazorzi, F., Crosilla, F., Fusiello, A., Gasparetto, A.: Preliminary comparison between handheld and mobile robotic mapping systems. In: International Workshop IFToMM for Sustainable Development Goals, pp. 290–298. Springer, Cham (2021).https://doi.org/10.1007/978-3-030-87383-7
19. Masuzawa, H., Miura, J., Oishi, S.: Development of a mobile robot for harvest support in greenhouse horticulture-person following and mapping. In: International Symposium on System Integration, pp. 541–546. IEEE (2017)
20. Matsuzaki, S., Masuzawa, H., Miura, J., Oishi, S.: 3D semantic mapping in greenhouses for agricultural mobile robots with robust object recognition using robots' trajectory. In: International Conference on Systems, Man, and Cybernetics, pp. 357–362 (2018)
21. Oliveira, L.F., Moreira, A.P., Silva, M.F.: Advances in agriculture robotics: a state-of-the-art review and challenges ahead. Robotics **10**(2), 52 (2021)
22. Post, M.A., Bianco, A., Yan, X.T.: Autonomous navigation with ROS for a mobile robot in agricultural fields. In: International Conference on International Conference on Informatics in Control, Automation and Robotics (2017)
23. Reina, G., Milella, A., Galati, R.: Terrain assessment for precision agriculture using vehicle dynamic modelling. Biosyst. Eng. **162**, 124–139 (2017)
24. Ristorto, G., Gallo, R., Gasparetto, A., Scalera, L., Vidoni, R., Mazzetto, F.: A mobile laboratory for orchard health status monitoring in precision farming. Chem. Eng. Trans. **58**, 661–666 (2017)
25. Santos, L., et al.: Path planning aware of robot's center of mass for steep slope vineyards. Robotica **38**(4), 684–698 (2020)
26. Santos, L.C., Santos, F.N., Pires, E.S., Valente, A., Costa, P., Magalhães, S.: Path planning for ground robots in agriculture: a short review. In: International Conference on Autonomous Robot Systems and Competitions, pp. 61–66. IEEE (2020)
27. Shu, F., Lesur, P., Xie, Y., Pagani, A., Stricker, D.: Slam in the field: an evaluation of monocular mapping and localization on challenging dynamic agricultural environment. In: International Conference on Applications of Computer Vision, pp. 1761–1771 (2021)
28. Umari, H., Mukhopadhyay, S.: Autonomous robotic exploration based on multiple rapidly-exploring randomized trees. In: IIEEE/RSJ International Conference on Intelligent Robots and Systems. pp. 1396–1402 (2017)
29. Vidoni, R., et al.: Byelab: an agricultural mobile robot prototype for proximal sensing and precision farming. In: The International Mechanical Engineering Congress and Exposition (2017)
30. Wang, K., Zhou, J., Zhang, W., Zhang, B.: Mobile lidar scanning system combined with canopy morphology extracting methods for tree crown parameters evaluation in orchards. Sensors **21**(2), 339 (2021)
31. Yang, C., Watson, R.M., Gross, J.N., Gu, Y.: Localization algorithm design and evaluation for an autonomous pollination robot. In: International Meeting of The Satellite Division of the Institute of Navigation, pp. 2702–2710 (2019)
32. Yang, Y., et al.: 3d point cloud on semantic information for wheat reconstruction. Agriculture **11**(5), 450 (2021)

Novel Approaches for Periodic Depth Enhancement in Visual SLAM

Stephan Sandfuchs[1(✉)], Marco Schmidt[2], and Jörg Frochte[1]

[1] Bochum University of Applied Sciences, Heiligenhaus, Germany
{stephan.sandfuchs,joerg.frochte}@hs-bochum.de
[2] University of Applied Sciences Würzburg-Schweinfurt, Schweinfurt, Germany
marco.schmidt@fhws.de

Abstract. Most visual simultaneous localization and mapping (SLAM) systems use either monocular or stereo sensor information. However, in some situations, depth information may not be available for each camera frame, for example, when two cameras with different frame rates are used or the computer used does not have the computational resources to calculate a depth map for each camera frame. This work presents two novel approaches by enhancing either each n_k-th keyframe or each n_f-th camera frame periodically with depth information. The experimental results on the KITTI visual odometry benchmark show that both approaches improve scale drift compared to monocular SLAM. In terms of trajectory accuracy, the periodic camera frame enhancement outperforms the other novel approach. Even if only a small number of camera frames is enhanced with depth information (10% to 20%), this approach achieves comparable or more accurate results than stereo trajectories. Therefore, this work introduces novel approaches for depth enhancement in visual SLAM systems that do not require depth information at every camera frame. This allows systems with limited access to measured depth or limited computational resources to improve on scale drift.

Keywords: Depth enhancement · Monocular SLAM · Stereo SLAM · ORB-SLAM2

1 Introduction

In recent years, monocular and stereo cameras have been the most commonly used sensors in visual SLAM research. A common issue faced by monocular systems is the scale drift of the map and the resulting robot trajectory. Monocular systems have to estimate the scale for every camera frame based solely on 2D geometric information. The first depth estimation methods were statistical models with specific filters or textures such as [15]. Other methods use image pyramids to check which features are observable and at what scale [1], and then estimate the distance to the camera [9]. However, these methods often fail in complex environments and are susceptible to scale variations in distance. The issue of

A. Müller and M. Brandstötter (Eds.): RAAD 2022, MMS 120, pp. 436–443, 2022.
https://doi.org/10.1007/978-3-031-04870-8_51

scale drift occurs when estimation errors accumulate and the estimated scale in a local area drifts with time.

Therefore, some monocular SLAM systems are adapted to support stereo cameras as sensors. Because of their architecture, stereo cameras can be used to calculate a depth map from the disparity between the left and right images. While this reliably solves the problem of scale drift, it also requires the use of a completely different sensor or an additional camera. Although the necessary disparity matching for each pixel can consume many computational resources, the easily available depth information in the stereo SLAM system leads to better performance compared to a monocular system, especially in terms of scale drift.

In SLAM systems, the combination of stereo and monocular information is often not considered [2–6,8,9,16]. Though, under certain conditions it may be necessary to process both stereo and monocular information during runtime. One reason could be that two fixed cameras with different frame rates are used as a stereo camera, or that the device controlling a robot does not have the computational resources to process a depth map for each camera frame. A second reason could be the increasingly widespread use of artificial neural networks to estimate depth information from a single image, which is a major challenge for mobile robots with limited computational resources and no GPU support. Although some research has been done for the latter use case, most of it focuses on either improving the real-time capabilities of the trained models on CPUs and embedded CPUs [10,11] or optimizing the architecture of the models to increase the accuracy of the trajectories [14,17]. Even though these research topics are important, to our knowledge there is no research on the approaches of introducing depth information into a monocular SLAM system.

What if depth information does not need to be available at each camera or keyframe to obtain an accurate trajectory and reduce scale drift? To answer this question, we present two different approaches to introducing depth information into a SLAM system, with the goal of analyzing and determining the effects of each approach on scale drift and trajectory accuracy.

2 Related Work

The focus of this work is not to compare multiple SLAM systems, but to analyze the impact that different approaches to introducing depth information may have on scale drift and trajectory accuracy in the scenario where limited depth information is available.

Therefore, we use the stereo system ORB-SLAM2 [8], because it is widely used and designed to handle monocular keypoints, even if it otherwise uses a stereo camera. But ultimately, any system such as [2–6,8,9,16] could have been used. ORB-SLAM2 also uses a global covisibility graph outside a local window, unlike, e.g. [3,6,16,17] and it's monocular predecessor ORB-SLAM has scale drift issues in certain situations [9], so the improvements in scale drift are easy to show.

ORB-SLAM2 and the monocular ORB-SLAM are both feature-based visual SLAM systems using ORB features as keypoints [8,9,12]. Depth information is encoded as stereo keypoints $\mathbf{x}_s = (u_L, v_L, u_R)$ which represent a correspondance between the left key point (u_L, v_L) and the horizontal coordinate of the right keypoint u_R. It also supports RGB-D cameras transforming the depth value into a virtual right coordinate [13]. Therefore, from this point on, only the monocular and the stereo ORB SLAM are distinguished.

During feature extraction, each key point is also classified as close or far depending on whether it is closer or farther than 40 times the base value. Close points provide accurate scale, translation and rotation information, while far points provide accurate rotation information but weak scale information. In stereo mode, monocular keypoints $\mathbf{x}_m = (u_L, v_L)$, which have no match in the right image, are used for rotation and translation estimation, but not for scale information. In monocular mode, the scale is estimated by an image pyramid and then further optimized with Sim(3) coordinates during loop closure [9].

During pose tracking the current camera pose $\left(\mathbf{t}_c \in \mathbb{R}^3, \mathbf{R}_c \in SO(3)\right)$ is optimized by motion-only bundle adjustment (BA) [9]. The optimization minimizes a reprojection error between the measured keypoints $\mathbf{x}^i_{(\cdot)}$ of the most recent image and the associated map point position $\mathbf{X}^i$, with the set of all matches $\mathcal{X}$, with the robust Huber cost function ρ and the covariance matrix Σ associated to the scale of the keypoint:

$$\{\mathbf{R}_c, \mathbf{t}_c\} = \underset{\mathbf{R}_c, \mathbf{t}_c}{\operatorname{argmin}} \sum_{i \in \mathcal{X}} \rho\left(E(c,i)\right), \quad E(c,i) = \left\| \mathbf{x}^i_{(\cdot)} - \pi_{(\cdot)}\left(\mathbf{R}_c \mathbf{X}^i + \mathbf{t}_c\right) \right\|^2_{\Sigma} \tag{1}$$

The stereo reprojection function $\pi_s \colon \mathbb{R}^3 \to \mathbb{R}^3$ and the monocular reprojection function $\pi_m \colon \mathbb{R}^3 \to \mathbb{R}^2$ project a map point $\mathbf{X}^i$ into the keypoint coordinates.

After optimizing the camera pose, a decision is made whether to insert the tracked camera frame as a new keyframe into the covisibility graph. A keyframe is inserted if more than 10% of the observed map points deviate from the reference keyframe or enough time has passed since the last insertion of a keyframe [8]. The undirected weighted covisibilty graph uses the keyframes as nodes connected by edges whose weight represents the number of shared observations, i.e., map points [13] The Essential Graph reduces the number of edges to the most important ones with more than 100 shared observations and introduces loop closure edges [9].

An optimization of this Essential Graph is performed over a set of covisible keyframes $\mathcal{K}_L$ and the map points $\mathcal{P}_L$ observed by them. The keyframes $\mathcal{K}_F$ outside $\mathcal{K}_L$ are fixed. With the set $\mathcal{X}_k$ containing the matches between the keypoints in frame k and the map points in $\mathcal{P}_L$, the optimization is defined as follows:

$$\{\mathbf{X}^i, \mathbf{R}_l, \mathbf{t}_l | i \in \mathcal{P}_L, l \in \mathcal{K}_L\} = \underset{\mathbf{X}^i, \mathbf{R}_l, \mathbf{t}_l}{\operatorname{argmin}} \sum_{k \in \mathcal{K}_L} \sum_{j \in \mathcal{X}_k} \rho\left(E(k,j)\right) \tag{2}$$

This optimization is performed in two cases [8,9]. During local map generation, the keyframes poses observing the same map points as the current camera frame, and these map points are optimized by a local BA over a factor of the Essential Graph. During loop closure, a global BA over the entire Essential Graph optimizes all keyframe poses and map point positions. In monocular ORB-SLAM the global BA is performed using Sim(3) constraints, which also optimize the scale [9].

3 Approach

The general approach chosen for this work is the configurable step wise reduction of depth information from a stereo-like case to a monocular SLAM. Therefore, the new approaches can be configured depending on the constraints of the system. The presented SLAM system uses camera images in the initial pose estimation and keyframes in the (covisibility) graph for optimization. Thus, depth information can be introduced during both operations, and two different approaches can be derived to enhance either keyframes or frames. For the purposes of this work, a keyframe or camera frame is defined as enhanced, if it has access to some kind of depth information.

In the first approach, every n_k-th keyframe is enhanced (see Fig. 1a). Therefore, a predefined number of enhancements per set of keyframes is usable for optimization. However, some variation is to be expected if a SLAM system uses a keyframe culling mechanism during runtime.

In the second approach each n_f-th camera frame is enhanced (s. Fig. 1b). Thus, depth information is regularly available during pose tracking and some keyframes are also enhanced. While an enhanced camera frame is periodically available during tracking, the number of enhanced keyframes can vary. The keyframe decision is mostly made based on the change in observations, so not every enhanced camera frame might be selected as keyframe.

In ORB-SLAM2 each optimization type is based on the reprojection error E in (1). Depending on the optimization type the target variables and the size of the local window change. Therefore, the number of covisible keyframes in the set $\mathcal{K}_L$ and the set of observable map points $\mathcal{X}_k$ change as well.

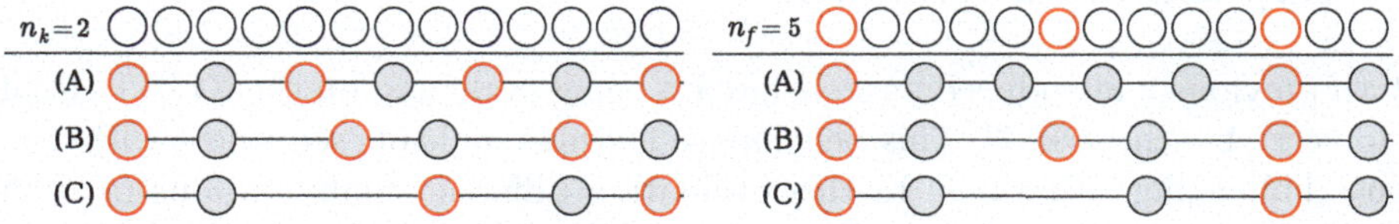

(a) Every 2nd keyframe ($n_k = 2$) has depth information

(b) Every 5th camera frame ($n_f = 5$) has depth information

Fig. 1. Examples for the distribution of enhanced nodes under different periodic keyframe selections using each 2nd camera frame (A) or the periodic sequences {2[nd], 3rd} (B) and {2[nd], 5th, 3rd} (C). The white circles are camera frames and the grey circles are keyframes. A thick red contoure indicates depth enhancement.

When using the first approach of enhancing each n_k-th keyframe, the motion-only BA is always done with a monocular reprojection and coordinates. The keypoint distances from the camera are all estimated. Therefore, the scale covariance Σ might not be very accurate. However, the keypoint coordinates $\mathbf{x}_m^i$ are not affected while the reprojection function π_m is partially affected by an inaccurate scale estimation.

The benefit this approach has towards pure monocular SLAM is that some distances between keyframes and map points are more accurate if the keyframe they are observed from is enhanced. Although no real constraint these distances create a rigid relationship between a keyframe pose $(\mathbf{R}_k, \mathbf{t}_k)$ and the observable map points $\mathcal{X}_k$. Thus, the relative distances between a keyframe and it's observable map points are more likely to be correct, reducing the scale uncertainty resulting in a more accurate the scale covariance Σ.

For motion-only BA these rigid relationships have an indirect benefit. The initial pose is optimized using the more scale accurate reprojection of a map point observable by an enhanced keyframe. In local and global BA these relationships result in a covisibility graph with periodically occurring enhanced nodes lowering the chance of scale drift to accumulate. Therefore, the enhancement of the covisibility graph indirectly affects the initial monocular pose estimation, reducing the chance of scale drift.

The second approach affects both the tracking as well as local and global optimization to some degree. It has the same benefits as the previous approach, when the current camera frame is not enhanced. However, if the current frame is enhanced, the measured distances are more accurate as well improving the estimation accuracy for the initial camera compared to the first approach. Given that each n_f-th frame has an accurate distance measurement not a lot of scale drift is accumulated during tracking.

The issue this approach faces is the scene dependant keyframe decision, which might lead to non-enhanced camera frames to be inserted as keyframes. Therefore, some depth information is not carried over to the graph rendering it unavailable for local or global BA.

4 Experimental Evaluation

Both previously introduced approaches are being evaluated on the KITTI visual odometry benchmark [7]. This benchmark features multiple sequences with various challenging situations. In this evaluation different configurations of both approaches are executed on each of the publicly available sequences (00-10). As references for comparison the groundtruth trajectory as well as recorded trajectories using stereo ORB-SLAM and monocular ORB-SLAM respectively are used. For this evaluation both approaches are implemented as a modified version of the ORB-SLAM2 ROS node [8]. For each approach multiple configurations are used. Each configuration is executed ten times on all sequences. For the final evaluation the results in each sequence are averaged to achieve a more representative result.

For each configuration the mean distance $\bar{d}$ between the keyframe poses and the groundtruth poses and it's standard deviation σ_d are calculated as an indication of the overall accuracy of each configuration. Additionally an average relative scale indicator $\bar{r}_s$ is used as a moving window average over the trajectory $\mathcal{X}$ calculating the relative distance difference quotient of n_r subsequent keyframes $[\mathbf{x}^i, \mathbf{x}^{i-n_r}]$ and the groundtruth in the same section $[\mathbf{x}_g^i, \mathbf{x}_g^{i-n_r}]$:

$$\bar{r}_s = \frac{1}{N} \sum_{i \in \mathcal{X}} \frac{\|\mathbf{x}^i - \mathbf{x}^{i-n_r}\|^2}{\|\mathbf{x}_g^i - \mathbf{x}_g^{i-n_r}\|^2} \tag{3}$$

5 Experimental Results

The experimental results show that the approach enhancing each n_f-th camera frame is an improvement of monocular ORB-SLAM, and is in certain sequences even comparable to stereo ORB-SLAM. Table 1 displays the average distance between the keyframe and the groundtruth poses $\bar{d}$ including it's standard deviation σ_d as well as the average relative scale indicator $\bar{r}_s$ with $n_r = 10$ for all KITTI sequences. Figure 2 displays selected sequences for visual comparison.

Based on these results the approach enhancing camera frames (n_f) performs significantly better than the approach enhancing only keyframes (n_k). This is also visible in Fig. 2 as the trajectories for $n_f = 5$ and $n_f = 10$ are much closer to the groundtruth than those from the other approach.

Both new approaches improve the relative scale drift $\bar{r}_s$ compared to monocular ORB-SLAM. Especially in the sequences where significant scale drift occurs (02, 08, 10) both approaches greatly improve scale accuracy. As expected, the inaccuracies mostly occur either at sharp corners or during straight sections with few observable close keypoints. Even though the approach enhancing only keyframes is more accurate than monocular ORB-SLAM in some sequences (02, 08, 10), where scale drift is especially problematic, it is still the worst approach in most other sequences. However, this approach might work better in other SLAM systems, but without modifications to the tracking and local optimization it is not suited to be used in ORB-SLAM.

Table 1. Experimental results. The most accurate approach is highlighted.

Seq.	$n_k = 1$			$n_k = 2$			$n_f = 5$			$n_f = 10$			monocular			stereo		
	$\bar{d}\,[m]$	$\sigma_d\,[m]$	$\bar{r}_s$	$\bar{d}\,[m]$	$\sigma_d\,[m]$	$\bar{r}_s$	$\bar{d}\,[m]$	$\sigma_d\,[m]$	$\bar{r}_s$	$\bar{d}\,[m]$	$\sigma_d\,[m]$	$\bar{r}_s$	$\bar{d}\,[m]$	$\sigma_d\,[m]$	$\bar{r}_s$	$\bar{d}\,[m]$	$\sigma_d\,[m]$	$\bar{r}_s$
00	12.92	10.98	0.94	14.07	10.24	0.95	7.17	3.22	0.97	**6.71**	2.93	0.96	10.69	6.01	0.96	7.81	4.77	0.97
01	340.18	308.32	1.56	197.12	216.82	1.43	**11.36**	9.35	1.01	127.49	209.28	1.01	599.08	469.40	1.60	75.04	74.21	1.01
02	66.26	37.57	0.96	64.80	44.45	0.97	**12.10**	5.31	1.01	15.88	8.89	1.01	164.17	129.93	1.87	13.77	8.09	1.01
03	20.01	13.47	1.00	22.58	17.09	1.00	<u>7.29</u>	4.88	1.00	8.04	5.15	1.01	**5.45**	3.08	0.98	6.89	4.75	1.00
04	3.40	2.51	1.01	2.74	2.43	1.01	**1.64**	0.97	1.01	1.77	1.41	1.01	2.09	2.23	1.01	1.81	1.09	1.01
05	30.34	6.75	1.00	31.42	6.56	1.00	<u>3.71</u>	2.66	1.01	6.37	5.30	1.01	19.82	7.69	1.02	**3.70**	3.36	1.00
06	22.88	17.39	1.06	34.47	37.55	1.03	**2.39**	1.36	1.01	4.55	3.42	1.02	16.80	13.50	1.08	3.02	1.93	1.01
07	5.34	3.68	0.97	4.54	2.99	0.97	<u>1.49</u>	1.00	0.99	1.87	1.59	0.99	2.92	1.75	0.98	**0.91**	0.46	0.99
08	127.12	92.08	0.91	132.28	99.96	0.92	<u>15.04</u>	4.38	0.99	16.68	4.58	0.99	206.94	149.99	0.85	**14.02**	4.86	0.99
09	17.09	25.10	1.03	28.39	51.95	1.02	**6.29**	3.25	1.01	6.76	4.53	1.00	19.72	26.62	1.00	12.50	12.52	1.01
10	9.34	7.05	0.97	10.06	8.34	0.97	<u>5.81</u>	3.68	1.01	6.05	3.90	1.01	66.12	53.90	1.98	**4.97**	3.44	1.01

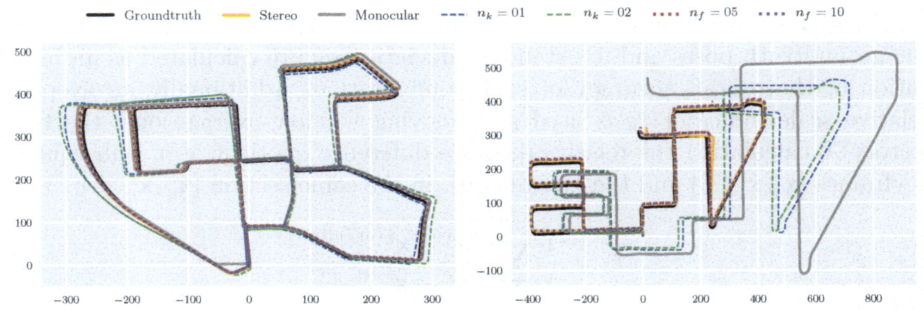

(a) Sequence 00 (b) Sequence 08

Fig. 2. Exemplary results displaying the scale drift and accuracy improvements

The two configurations for the approach enhancing camera frames perform very similar. As expected, in most sequences the trajectory using $n_f = 10$ is less accurate compared to $n_f = 5$, but in sequence 00 it is slightly better. In about half of the sequences the approach using $n_f = 5$ is even more accurate than the stereo trajectory. Overall the approach enhancing camera frames periodically performs very well compared to the stereo and groundtruth trajectory considering the depth information is only available in 10% to 20% of camera frames.

6 Conclusion

In this work, two new approaches to periodic depth enhancement in SLAM systems are presented. The experimental results show that the approach of enhancing each n_f-th camera frame can be used as an alternative to stereo ORB-SLAM. Systems or robots where either depth information is not available at each camera frame or computational resources do not allow the computation of a depth map at each camera frame can benefit from using this approach instead of pure stereo or monocular SLAM.

The other introduced approach of enhancing each n_k-th keyframe is able to improve the scale drift compared to monocular ORB-SLAM. However, the experimental results show that the approach of enhancing each n_f-th camera frame is more promising in terms of overall performance. Therefore, in future work, we plan to analyze whether different visual SLAM systems can take better advantage of the scale drift improvements from periodic keyframe enhancement than ORB-SLAM.

In summary, this work shows that the introduced periodic depth enhancement approaches improve the scale drift of monocular SLAM, and can even provide more accurate results than stereo SLAM in certain situations. This is especially beneficial for robots or other systems with limited computational resources or those that use cameras with different frame rates. Therefore, in future works we want to research the possibilities of these approaches even further.

References

1. Adelson, E.H., Anderson, C.H., Bergen, J.R., Burt, P.J., Ogden, J.M.: Pyramid methods in image processing. RCA Eng. **29**(6), 33–41 (1984)
2. Bescos, B., Fácil, J.M., Civera, J., Neira, J.: DynaSLAM: tracking, mapping, and inpainting in dynamic scenes. IEEE Robot. Autom. Lett. **3**(4), 4076–4083 (2018)
3. Engel, J., Koltun, V., Cremers, D.: Direct sparse odometry. IEEE Trans. Pattern Anal. Mach. Intell. **40**(3), 611–625 (2017)
4. Engel, J., Schöps, T., Cremers, D.: LSD-SLAM: large-scale direct monocular SLAM. In: Fleet, D., Pajdla, T., Schiele, B., Tuytelaars, T. (eds.) ECCV 2014. LNCS, vol. 8690, pp. 834–849. Springer, Cham (2014). https://doi.org/10.1007/978-3-319-10605-2_54
5. Engel, J., Stückler, J., Cremers, D.: Large-scale direct SLAM with stereo cameras. In: 2015 IEEE/RSJ International Conference on Intelligent Robots and Systems (IROS), pp. 1935–1942. IEEE (2015)
6. Gao, X., Wang, R., Demmel, N., Cremers, D.: LDSO: direct sparse odometry with Loop Closure. In: 2018 IEEE/RSJ International Conference on Intelligent Robots and Systems (IROS), pp. 2198–2204. IEEE, October 2018
7. Geiger, A., Lenz, P., Urtasun, R.: Are we ready for autonomous driving? The KITTI vision benchmark suite. In: Conference on Computer Vision and Pattern Recognition (CVPR) (2012)
8. Mur-Artal, R., Tardós, J.D.: ORB-SLAM2: an open-source SLAM system for monocular, stereo, and RGB-D cameras. IEEE Trans. Robot. **33**(5), 1255–1262 (2017)
9. Mur-Artal, R., Montiel, J.M.M., Tardós, J.D.: ORB-SLAM: a versatile and accurate monocular SLAM system. IEEE Trans. Robot. **31**(5), 1147–1163 (2015)
10. Peluso, V., et al.: Monocular depth perception on microcontrollers for edge applications. IEEE Trans. Circuits Syst. Video Technol. 1 (2021)
11. Poggi, M., Aleotti, F., Tosi, F., Mattoccia, S.: Towards real-time unsupervised monocular depth estimation on CPU. In: 2018 IEEE/RSJ International Conference on Intelligent Robots and Systems (IROS), pp. 5848–5854. IEEE, October 2018
12. Rublee, E., Rabaud, V., Konolige, K., Bradski, G.: ORB: an efficient alternative to SIFT or SURF. In: 2011 International Conference on Computer Vision, pp. 2564–2571. IEEE (2011)
13. Strasdat, H., Davison, A.J., Montiel, J.M., Konolige, K.: Double window optimisation for constant time visual SLAM. In: 2011 International Conference on Computer Vision, pp. 2352–2359. IEEE (2011)
14. Tateno, K., Tombari, F., Laina, I., Navab, N.: CNN-SLAM: real-time dense monocular SLAM with learned depth prediction. In: Proceedings of the IEEE Conference on Computer Vision and Pattern Recognition (CVPR), July 2017
15. Torralba, A., Oliva, A.: Depth estimation from image structure. IEEE Trans. Pattern Anal. Mach. Intell. **24**(9), 1226–1238 (2002)
16. Wang, R., Schwörer, M., Cremers, D.: Stereo DSO: large-scale direct sparse visual odometry with stereo cameras. In: International Conference on Computer Vision (ICCV), Venice, Italy, October 2017
17. Yang, N., Wang, R., Stuckler, J., Cremers, D.: Deep virtual stereo odometry: leveraging deep depth prediction for monocular direct sparse odometry. In: Proceedings of the European Conference on Computer Vision (ECCV), pp. 817–833 (2018)

Analysis of Object Detection Under Different Weather Conditions in Simulated and Real Environment

Pragati Jaiswal(✉), Axel Vierling, and Karsten Berns

Robotics Research Lab, Technische Universität Kaiserslautern, Kaiserslautern, Germany
{p_jaiswal19,vierling,berns}@cs.uni-kl.de

Abstract. In modern robotic applications, like autonomous driving in an outdoor environment, a very important task is robust object detection. The object detection model tends to perform poorly under weather conditions including rain. Therefore, in order to have good performance, the data required for the training of these models has to be from a wide variety of conditions, like different lighting, weather, locations, etc. Any model not trained on a different variety of data is more likely to overfit on the dataset it is trained on and hence have poor performance over other situations. To avoid this situation often simulation data is used. The main goal of this paper is to analyse the performance of the object detection model on the various simulated and real weather conditions in order to achieve good performance under the real rain condition with a model trained only on simulated rain data.

Keywords: Object detection · Weather noises · Faster R-CNN

1 Introduction

Most robotic applications require the perception of their environment in a detailed manner. As the most commonly used sensors are cameras, image processing and object detection plays a crucial role in robotic applications, like autonomous driving. It can become an arbitrarily difficult task to detect objects in the vicinity of the robot in complex environments. Nice object detection can be used in a multitude of ways like safety [1], mapping [2] and localization [3]. The problem of detecting objects in a robust manner becomes even more difficult if an outdoor environment is considered, as here not only the complexity of the environment needs to be considered, but additionally, weather affects the images. Even state of the art object detection solutions are strongly affected by high frequency, similar to e.g. rain, careful consideration of such effects needs to be done. The application behind the considerations of this paper is driving in a campus environment with a delivery robot as seen in Fig. 1.

Most often the object detection model for robotic applications is trained on a simulated environment which often does not reciprocate the same outcome and

A. Müller and M. Brandstötter (Eds.): RAAD 2022, MMS 120, pp. 444–451, 2022.
https://doi.org/10.1007/978-3-031-04870-8_52

Fig. 1. Robot used for real data collection

Table 1. Dataset

Environment	Weather	Number of images
Real	Sun	496
Real	Rain	386
Simulated	Sun	408
Simulated	Rain	218
Simulated	Rain lens	535

performance in the real-world environment, mostly due to very minor differences in the images rendered artificially, as the simulation cannot reproduce similar kinds of results as noises generated by it is different from the noises captured in real-world data. In some situations, weather and other kinds of noises like snow, rain, dust, etc. are incorporated in the simulation. But it is often not guaranteed that the performance would improve as simulation noises do not emulate well with the actual noises.

The best way to make the simulation data usable is to analyse the noises in the real-world data and create simulation data with such noises so that the models trained on simulation data can easily perform in real-world tests. In this paper, we analyse the model's performance on simulation and real data and try to bring the simulation data close to real. In order to analyse the source of differences in the data, the model's filters and feature maps are visualised.

2 Related Work

Object detection is a very common task in image processing and robotics, used in a multitude of applications and subtasks such as face detection [4], pedestrian detection [5], etc. It is able to provide valuable information for semantic understanding of images and videos and is related to many applications, including image classification [6,7], face recognition [8] and autonomous driving [9,10].

However, there is not a strong focus on object detection under specific weather conditions. A major reason might be, that collecting data under difficult weather conditions is inherently difficult, as not all robots and sensor setups are usable under all weather conditions and certain conditions, like snow, only occur seldom. While there exists some work in this direction [11] which uses image adaptive YOLO for adverse weather conditions but they do not use simulation data rather try to adaptively defog and process real environment data to increase the performance of the detection model. Also, recent research tries to make use of the improving simulation engines to generate sufficient realistic data for specific robotic setups [12], but still, weather conditions such as rain are not realistic. In this paper, we try to find the differences between such simulation generated data and real data so as to incorporate them to further improve the

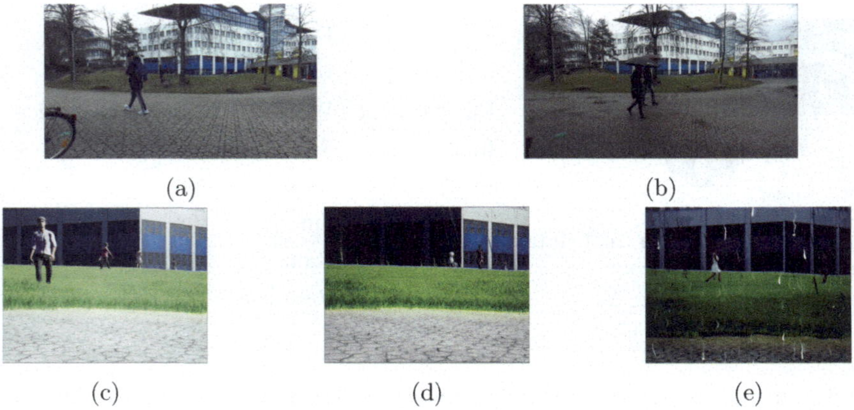

Fig. 2. **(a)** Sun [real environment] **(b)** Rain [real environment] **(c)** Sun [simulated environment] **(d)** Rain [simulated environment] **(e)** Rain lens [simulated environment]

simulation. In research, [13], authors demonstrate the effectiveness of using simulation data to address real-world challenges in the practice of object detection and achieved even higher performance than the model trained on other publicly available data. Since, they are using data from Grand Theft Auto V (GTA V) game environment, they are limiting their application to a single scenario.

3 Proposed Work

The main objective of this paper is to analyse the models trained on real-world data and models trained on different simulation data with the introduction of environment and weather noises especially rain. The paper explores the differences occurring due to changes in the environment. Furthermore, different feature maps and Region Proposal Network (RPN) outputs are visualised to bring forward the inherent changes occurring in the network.

4 Dataset

The dataset is prepared from scratch under different weather conditions in real and simulated environment. The real environment data was collected using a Zed Stereo camera[1] (Fig. 1) mounted on a prototype of autonomous delivery robot with camera fixed at the robot's height. The robot was driven around the campus to collect data under different weather conditions. Multiple video feeds were recorded and later frames were sampled from the video randomly and filtered based on people present in the frame after which they carefully annotated

[1] https://www.stereolabs.com/zed/.

using the CVAT [14] image annotation tool. During the collection of data, no external stabilizations were applied to level or avoid the jitters in the video feed.

For simulated environment data, a simulated model of the university campus made in Unreal Engine was used. Two different weather conditions sunshine and rain were generated for the data collection. Two different rain simulations, one with a clear lens and the other with water droplets on the lens, was done while data collection. Random human objects were added in the simulation walking around and the data was collected from a camera view placed at the robot's height. The camera also captures visible areas of people in the frame, the mask for objects in the frames were converted to bounding boxes [15]. The anomaly due to overlapping boxes was manually rectified using the same annotation tool and also the boxes with the very small visible area (only parts of head or shoes, etc.) were removed. The entire data was manually verified before using it for training the model. Using the above process, 5 different datasets were collected and stored as shown in Table 1. Examples of each are shown in Fig. 2.

5 Training

The training of the model was done on ResNet-50 [16] using different datasets separate from each other in order to keep the models individually optimized for only one type of dataset. As mentioned above, the dataset collected from the real environment and the simulated environment is in 5 different categories. The difference in the number of images among the datasets (Table 1) is small and therefore has a negligible effect on the model's performance. Table 2 summarizes the model name and the dataset used for its training.

Each model had the same training and hyper-parameters to converge to the final result. All the models were loaded with pre-trained weights of the model trained on the COCO-17 dataset [17]. The models were trained for a total of 50k steps with a variable learning rate. Initially, the rate increased linearly from 0 to 0.005 in 5k steps and then a cosine decay was applied till step 50k dropping the rate to zero (see Fig. 3). The linear increase in the learning rate, in the beginning, help the pre-trained model to fit the data smoothly while decaying it improves the learning of complex patterns. Along with this, multiple random augmentations of horizontal flip, hue, contrast, saturation and scale (0.9x to 1.1x) was applied to the image during training which helps the model independent of the counterfeit features of the data space.

Table 2. Model training

Model name	Training dataset used
Model real sun	Sun (real environment)
Model real rain	Rain (real environment)
Model sim sun	Sun (simulated environment)
Model sim rain	Rain (simulated environment)
Model sim rain lens	Rain lens (simulated environment)

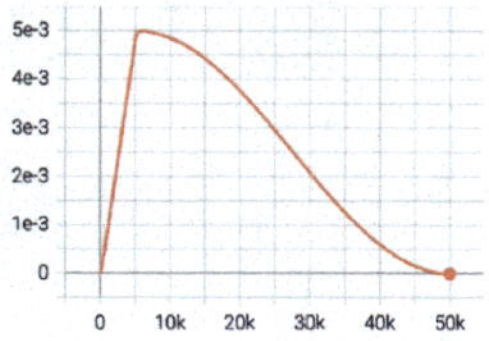

Fig. 3. Cosine decay learning rate

6 Evaluation and Visualisation

After training the models as in Table 2, each model was self-evaluated (evaluated on the same dataset as it was trained on) and cross-evaluated (evaluated on a dataset other than the one it was trained on). This kind of self and cross-evaluation give a clear picture of the performance of a model on different kinds of environment and weather datasets.

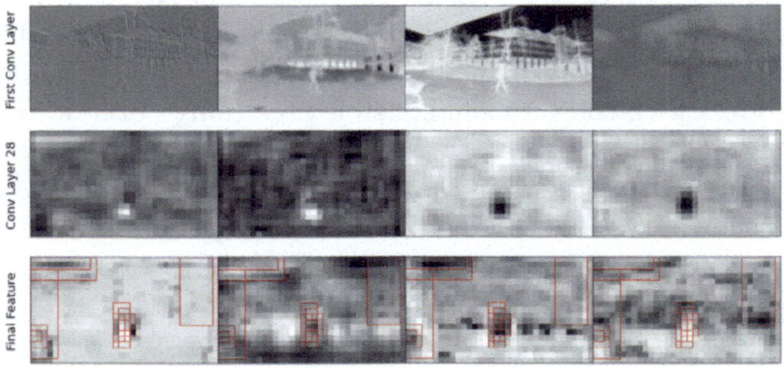

Fig. 4. Output of model real rain

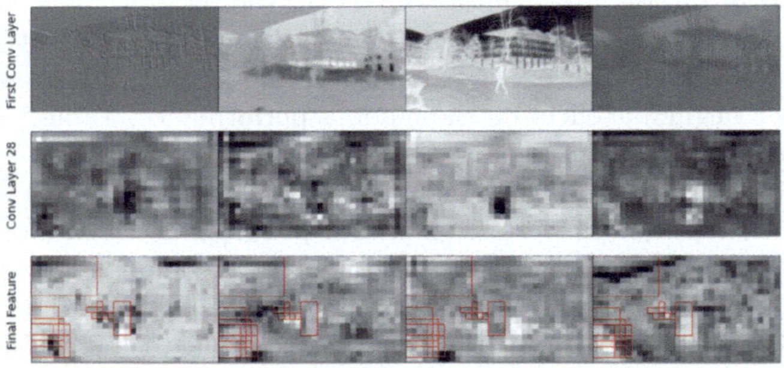

Fig. 5. Output of model sim rain

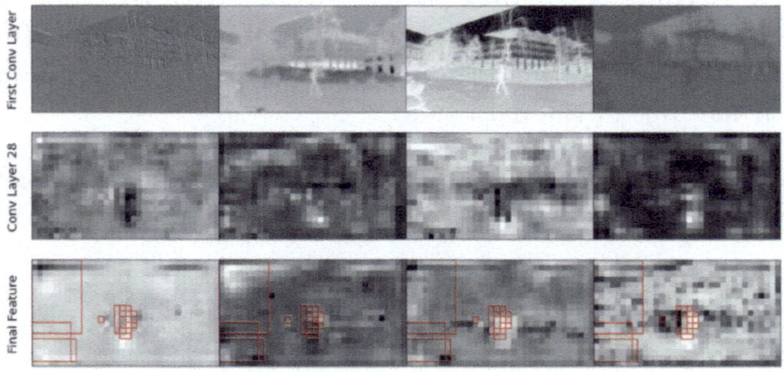

Fig. 6. Output of model sim rain lens

For evaluation, multiple methods were used. Firstly, coco mAP (mean average precision) was used to check the overall performance of a model on a particular dataset. In coco mAP, 10 IoU thresholds of 0.50: 0.05: 0.95 are used and the scores are averaged over these intersections. Secondly, visualisation of the model output at each layer was done to analyse the changes in the features generated by the ResNet50 model. The output of the RPN network was also overlayed with the feature map to see the proposal regions generated by the model and hence analyse if the model has learned to segregate objects from the background. This also helps us understand the features in the image which belongs to the negative space but the model identifies it as a positive space. The feature output from intermediate layers of the models is shown in Fig. 4, Fig. 5, and Fig. 6. In Fig. 7, the first 15 regions of the RPN is overlayed on white background with 50% transparency to analyse the concentration of ROIs and hence the performance of the model.

7 Results and Conclusion

Upon analysis of the mAP scores, it was found the performance of models were highest when evaluated on the data it was trained on. The model showed a good result when they were tested on the dataset from different weather conditions inside the same environment i.e. simulation environment model performed well on simulated environment data, similar holds for the real environment. The mAP scores of the models fell when they were tested against the data from the different environments with the minimum being when the model trained on Simulated Rain Lens was tested on the Real Rain dataset.

Upon further analysis of the RPN output of different models on some sample images, it can be seen that the models have a higher concentration of boxes in the positive space of the image i.e. the area containing the object when it is tested on the same environment (for say Fig. 7: Model Sim Rain on Sim Rain), but when tested against other environments, the boxes generated by the RPN is scattered in other regions of the image too, showing that the model is not able to differentiate positive space from the background (like Fig. 7 Model Real Rain on Sim Rain). This issue can be seen to have occurred the most when the model trained on Simulated Sun is tested on the Real Rain dataset (Fig. 7 Model Sim Sun on Real Rain), where the boxes generated by the RPN are scattered all over the image.

The analysis of models revealed that under the same environment changes in the weather conditions the performance deteriorates very slightly and it doesn't affect the model much. The performance falls in a cross-environment and is the worst for model sim sun tested on real rain. We can see very clearly from the RPN output that the boxes are scattered all over the image and the model is incapable of detecting an object (Fig. 7 Model Sim Sun on Real Rain). This clearly shows that the network didn't get any weather or environmental noise at the time of training and hence wasn't able to model any of it, but at the time of testing the model was strongly affected by high-frequency noise i.e. rain. To conclude it

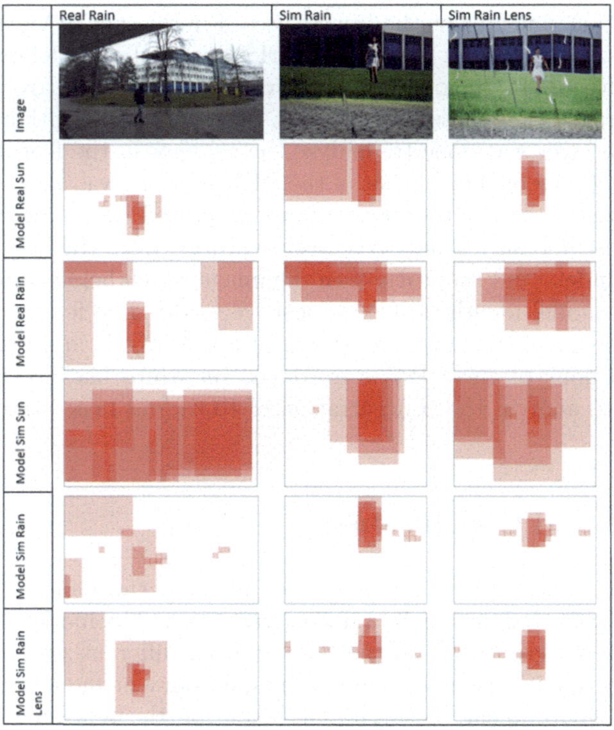

Fig. 7. RPN outputs

Models \ Data	Real Sun	Real Rain	Sim Sun	Sim Rain	Sim Rain Lens
Real Sun	0.499755	0.246596	0.558817	0.351102	0.338902
Real Rain	0.306468	0.49594	0.477295	0.319305	0.238021
Sim Sun	0.269804	0.097323	0.812577	0.633294	0.520916
Sim Rain	0.246674	0.05589	0.652116	0.808397	0.54567
Sim Rain Lens	0.202315	0.050856	0.688877	0.678745	0.752141

(a)

Models \ Data	Real Sun	Real Rain	Sim Sun	Sim Rain	Sim Rain Lens
Real Sun	0.275954	0.353729	0.368869	0.329921	0.321041
Real Rain	0.330032	0.263085	0.394389	0.353415	0.35202
Sim Sun	0.379	0.468697	0.194502	0.230034	0.265534
Sim Rain	0.407171	0.463907	0.362479	0.222772	0.284291
Sim Rain Lens	0.331426	0.335009	0.271933	0.209381	0.165252

(b)

Fig. 8. Model evaluation results : **(a)** mAP scores, **(b)** Losses

was found that a model solely trained on simulation data under varied weather conditions fails to perform well in the real environment. To overcome this, using some amount of real environment data along with simulated data can prove to improve the performance of the network.

References

1. Fekri, P., Z.M., Dargahi, J.: On the safety of autonomous driving: a dynamic deep object detection approach (2019). https://doi.org/10.4271/2019-01-1044
2. Ekvall, S., Kragic, D., Jensfelt, P.: Object detection and mapping for service robot tasks. Robotica **25**, 175–187 (2007). https://doi.org/10.1017/S0263574706003237
3. Kunz, M., Vierling, A., Wolf, P., Berns, K.: Localization using OSM landmarks. In: 2021 Latin American Robotics Symposium (LARS), 2021 Brazilian Symposium on Robotics (SBR), and 2021 Workshop on Robotics in Education (WRE), pp. 228–233 (2021). https://doi.org/10.1109/LARS/SBR/WRE54079.2021.9605475
4. Ebrahimpour, R., Kabir, E., Yousefi, M.R.: Face detection using mixture of MLP experts. Neural Process. Lett. **26**, 69–82 (2007)
5. Dollar, P., Wojek, C., Schiele, B., Perona, P.: Pedestrian detection: an evaluation of the state of the art. IEEE Trans. Pattern Anal. Mach. Intell. **34**(4), 743–761 (2012). https://doi.org/10.1109/TPAMI.2011.155
6. Jia, Y., et.al.: Caffe: convolutional architecture for fast feature embedding. In: Proceedings of the 22nd ACM International Conference on Multimedia, pp. 675–678 (2014)
7. Krizhevsky, A., Sutskever, I., Hinton, G.E.: Imagenet classification with deep convolutional neural networks. Adv. Neural Inf. Process. Syst. **25**, 1097–1105 (2012)
8. Yang, Z., Nevatia, R.: A multi-scale cascade fully convolutional network face detector. In: 2016 23rd International Conference on Pattern Recognition (ICPR), pp. 633–638. IEEE (2016)
9. Chen, C., Seff, A., Kornhauser, A., Xiao, J.: Deepdriving: learning affordance for direct perception in autonomous driving. In: Proceedings of the IEEE International Conference on Computer Vision, pp. 2722–2730 (2015)
10. Chen, X., Ma, H., Wan, J., Li, B., Xia, T.: Multi-view 3d object detection network for autonomous driving. In: Proceedings of the IEEE conference on Computer Vision and Pattern Recognition, pp. 1907–1915 (2017)
11. Liu, W., Ren, G., Yu, R., Guo, S., Zhu, J., Zhang, L.: Image-adaptive yolo for object detection in adverse weather conditions (2021)
12. Wolf, P., Groll, T., Hemer, S., Berns, K.: Evolution of robotic simulators: using UE 4 to enable real-world quality testing of complex autonomous robots in unstructured environments, July 2020. https://doi.org/10.5220/0009911502710278
13. Liu, D., Cui, Y., Cao, Z., Chen, Y.: A large-scale simulation dataset: Boost the detection accuracy for special weather conditions. In: 2020 International Joint Conference on Neural Networks (IJCNN), pp. 1–8 (2020). https://doi.org/10.1109/IJCNN48605.2020.9206716
14. Sekachev, B., et al.: OpenCV/CV at: v1.1.0, August 2020. https://doi.org/10.5281/zenodo.4009388
15. Vierling, A., Sutjaritvorakul, T., Berns, K.: Dataset generation using a simulated world. In: RAAD (2019)
16. He, K., Zhang, X., Ren, S., Sun, J.: Deep residual learning for image recognition. In: 2016 IEEE Conference on Computer Vision and Pattern Recognition (CVPR), pp. 770–778 (2016). https://doi.org/10.1109/CVPR.2016.90
17. Lin, T.Y., et.al.: Microsoft COCO: Common Objects in Context (2015)

Putting Mobile Robots into Industrial Warehouses

Dario Lodi Rizzini(✉) and Stefano Caselli

RIMLab, University of Parma, Parco Area delle Scienze, 181A, 43124 Parma, Italy
{dario.lodirizzini,stefano.caselli}@unipr.it
http://www.rimlab.ce.unipr.it

Abstract. This paper illustrates the effort to transfer advanced results of mobile robot research into the practical operation of Automated Guided Vehicles (AGVs) in industrial warehouses. We focus on three contributions: automatic calibration of AGVs to achieve accurate positioning, AGV localization based on landmark extracted from laser scans, and detection of moving objects for safe navigation.

Keywords: Industrial applications · Mobile robots

1 Introduction

The role of mobile robots in operation of industrial warehouses has constantly increased in the last three decades. In automation, they are commonly referred to as Automated Guided Vehicles (AGVs). Standard equipment of AGVs (Fig. 1) includes laser scanners for localization and obstacle detection as well as forks or other grasping devices used in transportation of goods, that are often packed on pallets. AGVs must be able to localize in the environment, to plan their paths, to safely navigate without collision with people and obstacles, and to precisely reach with the forklift the desired operation points. The execution of these tasks is often guaranteed under rigid assumptions on the operating environment, like presence of artificial landmarks that must be carefully mapped, demanding manual calibration and configuration, strict safety regions.

Recently, the need for more flexibility has required to overcome these constraints and led to the investigation of novel algorithms and solutions for AGVs, often transferred from scientific research. The setup time of a new warehouse is a significant fraction of production and installation costs. Placement and mapping of special markers in large warehouses contribute to increase such setup time, and there may be areas where it is actually impossible to put stable markers. Moreover, the accurate positioning of each vehicle is affected by estimation of intrinsic and extrinsic parameters used respectively in odometry and sensor-based localization. If manually done, AGV calibration is prone to error and can cause large delays or downtimes in warehouse operation which should be avoided. Finally, key issues for human operators and AGVs co-existence in the warehouse must

A. Müller and M. Brandstötter (Eds.): RAAD 2022, MMS 120, pp. 452–459, 2022.
https://doi.org/10.1007/978-3-031-04870-8_53

Fig. 1. Example of industrial AGV equipped with forklift as well as navigation and safety laser scanners (respectively, on the top and bottom part of the AGV). Also shown are reflective artificial landmarks used for localization and the outline of braking and halting safety areas (bottom-right).

be addressed. Safety laser scanners are placed around the AGVs at ground level to monitor the presence of obstacles in given areas around the vehicle. The size of these areas must be carefully chosen as trade-off between safety and efficiency, since the vehicle slows down or stops whenever the area is occupied.

More advanced perception and estimation methods can increase the autonomy of industrial AGVs and the flexibility of plant setup. Localization or obstacle detection based on raw sensor data can address the previously discussed limitations. Solutions to quickly achieve high technology readiness level must take into account requirements and constraints of industrial systems. For example, industrial control architectures are often based on PLC-like embedded systems, which guarantee robustness, real-time execution as well as compliance with safety regulations, but are also subject to strong limitations on available programming libraries for advanced data structures and on computation capabilities.

In this paper, we present an overview of our works aimed at improving the capabilities of AGVs in actual industrial warehouses during our collaboration with AGV manufacturers. The first contribution is a set of algorithms for automatic calibration of intrinsic and extrinsic parameters. The intrinsic parameters are related to the model of the robot and are essential for computing its odometry. Calibration algorithms have been developed for common AGV kinematic models with both three and four wheels, namely the tricycle [4], and the Ackermann and the Dual-Drive models [1]. The extrinsic parameters establish the relation between the sensors mounted on the AGV as well as their measurements and the robot reference frame. The working principle of calibration is the comparison of the expected paths given the input commands and of the real paths observed by one or more sensors.

The second contribution is the development of localization methods for AGV guidance in industrial plants without handcrafted landmarks. In spite of the extensive research on localization, there is a limited number of real industrial applications meeting with requirements in safety, robustness and technology readiness. We investigated several solutions designed for laser scanners including keypoint features like *FALKO* (Fast Adaptive Laser Keypoint Orientation-invariant) [3] and SKIP (SKeleton Interest Point) [8], as well as complete scan fingerprints like ARS (Angular Radon Spectrum) [6] or registration-based techniques. Both approaches have been integrated in the sensor models of the Extended Kalman Filter (EKF) localization previously implemented for artificial landmark localization.

The third contribution is the detection of moving objects like people and vehicles [8] or semi-static items [7] in the space around the AGV. The proposed algorithm exploits the laser scanners already mounted on AGVs for safety rules compliance. These sensors operate according to simple and rigid area-based policies that slow down or stop the AGV when an unexpected item is found. We extended their use to discriminate measurements belonging to dynamic parts of the scene in order to predict their motion and prevent collision.

The proposed technological advancements have been developed by taking into account actual industrial mobile robotic platforms. We claim that gradual integration of state-of-the-art solutions is an effective approach to achieve high technology readiness levels and bring the full potential of mobile robotics into operational warehouses.

2 Calibration of AGVs

Calibration parameters are divided into intrinsic and extrinsic. The intrinsic parameters depend on the specific kinematic model whereas the extrinsic ones represent the pose of the navigation sensor (usually a laser scanner) w.r.t. the robot frame. Since mobile robots move on the ground plane, the kinematic models are planar.

The robot configurations typically used in warehouse logistics have three or four wheels, like those displayed in Fig. 2. Intrinsic parameters are often related to the state of wheel. They include the *steering offset* of a wheel, which is the angle between the real wheel direction and the reference angle of steering encoder, and the *driving scale*, which is the length of the travelled path of a wheel by a single encoder tick. Steering offset and driving scale allow to relate the motion of each single wheel with the state of the mobile robot. In earlier work, we addressed the following models:

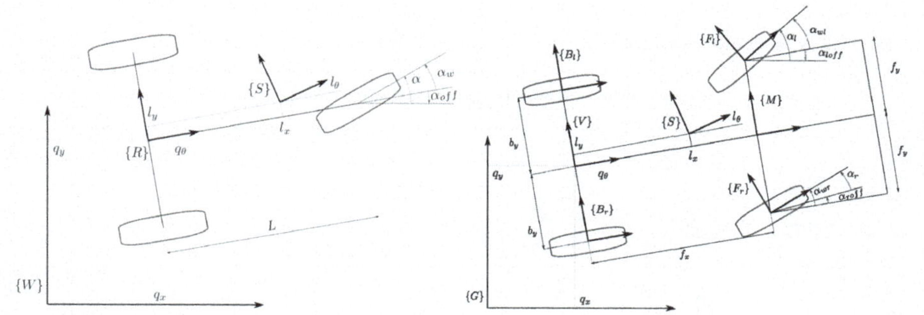

Fig. 2. Widespread AGV kinematics: the three-wheel model or *Tricycle* (left), and the four-wheel models, namely *Ackermann* and *Dual-Drive* (right).

- *Tricycle.* It has a single front actuated and steering wheel with steering offset α_{off} and driving scale s_w and two passive back wheels.
- *Ackermann.* It has two front actuated steering wheels and two passive back wheels. Both the front left and front right wheels are described by steering offsets (respectively α_{loff} and α_{roff}) and driving scales (respectively s_{wl} and s_{wr}).
- *Dual-Drive.* It is similar to Ackermann, but the two front wheels are steering whereas the back wheels are non-steering and actuated. The front steering wheels are defined by steering offsets α_{loff} and α_{roff}, the back wheels are described by their driving scales s_{wl} and s_{wr}.

Extrinsic parameters are the pose of the sensor frame $\{S\}$ w.r.t. the robot frame $\{R\}$, i.e. $\mathbf{l} = [l_x, l_y, l_\theta]^\top$.

The principle adopted for estimation of calibration parameters is the comparison between the trajectory measured by the on-board sensor and the expected trajectory of the AGV. We assume that at every time instant each wheel is aligned with the Instantaneous Center of Rotation (ICR) of the AGV rigid body and that there is no wheel slipping. Under such assumption the calibration parameters are observable. The calibration procedure collects data while the AGV iteratively moves along several paths segments. We choose path segments to be circular arcs obtained with constant input commands because the kinematic equations have straightforward and reliable solutions, although other path geometries could be used.

For each path segment k, the robot collects the nominal steering angles α_* measured by steering encoders and the number of encoder ticks n_* measured by the motion encoders (symbol $*$ is a placeholder for any wheel label). The AGV is equipped with a sensor, usually a laser scanner, that can be used to estimate the egomotion $\boldsymbol{\zeta}_k$ over path segment k. The egomotion, i.e. the relative motion of the sensor over the time, is estimated through registration or alignment of consecutive laser scans. The specific kinematic equations of the AGV constrain the relative motion of the robot $\boldsymbol{\delta}_k$ on the path segment with α_*, n_* and the intrinsic parameters to be estimated. The egomotions $\boldsymbol{\zeta}_k$ and the relative motion

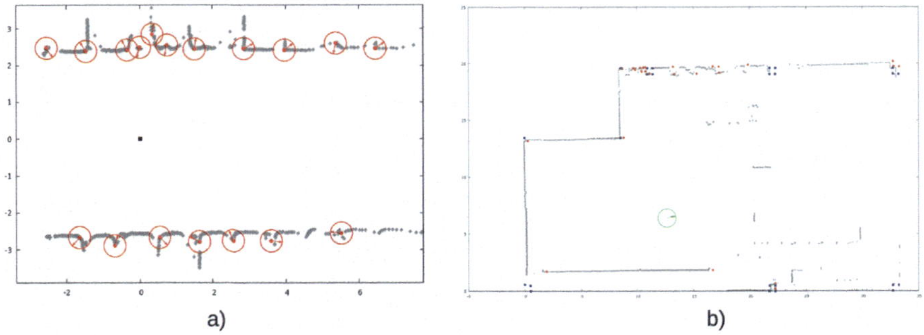

Fig. 3. Examples of keypoint features for robot localization: a) FALKO (red circles), b) SKIP (blue squares).

$\boldsymbol{\delta}_k$ are constrained by extrinsic parameters as $\mathbf{l} \oplus \boldsymbol{\delta}_k = \boldsymbol{\zeta}_k \oplus \mathbf{l}$ where $\oplus$ is the pose composition operator. The solutions for these constraints have been derived for each kinematic model [1,4].

Automatic calibration has many advantages over the manual procedures previously used to assess the values of parameters. First, the precision of the estimated AGV parameters in repeated calibration trials is at most $0.1°$ for angular parameters and typically less than 5 mm for position parameters. More importantly, the positioning error on operation points is about $10 \div 15$ mm. Second, a standard automatic calibration procedure leads to more uniform positioning of a fleet of AGVs than the one achieved by the manual expertise of human operators. Finally, calibration requires about 15 minutes for each AGV with limited human intervention and no special expertise of the operator, instead of about one hour per AGV which must be manually calibrated by an expert operator. Indeed, expert operators with the ability to perform such calibrations are a scarce resource, whose unavailability leads to delays in warehouse setup or extended AGV downtimes. In plants with tens of AGVs, automatic calibration significantly reduces the setup time and potentially allows frequent re-calibration.

3 AGV Localization Without Artificial Landmarks

The standard commercial solution for AGV localization is based on artificial markers made of reflective material observable by a laser scanner. The robot position and orientation are computed using EKF and landmark maps. Feature detection from raw laser scans avoids the reliance upon special landmarks for AGV localization.

The detection of stable and distinguishable interest points from laser measurements is a not trivial task. Scientific literature includes few keypoint features with descriptors like FLIRT [9]. In an effort to devise effective features suitable for industrial environments, we have developed two different algorithms: FALKO [2,3] and SKIP [8], illustrated in Fig. 3. Both FALKO and SKIP points are detected by associating to each point a saliency score. Such score is related to geometric parameters like the curvature of underlying obstacle surfaces. In the case of FALKO, features are extracted where there is a distinctive local distribution of neighbor points. The *cornerness* of a point $\mathbf{p}_i$ is computed by comparing the local points belonging to its neighborhood after a careful selection of valid points, e.g. avoiding those close to gaps due to occlusion. FALKO points allow the identification reasonably stable references for navigation and local mapping of AGVs. Experiments on AGVs have shown the effectiveness of FALKO keypoints as temporary landmarks for navigation in block storages, i.e. passages between lines of pallets (see Fig. 3(a)). Hence, a temporary map of FALKO features allows short-term navigation of AGVs. The main drawback of such interest points is that they do not faithfully describe the global shape of the free space in a laser scan, since the scoring depends only on a local neighborhood.

The novel feature SKIP [8] extracts interest points through simplification of the global contour defined by scan points. The general idea is brought from curve simplification techniques like Discrete Contour Evolution [5]. Instead of directly connecting the laser points adjacent according to their radial order, the scan is split into intervals in correspondence to gaps caused by occlusion. Gaps are pairs of consecutive points whose distance is above a given threshold and are strong discontinuities in the range values. Gaps are not included in the following assessment. A cornerness score is computed for each of the remaining points $\mathbf{p}_i$ based on its previous and next points, respectively $\mathbf{p}_{i-1}$ and $\mathbf{p}_{\underline{i+1}}$. The chosen score is equal to $score(\mathbf{p}_i) = |\overline{\mathbf{p}_i\mathbf{p}_{i-1}} + \overline{\mathbf{p}_i \mathrm{i}\underline{+}1} - \overline{\mathbf{p}_{i+1}\mathbf{p}_{i-1}}|$ where $\overline{\mathbf{ab}}$ is the length of the segment connecting points a and b. Then, the points are ordered according to their score and those with score value below a given threshold are iteratively removed. When a point is removed, the score of its adjacent points is updated according to their new neighbors. An example of the outcome of SKIP algorithm is illustrated in Fig. 3(b). SKIP points are the blue squares identified at the corners. Gaps and other discarded points are marked in red. The algorithm has been tested on datasets acquired in an industrial warehouse and has shown to detect meaningful environmental features.

4 Detection of Semi-static and Dynamic Objects

AGVs are usually equipped with laser scanners at ground level to detect potentially colliding objects like obstacles, people and other vehicles. Decisions on the robot behavior, e.g. to slow down or halt its motion, are taken according to simple area-based policies. While required by safety regulations, this approach may prove rigid. Safety laser scanner could be used to predict the motion and to replan the robot trajectory.

Parts of the scene can be classified into three categories: *dynamic*, *semi-static* and *static*. The classification depends on the frequency of location changes for the objects in the scene. For dynamic objects such change occurs while the AVG collects sensor data with the laser scans. When the AGV moves in a static or semi-static scene, the points acquired at two consecutive time instants are aligned by the same rigid motion that is the inverse of the robot motion. On the

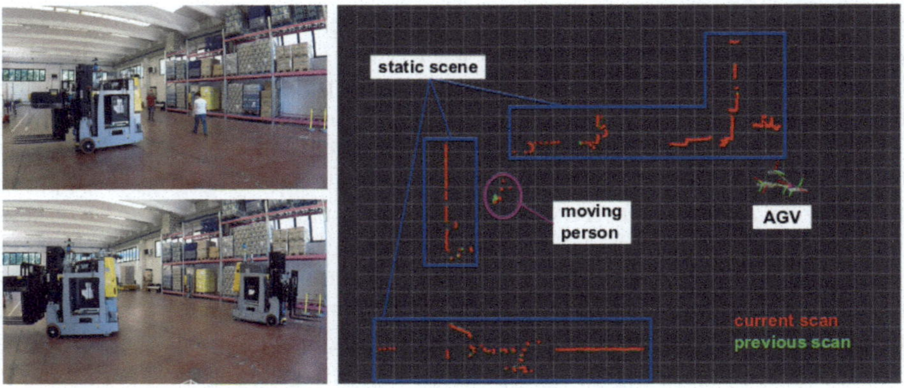

Fig. 4. An AGV coping with a moving person (top-left) and another AGV (bottom-left). Detection of dynamic objects and static scene is achieved by comparison of consecutive scans (right).

other hand, dynamically moving objects are aligned by a different transformation w.r.t. the background. The approach is illustrated in Fig. 4.

In [8] an algorithm is proposed for dynamic object detection based on registration and difference between scans. First, the relative pose between two consecutive AGV poses is estimated via scan matching. Hough spectrum correlation [6] allows robust estimation of robot rotation. Once rotation is estimated, clustering is performed only on translation. *Motion clustering* is achieved by iteratively performing k-means clustering on groups of points with coherent translation motion. In particular, the points of the current scan are associated to their nearest neighbors in the previous scan and the respective difference vectors are computed. The difference vectors are then assigned to a cluster using k-means algorithm. The number of clusters k should be equal to the number of moving objects, which is unknown. Thus, the algorithm initially overestimates $k = 15$ clusters and, when the centroids of two clusters are very close, the two corresponding clusters are merged. The achieved classification is improved under the hypothesis that points belonging to the same object are more likely neighbors. Spatial clustering is performed and the initial point labels are corrected by adopting the same classification for the points in the same spatial cluster. Errors in cluster identification are further avoided through their tracking.

The identification of semi-static objects is more complex, since their location does not change during the transit of the AGV. Examples of semi-static objects in warehouses include pallets deposited for hours or days in a given location where they stay until their transportation to the buyer. Semi-static objects usually do not represent risks for safety, but they may unnecessarily slow down the AGV. We proposed a semi-static detection algorithm [7] that compares maps of the same region visited at different time instants. The environment is represented in the form of polyline maps that have limited footprint. Polylines are extracted from laser scans, simplified and compared with proper procedures. Experiments have shown the potential of the approach to detect semi-static objects, although the procedure has not been validated yet in an operational warehouse.

5 Conclusion

This paper has presented several works for improving mobile robot (aka AGV) operation and navigation in industrial warehouses through fast and accurate calibration of vehicles, localization and tracking of moving objects. The leading criterion in our work has been gradual integration on the existing industrial platforms. The first two contributions have achieved high technology readiness level and have been successfully applied in real plants.

References

1. Galasso, F., Lodi Rizzini, D., Oleari, F., Caselli, S.: Efficient calibration of four wheel industrial AGVs. Robot. Comput. Integr. Manuf. (RCIM) **57**, 116–128 (2019). https://doi.org/10.1016/j.rcim.2018.11.005
2. Kallasi, F., Lodi Rizzini, D.: Efficient loop closure based on FALKO LIDAR features for online robot localization and mapping. In: Proceedings of the IEEE/RSJ International Conference on Intelligent Robots and Systems (IROS), pp. 1206–1213 (2016)
3. Kallasi, F., Lodi Rizzini, D., Caselli, S.: Fast keypoint features from laser scanner for robot localization and mapping. IEEE Robot. Autom. Lett. (RA-L) **1**(1), 176–183 (2016). https://doi.org/10.1109/LRA.2016.2517210
4. Kallasi, F., Lodi Rizzini, D., Oleari, F., Magnani, M., Caselli, S.: A novel calibration method for industrial AGVS. Robot. Auton. Syst. **94**, 75–88 (2017). https://doi.org/10.1016/j.robot.2017.04.019
5. Latecki, L., Lakamper, R.: Shape similarity measure based on correspondence of visual parts. IEEE Trans. Pattern Anal. Mach. Intell. **22**(10), 1185–1190 (2000). https://doi.org/10.1109/34.879802
6. Lodi Rizzini, D.: Angular radon spectrum for rotation estimation. Pattern Recogn. **84**, 182–196 (2018)
7. Lodi Rizzini, D., Boccalini, G., Caselli, S.: Semi-static object detection using polygonal maps for safe navigation of industrial robots. In: Proceedings of the International Conference on Informatics in Control, Automation and Robotics (ICINCO), pp. 191–198 (2012)
8. Lodi Rizzini, D., Galasso, F., Caselli, S.: Safe feature-based navigation for industrial AGVs. In: Proceedings of IROS Workshop on Robotics for Logistics in Warehouses and Environments Shared With Humans (2018)
9. Tipaldi, G.D., Arras, K.O.: Flirt-interest regions for 2d range data. In: Proceedings of the IEEE International Conference on Robotics and Automation (ICRA) pp. 3616–3622 (2010)

Passive Guidance for Offroad-Vehicles

Werner Bailer[1(✉)], Hermann Fürntratt[1], Manfred Klopschitz[1], Dominik Lampel[2], Roman Lesjak[1], Matthias Rüther[1], Susanne Schweitzer[1], and Gerald Steinbauer-Wagner[2]

[1] DIGITAL - Institute for Information and Communication Technologies, JOANNEUM RESEARCH Forschungsgesellschaft mbH, 8010 Graz, Austria
{werner.bailer,hermann.fuerntratt,Manfred.Klopschitz,Roman.Lesjak, Matthias.Ruether,Susanne.Schweitzer}@joanneum.at

[2] Institute of Software Technology, Graz University of Technology, 8010 Graz, Austria
lampel@student.tugraz.at

Abstract. In contrast to commercial automated driving, where systems are usually able to use infrastructure such as high-definition maps and GNSS as well as active sensors such as Lidar or Radar, in unstructured offroad environments for military or disaster response application such means are not available. In order to address a logistics tasks in such a setting we propose a vehicle guidance approach that is based on IMU pose tracking and visual image retrieval in an image database build up from a training trajectory. The approach has been implemented as a prototype and evaluated with IMU and image data recorded in several recording campaigns in military offroad areas.

Keywords: Robot navigation · Offroad environment · Passive sensors · GNSS-denied environment

1 Introduction

The automation of vehicles offers a number of advantages. For commercial automated driving the goal is to navigate automatically on streets in urban areas or on highways and to free users form the driving activity, improve safety, and reduce traffic jams. In military or disaster response situations the goal is different in removing the person from a vehicle in a dangerous situation. Besides the use cases there are also major differences in the environments the systems are deployed in, the infrastructure which is available to the systems, and the sensor technology that can be used. Commercial automated driving heavily relies on available infrastructure such as GNSS, pre-recorded high-definition maps, road markings, and road signs and on active sensor technology such as Lidar and Radar [17]. Moreover, the used road network is quite structured. In contrast in military and disaster response application the environments are unstructured

The authors are in alphabetical order.

A. Müller and M. Brandstötter (Eds.): RAAD 2022, MMS 120, pp. 460–468, 2022.
https://doi.org/10.1007/978-3-031-04870-8_54

(e.g. offroad scenarios, disaster sites), services and infrastructure are not available or cannot be used, and sensor technology is limited to passive systems like IMUs and cameras.

The goal of our research is to develop technology that allows vehicles to navigate automatically under the above restrictions in order to carry out tasks such as unmanned transport in hazard situations such as battle fields or when avalanches are possible. A long lasting competition to foster development of such automated systems is the European Land Robot Trial (ELROB) [12]. Within this competition in particular the mule challenge that resembles an automated logistics in an offroad environment. The MuCar system [5] is a good example for carrying over research results from automated driving to solve automated offroad logistics tasks. A drawback of this system is that it still depends on infrastructure such as GNSS and active sensors such as Lidars.

A major problem in the mule scenario is the guidance of the vehicle that is mainly concerned with the determination of the relation between the actual pose of the vehicle to a given goal or trajectory towards it. The problem can be addressed relatively and locally without a global localization or with a global environment representation and localization. There exists well performing visual-inertial odometry (ROVIO [2]) or SLAM (ORB-SLAM3 [3]). The former is able to track the pose of a vehicle while the latter is able to build up a map and localize a vehicle on that map.

In this paper we propose an approach for the guidance problem of offroad logistics tasks that also employs an IMU and stereo cameras but focuses stronger on the appearance of locations in the guidance than on geometrical relations in the environment. The core difference is that the environment representation is formed by an indexed image database rather than using a geometrical map.

2 Passive Guidance Concept

In order to solve the mule task in a setting without infrastructure and passive sensors only we propose an approach with two stages. In the first stage the vehicle will be manually driven on a suitable trajectory from the start to the end location. During this drive IMU and camera data are recorded, processed, and converted to an indexed image database. The index represents the spatial and temporal structure of the images recorded along the training trajectory. In stage two the relative pose of the vehicles is tracked using IMU measurements. The result is used as a prior for the retrieval of the image from the database that is closest in appearance to the actual camera image. This image pair is then used to determine the geometrical displacement of the actual vehicle pose and the desired pose on the recorded trajectory. This estimated displacement can be used in a trajectory following algorithm. The advantage of this approach is that the image database and the pose tracking needs only to be consistent locally to allow robust trajectory following. Moreover, the estimated displacement is also used to estimate the actual velocity of the vehicle that can be used to limit the drift of the IMU pose tracking. In the next sections we describe the core modules of the approach in more detail.

2.1 Training Trajectories

During a training run from the starting to the end location the raw IMU data and the images of the involved cameras are recorded including their timestamps. The training run contains also IMU calibration patterns (bias estimation) around the starting location. The recorded IMU data are used to estimate the vehicle pose during the run (see Sect. 2.2). The recorded images are sub-sampled spatially based on the pose estimation and stored in an image database together with further computed features (see Sect. 2.3). Please note that the pose estimation needs only be consistent in a local context (IMU data drift strongly on the long run) because the images serve as global anchor points.

2.2 Inertial Pose Tracking

An inertial navigation system is ideal for use in passive navigation systems as it cannot be disturbed by external influences and ensures a continuous navigation solution comprising position, velocity and attitude (PVA) [15].

An inertial measurement unit (IMU) provides angular rate and acceleration data. By means of the angular rate data, changes in the position of a vehicle can be determined which allows to determine its attitude by integration. Before this integration, the angular rates must be compensated for the contributions due to the Earth rotation rate. The attitude is then used to transform the measured accelerations from the body frame (which is related to the vehicle where the IMU is integrated) to a space-fixed reference system, which is oriented towards north and east, including a vertical dimension (up or down; called navigation frame). The velocity is then propagated from integration of the accelerations. Thereby, gravitational acceleration and Coriolis acceleration are compensated for, as the IMU measures the correlated forces as apparent accelerations (an upward acceleration in the case of gravitation). The position is finally propagated through integration of the velocity.

The whole calculation rule described above, for deriving the PVA solution from the measured angular rate and acceleration data is called strapdown algorithm. The logic of the strapdown is visualized in Fig. 1. A detailed description of the algorithms can be found in e.g. [7,11,15].

2.3 Vision-Based Passive Pose Estimation

To compute a 6DoF relative orientation based on a pair of images, one image from the map with known positions (the image database obtained in the previous stage) and one from the query sequence that is to be localized with respect to the map, similar images in the map have to be found and then detailed geometry can be computed on image pairs that view a similar scene content. Therefore, the problem of passive localization is decomposed into two parts, first an appearance based search is performed to find views on the map images that are similar to the query image, secondly the actual geometry of the motion is computed.

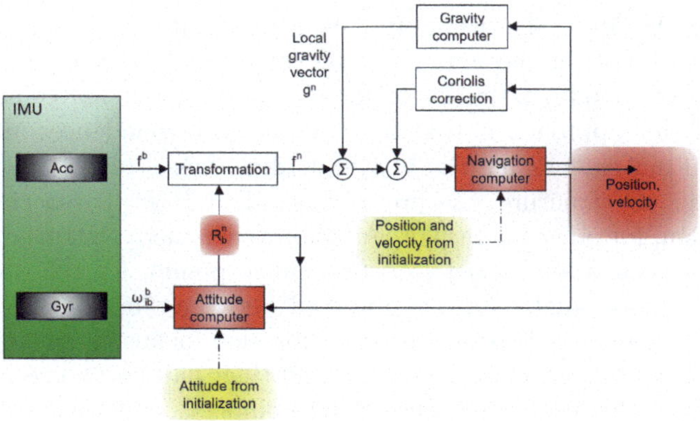

Fig. 1. Scheme of the strapdown algorithm [9].

Image-Based Localization. In computer vision, the problem of location recognition has been addressed in the past by a variety of approaches, the most successful methods rely on descriptor based matching techniques of invariant interest points and local and global image descriptors. The basic idea behind these methods is to compute the position of a query image with respect to a database of registered reference images, planar surfaces or 3D models. Assuming a static scene, geometric verification can be used to determine the actual pose of the camera with respect to the exemplar database. Different viewpoints or illumination changes are largely handled by robust feature detectors and descriptors that act as descriptors of local image patches and global appearance.

Additionally, there is more temporal redundancy available in video images of a scene. We propose to use a compressed descriptor for image sequences, which encodes a segment of video using a single descriptor. The proposed approach is a framework that can be used with different local descriptors, including compact descriptors. The details of our image based localization method can be found in [1]. Our method utilizes local and global image descriptors and the matching process is hierarchical, in the sense that matching of details is only performed if some degree of similarity is found on the coarser level. This matching is illustrated in Fig. 2. The most similar map images are then used for detailed relative orientation computation. It is also possible to reduce the search space by adding an IMU prior localization information.

Relative Pose Estimation. In contrast to the image based localization, which performs essentially 1D localization in an image sequence, we require a full 6DoF motion. Generally, image based camera motion can be computed by point correspondences between image pairs. One possibility to classify motion estimation is by discriminating between $2D-2D$ and $2D-3D$ point correspondences that are used to compute the motion. Given a set of $2D-2D$ point correspondences the relative motion can be computed up to scale, for example with a calibrated camera using the $5pt-$ algorithm [10]. Additional information is necessary to compute the scale of the motion. $2D - 3D$ correspondences require that depth information ($3D$ points) is available for one part of the matching pair. The advantage of this approach is that fewer point correspondences are necessary and the resulting camera pose is absolute (including scale of the motion) in regards to the $3D$ structure, the problem is called $P3P$ (perspective 3 point) pose estimation, for one camera pair. A calibrated camera (intrinsic parameters are known) measures angles between two image points, for the absolute pose problem, three such angles θ_{ij} are observed between three image points. For pinhole camera models a known camera calibration means that the image measurements m_i can be converted to rays v_i and their pairwise angle $\angle(v_i, v_j)$ can be measured. In this case three known 3D points X_i and their corresponding image measurements m_i give rise to 3 pairwise angle measurements. These are sufficient to compute a finite number of solutions for the camera location and orientation. For three observed 3D points X_i, the pairwise 3D point distances l_{ij} can be computed. Furthermore, the angles between pairs of image measurements θ_{ij} are known from the corresponding image measurements m_i. The unknowns are the distances x_i between the center of projection C and the 3D point X_i:

$$\begin{aligned} l_{ij} &= \|X_i - X_j\| \\ \theta_{ij} &= \angle(v_i, v_j) \\ x_i &= \|C - X_i\|. \end{aligned}$$

Using the law of cosines each of the three point pairs gives one equation:

$$l_{ij}^2 = x_i^2 + x_j^2 - 2x_i x_j cos\theta_{ij}$$

Different strategies for solving this equation system can be found, see the classical approaches in [6] for an introduction to the concept.

As we have multiple cameras mounted on the vehicle, we use a generalized approach. Generalized here means that a camera is not represented by one projection center but decomposed into multiple projection centers that are used simultaneously. Feature points are represented as rays with an origin, each physical projection center of a camera is represented by a ray origin, we use the generalized problem formulation and approach to solve for the motion of [8].

The set of $3D$ points is created for the image sequence of the map images by dense stereo matching [16]. The vehicle uses 3 stereo camera pairs, therefore $3D$ structure is generated by these three image pairs for mapping part of the sequence. Figure 3 shows a stereo example for one vehicle position. For the relative orientation part, the three corresponding stereo reference cameras are used to form one calibrated generalized camera. $2D$ image feature matching [14] is used to generate tentative $2D - 3D$ point correspondences, using the search result of the most similar location from the image based localization stage. Figure 4 shows a feature matching result from the best matching map image set obtained by image based localization and the three stereo key frames. The generalized three point pose estimation is used in a simple RANSAC [4] scheme to generate hypotheses for the pose of the vehicle.

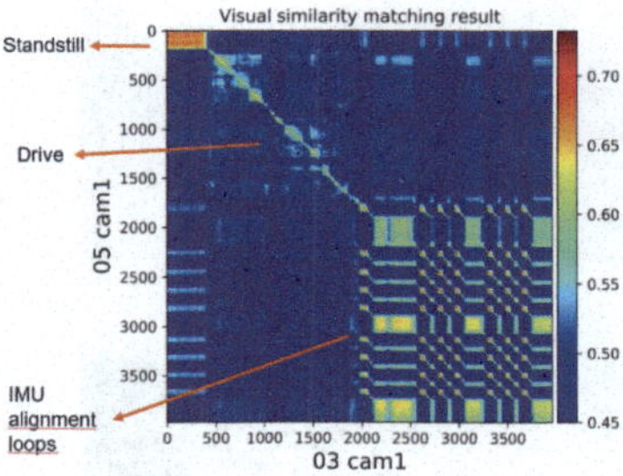

Fig. 2. Image based localization example: Two sequences of one camera are compared, the trajectory is similar, this can be seen in the structure of the similarity scores. The similarity is color coded, red means high similarity, dark blue low similarity. On the x–axis and y–axes image indices are shown. 3 different matching patterns are visible.

Fig. 3. Stereo matching: on the left are the key images of three stereo pairs looking front, left and right, on the right side the dense depth images generated by stereo matching to compute the $3D$ structure for the $P3P$ problem.

2.4 Vehicle Velocity Estimation

In order to estimate the actual vehicle velocity an Extended Kalman Filter is employed. The state vector contains the 3D position of the vehicle, its attitude, and its translational and rotational velocities. This Filter is tightly coupled with the IMU Pose Tracking and uses its result for the prediction step. The measurement update of the filter is based on the estimated displacement of the vehicle to the related image and its pose from the training run. Since the vision-based estimation runs at a rate much lower than the IMU (1 Hz versus 500 Hz) and is significantly delayed (appr. 1 s due to operations in the image database) an update procedure needs to be applied that allows delayed measurements [13]. The estimated velocity is fed back into the IMU pose tracking to limit the drifting.

3 Preliminary Results

In order to evaluate the proposed approach a sensor platform was developed, carrying eight cameras (two looking to the front in a stereo-setup, two looking diagonal to the front, and two further stereo-camera systems looking to the left and right side of the vehicle), an IMU, as well as a GNSS system for validation purposes. The platform was mounted on two different vehicles; a military truck by IVECO and an electrical utility vehicle developed by AVL (see Fig. 5).

Several data collection runs were carried out under offroad conditions in the military training areas Strass, Seetaler Alpe and Allentsteig (all located in

Fig. 4. Feature matching: on the left column are the three key stereo images of the map query result, on the right column the query key images. The vertical lines indicate feature correspondences used for the $P3P$ motion computation.

Austria). The runs included different scenarios, like flat- and hilly offroad conditions, different underground conditions (asphalt, gravel road, road with ditches and meadow), velocities between 10 km/h and 30 km/h, obstacles and convoy drives (sensor car following another vehicle). Drives took place in sunshine and rain. The runs comprise recordings allowing training the trajectory as well as test runs that are off the training trajectory for a meter to allow an evaluation of the displacement estimation.

The sensor data recorded during these drives are the basis for the algorithm development. The proposed algorithms were integrated in a Python-based software framework that allows to run the approach slower than in realtime but is able to maintain the proper temporal sequence of the processing steps. This is necessary because the individual processing steps run at different rates and the implementation is not yet realtime capable.

First evaluations of recorded runs show that the proposed image-based displacement estimation using the image database is able to report proper displacements for the test runs and is robust in terms of the IMU position prior.

Fig. 5. IVECO truck and AVL vehicle carrying the sensor platform during recording runs in Strass respectively Seetaler Alpe.

4 Conclusion

In this paper we presented a guidance approach for offroad vehicles in logistics applications without access to infrastructure like GNSS or the possibility to use emitting sensors. The approach follows two stages. In the first stage on a manually driven reference trajectory IMU and stereo vision data are recorded. The data are converted into a spatially and temporally ordered image database that allows to query reference images on the training trajectory from an actual set of camera images. In the second stage IMU data are used as a prior for the retrieval of reference images and the geometrical displacement from the actual vehicles pose to the reference on trajectory. The estimated displacements can be used in a trajectory following controller. The main advantage of the proposed method is that the IMU pose tracking and organization of the image database needs to be consistent only on the local scale. A first non-realtime capable prototype implementation of the approach has been implemented and tested using recorded data from various offroad areas. First results are promising and showed the robustness of the image-based pose estimation. Further steps will be a detailed metric evaluation of the approach, the implementation of a realtime version, and the integration of the approach into the navigation on a target vehicle.

Acknowledgement. This paper is a result of the PALONA project which is funded by the FORTE programme. It is a research, technology and innovation programme of the Austrian Ministry for Agriculture, Regions and Tourism. The Austrian Research Promotion Agency (FFG) is authorized for the programme management. We like also to acknowledge the support of the Austrian Ministry of Defense and the Austrian Armed Forces in conducting the research and the experiments.

References

1. Bailer, W., Wechtitsch, S., Thaler, M.: Compressing visual descriptors of image sequences. In: Amsaleg, L., Guðmundsson, G., Gurrin, C., Jónsson, B., Satoh, S. (eds.) MMM 2017. LNCS, vol. 10133, pp. 124–135. Springer, Cham (2017). https://doi.org/10.1007/978-3-319-51814-5_11
2. Bloesch, M., Burri, M., Omari, S., Hutter, M., Siegwart, R.: Iterated extended Kalman filter based visual-inertial odometry using direct photometric feedback. Int. J. Robot. Res. **36**(10), 1053–1072 (2017)
3. Campos, C., Elvira, R., Rodríguez, J.J.G., Montiel, J.M.M., Tardós, J.D.: ORB-SLAM3: an accurate open-source library for visual, visual-inertial and multi-map SLAM. IEEE Trans. Robot. **37**(6), 1874–1890 (2021)
4. Fischler, M.A., Bolles, R.C.: A paradigm for model fitting with applications to image analysis and automated cartography (reprinted in readings in computer vision, ed. MA Fischer). Comm. ACM **24**(6), 381–395 (1981)
5. Fries, C., Burger, P., Kallwies, J., Naujoks, B., Luettel, T., Wuensche, H.-J.: How MuCAR won the convoy scenario at ELROB 2016. In: 2017 IEEE 20th International Conference on Intelligent Transportation Systems (ITSC)
6. Haralick, B.M., Lee, C.-N., Ottenberg, K., Nölle, M.: Review and analysis of solutions of the three point perspective pose estimation problem. Int. J. Comput. Vis. **13**(3), 331–356 (1994)
7. Hofmann-Wellenhof, B., Legat, K., Wieser, M.: Principles of Positioning and Guidance. Springer, Wien (2003). https://doi.org/10.1007/978-3-7091-6078-7
8. Kukelova, Z., Heller, J., Fitzgibbon, A.: Efficient intersection of three quadrics and applications in computer vision. In: Proceedings of the IEEE Conference on Computer Vision and Pattern Recognition, pp. 1799–1808 (2016)
9. Lesjak, R.: Die rolle einer intertialen messeinheit in der anwendung der moving-base gravimetry. Austria (2010)
10. Nistér, D.: An efficient solution to the five-point relative pose problem. IEEE Trans. Pattern Anal. Mach. Intell. **26**(6), 756–770 (2004)
11. Noureldin, A., Karamat, T.B., Georgy, J.: Fundamentals of Inertial Navigation, Sattelite-Based Positioning and Their Integration. Springer, Heidelberg (2013). https://doi.org/10.1007/978-3-642-30466-8
12. Schneider, F.E., Wildermuth, D., Wolf, H.-L.: Professional ground robotic competitions from an educational perspective: a consideration using the example of the European Land Robot Trial (ELROB). In: 6th IEEE International Conference Intelligent Systems, pp. 399–405 (2012)
13. Tessier, C., Cariou, C., Debain, C., Chausse, F., Chapuis, R., Rousset, C.: A real-time, multi-sensor architecture for fusion of delayed observations: application to vehicle localization. IEEE Intell. Transp. Syst. Conf. **2006**, 1316–1321 (2006)
14. Vedaldi, A., Fulkerson, B.: Vlfea: an open and portable library of computer vision algorithms. In: Proceedings of the 18th ACM International Conference on Multimedia, pp. 1469–1472 (2010)
15. Wendel, J.: Integrierte Navigationssysteme. Oldenburg Verlag München, München (2011)
16. Yang, G., Manela, J., Happold, M., Ramanan, D.: Hierarchical deep stereo matching on high-resolution images. In: Proceedings of the IEEE/CVF Conference on Computer Vision and Pattern Recognition, pp. 5515–5524 (2019)
17. Yurtsever, E., Lambert, J., Carballo, A., Takeda, K.: A survey of autonomous driving: common practices and emerging technologies. IEEE Access **8**, 58443–58469 (2020). https://doi.org/10.1109/ACCESS.2020.2983149

Computer Vision Based Object Tracking for Multiple Robot Collaboration

Denis Chikurtev[1(✉)] and Kaloyan Yovchev[2]

[1] Institute of Information and Communication Technologies, Bulgarian Academy of Science, Sofia, Bulgaria
denis.chikurtev@iict.bas.bg
[2] Faculty of Mathematics and Informatics, Sofia University, Sofia, Bulgaria
k.yovchev@fmi.uni-sofia.bg

Abstract. In this research we will consider the case in which industrial robotic manipulators execute pick and place operations with dynamically moving objects. Objects are considered to move on a conveyor belt of the production line without being precisely positioned on it. This research will investigate how a simple color camera can be used for recognition using computer vision methods for background subtraction. The proposed algorithm detects objects located on the conveyor belt of a production line and plans the grasping process for the multiple manipulating robots. It uses methods such as Mixture of Gaussians, which appear to be suitable for application in industrial environments. The algorithm was experimentally tested in a simulation environment. This leads to a universal approach, which does not require usage of special tools and sensors for the production line. The approach could be applied for each conveyor belt of any industrial production line. The grasp planning can consider a larger number of robots and optimize the sequence of their movements.

Keywords: Robotic manipulator · Computer vision · Image processing · Object detection

1 Introduction

One of the main tasks of industrial robotics is to move objects [1]. There are different configurations of the industrial production line. In some cases, the industrial robotic manipulators pick an object from a table, rotary table, another robot, or a machine or from a conveyor belt and then place the object at a certain location [2]. In some operations, robots manipulate with static objects - not moving relative to the robot. In other types of operations, however, we may have moving objects. One of the most common cases of manipulating moving objects is when they move on a conveyor belt and in the process of movement the robot must pick and place these objects - dynamic picking and placing.

When manipulating static objects in the production processes, sensors are usually used to signal and indicate whether there is an object in a position, so the robot knows when to act. When working with moving objects, this approach using sensors is complex

A. Müller and M. Brandstötter (Eds.): RAAD 2022, MMS 120, pp. 469–476, 2022.
https://doi.org/10.1007/978-3-031-04870-8_55

and difficult to implement and cannot guarantee reliability [3]. When working with moving objects, we offer the application of a computer vision system as a more suitable option.

In this research we will consider the case of dynamically moving objects. Objects will move on a conveyor belt line without being precisely positioned on it - their position is arbitrary in two-dimensional space. When they reach the working area of the manipulation robots, the objects will have to be picked and placed from the production line in baskets. In this study we will rely on a simple color camera and object recognition using computer vision methods.

Computer vision is a method that is widely used in many areas of life, industry, and transport. Specifically for the industry it is applied in quality control operations, sorting of objects, categorization, and others [4, 5]. In the new trends related to Industry 4.0, computer vision is used for intelligent human-robot or machine-robot interaction [6, 7]. In this study, we consider the application of computer vision in the process of robotization. The task that will have to solve computer vision in our case is to recognize moving objects, to track their movement and to provide data on the position of each recognized object at a given time.

In the case of object recognition on a production line, we need algorithms that work well for tracking by processing the real-time video stream from a camera pointing towards the conveyor belt of the production line. The position of the camera is fixed relative to the conveyor belt. In this situation, the following methods for separating an object from a background can be used: Mixture of Gaussians (MOG), k-nearest neighbors (kNN) or the background subtractor of Godbehere-Matsukawa-Goldberg (GMG).

The article is structured in the following sections: Sect. 2 discusses in detail the principle of operation of computer vision algorithms and their application. In addition, in this paragraph we represent an experiment that compares the proposed methods. Section 3 presents the description of an algorithm for optimized pick and place of objects, through feedback from the computer vision system. Section 4 presents the experimental setup, the experiments performed and the obtained results.

2 Methods for Background Subtraction and Object Detection

Background subtraction (BS) is a common and widely used technique for generating a foreground mask (namely, a binary image containing the pixels belonging to moving objects in the scene) by using a static camera [8]. As the name suggests, BS calculates the foreground mask performing a subtraction between the current frame and a background model, containing the static part of the scene or, more in general, everything that can be considered as background given the characteristics of the observed scene. Background subtractors update the foreground/background model over time, typically by applying machine learning to a series of frames [9, 10].

In this research, we study three methods for background subtraction. These methods are implemented into the OpenCV library. Given that we are looking for a suitable method for application in industry, where the parameters of the environment are usually constant and do not change, we will compare these methods to what extent they are applicable. The studied methods are the following:

- Mixture of Gaussians models (MOG): The background model based on Mixture of Gaussians developed by Stauffer and Grimson [11] is the most common approach. This is a statistical model that is based on modeling the values of a pixel as a combination of Gaussian values. Based on the constancy and variance of each of the Gaussian combinations, it is determined which Gaussians can match the background colors. Pixel values that do not match the background distributions are in the foreground until a Gaussian one appears, which includes sufficient and consistent evidence to support this.
- MOG2, which is an improved version of MOG. This method is based on an adaptive algorithm using the Gaussian mixture probability density. Recursive equations are used to constantly update the parameters and at the same time to select the appropriate number of components for each pixel [12].
- k-nearest neighbors (kNN): this algorithm is a simple, supervised machine learning algorithm that can be used to solve both classification and regression problems. This method is represented by recursive equations, which are used to constantly update the parameters of a Gaussian mixture model and to simultaneously select the appropriate number of components for each pixel. In some studies, the KNN method has been proposed to better estimate core density [13].
- background subtractor of Godbehere-Matsukawa-Goldberg (GMG): The algorithm used in this method combines statistical evaluation of the background image, Bayesian pixel segmentation and an approximate solution to the problem of tracking multiple targets using a Kalman filter bank and Gail-Shepley matching. The heuristic trust model allows selective filtering of tracks based on dynamic data [14].

To choose a suitable method for recognizing and tracking objects on the production line, we conducted a brief experiment with each of them. According to a study [15], the KNN and MOG2 methods give the best results and separate objects from the background as well as possible. Because the environment is slightly changing and objects have simple shapes, we decided to compare whether there is a significant difference in the separation of the background. Figure 1 presents the results of the experiments performed, in the form of images from the different stages of operation of each method. We tested the MOG2, GMG and kNN methods. In Fig. 1a. is the result of the MOG2 method, Fig. 1b. is the result of the kNN method and Fig. 1c. is the result of the GMG method.

Because the background is a bit dynamic (the conveyor belt moves slowly and is patterned), all methods are successful in separating objects from the background and tracking them. Comparing the images and the results of the experiments, we can state that MOG2 and kNN are performing better than GMG. In our case the MOG2 gives the best coverage of the pixels related to the objects. Therefore, we will use it as a method to achieve the goal of this study.

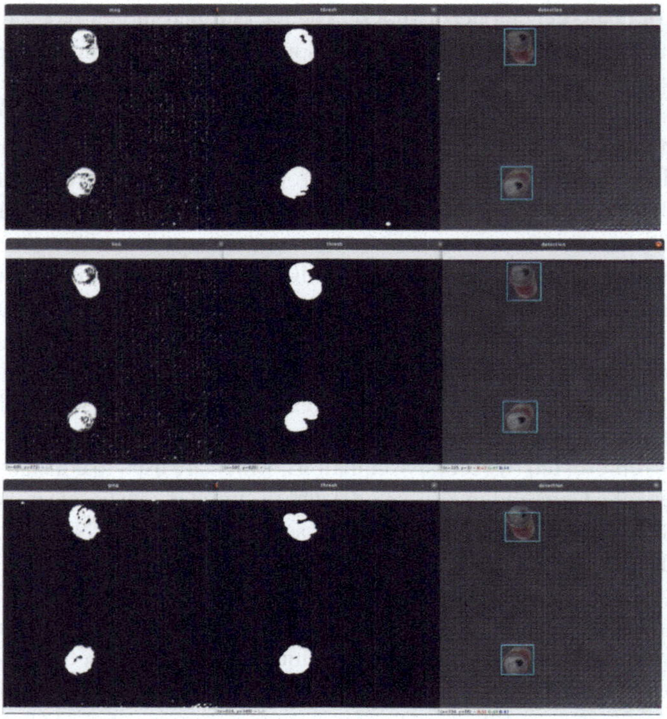

a.

b.

c.

Fig. 1. Results from the background subtraction methods: **a.** MOG2; **b.** kNN; **c.** GMG. The left image represents the result from the background subtractor, the middle image represents the result form the binarization, and the right image shows the marked object on the camera feed.

3 Algorithm for Automated Grasping in Case of Multiple Robot Collaboration

In production processes, several machines often must work together and sequentially. In this study, we consider a case of collaboration between two robots. The purpose of the algorithm is to improve the interaction between robots, to optimize the process of picking and placing objects using a visual guided system.

We will have a conveyor belt that runs at a constant speed. At the end of the conveyor belt there are two manipulation robots that will be able to grasp the objects from the conveyor belt and move them to the baskets. A camera, which captures the objects is placed in a fixed position above the conveyor belt. The experimental setup is shown in Fig. 2. The camera is positioned at the top right above the production line in the form of a blue cone. To the right of the robots is the production line, and to the left the cartridges. The three squares on the left of the cassettes show the data from the camera and the method used to identify and track the objects.

When an object appears under the camera, the algorithm plans which of the two robots to pick and place the object. The camera is in front of the robots so that the time required to move objects from the robots is considered. The algorithm will be able to change the decision of which robot to capture which object, depending on the current number of objects that have been intercepted by the camera and have not yet been moved by the robots. Thus, the algorithm will contribute to improving the interaction between the two robots. And in further research, this algorithm can be used to provide optimized human-robots interaction.

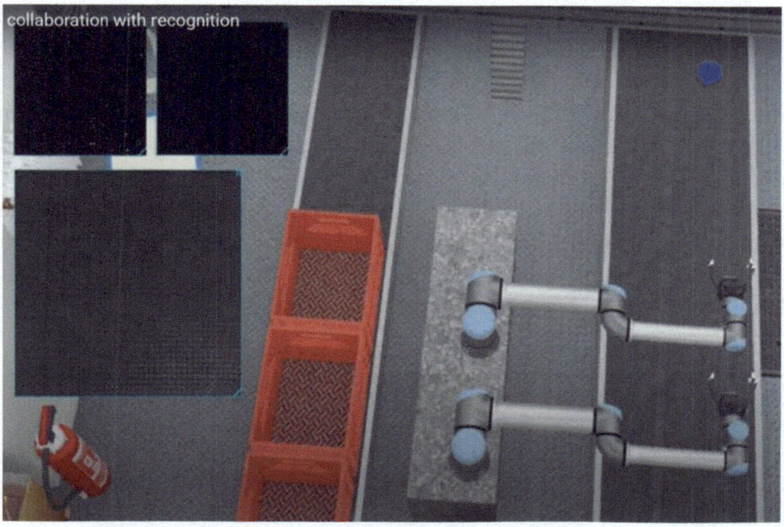

Fig. 2. Experimental setup if the production process.

3.1 Algorithm for Detecting Objects and Determination of Their Grasping Positions

1. Initialization of the background subtractor: in our case, this is achieved by capturing frames, when the conveyor belt is moving and empty.
2. Separating objects from the background: applying the MOG2 background subtractor.
3. Binarization of found pixels and subsequent morphological erosion and dilation operations to filter the noise and fill holes.
4. Finding the contours of the objects in the binarized image.
5. Calculating the mass centers of the contours, considering the position of the object relative to the conveyor belt, determining the time to reach the grasping point of each of the robots for the object and adding it to the list of active objects.
6. Analyze the list of active objects and set the next grasping position for each of the robots. In this analysis and positioning, it is important to align the movement of the robots with that of the objects and to avoid a collision of an object with any robot. Because an object may be more optimal to be grasped by a rear-standing robot.
7. Exclude objects from the active list when they are grasped by a robot. To avoid problems with time synchronization, the gripper of each robot is equipped with a distance sensor based on which the robot decides when to close the gripper and start moving the object in the basket.

This algorithm is experimentally validated in the next section.

4 Experiment and Results

For the experiments we use the Webots simulation environment with two UR10e robots and a conveyor belt. The conveyor belt line is 0.7 m wide. The centers of the robot bases are at 0.95 m from the center of the production line. The centers of the bases of the two robots are at 0.6 m from each other. The camera is located above the production line so that it can see the entire width of the line. The camera is at 1 m along the axis parallel to the conveyor belt of the first robot and 1.6 m relative to the second robot.

The production line moves at a constant speed of 0.2 m/s, which means that the time to reach the objects from the recognition and localization to the grasping position is 5 s for the first robot and 8 s for the second. The time required for robots to pick and place objects is measured at 7 s in the worst case. These are the times that are considered in the grasp/pick planning phase. The objects are cylindrical bodies with a base diameter of 0.065 m and a height of 0.125 m. Figure 3 shows a moment in the process when we already have objects on the production line, and they pass under the camera. Then the algorithm is executed and begins to control the robots.

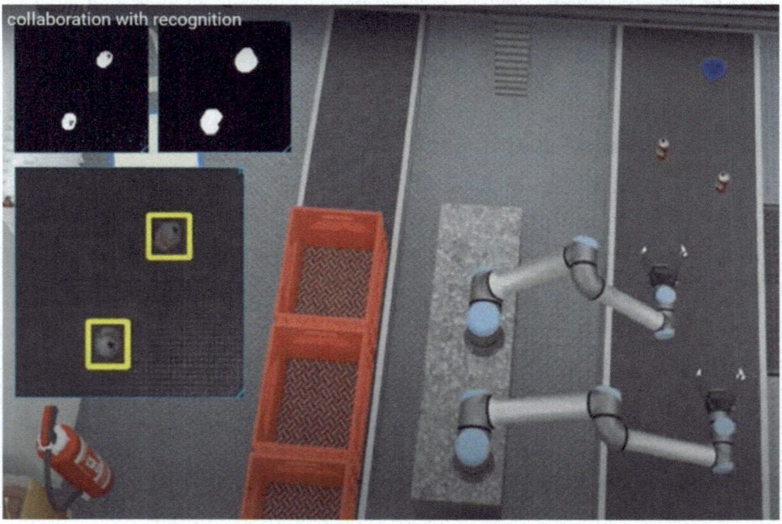

Fig. 3. Algorithm behavior during experimentation.

The video stream from the camera has been resized to 128 × 128 pixels. MOG2 considers both objects and their shadows. Shadows are removed from binarization by a threshold of 244. After binarization, noise and holes are filled by a sequence of morphological operations: two consecutive erosions with an elliptical structural element of size 3 × 3 and two consecutive dilations with an elliptical structural element of 7 × 7. Then all areas with at least 100 adjacent pixels are considered objects.

To demonstrate the operation of the algorithm, a video of the experiment was taken, which can be found at the following address: https://youtu.be/CPy-oq5xMB8.

As seen when the first object is detected, the algorithm sets it for the nearest robot. Then a second object appears, and the algorithm changes the job: it selects the second object for the first robot, and the first object is already assigned to the nearest robot.

From the frames of the video stream, it can be noticed that the conveyor belt is not entirely monochromatic due to the structure of its surface, as well as due to the not completely uniform illumination. However, the algorithm manages to successfully detect objects and plan their pick and place.

The results of the experiment confirm the applicability of the algorithm for industrial tasks.

5 Conclusion

The proposed algorithm detects objects located on the conveyor belt of a production line and plans the grasping process for the multiple manipulating robots. To achieve the goal a simple color camera is used, and an algorithm based on the area of image processing and computer vision. The proposed algorithm uses methods, which are suitable for industrial environments. This leads to a universal approach, which does not require usage of special tools and sensors for the production line. The approach could be applied for each conveyor belt of any industrial production line. The only requirement is to mount a camera on the appropriate place over the conveyor belt and to set the positioning parameters related to the conveyor belt and the robots. The grasp planning can consider a larger number of robots and optimize the sequence of their movements. The algorithm was experimentally tested in a simulation environment with two manipulation robots and successfully managed to plan the pick and place of all objects.

For future research, it is interesting to consider cases of a circular production line with densely located objects, then an optimization problem to grasp all available objects as quickly as possible, will have to be solved. It is possible to consider cases of a production line, which can control the speed of movement of the conveyor belt. Also, the planning of the objects grasping process when collaborative interaction with humans is considered.

Acknowledgement. The research presented in this paper was supported by the Bulgarian National Science Fund under the contract # KP-06-M47/5 – 27.11.2020.

This research is carried out as part of the project funded from the Bulgarian Ministry of Education and Science under the National science program "INTELLIGENT ANIMAL HUSBANDRY", grant agreement # D01–62/18.03.2021.

References

1. Huang, P.-C., Mok, A.K.: A case study of cyber-physical system design: autonomous pick-and-place robot. In: 2018 IEEE 24th international conference on embedded and real-time computing systems and applications (RTCSA), pp. 22–31. IEEE (2018)
2. Smys, S., Ranganathan, G.: Robot assisted sensing control and manufacture in automobile industry. J. ISMAC **1**(03), 180–187 (2019)
3. Han, S.D., Si, W.F., Yu, J.: Toward fast and optimal robotic pick-and-place on a moving conveyor. IEEE Robot. Autom. Lett. **5**(2), 446–453 (2019)

4. Hocenski, Ž., Matić, T., Vidović, I.: Technology transfer of computer vision defect detection to ceramic tiles industry. In: 2016 International Conference on Smart Systems and Technologies (SST), pp. 301–305. IEEE (2016)
5. Jain, T.: Industrial objects recognition in intelligent manufacturing for computer vision. Int. J. Intell. Unmanned Syst. (2021)
6. Xu, Y., Fang, G., Lv, N., Chen, S., Zou, J.J.: Computer vision technology for seam tracking in robotic GTAW and GMAW. Robot. Comput. Integr. Manuf. **32**, 25–36 (2015). https://doi.org/10.1016/j.rcim.2014.09.002. (ISSN: 0736–5845)
7. Suma, V.: Computer vision for human-machine interaction-review. J. Trends Comput. Sci. Smart Technol. (TCSST) **1**(02), 131–139 (2019)
8. Goyal, K., Singhai, J.: Review of background subtraction methods using Gaussian mixture model for video surveillance systems. Artif. Intell. Rev. **50**(2), 241–259 (2017). https://doi.org/10.1007/s10462-017-9542-x
9. Supreeth, H.S.G., Patil, C.M.: Efficient multiple moving object detection and tracking using combined background subtraction and clustering. SIViP **12**(6), 1097–1105 (2018). https://doi.org/10.1007/s11760-018-1259-z
10. Kini, S.: Real time moving vehicle congestion detection and tracking using OpenCV. Turkish J. Comput. Math. Educ. (TURCOMAT) **12**(10), 273–279 (2021)
11. Stauffer, C., Grimson, E.: Adaptive background mixture models for real-time tracking. In: IEEE Conference on Computer Vision and Pattern Recognition, CVPR, pp. 246–252 (1999)
12. Zivkovic, Z.: Improved adaptive gaussian mixture model for background subtraction. In: 2004 Proceedings of the 17th International Conference on Pattern Recognition ICPR 2004, vol. 2, pp. 28–31. IEEE (2004)
13. Zivkovic, Z., van der Heijden, F.: Efficient adaptive density estimation per image pixel for the task of background subtraction. Pattern. Recogn. Lett. **27**(7), 773–780 (2006)
14. Godbehere, A.B., Matsukawa, A., Goldberg, K.: Visual tracking of human visitors under variable-lighting conditions for a responsive audio art installation. In: American Control Conference (ACC), 2012, pp. 4305–4312. IEEE (2012)
15. Trnovszký, T., Sýkora, P., Hudec, R.: Comparison of background subtraction methods on near infra-red spectrum video sequences. In: Procedia Engineering, vol. 192, pp. 887–892 (2017). https://doi.org/10.1016/j.proeng.2017.06.153. (ISSN: 1877–7058)

The AMADEE-20 Robotic Exploration Cascade: An Experience Report

Hamid Didari[1], Matthias Eder[1], Gernot Grömer[3], Richard Halatschek[1], Seda Özdemir-Fritz[3], Raphael Prinz[2], Johannes Scholz[2], and Gerald Steinbauer-Wagner[1](✉)

[1] Institute of Software Technology, Graz University of Technology, 8010 Graz, Austria
{hamid.didari,matthias.eder,richard.halatschek,steinbauer}@ist.tugraz.at
[2] Institute of Geodesy, Graz University of Technology, 8010 Graz, Austria
{raphael.prinz,johannes.scholz}@tugraz.at
[3] Austrian Space Forum (ÖWF), 6020 Innsbruck, Austria
{gernot.groemer,seda.ozdemirfritz}@oewf.org

Abstract. As part of the AMADEE-20 Mars analog mission, an exploration cascade for the remote sensing of extraterrestrial terrain was implemented. For this purpose autonomous mobile as well as aerial robots were conceptualized and used in an iterative exploration cascade. First, satellite imagery was analyzed using Deep Learning methods. This large but low resolution image was used as a basis for mission planning as well as for an initial traversability analysis for the ground robot. High resolution low orbit aerial images of a defined area of interest were then recorded, and utilized together with the low resolution map were in a path planning tool for the mobile robot. The mobile robot uses a navigation concept of global and local planning as well as local perception to navigate to the area of interest. Experiments during the AMADEE-20 mission in the Israeli Negev Desert validated the exploration cascade.

Keywords: Field robotics · Space exploration · Robot navigation

1 Introduction

The first human-robotic Mars missions are envisioned to materialize no earlier than the late 2030s, aiming for surface exploration and sojourns lasting at least one month [1]. To prepare these missions and improve the probability of success, analog planetary research missions are conducted on Earth to investigate the potential of instruments, workflows, materials, and human factors [5]. In the latest Mars Analog Research mission (AMADEE-20) organized by the Austrian Space Forum (OeWF) in cooperation with the Israel Space Agency, an operational workflow for an exploration cascade (EC) using robotic systems was

The authors are in alphabetical order. This work was partially supported by the Austrian Research Promotion Agency (FFG) with the project RoboNav.

A. Müller and M. Brandstötter (Eds.): RAAD 2022, MMS 120, pp. 477–484, 2022.
https://doi.org/10.1007/978-3-031-04870-8_56

implemented which is presented in this work. For this, autonomous mobile robots as well as aerial robots were conceptualized for use in an iterative exploration cascade pertinent to a human robotic exploration scenario on Mars: Satellite remote sensing data were analyzed for geographical features using convolutional neural networks. This large but low resolution image was used as a basis for mission planning as well as an initial navigation map for the ground robot. High resolution low orbit aerial imagery of defined areas of interest was obtained, which together with the low resolution map serves as input for the mobile robot's path planning tools. The mobile robot platform "Mercator" utilized a navigation concept of global and local planning and local perception to navigate through the area of interest. The mission took place in October 2021 and was conducted in the Ramon Crater, Negev Desert, Israel.

2 Mars Analog Research

Starting in the Apollo era, analog planetary field sites were used to train astronauts. Analog research offers an effective tool to prepare for such surface missions, including future Mars missions. They facilitate and catalyze the field testing of equipment, scientific strategies, and robotic operations in a representative environment, often characterized by a steep operational learning curve.

Analog missions such as NASA's D-RATS campaigns, its HI-SEAS missions, or the European MOONWALK or PANGEA projects have supported the preparations for future Mars exploration missions. These field campaigns focus on high fidelity surface operations by using research-grade spacesuit simulators and on conducting experiments in the domains of geoscience, human factors, and astrobiology. The analog missions also enable the validation of robotic systems. It also includes the establishment of an exploration cascade that optimizes instrument workflows, required resources, environmental and flight planning border conditions, as well as the ground segment data processing pipeline [5].

3 AMADEE-20 Exploration Cascade

To maximize the efficiency and scientific output of Mars exploration missions, they need to be properly planned both from a scientific as well as from an operational perspective. The AMADEE missions' exploration cascade (EC) was developed to facilitate the inclusion of those key aspects in planetary surface operations and is an OeWF-coined term for optimizing the tactical sequence of measurements addressing pre-defined scientific questions [5]. It helps to manage multiple assets to optimize the scientific return of planetary surface missions, to search for extinct and/or extant traces of life, and to characterize the geoscientific context of sites of interest. Hence, during AMADEE-20, the EC with the experiments therein aims to study the geology of the analog site by investigating geological samples and their composition. Those results shall provide insights into geomorphological processes, e.g., whether aqueous formations have been present and what forms of life could have dominated throughout different periods, exploring the test sites' astrobiological potential. Incorporating those aspects, the EC represents the framework for the robotic experiments outlined in the following sections. The order represents the deployment sequence from first obtaining high altitude aerial imaging over refining situation awareness using UAVS and leading to in-situ investigations by rovers in the end.

3.1 Flight Control and Operation

The planning and control of the AMADEE-20 missions, including the exploration cascade, was done by the Remote Science Support (RSS) Team of the OeWF as part of the Mission Support Center (MSC), located in Innsbruck/Austria. The RSS tasks included the tactical sequence of what instrument to deploy by which mean at which time and at which location in pre-defined regions of interest. The deployment, preparatory activities, and additional experiments (e.g. geological sampling) performed by human analog astronauts are all scheduled in the Daily Activity Plan. The MSC is also the focal point of scientific data, mission planning, and mission support activities. The science data is archived in the Science Data Archive of the OeWF available to the scientific community.

3.2 Mission Preparation Using Satellite Images

Exploration of the Martian surface requires identifying interesting locations based upon satellite reconnaissance for which the overall resolution remains low. This is the basis for surface mission planning, including navigational context information for the mobile robot platform.

Therefore, convolutions neural networks (CNN) with an U-NET architecture [6] are trained on a high resolution orthoscopic satellite image of the mission environment to assess the traversability for the rover on the ground. The parameters for this analysis include obstacles as well as rims, tracks, and roads. For the AMADEE-20 mission area, bushes, buildings, boulders, and bodies of water are identified and classified as non-traversable by the rover, whereas existing roads and homogeneous terrain are suitable for rover movement. The objective is to generate a map of the area, that provides navigational information and slope as well as a surface model of the area, that provides accurate terrain information concerning the suitability for rover navigation. The AMADEE-20 mission area covers an area of 5×5 km around the base station having a resolution of 0.5m per pixel. Two models are used for the segmentation. The first model was trained on larger features (macro-features) , to extract larger rock formations and river beds. For the second one (micro-features), samples were manually labeled to train the CNN for the semantic segmentation. These samples include bushes, boulders, bodies of water as well as gravel and asphalt roads (see Fig. 1).

The separation of gravel roads from the barren land of the Ramon Crater proves difficult for the CNN, as the colors and textures of roads exhibit a low variability challenging detection by the model in that specific geographic region. In addition to semantic segmentation, a standard-deviation kernel-filter with a kernel size of 5×5 pixels was applied to the existing orthoscopic image.

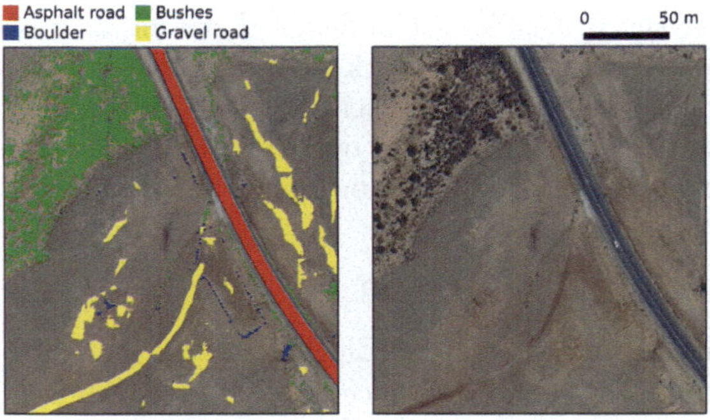

Fig. 1. In the left graphic a detailed view of the classified micro-features of surface - generated with the CNN approach - is depicted. The picture to the right is the corresponding orthosopic image that serves as input data source for the CNN.

This method estimates the heterogeneity of an area. The reasoning behind this is, that areas with high variance in the image provide evidence, that obstacles like groups of boulders and rocky terrain exist. While this method does not identify the objects on the ground, it still provides useful information and proved successful in identifying areas that should be avoided by the rover - due to their low suitability for navigation. In the slope raster image, pixels with a value larger than 20°C of inclination are flagged as not traversable by the rover. The resulting navigation map (i.e. cost map) has values ranging from 0 to 255 (see Fig. 3).

3.3 Aerial High Resolution Exploration

Another critical part of the EC toolchain is the deployment of UAVs in lower altitudes, to obtain more details of areas of interest. While higher altitude missions yield an overview of the entire area at a low resolution, very high-resolution is sought with the low-altitude flights. In addition, specific zones deemed to be of special interest require ego-motion-free images and close-up views. For the exploration cascade in AMADEE-20, a UAV was chosen which features a high-resolution camera in the visual spectrum, and an Inertial Measurement Unit (IMU). Both sensors are used as input for visual-inertia algorithms to guide the aerial vehicle autonomously from A to B including take-off and safe landing following the manually provided exploration path.

For a specific area of interest, the UAV was used to take 230 images of an approximately 450 by 500 m area. The images are used to reconstruct the surface using structure from motion. For processing, the software Agisoft Metashape is used to calculate a point cloud, a high-resolution surface model, and orthographic imagery from the drone footage. The Progressive Morphological Filter developed by [8] is used to classify the point cloud into ground and none ground, allowing for further obstacle detection. Although this filter was developed with LIDAR data in mind, descent results could be achieved with the structure from motion data. After filtering, the ground points are used to compute an elevation model.

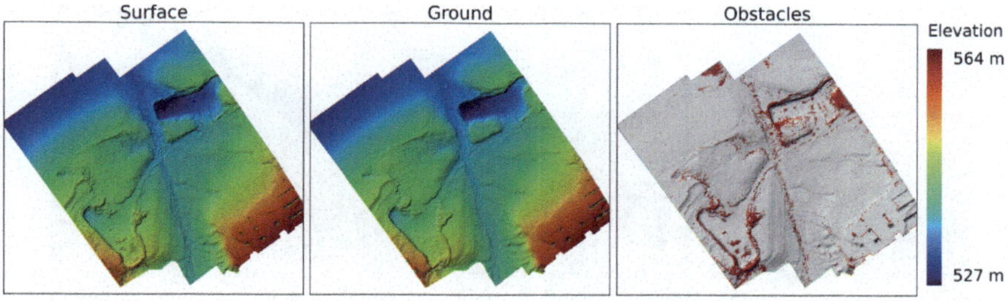

Fig. 2. Three different maps created with the GeoAI approach - the digital surface model (denoted as *Surface*), the digital terrain model (denoted as *Ground*), and the obstacles detected by the approach with CNN - denoted as *Obstacles*.

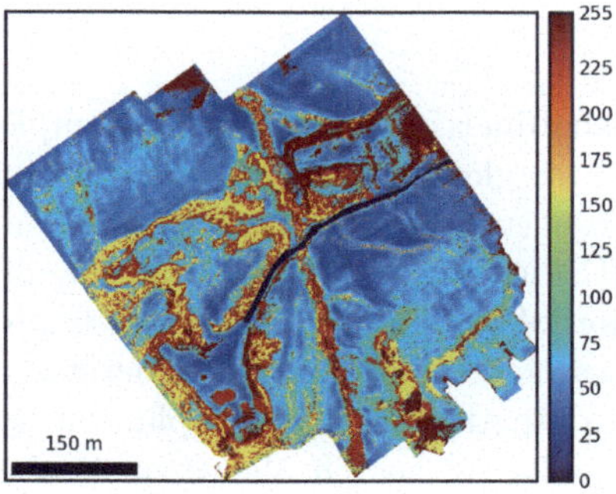

Fig. 3. The costmap for the AMADEE mission in Ramon Crater, Israel. The higher the value of a pixel, the less suitable is the area for the rover of the mission.

The surface and the elevation model are subtracted from each other to obtain a raster map that represents obstacles for the rover. The elevation model was further used to compute the slope of the terrain. The results are depicted in Fig. 2. Additionally, the micro-feature CNN was tested on the orthoscopic image generated from the drone footage. Numerous obstacles could be identified, and a combination with the structure from motion data resulted in a detailed map of obstacles. These obstacles are combined with the slope and standard deviation of the area in the same fashion as the higher altitude map described above. The resulting raster provides a high-resolution map of the traversability for the rover in the area. The result is depicted in Fig. 3.

3.4 Autonomous Robot Navigation

To provide detailed maps of an area of interest or to conduct close-up inspection (e.g. visual or spectroscopy) of a spot, an autonomous rover named "Mercator" (see Fig. 4 left) is added to the exploration cascade. Due to experiences of AMADEE-18 mission where ground robot exploration was limited to a few hundred meters a day [7] the goal was to extend the reach of the rover to several kilometers per day to reach interesting remote areas. Although the localization of

Fig. 4. Rover Mercator during a deployment in the exploration cascade (left). AMADEE-20 field site in Ramon Crater with storage container in the center, the astronauts' habitat on the right, and the solar-panels on the bottom (right).

Mars rovers employs visual-inertial methods with complex terrain matching [3] we decided to use GNSS for global position estimation for practical reasons. As GNSS offers only limited accuracy ($\geq$0.5 m), the position data is fused with the data of an onboard IMU for local and precise movement measurements. Additionally, a transformation of the global GNSS coordinates into the global map coordinates has to be conducted to overlay the generated map. To autonomously navigate a robot to the desired pose, a rough plan on how to reach the goal is needed. To achieve this, the A* search algorithm is used, as it is guaranteed to find the optimal path to the goal using the retrieved 2D cost map which was created by the fusion of the high and low orbital images (see Sect. 3.1 and 3.3). The result of the planning component is a low-resolution path, consisting of multiple poses providing a coarse direction to the desired goal position. The global plan is not continuously updated but only replanned on request, as the calculation of long paths (up to 2 km on a 25 km^2 map) is quite resource intense. Local navigation aims to follow the global path while considering the robot's surroundings and its kinematic constraints. As the resolution of the 2D map is rather poor for precise navigation an internal 3D representation of the environment with a higher resolution (0.05 m) is created by using 3D LiDAR scanners and stereo cameras. These sensors constantly map the robots surrounding to identify traversable areas to create an elevation map, as proposed in [2]. To classify obstacles as traversable or non-traversable the shape and kinematic constraints of the robot have to be considered. As the robot navigates locally in a three-dimensional space and the global path planning is done in 2D, a projection of the 2D path into the robot's local environment representation needs to be done. This information is then used as basis for planning motion primitives which take obstacles and kinematic constraints into account. In the experiments, the dynamic window approach (DWA) is used, which calculates a series of motion primitives to create a smooth kinematically feasible path for the robot to follow, while considering dynamic obstacles [4]. Please note that the robot employs a double Ackermann-steering.

4 Evaluation in Field Trial

The proposed EC using robotic vehicles was evaluated during the AMADEE-20 Mars analog mission which took place during October 2021 in the Ramon Crater, Negev desert, Israel (see Fig. 4 right). The mission comprises a one-week bridgehead phase, where the experimenter and field crew set up the field infrastructure, different scientific systems, and experiments as well as a three-week isolation phase, where the analog astronauts performed their duties with only delayed contact to the control center in Austria. Post-mission analysis of data and results is still ongoing. The EC was built up by three experiment teams: GEOS, AEROSCAN, and EXOSCOT. Based on the procedures of the EC, these experiments worked cooperatively before, during, and after the mission phases. Before the field deployment, the GEOS experiment used satellite and aerial images to define large regions of interest (ROIs), with satisfactory results from the geoscientific perspective (see Sect. 3.2). However, the RSS team proposed that future missions should utilize more sophisticated remote sensing data. Therefore, the GEOS team aimed to minimize these areas to points of interest (POI) by using detailed data (e.g. 3D point clouds) and images obtained by the ground robot as well as lower altitude images. The lower altitude imagery was done by the AEROSCAN drone experiment which aims to design, manufacture, and test an autonomous (dawn-sunset flight time) solar-powered, convertible Vertical Take-Off and Landing (VTOL) drone (see Sect. 3.3). Its main purpose is to autonomously survey POIs selected from satellite data to obtain detailed maps. These results are useful to plan site-specific operations like Extra-Vehicular Activities (EVAs) or surveys with ground robots. In the frame of the AMADEE-20 exploration cascade, the drone was launched by astronauts during EVAs and supervised by the remaining crew in the habitat. From deployment to landing, the drone performed its tasks autonomously, following a pre-loaded mission. In the meantime, the habitat crew monitored the drone's telemetry to ensure that the path is followed and correct for non-nominal events. Several areas of interest around the AMADEE-20 station were photographed with the drone's camera. To allow the in-situ exploration of the EC by the EXOSCOT rover team, the recorded images were processed into geo-referenced maps using the process described in Sect. 3.3. This detailed terrain data allows evaluating which areas the rover can traverse. On one hand, this information is used by the RSS to select Regions of Interest (ROI) and POIs reachable by the rover, to provide the best research areas to address geology and astrobiology questions. On the other hand, this information is used by Mercator's onboard navigation system to plan a suitable path to the selected region. Once the region was selected and a plan was obtained, the navigation system guides the robot to the selected region. During the rover's traverse, camera images and 3D point clouds (stereo camera, 3D LIDAR) together with a GNSS ground truth pose are recorded. The recording is subsampled in a way to provide a suitable geometrical and temporal coverage of the region. These data are later used by RSS to better plan EVAs. During AMADEE-20 mission, 17 EVA runs were performed with about 80 h of EVA operations. The planning of these EVAs was supported by the following deployment of modules of the EC experiments. The GEOS experiment ran 4 times. Reconnaissance with AEROSCAN was performed 5 times. The EXOSCOT was deployed 7 times with a maximum distance of 2 km in one run.

5 Conclusion

The AMADEE-20 Analog Mars Mission provided a detailed and sophisticated framework for research groups focused on planetary robots validating their systems in Mars-like realistic environments, but more important in a spaceflight representative operational setup given by the mission architecture. The experiments for the proposed Robotic Exploration Cascade were conceived as a success, as the decision making pipeline based upon 1) high altitude image processing derived from satellites via the 2) high resolution images from a UAV to the 3) navigation of a mobile robot using the processed image data was successfully implemented.

References

1. Drake, B.G., Hoffmann, S.J., Beaty, D.W.: Human exploration of mars, design reference architecture 5.0. In: IEEE Aerospace Conference (2010)
2. Fankhauser, P., Bloesch, M., Hutter, M.: Probabilistic terrain mapping for mobile robots with uncertain localization. IEEE Robot. Autom. Lett. (RA-L) **3**(4), 3019–3026 (2018). https://doi.org/10.1109/LRA.2018.2849506
3. Farley, K., Williford, K., et al.: Mars 2020 mission overview. Space Sci. Rev. **216**, 1–41 (2020)
4. Fox, D., Burgard, W., Thrun, S.: The dynamic window approach to collision avoidance. IEEE Robot. Autom. Mag. **4**(1), 23–33 (1997). https://doi.org/10.1109/100.580977
5. Groemer, G., Ozdemir, S.: Planetary analog field operations as a learning tool. Front. Astron. Space Sci. **7**, 32 (2020). https://doi.org/10.3389/fspas.2020.00032
6. Ronneberger, O., Fischer, P., Brox, T.: U-net: convolutional networks for biomedical image segmentation. In: Navab, N., Hornegger, J., Wells, W.M., Frangi, A.F. (eds.) Medical Image Computing and Computer-Assisted Intervention - MICCAI 2015, pp. 234–241. Springer International Publishing, Cham (2015)
7. Stradner, M., Steinbauer, G.: Lifting robot exploration to 3d environments. In: 2019 IEEE International Symposium on Safety, Security, and Rescue Robotics (SSRR), pp. 1–8 (2019). https://doi.org/10.1109/SSRR.2019.8848942
8. Zhang, K., et al.: A progressive morphological filter for removing nonground measurements from airborne lidar data. IEEE Trans. Geosci. Remote Sens. **41**, 872–882 (2003)

Humanoids, Legged, and Bioinspired Robots

Garment Type Agnostic Robotic Unfolding of Garments from Random Configuration

Dimitra Triantafyllou(✉) and Nikolaos Aspragathos

University of Patras, Rio, Greece
{dimtria,aspragat}@upatras.gr

Abstract. This paper presents a method for the robotic unfolding of garments from random initial configurations. In order to handle various types of garments and fabrics, features common in all garments are explored and identified, while manipulations mimicking the human unfolding movements are derived by a learning procedure. The proposed approach involves the transformation of the garment from random to half-folded state so that its configuration can be analyzed through a visual hierarchical architecture. Then, the two layers of the garment and the folding axis that unites them are extracted. The upper layer is finally unfolded by the application of a path extracted based on learning by demonstration. The proposed framework is experimentally evaluated on a variety of garments and the produced results indicate the feasibility of the method.

Keywords: Garment · Unfolding · Depth sensors · Robotic manipulations

1 Introduction

Service robots have attracted the interest of robotic researchers who explore their use in various everyday life scenarios to alleviate humans from time-consuming and tedious tasks, such as housework. In this scope, the robotic handling of garments has been investigated. Especially the robotic unfolding of garments from random initial state constitutes a challenging task due to the difficulties in the recognition of their configuration, the detection of handling points and the large deformations caused by their highly non-rigid nature. Moreover, the large variety of garments' types and fabrics makes the robotic unfolding an even more demanding task.

Although most robotic garment unfolding methods require the type classification of the garment (shorts, shirt etch) to unfold it [1, 2], there are few authors that proposed garment agnostic approaches. In their approach, the garment is transformed from a random to a half-folded state by hanging it in the air and either iteratively grasp its lowest point [3] or detect its folds through the created shadows and grasp them [4]. Then, the garment is placed onto a table and its upper layer is extracted [5, 6] by clustering the garment into areas based on their depth values. Finally, in [5] the garment is fully unfolded by applying a triangular path to its upper layer.

In this paper, a garment type agnostic unfolding method is proposed based on previous works of the authors [7–9]. Our main goal is to provide for the first time an overall

A. Müller and M. Brandstötter (Eds.): RAAD 2022, MMS 120, pp. 487–495, 2022.
https://doi.org/10.1007/978-3-031-04870-8_57

description of the pipeline that we propose for the robotic unfolding of crumbled garments. Although the approach follows the aforementioned sequence of steps (garment handling in the air to bring it to a half-folded state [7], upper layer extraction [8], complete unfolding [9]) there are significant differences between our work and the related work of other authors. Our method depends on the identification of local features extracted from depth data, which are independent from colors and lighting conditions. Folds and hemlines are detected while handling the garment in the air and evaluated based on their graspability providing grasping point alternatives. Moreover, folds, overlaps, corners, edges and wrinkles of the half-folded garment are identified using the proposed junctions' "dictionary" and are combined through a visual architecture to high-level features, such as the folding axis, the upper and lower level. Finally, a method for the extraction of an unfolding path mimicking the human movements is derived without requiring any a priori knowledge on the garment's fabric and depending only on depth data, such as the folding curve's height.

2 Method

The proposed approach refers to the robotic unfolding of garments through visual analysis and robotic manipulation processes that transform garments from random to fully unfolded configurations. The steps of the unfolding process comprise: 1) the initial transformation of a garment from a random state (Fig. 1a) to a half-folded configuration (Fig. 1b), 2) the analysis of this configuration so that the layers of the folded garment are identified (Fig. 1c), and 3) the complete garment unfolding (Fig. 1d–e).

2.1 Garment Transformation from Random to Half-Folded Configuration

For the transformation of the garment to a half-folded configuration, a sequence of actions is applied. Initially, the crumbled garment (Fig. 1a – the method does not currently evaluate the initial state of the garment to verify if it is crumbled or unfolded) is lifted in the air and is rotated in front of a depth camera so that points of its outline are detected. Based on a voting procedure and a graspability check, one of the detected outline points is selected as grasp point. Then, the garment is grasped by the provided grasp point and the procedure is iterated for the selection of a second grasp point. In this way, the garment results in a half-folded configuration grasped by two points of its outline (Fig. 1b).

An important aspect of this method is that the detection of the outline points is based on features common to all garments, remaining independent of the garment type or fabric. Therefore, folds and hemlines with small inclination regarding the horizontal surface are considered as candidate grasp points (in Fig. 1b, folds and hemlines are marked on the hanging garment with blue and yellow colors respectively). Their detection is based on the extraction of the garment's edges as they occur from the depth differences of its surface, while the 2D geometry and the depth differences between these edges constitute the features' identification criteria.

As the garment is rotated in the air, multiple depth images are examined for outline points while simultaneously its 3D model is extracted. The detected outline points are mapped on the 3D model. To avoid false positives, only points with the highest number

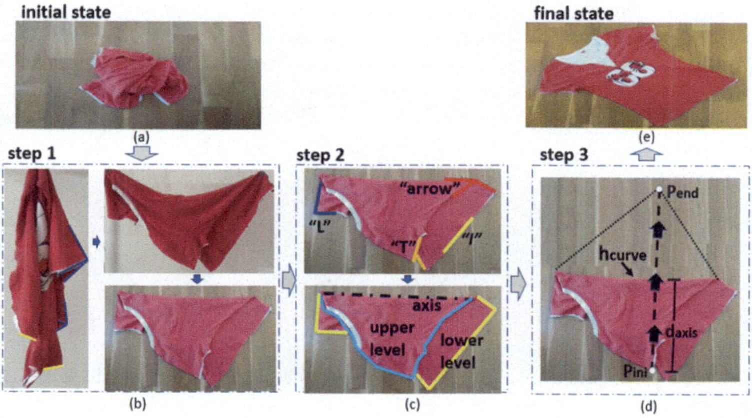

Fig. 1. Unfolding method outline: (a) initial state, (b–d) step 1–3, (e) unfolded garment.

of appearances in depth images are considered as grasp candidates and are examined regarding their graspability. In this scope, various orientations of the gripper are explored in simulation aiming to find an orientation that ensures successful grasping (e.g., avoid collision with the garment before the completion of the grasping).

Finally, the candidate grasp point that appears in most images and satisfies the graspability conditions is selected and grasped. The same procedure is repeated so that a second point is grasped, and the garment ends up to a half-folded configuration.

2.2 Analysis of the Half-Folded Configuration

Once the garment is half-folded, it is placed onto a table so that its configuration is analyzed, in order to proceed to its full unfolding. To achieve this, a hierarchical visual architecture that decomposes the folded configuration into low and high-level features is proposed. The low-level features comprise junctions of edges, such as "arrow", "T", "L" and "I" junctions (the main junction types depicted in Fig. 2.a), which are translated through a proposed "dictionary" to localized indicators of the garment's state, such as folds, overlaps, corners, and edges respectively (Fig. 1c). Moreover, combinations of "L" or "I" junctions with "noise" edges, i.e. edges that occur from a large seam, pockets or other 3D features of the garment, are also taken into account (Fig. 2.b). On the other hand, the high-level components refer to the two layers of the half-folded garment, i.e., the upper and lower layer, and the axis that unites them (Fig. 1c). The proposed configuration analysis method detects and classifies the low-level features into the "dictionary's" elements and combines them to reconstruct the high-level hierarchical features.

The heart of the proposed approach is the junctions' "dictionary" that attributes meaning to junctions of two or three edges. The junction's type depends on the 2D geometry and the depth levels formed between the junction's edges. A summary of the "dictionary's" junctions and their meaning is provided in Fig. 2 while a detailed analysis can be found in [8].

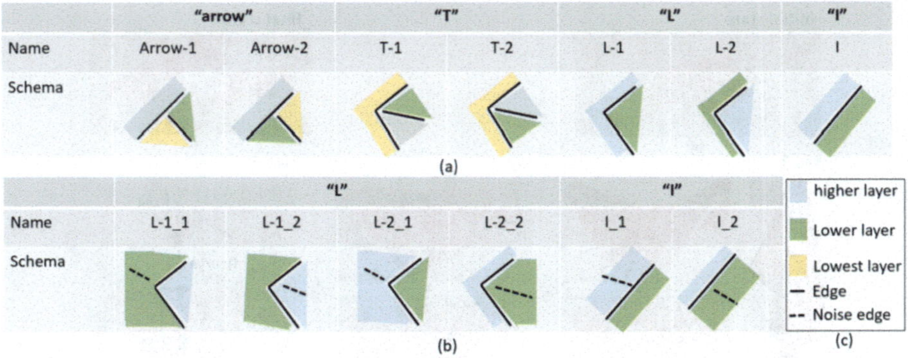

Fig. 2. "Dictionary" of junctions: (a) main junction types, (b) junctions with noise edges, (c) explanatory table.

The first step of the configuration analysis involves the extraction of the garment's edges based on their depth differences. Then, the edges are combined into junctions of two or three edges based on the Euclidean distance between their ends. The junctions are classified into elements of the "dictionary" through the Random Forest classifier. In particular, the classifier uses input data that refer to the distances, the angles and the depth differences formed between the edges of the junction to be classified as well as information that shows if an edge is located on the outline of the half-folded garment. The output provides the classification probabilities for each element of the dictionary and the one with the higher value corresponds to the junction's type.

Once the junction types are identified, the reconstruction of the architecture's higher-level elements begins, starting with the extraction of the garment's folding axis. The axis identification is achieved in a two-step process. Initially, all the external edges of the half-folded garment are evaluated and only the ones whose ends refer to combinations of two "arrow", two "L" or an "arrow" and an "L" junction are considered as axis candidates. In this way, the number of possible axes is reduced excluding cases that could not refer to an axis, e.g., an edge resulting in a "T"-junction, meaning an overlap. Then, the candidate edges are individually evaluated by an SVM classifier that provides the probability for an edge to be an axis. The inputs of the classifier refer to: 1) the number of "arrow"-junctions that lie on a candidate edge, since these junctions are strong indicators of folds, 2) the relative width of each edge in comparison to the other edge candidates, 3) the edge's length, 4) the number of arrow junctions that belong to other candidate edges. Once all the edges are evaluated, the candidate with the highest probability is the garment's axis.

The identification of the axis provides valuable information for the extraction of the upper layer. Specifically, the junctions that lie on its ends are the starting and ending point for the proposed algorithm that seeks the sequence of edges that form the upper layer. The algorithm starts from one of the junctions lying at the axis, it classifies it into one of the "dictionary's" elements and, based on the classification, the junction's edge that belongs to the upper edge is identified. Then, the algorithm searches for new junctions that include this edge, so that they are classified and a new edge that belongs to the garment's upper layer is found. This procedure is iterated until it ends up to the second end of the axis. In cases that the upper layer search fails to reach the axis's end resulting to a dead-end situation, a depth first search algorithm is applied. The algorithm returns to the previous junctions and re-evaluates them while excluding from the classification procedure the junction type that led to the erroneous result.

Once the upper layer of the garment is found, the extraction of the lower layer is straightforward. Since the garment's outline edges constitute a combination of all the visible lower layer's edges with edges that correspond to upper layer, the edges that are not mapped to upper layer are regarded as edges of the lower layer.

2.3 Unfolding of the Half-Folded Garment

This section presents the final phase of the unfolding procedure, namely the robotic manipulation of the garment's upper layer in order to bring the garment in its fully unfolded state. Therefore, a method is proposed for the extraction of the unfolding path applied on the garment's upper layer without lifting it from the table. The goal of the method is to extract a path that can be applied on garments based only on features that can be extracted by an RGB-D sensor, remaining independent from the physical characteristics of the garment's fabric.

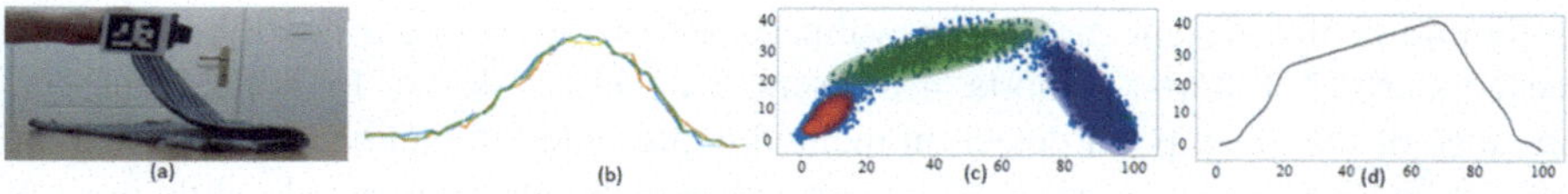

Fig. 3. Path extraction using learning by demonstration: (a) recording of unfolding paths, (b) paths' normalization and alignment, (c) modelling paths to segments using Gaussian Mixtures Models, (d) path extraction using Gaussian Mixture Regression.

The proposed approach is based on learning by demonstration methods to mimic the human's unfolding movement. To achieve it, the unfolding paths applied on the upper layer of the garment by humans are recorded using Aruco markers placed at their hands, near the point they grasp the garment to handle it (Fig. 3a). The unfolding procedure is applied on a variety of folds, garments, and fabrics. The recorded paths are normalized and aligned through the Dynamic Time Warping method (Fig. 3b). Since the normalized paths of different folds resemble to a large extend, the extraction of a generic path is proposed. To achieve it, Gaussian Mixture Models of the paths are used (Fig. 3c) to extract the generic path with Gaussian Mixture Regression (Fig. 3d).

For the application of the extracted path to a specific fold, its starting and ending point must be determined (Fig. 1d). To achieve it, the configuration analysis described in Sect. 2.2 is used. Since the upper layer of the garment is known, the initial point is selected based on the approach presented in [10], i.e., we select the convex vertex of the upper layer's outline that guarantees that the movement of the upper layer will not lead to new folds. In this specific case, since a single robotic arm is used, we mainly refer to triangular upper layers and we select the vertex that does not correspond to the unfolding axis (Fig. 1d). On the other hand, the final point P_{end} is detected through calculations that consider its symmetry to the starting point in reference to the folding axis. Therefore, due to the symmetry, the direction of the path is defined by the perpendicular to the unfolding axis, while the distance between the starting and ending point is calculated by the following equation:

$$d_{traj} = 2 \cdot d_{axis} + h_{curve} \tag{1}$$

where d_{axis} is the vertical distance of the initial point P_{ini} from the axis and h_{curve} is the height of the garment's fold curve. In particular, h_{curve} is used to approximate the length of the fold's curve, since its exact calculation is challenging and depends on various parameters such as the garment's fabric, knitting, and the existence of features that create local inhomogeneities (e.g., pockets and zippers).

3 Experiments

In this section, the results of the experiments that evaluate the phases of the proposed method are provided. Each stage of the unfolding method is evaluated separately, while more experiments can be found in [7–9]. For the garment transformation the robot holds and rotates the garment in a 1m distance from a RGB-D sensor while for the next phases the set-up of the camera is 0.8–1 m from the working table.

For the evaluation of the garment transformation from random state to half-folded configuration 50 different cases were evaluated with a variety of garment types (5 long-sleeved shirts, 5 T-shirts, 4 shorts, 4 trousers, 3 towels, 2 skirts). In Table 1, the performance of the grasp point detection method is presented. It can be observed that the 97% of the detected grasp points were correctly identified and grasped, while most of the detected points corresponded to folds. Regarding the erroneous results, a fold was correctly detected but the robot erroneously grasped a neighbor area, while two edges of large shorts' pockets were mistaken as hemlines of the garments.

Table 1. Performance of the grasp point detection and grasp evaluation method (%)

Type of grasp area	True positives	False positives
Fold	75	1
Edge	22	2
Total	97	3

The performance of the configuration analysis method is described by the evaluation of the junctions' classification, the axis and upper layer detection methods. Specifically, for the junction classification method 200 junctions per junction category were used. For the training of the Random Forests 160 trees were utilized with max depth equal to 20 while 80% of the samples were used for the training of the classifier and the remaining 20% for its testing. The classifier managed to recognize correctly 71.5% of the cases. Nevertheless, most of the misclassifications did not fail to map the edges correctly to the upper and lower layer and mainly referred to their relative inclination (e.g., misclassified an "L-2_2" to an "I-2" junction). Under this perspective, the correct layer mapping of the junctions reached the 91.5% of the tested cases.

For the evaluation of the axis detection method 375 different garment configurations were used. For the SVM training 60% of the cases were used and the remaining 40% were left for testing while k-fold cross validation (k = 5) was applied to achieve generalization. In 91% of the cases, the axis was detected correctly. Specifically, 5% of the cases were not considered as axis candidates due to misclassifications of the "L" and "arrow" junctions that laid to the axis ends and the remaining erroneous 4% occurred due to the SVM's mistaken output.

The upper detection method was applied in 112 different configurations of garments of various types and its performance was 83.04%. The main reasons of failure were misclassifications of the "dictionary's" elements or missing edges (i.e., not extracted from the depth differences) that hampered the layer extraction procedure (Table 2).

Table 2. Performance of the garments' configuration analysis.

	Junction classification	Junction layer mapping	Axis detection	Upper layer detection
Performance	71.5%	91.5%	91%	83.04%

For the evaluation of the unfolding path extraction method, the path is compared with other two folding/unfolding paths that do not require a priori knowledge of the garment's physical characteristics, i.e., the triangular [5] and the circular path [11]. For the evaluation and comparison of the results between the paths, we calculate the garment's deformation based on the area E_{traj} surrounded by its outline, as it occurs after the unfolding, and its area $E_{full,}$ when it is in the fully unfolded planar state. The deformation error of an unfolded garment is given by the following equation:

$$P_e = 1 - \left(E_{traj}/E_{full}\right) \tag{2}$$

The unfolding paths were applied by the robotic arm on various garments of four different fabrics (Table 3) and on folds with lengths ranging from 5–20 cm. For each type of fabric 10 folds were examined, i.e., 40 different folds for each type of path (proposed, triangular, and circular). The results show that the paths of the proposed method present smaller errors in all the fabric categories, the triangular is slightly better than the circular in highly flexible garments while the circular is better in more stiff cases. Moreover, although the average errors of more rigid fabrics do not vary a lot, the difference between the maximum errors is apparent. Due to their stiffness, these garments tend to resist to deformations and in most cases avoid them by sliding onto the table. Nevertheless, this did not occur to the cases that provided the maximum error.

Table 3. Deformation errors (max and average) for garments of various fabric categories.

	Proposed		Triangular		Circular	
	Average	Max	Average	Max	Average	Max
Denim	0.022	0.037	0.034	0.102	0.030	0.099
Towel	0.035	0.042	0.041	0.090	0.038	0.082
Thin cotton	0.019	0.024	0.020	0.027	0.021	0.029
Polyester	0.018	0.024	0.019	0.028	0.021	0.028

4 Conclusions

This paper presents a method for robotic fabric and type agnostic garment unfolding. By summarizing previous works of the authors that focused individually on each step of the unfolding method, the aim of this paper is to provide the description of the overall method and present its full pipeline by highlighting its most important contributions.

The proposed methodology comprises of approaches based on features common in all garments and manipulations mimicking the human unfolding movements, achieving in this way independency from specific garment types and fabrics. The proposed features are mapped to local configurations serving as handling points for garment manipulation and as low-level features for the extraction of the half-folded garment's state through a visual architecture and a junctions' dictionary. Moreover, the upper layer unfolding approach utilizes local features, such as the curve of the folding axis, to customize the unfolding path to each fold while applying a human-like movement.

Further research efforts are considered for the development of supplementary approaches used for the improvement of the proposed method. Such example could be a wrinkle removal approach used to eliminate any remaining wrinkles on the garment's surface after the application of the proposed unfolding method, leading to a completely planar garment configuration.

References

1. Corona, E., Alenya, G., Gabas, A., Torras, C.: Active recognition and target grasping point detection using deep learning. Pattern Recogn. **74**, 629–641 (2018)
2. Doumanoglou, A., Kim, T.-K., Zhao, X., Malassiotis, S.: Active random forests: an application to autonomous unfolding of clothes. In: Fleet, D., Pajdla, T., Schiele, B., Tuytelaars, T. (eds.) ECCV 2014. LNCS, vol. 8693, pp. 644–658. Springer, Cham (2014). https://doi.org/10.1007/978-3-319-10602-1_42
3. Osawa, F., Seki, H., Kamiya, Y.: Unfolding of massive laundry and classification types by dual manipulator. JACIII **11**(5), 457–463 (2007)
4. Hamajima, K., Kakikura, M.: Planning strategy for task of unfolding cloths. J. Robot Mechatron **12**, 557–584 (2000)
5. Stria, J., Petrik, V., Hlavac, V.: Model free approach to Garments Unfolding Based on detection of folded layers. In: Intelligent Robots and Systems, pp. 3274–3280 (2017)
6. Estevez, D., Victores, J.D., Fernandez, R., et al.: Enabling garment agnostic laundry tasks for a robot household companion. Robot. Auton. Syst. **123**, 134–139 (2020)

7. Triantafyllou, D., Mariolis, I., Kargakos, A., et al.: A geometric approach to robotic unfolding of garments. Robot. Auton. Syst. **75**, 233–243 (2016)
8. Triantafyllou, D., Koustoumpardis, P., Aspragathos, N.: Type independent hierarchical analysis for the recognition of folded garments' configuration. Intel. Serv. Robot. **14**(5), 427–444 (2021)
9. Triantafyllou, D., Koustoumpardis, P., Aspragathos, N.: Type and fabric agnostic methods for robotic unfolding of folded garments, submitted
10. van den Berg, J., Miller, S., Goldberg, K., Abbeel, P.: Gravity-based robotic cloth folding. In: Hsu, D., Isler, V., Latombe, J.C., Lin, M.C. (eds.) Algorithmic Foundations of Robotics IX. Springer Tracts in Advanced Robotics, vol. 68, pp. 409–424. Springer, Heidelberg (2010). https://doi.org/10.1007/978-3-642-17452-0_24
11. Pau, F.: Acquisition, placement and folding of fabric materials. Cloth. Sci. Technol. **16**(1), 227–237 (2004)

Phase State System for Generating Interactive Behaviors for Humanoid Robots

Tadej Petrič(✉) and Leon Žlajpah

Department for Automation, Biocybernetics and Robotics, Jožef Stefan Institute, Ljubljana, Slovenia
{tadej.petric,leon.zlajpah}@ijs.si

Abstract. When it comes to physical collaboration between humans and robots, robots currently have a shortcoming: their ability to observe and adapt to human dynamics is limited. This leads to inefficient collaboration and unergonomic interaction. In this work, we combine a dynamic phase state system (PSS) based on a network of stable heteroclinic channels (SHC) with Compliant Movement Primitives (CMP). The combination of PSS and CMP enables intuitive human interaction in tasks where humans and robots physically cooperate. The capabilities of the control system were demonstrated in simulations involving helping a humanoid bipedal robot Talos stand up from a squat position by pulling on its hands.

Keywords: Humanoids · Compliant movement primitives · Phase state system

1 Introduction

Humans can perform dynamic and dexterous movements with relative ease: walking, running and jumping, or even dancing are just a few examples. Dancing or getting up from a chair with someone's help are typical everyday tasks that seem easy but are quite complex and dynamic. They also require full body coordination and perfect synchronization of both partners to accomplish the task and ensure that both partners are safe and remain stable. In humanoid robotics, we are interested in physical, cooperative movements such as standing up with the help of another person better to understand the complexity and neuromechanical mechanics of humans, as this will allow us to improve and evolve robotic systems. Conventional approaches to humanoid robots require special programming skills, making humanoid robots inaccessible to the general public [11]. On the other hand, future generations of humanoid robots are expected to perform tasks in our natural environment. Therefore, we must study the natural human

This work was supported by Slovenian Research Agency grant N2-0153.

A. Müller and M. Brandstötter (Eds.): RAAD 2022, MMS 120, pp. 496–503, 2022.
https://doi.org/10.1007/978-3-031-04870-8_58

motion and physical interaction to gain knowledge that can be used to improve control systems for humanoid robots.

Because humanoid robots must remain stable all the time, it is critical to developing control systems that can respond instantly to human interactions. This makes the development of the control system challenging, as it must be able to recognize and respond to the user's intentions from the beginning. Since physical interaction between humans and humanoids is still in its infancy and the literature is sparse, exoskeleton controls can be used as inspiration for the development of the control system, as they too must first provide stability and then adapt to the user's intentions.

In this paper, we investigate an approach to control humanoid robots in physical cooperation with humans by merging a phase state system (PSS) based on a network of stable heteroclinic channels (SHC) [5] with compliant motion primitives (CMP) [9]. The novelty of our approach is the use of PSS phase output to drive the CMPs based on human physical interaction. This simply means a continuous path through the state space of the CMPs that can go back and forth instantaneously. The PSS was previously presented in [2] as a successful tool for human interaction. There, the ability of the control system to reverse and hold transitions to the next successor state was presented. The novelty of this work is the combination of the PSS with the CMPs to generate reactive behaviors for physical humanoid human collaboration that can provide stability while performing a specific task. The control concept is shown in Fig. 1.

2 Control Framework

The main building blocks for the proposed controller are the PSS based on the network of SHC [2,5,7,10] and the CMPs [9]. The main feature of the PSS is its dynamical system, where the path in the state space amounts to a sequence of states. In general, the PSS can have an arbitrary number of states. In this work, we use the PSS phase output to control the CMPs. This essentially leads to a hybrid state machine system where the transitions between states are continuous, non-exclusive, and reversible and thus used to generate CMP trajectories.

2.1 Phase State System (PSS)

The phase state system (PSS) is based on the network of SHC as presented in [2]. For completeness, a brief recapitulation is given here. The dynamics of the PSS system is determined by

$$\dot{\boldsymbol{x}} = \boldsymbol{x} \circ (\boldsymbol{\alpha} + \boldsymbol{x}^{\gamma} (\boldsymbol{\rho}_0 + \boldsymbol{\rho}_{\Delta} \circ (\mathbf{T} + \mathbf{G}))) \, \eta(t) + \dot{\boldsymbol{\delta}}(t), \tag{1}$$

where $\boldsymbol{x}$ is a n-dimensional state vector, γ is a positive scalar exponent, $\circ$ is element-wise multiplication, $\eta(t)$ is a scalar multiplier that adjusts the evolution rate of the PSS, and $\dot{\boldsymbol{\delta}}(t)$ is used to push the system away from saddle points and modulate transitions. The state transitions of the PSS are defined with a

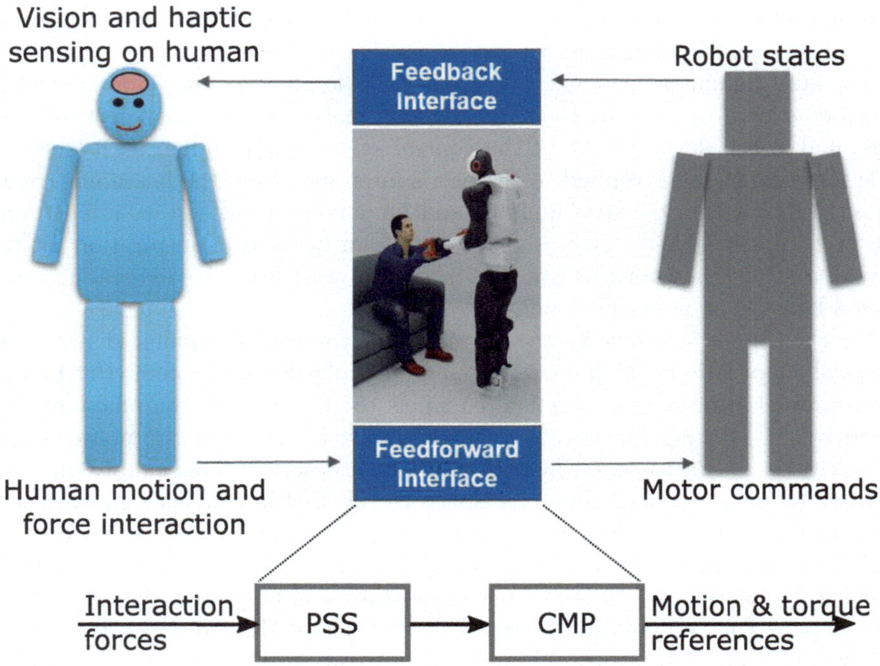

Fig. 1. Schematic representation of the control concept for a reactive physical interaction between human and humanoid robot. The two-layer control framework combines PSS with CMPs.

transition matrix $\mathbf{T}$, where transitions from i to j exist if $T_{j,i} = 1$, otherwise $T_{j,i} = 0$, and the matrix $\mathbf{G}$, which weights competing transitions. The $n \times n$ matrix $\boldsymbol{\rho}_0$ is defined such that the PSS has the n mutually exclusive states placed on its exclusive coordinate axis. The PSS is created with the $n \times n$ matrix $\boldsymbol{\rho}_\Delta$ multiplied by $\mathbf{T}$. Given the n-dimensional parameter vectors [5], the state locations $\boldsymbol{\beta}$, the shapes of the state saddle point shapes $\boldsymbol{\nu}$ and the growth rate $\boldsymbol{\alpha}$, we can define the matrices $\boldsymbol{\rho}_0$ and $\boldsymbol{\rho}_\Delta$ as

$$\boldsymbol{\rho}_0 = \left(\boldsymbol{\alpha}(\boldsymbol{\beta}^{-1})^T\right) \circ \left(\mathbf{I} - 1 - (\boldsymbol{\alpha}\boldsymbol{\alpha}^T)^{-1}\right), \tag{2}$$

$$\boldsymbol{\rho}_\Delta = \left(\boldsymbol{\alpha} \circ (1 + \boldsymbol{\nu}^{-1})\right)(\boldsymbol{\beta}^{-1})^T. \tag{3}$$

The main feature of this formulation is that $\boldsymbol{\rho}_0$ and $\boldsymbol{\rho}_\Delta$ remain constant, while the transition matrix $\mathbf{T}$ and the matrix $\mathbf{G}$ can change arbitrarily. To define all states and transitions we can use activation as in [2] given by

$$\boldsymbol{\Lambda}^t = \frac{16\boldsymbol{x}\boldsymbol{x}^T|\boldsymbol{x}^2|}{(\boldsymbol{x}\mathbf{1}^T + \mathbf{1}\boldsymbol{x}^T)^2 + |\boldsymbol{x}|^2} \circ \mathbf{T}. \tag{4}$$

Here the $\mathbf{1}$ is an n-dimensional vector of ones. Basically, this function makes the transitions invariant to the scaling of $\boldsymbol{x}$ and thus bounded to the interval $[0, 1]$. The residue of the transition matrix $\boldsymbol{\Lambda}^t$ can also be used to compute the states by

$$\boldsymbol{\lambda}^s = \frac{\boldsymbol{x}^2}{\sum \boldsymbol{x}^2}\left(\mathbf{1} - \sum \boldsymbol{\Lambda}^t\right). \tag{5}$$

Here the diagonal values of Λ^t have no meaning. Therefore, as proposed by [2], the transitions and state activations can be merged into

$$\boldsymbol{\Lambda}_{ji} = \begin{cases} \boldsymbol{\Lambda}^t_{ji}, & j \neq i \\ \boldsymbol{\lambda}^s_i, & j = i \end{cases}. \tag{6}$$

Since all PSS states are defined on a two-dimensional plane spanned between the coordinate axes of their predecessor and successor states, it is possible to calculate the phase variable. The phase signal from state i to state j is given by

$$\phi_{j,i} = \frac{|\boldsymbol{x}_j|}{|\boldsymbol{x}_j| + |\boldsymbol{x}_i|}. \tag{7}$$

Typically, the transitions in the state machine are discrete, while here, the transitions now have an associated phase signal that can be used to drive the CMPs [3,9]. Note that the system can be used to control any phase-based trajectory generator such as DMPs as in [2,6].

The system (1) has three main inputs that can be used to modulate PSS. The matrix $\mathbf{T}$ that defines the existing transitions, the greediness matrix $\mathbf{G}$ that can modify the direction of the transitions, and the vector $\boldsymbol{\delta}$ that modulates the time spent in a state and determines which transitions are active to leave a state. It is given by

$$\dot{\boldsymbol{\delta}} = (\boldsymbol{\Lambda} \circ \mathbf{B}) + \epsilon \boldsymbol{W}(t). \tag{8}$$

Here the elements $\mathbf{B}_{ji}$ are set as bias for state j in state i. In PSS, it is typically used to determine the next successor state from multiple states when transitions exist.

2.2 Compliant Movement Primitives (CMP)

Here we give a brief summary of CMP $h(t)$ consisting of kinematic trajectories encoded in Dynamic Movement Primitives (DMPs) [6] and corresponding torque profiles encoded in Torque Primitives (TPs) [3]

$$\boldsymbol{h}(t) = [\ddot{\boldsymbol{x}}_d(t), \dot{\boldsymbol{x}}_d(t), \boldsymbol{x}_d(t),\ \boldsymbol{\tau}_f(t)]. \tag{9}$$

Here $\ddot{\boldsymbol{x}}_d(t), \dot{\boldsymbol{x}}_d(t), \boldsymbol{x}_d(t)$ are the desired acceleration, velocity, and position trajectories in the task space encoded in the DMPs, and $\boldsymbol{\tau}_f$ are the corresponding torque profiles encoded in the TPs.

In the CMP framework, DMPs can be encoded in either joint or task space, and each degree of freedom is described by its own dynamical system. The details of Cartesian DMPs including rotations can be found in [13].

A short recap of the DMPs [6] for encoding the motion is given first. For details on Cartesian DMPs including rotations see [13]. Within this framework the trajectories can be given in either joint or task space, and every degree of freedom is described by its own dynamic system. The movement trajectory for each DOF is given by

$$\dot{z} = \Omega \left(\alpha_z \left(\beta_z \left(g - y\right) - z\right) + f(\phi)\right), \tag{10}$$

$$\dot{y} = \Omega z. \tag{11}$$

Here the nonlinear part $f(\phi)$ is defined as a linear combination of the radial basis function $\Psi_i(\phi)$, which is defined as follows

$$\boldsymbol{f}(\phi) = \frac{\sum_{i=1}^{N} w_i \Psi_i(\phi)}{\sum_{i=1}^{N} \Psi_i(\phi)}, \tag{12}$$

$$\Psi_i(\phi) = \exp(h(\cos(\phi - c_i) - 1)), \tag{13}$$

Here N is the number of $\mathbf{w}$ weighted kernels defining the actual shape of the encoded trajectory, h defines the width, and c_i are uniformly distributed along the phase ϕ. Here c_i is defined in N steps from 0 to 2π for periodic motions and from 0 to 1 for discrete motions. Note that instead of this formulation a more typical DMP formulation for discrete motions as given in [6] can also be used.

Unlike DMPs, which can be encoded in joint or Cartesian space, TPs can only be encoded in joint space. For a single joint, the TPs are defined as a linear combination of kernels defined by

$$\tau_{f,j}(\phi) = \frac{\sum_{i=1}^{N} \psi_i(\phi) w_{i,j}}{\sum_{i=1}^{N} \psi_i(\phi)}, \tag{14}$$

Note that the phase ϕ used for TPs is common with the DMPs and it is defined within the PSS system in Eq. (7). Also the phase ϕ is common for TPs and DMPs. To run the CMPs on a robot we used the impedance controller with low feedback gains given by

$$\boldsymbol{\tau}_u = \mathbf{J}^T(\mathbf{K}_p \boldsymbol{e} + \mathbf{K}_d \dot{\boldsymbol{e}} + \mathbf{K}_i \ddot{\boldsymbol{e}}) + \mathbf{N}\mathbf{K}_n \dot{\boldsymbol{q}} + \boldsymbol{\tau}_f(s), \tag{15}$$

where $\mathbf{J}^T$ is the Jacobi transpose, $\mathbf{N}$ is the null space matrix, $\boldsymbol{e}$, $\dot{\boldsymbol{e}}$ and $\ddot{\boldsymbol{e}}$ are the differences between the desired and actual position $\boldsymbol{x}$, velocity $\dot{\boldsymbol{x}}$ and acceleration $\ddot{\boldsymbol{x}}$, respectively. The $\mathbf{K}_p$, $\mathbf{K}_d$, $\mathbf{K}_i$ and $\mathbf{K}_n$ are the constant gain matrices chosen so that the robot is stable and compliant.

Unlike DMPs, TPs are not time invariant and cannot be paused because torques do not scale linearly. Therefore, an execution time must be specified in advance. Consequently, a new TP has to be created for each movement variation, including movement speed. By storing TPs, a library of motions can be created. In addition, statistical generalization or graph search can be used to create new TPs that have not been explicitly learned before, as in [3,4].

3 Simulation Example

The robot used in this study was the Talos bipedal humanoid robot from PAL-Robotics. It has 32 degrees of freedom, including two in the head, seven in each arm, one in hand, two in the waist, and six in each leg. TALOS is fully torque controllable due to the torque sensor in all joints, which allows powerful sensing and multi-contact movements. Closed-loop torque control can be used to enable safe interaction with the environment or humans [12].

To define the robot motion and the kinematic part of the CMPs, we use the motion dataset of human subjects moving from sitting to standing and back. [1]. For this purpose, we defined three states: standing without holding hands, standing with hand support, and squatting. From the dataset in [1], we also extracted six main angles tied to the sagittal plane, e.g., ankle, knee, hip, torso, shoulder, and elbow. The angles were further adapted to the kinematics of the robot to mimic as closely as possible the natural human motion during stand-up, using the optimization approach described in [8].

To trigger the reactive responses based on the physical interaction between humans and humanoids, the forces were measured at the end-effector of the Talos robot, i.e., the hands. The measured forces were used as input to the bias matrix in Eq. (8), such that if a pulling force was present at the hands, a positive bias was added for the transitions to standing with support. If no force was measured,

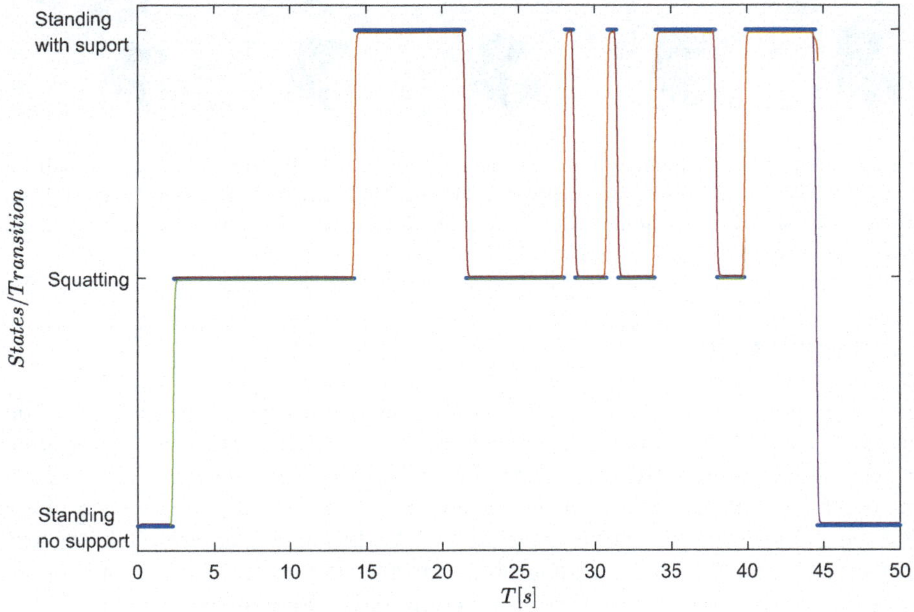

Fig. 2. State-space with several transitions between squatting, standing and standing without support are shown.

the positive bias was added for the transitions to standing, and if a pushing force was measured, a positive bias was added for the transition to sitting.

Simulation results showing the reactive transitions between three states as a function of force input are shown in Fig. 2 and the sequence for the transitions are shown in Fig. 3. Here, taking advantage of the internal state $\boldsymbol{x}$, the PSS system generated the phase signal ϕ for an active transition, which was used to drive the corresponding CMPs.

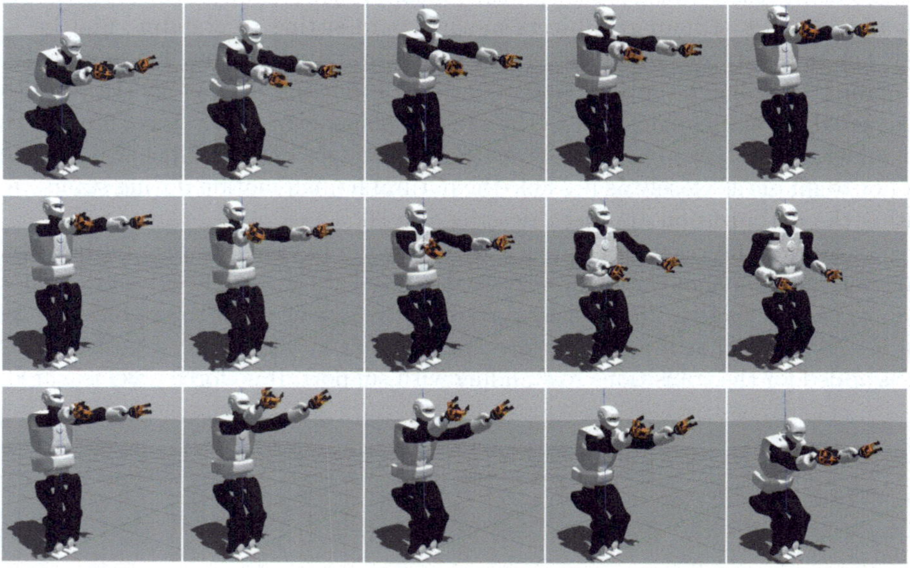

Fig. 3. Movement sequences for transitions. The upper figures show the transition from squatting to standing with support, the middle figures show the transition from standing to standing without support, and the lower figures show the transition from standing to squatting with support.

4 Conclusion

In this paper, we presented a novel control algorithm for physical interaction between human and humanoid robot based on a PSS. This allows the system to respond autonomously and intuitively, and predict the user's behavior, which is essential for physical interaction. In addition, the internal states of the phase state machine can be used to control the CMP, which can generate the kinematic trajectory and associated torque profiles. Overall, this results in fast and robust performance for reactive physical interaction with a humanoid.

Another important feature is that the proposed approach is general and can be used for any system that requires a state machine. We have shown that such a system has great potential for reactive control of humanoids. Although we have only demonstrated the use of the internal state space of PSS to control CMPs in simulations, we believe that the system can be effectively applied to a real humanoid robotic system, which is the goal of our further research.

References

1. Brecelj, T., Petrič, T.: Angular dependency of the zero moment point. In: Zeghloul, S., Laribi, M.A., Sandoval, J. (eds.) Advances in Service and Industrial Robotics, pp. 135–144. Springer, Cham (2021). https://doi.org/10.1007/978-3-030-75259-0-15
2. Deimel, R.: Reactive interaction through body motion and the phase-state-machine. In: 2019 IEEE/RSJ International Conference on Intelligent Robots and Systems (IROS), November 2019
3. Denisa, M., Gams, A., Ude, A., Petric, T.: Learning compliant movement primitives through demonstration and statistical generalization. IEEE/ASME Trans. Mechatron. (2016). https://doi.org/10.1109/TMECH.2015.2510165
4. Denisa, M., Petrič, T., Asfour, T., Ude, A.: Synthesizing compliant reaching movements by searching a database of example trajectories. In: International Conference on Humanoid Robots, pp. 540–546 (2013)
5. Horchler, A.D., Daltorio, K.A., Chiel, H.J., Quinn, R.D.: Designing responsive pattern generators: stable heteroclinic channel cycles for modeling and control. Bioinspir. Biom. **10**(2), 1–16 (2015). https://doi.org/10.1088/1748-3190/10/2/026001
6. Ijspeert, A.J., Nakanishi, J., Hoffmann, H., Pastor, P., Schaal, S.: Dynamical movement primitives: learning attractor models for motor behaviors. Neural Comput. **25**(2), 328–373 (2013)
7. Laurent, G., Stopfer, M., Friedrich, R.W., Rabinovich, M.I., Volkovskii, A., Abarbanel, H.D.: Odor encoding as an active, dynamical process: experiments, computation, and theory. Ann. Rev. Neurosci. **24**(1), 263–297 (2001). https://doi.org/10.1146/annurev.neuro.24.1.263
8. Miškovič, L., Koprivšek Leskovar, R., Gams, A., Petrič, T.: Optimizing end-effector force during the sit-stand task on the TALOS humanoid bipedal robot. In: Žemva, A., Trost, A. (eds.) Proceedings of the 30th International Electrotechnical and Computer Science Conference ERK 2021, pp. 175–178. Slovenska sekcija IEEE: Fakulteta za elektrotehniko, Ljubljana (2021)
9. Petrič, T., Gams, A., Colasanto, L., Ijspeert, A.J., Ude, A.: Accelerated sensorimotor learning of compliant movement primitives. IEEE Trans. Robot. **34**(6), 1636–1642 (2018). https://doi.org/10.1109/TRO.2018.2861921
10. Rabinovich, M.I., Huerta, R., Varona, P., Afraimovich, V.S.: Transient cognitive dynamics, metastability, and decision making. PLoS Comput. Biol. **4**(5), 25–30 (2008). https://doi.org/10.1371/journal.pcbi.1000072
11. Saeedvand, S., Jafari, M., Aghdasi, H.S., Baltes, J.: A comprehensive survey on humanoid robot development. Knowl. Eng. Rev. **34**, e20 (2019). https://doi.org/10.1017/S0269888919000158
12. Stasse, O., et al.: TALOS: A new humanoid research platform targeted for industrial applications. In: IEEE-RAS International Conference on Humanoid Robots, pp. 689–695 (2017). https://doi.org/10.1109/HUMANOIDS.2017.8246947
13. Ude, A., Nemec, B., Petrič, T., Morimoto, J.: Orientation in cartesian space dynamic movement primitives. In: 2014 IEEE International Conference on Robotics and Automation (ICRA), pp. 2997–3004, May 2014

Multi-contact Stability of Multi-legged Robots Operating in Unstructured Terrains

Parastoo Dastangoo(✉) and Alejandro Ramirez-Serrano

Department of Mechanical and Manufacturing Engineering, University of Calgary, 2500 University Drive N.W., Calgary, AB, Canada
{parastoo.dastangoo,aramirez}@ucalgary.ca

Abstract. This paper presents a practical solution to the problem of multi-legged robot's stability using a multi-contact approach applicable to unstructured terrains. The solution is based on a modified version of the Contact Wrench Set method which utilizes three arbitrary contact points to represent the physics of contact to test for stability. The developed method targets a variety of terrains including heterogenous surfaces extending the capabilities of the traditional Contact Wrench Set formulation. In this paper, the corresponding theoretical formulations and a novel visualization tool representing the robot's stability level are presented. Thorough verification of the new formulations is investigated via multi-body dynamic simulations. Experimental tests using a life-size humanoid robot show that the approach effectively quantifies the robot's stability on a wide range of surfaces.

Keywords: Multi-legged robotics · Humanoid stability · CWS method

1 Introduction

A viable approach to keep humans out of harm's way is to provide them with robotic artifacts that can assist them in performing their job in high-risk unstructured environments. Specifically, multi-legged robots including humanoids are promising devices with the capabilities to penetrate and perform tasks in complex chaotic unstructured (potentially compliant) environments. However, there are significant challenges that need to be resolved related to autonomy, locomotion, and adaptability before humanoids can be effectively used in such spaces.

Among all the mentioned challenges, stability is of significant importance for humanoids. The most well-known and widely used method in the body of humanoids' stability literature is the Zero Moment Point (ZMP) method [1] proposed to measure and assess the humanoids' dynamic balance. Although the ZMP method guarantees the stability of robots walking on flat horizontal surfaces where a large enough friction exists between the robot's foot and the ground it cannot be applied in unstructured compliant terrains. To overcome these issues, numerous works have been conducted to formulate a method capable of capturing the humanoid's whole dynamics including friction constraints [2]. Among these studies, one of the most generic approaches is the Contact Wrench Set (CWS) method capable of measuring the robots' stability when in contact

A. Müller and M. Brandstötter (Eds.): RAAD 2022, MMS 120, pp. 504–512, 2022.
https://doi.org/10.1007/978-3-031-04870-8_59

with multiple non-coplanar contacts where friction cone constraints hold [3]. The CWS method has been enhanced by investigating the extent to which the method is valid and its application in different operation scenarios.

Although the above-mentioned studies provide important advances, they fail to provide a comprehensive methodology for continuous surface contacts where the contacts are perceived as a surface having infinite contact points. To solve this issue, many approaches have used a large finite number of individual contact points and contact forces to capture the physics of the continuous surface [4]. However, such approaches are computationally expensive and cannot be effectively employed in real-time. In an attempt to address this problem, a 2015 study proposed the use of a closed-form formula that can compute the CWS stability explicitly, without reference to any individual contact points [4]. The method, however, is only applicable on rectangular surface areas. Such a limitation restricts its application to specific foot shapes and in cases where the sole of the robot does not make full contact with the ground.

2 Proposed Methodology

In this paper, we extend the scope of the CWS method by introducing a modified formulation that is applicable to more general cases such as any arbitrary contact surface defined by three contact forces. By doing so, the stability of the robot can be measured on practically any unstructured surface where the sole of the robot might not be in full contact with the ground (e.g., on rough, rocky terrains). This, in turn, enables to employ the solution on arbitrary terrain topographies provided there are at least three contact points between the robot and the environment.

Besides providing an effective tool for assessing the feasibility of the robots' stability, the CWS method can also be utilized to measure the robustness of the stability by defining a margin for the acceptable contact wrenches [5]. In this paper, a novel visualization method is proposed using radar charts to visualize the CWS stability condition.

2.1 Contact of Surface Areas

In order to determine whether each foot of the robot is in the fixed contact mode or not, the exerted force on the system is checked [6] using the concept of the Coulomb friction cone. However, the physics of a single contact point needs to be extended to consider the case where two surfaces (as opposed to two points) are in contact with each other (i.e., contact is continuous over a given area). For such cases, there exist infinite contact points each having a specific contact force between the corresponding objects/surfaces. Under such conditions, it would take an infinite amount of time to calculate the friction cone for each of these points and find the corresponding fix mode boundaries. In this regard, the authors in [4] proved that, under the Coulomb friction assumption, rather than dealing with an infinite number of reaction force vectors, a discrete number of friction cones can be defined at diverse points of the contact surface area adding up to the same wrench.

2.2 Triangular Surfaces and Corresponding Wrench Cone

With the virtual triangular area ABC, a novel analytical formulation can be developed for calculating the wrench cone for any surface represented by three points using the Fourier-Motzkin method [7]. Without loss of generality, the frame of reference defining the virtual or real (in the case of flat rigid terrains) triangular plane can be formed by aligning the x-axis alongside the BC edge of the contact triangular surface, with the orthogonal y-axis defined in the same plane. Based on the location of the contact points A, B, and C each having coordinates (xi, yi) and the geometric characteristic of the formed triangular contact plane, the following properties are computed (Fig. 1).

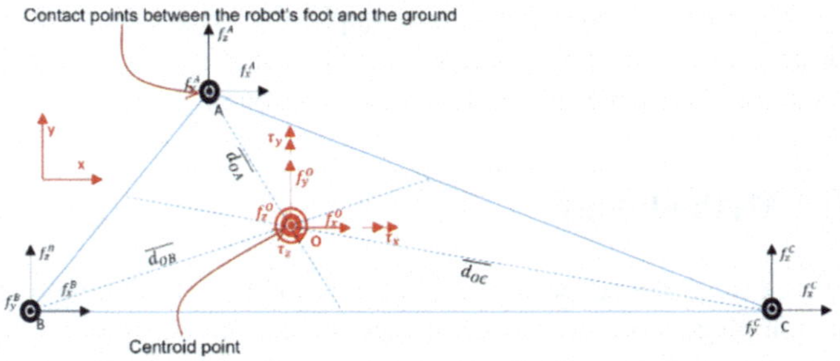

Fig. 1. A general contact triangle and the location and orientation of forces with respect to the frame of reference located on the triangle's centroid.

For a given triangle, the coordinates of its centroid (x_O, y_O) are calculated using Eq. (1).

$$x_O = \frac{x_A + x_B + x_C}{3};\ y_O = \frac{y_A + y_B + y_C}{3} \tag{1}$$

Under Coulomb friction rules, it can be shown that the contact forces on the triangular contact surface are equivalent to a set of forces at the triangular contact vertices. Therefore, in what follows, 3 contact forces at the vertices of the triangle and torque at the triangle's centroid are used to determine the physical interaction that exists between the foot and the ground contact area. For this, a set of mathematical formulations developed as part of this research work are used. Thus, the wrench cone of the 3 contact-point system shown in Fig. 1 can be described by Theorem 1 (proof of this theorem is provided in [8]).

Theorem 1: A system with 3 contact points f^A, f^B, f^C obeying the Coulomb friction rules (Eq. 1) exist, if and only if there exists a Contact Wrench Cone (CWC) with $wrench = \left(f_x, f_y, f_z, \tau_x, \tau_y, \tau_z\right)$ at the centroid, O(x0,y0), of the virtual contact triangle satisfying Eq. (2).

$$|f_x| \leq \frac{y_{OA}}{y_{sum}}.\mu f_z$$

$$|f_y| \leq \frac{x_{OC}}{x_{sum}}.\mu f_z$$

$$f_z > 0$$

$$-y_{OB}.f_z \leq \tau_x \leq y_{OA}.f_z \tag{2}$$

$$-x_{OC}.f_z \leq \tau_y \leq x_{OB}.f_z$$

$$\tau_{min} \leq \tau_z \leq \tau_{max}$$

where,

$$x_{Oi} = |x_i - x_O|,\ y_{Oi} = |y_i - y_O|\ i = A, B, C$$
$$x_{sum} = x_{OA} + x_{OB} + x_{OC},\ y_{sum} = y_{OA} + y_{OB} + y_{OC}$$
$$\tau_{min} = -\mu(x_{OC} + y_{OA})f_z + |y_{sum} \cdot f_x - \mu\tau_x| + |x_{sum} \cdot f_y - \mu\tau_y|$$
$$\tau_{max} = \mu(x_{OC} + y_{OA})f_z - |y_{sum} \cdot f_x + \mu\tau_x| - |x_{sum} \cdot f_y + \mu\tau_y|$$

Theorem 1 guarantees that if the CWC obeys Eq. (2) there exists a contact force at each of the 3 contact points that satisfy the Coulomb friction rules. The importance of this theorem is twofold:

Theorem 1 connects the physics of continuous surface contact to the physics of a wrench cone in the centroid of the surface.

It simplifies the stability condition check by examining the wrench only at the centroid of the surface rather than the need to check the entire surface.

Stability Margin and Visualizing Stability Condition.

A stability margin based on the five parameters (i.e., $f_x, f_y, f_z, \tau_x, \tau_y$, and τ_z) was used and developed to enable to know how close the robot is to become unstable. To this aim, a radar chart is proposed as a practical solution to visualize multivariate data on a 2-dimensional surface. On each axis, the acceptable range of each variable is represented. The maximum values of the five parameters form a polygon called stability polygon. While the values of the five parameters computed based on the state of the robot are used to form another polygon which is referred to in this paper as the data polygon in Fig. 2. As long as the data polygon lies inside the stability polygon, as shown in Fig. 2, all inequalities are satisfied, and the robot is said to be stable. In contrast, if one or more of the vertices of the data polygon crosses the boundaries of the stability polygon the robot becomes unstable.

The proposed radar graph can be divided into different regions each indicating how stable the robot is where each region can be defined based on user preferences or robot characteristics. As an illustration, in this paper, the allowable range for each of the five variables is divided into five severity regions each defined based on the stability characteristics of a life-size humanoid robot having 29 DoF. These stability levels are expressed as fuzzy variables and their ranges. If the values of the five parameters are inside the region closer to the center of the radar graph it indicates that the robot is more stable when compared to a case where the parameter values are in regions closer to the borders of polygons which would indicate that the robot is approaching instability.

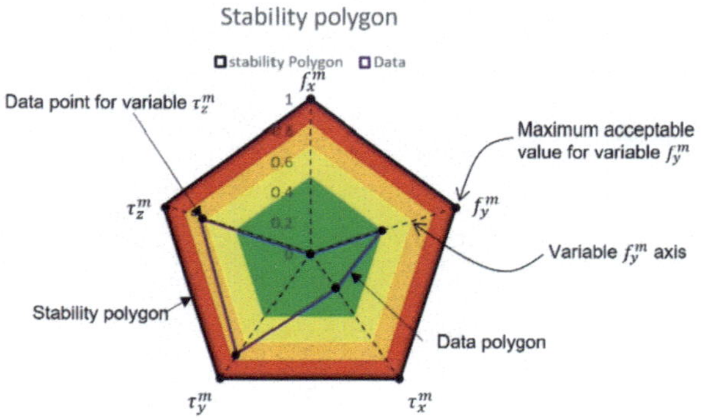

Fig. 2. Stability polygon on a radar graph

To build the radar graph the five parameters f_x, f_y, τ_x, τ_y, and τ_z are normalized based on the six variables f_x, f_y, f_z, τ_x, τ_y, and τ_z to obtain f_x^m, f_y^m, τ_x^m, τ_y^m, and τ_z^m. Among all the six variables, f_z has a direct impact on each of the five normalized variables and as a result does not appear as a normalized variable itself. The formulations to obtain f_x^m, f_y^m, τ_x^m, τ_y^m, τ_z^m are presented in [8].

3 Results

The proposed methodology was analyzed in simulation and experimentally test using Transcend, a life-size humanoid robot developed at 4Front Robotics Inc. at the University of Calgary.

3.1 Simulation Results

To conduct the simulations multi-body dynamics simulation software was used where the robot's dynamic equations were solved. Then the simulation results were compared with the results of the proposed method and the consistency between the results was used to determine the validity and accuracy of the proposed method. The performance of the method was evaluated according to two complementary measures of Accuracy and Reliability (definition for these terms are provided in [8]).

To be able to generate realistic results from the simulation that can be experimentally tested, the following four terrain characteristics were used: i) coefficient of friction, ii) damping ratio, iii) terrain stiffness, and iv) penetration depth. The value of each of these parameters is provided in [8].

During the tests, the robot was placed on a specific terrain in a given state (e.g., standing, walking) and the simulation was performed three times using three different penetration depth coefficients. The reason for this was to investigate the extent to which the method can be used to determine the robot's stability when standing on compliant (deformable) surfaces and not just solid flat terrains. Although numerous surface conditions (e.g., slope, friction, etc.) were used, in the context of this section three horizontal flat surface conditions were considered: *Surface1, Surface 2,* and *Surface 3.*

Surface 1 refers to a solid surface with a penetration depth of zero, *Surface 2* refers to a semi-solid surface with a penetration depth of 10^{-4}, and *Surface 3* refers to a compliant surface with a penetration depth of 2.1×10^{-3}. The values of the contact forces in the x and y axes, f_x^i and f_y^i at the contact point C_i were varied from 0 to 100 N between tests. Figure 3 shows the location of the three contact points on the robot's sole used during the tests, and the reference coordinate system used to specify the location of each contact point.

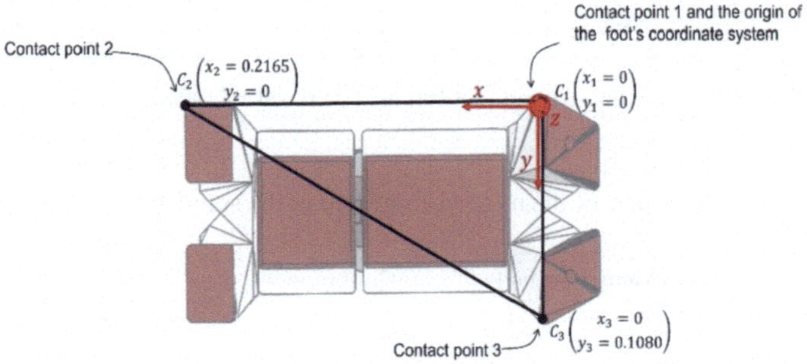

Fig. 3. Location of the pre-defined three contact points (C_1, C_2, *and* C_3) on the robot's sole.

Figure 4 is an example of the results which illustrates the robot's stability on different values for f_y^2 on the surface 1. The bars denote the binary stability condition obtained from the proposed method (orange) and the simulation (blue bars).

From Fig. 4 it is observed that the proposed methodology agrees 100% with the full-body dynamic simulation for that specific test.

The same procedure was also performed for the three components of the three contact points, $f_x^2, f_x^3, f_y^1, f_y^2$, and f_y^3 while varying the corresponding force caused by gravity at each simulation on a specific surface. Such tests produced similar results to the results shown in Fig. 4. The results of these tests in terms of their accuracy and reliability are calculated and Table 1 is an example of their result.

The obtained results show that the proposed method is successful in estimating the stability of the robot on all surfaces. With almost the same accuracy for the solid and semi-solid surfaces. Also, the method shows reliable results in all cases tested which suggests that the proposed method is able to accurately predict all unstable conditions.

The proposed approach was also tested on slope surfaces. To this aim, three surfaces having slopes of 10, 20, and 30° were utilized using the simulation settings same as provided in Fig. 3.

The results of the stability on sloped surfaces having an angle of 10 or 20° produced a reasonable performance with an accuracy value between 72% to 100%. Like the tests on flat horizontal solid and semi-solid surfaces, the results were reliable, but some results for the compliant surface were unreliable (detailed results are provided in [8]). When increasing the surface slope to 30°, it was observed that under the same setting used for a 20° sloped terrain, both the simulation and the proposed method show that the robot falls. This is due to the fact that the 30° slope is too steep for the robot to keep balance without the help of external forces (e.g., holding a banister with its hand).

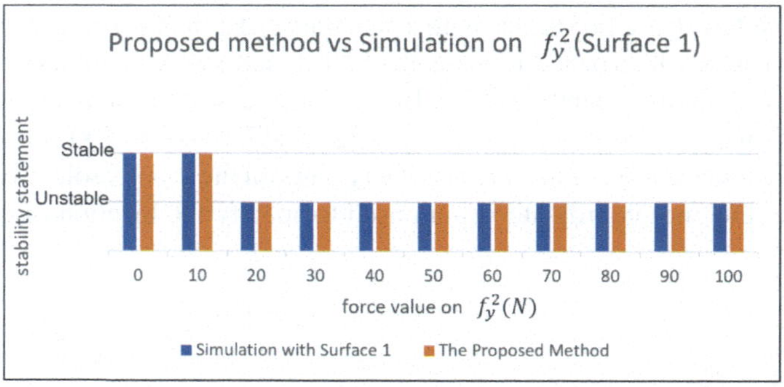

Fig. 4. Stability results when the contact forces (f_y^2) is varied on a flat horizontal surface.

Table 1. Performance of the method compared to the simulation on Surface 1.

Surface #1	Contact point 1		Contact point 2		Contact point 3	
	f_x^1	f_y^1	f_x^2	f_y^2	f_x^3	f_y^3
Accuracy (%)	72.7	100	72.7	100	81.8	100
Reliability (Y/N)	Y	Y	Y	Y	Y	Y

3.2 Experimental Tests

After validating the proposed method in simulation, the method was experimentally tested on real hardware under real-world conditions that are difficult to simulate (e.g., latency in the F/T sensor data collected). To implement the method on the humanoid robot, a ROS (Robotic Operation System) package was developed encoding the proposed method. The corresponding ROS package, written in Python, was configured to read the data from the robot's F/T sensors and use it to compute and publish the robot's stability in real-time. Table 2 shows the pseudo-code for the developed ROS package.

To check the stability when standing on a sloped surface the robot was set up to stand on one foot while the stability ROS package was executed. After numerous tests on various surfaces, it was concluded that the method is successful in predicting the true stability conditions of the robot on various flat and sloped surfaces having different friction coefficients and levels of surface discontinuities (i.e., bumpiness).

Table 2. The pseudo-code of the developed ROS package.

<table>
<tr><th>Inputs</th><th>Algorithm</th></tr>
<tr><td>- Wrench at the location of the F/T sensor for each timestamp comprising 3 force values and 3 torque values:
$w_{inst} = \begin{bmatrix} f_x \; f_y \; f_z \\ \tau_x \; \tau_y \; \tau_z \end{bmatrix}$
- Friction coefficient value (μ)
- The location of the contact points</td><td rowspan="5">Step 1: Read the wrench data published by F/T sensors (w_{inst})
Step 2: Calculate the upper and lower boundaries of the 6 variables ($f_x, f_y, f_z, \tau_x, \tau_y$, and τ_z) based on the given friction coefficient and pre-defined location of the contact points
Step 3: Check whether the values of w_{inst} lies inside the specified boundaries of variables in step 2
- If all values of w_{inst} lies inside their related boundaries, then the foot is stable
- If one of w_{inst} values do not lie inside the related boundaries, then the foot is unstable
Step 4: Publish the results of Step 3</td></tr>
<tr><td>Outputs</td></tr>
<tr><td>The stability condition for each timestamp</td></tr>
<tr><td>Parameters used in the algorithm:</td></tr>
<tr><td>- f_x, f_y, f_z: The force values in the x-axis, y-axis, and z-axis, respectively, at the origin of the F/T sensor's coordinate system
- τ_x, τ_y, τ_z: The Torque values at the origin of the F/T sensor's coordinate system</td></tr>
</table>

4 Conclusions

The approach provides an efficient stability computation for multi-legged robots operating in a broad spectrum of applications. Hence, it enables to deploy humanoids and other types of robots in complex real-world environments as those encountered in hazardous urban search and rescue operations. The results show that the method is capable of computing stability in a very fast way with reasonable accuracy and reliability. The method, indirectly, is also able to identify when the robot's foot slides providing a tool for the robot's controller to prevent falls. As a result, the proposed approach provides useful/relevant information to control the movement of robots so that their motions do not compromise their stability.

References

1. Vukobratović, M., Borovac, B.: Zero-moment point—thirty five years of its life. Int. J. Humanoid Robot. **1**(01), 157–173 (2004)
2. Takao, S., Yokokohji, Y., Yoshikawa, T.: FSW (feasible solution of wrench) for multi-legged robots. In: 2003 IEEE International Conf. on Robotics and Automation, vol. 3, pp. 3815–3820. IEEE (2003)
3. Dai, H.: Robust multi-contact dynamical motion planning using contact wrench set, Massachusetts Inst. of Tech., Thesis (2016)
4. Caron, S., Pham, Q.-C., Nakamura, Y.: Stability of surface contacts for humanoid robots: closed-form formulae of the contact wrench cone for rectangular support areas. In: IEEE International Conference on Robotics and Automation (ICRA), pp. 5107–5112. IEEE (2015)

5. Kagami, S., Kitagawa, T., Nishiwaki, K., Sugihara, T., Inaba, M., Inoue, H.: A fast dynamically equilibrated walking trajectory generation method of humanoid robot. Auton. Robot. **12**(1), 71-82. (2002). https://doi.org/10.1023/A:1013210909840
6. Caron, S.: Computational foundation for planner-in-the-loop multi-contact whole-body control of humanoid robots, University of Tokyo, Thesis (2016)
7. Stehr, G., Graeb, H.E., Antreich, K.J.: Analog performance space exploration by normal-boundary intersection and by Fourier-Motzkin elimination. IEEE Trans. Comput.-Aided Des. Integr. Circ. Syst. **26**(10), 1733–1748 (2007)
8. Dastangoo, P.: Multi-contact Stability of Multi-Legged Robots Operating in Unstructured Spaces, MSc thesis. U of Calgary (2021)

SnakeTrack, A Bio-inspired, Single Track Mobile Robot with Compliant Vertebral Column for Surveillance and Inspection

Shahab E. Nodehi, Luca Bruzzone(✉), and Pietro Fanghella

University of Genoa, 16145 Genoa, GE, Italy
shahabedin.nodehi@edu.unige.it, {luca.bruzzone, pietro.fanghella}@unige.it

Abstract. This paper presents the conceptual and embodiment design of a bio-inspired single track ground mobile robot, named SnakeTrack, designed for surveillance and inspection tasks in unstructured environments with narrow spaces. Its main components are the vertebral column, characterized by two end modules and a variable number of vertebrae connected by compliant joints, and the peripherical track. The robot is actuated by four motors, two for the track motion and two to command the lateral flexion of the vertebral column for steering, while limited passive torsion and retroflexion of the vertebral column are allowed by the compliant joints to adapt to terrain unevenness, improving traction. Vision is provided by two cameras placed on the end modules, behind the tracks, which are characterized by dedicated openings. The pitch of vertebrae and tracks is equal, then the robot is modular, and its length can be changed on the basis of the environment features and the required onboard equipment.

Keywords: Ground mobile robot · Tracked locomotion · Single track robot · Snake robot · Worm robot

1 Introduction

In the present technological scenario, service robotics is one of the fastest expanding research areas. In particular, ground mobile robots can substitute human beings in a wide variety of hazardous and unsafe applications, as well as surveillance, inspection of sites with chemical or radioactive contamination, intervention in extreme environments, rescue operations, homeland security.

There are many different kinds of calamities which originate from natural and man-made causes, such as earthquakes, floods, hurricanes, fires. It is crucial to reach the victims and casualties within the first 48 h after the disaster. To this aim, technological studies are concentrated on the development of search and rescue robots. The biggest contribution of these robots will be keeping the search and rescue personnel out of the disaster region and gathering and processing more information from the sites via several sensors and accordingly, preventing further casualties and losses of lives.

A. Müller and M. Brandstötter (Eds.): RAAD 2022, MMS 120, pp. 513–520, 2022.
https://doi.org/10.1007/978-3-031-04870-8_60

Independently of the payload, which depends on the specific task, a ground mobile robot is primarily characterized by its locomotion system. Ground mobile robots are often required to move in unstructured environments, and the selection of the locomotion architecture is based on the expected operative conditions.

Locomotion systems can be classified into three main categories: wheeled robots (W) [1, 2], tracked robots (T) [3, 4], legged robots (L) [5, 6]; moreover, hybrid robots exploit the four possible combinations of these systems: Legged-Wheeled (LW) [7–9], Legged-Tracked (LT) [10, 11], Wheeled-Tracked (WT) [12, 13], Legged-Wheeled-Tracked (LWT) [14]. A comparison of the operative features of these locomotion principles is outlined in [15].

In particular, tracked robots are characterized by excellent motion capabilities on yielding and uneven terrains, due to their large contact surface between tracks and ground. Despite such advantages, they are slower and less efficient than wheeled robots. For inspection of narrow spaces, some researchers have proposed snake-like tracked robots, with tracked modules in series [16]. A similar approach to design robots for inspection of narrow spaces is to adopt a single peripheral track. In [17] a flexible mobile mono-track named FMT is proposed and compared to classical differential steering tracked robots. In [18], another similar robot with peripherical single track is proposed, named Reconfigurable Continuous Track Robot (RCTR).

In the following a novel snake-like, single track robot with steering capability is proposed, named SnakeTrack, designed for surveillance and inspection of narrow spaces.

2 Functional Design of SnakeTrack

Figure 1 represents the external view of the proposed SnakeTrack robot. Its overall dimensions are 640 mm (length) × 150 mm (width) × 120 mm (height). Its main structural components are the vertebral column and the track.

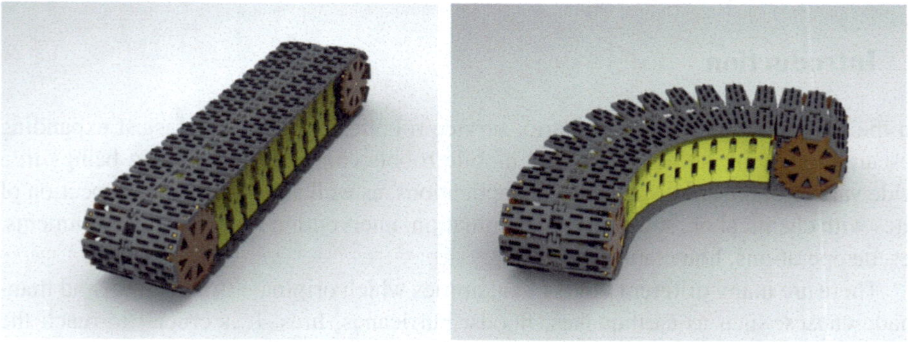

Fig. 1. External view of SnakeTrack, in straight position (left) and in steering position with minimum turning radius (right).

The vertebral column of SnakeTrack is composed of a variable number of vertebrae (ten in Fig. 2, V) and two end modules (Fig. 2, E). The length of the robot can be changed by varying the number of the vertebrae; they are connected by compliant joints, with low stiffness for the yaw rotation, to allow lateral flexion for steering (Fig. 1, right) with low energy consumption.

The robot is symmetric both in front-rear and up-down planes, so the end modules are equal, and each one hosts two actuators, one for the track motion (Fig. 2, TA) and one to command lateral flexion (Fig. 2, SA). Each SA actuator contains a gearmotor connected to a winch (Fig. 2, W). The winch can pull a rope (Fig. 2, R) which passes through lateral holes in the vertebrae. One end of the rope is fixed to the winch, and the other end is fixed to the opposite end module. Therefore, lateral steering is obtained by pulling the rope on the side of the center of rotation of the desired turn. Consequently, the motions of the two TA actuators are not independent, as well as the motions of the two SA actuators. As a matter of fact, the two TA actuators, connected to the track sprockets, must rotate at the same speed, and when one SA actuator pulls a rope, the other SA actuator must loosen the opposite rope.

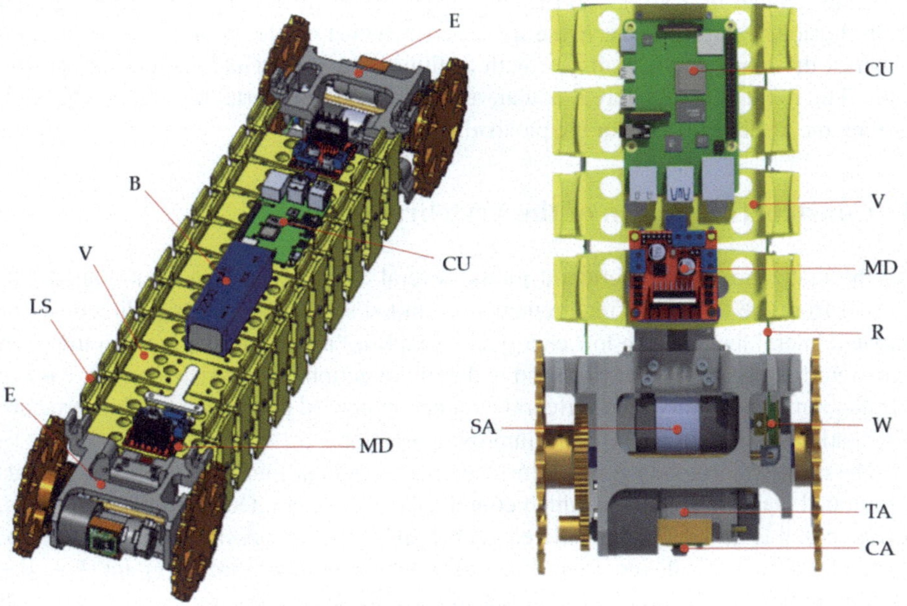

Fig. 2. Vertebral column of SnakeTrack.

A detail of the track modules is shown in Fig. 3. They are characterized by central openings (Fig. 3, O) that allow, during the track motion, an intermittent view by means of the two cameras mounted on the end modules (Fig. 3, CA). The track modules are guided by lateral supports (Fig. 3, LS), attached to each vertebra, in which the track guiding pegs (Fig. 3, GP) pass during the track motion.

Camera vision is active during the track motion only in the phases in which the camera faces the track opening. Considering a robot speed of 0.1 m/s, with a track pitch of 40 mm, the camera view can be updated 2.5 times per second; even if the vision is not continuous, this rate is sufficient to guide the robot at this speed. However, when

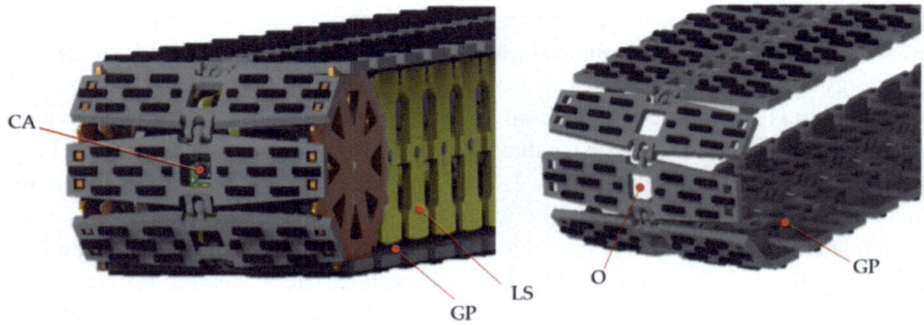

Fig. 3. Detail of the track modules with central opening for camera vision.

continuous camera vision is required for monitoring, the robot can stop with the track opening aligned with the camera.

Control unit (Fig. 2, CU), batteries (Fig. 2, B), motor drivers (Fig. 2, MD), and sensors can be hosted by the vertebrae, in the space between the lateral supports. Besides the two cameras, the robot can be equipped with additional environmental sensors for specific tasks. The robot is completely modular, and the number of vertebrae can be changed, varying the available space for the payload.

3 Constructive Design of the Vertebral Joints

For the realization of the vertebral joints, several alternatives have been considered, starting from the functional requirements of limited stiffness in the yaw direction (for actuated lateral flexion with low energy consumption) and higher stiffness in the pitch direction (for passive retroflexion) and in the roll direction (for passive torsion). Passive retroflexion and torsion of the vertebral column are needed to obtain a flexible structure which adapts to terrain unevenness, improving traction.

The most widespread approach for realizing compliant joints in robotics is to realize lumped elastic elements in which compliance is concentrated. For universal joints, several possible designs are considered in [19]. In Fig. 4, two possible solutions are represented: the two-axis flexure joint (Fig. 4a) and the compliant cardan U joint (Fig. 4b). Such compliant realizations of universal joints also allow a compliance along x (roll torsion), therefore they can be used as vertebral joints for SnakeTrack.

The two-axis flexure joint is the extension to two axes of the revolute flexure hinge, and has the same main drawback: the stress is highly concentrated in the central section, and this lowers its mechanical resistance. Different stiffness along y and z can be obtained by adopting a rectangular central section (Fig. 4a). The compliant cardan U-joint of Fig. 4b reduces the stress concentration issue, since strains and stresses are evenly distributed along the flexure plates fy and fz, that usually have constant thicknesses. Different stiffnesses along y and z can be obtained by adopting different thicknesses and lengths for the fy and fx flexure plates. The mechanical resistance of this joint can be further improved by replacing the flexure plates with inserts in superelastic materials (Fig. 4c), as proposed in [20, 21], but this greatly increases the joint size and weight

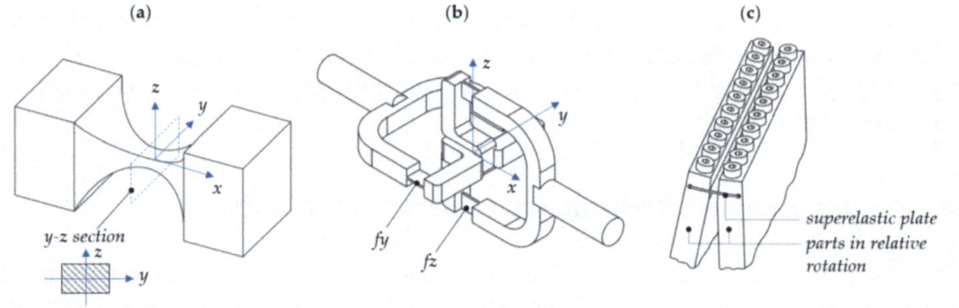

Fig. 4. Possible realizations of compliant universal joints: two-axis flexure joint (**a**) and compliant cardan U joint (**b**); revolute joint with superelastic insert (**c**).

because it can't be realized in a single part. On the other hand, the practical realization of a flexible cardan U joint is rather complex even with additive manufacturing techniques. An alternative approach to design a 3-DoF compliant joint is represented by beam-based compliant mechanisms [22], but they not being very compact, are not suited for the present application.

An alternative solution, which greatly simplifies the design of the vertebral joints, is to use the soft robotic paradigm, adopting a compliant joint without shrinkages, exploiting the hyperelasticity of materials such as thermoplastic polyurethane (TPU).

Figure 5 represents the design of the proposed compliant joints. Each vertebra is composed of a central core (C) and two external plates (P), fixed by bolts. A central TPU elastic element (EE) connects two adjacent vertebrae; the two external plates P lock the elastic element EE. Moreover, the contact of the profiles of the plates P of two adjacent vertebrae, with limited gap when EE is undeformed (0.13 mm), limits the relative pitch and roll motion and consequently the passive retroflexion and torsion, while lateral flexion and is almost not affected by the plates P.

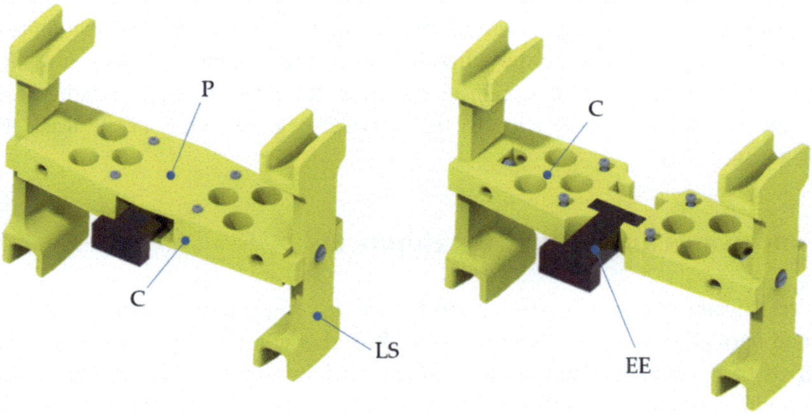

Fig. 5. Design of the compliant vertebral joints: core (C), external plates (P), lateral supports (LS) TPU elastic element (EE).

This design has the following advantages: low cost, resilience to shocks much higher than conventional compliant joints, possibility of easily tuning the range of retroflexion by changing the shape of the plates P and consequently the gap between the plates of two adjacent vertebrae.

4 Prototyping and Preliminary Experimental Tests

The first prototype of the SnakeTrack has been realized and the first experimental tests are in progress, at present with the robot radio controlled by an operator. These tests have confirmed the overall functionality of the proposed mechanical design, and in particular the capability to steer using the lateral flexion (Fig. 6, left) and to move adapting to terrain unevenness by means of the passive retroflexion and torsion of the vertebral joints (Fig. 6, right).

Fig. 6. Preliminary experimental tests on the SnakeTrack prototype: steering (left) and locomotion on uneven grounds (right).

These tests are fundamental to refine the shape of the components which are in relative motion during the track rotation. In particular, the shape of these elements must be carefully revised to increase the reliability of the track functionality even in steered position and/or in presence of terrain unevenness: the pegs of the track sprocket, the holes of the track modules which host the sprocket pegs, the ends of the track lateral supports (Fig. 5, LS).

5 Conclusions and Future Developments

The overall design of a small size tracked robot for surveillance and inspection of unstructured and narrow spaces has been presented. It is characterized by a modular structure composed of a vertebral column and a single peripherical track revolving around it. The vertebral column can perform actuated lateral flexion for steering, while passive retroflexion and torsion are allowed by the compliant vertebral joints to adapt to terrain unevenness. Thanks to the modular architecture, the length of the robot can be adapted to the payload necessary for the specific task.

The robot is fully symmetric and can continue operating after a capsize; moreover, if it falls on its flank, it can restore the correct position by combining lateral flexion, to lift on the track ends, and track motion. The conceptual design of the robot has been presented and the embodiment design of the vertebral joints has been discussed. The preliminary tests on the first prototype are in progress. The first experimental results have confirmed the robot capability to walk and steer while adapting to uneven grounds and suggested possible refinements in the detailed design of the tracks and of the steering system. In the next steps of the research, these modifications will be implemented to obtain improved prototypes.

Author Contributions. S.N. conceived the robot architecture and the locomotion system; S.N. and L.B. developed the detailed embodiment design, performed the experimental characterization of the compliant vertebral joints and prepared the manuscript; P.F. supervised the scientific methodology.

References

1. Stückler, J., Schwarz, M., Schadler, M., Topalidou-Kyniazopoulou, A., Behnke, S.: Nimbro explorer: Semiautonomous exploration and mobile manipulation in rough terrain. J. Field Robot. **33**(4), 411–430 (2016)
2. Arvidson, R.E., et al.: Mars science laboratory curiosity rover megaripple crossings up to Sol 710 in Gale crater. J. Field Robot. **34**(3), 495–518 (2017)
3. Tao, W., Ou, Y., Feng, H.: Research on dynamics and stability in the stairs-climbing of a tracked mobile robot. Int. J. Adv. Rob. Syst. **9**(4), 146 (2012)
4. Ottonello, G., Berselli, G., Bruzzone, L., Fanghella, P.: Functional design of Elloboat, a tracked vehicle for launching and beaching of watercrafts and small boats. In: Proceedings of the 14th IEEE/ASME International Conference on Mechatronic and Embedded Systems and Applications, MESA 2018. IEEE, Oulu (2018). https://doi.org/10.1109/MESA.2018.8449173
5. Rodinò, S., Matteo Curcio, E., Di Bella, A., Persampieri, M., Funaro, M., Carbone, G.: Design, simulation, and preliminary validation of a four-legged robot. Machines **8**(4), 82 (2020)
6. Mahapatra, A., Roy, S.S., Pratihar, D.K.: Multi-legged robots—a review. In: Multi-body Dynamic Modeling of Multi-legged Robots. CIR, pp. 11–32. Springer, Singapore (2020). https://doi.org/10.1007/978-981-15-2953-5_2
7. Quaglia, G., Bruzzone, L., Oderio, R., Razzoli, R.P.: EPI.Q mobile robots family. In: Proceedings of the ASME 2011 International Mechanical Engineering Congress and Exposition, IMECE 2011, vol. 7 (Parts A and B), pp. 1165–1172. ASME, Denver (2011)
8. Quaglia, G., Butera, L.G., Chiapello, E., Bruzzone, L.: UGV Epi.q-Mod. In: 20th CISM-IFToMM Symposium on Theory and Practice of Robots and Manipulators, ROMANSY 2014. Mechanisms and Machine Science, vol. 22, pp. 331–339 (2014)
9. Bruzzone, L., Fanghella, P.: Functional redesign of Mantis 2.0, a hybrid leg-wheel robot for surveillance and inspection. J. Intell. Rob. Syst. **81**(2), 215–230 (2015). https://doi.org/10.1007/s10846-015-0240-0
10. Hirose, S., et al.: Quadruped walking robots at tokyo institute of technology. IEEE Robot. Autom. Mag. **16**, 104–114 (2009)
11. Ben-Tzvi, P., Goldenberg, A.A., Zu, J.W.: Design, simulations and optimization of a tracked mobile robot manipulator with hybrid locomotion and manipulation capabilities. In: Proceedings of the 2008 IEEE International Conference on Robotics and Automation, pp. 2307–2312. IEEE, Pasadena (2008)

12. Gao, X., Cui, D., Guo, W., Mu, Y., Li, B.: Dynamics and stability analysis on stairs climbing of wheel–track mobile robot. Int. J. Adv. Rob. Syst. **14**(4), 1–13 (2017)
13. Moskvin, I., Lavrenov, R., Magid, E., Svinin, M.: Modelling a crawler robot using wheels as pseudo-tracks: model complexity vs performance. In: Proceedings of the 7th IEEE International Conference on Industrial Engineering and Applications (ICIEA 2020), pp. 1–5. IEEE, Bangkok (2020)
14. Bruzzone, L., Baggetta, M., Nodehi, S.E., Bilancia, P., Fanghella, P.: Functional design of a hybrid leg-wheel-track ground mobile robot. Machines **9**(1), 10, 1–11 (2021)
15. Bruzzone, L., Quaglia, G.: Review article: locomotion systems for ground mobile robots in unstructured environments. Mech. Sci. **3**(2), 49–62 (2012)
16. Neumann, M., Predki, T., Heckes, L., Labenda, P.: Snakelike, tracked, mobile robot with active flippers for urban search-and-rescue tasks. Ind. Robot. **40**(3), 246–250 (2013)
17. Haji, T., Kinugasa, T., Yoshida, K.: Experiment of maneuverability of flexible mono-tread mobile track and differential-type tracked vehicle. Ind. Robot. **37**(3), 263–272 (2010)
18. Kislassi, T., Zarrouk, D.: A minimally actuated reconfigurable continuous track robot. IEEE Robot. Autom. Lett. **5**(2), 652–659 (2020)
19. Farhadi Machekposhti, D., Tolou, N., Herder, J.L.: A review on compliant joints and rigid-body constant velocity universal joints toward the design of compliant homokinetic couplings. J. Mech. Des. **137**(3), 032301(2015)
20. Bruzzone, L., Molfino, R.M.: A novel parallel robot for current microassembly applications. Assembly Autom. **26**(4), 299–306 (2006)
21. Bruzzone, L., Bozzini, G.: A flexible joints microassembly robot with metamorphic gripper. Assembly Autom. **30**(3), 240–247 (2010)
22. Bilancia, P., Berselli, G.: An overview of procedures and tools for designing nonstandard beam-based compliant mechanisms. Comput. Aided Des. **134**(6),103001 (2021)

The Use of Robots in Aquatic Biomonitoring with Special Focus on Biohybrid Entities

Wiktoria Rajewicz(✉), Thomas Schmickl, and Ronald Thenius

University of Graz, Universitätplatz 2, Graz, Austria
wiktoria.rajewicz@uni-graz.at

Abstract. Aquatic monitoring is an essential part of battling the rising ecological crisis. Classical methods involving extensive sampling and sensor measurements are precise, however, time and money consuming. For these reasons, they are unsuitable for long-term continuous data collection. With the increasing water pollution, there is a need to monitor the environment in new, more efficient ways over a long period of time. Project Robocoenosis introduces a novel concept of autonomous, long-term aquatic monitoring with the use of biohybrids. By linking technological parts with living organisms, a more well-balanced information on the state of the environment can be obtained. This will be done by using organisms such as mussels and Daphnia as live biosensors and combining them with low-power robotics. The autonomous biohybrid entity will use Microbial Fuel Cells (MFC) as a natural power source through electricity harvesting. The fields of operation are focused on various Austrian lakes including the lakes Lake Hallstatt and Lake Millstatt in the Alpine Region as well as Lake Neusiedler.

Keywords: Animal-robot interaction · Biohybrid · Environmental monitoring

1 Introduction

Planet Earth has been facing an immense ecological crisis for decades. The environmental decay has both economic and environmental effects on its inhabitants. One of the majorly affected ecosystems that are crucial for the proper functioning of the planet is the aquatic ecosystem. Climate change, pollution, farming and other anthropogenic factors have caused the rapid worsening of the water quality over the recent years [7]. This trend became a threat to many organisms inhabiting the aquatic environments but also to humans who are almost entirely dependant on the natural fresh water supplies. An efficient and robust monitoring and global sharing of the water quality data is the most effective method of preventing the ecosystem collapse. Aquatic monitoring has been practiced for a long time with the use of various sensors and sample taking. While these methods are most accurate and reliable in terms of the predictability of the outcome,

The original version of this chapter was revised: This chapter has been changed to Open Access, licensed under the terms of the Creative Commons Attribution 4.0. The correction to this chapter is available at https://doi.org/10.1007/978-3-031-04870-8_72

A. Müller and M. Brandstötter (Eds.): RAAD 2022, MMS 120, pp. 521–527, 2022.
https://doi.org/10.1007/978-3-031-04870-8_61

they are time and money consuming [22]. Classical sensors can be costly and require frequent calibration and maintenance, therefore a continuous monitoring is close to impossible. An urgent need for more efficient methods of aquatic monitoring has arisen. The door to long-term environmental monitoring opens with the use of robots. Robotic entities are able to reach not easily accessible habitats and remain there for long periods of time continuously gathering large amounts of data. One way to achieve a long-term energy-efficient monitoring is combining the electronic parts with living organisms to create so-called "biohybrids" [17]. Initiatives surrounding this topic have been carried out in the recent years [10,18,20]. Biohybrids for long-term aquatic monitoring were implemented in a European Union funded project "Subcultron" [20]. Through the use of MFCs as a natural power source, bacterial activity was used to gather data on the environment by tracking their electricity patterns. By using the swarm strategy, the robots were able to collect large amounts of data from many places simultaneously, providing a better overview of the investigated place. This project created three types of robots: aMussel, aFish, aPads, each with their own purpose in the robotic swarm.

This innovative project succeeded on performing various unsupervised marine monitoring missions. MFC was used as a power source for the aMussels which were embedded in the sediment. The microbial activity enabled the aMussel to operate autonomously and also provided an insight into the benthic processes by monitoring the power output [20]. The concept of using a living organism as a sensor in biohybrid entities was extended with the project Robocoenosis [15–17]. Project Robocoenosis introduced here, aims to take a new perspective on this methodology and expand it with the use of multiple locally found lifeforms, such as mussels, *Daphnia* spp., *Hydra* spp., various larvae etc. [15]. Here, is where Robocoenosis uses the extreme complexity of the aquatic environments to its advantage.

The intricacy of the aquatic environments makes investigating one parameter at a time, which results in a very precise data, quite inefficient long-term. Ultimately, it might not accurately portray the overall condition of the water body. Lifeforms are in constant contact with the surrounding environment and respond to the overall of factors and changes occurring within it. Moreover, lifeforms do not require frequent maintenance as under normal conditions, they sustain themselves in their natural habitats. Lifeforms employed as live biosensors can be mussels, Daphnia, benthic community structure and many others [15]. Here, core features of the project Robocoenosis and its various applications are presented [15–17].

1.1 Robocoenosis Biohybrid Entities

Project Robocoenosis aims to develop novel methodology of utilising lifeforms as live sensors in place of the classical sensors with the use of biohybrids [17]. By combining the technical and electrical parts with the live sensors provided by organisms a complex biohybrid entity is able to hack into the ecosystem and obtain reliable data. The entities developed within this project will allow the scientific community to obtain long-term data on the status of the selected lake. This will be achieved by the low-power electronics and self-charging module with the use of Microbial Fuel Cells (MFC).

Autonomy. The aim for these prototypes is for them to be self-powering. To ensure the power autonomy the robot is supplied with MFCs. MFCs are one of many types of fuel cells [1]. They rely on the anaerobic breathing processes of the bacteria present in anoxic sediments. They use locally available organic matter to produce low-current electricity which can be accumulated and used by the biohybrid entity [1].

Low-Power. Due to the low currents obtained by the MFC, this method will only be used for short periods of time. To overcome the power shortage, the internal electrical parts (such as a single board computer "Raspberry Pi") are turned off or put into a sleep mode between the data collection periods. Low-power micro-controllers and other micro devices will be used.

Biodegradability. On top of the features mentioned above, the biohybrid entities designed by Robocoenosis are meant to be semi-biodegradable. This is accomplished by constructing two segments: 1) non-biodegradable core containing all the necessary electronic parts and 2) biodegradable platforms hosting organisms and the lifeforms themselves. After the monitoring mission is terminated, the non-biodegradable core will return to the surface and will be collected by the researchers while the biodegradable parts will remain in the environment.

2 Methods

2.1 Fields of Work

Various Austrian lakes have been chosen as the field of work. Each of them represents a vastly different habitat to make the biohybrid testing relevant to the a broader variety of surroundings. Firstly, Lake Millstatt (N46°47′ E13°34′) was chosen as a classic example of an oligotrophic lake, that is, a lake with a low amount of nutrients and clear, well-oxygenated waters. Glacial lakes, such as Lake Millstatt, are the ones created due to the glacial activity, whether being carved out by the glacier or supplied by the melt-water [4].

Lake Neusiedl (N47°56′ E16°50′) represents a completely different habitat. It is highly eutrophic, i.e. contains a high content of nutrients and algae. It is also likely to be met with anoxic conditions near the bottom sediments [19]. While being the largest lake in Austria surface-wise, it is also extremely shallow, being less than 2 m deep at the deepest point. These characteristics make this body of water extremely vulnerable to changing temperature and nutrient inflow.

Another example of a glacial lake that will be used for the field experiments is Lake Hallstatt (N47°34′ E13°39′) located in Upper Austria. The nutrient content is slightly higher than in Lake Millstatt and thus, this lake is classified as oligo-mesotrophic meaning that its water quality parameters fall in between oligo- and eutrophic conditions and leaning towards the former [8,9]. All of the lakes mentioned above are protected and highly sensitive areas. Biohybrid entities allow for an extensive and long-term environmental monitoring of these water bodies with minimal human involvement.

2.2 Lifeforms

Daphnia. When using a lifeform's physiological or behavioural responses as an indicator, it is essential to choose a well-documented species in order to obtain a reliable result [15]. One of the invertebrates most widely used as bioindicator is *Daphnia* sp. Because of its sensitivity to various chemicals and changes in the environment, it has become a popular choice for toxicological studies, behavioural and pharmaceutical research [6]. In case of a stressor present, certain reactions, such as swimming speed, vertical distribution or heart rate, can be used to identify its presence and intensity [14]. Daphnia's swimming behaviour is a good indicator of its health as affected individuals may display decreased swimming speed, increased sinking and other behaviours. By closely monitoring their swimming patterns there is a high chance of detecting a stressor in the environment. In Robocoenosis, this is achieved by constructing a flow-through chamber in which Daphnia can swim freely in the enclosed layer of water (Fig. 1). Thanks to an appropriately-sized mesh build into the setup, animals can still easily feed on the algae naturally occurring in the water. Daphnia specimens were sampled from the respective lakes to avoid introducing alien species to the environment. A camera piece controlled by a single board computer (Raspberry Pi) included in the setup is facing the chamber. The setup is lowered with suspension cables the bottom of the respective lakes at various depths.

Fig. 1. Early biohybrid prototype during field test in the Lake Neusiedl. The prototype includes A) a chamber containing swimming Daphnia, B) a Raspberry Pi camera module and C) a Raspberry pi contained within the plexiglass tube.

Zebra Mussel. Another species of focus is the zebra mussel *Dreissena polymorpha.* It is a relatively small species with the size ranging between 2–2.5 cm, and most commonly lives in large colonies. It thrives in well-oxygenated waters and thanks to its resilience it is now widely spread across the world and was labeled as highly invasive by the Invasive Species Compendium [3,5]. As zebra mussel feeds and breaths through filtrating the surrounding waters it responds quickly to any changes in the environment or the presence of toxic substances with certain stress behaviours [11]. A sudden appearance of an undesirable factor such as low oxygen, temperature spike or a toxic substance will result in an instant closing or fluttering movements of the valves. When the observed specimen presents any of those behaviours, it is a sign of a stressor present in the environment. In Robocoenosis, data from this behaviour is harvested by combining visual observation and automated movement tracking (Fig. 2). A mussel is mounted to the setup with a tag on the top valve to augment the valve movements. The mussel is recorded at a set interval to observe the movements and any potential signs of stress.

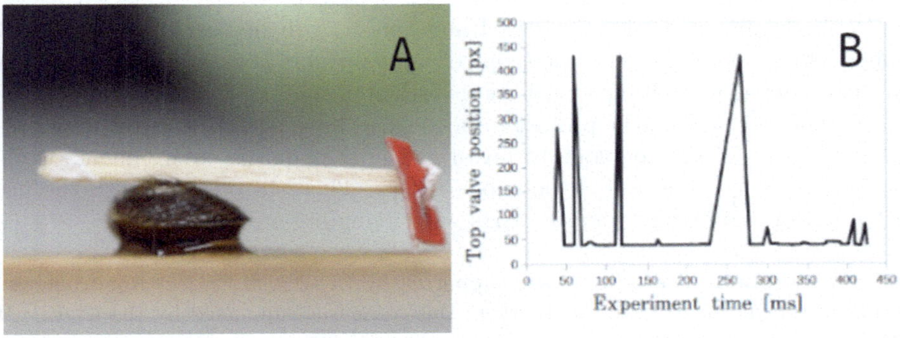

Fig. 2. A: Zebra mussel *Dreissena polymorpha* with a red tag mounted on the top valve in order to facilitate the movement detection. B: Preliminary results of the object tracking done in Lake Millstatt.

3 Discussion

The project Robocoenosis introduces a new approach to environmental monitoring. Robocoenosis aims to develop novel methodologies to using living organisms as natural biosensors. This opens the door to an efficient long-term collection of large amounts of data with minimal human impact on the environment. The outcomes of this project will find many applications in environmental management and biological research, as well as exploring the interactions between lifeforms and the environment. The use of biohybrids will allow to monitor bodies of water that would otherwise be inaccessible or where sampling surveys are strictly limited. It will also provide a new methodology for ethological and ecological research. Through the robotic setup it is possible to observe the animals in their natural habitat without the usual disruptions of in-person observation, such as gas bubbles or movements while handling or maintenance.

Over the years, various lifeforms were used as biosensors. Daphnia has been extensively used, for example, by pharmaceutical companies for testing the effects on cardioactive drugs on the heart rate [21]. Daphnia's light responses were also successfully employed to detect small quantities of toxic chemicals in the water [13]. A disruption in the overall swimming behaviour including clustering, swimming speed and distribution, was treated as an early-warning signal by Daphnia Toximeter II [12]. When the behaviour of enough individuals deviates from normal, an alarm is triggered. With these laboratory-based uses of Daphnia's sensitivity, it is of merit to investigate its uses in the field. Robocoenosis will move the toxicological studies directly into the water body of interest decreasing significantly the time of detection and sample processing period [2].

As mentioned before, this was successfully achieved by the project Subcultron based in Adria region [20]. With the use of a robotic swarm equipped with various sensors, the researchers were able to gather data not only in a continuous manner but also cover large areas at the same time [20]. This initiative will be taken a step further with the project Robocoenosis.

With regards to ethics, project Robocoenosis will use exclusively invertebrates which do not fall into the scope of animal rights laws. Regardless of this, minimally invasive techniques will be investigated.

Here, major aims of the project Robocoenosis have been presented. Its aim is designing a novel methodology to investigating the aquatic environments while focusing on long-term and autonomous setups. This will be done with the use of biohybrids, combining robotic parts with living organisms.

Acknowledgments. This work was supported by EU-H2020 Project Robocoenosis, grant agreement No 899520. Furthermore this work was supported by the COLIBRI initiative at the University of Graz. We further want to thank the "Österreichische Bundesforste AG", the "Burgenländische Landesregierung, Abt. 4 (Natur- und Klimaschutz)" and the "Biologische Station Illmitz" for their immense support.

References

1. Afroz, A.S., Romano, D., Inglese, F., Stefanini, C.: Towards bio-hybrid energy harvesting in the real-wolakerld: pushing the boundaries of technologies and strategies using bio-electrochemical and bio-mechanical processes. Appl. Sci. **11**(5), 2220 (2021)
2. AquaDect. Mosselmonitor - the biological early warning system (2021). http://www.mosselmonitor.nl/html/Engels/functional.html. Accessed on 24 Feb 2021
3. Benson, A., Raikow, D., Larson, J., Fusaro, A., Bogdanoff, A., Elgin, A.: Dreissena polymorpha (Pallas, 1771): U.S. Geological Survey, Nonindigenous Aquatic Species Database, Gainesville, FL (2021). https://nas.er.usgs.gov/queries/FactSheet.aspx?speciesID=5. Accessed 15 Feb 2021
4. Buckel, J., Otto, J., Prasicek, G., Keuschnig, M.: Glacial lakes in Austria - distribution and formation since the little ice age. Global Plan. Change **164**, 39–51 (2018)

5. CABI. Invasive Species Compendium, Wallingford, UK: CAB International (2022). www.cabi.org/isc. Accessed 18 Jan 2022
6. Choi, J., et al.: An evaluation of acute toxicity using the phototactic behavior of Daphnia magna. J. Pharmacol. Toxicol. Methods. **68**(1), 50–51 (2013). (10th Annual Focused Issue on Methods in Safety Pharmacology)
7. Damania, R., Desbureaux, S., Rodella, A.-S., Russ, J., Zaveri, E.: Quality Unknown: The Invisible Water Crisis. World Bank Group (2019)
8. Dokulil, M.: Contribution of green algae to the phytoplankton assemblage in a Mesotrophic Lake, Mondsee, Austria. Archiv für Protistenkunde **139**(1), 213–223 (1991)
9. Ernst, B., Hoeger, S.J., O'Brien, E., Dietrich, D.R.: Abundance and toxicity of Planktothrix rubescens in the prealpine Lake Ammersee, Germany. Harmful Algae **8**(2), 329–342 (2009)
10. García-Carmona, L., et al.: Biohybrid systems for environmental intelligence on living plants: watchplant project. In: Proceedings of the Conference on Information Technology for Social Good 2021, pp. 210–215. Association for Computing Machinery, New York, NY, USA (2021)
11. Grabarkiewicz, J., Davis, W.: An introduction to freshwater mussels as biological indicators. EPA-260-R-08-015. U.S. Environmental Protection Agency, Office of Environmental Information, Washington, DC (2008)
12. Green, U., Kremer, J.H., Zillmer, M., Moldaenke, C.: Detection of chemical threat agents in drinking water by an early warning real-time biomonitor. Environ. Toxicol. **18**(6), 368–374 (2003)
13. Martins, J., Soares, M., Saker, M., Oliva Teles, L., Vasconcelos, V.: Phototactic behavior in Daphnia magna Straus as an indicator of toxicants in the aquatic environment. Ecotoxicol. Environ. Safety **67**, 417–22 (2007)
14. Nikitin, O.: Effect of various temperature and light intensity regimes on Daphnia magna swimming behaviour. In: 19th SGEM International Multidisciplinary Scientific GeoConference EXPO (2019)
15. Rajewicz, W., Romano, D., Varughese, J.C., Vuuren, G.J.V., Campo, A., Thenius, R., Schmickl, T.: Freshwater organisms potentially useful as biosensors and power-generation mediators in biohybrid robotics. Biol. Cybern. **115**(6), 615–628 (2021)
16. Robocoenosis. Robocoenosis- robots in cooperation with a biocoenosis (2022). https://www.robocoenosis.com/. Accessed 26 Jan 2022
17. Thenius, R.: Biohybrid entities for environmental monitoring. In: ALIFE 2021: The 2021 Conference on Artificial Life. MIT Press (2021)
18. Schmickl, T., et al.: Cocoro-theself-aware underwater swarm. In: 2011 Fifth IEEE Conference on Self-Adaptive and Self-Organizing Systems Workshops, pp. 120-126. IEEE (2011)
19. Soja, G., Züger, J., Knoflacher, M., Kinner, P., Soja, A.M.: Climate impacts on water balance of a shallow steppe lake in Eastern Austria (Lake Neusiedl). J. Hydrol. **480**, 115–124 (2013)
20. Thenius, R., et al.: subCULTron - cultural development as a tool in underwater robotics. In: Lewis, P.R., Headleand, C.J., Battle, S., Ritsos, P.D. (eds.) ALIA 2016. CCIS, vol. 732, pp. 27–41. Springer, Cham (2018). https://doi.org/10.1007/978-3-319-90418-4_3
21. Villegas-Navarro, A., Rosas-L, E., Reyes, J.L.: The heart of Daphnia magna: effects of four cardioactive drugs. Comp. Biochem. Physiol. Part C Toxicol. Pharmacol. **136**(2), 127–134 (2003)
22. Volkov, A., Ranatunga, D.: Plants as environmental biosensors. Plant Signal. Behav. **1**, 105–15 (2006)

Four-Legged Walking Robot, Steering and Turning Using Phase Control

Tomer Maizel[1], Izhak Bucher[1,2(✉)], Eyal Baruch[1], Yoav Vered[2], and Harel Plat[1]

[1] Dynamics Laboratory, Mechanical Engineering, Technion, Haifa, Israel
bucher@technion.ac.il

[2] Mechanical Engineering, Technion, Technion, Haifa 3200003, Israel

Abstract. Presented is a 4-legged walking robot. Unlike classical models, which require extensive actuation to walk, this robot achieves precise steering using 2 motors. Each motor controls a Theo-Jansen-like mechanism consisting of 2 legs. It is shown that the relative phase between the 2 motors determines the turning rate and direction of movement of the robot.

The legs of the robot were designed using a model optimization algorithm, based on the desired trajectory of the legs. A quasi-static kinematic model was developed to predict the trajectory of robot and validate the control process, which is based on the relative phase between the motors. The model is verified via simulations, and a robot was constructed to demonstrate its movement and control process experimentally.

Keywords: Walking robot · Theo-Jansen mechanism · Compliant mechanism · Phase control

1 Introduction and Motivation

Walking mobile robots have several advantages when compared to wheeled mobile robots, especially while moving across uneven rough and natural terrains [1]. Therefore, various walking mechanisms have been developed for legged robots [2, 3]. However, these mechanisms tend to be heavily actuated, and require complex control algorithms in order to walk successfully [1–3]. The present paper focuses on a restricted version of a four-legged walking robot, inspired by walking and running animals [4, 5], which is actuated using 2 motors. It has been described (e.g. [8]) that mammals' locomotion is controlled by a network of coupled oscillators. A flexible combination of unit oscillators gives rise to different forms of locomotion. Inter-oscillator coupling can be modified by changing the synaptic strength (or polarity) of the relative spinal connections. As a result, unit oscillators can be coupled in phase, out of phase, or with a variable phase, giving rise to different behaviours, such as speed increments or reversal of gait direction (from forward to backward). We adopt the notion of relative phase only, keeping the gait unchanged.

The legs were designed based on a compliant version of the Theo-Jansen Mechanism [6, 7]. The Jansen mechanisms on each side are both driven by a dedicated motor with

A. Müller and M. Brandstötter (Eds.): RAAD 2022, MMS 120, pp. 528–535, 2022.
https://doi.org/10.1007/978-3-031-04870-8_62

a fixed-amplitude excentre. Therefore, the robot cannot change its gait, but the relative phase between the left and the right side is digitally controlled by a central microprocessor and suitable sensors and electronics.

This arrangement is considerably simpler than what most quadruple robots employ (see [5, 9–11]). When each leg is controlled independently, the design space is large and genetic algorithms [12] or other optimization methods are used find the best gait for optimal mobility.

This paper is structured as following: Sect. 2.1 describes the mechanical design and optimization of the mechanism. Section 2.2 puts forward a quasi-static analysis of the robot. In Sect. 2.3 the control method of the motors is presented. Finally, Sect. 2.4 presents the numerical simulations and experimental results carried out on a constructed mechanism. The results are compared to the analytical model, validating it and demonstrating the ability to successfully control the trajectory of a 4-legged mechanism.

2 Robot Model, Kinematics and Realization

In this section, the geometry, configuration and realization of the proposed walking robot is described.

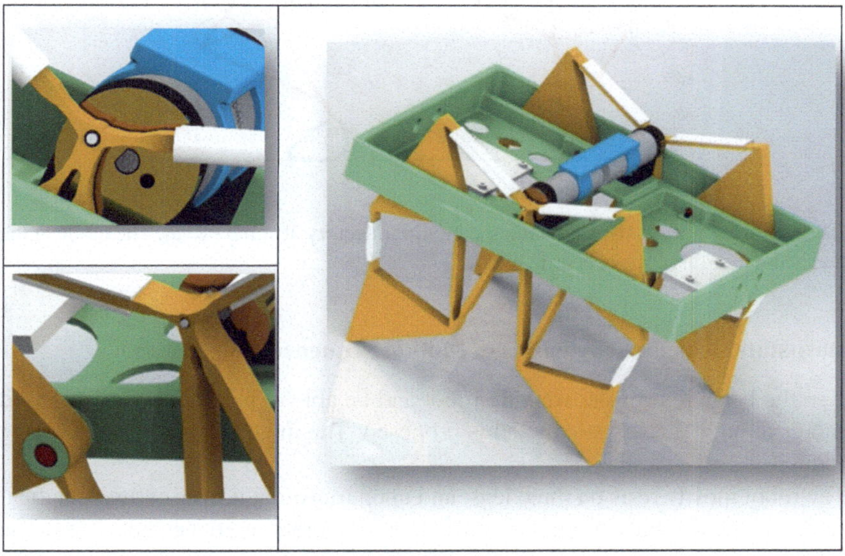

Fig. 1. The walking robot model, consisting of 2 compliant Theo-Jansen mechanisms. Each mechanism is driven at its central point in a circular manner. Top left: eccentric driving actuator producing the circular motion. Bottom left: excentre and one hinge (red). (Color figure online)

Figures 1 and 2 present the driving mechanism and the overall construction of the walking robot. It is clear that not all 4 tips are in contact with the floor simultaneously, since 3 points define a plane.

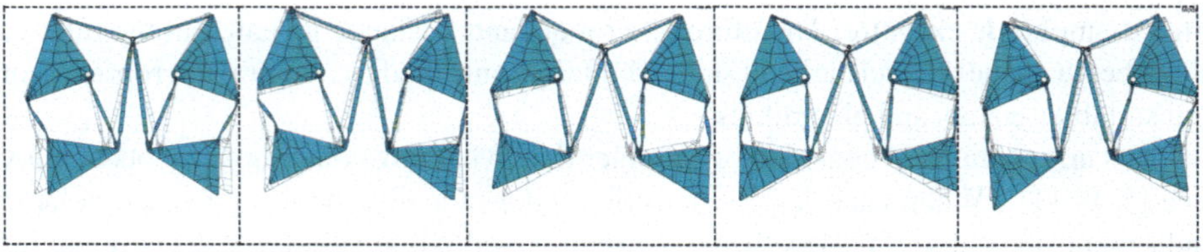

Fig. 2. A sequence of the compliant mechanism states (left legs), showing the state of legs within a cycle and the compliant links undergoing bending.

2.1 Walking Mechanism Geometry Design and Optimization

The leg mechanisms were optimized in terms of dimensions (similarly to [7, 13]) to perform a tip motion exhibiting a non-intersecting closed-curve. The curve (Fig. 3) is designed to produce propulsion through friction with the floor and raise the contacting tip upon completion of the movement.

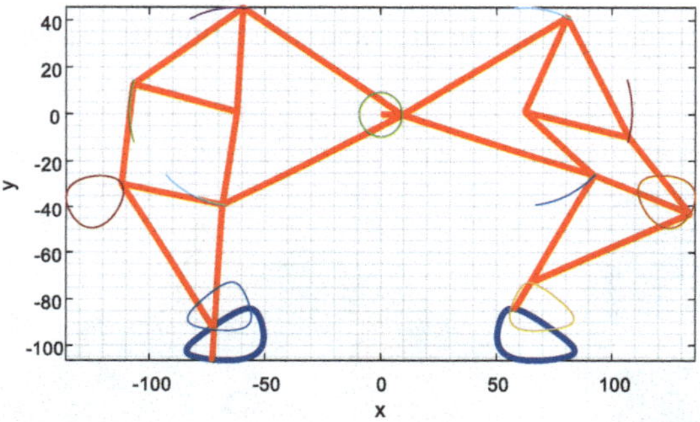

Fig. 3. The optimized mechanism showing the leg-tip trajectory at some $t = t_0$. The circle at the middle traces the motion of the eccentric motor.

2.2 Quasi-static Kinematic Analysis of Motion Sequence

To analyse the behaviour of the walking robot and be able to predict its motion based on the motor input, a quasi-static model is proposed. The model follows the following assumptions:

(i) The robot mostly rests on three legs, and upon transition it rests on two diagonal legs for a short time (ii) Since motion is rather slow, accelerations are neglected. (iii) The leg with largest normal force does not slip. (iv) The centre gravity is offset towards the rear legs, relative to the geometrical centre. therefore, the normal force due to self-weight is largest on these legs and during stepping motion, one of these legs (leg #1 in Fig. 4) remains immovable.

As a result of these assumptions, since all 4 legs are driven kinematically, the location of their tips can be computed. Consequently, we are able to find the plane defined by the 3 legs contacting the floor. During a complete revolution of the eccentric motors, the two front legs change their role and the one held in the air becomes in contact with the floor while the other is lifted up.

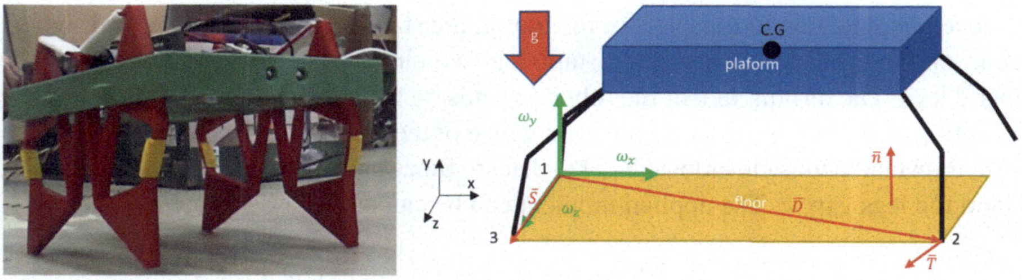

Fig. 4. Left: the robot resting on 3 legs during a walking sequence. Right: a simplified quadruped robot resting on 3 legs with its local and global systems of coordinates.

Using these assumptions, force and moment balance of the 4 legs can be derived (Fig. 5) and represented by (1)–(4).

$$\left((\bar{r}_{22} - \bar{r}_{11}) \times \left(Fy_{22} \cdot \hat{n}\right)\right) \cdot (\bar{r}_{21} - \bar{r}_{11}) - \left((\bar{r}_{cm} - \bar{r}_{11}) \times \left(mg \cdot \hat{n}\right)\right) \cdot (\bar{r}_{21} - \bar{r}_{11}) = 0 \tag{1}$$

$$\left((\bar{r}_{21} - \bar{r}_{11}) \times \left(Fy_{21} \cdot \hat{n}\right)\right) \cdot (\bar{r}_{22} - \bar{r}_{11}) - \left((\bar{r}_{cm} - \bar{r}_{11}) \times \left(mg \cdot \hat{n}\right)\right) \cdot (\bar{r}_{22} - \bar{r}_{11}) = 0 \tag{2}$$

$$Fy_{11} + Fy_{21} + Fy_{22} = mg \tag{3}$$

$$Fx_{22} + Fx_{21} + Fx_{11} = 0 \tag{4}$$

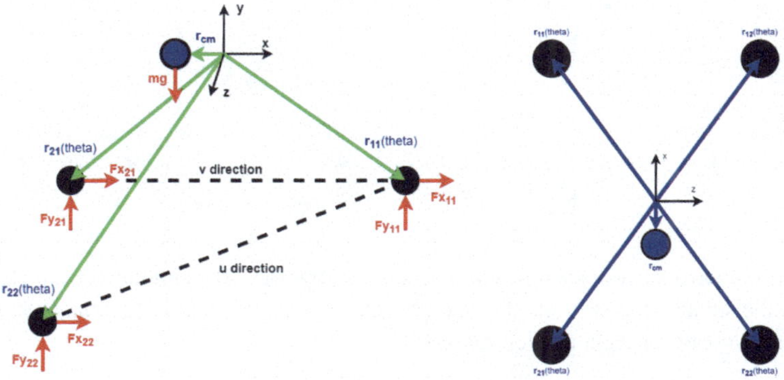

Fig. 5. Left: force diagram of 3 supporting legs during walking sequence. Right: top view, showing the locations of the 4 legs vs the centre of mass and local coordinates.

During a complete revolution of the eccentric motors, the two front legs change their role and the one held in the air becomes in contact with the floor while the other is lifted up. Using static analysis, it can be shown that leg #2 force F_{y22} (see Fig. 5) is larger than F_{y21} and F_{y11}. Therefore, it is assumed that it does not slip, while the 2 other legs must slip to allow for the kinematic motion. The friction forces will be:

$$Fx_{11} = -\mu Fy_{11} sign(\dot{x}_{11}), \quad Fx_{21} = -\mu Fy_{21} sign(\dot{x}_{21}) \tag{5}$$

In a complete revolution cycle of the motor, the static legs alternate and the moment acting on the robot frame about the static leg (which acts as an axis) is caused by the other 2 legs. The turning rate of the robot depends on the average moment in a complete revolution of the motor, which depends on which of the front legs is in the air, determined by their instantaneous elevations (Y_1, Y_2) that are functions of the motors' relative phase, ϕ, and the legs gap, h. The applied moment can be calculated as:

$$M_y(t,\phi) = \mu h \cdot \begin{cases} Fy_{11}\text{sign}(\dot{x}_{11}) + Fy_{21}\text{sign}(\dot{x}_{21}) \; Y_1 > Y_2 \\ Fy_{22}\text{sign}(\dot{x}_{22}) + Fy_{12}\text{sign}(\dot{x}_{12}) \; Y_2 > Y_1 \end{cases} \tag{6}$$

The average moment per cycle or revolution of the motors is thus:

$$M_{avg,cycle}(\phi) = \frac{\Omega\mu h}{2\pi}\begin{pmatrix} \int (Fy_{11}\text{sign}(\dot{x}_{11}) + Fy_{21}\text{sign}(\dot{x}_{21}))dt \; Y_1 > Y_2 \\ \int (Fy_{22}\text{sign}(\dot{x}_{22}) + Fy_{12}\text{sign}(\dot{x}_{12}))dt \; Y_2 > Y_1 \end{pmatrix} \tag{7}$$

2.3 The Dual Motor Control Algorithm

A closed-loop PI controller was constructed for each of the 2 motors independently, to control their respective angular position. Enforcing a constant motor speed of rotation $\dot{\theta} = \Omega$, the instantaneous reference angle fed to the right motor's controller is thus $\theta_{ref} = \Omega t$. The left motor follows the rate of the right one at a given phase lag ϕ (Fig. 6).

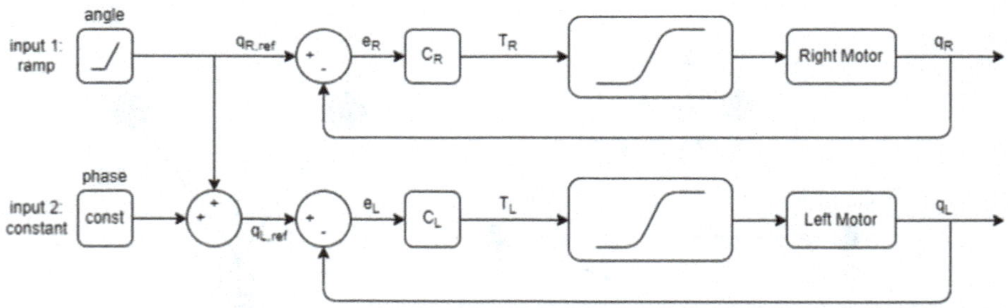

Fig. 6. The system control algorithm. CR, CL, are the PID controllers for the right and left motors generating the suitable torque to the DC motors. The inputs are $angle = \Omega t$, $phase = \phi$-relative phase reference between right and left motors.

2.4 Numerical Simulations and Experimental Verification

In this section, a comparison between the analytical model presented in Sect. 2.2, a simulated model of the robot, and an experimental constructed mechanism is presented.

Numerical Simulations

The robot's dynamics were simulated using SimMechanics™, assuming massless legs and a rigid main body. The motors were controlled by the control algorithm illustrated above. The simulated model is depicted in Fig. 7.

Fig. 7. Matlab SimMechanics™ based simulation at 2 time instances, see Link

Experimental System

The robot was manufactured using standard 3D printing. The main body was made of polymeric (PLA) material while the compliant parts of the legs used rubber-like material (TPU). Two geared DC motors rotating at a maximal speed of 130 RPM, with digital shaft encoders were implemented to actuate the legs. The sequencing and control algorithm was realized on a microcontroller. The measurements were executed via video tracking of the robot. Several videos taken during the experiment, with different relative phase values set as input. The tested values are 90, 135, 180, 225 and 270°, and we examined the trajectory the robot makes at each and its turning speed. The motors rotated at 60 RPM during the experiment. Recorded videos from these experiments are shown in Table 1 via web links (Fig. 8).

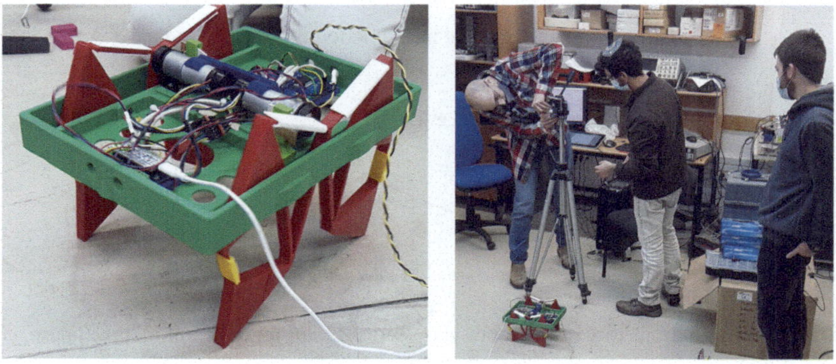

Fig. 8. Left: Walking robot during experiments, showing the motors and the electronics. Right: Top video camera set-up, used to reconstruct the trajectories in Figs. 9 and 10.

Table 1. Links to online videos of the experiment at various phase lags - ϕ.

Phase ϕ	Link 180°	Link 90°	Link 270°	Link 45°

Results

The turning rate of the robot vs relative phase, ϕ, was compared for the analytic, numerically simulated and experimental models given in Fig. 9. The analytic values were calculated with the integration of the moment twice over a cycle,

$$\omega_{avg,cycle}(\phi) = \frac{\Omega}{2\pi} \int_0^{\frac{2\pi}{\Omega}} \left(\int_0^t M_y d\tau \right) dt, \quad \theta = \Omega t, \tag{8}$$

where Ω is the motors speed of revolution. It is clear that the analysis, simulation and experiment show some resemblance and that for $\phi = \pi$ the robot walks straight, while around $\phi = \pm\pi/2$ the maximal turn rate occurs. The simulation results are largely affected by the motors' speed and the coefficient of friction giving rise to moderate deviations of the simulated and experimental results.

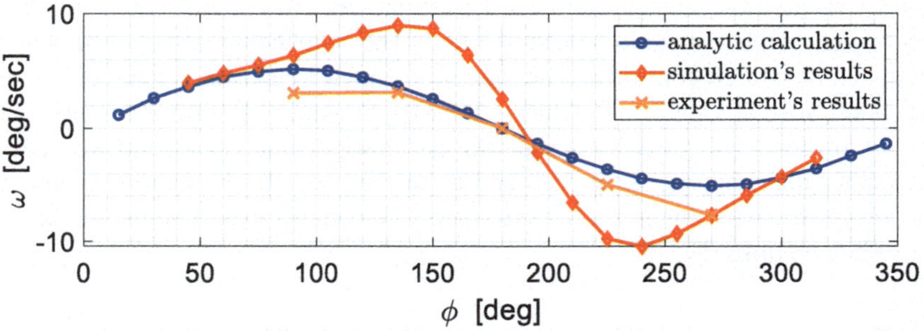

Fig. 9. Turning rate of the robot as a function of the relative phase, ϕ

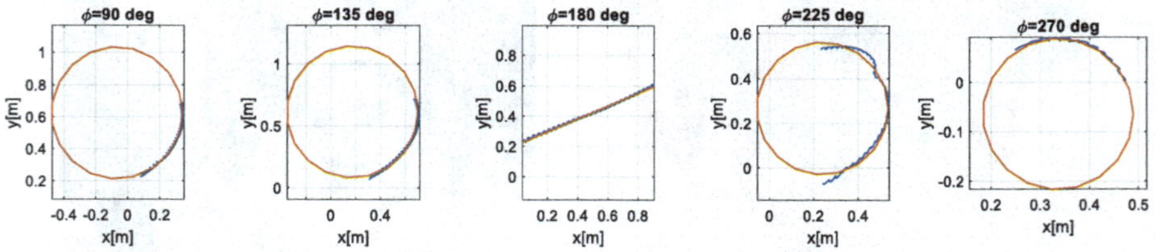

Fig. 10. Motion trajectory of the robot via video tracking for different relative phase, ϕ

As can be seen in Table 1, and in Figs. 9 and 10, the radius of the trajectory is set by, ϕ, and the speed of progress along the trajectory is set by Ω.

3 Conclusions

It was shown that a quadruped animal inspired robot's turning rate can be actuated using two motors, and controlled by a single parameter, which is the relative phase between the left and right-side legs. A simplified, quasi-static model was able to predict this behaviour, which was verified in an elaborate multi-body numerical simulation (SimMechanics™) and by a real walking robot produced using 3D printing.

This conclusion is obviously affected by the unique geometry used to model the robot's legs, which can create an optimized step trajectory curve. The speed of the motors, flexibility, inaccurate dimensions, inertia and friction all introduce some uncertainty which did not affect the main conclusion greatly and the quasi-static model was able to explain what is controlling the turn rate.

References

1. Silva, M.F., Machado, J.A.T.: An overview of legged robots. In: Conference: MME– International Symposium on Mathematical Methods in Engineering: Cankaya, Ankara, Turkey, April 2006
2. Kajita, S., Espiau, B.: Legged robots BT. In: Siciliano, B., Khatib, O. (Eds.) Springer Handbook of Robotics, pp. 361–389. Springer, Heidelberg (2008)
3. Gregorio, P., Ahmadi, M., Buehler, M.: Design, control, and energetics of an electrically actuated legged robot. IEEE Trans. Syst. Man Cybern. Part B. Cybern. **27**(4), 626–634 (1997)
4. Parkes, R.S.V., Pfau, T., Weller, R., Witte, T.H.: The effect of curve running on distal limb kinematics in the Thoroughbred racehorse. PLoS One. **15**(12), 1–15 (2020)
5. Gan, Z., Wiestner, T., Weishaupt, M.A., Waldern, N.M., David Remy, C.: Passive dynamics explain quadrupedal walking trotting and tölting. J. Comput. Nonlinear Dyn. **11**(2), 1–12 (2016)
6. Patnaik, S.: Analysis of Theo Jansen mechanism (Strandbeest) and its comparative advantages over wheel based mine excavation system. IOSR J. Eng. **05**(07), 43–52 (2015)
7. Nansai, S., Elara, M.R., Iwase, M.: Dynamic analysis and modeling of Jansen mechanism. Proc. Eng. **64**, 1562–1571 (2013)
8. Lacquaniti, F., Ivanenko, Y.P., Zago, M.: Kinematic control of walking. Arch. Ital. Biol. **140**(4), 263–272 (2002)
9. Hancher, M.D., Hornby, G.S.: Evolving quadruped gaits with a heterogeneous modular robotic system. In: IEEE Congress Evolutionary Computation CEC 2007, pp. 3631–3638 (2007)
10. Chang, S., Du, H., Cong, Y., Xie, F., Zhang, J.: Gait planning of quadruped robot based on ROS. In: 2020 7th International Conference on Information, Cybernetics Computational Social Systems ICCSS 2020, pp. 761–766 (2020)
11. McClain, E.W., Filsoofi, A.S., Meek, S.: Stability and control of a pacing quadruped using stabilizing switching design. IEEE Robot. Autom. Lett. **6**(4), 8466–8473 (2021)
12. Hornby, G.S., Takamura, S., Yamamoto, T., Fujita, M.: With two quadruped robots. IEEE Trans. Robot. **21**(3), 402–410 (2005)
13. Morita, K., Ishihara, H.: 4-legged mechanism of realizing dynamic running-a study on change of locomotion pattern by phase declination. In: IEEE International Conference Mechatronics and Automation, 2005, vol. 4, pp. 2084–2089 (2005)

Finding Optimal Placement of the Almost Spherical Parallel Mechanism in the Recupera-Reha Lower Extremity Exoskeleton for Enhanced Workspace

Ibrahim Tijjani(✉)

Robotics Innovation Center, German Research Center for Artificial Intelligence (DFKI GmbH), Robert-Hooke-Str. 1, 28359 Bremen, Germany
ibrahim.tijjani@dfki.de

Abstract. The ankle joint of an exoskeleton plays a vital role in maintaining balance and posture during locomotion by grounding the upper body's weight. Parallel designs are advantageous for constructing ankle joints in exoskeleton but challenges include complex workspace analysis and finding the optimal placement in the overall structure of the exoskeleton. This paper presents a strategy for finding the optimal placement of the prototype Active Ankle 3[R2[US]] mechanism in the Recupera-Reha lower extremity exoskeleton for enhanced workspace. To this end, rotative inverse geometric model of the Active Ankle is exploited to find the alignment between principle human joint axes and axes along which the range of motion is optimum. It is demonstrated that by rotating the mechanism along the adduction-abduction axis and by selecting appropriate ball and socket joints, it is possible to enhance the usable workspace of the mechanism for the human wearing the exoskeleton.

Keywords: Spherical-parallel manipulators · Kinematic analysis · Optimal placement point · Exoskeletons and prosthesis devices

1 Introduction

There has been an increasing societal need for procuring modular ankle exoskeleton devices for medical rehabilitation, walking assistance, and augmenting locomotion strength for the physically disabled, aged persons, and athletes. A parallel manipulator (PM) is a robot system design that offers high stiffness, payload capacity, speed, and accuracy due to combination of two or more kinematic chains. In particular, spherical parallel manipulator (SPM) is a PM which allows only rotational motions of its end-effector platform. For instance, the prominent three (3) degree of freedom (DoF) Agile Eye [3] is a type of SPM and finds application in camera orienting devices. The PKAnkle [11] is a redesigned prototype of Agile Eye that fits in for ankle neuro-rehabilitation and it's capable of providing wide ankle motion ranges but has problems associated to alignment with the

A. Müller and M. Brandstötter (Eds.): RAAD 2022, MMS 120, pp. 536–544, 2022.
https://doi.org/10.1007/978-3-031-04870-8_63

human ankle joint complex. Majority of the researchers in the exoskeleton application domain for gait rehabilitation [1,2,4,10,12,14,16], exploits the advantages of an SPM design approach (see [8] for a recent survey), with emphasis on increasing the workspace, and finding suitable alignment configuration with the human ankle joint.

The novel Active Ankle (see Fig. 1) first introduced in [12] is a wearable ankle exoskeleton device that exploits the advantages of a PM. It is an almost spherical parallel mechanism (ASPM) similar to the SPM in structural design, with a 3 DoF and 3 actuators capable of three principal movements; dorsiflexion-plantarflexion (DF-PF), eversion-inversion (EV-IN), and adduction-abduction (AD-AB) motions. The design of the Active Ankle mechanism represents a module for the ankle and hip joints shown in Fig. 2 of the Recupera-Reha full-body exoskeleton [9] in Fig. 3. An extensive kinematic analysis of the mechanism is available in [5–7]. The current mounting of the active ankle mechanism in the Recupera-Reha exoskeleton has some limitations. The orientation of its joint axes is not properly aligned with that of the human ankle joint complex which leads to a reduced workspace. Hence, it is essential to find an optimal placement point of the ankle mechanism in the exoskeleton leg design. Another challenge with the device is a passive ball and socket joint limit on the motor actuator that leads to a small usable workspace available for the movement trajectories if it's to be utilized, as a use-case for dynamic walking or gait rehabilitation, proper alignment with the human ankle joint complex with increased motion ranges is necessary.

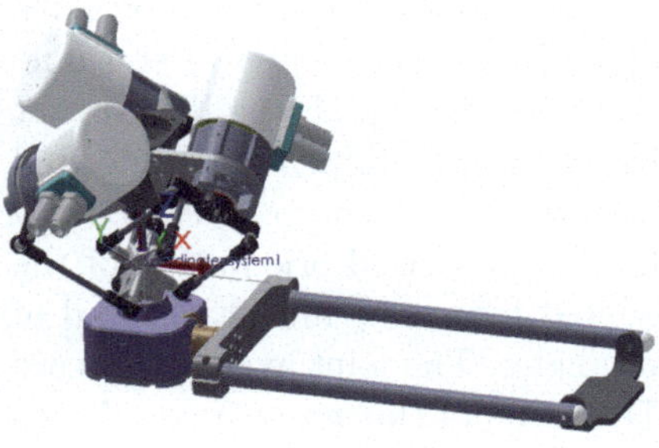

Fig. 1. Active ankle

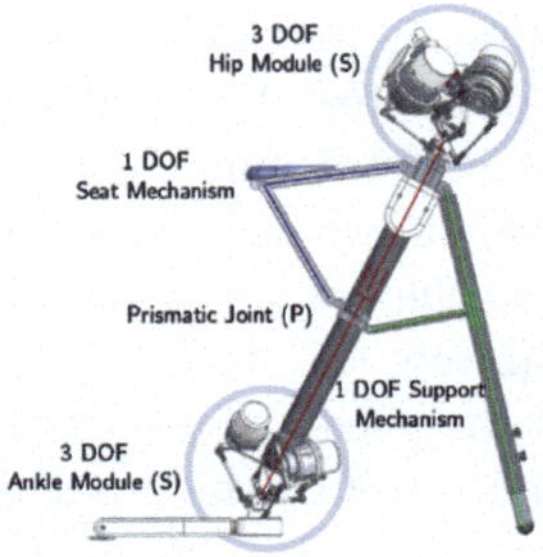

Fig. 2. Joints module

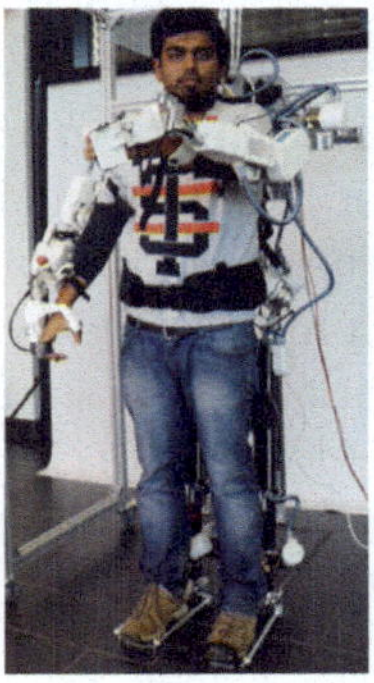

Fig. 3. Full-body exoskeleton [9].

Contribution. Since the desire to lift the upper body's weight and perform some dynamic walking with the active ankle mechanism, this paper has exploited the rotative inverse geometric model (RIGM) to produce a larger usable workspace and found an optimal placement configuration point for the ASPM ankle joint which fits better to the human ankle joint complex. Further, a proof that by modifying the opening angle of the ball and socket joints, it is possible to further extend the workspace without making any big structural changes in the system design.

2 Rotative Inverse Geometric Model of the ASPM Ankle

This section highlights the analysis of the inverse kinematic problem of the active ankle mechanism adapted from [6]. As earlier stated, the device has a feature of an ASPM, designed to drive a spatial quadrilateral that intersects its three rotative actuators perpendicularly at its end-effector (EE) point. The opening angle of the ball and socket joint for the mechanism is limited by $\pm 25°$ motor angle constraint (q_{min}, q_{max}), while the available range of motion (RoM) for the task space angles $(\theta_{min}, \theta_{max})$, is constrained by a $\pm 90°$ physical limit (θ). The fascinating feature in the mechanism design structure is that the physical configuration of its joint axes is capable of bearing any force applied without external torque from the motor actuators. In essence, a design structure of the joint axes different from the previous configuration will require additional torques from the motors to drive the device, thus, increased cost. Due to the ASPM design nature of the device, the existence of translation shifts in the EE point is neglected due to its small value when the mechanism is used in the application point of view. Hence, only the rotative universal joints with the spherical cut-joints were utilized in the mechanical design structure, but in the analytical kinematic formulation, the translation shifts were included. The forward kinematic problem that gives the position of the EE as a function of its joint variables is difficult to compute analytically for a complex PM like the ankle joint since it has a limited usable workspace. Therefore, you can not ascertain the feasible workspace that fits into a specific joint configuration. For example, the available workspace in the rotative domain shown in Fig. 4 as reported in the previous variant of the active ankle mechanism. We can not discern which one within the available workspace has the exact configuration for the given set of joint angles. Hence, the knowledge of the positional coordinates of the device alone is not sufficient to determine the placement of the mechanism in the leg. Therefore, we exploit the RIGM that computes the rotative joint space angles and EE positional shifts from the desired orientation, similar to the human ankle joint. However, to achieve a desired orientation of the EE-frame, the RIGM uses the parameterized orientation of the principal movement joint axes vectors represented by u_x, u_y and u_z instead of the position coordinates to find the input joint angles. The joint axes vectors are obtained from a computer-aided design (CAD) model of the active ankle device and adapted to the RIGM.

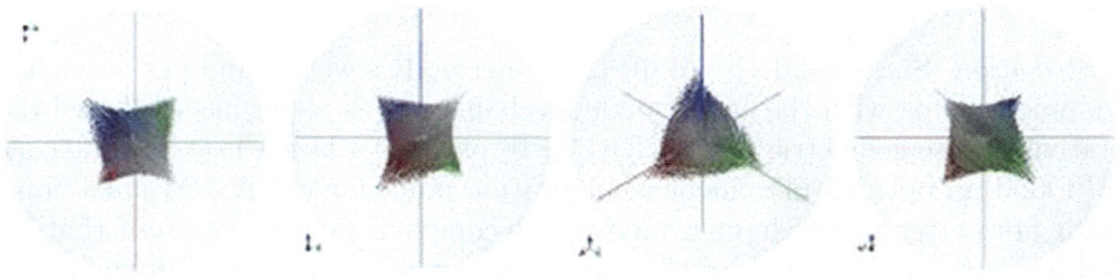

Fig. 4. Feasible workspace configurations adapted in [6].

To find an optimal placement point of the ankle mechanism in the leg, with a similar orientation to the human ankle joint complex in Fig. 5, three different coordinate frame systems were considered at the EE point shown in Fig. 6. The first coordinate frame is the EE-frame (joint axes encoded in yellow color) with a different orientation, a second coordinate frame is the task-space frame located at the EE-point whose joint axes vectors (u_x, u_y, u_z) are parallel to the global coordinate frame attached to the leg mount point. Basically, the constraint that mounting frames at the leg mount point and the EE-point should be parallel is the initial choice of optimizing the placement angle (α). The third coordinate frame (joint axes encoded in black color) is called the rotated frame. Considering the rotated frame with the joint axes (du_x, du_y, du_z) at zero-configuration, only the du_z joint vector is fixed in the same direction with the z-axis (blue color) of the global frame, unlike du_x and du_y which are perpendicular to each other. The ASPM mechanism placement angle α is the angle between the joint vectors of the task-space frame and the rotated frame. We want to parameterize the rotated frame with respect to the task-space frame and map the generated joint angles which are within the actuator motor limit to the EE-frame.

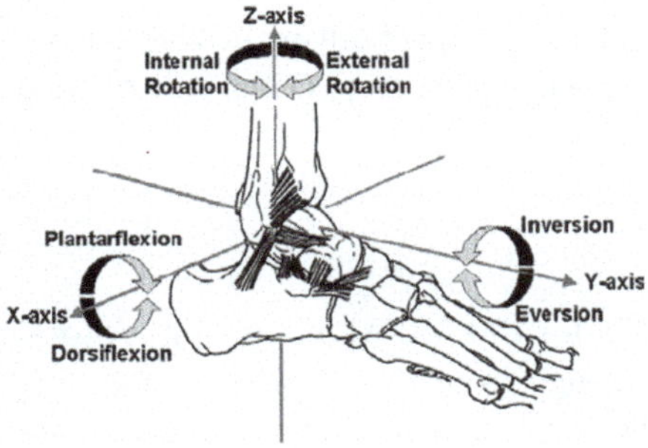

Fig. 5. Human ankle complex [15]

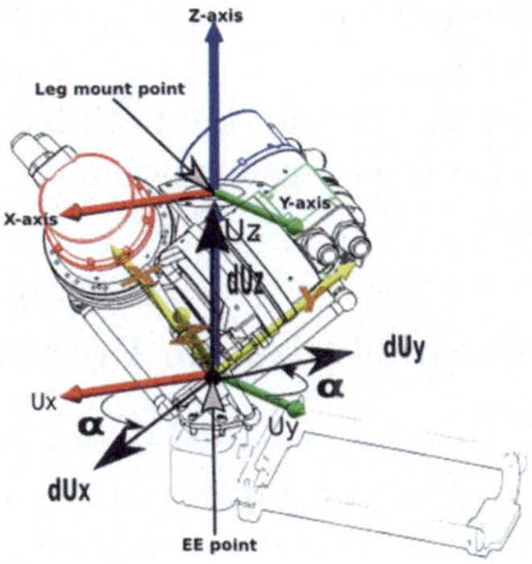

Fig. 6. ASPM ankle frames

The following Table 1 presents the RoM for the movement trajectories. In comparison with the human ankle joint motions reported in [6]. At zero-configuration, you can discern the differences between the human ankle motion ranges and the three other mechanisms from their absolute (Abs.) sum of the minimum (Min.) and maximum (Max.) angles. The active ankle has limited RoM in the DF-PF motion of the previous work [6], which is an essential motion trajectory for the ankle joint. However, this could be insufficient for dynamic walking towards rehabilitation of gait. SPKM and PKAnkle devices have limited RoM in their respective EV-IN and AD-AB motion types, which could be sufficient motion range for human normal daily walking activities. However, for stair ascending and descending, a minimum RoM of 37° and 56° respectively is required according to the authors in [13] for the DF-PF motion trajectory, producing an absolute sum of 93°. Therefore, an increased RoM is imperative for enhanced dynamic walking for the most frequent human daily activities.

Table 1. Comparison between human and existing ankle joint mechanisms RoM.

Motion type	Human ankle			Active ankle [6]			SPKM [1]			PKAnkle [11]		
	Min.	Max.	Abs.	Min.	Max.	Abs.	Min.	Max.	Abs.	Min.	Max.	Abs.
DF-PF	−20°	50°	70°	−19.83°	37.23°	57.06°	29.8°	45.8°	75.6°	−40°	35°	75°
EV-IN	−15°	35°	50°	−15.00°	35.00°	50.00°	17°	22°	39°	−25°	20°	45°
AD-AB	−30°	45°	75°	−29.20°	36.96°	66.16°	25.9°	36°	69.10°	−20°	10°	30°

3 Optimal Placement of the ASPM Ankle in the Leg

It is of interest to know the criteria for modification of the current design even though the results obtained from the kinematic model analysis of the mechanism in [6] demonstrated the DoFs for the three rotative actuators with a workspace modality that shows three principal motion trajectories for application as an ankle joint. However, for use-case as an ankle rehabilitation device towards enhancing dynamic walking, the following two cases could be used to increase the usable workspace for the active ankle:

1. **Case 1:** Increasing the opening angle for the ball and socket joint to have wider RoM.
2. **Case 2:** For the optimal placement point of the mechanism in the leg, we require the transformation of the task space frame respecting the rotated frame.

3.1 Parameterization Procedure

Matlab simulation is used to improve the RoM for the ASPM ankle joint. A total number of 1000 samples (n) for a time (t) range from 0 to 1 with a sampling rate of 0.001 is used for the simulation. The $\pm 25°$ motor angle constraints are initially set for the simulation to compare the previous experimental results obtained in [6] and later increased to $\pm 27°$. The parameterization procedure is described in Algorithm 1, for a range of α in Line 1 which is limited by θ, the transformation matrix (T) in Line 2 is built from a Matlab command *makehgtform*, that rotates the vector u_z by α in radians. Then, Line 3 extracts the rotation matrix (R) part of T. The vectors u_x and u_y are now parameterized by R in Line 4 and Line 5 respectively as the orientation vectors, such that they are orthogonal to each other and lie on the same plane perpendicular to the u_z. The *RIGM* function in Line 6 takes into account the computed orientation vectors from the parameterized vectors, and the motor constraints as input while Line 7 produces the joint space angles which are within the motor constraints as outputs.

In Algorithm 2, the function routine *TaskSpaceAngles* in Line 1 of Algorithm 2 maps out the joint space angles to α in Lines 2, 3, and 4 respectively, to find the relationship between them. Line 7 returns the boundary limits on the task space angle curves computed from Lines 5 and 6. The function routine

IntersectCurves in Line 8 uses the Mathworks *InternX* function to compute the minimum and maximum intersection points on the three curves. The intersection points within the curves that are less than or greater than zero are computed in Line 9 and Line 10 respectively. These are the trade-off points within the actuator motor limit and the physical limit.

Algorithm 1: Parameterization procedure

Input: Joint axes vectors:
$\overrightarrow{u_x} = [-0.8165; 0.4082; 0.4082]$,
$\overrightarrow{u_y} = [0; -0.7071; 0.7071]$, and
$\overrightarrow{u_z} = [1; 1; 1]$
Output: Joint angles:
q_x, q_y, *and* q_z

1 **for** $\alpha = -\pi : 2\frac{\pi}{n} : \pi$; **do**
2   **T** ←
$$\begin{bmatrix} -0.3333 & 0.6667 & 0.6667 & 0 \\ 0.6667 & -0.3333 & 0.6667 & 0 \\ 0.6667 & 0.6667 & -0.3333 & 0 \\ 0 & 0 & 0 & 1 \end{bmatrix};$$
3   **R** ← **T**$(1:3, 1:3)$;
4     $du_x \leftarrow (R \cdot u_x)$;
5     $du_y \leftarrow (R \cdot u_y)$;
6   **Function** RIGM(du_x, du_y, q_{min},q_{max}):
7         $[q_x, q_y, q_z] \leftarrow (-25° \leq q \leq 25°)$;
8   **return** q_x, q_y, q_z
9 **End**

Algorithm 2: Computation of task space angles

Input: $(\alpha, q_{min}, q_{max}, q_x, q_y, q_z)$
Output: Task space angles $(\theta_{min}, \theta_{max})$

1 **Function** TaskSpaceAngles(α,q_x,q_y,q_z,q_{min},q_{max}):
2     $L_x \leftarrow [\alpha; q'_x]$
3     $L_y \leftarrow [\alpha; q'_y]$
4     $L_z \leftarrow [\alpha; q'_z]$
5     $L_{min} \leftarrow [\alpha; q_{min} \cdot length(\alpha)]$
6     $L_{max} \leftarrow [\alpha; q_{max} \cdot length(\alpha)]$
7   **return** L_{min}, L_{max}
8   **Function** IntersectCurves(L_{min}, L_{max}, L_x, L_y,L_z):
9         $\theta_{min} \leftarrow max(\theta < 0)$
10         $\theta_{max} \leftarrow min(\theta > 0)$
11   **return** $\theta_{min}, \theta_{max}$
12 **End Function**

4 Simulation Result

The Matlab simulation in Fig. 7 shows the graphical plot of the $(\theta_{min}, \theta_{max})$ in (radian) against (q_x, q_y, and q_z) also in (radian) that are limited by the same motor constraints. The graph describes the computation of the intersection points on the rotated curves. The upper and lower horizontal lines (maximum and minimum) and the vertical lines are created to show all the intersection points, the point with the highest RoM closer to the zero-configuration is selected as the optimal placement point of the mechanism in the leg. The graphical representation for the variation of $(\theta_{min}, \theta_{max})$ and α, respecting the ±25° motor constraint on the ball and socket joint opening angle is depicted in Fig. 8, for 0° configuration point. The sum of the intersection points on the lower and upper boundaries of each curve produced the absolute sum of 63.46° DF-PF, 74.35° EV-IN, and 71.29° AD-AB RoM for the three principal motion trajectories of the ASPM ankle module. However, modifying the opening angle of the ball and socket joint which is constrained by a motor constraint of ±27°, the result in Table 2 is obtained for the following configuration points (0°, ±15.12°, ±30.24°, 45°). The green shaded part is the chosen optimal placement point of the ankle mechanism in the leg. We can see a symmetric movement in both directions for the chosen point producing a wide RoM for the motion trajectories.

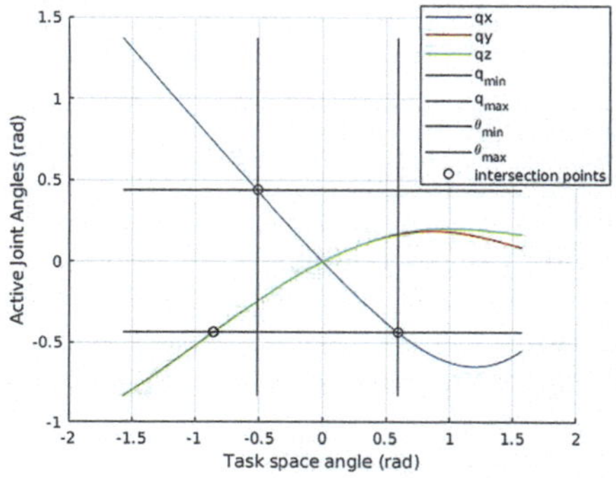

Fig. 7. Intersection points on the curves

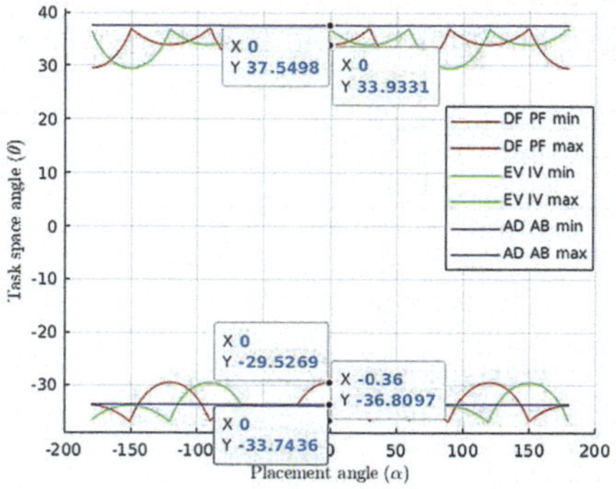

Fig. 8. 0° configuration point

Table 2. Ankle Motion Ranges with ±27° motor constraint.

Placement angle(α)	DF-PF			EV-IN			AD-AB		
	Min.	Max.	Absolute	Min.	Max.	Absolute	Min.	Max.	Absolute
0°	−31.83	37.28	69.11	−39.77	40.14	79.91	−33.74	40.26	74
−15.12°	−33.74	37.87	71.61	−38.16	33.61	71.77	−33.74	40.26	74
15.12°	−33.74	37.85	71.59	−33.74	37.85	71.59	−33.74	40.26	74
−30.24°	−39.94	40.26	80.20	−37.28	31.83	69.11	−33.74	40.26	74
30.24°	−40.09	40.26	80.35	−31.83	37.28	69.11	−33.74	40.26	74
45°	−37.86	33.41	71.27	−33.64	38.17	71.81	−33.74	40.26	74

5 Conclusion

This paper presents the optimal placement of the ASPM ankle joint in the leg. The previous prototype has an opening angle on the ball and socket joints that leads to the motor constraint of ±25°, limiting the workspace of the mechanism principal movement trajectories. Likewise, there is difficulty finding an optimal placement point for the mechanism in the leg. However, employing the RIGM to the mechanism, an optimal placement point was obtainable with an enhanced workspace for the motion ranges. Since the ASPM ankle represents a module for the ankle and hip joint, only the ankle joint module is presented in this work with an increased opening angle of the ball and socket joints to ±27°. Forthwith, a large workspace was produced at 30.24° configuration point. As a prospect for the future, the dynamic model of the ASPM ankle joint will be used to control the device and enhance dynamic walking towards rehabilitation of gait, as a proof of concept to the Recupera-Reha full-body exoskeleton.

Acknowledgement. This research work is conducted at the Robotics Innovation Center, DFKI GmbH, Bremen, Germany in the M-RoCK project funded by the Federal Ministry of Education and Research with grant no. 01IW21002. Additionally, the author acknowledges the support of the Nigerian-German Academic Exchange Programme, sponsored by the DAAD.

References

1. Bi, Z.M.: Design of a spherical parallel kinematic machine for ankle rehabilitation. Adv. Robot. **27**(2), 121–132 (2013)
2. Gallardo-Alvarado, J., Rodriguez-Castro, R., Perez-Gonzalez, L., Aguilar-Najera, C.R.: Kinematics of the 3(RPSP)-S fully spherical parallel manipulator by means of screw theory. Robotics **7**(2), 29 (2018). https://doi.org/10.3390/robotics7020029
3. Gosselin, C., St.-Pierre, É., Gagné, J.A.M.: On the development of the agile eye. IEEE Robot. Autom. Mag. **3**(4), 29–37 (1996). https://doi.org/10.1109/100.556480
4. Kirchner, E.A., et al.: Recupera-Reha: Exoskeleton technology with integrated biosignal analysis for sensorimotor rehabilitation (2016)
5. Kumar, S., Bongardt, B., Simnofske, M., Kirchner, F.: Task space controller for the novel active ankle mechanism (2016). https://doi.org/10.13140/RG.2.2.30022.11845
6. Kumar, S., Bongardt, B., Simnofske, M., Kirchner, F.: Design and kinematic analysis of the novel almost spherical parallel mechanism active ankle. J. Intell. Robot. Syst. **94**, 303–325 (2019)
7. Kumar, S., Nayak, A., Bongardt, B., Mueller, A., Kirchner, F.: Kinematic analysis of active ankle using computational algebraic geometry. In: Zeghloul, S., Romdhane, L., Laribi, M.A. (eds.) Computational Kinematics. MMS, vol. 50, pp. 117–125. Springer, Cham (2018). https://doi.org/10.1007/978-3-319-60867-9_14
8. Kumar, S., Wöhrle, H., de Gea Fernández, J., Müller, A., Kirchner, F.: A survey on modularity and distributivity in series-parallel hybrid robots **68**, 102–367 (2020). https://doi.org/10.1016/j.mechatronics.2020.102367. ISSN: 0957-4158
9. Kumar, S., et al.: Modular design and decentralized control of the Recupera exoskeleton for stroke rehabilitation. Appl. Sci. **9**(4), 626 (2019)
10. Leiyu, Z., et al.: Design and workspace analysis of a parallel ankle rehabilitation robot (PARR). J. Healthcare Eng. **2019**, 1–10 (2019). https://doi.org/10.1155/2019/4164790
11. Malosio, M., Negri, S., Pedrocchi, N., Vicentini, F., Caimmi, M., Molinari Tosatti, L.: A spherical parallel 3 degrees-of-freedom robot for ankle-foot neuro-rehabilitation. In: IEEE Engineering in Medicine and Biology Society Conference, pp. 3356–3359 (2012)
12. Marc, S., Shivesh, K., Bertold, B., Frank, K.: Active ankle - an almost-spherical parallel mechanism. In: Proceedings of ISR 2016: 47st International Symposium on Robotics, pp. 1–6 (2016)
13. Stauffer, R.N., Chao, E.Y., Brewster, R.C.: Force and motion analysis of the normal, diseased, and prosthetic ankle joint, vol. 127, pp. 189–196 (1977)

14. Wang, C., Fang, Y., Guo, S., Chen, Y.: Design and kinematical performance analysis of a 3-RUS/RRR redundantly actuated parallel mechanism for ankle rehabilitation. J. Mech. Robot. **5**(4), 041003 (2013). https://doi.org/10.1115/1.4024736
15. Yuki, T.: Effect of arch supports on ankle-subtalar complex instability: a biomechanical experimental study. Foot Ankle Int. **24**(8), 634–639 (2003). https://doi.org/10.1177/107110070302400811
16. Zuo, S., Li, J., Dong, M., Zhou, X., Fan, W., Kong, Y.: Design and performance evaluation of a novel wearable parallel mechanism for ankle rehabilitation. Front. Neurorobotics **14**, 9 (2020). https://doi.org/10.3389/fnbot.2020.00009

Collaborative Robotics

An Investigation of the Measurement of Transient Contacts in Human-Robot Interaction

Clara Fischer[1](✉), Martin Steiner[2], Michael Neuhold[2], Maximilian Papa[1], Alexandra Markis[3], and Sebastian Schlund[1]

[1] Institute of Management Science, TU Wien, Theresianumgasse 27, 1040 Vienna, Austria
clara.fischer@tuwien.ac.at
[2] TÜV AUSTRIA Service GmbH, Deutschstraße 10, 1230 Vienna, Austria
[3] TÜV AUSTRIA Holding AG, TÜV AUSTRIA-Platz 1, 2345 Brunn am Gebirge, Austria

Abstract. The collaborative robot safety mode 'power and force limiting' requires the compliance of biomechanical limits to ensure human safety. As part of the risk assessment, it is common to test possible contact points of a collaborative application for a) quasi-static contact (e.g. squeezing or clamping) and b) transient contact (collision with free impact). Although, there are standardized power and force measuring devices (PFMD), which are offered by different companies on the market, multiple, partly differing measuring methods exist. Especially for the transient contact, the respective measuring setup is not consistently defined. Therefore, we carried out an investigation of three state-of-the-art measurement approaches for transient contacts: i) fixed measuring device, ii) linear moveable device on a sledge and iii) device on a pendulum. For a reproducible comparison, we first compared them on an analytical and an experimental perspective. Furthermore, we addressed the specific requirements of cobot applications within flexible working systems. Finally, we analyzed and interpreted the results to derive recommendations for the selection of the measurement setup of the transient contact.

Keywords: Collaborative robots (cobots) · Human-robot collision · Robot safety · Power and force limiting · Biomechanical measurement · Transient contact

1 Introduction

Safety requirements for the most integrated type of human-robot interaction, called collaboration, are usually realized by limiting power and forces of the respective machine, the robot (a.k.a collaborative robot or cobot). In the power and force limiting (PFL) mode, the robot application is regulated in such a way, that in case of a collision the impact force and pressure may not extend biomechanical limits. The ISO/TS 15066:2016 defines these limits for 29 regions of the human body for two different contact scenarios: a) quasi-static contact (squeezing or clamping of body region) and b) transient contact

A. Müller and M. Brandstötter (Eds.): RAAD 2022, MMS 120, pp. 547–555, 2022.
https://doi.org/10.1007/978-3-031-04870-8_64

(free impact from which the person can move back). The limits for quasi-static contact result from studies of pain sensitivity and have been doubled for the transient contact [1].

The verification whether a human-robot application complies with the biomechanical limits has to be carried after a risk assessment, which is necessary for the conformity procedure [2]. According to the Machinery Directive 2006/42/EC [3] a risk assessment is required not only for the Cobot but the entire application. As defined in the ISO 12100:2010 [4] the assessment should identify all hazards and determine actions. In case of a collaborative application, measurements of the impact forces and pressures for possible collision points can be used as verification of the effectiveness of the actions taken to comply to the corresponding biomechanical limits [5].

There exist some approaches for robot collision tests with crash test dummies [6, 7], however, power and force measuring devices (PFMDs) which illustrate the body region as a mass spring-damper model have been widely established and are recommended according to the recent draft of the ISO 10218-2:2020 [8] and an information sheet of the DGUV (Deutsche Gesetzliche Unfallversicherung) [9]. Figure 1a illustrates a conventional PFMD, that consists of a movable platen (A) supported by a spring (B), which the robot lowers during an impact and an integrated force sensor (C) detects the value of the collision force. On the platen, a damping element (D) and a pressure foil (E) are placed to generate a pressure profile during the impact. The spring simulates the mechanical properties of the human body and the damping element the behavior of the tissue. The corresponding spring stiffnesses are specified in the ISO/TS 15066:2016 [1] and damping materials for the body regions in the information sheet of the DGUV [9].

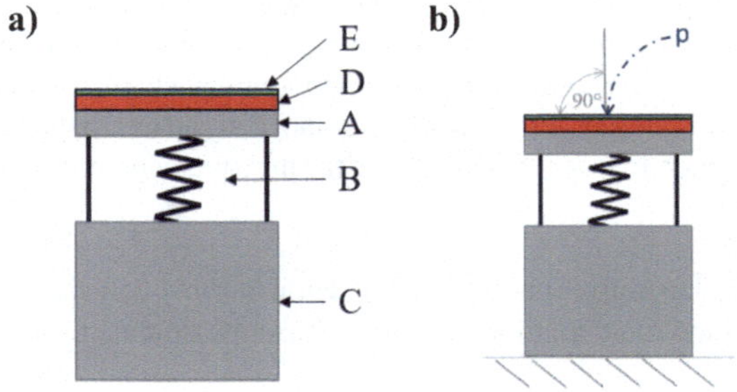

Fig. 1. Schematic sketch of **a)** power and force measuring device (PFMD) and **b)** measurement setup for quasi-static contact, with the robot path p.

For the measurement of a quasi-static contact, the measuring device is mounted on one side, e.g. the table, and the robot is moved at an angle of 90° towards it (see Fig. 1b) [10]. For a transient contact, however, the setup and the performance of the measurement is more difficult, as the person can move back during the impact. Although this simplification and its consequences are known, no standardized regulation takes it into account. Therefore, we have carried out an investigation of different measurement setups for transient contacts in a human-robot collision.

First, we describe how the measurement of the transient contact is represented in different guidelines and the way it has already been measured in Sect. 2. In Sect. 3 we investigate three different measurement setups, analytically and in an experimental setup to compare the resulting impact forces. Furthermore, the effects of the measurements for flexible work systems are described in Sect. 4. The discussion of our results can then be found in Sect. 5, and our conclusion and outlook in Sect. 6.

2 Measuring Approaches of a Transient Contact

A relevant description for the measurement of a human-robot collision with a PFMD was published by the DGUV in 2017 [9]. According to this document the measuring device shall be fixed for the transient contact, which leads to a similar shape of the curve of the measured contact force or pressure as for the quasi-static contact: An increase from zero to a maximum value at the time of the first contact, which then decreases and finally settles to a constant value. The maximum value is the relevant one, which has to be considered for the transient contact and only lowers to a constant value due to the fixed measuring device. In addition, the constant force or pressure is the one to be considered for the quasi-static contact [9]. Measurements carried out this way can be found in the work from Valori et al. [11] and Hofbaur and Rathmair [12].

However, the draft of the ISO 10218-2:2020, Annex N of which will replace the ISO/TS 15066:2016, describes that for the transient contact, the PFMD should be freely moveable in the contact direction to compensate the effective mass of the affected body region. This can be realized by mounting the measuring device on a sledge with low friction and additional weights [8]. The corresponding American technical report, ANSI RIA TR R15.806:2018 [13] prescribes the same procedure. Even before these two standards, a prototype of a moveable measuring device, the NIST PFL was presented 2012 by Falco et al. [14]. Another collision test bed setup for a human-robot collision is described by Matthias et al. [15], where a sledge equipped with a force-torque sensor and a spring, similar to a PFMD, is mounted on linear guides and able to move for tests of the transient contact force. Additionally, high speed detector foils, fixed on the sensor, are used to detect the pressure during the contact.

A different approach is presented by Schneider et al. [16], where a PFMD is fixed on a pendulum and guarantees a free oscillation to realistically simulate the recoil of the human during impact. The pendulum was also adjustable in weight to regulate the mass properties of the effected body region.

According to the described state of the art, the following three types of measurement for transient impacts have become established: i) fixed measuring device, ii) linear moveable device on a sledge and iii) device on a pendulum.

3 Investigation of the Different Types of Measurement

In this section, the three established measuring setup types will be investigated and compared, in terms of which is the best to be applied in practice.

3.1 Analytical Investigation

First an analytical investigation with simplified mechanical models was done to compare the impact forces of the three types (see Fig. 2). In each case the robot collides with its effective mass m_R and relative velocity $v_R = v_{rel}$ against the human body part with the effective mass of m_H. We assumed that the human is not moving, therefore its velocity v_H is always zero. In case of the fixed device i), the velocity of the human body part after the collision $v_H{}'$ is zero as well, and the ones of the robot is $v_R{}'$. Case ii) represents an inelastic impact, both masses are moving on together with v'. In case of the pendulum with the length l, m_H has a velocity of $v' = \omega' l$, ω' is the angular velocity.

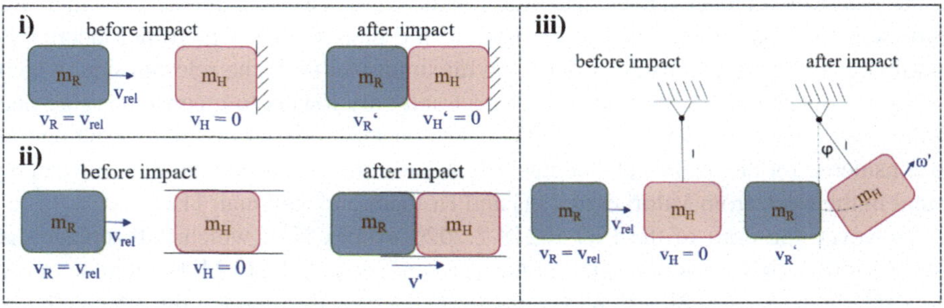

Fig. 2. Mechanical models for **i)** fixed device, **ii)** linear moveable device and **iii)** device on a pendulum.

For all three types, we assumed that the energy difference ΔE during the impact, is completely transferred to the affected body region and can therefore be equated with the potential energy of the spring V_F, which is defined in Eq. (1). F is the impact force, k the effective stiffness, and x the lowering of the spring.

$$\Delta \mathrm{E} = V_F = \frac{1}{2} k\, x^2 = \frac{F^2}{2k} \tag{1}$$

The contact forces of each of the three measurement types derive from defining the energy balance of each of the three systems and can be seen in Eqs. (2), (3) and (4). The impact number e ranges from zero to one for case i) and iii), depending on the robot's material and the effected body region, and is zero for the inelastic impact ii).

$$F(i) = v_{rel}\sqrt{k\ m_R(1+e)} \tag{2}$$

The contact force of Eq. (3) can be found in the ISO/TS 15066:2016 [1]. This document also defines the reduced mass μ according Eq. (5), and an equation for m_R.

$$F(ii) = v_{rel}\sqrt{k\mu} \tag{3}$$

In iii) the human's kinetic energy has a rotatory component, which includes i_s, the radius of inertia. The term with gravity g results from the potential energy difference.

$$F(iii) = \sqrt{k[v_{rel}^2 \frac{\mu(1+e)}{(m_H + m_R)}[m_H(1-e) + m_R(2 - (\frac{i_s}{l})^2 (1+e))] - 2m_H gl(1-\cos\varphi)]} \quad (4)$$

$$\mu = \left(\frac{1}{m_H} + \frac{1}{m_R}\right)^{-1} \quad (5)$$

A comparison of Eqs. (2), (3) and (4) results that the fixed device i) derives a bigger contact force than the other measurement setups, see Eqs. (6) and (7).

$$F(i) > F(ii) \quad (6)$$

$$F(i) > F(iii) \quad (7)$$

3.2 Experimental Collision Test

For the experimental collision test we used the Universal Robot UR3e and the GTE CBSF-35 PFMD. The measuring device was calibrated beforehand by the manufacturer. A robot movement was teached to simulate a transient collision with the upper arm, of a 176 cm tall person and we measured the impact force for: i) fixed measuring device, ii) linear moveable device and iii) device on pendulum. Since impact force and pressure are physically related, an investigation of the force was sufficient for our analysis. The ISO/TS 15066:2016 defines a spring constant of 30 N/mm for the upper arm, limited by our measuring equipment, we used 35 N/mm, as suggested by GTE. According to DGUV [9] we utilized a damping element with a shore hardness of 30. We used an effective mass of 3 kg for the upper arm, as defined in [1].

Measurement Setup. For each scenario we mounted the PFMD on the effective mass, which was fixed on the table for i), linear moveable on rails for ii), as defined in [8] and suspended by ropes on a gallows in case iii), see Fig. 3. We marked the positions of the effective mass and the collision point on the table, in order to always achieve the same.

For each case we assumed that the human is not moving. We programmed a linear robot movement, that generates an impact 90° against the PFMD. According to the collaborative safety mechanism of the robot, it stopped shortly after the collision was detected, and the PFMD recorded the transmitted impact force. The friction of the sledge in case ii) can be neglected, as it requires a maximum force of only 2 N to set it in motion. In case of the pendulum iii) the measuring head was held manually after the collision, in order not to measure another impact.

Measurement Results. In order to achieve significant measuring results, we performed ten measuring tests for each setup. According to GTE, the PFMD has a max. measuring uncertainty of ±3%, which is reflected in the results, except for one value of case ii). This may be caused by further uncertainties of the measuring setup, such as the friction of the sledge. Table 1, we received the highest impact forces for i) with an average value of 200.5 N. This also corresponds with Eqs. (6) and (7) of Sect. 3.1. The values for ii) with

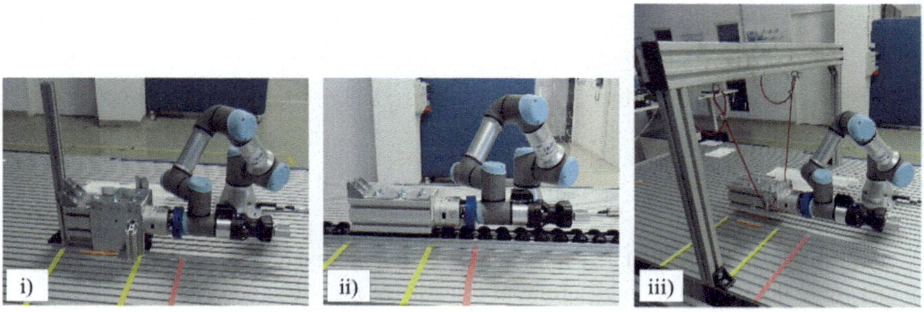

Fig. 3. Measurement setup of **i)** fixed device, **ii)** linear moveable device and **iii)** device on a pendulum.

an average of 100.4 N, and iii) with an average of 100.3 N, are quite similar and about the half of the ones of i). However, this only relates to the used robot and its generated movement. In order to receive a meaningful relation between the impact force of i) and ii) or iii), experimental tests with different cobots, as they recognize the impact with different accuracies, and several types of movements are required. The same applies for the relation between the forces of the linear moveable device and the pendulum. As the deflection of the pendulum is so small, the movement of the measuring head is almost linear and derives quite the same results as ii).

A comparison with the model-based estimations of Sect. 3.1 showed 12.4% and 34.3% higher calculated impact forces ($F_c(i) = 225.63$ N, $F_c(ii) = 134.84$ N) than the measured ones. The reason for this deviations are the simplified mechanical models, especially the calculation of effective mass m_R according to ISO/TS 15066:2016 [1] is too inaccurate, which was already proven by Kirschner et al. [17].

Table 1. Results of the measurement experiment.

Measurement type	Measurement (Impact force in Newton)									
	1	2	3	4	5	6	7	8	9	10
i) Fixed device	197	203	202	201	202	201	199	199	200	201
ii) Linear moveable device	101	102	100	95	101	101	101	99	102	102
iii) Device on pendulum	101	99	102	99	100	101	101	100	100	100

4 Impact on Flexible Work Systems and Their Reconfiguration

The measurements only represent one part of the whole risk assessment by estimating the severity of harm for corresponding risks, while the foundation of the risk assessment is represented by the determination of the limits of machinery. Therefore, a modification in the machine's use, space or time definition may result in new identified hazards and the need of new power and force measurement setups.

This challenge becomes especially bearing in the present time, as work systems are often not anymore planned for a lifetime use without change. Flexible work systems communicating with each other to combine a suitable production outcome are getting widely discussed and established within the fourth industrial revolution [18]. These systems will reconfigure in different ways to archive the requirements given. A distinction can be made between planned and new work system reconfigurations. The planned reconfigurations can already be planned during the first conformity process (e.g., if the development of two different products is already planned and completely defined on the same production line). Thereby the dangers for these reconfiguration possibilities are already identified and suitably minimized. On the contrary, however, new reconfigurations can occur that were not planned in the conformity process before. For example, if a production facility switches its outcome briefly to much more needed respiratory masks due to a pandemic. Here, new unidentified hazards can occur and new power and force measurement setups will be needed. This creates a tremendous conformity checking effort for any planned new reconfiguration. Especially since one can imagine that the PMFD cannot always be set up optimally in the corresponding pinch point and considerable preliminary set up work is necessary.

5 Recommendation of Measuring Types Based on Our Results

In order to give a recommendation of the measurement setup to use, more tests with several cobots and movement are required. However, based on our results it can be said, that if a measurement with a fixed device complies the biomechanical limits, the application is safe according to the ISO/TS 15066:2016. Furthermore, it is the easiest to realize, especially for applications where slides or pendulums are difficult to implement. Therefore, we would recommend to use the fixed device. This is often the case with flexible systems. If the limit values are not complied with a fixed device, it can be checked whether they are at least met with a movable device. The choice between slide and pendulum depends on movement, application and installation space. No general recommendation can be done here as to which setup is better. However, it should be mentioned, that the linear moveable device and the pendulum do not represent the real human movement, since in both cases the PFMD is not freely moveable and therefore the forces are limited in directions or length of the pendulum.

6 Conclusion and Outlook

This paper investigated the measurement of the transient contact in human-robot interaction and compared three different measurement setups: i) fixed measuring device, ii) linear moveable device on a sledge and iii) device on a pendulum. Our experimental collision test derived nearly equal impact forces for ii) and iii) and the double values for i). However, this only applies to the used robot and it's programmed movement. At the moment, the choice of the measurement setup depends on the robot's movement, application and installation space. With the fixed device you are always on the safe side, as also its analytical investigation derives the highest impact forces.

However, no matter which type of setup is used, the measurement itself, as well as the identification of possible collision points beforehand, is a very time consuming and expensive process, which requires physical laboratory setups. Therefore, further research within the #SafeSecLab doctoral project [19] deals with this problem, in order to create a model for the automatic check of biomechanical limits. The results of this paper showed that the mathematical models of Sect. 3.1 needs further development, like calculating m_R according to [17], in order to receive meaningful values that can be compared to the measurement results. For this validation further measuring experiments are required and more data has to be collected to calculate m_R and the contact force of case iii.

Acknowledgments. This research paper is part of the #SafeSecLab funded by TÜV AUSTRIA and was also supported by the BMK endowed professorship HCCPPAS (No.852789).

References

1. ISO/TS 15066:2016: Robots and robotic devices. Collaborative robots (2017)
2. DIN EN ISO 10218-2:2011. Robots and robotic devices – Safety requirements for industrial robots – Part 2: Robot systems and integration (2012)
3. DIRECTIVE 2006/42/EC: of the European Parliament and of the Council of the European Union. Machinery Directive (2006)
4. DIN EN ISO 12100:2010: Safety of machinery – General principles for design – Risk assessment and risk reduction (2011)
5. Bauer, W., Bender, M., Braun, M., Rally, P., Scholtz, O.: Leichtbauroboter in der manuellen Montage. IRB Mediendienstleistungen, Stuttgart (2016)
6. Oberer-Treitz, S.: Abschätzung der Kollisionsfolgen von Robotern zur Bewertung des sicheren Einsatzes in der Mensch-Roboter-Kooperation. Fraunhofer Verlag (2018)
7. Haddadin, S., Albu-Schaeffer, A., Hirzinger, G.: Safety Evaluation of Physical Human-Robot Interaction via Crash-Testing. In: Robotics: Science and systems (2007)
8. prEN ISO 10218-2:2020: Robotics – Safety requirements for robot systems in an industrial environment – Part 2: Robot systems, robot applications and robot cells integration (2021)
9. DGUV-Information FB HM 080: Collaborative robot systems (2017)
10. Kirschner, R.J., Mansfeld, N., Abdolshah, S., Haddadin, S.: Experimental Analysis of Impact Forces in Constrained Collisions According to ISO/TS 15066. IEEE ISR (2021)
11. Valori, M., et al.: Validating safety in human-robot collaboration: standards and new perspectives. Robotics **10**, 65 (2021)
12. Hofbaur, M., Rathmair, M.: Physische Sicherheit in der Mensch-Roboter Kollaboration. e & i Elektrotechnik und Informationstechnik (2019)
13. ANSI RIA TR R15.806-2018: Technical Report for Industrial Robots and Robot Systems - Technical Report for Industrial Robots and Robot Systems - Testing Methods for Power & Force Limited Collaborative Applications (2018)
14. Falco, J., Marvel, J., Norcross, R.: Collaborative robotics: measuring blunt force impacts on humans. Chest **140**, 45 (2012)
15. Matthias, B., Oberer-Treitz, S., Ding, H.: Experimental Characterization of Collaborative Robot Collisions. In: IRS Robotik (2014)
16. Schneider, C., Seizmeir, M., Suchanek, T., Hutter-Mironovová, M., Bdiwi, M., Putz, M.: Empirical Analysis of the Impact of Additional Padding on the Collaborative Robot Velocity Behavior in Transient Contact Cases. SCITEPRESS (2021)

17. Kirschner, R.J., Mansfeld, N., Pena, G.G., Abdolshah, S., Haddadin, S.: Notion on the Correct Use of the Robot Effective Mass in the Safety Context and Comments on ISO/TS 15066. IEEE ISR (2021)
18. Vogel-Heuser, B., Bauernhansl, T., Hompel, M.T.: Handbuch Industrie 4.0 Bd. 2, 2nd edn. Springer Reference Technik. Springer Vieweg, Heidelberg (2017). https://doi.org/10.1007/978-3-662-53248-5
19. TÜV AUSTRIA #SafeSecLab. https://safeseclab.tuwien.ac.at/

Development of a Human-Robot Interface for a Safe and Intuitive Telecontrol of Collaborative Robots in Industrial Applications

Andrea Raviola(✉), Alberto Coccia, Andrea De Martin, Antonio Carlo Bertolino, Stefano Mauro, and Massimo Sorli

Department of Mechanical and Aerospace Engineering, Politecnico di Torino, Corso Duca degli Abruzzi 24, 10129 Turin, Italy
andrea.raviola@polito.it

Abstract. Teleoperated robots have been used to substitute the human presence in dangerous applications and environments. The present paper, on the contrary, describes a user interface aimed to intuitively support the operator controlling a collaborative robot in a shared workspace. Such a result is achieved by commanding the robot arm according to a reference frame integral with the user. This approach allows maintaining the same control strategy no matter the relative orientation of the operator with respect to the manipulator and reducing the mental effort required for the teleoperation. The research campaign has been carried out using a UR5 collaborative robot from Universal Robots controlled through a user-friendly interface and an off-the-shelf gaming joypad. A virtual environment, which replicates the experimental setup, has also been designed to act as a training platform for the user who intends to telecontrol a robot arm. In addition, to assure the safety of the operator sharing his workspace with the robot, possible malfunctions of the teleoperation have been properly analyzed and addressed.

Keywords: Collaborative robotics · Industrial robots · Teleoperation · Telerobotics · Human-robot collaboration · Robot control · Industry 4.0

1 Introduction

Telecontrolled robots were created to substitute human operators in hazardous tasks and environments such as nuclear power plants [1] and mines [2]. According to [3], when the levels of uncertainty and complexity are too high to be autonomously handled by a machine, like in medical applications [4], remotely controlled robots should be preferred over autonomous ones. Here haptic interfaces, like the one described in [5], have also been developed to provide a more immersive and reliable telecontrol experience. Recent studies, like [6], also adopted force feedback to enhance new teleoperation algorithms based on impedance control. Other types of telecontrol approaches are based on making a manipulator replicate the operator's movements detected by one or more cameras [7].

A. Müller and M. Brandstötter (Eds.): RAAD 2022, MMS 120, pp. 556–563, 2022.
https://doi.org/10.1007/978-3-031-04870-8_65

However, even though this solution has been proved to be effective in pick-and-place or grasping applications, it might not be sufficiently accurate in the ones where a higher precision of the robot movements is required, as in [8].

Within this framework, the present research provides an intuitive [9] and flexible user interface to make teleoperation an added value for human-robot collaboration in industrial applications. The novelty of this work relies on the fact that, even though the user changes his relative orientation with respect to the robot, the control logic remains the same. So, since the robot movements are always intuitively defined according to the user, the mental effort required to properly command the robot, together with the risk of human error, is highly reduced. This is achieved by setting a reference frame, integral with the operator who controls the robot, according to which all the commands are given to the machine. A similar approach is used in [10], where, to achieve an intuitive control of an assistive robot, a new reference frame is continuously determined as a function of the orientation of the robot Tool Center Point (TCP). However, here the operator is always aligned with the robot base reference frame, an assumption that is not acceptable for a dynamic environment like the industrial one. A further added value is represented by the virtual copy of the experimental environment that can be used not only by neophytes as a first approach to telerobotics, but also to evaluate the feasibility of specific applications without any need of a physical setup.

In addition, to assure a safe and reliable human-robot cooperation, possible malfunctions in the teleoperation have been analyzed and taken into account.

2 System Description

The proposed control algorithm, developed in LabVIEW, has been tested with a reconfigurable test bench designed for maintenance and testing of different components. The manipulator adopted in the research campaign is a UR5 collaborative robot (cobot) from Universal Robots controlled through a common gaming joypad. In addition, a simulated environment, containing a digital twin of the robot, has also been developed. Besides being used as a training platform by the user, such a simulated environment allows reducing the number of trials required for debugging, testing, and commands calibration.

2.1 Experimental Setup

Figure 1 shows an overview of the test bench used for the proposed study and a schematic representation of the interactions among the main elements of the teleoperation. The UR5 robot, equipped with an FT300 force-torque sensor from Robotiq, is mounted on a movable cart together with its control unit and teach pendant. The ability of the manipulator to replicate different tasks, like the ones reported in [8], is guaranteed by several tools stored inside a modular warehouse mounted on the work cart.

The external computer, on which the proposed control interface runs, acts as the Local Computer (LC) and converts the user inputs into signals understandable by the Remote Computer (RC), identified with the UR5 control unit. In addition, it provides a Graphical User Interface (GUI) on which the operator can monitor robot feedback

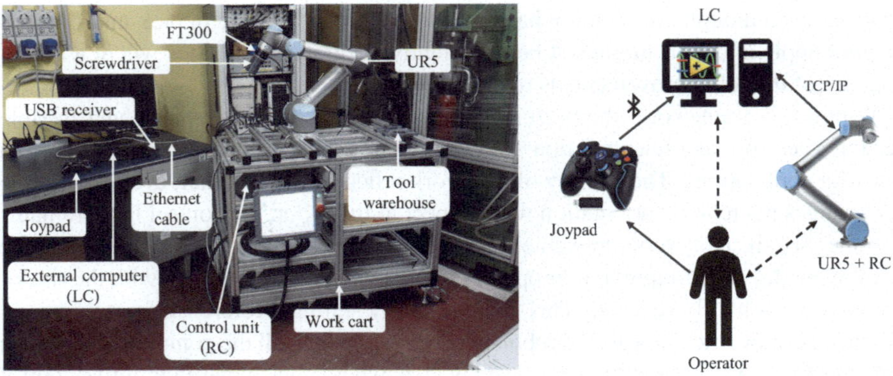

Fig. 1. (**a**) Experimental setup used to telecontrol the UR5 collaborative robot; (**b**) Schematic overview of the connections among the fundamental elements of the proposed system.

signals and potential warnings messages. To control the robot arm, the operator uses a joypad that sends signals to a USB receiver, plugged inside the LC, via a 2.4 GHz wireless connection. On the contrary, a TCP/IP connection is established between the LC and the RC via an Ethernet cable allowing commands and feedback to be exchanged via bidirectional communication. In Fig. 1(b), the dashed arrows represent the visual monitoring of the cobot and the GUI by the operator, while the solid lines stand for physical connections.

2.2 Simulated Environment and UR5 Digital Twin

The proposed interface has also been designed for neophytes in telecontrol and robotics. To allow new users to rapidly become familiar with the telecontrol algorithm, a virtual environment, depicted in Fig. 2(a), has been developed.

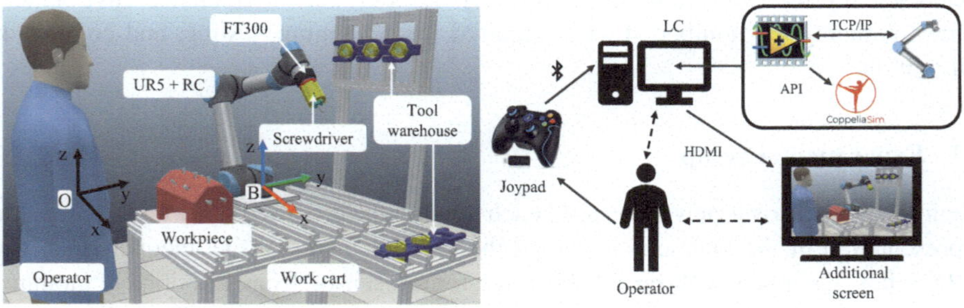

Fig. 2. (**a**) Virtual environment which reproduces the experimental setup for telecontrolling the UR5 robot; (**b**) Schematic overview of the connections among the fundamental elements of the teleoperation in the simulated setup.

As with the experimental setup, the proposed control algorithm commands the robot via a TCP/IP connection. However, here, the manipulator is a digital twin of the UR5, created using URSim v3.13.0, the official offline simulator developed by Universal Robots. By contrast, dedicated application programming interfaces have been adopted to communicate with the digital scene built using the free educational version of CoppeliaSim (release 4.0.0) by Coppelia Robotics GmbH. The main goal of such a virtual environment is to provide an immersive and realistic simulation of the teleoperation by recreating the test bench reported in Fig. 1(a), instead of using the simple and not customizable workspace of the URSim.

Moreover, since a single PC acts as both the local and the remote computer, the setup can be simplified with respect to the experimental one, while maintaining unaltered the relationships among the fundamental elements of the teleoperation. To do so, the URSim runs on a Linux Mint Rebecca 17.1 operating system installed on the Oracle VM VirtualBox v6.0.20 virtual machine. This allows simulating the TCP/IP communication between the external computer and the UR5 control unit.

3 Telecontrol of the UR5 Collaborative Robot

The intuitive telecontrol of the robot arm is achieved by commanding the manipulator with a set of signals provided according to the operator's reference frame (O) reported in Fig. 2(a) and defined as:

- X axis: from the operator's chest to his right arm;
- Y axis: from the operator's chest to the robot base joint;
- Z axis: from the operator's chest to his head.

With this approach, no matter the relative orientation between the user and the manipulator, the TCP will move according to the person's left/right (±X), back/front (±Y), and bottom/top (±Z). This is done by converting the operator's commands of the TCP linear and angular velocities (${}^{O}\boldsymbol{v}_{cmd} \in \mathbb{R}^6$), into the ones (${}^{B}\boldsymbol{v}_{cmd} \in \mathbb{R}^6$) defined with respect to the robot base reference frame (B), according to:

$$ {}^{B}\boldsymbol{v}_{cmd} = {}^{B}\boldsymbol{A}_{O} \cdot {}^{O}\boldsymbol{v}_{cmd} \tag{1} $$

where ${}^{B}\boldsymbol{A}_{O} \in \mathbb{R}^{6x6}$ is the orientation matrix built as ${}^{B}\boldsymbol{A}_{O} = \text{diag}(\boldsymbol{Rz}(\theta), \boldsymbol{Rz}(\theta))$, being $\mathbf{0} \in \mathbb{R}^{3x3}$ a null matrix, and $\boldsymbol{Rz}(\theta) \in \mathbb{R}^{3x3}$ the rotation matrix around the user's Z axis of an angle θ which represents the relative orientation between the robot and the operator. In case the two reference frames are aligned, as in Fig. 2(a), θ is set to zero, leading to ${}^{B}\boldsymbol{A}_{O} = \mathbf{I}$, being $\mathbf{I} \in \mathbb{R}^{6x6}$ the identity matrix, and ${}^{B}\boldsymbol{v}_{cmd} = {}^{O}\boldsymbol{v}_{cmd}$. Since, as depicted in Fig. 3, the information about the angle θ is manually provided by the operator through the joypad, it has been chosen to limit its possible values to five. Four of them describe the relative orientations between the user and the robot (front, left, behind, and right), while the fifth one is integral with the TCP reference frame, and it is used for precision movements such as aligning the tool with a screw. This solution represents a compromise between a flexible telecontrol of the UR5 and a high complexity of the user interface which could negatively affect the telecontrol experience [11]. However, an ongoing research campaign is proving the feasibility of adopting a camera for the identification of the operator around the robot, leading to a higher discretization of the angle θ. This approach also opens to the possibility of merging telecontrol with collision avoidance [12, 13], and hand-over [14] algorithms, that could be used to improve the safety and the intuitiveness of teleoperations.

For a precise and adjustable telecontrol of the robot arm, the analog signals of sticks and triggers of the joypad have been converted, according to a proportional law, into acceptable values of linear and angular velocities to be sent to the robot. Moreover, to prevent undesired movements of the manipulator caused by a non-ideal return of a joypad lever to the default null position, a safety threshold, under which ${}^{O}\boldsymbol{v}_{cmd} = \mathbf{0} \in \mathbb{R}^6$, has been defined. After several tests, performed by people with different backgrounds using both the simulated and the experimental setups, TCP linear and angular velocities have been limited, respectively, between 5 and 100 mm/s and 5 and 45°/s, while the leading axis acceleration [15] has been set to 1 rad/s^2.

A more extensive study, similar to the one described in [10], is ongoing. This will allow adjusting and optimizing the control strategy and to properly define the delay between the actual movements of the robot and the commands sent by the operator through the proposed interface.

3.1 Human-Robot Control Interface

Both in the experimental and the simulated setup, the operator can control the cobot through the intuitive human-robot interface developed in LabVIEW and reported in Fig. 3. Here, in addition to the real manipulator, the user constantly receives feedback on the current status of the robot and selects different features of the control algorithm to perform the desired tasks.

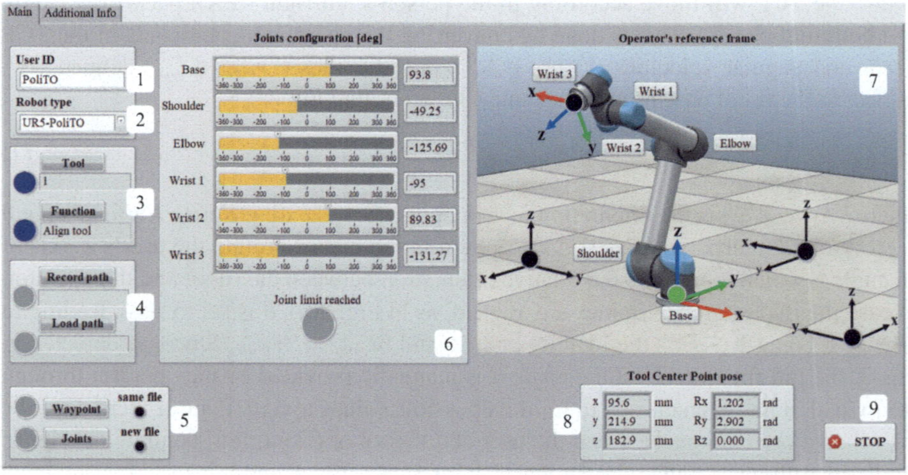

Fig. 3. Graphical user interface used to telecontrol the UR5 collaborative robot.

The proposed GUI is divided into nine main sections:

1. User ID: it is possible to specify the operator's name or identification number to keep track of who controls the robot.
2. Robot type: a drop-down menu that allows the user to choose which robot to control by selecting the static IP address of a specific manipulator.
3. Tool/function: it allows selecting different tools (i.e., wrenches and screwdrivers) or pre-programmed functions (e.g., autonomous tool change or unfastening a screw). For high and fast customization, these data are stored inside text files and, since the control interface automatically adapts itself to such changes, it is not necessary to make any changes to the LabVIEW software.
4. Record/load path: at any time, at the robot frequency of 125 Hz, it is possible to record the TCP trajectory which can then be replicated by the robot. This feature is useful in applications, like welding, in which it is easier and faster to teach the robot a desired path than programming specific and complex movements.
5. Waypoint/joints storage: when selected, the information related to a waypoint and the robot joints angular positions is stored in a text file for future usage.
6. Joints configuration: representation of the current angular positions of each joint of the robot to inform the operator of possible proximity of one or more joints to their rotation limits of $\pm 360°$, which would lead to a protective stop. To avoid it, when one or more joints exceed the virtual position limit of $\pm 355°$, the *joint mode* is automatically activated. In this scenario, all the joypad commands are disabled except for the triggers, which must be used to directly control the robot joints and restore the nominal operating conditions.
7. Operator's reference frame: this tab gives the operator constant feedback of the reference frame according to which the robot is being commanded. By default, when the telecontrol starts, the operator's reference frame is aligned with the one of the cobot base (green led in Fig. 3). When the user changes his relative orientation according to the manipulator, a different reference frame can be manually selected to preserve an intuitive control experience.
8. TCP pose: an indicator of the current pose of the tool center point.
9. STOP: indicator, set to *false* by default, that is turned to *true* when the telecontrol ends.

The present GUI should be considered as an extension of the teach pendant with which the robot is normally programmed. Together with the joypad, it can be used to interact with external devices, such as a camera that can be adopted to automatically detect the operator's orientation relative to the robot. The control algorithm could also be adapted to command, for example, an Unmanned Ground Vehicle (UGV) on which the manipulator could be mounted. In this scenario, the operator could easily select which robot to control without needing to change the control device.

3.2 Error Handling

To guarantee the operator's safety in case of malfunction, three main unexpected events have been addressed:

1. Loss of connection between the local computer and the joypad: this event can be caused by a physical disconnection of the USB receiver from the external computer (LC) or when the joypad is out of power. When this happens, the control algorithm generates an error message, and it automatically stops the robot.
2. Joypad jamming: this considers the possibility of a mechanical failure of one or more components of the control device as a consequence of jamming. Since the outputs of the joypad buttons are mainly used to select options on the GUI, they have not been labeled as safety-critical. On the contrary, a blocked stick or trigger could lead the manipulator to keep moving uncontrollably in one direction. Since this event is interpreted by the proposed control algorithm as a command given by the operator, the robot can only be stopped by pressing the STOP button on the joypad or the emergency one on the robot teach pendant.
3. Loss of TCP/IP connection: it takes place when the LabVIEW interface, the robot, or both, leave the communication channel. Since such bidirectional connection is the only way for the manipulator and an external device to interact, the cobot cannot be stopped using the joypad. To avoid undesired movements of the robot arm, the functions used to move the TCP have been defined to return after 0.2 s [15]. If the UR5 control unit does not receive any external input within such time interval, the robot autonomously stops.

4 Conclusions

The present paper describes a human-machine interface for intuitively teleoperating industrial manipulators when sharing their workspace with a human operator. The intuitiveness of the teleoperations is guaranteed by the possibility to control the robot according to the user's reference frame despite his relative orientation with respect to the manipulator. In addition, a virtual environment, able to replicate the experimental setup, has been developed for both training and testing purposes. Since the adopted software are commonly available at universities, or even entirely open source, the proposed simulated setup could be effectively used to make students come closer to robotics, leading to a reduction of the gap between the actual knowledge and the skills required from a market oriented towards Industry 4.0.

At present, several people with different backgrounds, both within and outside robotics, provided positive feedback on the ease of use of the proposed control interface. Nevertheless, a larger study, which will allow measuring proper metrics representative of the intuitiveness of the teleoperation both in the real and the simulated environment, is currently ongoing.

Future work will be devoted to make the proposed interface automatically adapt in controlling not only well-known industrial and collaborative robots, but also innovative structures as the one proposed in [16].

References

1. Nagatani, K., et al.: Emergency response to the nuclear accident at the Fukushima Daiichi nuclear power plants using mobile rescue robots. J. Field Robot. **33**(1), 1–17 (2014)

2. Szrek, J., Wodecki, J., Błazej, R., Zimroz, R.: An inspection robot for belt conveyor maintenance in underground mine-infrared thermography for overheated idlers detection. Appl. Sci. **10**(14), 1–17 (2020)
3. Parker, L.E., Draper, J.V.: Robotics Applications in Maintenance and Repair. Handbook of Industrial Robotics. Wiley, New York (1999)
4. Avgousti, S., et al.: Medical telerobotic systems: current status and future trends. Biomed. Eng. Online **15**(1), 1–44 (2016)
5. Song, G., Guo, S.: A novel force feedback-based teleoperation system for medical application. Int. J. Innov. Comput. Inf. Control **3**(3), 737–750 (2007)
6. Laghi, M., Ajoudani, A., Catalano, M., Bicchi, A.: Tele-impedance with force feedback under communication time delay. In: IEEE International Conference on Intelligent Robots and Systems (IROS), vol. 2017, pp. 2564–2571, Vancouver, BC, Canada, September 2017
7. Kofman, J., Wu, X., Luu, T., Verma, S.: Teleoperation of a robot manipulator using a vision-based human-robot interface. IEEE Trans. Ind. Electron. **52**(5), 1206–1219 (2005)
8. Raviola, A., Antonacci, M., Marino, F., Jacazio, G., Sorli, M., Wende, G.: Collaborative robotics: enhance maintenance procedures on primary flight control servo-actuators. Appl. Sci. **11**(11), 4929 (2021)
9. Naumann, A., et al.: Intuitive use of user interfaces: defining a vague concept. In: Harris, D. (ed.) EPCE 2007. LNCS (LNAI), vol. 4562, pp. 128–136. Springer, Heidelberg (2007). https://doi.org/10.1007/978-3-540-73331-7_14
10. Campeau-Lecours, A., Cote-Allard, U., Vu, D.S., Routhier, F., Gosselin, B., Gosselin, C.: Intuitive adaptive orientation control for enhanced human-robot interaction. IEEE Trans. Robot. **35**(2), 509–520 (2019)
11. De Barros, P.G., Lindeman, R.W.: A Survey of User Interfaces for Robot Teleoperation, Worcester, Massachussets, USA (2008)
12. Mauro, S., Scimmi, L.S., Pastorelli, S.: Collision avoidance system for collaborative robotics. In: Ferraresi, C., Quaglia, G. (eds.) RAAD 2017. MMS, vol. 49, pp. 344–352. Springer, Cham (2018). https://doi.org/10.1007/978-3-319-61276-8_38
13. Melchiorre, M., Scimmi, L.S., Pastorelli, S.P., Mauro, S.: Collison avoidance using point cloud data fusion from multiple depth sensors: a practical approach. In: 23rd International Conference on Mechatronics Technology (ICMT 2019), Salerno, Italy (2019)
14. Melchiorre, M., Scimmi, L.S., Mauro, S., Pastorelli, S.P.: Vision-based control architecture for human–robot hand-over applications. Asian J. Control **23**(1), 1–13 (2020)
15. Universal Robots: The URScript Programming Language. Universal Robots A/S, Odense, Denmark (2019)
16. Palmieri, P., Gaidano, M., Ruggeri, A., Salamina, L.: An inflatable robotic assistant for onboard applications. In: 72nd International Astronautical Congress (IAC 2021), Dubai, UAE (2021)

Assessment of the Human-Robot Collaborative Polishing Task by Using EMG Sensors and 3D Pose Estimation

Milos Petrovic[1], Arso M. Vukicevic[2], Branko Lukic[1], and Kosta Jovanovic[1](✉)

[1] School of Electrical Engineering, University of Belgrade, Belgrade, Serbia
{petrovic.milos,kostaj}@etf.bg.ac.rs

[2] Faculty of Engineering, University of Kragujevac, Kragujevac, Serbia

Abstract. Safe human-robot collaboration, especially considering ergonomics, still represents a challenging problem. The objective of this paper is to develop the procedure for assessing human-robot collaboration with the help of computer vision system and electromyography (EMG) data. The conducted study involved three participants performing the polishing task in four different configurations interacting with Franka Emika's Panda Robot. The robot is controlled in the impedance mode to achieve safe human-robot interaction and, at the same time, to achieve desired robot impedance in the translational and the rotational axis. In order to study human kinematics, the experiment was recorded using four IP cameras, while muscle activity was measured by Trigno Avanti Duo sensors by Delsys to account for human dynamics in the contact task. The collected EMG signals were processed using a bandpass filter, a notch filter, envelope calculation, and normalization. The reconstruction of a human pose from the collected videos was performed using the VIBE algorithm, while body pose parameters (characteristic postural angles) were computed from the output SMPL parametric human model. The obtained results showed significant variations among different body configurations, as verified by the trend of the EMG signals. The proposed approach demonstrates the potential to be an effective tool for enhancing ergonomics in an industrial environment.

Keywords: Human-robot collaboration · Pose estimation · Computer vision · Ergonomics · Artificial intelligence · Machine learning

1 Introduction

According to the report from the International Federation of Robotics, the use of robotic technology records a constant increase with almost 400.000 installed devices worldwide [1], whereas an increase of collaborative robots share is expected to reach almost 30% of all installed robots compared to meager 3.24% in 2018. Thus, it is expected that in the near future human-robot collaboration (HRC) will become a standard way of performing tasks in many industries, including manufacturing. The latest generation of collaborative robots has been developed to work alongside humans and can be fully integrated

A. Müller and M. Brandstötter (Eds.): RAAD 2022, MMS 120, pp. 564–570, 2022.
https://doi.org/10.1007/978-3-031-04870-8_66

into workbenches, creating efficient hybrid workplaces. Contemporary collaborative robots still have rigid connections between their actuators and joints/links while safety is increased by sensors - collisions can be promptly and accurately identified, and the robot stopped to prevent damage. Recommendations for collaborative robots are summarized in the technical specification ISO/TS 15066 [2]. Besides automation and increase of productivity, the HRC has a high potential to reduce the physical load of workers and thus the occurrence of work-related musculoskeletal disorders (WMSDs). The first step in providing a robust collaboration of robots with human operators is enabling them to perceive human actions. However, the practice has shown that there are a series of challenges in terms of planning and managing HRC, as it is not trivial to predict or monitor human behavior. Therefore, monitoring and ensuring the safety and ergonomics of operator actions during the HRC has gained a lot of attention. In literature, this problem has been studied by utilizing wearables (commonly electromyography) and a motion capture system, with the aim to analyze or optimize ergonomics [3]. Apart from detailed human model and overloading joint torques, compressive forces and fatigue index are considered to identify physical risk factors [4, 5]. Although the motion capture system provides accurate results in laboratory conditions, the price and the complexity of use limit its applicability in industrial environments. Accordingly, there is a need for studying and further development of procedures based on using conventional devices and cameras. The human pose estimation could be considered as a 2D and 3D problem. Even though 2D pose estimation appears to be more stable and frequently used in literature, 3D pose estimation is more suited for advancing HRC. This study is focused on 3D pose estimation from monocular videos [6–9]. Particularly, the assumption was that by using deep learning algorithms, which deliver full 3D human body shape from conventional cameras, there is a possibility to extract information related to both human actions and ergonomics - which are key information needed for improving the efficiency and safety of HRC.

Accordingly, in this study, we present an experimental setup and the concept for the assessment of the polishing ergonomics by using Deep learning 3D pose estimation and wearables for measuring muscle activities. The rest of the paper is organized as follows. Section 2 gives an overview of the proposed method and the experimental setup. In Sect. 3, the obtained results are discussed. Lastly, the article ends with some concluding remarks and future work prospective in Sect. 4.

2 Materials and Methods for Human Motion Acquisition

The graphical illustration of the proposed concept for human ergonomics monitoring and improvement in HRC is given in Fig. 1. As the aim to develop a data-driven decision support system, the first step consists of collecting the measurements of the human movement and of the forces the operator exchanges with the external environment. Accordingly, the laboratory study included three participants (age: 29.7 $\pm$ 4.7 years, mass: 80.1 $\pm$ 10.2 kg, height: 1.77 $\pm$ 0.1 m) performing the polishing task with the support of a collaborative robot. We aimed to investigate the trend of human muscular activity and its correlation with body posture configurations and acting force during the collaborative polishing task. The subjects were provided with a graphic interface showing the intensity of the force exerted on the board, which is the subject of the polishing, enabling a relatively uniform procedure suitable for comparison. Each subject was instructed to adopt four different body postures and perform the polishing task in each experimental configuration for 2 min exerting a constant force of 10 N and 20 N, respectively. The data regarding the human whole-body motion, the force exerted on the working piece as well as the human muscle activity were recorded throughout all the experimental sessions. Moreover, human arm manipulability was taken into account. The manipulability measure, defined as $w = \sqrt{det(J(q)J(q)^T)}$ [10] is the kinematic parameter that reflects the capacity to change the end effector's position as a function of the joint configuration. It is a quality measure for redundant manipulators, which describes the distance to singular configurations based on analyzing the manipulability ellipsoid that is spanned by the singular vectors of the Jacobian.

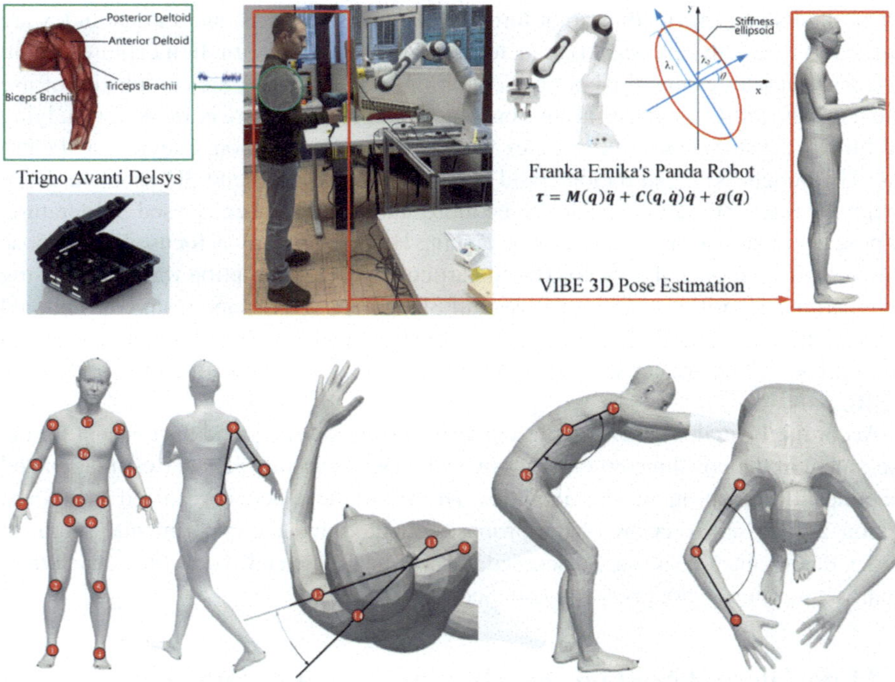

Fig. 1. The concept of the proposed solution and its components: wearable sensors setup for human motion and muscle effort tracking; polishing task – an experiment setup; representation of robot impedance for contact tasks; SMPL 3D model of the human body

2.1 EMG Measurements

For the analysis of human muscle activity in the four predefined configurations, four electromyography (EMG) surface electrodes (Trigno Avanti Duo sensors by Delsys) were placed on the subjects' skin. For the task, dominant shoulder and elbow muscles were selected: Biceps Brachii (BB), Triceps Brachii (TB), Anterior Deltoid (AD), Posterior Deltoid (PD) (Fig. 1). Trigno Avanti™ sensors support a low noise, high fidelity sensing circuit for detecting EMG signals from the surface of the skin when muscles contract. The Trigno EMG System is supported by EMGworks Acquisition software for standardized and simple data acquisition and EMGworks Analysis software for analysis of the record signals. Before starting any experiment, the maximal voluntary contraction (MVC) for each muscle was extracted according to the protocol for the MVC test positions [11]. The EMG signals were processed using a bandpass filter, a notch filter, envelope calculation, and normalized to the MVC to account for muscular activities following the methodology described in [12]. It is important to note that the processed EMG signals were used to provide additional means of verifying the results of this study.

2.2 Execution of Robot Behavior

The task of the robot was to bring the board to the human, whereas the task of the human was to polish it using a 1.2 kg polisher. The board was provided with a force torque sensor to measure the interaction force between the tool and the working piece. During each experimental session, the board was placed in four different positions and with different orientations in the workspace. In order to achieve ergonomic and adaptive interaction between the human co-worker and robot, the robot end effector stiffness was set to 1500 N/m in the all translational axis and 100 Nm/rad in the all rotational axis, providing a reasonable trade-off between the end effector compliance and the trajectory tracking performance (Fig. 1) [13].

2.3 Visual Monitoring and Analysis of Subjects Motion

In this study, we used four DAHUA IPC-HFW2831TP-ZS 8 MP WDR IR Bullet IP Cameras, with DAHUA PFS3010-8ET-96 8 port Fast Ethernet PoE switch. Host PC had CPU 1151 INTEL Core i3-8100 3.6 GHz 6 MB BOX. The 3D pose reconstructions were obtained by using the VIBE deep learning architecture [14]. Briefly, it considers pose reconstruction in an adversarial manner, which assumes that it estimates 3D human actions from previously detected 2D landmark points - so that the discriminator part cannot spot the difference between the obtained motion and motions(s) learned from the AMASS database that contains ~1000 movements of ~300 subjects [15]. The obtained poses were represented with the Skinned Multi-Person Linear (SMPL) parametric body model (composed of 6890 Quad4 elements). Following the previous study, we extracted a series of parameters that are considered ergonomic indicators (Fig. 1) [16].

3 Results and Discussion

Four different task configurations adopted by a human co-worker during the collaborative task are illustrated in Fig. 2.

The experimental results are shown in Fig. 3, where the normalized muscle activity is reported for each configuration averaged across subjects for the exerted force of 10 N and 20 N, respectively.

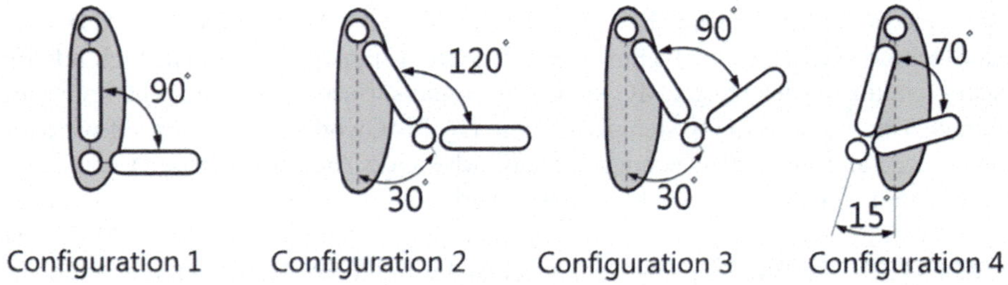

Fig. 2. Four different arm configurations for polishing task adopted by human co-worker

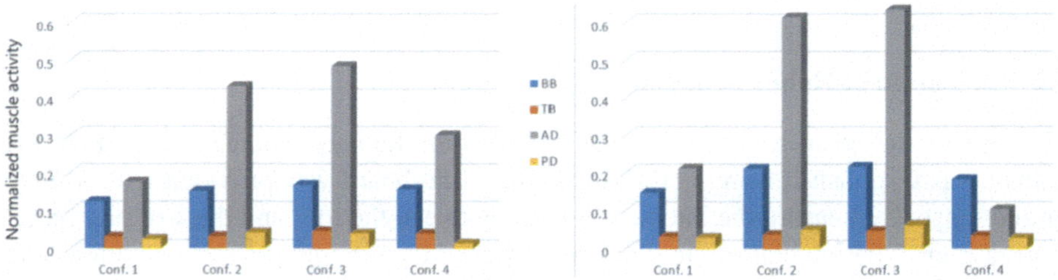

Fig. 3. Normalized average muscle activity (Biceps Brachii – BB, Triceps Brachii – TB, Anterior Deltoid – AD, Posterior Deltoid - TD) across the subjects for each configuration, exerted force 10 N (left) and 20 N (right)

From the average muscle activity in different configurations, we can observe that configurations 1 and 4 have overall lower muscle activity than configurations 2 and 3. Furthermore, for this collaborative task, dominant muscles are the Anterior Deltoid and Biceps Brachii. The activity of Triceps Brachii and Posterior Deltoid is almost negligible. With the increase of exerted force, from 10 N to 20 N, a significant increase, especially in Anterior Deltoid, and to some extent in Biceps Brachii muscle activity is observed. Compared to other configurations, configuration 2 has a significantly lower manipulability capacity of the arm, which could affect task production.

Furthermore, the contribution of this study is the creation of a dataset that contains motions with labels relevant for ergonomics studies for collaborative robotics. Aside from its unique activities and labels, the dataset contains EMG signals (Fig. 4) in order to understand and analyze human dynamic parameters, as well as videos for future human pose estimation and creation of the robot control strategies.

The potential of successful pose differentiation by the computer vision system (Fig. 5) opens the possibility to perform pose classification. Postural angles - elbow and armpit angle are defined on the SMPL model with keypoints 7-8-9 and 8-9-13, respectively (Fig. 1). Pose classification can lead to adaptation of the robot behavior to assist and guide the human partner to perform the collaborative task efficiently, with less effort, and in a more ergonomic way.

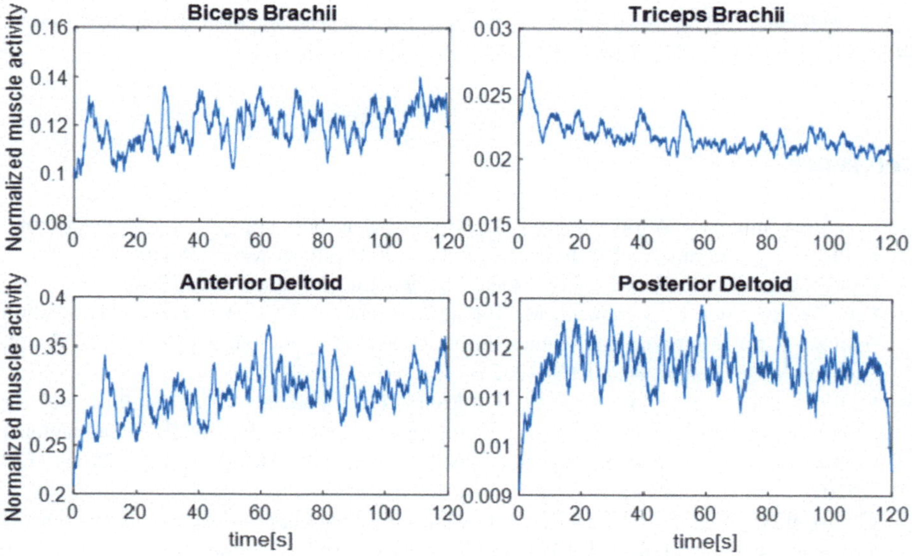

Fig. 4. Normalized muscle activity, configuration 1, exerted force 20 N

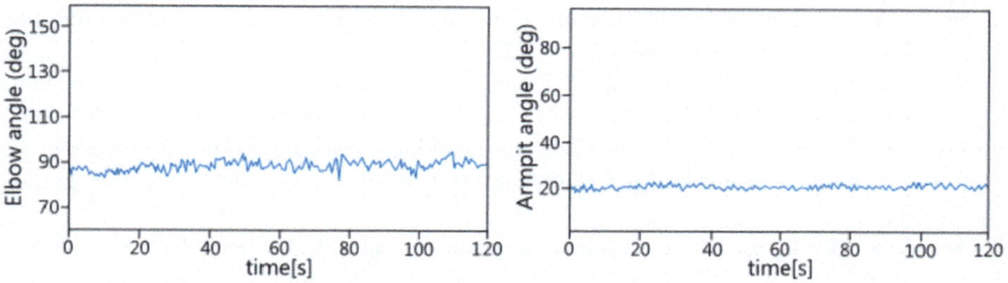

Fig. 5. Estimated postural angles by the computer vision system, configuration 1, exerted force 20 N

4 Conclusion

In this paper, we introduced an approach to improve human-robot collaboration while carrying out collaborative tasks in the industrial environment. The proposed method has twofold potential: to be an effective tool to enhance ergonomics during complex dynamic interactions between human and robot, and to enable the replacement of the worker in the laborious tasks by collaborative robot able to achieve the worker performance. Some significant differences were found among different body configurations, as verified by the trend of the EMG signals. Our final goal is to create a real-time comprehensive computer vision system for the monitoring and assessment of human ergonomics able to account for different risk factors in the workplace providing a collaborative robot with sufficient information giving it the ability to adapt and lower that risk.

Acknowledgement. This research was supported by the Science Fund of the Republic of Serbia, PROMIS, #6062528, ForNextCobot and #6524219, AI4WorkplaceSafety.

References

1. Executive Summary World Robotics 2021 Industrial Robots https://ifr.org/img/worldrobotics/Executive_Summary_WR_Industrial_Robots_2021.pdf. Accessed 26 Jan 2022
2. ISO/TS 15066 https://www.iso.org/standard/62996.html. Accessed 26 Jan 2022
3. Kim, W., Peternel, L., Lorenzini, M., Babic, J., Ajoudani A.: Human-robot collaboration framework for improving ergonomics during dexterous operation of power tools. Robot. Comput. Integr. Manuf. **68**, 102084 (2021)
4. Fortini, L., Lorenzini, M., Kim, W., De Momi, E., Ajoudani, A.: A real-time tool for human ergonomics assessment based on joint compressive forces. In: 29th IEEE International Conference on Robot and Human Interactive Communication (RO-MAN), pp. 1164--1170 (2020)
5. Lorenzini, M., Kim, W., Ajoudani, A.: An online multi-index approach to human ergonomics assessment in the workplace. IEEE Trans. Hum. Mach. Syst. 1–12 (2022)
6. Bogo, F., Kanazawa, A., Lassner, C., Gehler, P., Romero, J., Black M.J.: Keep it SMPL: automatic estimation of 3D human pose and shape from a single image. In: European Conference on Computer Vision (2016)
7. Huang, Y., et al.: Towards accurate marker-less human shape and pose estimation over time. In: International Conference on 3DVision (2017)
8. Kocabas, M., Karagoz, S., Akbas, E.: Selfsupervised learning of 3D human pose using multi-view geometry. In: IEEE Conference on Computer Vision and Pattern Recognition (2019)
9. Pavlakos, G., Zhu, L., Zhou, X., Daniilidis, K.: Learning to estimate 3D human pose and shape from a single color image. In: IEEE Conference on Computer Vision and Pattern Recognition (2018)
10. Yoshikawa, T.: Manipulability of robotic mechanisms, Int. J. Robot. Res. **4**(2), 3–9 (1985)
11. Konrad, P. The ABC of EMG: A Practical Introduction to Kinesiological Electromyography, pp. 30--35, Noraxon Inc, (2005)
12. Radmilovic, M., Urukalo, D., Petrovic, M., Becanovic, F., Jovanovic, K.: Influence of muscle co-contraction indicators for different task conditions, In: ICEtran (2021)
13. Lukić, B., Jovanović, K., Knežević, N., Žlajpah, L., Petrič, T.: Maximizing the end-effector Cartesian stiffness range for kinematic redundant robot with compliance. In: Zeghloul, S., Laribi, M.A., Sandoval Arevalo, J.S. (eds.) RAAD 2020. MMS, vol. 84, pp. 208–217. Springer, Cham (2020). https://doi.org/10.1007/978-3-030-48989-2_23
14. Kocabas, M., Athanasiou, N., Black, M.J.: Vibe: video inference for human body pose and shape estimation. In: Proceedings of the IEEE/CVF Conference on Computer Vision and Pattern Recognition, pp. 5253–5263 (2020)
15. Loper, M., Mahmood, N., Romero, J., Pons-Moll, G., Black, M.J.: SMPL: a skinned multi-person linear model. ACM Trans. Graph. **34**(6), 1–16 (2015)
16. Vukicevic, A.M., Macuzic, I., Mijailovic, N., Peulic, A., Radovic, M.: Assessment of the handcart pushing and pulling safety by using deep learning 3D pose estimation and IoT force sensors. Exp. Syst. Appl. **183**, 115371 (2021)

Application of Model Predictive Control in Physical Human-Machine Interaction

Maria Paterna, Daniel Pacheco Quiñones, Carlo De Benedictis(✉), Daniela Maffiodo, Walter Franco, and Carlo Ferraresi

Department of Mechanical and Aerospace Engineering, Politecnico di Turin, Turin, Italy
carlo.debenedictis@polito.it

Abstract. In this paper, an application of Model Predictive Control (MPC) to physical human-machine interaction is presented. In particular, the study focuses on the development of a mechatronic device able to apply a well-controlled mechanical impulsive force on the human body, for clinical investigation of postural control. The need for high accuracy and repeatability led to the MPC design, which is able to manage the non-linearities related to the human-machine interaction. The hardware architecture design of the prototype, the development of the control system (based on motor current saturation) and its optimization are presented. The results of experimental trials carried out in the laboratory and on healthy subjects show that the MPC algorithm is able to provide the accuracy and robustness requested by the application.

Keywords: Model Predictive Control · Linear electric actuator · Human-machine interaction · Force control · Speed control · Automated posturography

1 Introduction

Human-machine interaction (HMi) has always been a fundamental part of robotics design and mechatronics in general aiming to avoid dangerous scenarios which could potentially involve serious human injuries and device damage. The principles of Industry 4.0 are starting to spread in industrial and service robotics. HMi has begun to be the target of human-machine cooperation-oriented design and involves many new application domains, such as smart industry [1], smart-home design [2], bioengineering [3, 4], or driver assistance systems [5].

As a function of a suitable hazard and risk assessment analysis, HMi control design and testing play a crucial role in system definition. Modern control techniques involving adaptive [6] or robust [7] strategies are generally preferred over simpler but less performing design solutions. To this extent, Model Predictive Control (MPC) [8–10] poses a viable design option if the system is totally or partially known a priori, such in grey-box models describing the general range of possible behaviors of the HMi. As the control algorithm can predict in an optimized way the interaction behavior thanks to an input model of the system, derived MPC architectures are indeed a pertinent solution

A. Müller and M. Brandstötter (Eds.): RAAD 2022, MMS 120, pp. 571–579, 2022.
https://doi.org/10.1007/978-3-031-04870-8_67

to overcome uncertainty and disturbance factors involving test environment, versatility, and fallibility.

Previous work from the authors [11] showed how the MPC strategy could manage complex HMi interfacing issues on a hand-held mechatronic device named automatic perturbator, designed to provide, through a linear electric actuator, a controlled pushing force to the upper body of a patient in clinical postural analyses. In such environment, the contact force must be properly controlled both in terms of amplitude, to avoid overshoot potentially dangerous for the patient, and duration, to minimize postural adaptation. Based on preliminary experiments [12, 13] the force impulse (i.e., the time integral of the force signal) of the applied perturbation should be comprised between 2 Ns and 10 Ns, while the duration of the stimulus should be less than 300 ms, or even less than the latency of the postural reflex (about 75–100 ms) for specific analyses. Pacheco Quiñones et al. [11] implemented a mixed speed-force control, that switched between the two references depending on the detection of the contact and on the desired motion pattern of the rod. Although the dynamics of the system was considered appropriate for the preliminary testing application, the switching behavior negatively affected the MPC strategy, due to high nonlinearities involved, and resulted in a loss of tracking accuracy.

This paper presents a renewed design based on MPC for the control of impulsive mechanical interactions directed to the human body. The hardware architecture of an automatic perturbation device is briefly presented, then the control system and its optimization are carried out by performing experimental analyses with a prototype of the perturbator. Finally, the device flexibility and robustness are also evaluated.

2 Model Predictive Control Design

To fulfill the project specifications, the automatic perturbator (AP) is conceived to be operated in a hand-held arrangement as shown in Fig. 1 and Fig. 2.

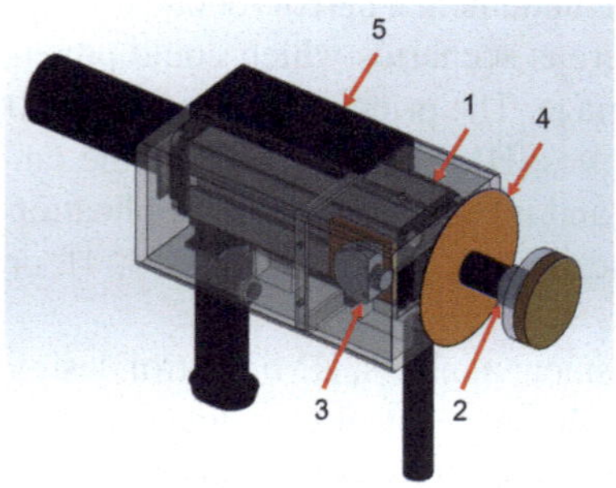

Fig. 1. CAD model of the automatic perturbator

Table 1. List of automatic perturbator components.

N°	Component
1	GD160Q motor (NiLAB GmbH)
2	UMM 50 kgf-ranged load cell (Dacell Co. Ltd.)
3	Q4XTULAF400-Q8 optical sensor (Banner Engineering Corp.)
4	Optical sensor target disk
5	3D-printed chassis

Referring to Fig. 1 and Table 1, a 3D-printed chassis (5) surrounds a linear electric motor (1), provided with an embedded encoder and controlled through a SLVD1N driver (Parker Hannifin Corp.) and a Baseline Real-Time target machine (Speedgoat Inc.). A load cell (2) senses the impulsive force during contact, while an optical sensor (3) monitors the piston stroke through the aid of a 3D-printed target disk (4) and manages limit-switch control. Simulink® environment (The MathWorks Inc.) is used to design the control architecture of the system and to interact with the driver's software, MotionWiz (Parker Hannifin Corp.). The device's control logic is based on the Finite State Machine (FSM) design, involving the following fundamental phases:

1. idle: in which the actuator's rod is retracted, still, and ready for operation;
2. approach: button-triggered, the rod approaches the target body (Fig. 2*a*);
3. strike: the rod imparts on the target body the predefined impulsive force. The switch to this phase only occurs if the load cell output exceeds 3 N (Fig. 2*b*);
4. retraction: the rod pulls back from the target (Fig. 2*c*).

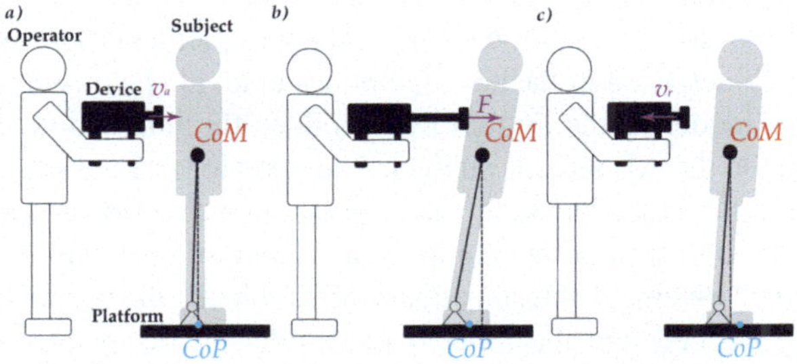

Fig. 2. Perturbator-subject interaction during approach (a), strike (b), and retraction (c) phases. CoM is the subject's center of mass, CoP is the center of pressure on the platform. In magenta, the control input signals: F is the impulsive force applied on the subject's body; v_a and v_r are, respectively, the piston approach and retraction velocities of the piston.

The operating mode of the driver's software, shown in Fig. 3, features the following elements:

- a closed-loop speed control system, based on a PI controller, working with the main driver's analog input as speed reference and using the motor's embedded encoder for feedback;
- a force reference generator, working as an alternative to the speed controller with the same main analog input, translating the voltage signal into force;
- a pico-PLC code, responsible to control the driver's software architecture during online operation;
- a current (i.e., force) saturation block, imposing the actuation current to be subjected to a saturation value equal to the minimum value among the following: peak current during normal operation; nominal current during i^2t protection; a custom constant value; an auxiliary analog input signal.

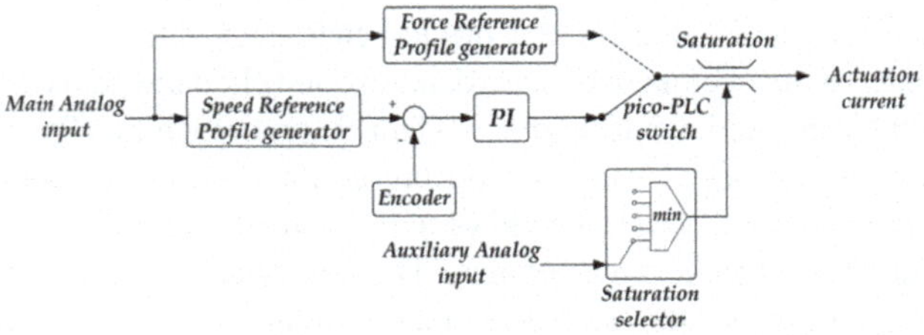

Fig. 3. Scheme of the driver's operating mode.

The presented control system focuses on the auxiliary analog input entering the saturation block (Fig. 3) to control the contact force entity. The actuator is continuously motion-controlled with a speed reference acting as the main analog input of the driver, while the auxiliary input is used to saturate the output force to achieve the desired behavior. During phases 1, 2, and 4, the force saturation is set high enough to not interfere with the inertia of the piston while, in phase 3, the current saturation is modulated by the MPC action, and the speed reference is appropriately increased to achieve the desired saturated force as fast as possible. While phases 2 and 4 can take advantage of the PI closed-loop speed control, the control of force saturation during impact implements a derived architecture of the linear MPC block used in Pacheco Quiñones et al. [11], considering a linearized lumped-parameter model of the system to reproduce the interaction shown in Fig. 2 between the operator, the AP, and the subject. The operator and the subject are modeled as lumped masses which translate in the antero-posterior direction and are connected to the environment through springs and dampers; a viscoelastic behavior is also associated to the mechanical AP-subject interface. The MPC block, thanks to the force signal provided by the load cell and the a-priori knowledge of the plant coming from the model, is able to compute an optimized step-by-step solution, predicting the system's behavior and adjusting the auxiliary input accordingly. The MPC bases its optimization on the minimization of the cost function J described in Eq. (1):

$$J(U(k|k)) = \sum_{i=0}^{H_p-1} e_y^T(k+i|k) Q e_y(k+i|k) + e_u^T(k+i|k) R_u e_u(k+i|k) + du^T(k+i|k) R_{du} du(k+i|k) + \rho_\varepsilon \varepsilon_k^2 \qquad (1)$$

in which: "i|j" means that the prediction of the ith time step is computed during the jth time step ($j < i$); H_p is the prediction horizon, i.e., the number of future control steps that the MPC controller must evaluate by prediction; e_y, e_u are the predicted error of the output and input variables, respectively, and du is the control input rate; while Q, R_u, R_{du} are their respective weight matrices. Finally, $\rho_\varepsilon {\varepsilon_k}^2$ is the constraint violation cost function. U(k|k) is the input macro vector, further details are given in Pacheco Quiñones et al. [11]. The force constraints of the actuation system depend on the application, i.e., the MPC controller behavior is saturated to $u_{max} = 52$ N (maximum force allowable), during a tunable Saturation time interval (Sdt) happening at the beginning of phase 3.

3 Experimental Trials and Control Logic Optimization

Experimental tests are designed to optimize the performance of the AP control logic. The analysis focuses on the MPC cost function, the operator stiffness (k_a) within the model, and the Sdt parameter, that are adjusted through a trial-and-error optimization process. In these tests, a rectangular pulse with an amplitude of 40 N and a duration of 250 ms is selected as the reference force signal, and the AP is handled by one operator and used to hit a rigid, fixed target. For each parameter configuration, 6 distinct stimuli are delivered.

To evaluate the device's ability to follow the desired force profile, the Tracking Accuracy Error (TAE) (Eq. (2)) and the force impulse deviation (FID) (Eq. (3)) are calculated over the time interval in which the contact force is higher than 3 N:

$$\mathrm{TAE} = 100 \cdot \frac{\int_{\Delta t} |\text{measured force} - \text{reference force}|}{\text{reference impulse value}}, \quad \Delta t = \text{contact time interval} \tag{2}$$

$$\mathrm{FID} = 100 \cdot \frac{\text{measured impulse value} - \text{reference impulse value}}{\text{reference impulse value}} \tag{3}$$

The force profiles obtained by varying R_u, Q, H_p, k_a, and Sdt are shown in Fig. 4 and Table 2, while R_{du} is kept constant and equal to 10. As R_u increases, the contact force oscillates around a lower mean value; decreasing Q causes a decrease in the amplitude and frequency of the force oscillations; despite the controller based its computation on a simplified model, increasing its power of prediction H_p from 10 to 20 steps allowed it to obtain better performance. For even higher Hp values, the system fails to calculate the optimal solution in real-time due to the high computational cost. Increasing k_a reduces ringing and allows for more stable force profiles. Sdt higher than 0 is necessary to compensate for the nonlinearities involved in the impact phase, which are not modeled in the plant due to the unknown deformation of the patient's body tissues. Sdt should be properly set and adapted to the magnitude of the reference force value to reduce the initial overshoot by maintaining a limited rising time (17.5 ± 5.3 ms). The final configuration of the parameters (*k* profile) highlights good performance: the contact force is almost constant and close to the reference throughout the considered time interval. Moreover, FID and TAE are less than 5% and 15%, respectively.

The AP flexibility and robustness are evaluated by providing perturbations at different magnitudes (20, 30, 40, 50 N) in two series, carried out by two distinct operators, on a fixed target and on a healthy young subject, respectively. The results presented in Fig. 5 and Table 3 show that the tracking accuracy is not affected either by the mechanical impedance of the subject or by that of the operator. Only slight differences in fact are noticeable among the force profiles applied on a rigid, fixed target (Fig. 5, left-top) and on a human (Fig. 5, left-bottom), regardless of the force amplitude. This result supports the robustness of the device and of the control logic detailed in the current work. Moreover, as highlighted in Fig. 5, on the right, the produced force profiles show a reasonable degree of repeatability despite the variability introduced by the interaction between the AP and both the operator and the subject. The repeatability is confirmed by a FI coefficient of variation equal to 2.51%, on average.

Table 2. Control parameters tuning. Mean ± standard deviation are given for force impulse (FI), force impulse deviation (FID) and tracking accuracy error (TAE). Data refer to the curves shown in Fig. 4, that have been labeled accordingly.

Profile	R_u	Q	H_p	k_a (N/m)	Sdt (ms)	FI (Ns)	FID (%)	TAE (%)
a	0.2	5	10	1600	30	9.29 ± 0.44	−7.11 ± 4.44	22.3 ± 2.66
b	0.5	5	10	1600	30	8.80 ± 0.45	−12.0 ± 4.52	21.8 ± 1.40
c	1	5	10	1600	30	7.71 ± 0.41	−22.9 ± 4.12	28.4 ± 3.81
d	0.2	2	10	1600	30	9.52 ± 0.49	−4.76 ± 4.89	18.2 ± 1.04
e	0.2	3	10	1600	30	9.61 ± 0.27	−3.91 ± 2.68	17.6 ± 1.40
f	0.2	3	15	1600	30	10.5 ± 0.92	5.06 ± 9.19	20.9 ± 4.27
g	0.2	3	20	1600	30	10.4 ± 0.18	3.60 ± 1.76	17.7 ± 1.58
h	0.2	3	20	3000	30	11.5 ± 0.49	14.7 ± 4.91	21.9 ± 4.21
i	0.2	3	20	15000	30	10.7 ± 0.44	6.98 ± 4.45	18.5 ± 1.80
j	0.2	3	20	15000	0	9.43 ± 0.50	−5.72 ± 4.97	14.5 ± 2.77
k	0.2	3	20	15000	15	10.4 ± 0.56	3.68 ± 5.61	13.6 ± 2.30

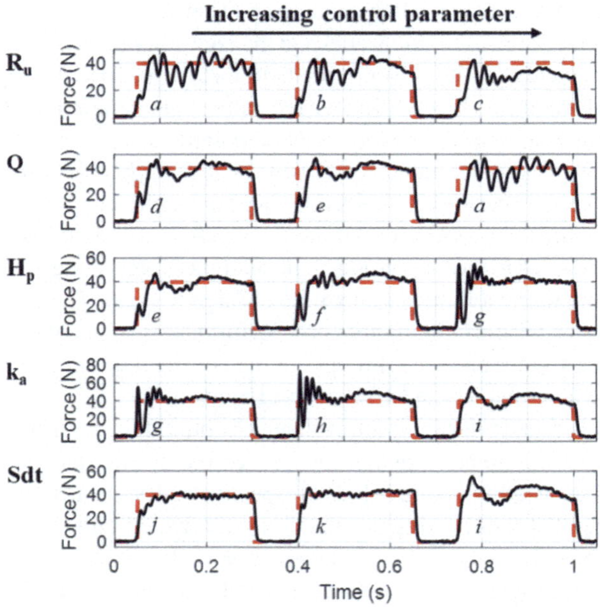

Fig. 4. Force tracking of the model predictive control algorithm for different value of R_u (control input rate), Q (tracking error), H_p (tracking error), k_a (simulated stiffness operator) and Sdt (saturation time interval). Each row of the plot refers to a specific control parameter, that has been increased from left to right side. The red dashed lines represent the reference force signal, while the averaged contact force signals are shown in black and are labeled as reported in Table 2.

In comparison with the previous software architecture outlined in Pacheco Quiñones et al. [11], or other analyzed devices, such as in Paterna et al. [13], the proposed control algorithm configuration shows substantial dynamic performance improvements. TAE, overshoot, and rise time of the force profiles obtained in Paterna et al. [13], in Pacheco Quinones et al. [11], and in the current work are respectively: [21.4; 24.9; 15.7]%, [44.5; 21.8; 22.6]% and [23.4; 25.4; 18.0] ms.

Table 3. Automatic perturbator performance for different force amplitudes and targets. Mean ± standard deviation are given for force impulse (FI), force impulse deviation (FID) and tracking accuracy error (TAE). Data refer to the corresponding curves shown in Fig. 5.

Force (N)	Target	Sdt (ms)	FI (Ns)	FID (%)	TAE (%)
20	Fixed	0	5.49 ± 0.38	9.81 ± 7.57	17.8 ± 3.73
30	Fixed	0	7.34 ± 0.35	−2.08 ± 4.67	13.9 ± 2.09
40	Fixed	15	10.4 ± 0.56	3.68 ± 5.61	13.6 ± 2.30
50	Fixed	20	12.8 ± 0.28	2.36 ± 2.28	11.5 ± 1.73
20	Subject	0	4.94 ± 0.18	−1.20 ± 3.69	17.5 ± 1.65
30	Subject	0	6.90 ± 0.24	−7.94 ± 3.22	18.1 ± 2.22
40	Subject	15	9.63 ± 0.08	−3.66 ± 0.82	15.7 ± 0.63
50	Subject	30	11.5 ± 0.22	−7.61 ± 1.80	15.1 ± 2.45

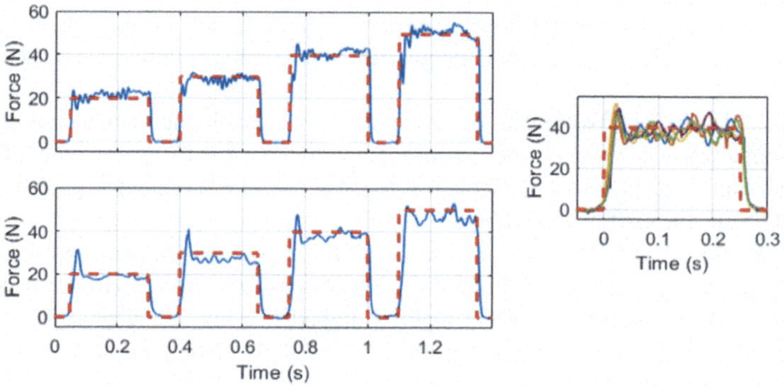

Fig. 5. Perturbation applied on a fixed target (top, left) and on a healthy subject (bottom, left) by two different operators. The red dashed lines represent the reference force profile, while the actual contact force, averaged over 6 successive stimuli, is in blue. On the right, six consecutive force signals obtained on the subject for the 40 N amplitude are shown.

4 Conclusion

The effectiveness of MPC algorithm in the control of impulsive mechanical interactions occurring between an electromechanical device and the human body has been demonstrated. The architecture of the device has been outlined and its accuracy, robustness and flexibility have been proved.

The automatic perturbator can be successfully employed in clinical dynamic posturography analysis, in which an unpredictable, short-lasting, controlled perturbation must be generated to challenge the patient's balance. However, further testing is needed to assess the robustness of the control system in a wider scenario, considering different subjects and operators.

References

1. Ronzoni, M., Accorsi, R., Botti, L., Manzini, R.: A support-design framework for cooperative robots systems in labor-intensive manufacturing processes. J. Manuf. Syst. **61**, 646–657 (2021). https://doi.org/10.1016/j.jmsy.2021.10.008
2. Park, K., Lee, H., Kim, Y., Bien, Z.Z.: A steward robot for human-frindly human-machine interaction in a smart house environment. IEEE Trans. Autom. Sci. Eng. **5**, 21–25 (2008). https://doi.org/10.1109/TASE.2007.911674
3. Kyrarini, M., et al.: A survey of robots in healthcare. Technologies **9**, 8 (2021). https://doi.org/10.3390/technologies9010008
4. Guang, H., Ji, L., Shi, Y., Misgeld, B.J.E.: Dynamic modeling and interactive performance of ParM: a parallel upper-limb rehabilitation robot using impedance control for patients after stroke. J. Healthc. Eng. **2018**, 8647591 (2018). https://doi.org/10.1155/2018/8647591
5. Clement, P., et al.: Enhancing acceptance and trust in automated driving through virtual experience on a driving simulator. Energies **15**(3), 781 (2022). https://doi.org/10.3390/en15030781
6. Bower, C., Taheri, H., Wolbrecht, E.: Adaptive control with state-dependent modeling of patient impairment for robotic movement therapy. In: 2013 IEEE International Conference on Rehabilitation Robotics (ICORR), pp. 1–6. IEEE, Seattle, Washington (2013). https://doi.org/10.1109/ICORR.2013.6650460
7. Krüger, J., Surdilovic, D.: Robust control of force-coupled human-robot-interaction in assembly processes. CIRP Ann. **57**, 41–44 (2008). https://doi.org/10.1016/j.cirp.2008.03.005
8. dos Santos, W.M., Siqueira, A.A.G.: Optimal impedance via model predictive control for robot-aided rehabilitation. Control Eng. Pract. 93, 104177 (2019). https://doi.org/10.1016/j.conengprac.2019.104177
9. Erickson, Z., Clever, H.M., Turk, G., Liu, C.K., Kemp, C.C.: Deep haptic model predictive control for robot-assisted dressing. In: 2018 IEEE International Conference on Robotics and Automation (ICRA), pp. 4437–4444. IEEE, Brisbane (2018). https://doi.org/10.1109/ICRA.2018.8460656
10. Teramae, T., Noda, T., Morimoto, J.: EMG-based model predictive control for physical human-robot interaction: application for assist-as-needed control. IEEE Robot. Autom. Lett. **3**, 210–217 (2018). https://doi.org/10.1109/LRA.2017.2737478
11. Pacheco Quiñones, D., Paterna, M., De Benedictis, C.: Automatic electromechanical perturbator for postural control analysis based on model predictive control. Appl. Sci. **11**, 4090 (2021). https://doi.org/10.3390/app11094090

12. Dvir, Z., et al.: Linearity and repeatability of postural responses in relation to peak force and impulse of manually delivered perturbations: a preliminary study. Eur. J. Appl. Physiol. **120**(6), 1319–1330 (2020). https://doi.org/10.1007/s00421-020-04364-y
13. Paterna, M., et al.: Center of pressure displacement due to graded controlled perturbations to the trunk in standing subjects: the force–impulse paradigm. Eur. J. Appl. Physiol. **122**, 425–435 (2022). https://doi.org/10.1007/s00421-021-04844-9

Evaluation of FMCW Radar for Potential Use in SSM

Peter Nimac[1,2(✉)], Tadej Petrič[1], Andrej Krpič[3], and Andrej Gams[1]

[1] Jožef Stefan Institute, Jamova cesta 39, 1000 Ljubljana, Slovenia
[2] Jožef Stefan International Postgraduate School, Jamova cesta 39, 1000 Ljubljana, Slovenia
peter.nimac@ijs.si
[3] Smart Com d.o.o., Brnčičeva ulica 45, 1231 Ljubljana-Črnuče, Slovenia

Abstract. Speed and separation monitoring (SSM), as one of the possible modes of human-robot collaboration operations, relies on appropriate sensors for proximity detection. While cameras and laser scanners suffer from reduced visual conditions and occlusions and may be unwelcome due to privacy concerns, radars do not have these problems. Furthermore, radars can offer direct target velocity estimation in order to comply with safety standards. However, in general, radars offer reduced spatial resolution. In this paper, we evaluate the applicability of an off-the-shelf Frequency-Modulated Continuous-Wave (FMCW) radar for SSM from the perspective of the accuracy of detection and tracking of people. The paper presents the FMCW radar and tracking system and a comparison of measurements of an off-the-shelf FMCW radar using a basic tracking algorithm with a highly accurate motion capture system. Results show relatively high accuracy of the radar in the radial direction but less accurate tracking in the tangential direction. Careful placement of the radar is thus crucial for use in SSM.

Keywords: Speed and separation monitoring · FMCW radar · Collaborative robot

1 Introduction

Collaborative robotics, where humans and robots work without explicit separation by fences, has gained attention, driven primarily by the demand for increased flexibility in production environments [10,15]. Standard ISO 10218 [8], and specifically ISO/TS 15066 [9], define four possible modes of operation: Safety-Rated Monitored Stop (SRMS), Hand Guiding, Speed and Separation Monitoring (SSM) and Power and Force Limiting (PFL). This paper focuses on the possible use of a radar sensor for SSM.

Supported by young researcher grant (PR-11324), program group Automation, robotics, and biocybernetics (P2-0076) by the Slovenian Research Agency.

A. Müller and M. Brandstötter (Eds.): RAAD 2022, MMS 120, pp. 580–588, 2022.
https://doi.org/10.1007/978-3-031-04870-8_68

In Speed and Separation Monitoring, as the name implies, the proximity of the robot is constantly observed, and the robot's behaviour is modified with respect to the distance, speed and direction of motion of a person [9]. If the human is far, the robot works at normal operating speed. If the human gets closer, the robot slows down, and finally, if the human gets too close, the robot initiates a protective stop. The distances to the robot are defined in the standards [9]. However, these can even change based on the direction and speed of the human dynamically, leading to dynamic SSM [1]. The monitoring of the robot's proximity and the tracking in collaborative tasks requires appropriate sensors. Cameras [22] and laser sensors [11,14] are typically applied [5]. As described in [21], both laser trackers and camera sensors suffer from light conditions, dirt, staining, visual occlusions and reflections. One issue is also the invasion of privacy and constant monitoring possible with the use of cameras [17]. Lately, radars have been proposed for SSM because they do not have these problems [21,26]. However, their biggest drawback is spatial resolution. Human tracking and even identifying has been shown with radars in [25] and [13]. In [26], the authors propose the use of a K-band FMCW radar for machine guarding and have declared it a possible candidate for SSM applications based on range and angular resolution and reaction time. Reaction time, determined by the update rate, has been shown to have a great effect on the determination of the separation distance [15]. In [21] it is shown how a two radars system can distinguish between a person and a mobile robot approaching the work cell and act differently in each case.

In this paper, we present and experimentally evaluate an of-the-shelf IWR6843AOP-EVM Frequency-Modulated Continuous-Wave (FMCW) radar module and GTRACK tracking algorithm based on clustering of point-cloud data, provided and applicable with out-of-the-box firmware, both offered by Texas Instruments. We used the FMCW radar to measure the location and velocity of a person and compare the measurements with an accurate motion capture (MoCap) system. Results show that tracking with the used FMCW radar is not equally accurate in all directions, meaning that proper radar placement can affect its usefulness for SSM. Figure 1 depicts the experimental setup.

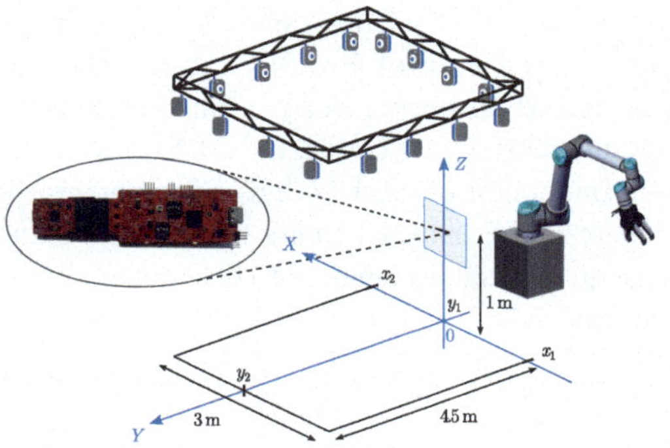

Fig. 1. Experimental setup for FMCW radar evaluation. The rectangle on the ground represents the radar's observation area. The same area was covered with a MoCap system. IWR6843AOPEVM FMCW radar module is in the magnified area.

2 FMCW Radar for Worker Detection and Tracking

There exist multiple approaches in radar design, e.g., Continuous-Wave Doppler radar (CW radar), pulse-Doppler radar, or the Frequency-Modulated Continuous-Wave (FMCW) radar. The advantage of using an FMCW as opposed to CW radar or pulse-Doppler radar is that it can determine both: the target's distance from the radar and its movement velocity in a radial direction, directly towards or away from the radar. FMCW radar also has the advantages to detect targets on very short distances r (minimal range is comparable to the average wavelength of the transmitted signal), as well as high accuracy in range estimation [24]. We have chosen the FMCW radar for SSM.

Similar use of FMCW radar for person detection and tracking has been demonstrated by Rizik et al. in [18], where they used a 24 GHz FMCW radar for security gate monitoring. They used the radar in conjunction with a Raspberry Pi computer for data acquisition, which was then forwarded to a remote personal computer (PC) for detection, recognition and target tracking. In this paper, we propose using a 60 GHz FMCW radar track workers near a robot cell to enable dynamic SSM with signal processing on a remote PC.

2.1 FMCW Radar

The basic operating principle of a generic FMCW radar is as follows. First, the frequency synthesizer generates a chirp signal TX(t) (TX stands for transmitted signal). Reflections of this signal are picked up by a receiver (RX) RX(t) $= x(t + \tau)$ with a delay τ which directly correlates to the target range r [12]. In the case of FMCW radars, the delay is not measured directly, but rather it is determined from an intermediate frequency IF(t), which results from frequency mixing of the signals TX(t) and its own delayed copy RX($t + \tau$). In Fig. 2 the height of intermediate frequency is directly correlated to the time delay τ, where more distant targets with longer time delays produce higher intermediate frequencies.The height of the intermediate frequency f_{IF} is described by the equation

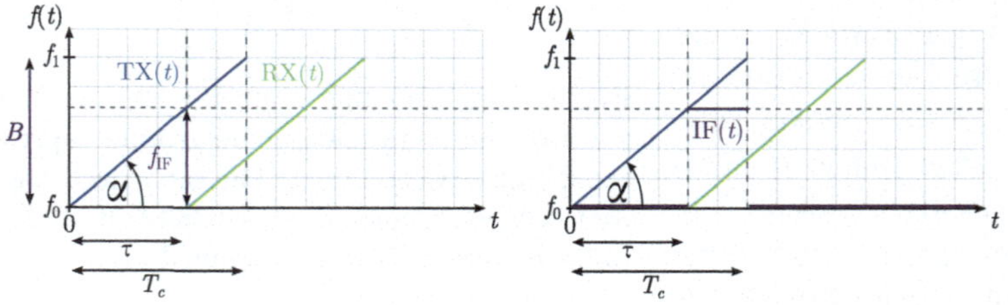

Fig. 2. The intermediate frequency is the difference of frequencies of signals TX(t) and RX(t). T_c marks the duration of a chirp, f_0 is the starting frequency, and f_1 is the stop frequency.

$$f_{\mathrm{IF}}(\tau) = \tan\alpha \cdot \tau = \tan\alpha \cdot 2\,\frac{r}{c_0}. \tag{1}$$

Here $\tan\alpha$, with units [Hz/s], is the slope of a chirp, defined as

$$\tan\alpha = \frac{f_1 - f_0}{T_c} = \frac{B}{T_c}, \tag{2}$$

where B is the chirp's bandwidth.

The FMCW radar's range is limited by chirp's duration T_c, as targets out of FMCW radar's unambiguous range cause reflections with a delay of $\tau > T_c$.

2.2 Chirp Configuration Parameters

For worker detection, the radar was configured for a maximum detectable range of 9.01 m with a range resolution of 8.8 cm. For better detection, it is better to have a more precise radial velocity resolution as different workers will have different movement velocities and are thus easier to separate from one another. Furthermore, most targets will not approach the radar in the radial direction but under different angles of approach, which reduces their radial velocity compared to their actual movement velocity. Because of this, it is favoured to have a better resolution. The radar was thus configured for the maximum radial velocity of 2.56 m/s and radial velocity resolution of 0.08 m/s. Parameters that we used in our setup are listed in Table 1.

Table 1. FMCW radar configuration parameters.

Parameter	Value
Starting frequency	60 GHz
Bandwidth	4 GHz
Slope	100 MHz/µs
Samples per chirp	128
Samples per frame	384
Max unambiguous range	9.01 m
Range resolution	8.8 cm
Max radial velocity	2.56 m/s
Radial velocity resolution	0.08 m/s
Measurement rate	15.00 Hz

3 GTRACK Based Worker Tracking

The detection and tracking of workers in the immediate vicinity of the robot cell are performed by the group tracking algorithm GTRACK [19]. GTRACK was initially developed by Texas instruments to be used with their line of mmWave sensors. Since the algorithm was designed to run on the sensors' integrated processor, we modified it into a python module to run on any external CPU capable of running python scripts. We took this approach to off-load the detection and tracking off of the integrated processor. Off-loading the detection and tracking from the integrated processor allows the sensor to process more reflection points while still meeting high real-time requirements. We modified GTRACK with the idea that it can be used directly alongside the out-of-the-box firmware, which comes pre-flashed on off-the-shelf mmWave FMCW radar.

To correctly recognise and track workers, we need to configure GTRACK with an appropriate worker's description. The description specifies how workers are visible by the radar in the shape of point-cloud clusters. We initialised a GTRACK instance with parameters described in Table 2 and Table 3. The parameters were determined empirically by scaling typical human dimensions and space requirements in Table 2. For depth limit and width limit, a space requirement for a person with luggage was taken [16] to approximate a worker who is carrying a load. We equated both depth and width measurements under the assumption that a point-cloud cluster of a worker carrying a load is of a roughly cylindrical shape, which we also observed during our experiments. For the height limit, we considered the average height of an adult male (1.87 m) [2], which we empirically scaled to 2 m. This height is also closer to 1.92 m, as listed in [20].

These parameters represent the upper dimensional boundaries for detection, so any worker within these boundaries will be recognised by GTRACK all the same.

Table 2. Typical human dimensions and space requirements [16] with corresponding GTRACK parameters.

Measurements	Value	GTRACK gating parameters	Value
Depth	1.0 m	Depth limit	1.125 m
Width	1.0 m	Depth limit	1.125 m
Height	1.87 m	Height limit	2.0 m
		Doppler spread limit	1.5 m/s
		Gain	3

Table 3. GTRACK allocation and state transition parameters.

Allocation parameters	Value	State transition parameters	Value
Min SNR	36.0 dB	det2actThre	2
Min SNR obscured	120.0 dB	det2freeThre	5
Min Velocity threshold	0.051 m/s	active2freeThre	5
Min Points threshold	9	static2freeThre	50
Max distance threshold	1.1 m^2	exit2freeThre	10
Max velocity threshold	3.0 m/s	sleep2freeThre	600

4 Experimental Evaluation

In our evaluations, we compared measurements using the FMCW radar and a sub-millimetre accurate motion capture system. The schematic of the experimental setup is depicted in Fig. 1.

We used the IWR6843AOPEVM FMCW radar by Texas Instruments, described in detail in Sect. 2. For the MoCap system, we used a passive marker-based system that can localise the positions of the markers and track their motion in space in real-time. Our MoCap system consists of 16 IR cameras and, when properly calibrated, achieves a marker position measurement accuracy of ±0.5 mm in the measurement volume. The acquisition frequency was set 120 Hz. To track the person, we defined a virtual rigid body consisting of a set of three markers that were attached to the person.

Results of measurements with the MoCap and the FMCW are shown in Fig. 3. Left plots in Fig. 3 show the positions. We can see that position is more accurately determined in the y direction, i. e., radial to the radar (to and from the radar). The center plots in Fig. 3 show the velocities. For the measurement error, in the bottom plots, we used MoCap values as reference. Direct velocity measurements, as acquired by the FMCW radar module are shown, while for the MoCap the velocity was determined through differentiation. Again, we can see that the resolution in the y direction is much higher, and therefore the error in this direction is also much smaller. A large error in the tangential direction x is present primarily due to the fact that radar can not measure velocity in this direction and is extrapolated by GTRACK. In the y direction, velocity is directly measured through the Doppler shift of the received signal $RX(t)$ and is far more accurate. One possibility to improve GTRACK extrapolation accuracy is to increase radar's measurement rate, for which radar's onboard firmware has to be modified.

The rightmost plot in Fig. 3 shows an $x-y$ plot of the motion of the detected person for the same measurements. The difference in accuracy between x and y directions can be observed.

5 Discussion

Even though radar's tracking performance in the x direction is not highly accurate, this does not necessarily mean it cannot be used in SSM use-cases, where it is more important to measure motion towards the robot. Correct radar placement is thus important so that the measurements are accurate for approaching the robot. One way to improve this is to set up two separate FMCW radars, e.g., as in [3]. A simple solution is to set two radars, where one radially covers the y direction, and one radially covers the x direction (for directions, see Fig. 1).

Another issue with radars is that this is an active, emitting device. As all Radio-Frequency (RF) emitting devices, radars must comply with standard EN 62232:2017 [7], to satisfy work-space safety requirements. The European Commission stipulates that electronic devices with effective isotropic radiative

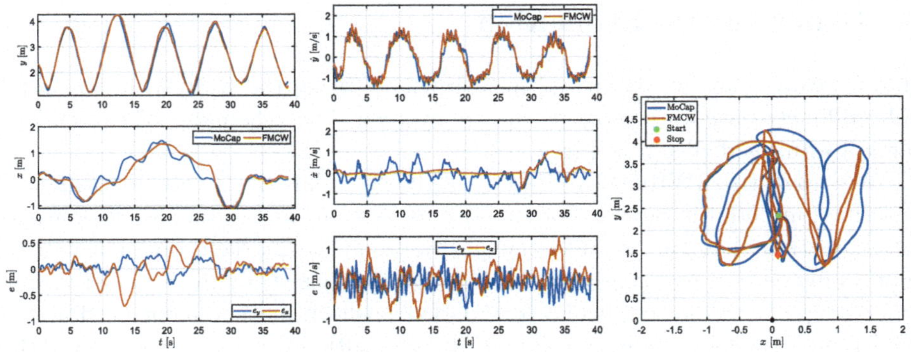

Fig. 3. Measured positions (left) and velocities (centre). In the top two rows, MoCap data is in blue and FMCW data is in orange. In the bottom row, errors in y and x directions are given in blue and orange, respectively. Motion in front of the radar in $x - y$ plane (right).

power $P_{\text{EIRP}} \leq 10$ W should be mounted with a distance of at least $l \geq 20$ cm between the main antenna lobe and a person [4]. Currently, the International Commission on Non-Ionizing Radiation Protection (ICNIRP) also suggests a minimum safe distance l for the general public [6] as a distance at which effective electric field falls below $|S_{\text{eff}}| \leq 10\ \text{W}_{\text{eff}}/\text{m}^2$ or below $|E_{\text{eff}}| \leq 61\ \text{V}_{\text{eff}}/\text{m}$, where S is the surface power density and E is the electric field of the emitted RF signal. According to [23], the minimum safety distance l is calculated with

$$l \geq \sqrt{\frac{Z_0\, P_{\text{EIRP}}}{4\pi |E_{\text{eff}}|}}, \tag{3}$$

where $Z_0 = 377\ \Omega$ (free space impedance), which for IWR6843AOP FMCW radar results in $l = 12.5$ cm.

6 Conclusion

The used, off-the-shelf FMCW radar, coupled with the provided GTRACK tracking algorithm, did not perform as accurately as the MoCap system. However, its performance in the radial, y direction was good. To observe the worker's proximity to a robot, we are mainly interested in observing the radial direction, where the radar's performance was excellent, considering its configured radial direction resolution of 8.8 cm. Thus, the correct placement of the radar is of crucial importance for effective use in SSM applications. Even if two radars have to be utilised to achieve better accuracy, this solution is still less expensive than protective fences and, at the same time, saves a lot of space. The added benefit is that radars do not present a risk for privacy invasion when used for tracking workers. Alternatively, modifying radar firmware to allow a higher measurement rate would improve GTRACK's extrapolation accuracy of tangential velocity $\dot{x}$ and, consequently, the accuracy in the x direction.

References

1. Byner, C., Matthias, B., Ding, H.: Dynamic speed and separation monitoring for collaborative robot applications - concepts and performance. Robot. Comput.-Integr. Manuf. **58**, 239–252 (2019)
2. Center of disease control and prevention: vital and health statistics (2021). https://www.cdc.gov/nchs/data/series/sr_03/sr03-046-508.pdf
3. Chiani, M., Giorgetti, A., Paolini, E.: Sensor radar for object tracking. Proc. IEEE **106**(6), 1022–1041 (2018)
4. European Union: Commission implementing regulation (EU) 2020/1070 of 20 July 2020 on specifying the characteristics of small-area wireless access points. Off. J. Eur. Union **63**(L 234), 11–15 (2020). https://eur-lex.europa.eu/eli/reg_impl/2020/1070/oj
5. Halme, R.J., Lanz, M., Kämäräinen, J., Pieters, R., Latokartano, J., Hietanen, A.: Review of vision-based safety systems for human-robot collaboration. Procedia CIRP **72**, 111–116 (2018). 51st CIRP Conference on Manufacturing Systems
6. International Commission on Non-Ionizing Radiation Protection (ICNIRP): Guidelines for limiting exposure to electromagnetic fields (100 kHz to 300 GHz). Health Phys. **118**(5), 483–524 (2020). https://doi.org/10.1097/HP.0000000000001210. https://www.icnirp.org/cms/upload/publications/ICNIRPrfgdl2020.pdf
7. International Electrotechnical Commission (IEC): Determination of RF field strength, power density and SAR in the vicinity of radiocommunication base stations for the purpose of evaluating human exposure. International Standard EN 62232:2017, Geneva, CH, August 2017
8. ISO 10218-2:2011-7 robots and robotic devices - safety requirements for industrial robots - part 2: robot systems and integration. Standard, International Organization for Standardization, Geneva, CH (2011)
9. ISO/TS 115066:2016-2 robots and robotic devices - collaborative robots. Standard, International Organization for Standardization, Geneva, CH (2016)
10. Kim, E., Yamada, Y., Okamoto, S., Sennin, M., Kito, H.: Considerations of potential runaway motion and physical interaction for speed and separation monitoring. Robot. Comput.-Integr. Manuf. **67**, 102034 (2021)
11. Kumar, S., Arora, S., Sahin, F.: Speed and separation monitoring using on-robot time-of-flight laser-ranging sensor arrays. In: 2019 IEEE 15th International Conference on Automation Science and Engineering (CASE), pp. 1684–1691 (2019)
12. Luck, D.G.C.: Frequency Modulated Radar, 1st edn. McGraw-Hill, New York (1949)
13. Ma, R., Liu, W., Miao, C., Yang, Z., Shao, H.: The simulation of human walking micro-Doppler echo and comparison of time-frequency analysis method. In: First International Conference on Electronics Instrumentation Information Systems (EIIS), pp. 1–6 (2017)
14. Marvel, J.A.: Performance metrics of speed and separation monitoring in shared workspaces. IEEE Trans. Autom. Sci. Eng. **10**(2), 405–414 (2013)
15. Marvel, J.A., Norcross, R.: Implementing speed and separation monitoring in collaborative robot workcells. Robot. Comput.-Integr. Manuf. **44**, 144–155 (2017)
16. Neufret, E., Neufret, P.: Man: dimensions and space requirements. In: Space Requirements, 3rd edn, pp. 16–18. Wiley-Blackwell (2000)
17. Rajpoot, Q.M., Jensen, C.D.: Video surveillance: privacy issues and legal compliance. IGI Global (2015)

18. Rizik, A., Tavanti, E., Chible, H., Caviglia, D.D., Randazzo, A.: Cost-efficient FMCW radar for multi-target classification in security gate monitoring. IEEE Sens. J. **21**(18), 20447–20461 (2021)
19. Texas Instruments Inc.: Tracking radar targets with multiple reflection points (2021). Available in SDK directory \mmwave_sdk_03_05_00_04\packages\ti\alg
20. Tilley, A.R., Henry Dreyfuss Associates: The Measure of Man and Woman: Human Factors in Design. Wiley (2002)
21. Ubezio, B., Schöffmann, C., Wohlhart, L., Mülbacher-Karrer, S., Zangl, H., Hofbaur, M.: Radar based target tracking and classification for efficient robot speed control in fenceless environments. In: 2021 IEEE/RSJ International Conference on Intelligent Robots and Systems (IROS), pp. 799–806 (2021)
22. Vicentini, F., Giussani, M., Tosatti, L.M.: Trajectory-dependent safe distances in human-robot interaction. In: Proceedings of the 2014 IEEE Emerging Technology and Factory Automation (ETFA), pp. 1–4 (2014)
23. Vidmar, M.: Antene in razširjanje valov, 1st edn. Založba FE, Ljubljana (2020)
24. Wolff, C.: Frequency-modulated continuous-wave radar (FMCW radar). https://www.radartutorial.eu/02.basics/FrequencyModulatedContinuousWaveRadar.en.html
25. Zhao, P., et al.: mID: tracking and identifying people with millimeter wave radar. In: 15th International Conference on Distributed Computing in Sensor Systems (DCOSS), pp. 33–40 (2019)
26. Zlatanski, M., Sommer, P., Zurfluh, F., Madonna, G.L.: Radar sensor for fenceless machine guarding and collaborative robotics. In: 2018 IEEE International Conference on Intelligence and Safety for Robotics (ISR), pp. 19–25 (2018)

On the Methodologies to Compute Minimum Jerk Trajectories and Their Application in Collaborative Robotics

Rafael A. Rojas and Renato Vidoni(✉)

Faculty of Science and Technology, Free University of Bozen-Bolzano, Bolzano, Italy
{rafael.rojas,renato.vidoni}@unibz.it

Abstract. Industrial collaborative robots have safety features that, if tuned correctly, make possible their physical interaction with humans. However, in addition to mechanical safety, reducing any kind of psychological discomfort or stress that the robot may induce is also crucial for Human-Robot-Interaction (HRI). On this regard, the term psychological safety has been coined to identify the absence of such unpleasant effects during the HRI experience. Several approaches propose to ensure safety by planning a motion subject to safety constraints. Among these, minimum jerk trajectories have drawn the interest of the robotics community for their advantages from the mechanical point of view and their similarity to human motions. In this work, we revise and comment the most common approaches that have been applied to compute this kind of trajectories through a common framework and discuss them with respect to the authors' recently introduced method based on the variational formalism.

1 Introduction

Collaborative robots are characterized by a design focused on Physical Human-Robot Interaction (pHRI), which facilitates the programming of tasks in compliance with the standards ISO 10218-1 [6], ISO 10218-2 [7] and TS-15066 [8]. In particular, these standards define four modalities of human-robot collaboration: hand guiding, safety-rated monitored stop, speed and separation monitoring, and power and force limiting.

The hand guiding modality provides an intuitive and simple way for teaching to robots. It allows to construct a motion by manually bringing the robot into a sequence of desired positions called via-points that are then used by a motion planner to compute the trajectory. Thus, this delegates the path planning to the robot's operator. In the power and force limiting modality, physical contact between the robot and the operator is allowed under certain limits specified by technical specification TS-15066 [8]. This specification provides velocity limits for the robot's moving parts that ensure mechanically safe contacts.

It follows that trajectory planning along via-points with velocity constraints can compute mechanically safe motions under the absence of quasi-static

A. Müller and M. Brandstötter (Eds.): RAAD 2022, MMS 120, pp. 589–596, 2022.
https://doi.org/10.1007/978-3-031-04870-8_69

contacts. Commercially available robots implement such a method through a trapezoidal velocity profile on the straight lines joining the given via-points in either the operational or joint space. However, experimental evidence shows that these motions could represent a source of psychological risk [1,13,16].

In order to reduce the psychological risk, authors in [9,19,22] propose the use trajectories that minimize the L^2 norm of the jerk, i.e. minimum-jerk trajectories. Indeed, since these trajectories have successfully modelled human arm/hand motions [4,14,15], it is also supposed that they are familiar to the operator and the degree of acceptance is higher. On this regard, the impact of such trajectories on the induced stress has been investigated in [10], showing that they significantly reduce heart rate in humans, thus suggesting an improvement of robot's acceptance and an increment of the psychological safety.

Studies on minimum jerk trajectories go back to [12] and [4], where the point-to-point and via-points problem have been addressed and solved as an optimal control problem (OCP). Other works like [5,17,23] use different kinds of approximations to construct a Constrained Non Linear Program. In this paper, we revise the theoretical background of the methodologies presented by [4,5,12, 17,23] thorough a unified framework and discuss them with the more recently approach based on the variational formalism presented in [19]. In Sect. 2, the general formulation of the problem is presented. In Sect. 3, the main methods are recalled and presented under the general framework. In Sect. 4 we present the variational formalism and discuss its advantages and drawbacks with respect to the other methods. In Sect. 5 we present some drawbacks of minimum jerk trajectories as well as proposed solutions.

2 General Problem Formulation

We consider a trajectory of a robot as an arbitrary curve $\mathbf{q}$ with execution interval $J = [0, t_f]$ in $\mathbb{R}^n$ without making specific considerations whether the motion is given in the configuration space of the robot or on a specific task.

Suppose that the operator desires to execute a task through with via-points $VP = \{\mathbf{w}_0, \mathbf{w}_1, \ldots, \mathbf{w}_N\} \subset \mathbb{R}^n$. The problem of trajectory planning along VP is that of finding a curve where via-point $\mathbf{w}_i$ is attained at the time t_i with $t_i < t_{i+1}$ during a given execution time interval $J = [0, t_f]$. Velocity and acceleration constraints can be added to the problem, deriving in the following constraints:

$$\begin{aligned} &\mathbf{q}(t_i) = \mathbf{w}_i \\ &\dot{\mathbf{q}}(t_0) = \dot{\mathbf{w}}_0 \quad \dot{\mathbf{q}}(t_N) = \dot{\mathbf{w}}_N \\ &\ddot{\mathbf{q}}(t_0) = \ddot{\mathbf{w}}_0 \quad \ddot{\mathbf{q}}(t_N) = \ddot{\mathbf{w}}_N. \end{aligned} \tag{1}$$

with $t_0 = 0$ and $t_N = t_f$.

We define the minimum jerk trajectory as the one that minimizes the following cost function

$$I(\mathbf{q}) = \int_0^{t_f} \|\dddot{\mathbf{q}}\|^2 \mathrm{d}t. \tag{2}$$

In order to formalize the proposed problem, we introduce a set of feasible solutions of (2) subject to (1). On the one hand, such a set is constituted by curves which are continuous up to some k-th derivative, i.e. $\mathbf{q} \in C^k(J, \mathbb{R}^n)$, where k may be found from the optimization problem or selected by a given criteria depending on the method. On the other hand, the evaluation (2) should be finite. However, in order to obtain a Hilbert space structure, we require that these solutions belong to the Sobolev space [3] $\mathbb{W}^{2,k}$, which is the set of functions with square integrable k-th derivative. Thus, we define the set of admissible trajectories as:

$$\mathcal{Q} = \left\{ \mathbf{q} \in \mathbb{W}^{2,k}(J, \mathbb{R}^n) \cap C^k(J, \mathbb{R}^n) \middle| \exists \{t_i\}_{i=0}^N \subset J \text{s.t.}(1) \right\}. \tag{3}$$

In addition to the minimization of (2) it is also important to fulfil other constraints, such as the joint maximum velocities, accelerations, torques related to collaborative safety constraints. We introduce these constrains as a map $\mathbf{g} : \mathcal{Q} \longrightarrow \mathbb{R}^m$, such that:

$$\begin{aligned} \min_{\mathbf{q} \in \mathcal{Q}} &= I(\mathbf{q}) \\ \mathbf{g}(\mathbf{q}) &\leq \mathbf{g}_{\max} \end{aligned} \tag{4}$$

In order to incorporate the safety constraints related to a collaborative scenario, the concept of the Danger Field is used [11]. It is defined as a map $DF : \mathbb{R}^n \times \mathbb{R}^n \times \mathbb{R}^3 \longrightarrow \mathbb{R}$ that associates the danger induced by a robot with joint positions $\mathbf{q}$ and velocities $\dot{\mathbf{q}}$ at the point $\mathbf{x}_0$ to a number $DF(\mathbf{q}, \dot{\mathbf{q}}, \mathbf{x}_0)$.

2.1 Invariance with Respect to Linear Dilations and Subsidiary Constraints

As it has been proven in [4,19], the integral (2) has the remarkable property that its extremals are invariant with respect to the linear scaling of the execution time. In other words, if $\mathbf{q}$ is an optimal in the interval $[0, t_f]$, the new trajectory with time scaling factor σ, $\mathbf{q}'(t) = \mathbf{q}(\sigma t)$ in the interval $[0, t_f/\sigma]$ will also be an optimal of the integral (2) in the new interval. As a consequence, the minimizers of (2) are characterized by a unique path which is independent of the execution time. This property may be exploited to find solutions of (4) when the constrains have the property that $\mathbf{g}(\mathbf{q}') = \sigma^\beta \mathbf{g}(\mathbf{q})$ with $\beta \in \mathbb{R}$. In fact, in such a case we can first find a minimizer of (2) $\mathbf{q}$ and then choose σ such that $\mathbf{g}(\mathbf{q}') = \mathbf{g}_{\max}$.

3 Main Methods to Find Minimum Jerk Trajectories in Collaborative Workspaces

3.1 Direct Approximation Methods

Authors like [5,23] and [17] choose an *a priory* set of feasible motions. These authors choose a fixed execution time and a set of n_c scalar functions $\phi \in C^\infty(J_0, \mathbb{R})$ in a canonical interval J_0. In order to simplify the computations, we choose $J_0 = [-1, 1]$. Then, their restrict the set of admissible motions such that they can be written as

$$q_i(t) = \sum_{i=1}^{n_c} y_i^j \phi_i \circ s_j(t) \text{if} t \in J_j. \tag{5}$$

where $\mathbf{y}_j \in \mathbb{R}^n$ and s_i is defined by

$$s_j(t) = 2\frac{t - t_j}{\tau_j} - 1, \tag{6}$$

where $\tau_i = t_{i+1} - t_i$.

The assumption (5) describes each trajectory with n_c vectors $\mathbf{y}_j$ and a vector $\boldsymbol{\tau} \in \mathbb{R}^N$ which components are τ_i. Following this idea, we can substitute (5) in (4) and integrate it with respect to time to obtain an optimization problem in real variables. In addition, continuity constraints must be added up to some desired degree $k > 1$ as well as safety constraints. It follows that the new set of admissible trajectories is

$$\begin{aligned} &\min_{\mathbf{y},\boldsymbol{\tau}} I'(\mathbf{y}, \boldsymbol{\tau}) \\ &\mathbf{g}(\mathbf{q}) \leq \mathbf{g}_{\max} \end{aligned} \tag{7}$$

The problem (7) is a Non Linear Program that can be solved with off the shelf optimizers.

3.2 Optimal Control Approaches

Works like [12] and [4], formulate the problem as an OCP. This approach requires to introduce the new variables

$$\begin{aligned} \mathbf{x} &= \left[\mathbf{q}, \dot{\mathbf{q}}, \ddot{\mathbf{q}}\right]^\top \\ \mathbf{u} &= \dddot{\mathbf{q}}. \end{aligned} \tag{8}$$

In the optimal control terminology $\mathbf{x}$ is called the state variable and $\mathbf{u}$ is the control variable. After the introduction of the variables (8), the (4) becomes:

$$\begin{aligned} &\min_{\mathbf{u} \in L^2(J,\mathbb{R}^n)} \int_0^{t_f} \|\mathbf{u}\|^2 \mathrm{d}t \\ &\dot{\mathbf{x}} = \mathbf{A}^\top \mathbf{x} + \mathbf{B}\mathbf{u}, \end{aligned} \tag{9}$$

where $L^2(J, \mathbb{R}^n)$ is the space of square integrable functions.

The Pontryagin Maximum Principle [4] states the following necessary optimality conditions:

$$\begin{aligned} \mathbf{u} &= \max_{\mathbf{u} \in L^2(J,\mathbb{R}^n)} \boldsymbol{\lambda}^\top (\mathbf{A}\mathbf{x} + \mathbf{B}u) - \|\mathbf{u}\|^2 \\ \dot{\boldsymbol{\lambda}} &= \mathbf{A}^\top \mathbf{x} \\ \boldsymbol{\lambda}(t_f) &= 0, \end{aligned} \tag{10}$$

where $\boldsymbol{\lambda}$ is called the costate, also referred as the Lagrange multipliers. The solution of (10) subject to (1) is stated in [4] and the safety constraints are added using the invariance argument.

4 Variational Method

The problem of the minimization of the jerk subject to a constraint of type (1) was firstly solved using the variational formalism in [19]. In order to do so, it is necessary to introduce the variation of $\mathbf{q}$ as by the family of maps $\mathbf{p} = \mathbf{q}(t) + \varepsilon\boldsymbol{\eta}(t)$, where $\boldsymbol{\eta} \in C^\infty(J, \mathbb{R}^n)$ is an arbitrary map called variation and ε is an arbitrarily small scalar. Then we search for the conditions that ensure

$$I(\mathbf{p}) - I(\mathbf{q}) = \mathcal{O}(\varepsilon^2), \tag{11}$$

where $\mathcal{O}(\varepsilon^2)$ represents a negligible quantity proportional to ε^2 called big-O notation. However, each trajectory in the neighbourhood under consideration has its own set of times $r_0, \ldots, r_N$ such that $\mathbf{p}(r_i) = q_i$. It follows that we also consider a variation δt_i of t_i, with $r_i = t_i + \varepsilon\delta t_i$. Furthermore, we also consider the variation of $\mathbf{q}$ and its derivatives at the prescribed points given by $\delta\dot{\mathbf{q}}_i = \dot{\mathbf{p}}(r_i) - \dot{\mathbf{q}}(t_i)$ and $\delta\ddot{\mathbf{q}}_i = \ddot{\mathbf{p}}(r_i) - \ddot{\mathbf{q}}(t_i)$. We can prove that these variations have the following relations [2]:

$$\boldsymbol{\eta}(t_i) = -\dot{\mathbf{q}}(t_i)\delta t_i; \quad \dot{\boldsymbol{\eta}}(t_i) = \delta\dot{\mathbf{q}}_i - \ddot{\mathbf{q}}(t_i)\delta t_i; \quad \ddot{\boldsymbol{\eta}}(t_i) = \delta\ddot{\mathbf{q}}_i - \dddot{\mathbf{q}}(t_i)\delta t_i. \tag{12}$$

As a consequence, the value of (11) can be written as:

$$\begin{aligned} I(\mathbf{p}) - I(\mathbf{q}) = \Bigg\{ & 2\int_0^T \left(\frac{d^6\mathbf{q}}{dt^6}\right)^\top \boldsymbol{\eta}\,\mathrm{d}t + \sum_{i=1}^{N-1}\left[F(t_i^+) - F(t_i^-)\right]\delta t_i \\ & + \sum_{i=1}^{N-1} 2\left[\frac{d^4\mathbf{q}}{dt^4}(t_i^+) - \frac{d^4\mathbf{q}}{dt^4}(t_i^-)\right]^\top \delta\dot{\mathbf{q}}_i \\ & + \sum_{i=1}^{N-1} 2\left[\dddot{\mathbf{q}}(t_i^+) - \dddot{\mathbf{q}}(t_i^-)\right]^\top \delta\ddot{\mathbf{q}}_i \Bigg\}\varepsilon + \mathcal{O}(\varepsilon^2), \end{aligned} \tag{13}$$

where F can be computed following the indicated procedure and t_i^- and t_i^+ represent the limits of t when approaching t_i from the left and the right, respectively. Since the variations $\boldsymbol{\eta}$, δt_i, $\delta\dot{\mathbf{q}}_i$, $\delta\ddot{\mathbf{q}}_i$ and their components are linearly independent, the following conditions to ensure the stationarity of I at $\mathbf{q}$ hold:

$$\frac{\mathrm{d}^6\mathbf{q}}{\mathrm{d}t^6} = 0 \ a.e. \tag{14}$$

$$\mathbf{q} \in C^4(J, \mathbb{R}^n) \tag{15}$$

$$\left[\frac{\mathrm{d}^5\mathbf{q}}{\mathrm{d}t^5}(t_i^+) - \frac{\mathrm{d}^5\mathbf{q}}{\mathrm{d}t^5}(t_i^-)\right]^\top \dot{\mathbf{q}}(t_i) = 0. \tag{16}$$

From (14), it derives that any extremal belongs to the set of piece-wise vector-valued fifth-order polynomials. Moreover, conditions (14)–(16) are preserved under linear dilations.

With respect to the approaches based on optimal control theory, this approach is simpler in the sense that it does not require to introduce further differential constraints (avoiding the difficulties of treating the costate) and artificial control actions. With respect to direct methods, this approach has the advantage of providing the exact solutions to the problem and a simple prove of the invariance with respect to linear parametrizations. However, direct methods allow a more accurate approach to complex constraints.

5 Limits of the Min-Jerk Trajectories and Proposed Solutions

Beyond the highlighted advantages, minimum jerk motions could have at least two potential drawbacks that degrade the robot's behaviour. On the one hand, they could be much more slower than motions like trapezoidal velocity profiles. On the other hand, they can show large deviations from the minimum-length path thus creating possible unwanted collisions or dangers. To overcome these issues, the authors of this work have proposed two main approaches [18,20,21]. Firstly, authors in [18,21] propose to minimize the convex combination of the smoothness and a cost proportional to the arc-length given a parameter $\alpha \in (0, 1)$,

$$I(\mathbf{q}) = \int_0^{t_f} \alpha\|\dot{\mathbf{q}}\|^2 + (1-\alpha)\|\dddot{\mathbf{q}}\|^2 \mathrm{d}t. \qquad (17)$$

In this manner, a trade off between the two extremals can be chosen. This can be appreciated in Fig. 1a where the same collaborative work-cell with a UR3 collaborative robot from Universal Robots adopted by the authors in [21] has been used. The trapezoidal speed profile, the minimum-jerk and the proposed alternatives are compared both in the joint space (position and velocity of the first joint is reported) and in the operative space (projection on the x-y and y-z planed). As it can be observed, in particular in the plane y, z, the minimization of (17) reduces the overshoots with respect to the minimum jerk trajectories and reduces the execution time as well.

As a second and alternative approach, the authors of this work in [20], leveraging on the formalisms of calculus of variations, propose a novel minimum-time parametrization to find the fastest possible re-parametrization of minimum jerk trajectories. Considering a minimum-jerk trajectory $\mathbf{q}_0$ and a small parametrization of the path ξ as a variation of the identity $\xi_\varepsilon(t) = t + \varepsilon\mu$, following a procedure similar to that in Sect. 4, a linear approximation of the robot's torque constraints and the execution time with respect to the variation μ is found. After solving a linear optimization problem, the optimal variation μ^* is obtained and then used to compute a small parametrization ξ_1 and a new trajectory $\mathbf{q}_1$. This procedure is iterated until specific optimality conditions are met. The effectiveness of the approach can be appreciated in Fig. 1b: it can be seen that under time optimal parametrizations minimum jerk trajectories can achieve execution times comparable with trapezoidal velocity profiles.

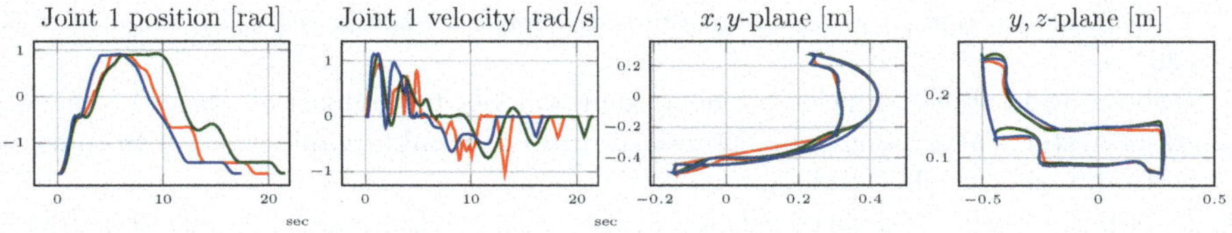

a Comparison with alternative cost functions (17) (blue).

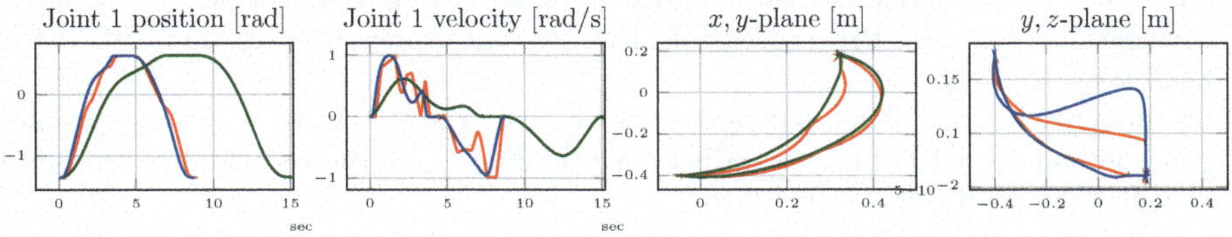

b Comparison with minimum time parametrization (blue).

Fig. 1. Comparison between trapezoidal velocity profile (red), minimum jerk (green) and proposed alternatives (blue).

6 Conclusions

In this work, the most common approaches to plan minimum jerk trajectories are revised and discussed under a collaborative robotics perspective considering their suitability as movements that are both mechanically "gentle" and human-friendly. After that, the authors' recently introduced method based on the variational formalism is recalled and the main drawbacks of min-jerk trajectories as well as two possible methods that avoid them shown also through experimental evidence.

Acknowledgements. This work was supported by the Unibz ID project "Automated Process Planning in Cyber Physical Production Systems of Smart Factories" (Acronym: SMART-APP - Cost centre: TN201I).

References

1. Arai, T., Kato, R., Fujita, M.: Assessment of operator stress induced by robot collaboration in assembly. CIRP Ann. **59**(1), 5–8 (2010)
2. van Brunt, B.: The Calculus of Variations. Universitext, Springer, New York (2006). https://doi.org/10.1007/b97436
3. Evans, L.C.: Partial Differential Equations, vol. 19. American Mathematical Society, Providence (2010)
4. Flash, T., Hogan, N.: The coordination of arm movements: an experimentally confirmed mathematical model. J. Neurosci. **5**(7), 1688–1703 (1985)

5. Gasparetto, A., Lanzutti, A., Vidoni, R., Zanotto, V.: Validation of minimum time-jerk algorithms for trajectory planning of industrial robots. J. Mech. Robot. **3**(3), 031003 (2011)
6. Robots and robotic devices – safety requirements for industrial robots – part 1. Technical report, International Organization for Standardization, Geneva, CH (2011)
7. Robots and robotic devices – safety requirements for industrial robots – part 2: robot systems and integration. Standard, International Organization for Standardization, Geneva, CH (2011)
8. Robots and robotic devices – collaborative robots. Technical specification, International Organization for Standardization, Geneva, CH (2016)
9. Kokabe, M.: Modeling of handling motion reflecting emotional state and its application to robots. In: Proceedings of SICE Annual Conference, pp. 495–501. IEEE (2008)
10. Kühnlenz, B., Kühnlenz, K.: Reduction of heart rate by robot trajectory profiles in cooperative HRI. In: Proceedings of ISR 2016: 47st International Symposium on Robotics, pp. 1–6. VDE (2016)
11. Kulić, D., Croft, E.A.: Real-time safety for human-robot interaction. Robot. Auton. Syst. **54**(1), 1–12 (2006)
12. Kyriakopoulos, K.: Minimum jerk path generation. In: Proceedings of the International Conference on Robotics and Automation (ICRA), pp. 364–369. IEEE (1988)
13. Lasota, P.A., Shah, J.A.: Analyzing the effects of human-aware motion planning on close-proximity human-robot collaboration. Hum. Factors **57**(1), 21–33 (2015)
14. Meirovitch, Y.: Geometrical invariance and smoothness maximization for task-space movement generation. IEEE Trans. Robot. **32**(4), 837–853 (2016)
15. Oguz, O.S., Zhou, Z., Glasauer, S., Wollherr, D.: An inverse optimal control approach to explain human arm reaching control based on multiple internal models. Sci. Rep. **8**(1), 5583 (2018)
16. Or, C.K., Duffy, V.G., Cheung, C.C.: Perception of safe robot idle time in virtual reality and real industrial environments. Int. J. Ind. Ergon. **39**(5), 807–812 (2009)
17. Piazzi, A., Visioli, A.: Global minimum-jerk trajectory planning of robot manipulators. IEEE Trans. Ind. Electron. **47**(1), 140–149 (2000)
18. Rojas, R.A., Garcia, M.A.R., Gualtieri, L., Rauch, E.: Combining safety and speed in collaborative assembly systems-an approach to time optimal trajectories for collaborative robots. Procedia CIRP **97**, 308–312 (2021)
19. Rojas, R.A., Garcia, M.A.R., Wehrle, E., Vidoni, R.: A variational approach to minimum-jerk trajectories for psychological safety in collaborative assembly stations. IEEE Robot. Autom. Lett. **4**(2), 823–829 (2019)
20. Rojas, R.A., Vidoni, R.: Designing fast and smooth trajectories in collaborative workstations. IEEE Robot. Autom. Lett. **6**(2), 1700–1706 (2021)
21. Rojas, R.A., Wehrle, E., Vidoni, R.: A multicriteria motion planning approach for combining smoothness and speed in collaborative assembly systems. Appl. Sci. **10**(15), 5086 (2020)
22. Sidobre, D., Desormeaux, K.: Smooth cubic polynomial trajectories for human-robot interactions. J. Intell. Robot. Syst. **95**(3–4), 851–869 (2019). https://doi.org/10.1007/s10846-018-0936-z
23. Simon, D., Isik, C.: A trigonometric trajectory generator for robotic arms. Int. J. Control **57**(3), 505–517 (1993)

Development of a Virtual Reality Application for the Assessment of Human-Robot Collaboration Tasks

Giacomo Palmieri(✉), Cecilia Scoccia, Daniele Costa, and Massimo Callegari

Department of Industrial Engineering and Mathematical Sciences, Polytechnic University of Marche, Ancona, Italy
{g.palmieri,c.scoccia,d.costa,m.callegari}@univpm.it

Abstract. The combination of Virtual Reality (VR) systems and advanced control strategies represent a winning approach to improve the Human Robot Collaboration (HRC) in industrial applications: both safety and ease of development and tuning of control parameters are achievable, responding to recent industrial needs of productivity and flexibility, and at the same time ergonomics for operators. In this paper it is described a VR application used to asses a control strategy for collision avoidance developed by the authors for the redundant collaborative robot KUKA iiwa LBR. The robot kinematics is virtually modelled, whereas the obstacle is a virtual copy of the controller which is manually moved by the operator. Spatial coordinates of the controller, i.e. the obstacle, are available in real time and can be used to update the motion of the robot in order to avoid a potential collision. The effectiveness of the method, demonstrated by the presented test case, suggests further developments in order to simulate real industrial applications and integrate other types of sensors to improve the system capabilities.

Keywords: Virtual reality · Human-robot collaboration · Collaborative robotics · Collision avoidance

1 Introduction

In recent years, production flexibility and short time-to-market are among the most urgent needs of the industries. Progress in robotics and Information and Communication Technologies (ICT) has given a strong contribution to the transformation of the production systems from mass production to small lot sizes and product mixes.

Traditionally industrial robots are able to guarantee a high quality and fast production; recently, with the advent of collaborative robotics, the development of techniques for Human-Robot Collaboration (HRC) allows to combine the repeatability and accuracy of robots with flexibility, adaptability ad dexterity of humans [3,13,16].

A. Müller and M. Brandstötter (Eds.): RAAD 2022, MMS 120, pp. 597–604, 2022.
https://doi.org/10.1007/978-3-031-04870-8_70

At the same time the digitization of manufacturing systems is becoming increasingly important in the field of production management: digital twins of the production systems can be exploited to implement various types of predictive analysis, simulation and maintenance. Virtual Reality (VR) is one of the pillars falling within this technological field. Among various applications, VR can be used to simulate a manufacturing cell before it is realized, eliminating or limiting the impact of the prototyping stage, thus reducing the time-to-market and costs.

Several examples can be found where VR techniques are used to simulate and test robotic systems in a safe user environment, giving useful insights for the off-line programming of the robots. Recent researches are focused on solutions based on VR for HRC, where the interaction between the humans and the machines can be assessed [6,9].

A virtual environment where an industrial robot is modelled by UNITY software is presented in [5]; the operator can observe the environment with VR Head Mounted Display (HMD) and interact with the robot by using some virtual control buttons.

A digital model of a redundant manipulator (KUKA LBR iiwa) is presented in [8] starting with kinematic modeling and then coupled with the real robot; the aim is to simulate HRC in several scenarios, to reproduce the safe behavior on the real robot, as well as to train operators.

In [10] it is proposed the use of a VR digital twin of a physical layout as a mechanism to understand human reactions to both predictable and unpredictable robot motions; a set of established metrics as well as a newly developed Kinetic Energy Ratio metric are used to analyse human reactions.

The objective of the study presented in [7] is a concept of Virtual Safety Training System (VSTS) in Human-Robot Collaboration. In this systems the virtual environment and its objects are modeled in CAD software and exported to game engines. Additionally, forward and inverse kinematics of the robot are simulated and scripted for visualization of robot tasks and movements. Furthermore, the collision of the system's components (robot, user, parts) can be simulated. Similarly, a VR and depth camera based system where vibration signals are used to alert the user of a probable collision with a robot is presented in [12].

In [18] a VR application is used to define and test the task sharing in a collaborative robotic cell for nut screwing.

Based on established research that trains Artificial Intelligence (AI) systems within simulated environments, a machine learning implementation that enables the training of a dual-arm robot within a VR application is proposed in [2].

Algorithms for collision avoidance have been presented by the authors in previous works. They have been developed for serial collaborative manipulators, in particular for the UR family [4], which is characterized by 6 Degrees Of Freedom (DOF), and for the KUKA iiwa family [11,17] which has 7 DOFs, thus has one redundant degree of freedom. This paper presents a VR application aimed at evaluating collision avoidance algorithms on virtual robots, while the motion of the operator, therefore his interaction with the robot, is derived from the physical sensors of the VR system. The goal is to insert the real behavior of the human in the simulation loop in order to fine-tune the parameters of the algorithms and reach an acceptable response from the robot in terms of physical and cognitive ergonomics for the operator.

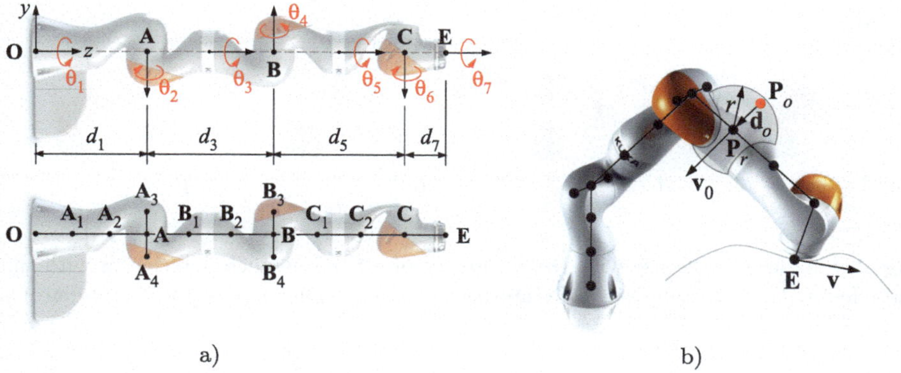

Fig. 1. a) KUKA LBR iiwa robot: kinematic parameters and collision control points; b) Linear velocity of the end-effector **E** and of the internal control point $\mathbf{P}_r$ closest to the obstacle.

2 Obstacle Avoidance Algorithms for HRC

At the basis of a synergistic collaboration between humans and robots, it is of fundamental importance to develop techniques for controlling the robot in a non structured workspace where collision with objects or humans are possible and must be avoided in real-time [14,15]. A series of algorithm for obstacle avoidance have been developed by the authors for some commercial collaborative robots. In this case the KUKA LBR iiwa 14 R820 is considered. The adopted strategy consists in a combination of an off-line path planning procedure and an on-line motion control algorithm: the trajectory planning is managed in order to define a path that avoids obstacles inside the workspace considering their positions at the initial step of the motion; the on-line motion control algorithm is then exploited for two effects, namely updating the motion of the end-effector to compensate for moving obstacles and also to generate internal movements of the robot arm in order to avoid collisions between obstacles and any point of the kinematic chain of the robot in addition to the end-effector.

The off-line trajectory planning strategy exploits the potential fields method. It is based on the definition of a potential field that drives the robot through the workspace until the target is reached while avoiding obstacles: a global minimum of the potential is set at the target point, whereas local maximums are set in correspondence of obstacles. The effect of the potential field on the robot end-effector results in a combination of forces which are proportionally converted in velocities, attractive towards the target and repulsive from obstacles.

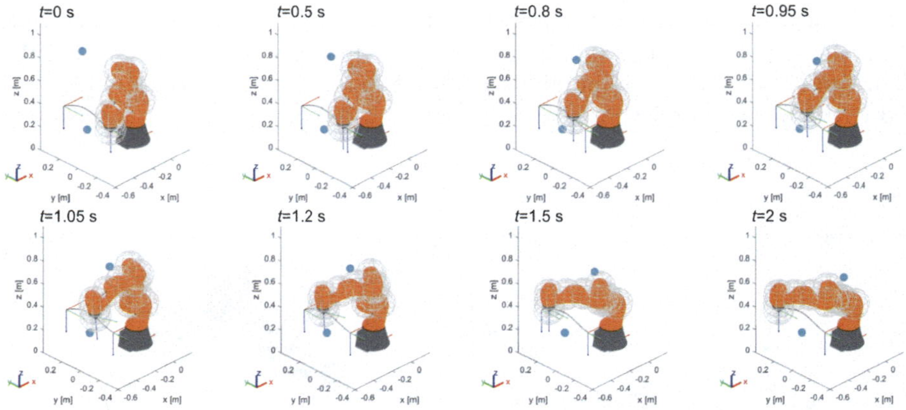

Fig. 2. Simulated test-case with the avoidance of a fixed obstacle interfering with the end-effector and a moving obstacle interfering with an inner link of the manipulator.

The trajectory resulting from the potential field method can often give problems due to low radius curves that generate high accelerations. A possible approach to solve such problem is to interpolate the trajectory using a type of parametric curve, as for example a third order Bezier curve. A best fit procedure con be executed keeping fixed the initial and final points of the trajectory, whereas the other parameters of the curve are found by least-square optimization. After that a timing law is assigned to the obtained path in order to generate the final motion law.

The on-line motion control is managed by an algorithm that is able to prevent collisions with internal points of the kinematic chain of the manipulator and allows also to update the trajectory of the end-effector to avoid moving obstacles. Basically, the redundancy of the manipulator is exploited to assume configurations that prevent for collisions without modify the path of the end-effector, unless a moving obstacles is going to collide with the end-effector, in which case a real time modification of the trajectory is planned.

In order to avoid collisions in all parts of the manipulator, 13 control points of the kinematic chain are considered in addition to the end-effector $\mathbf{E}$ and the base point $\mathbf{O}$ (Fig. 1a).

In addition to the basic control law for redundant manipulators, authors have developed a strategy for obstacle avoidance exploiting the null space method [11]: an additional velocity task is assigned to the control point of the robot which is the closest to one of the obstacles within the workspace, so that the collision with such obstacle can be avoided, without affecting the motion of the end-effector. Referring to the Fig. 1b, the couple of points $\mathbf{P}_r$ and $\mathbf{P}_o$ at the minimum distance d_o is identified at each time step. Each control point is shielded by a spherical volume of interest of radius r. If the condition $d_o < r$ is verified at a time step of the motion, a repulsive velocity $\dot{\mathbf{x}}_0$ along the direction of $\mathbf{d}_o$ is assigned to the relative control point. Due to the unitary degree of redundancy of the

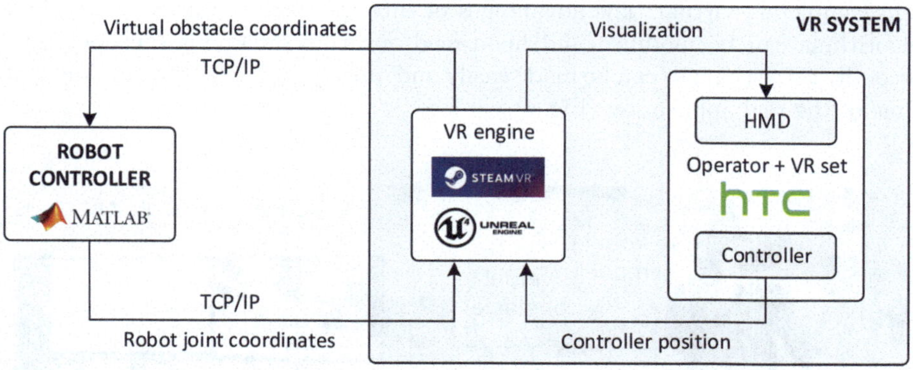

Fig. 3. Architecture of the control loop for the VR application

manipulator, only one repulsive velocity vector can be assigned at each time step in addition to the velocity of the end-effector, thus the point to which it is assigned may change over time with the criterion of minimum distance from one of the obstacles. Such a point could be the end-effector itself: the repulsive velocity vector can be applied to **E** if the position of an obstacle changes from its initial state or a new obstacle enters the workspace, interfering with the motion of the end-effector that was assigned by the off-line motion planner at the initial step of the motion. The resulting position drift from the planned trajectory can be subsequently recovered by a proportional corrective term, as typically done with Closed Loop Inverse Kinematics (CLIK) control laws.

An example is presented in the following of this section in order to understand the behaviour of the robot under the above described control algorithms. The simulated case study presents two obstacles (Fig. 2). The first, fixed, is located under the linear path of the robot. Since its proximity to the path is less than the safety distance for the end-effector, the off-line trajectory planner generates a Bézier curve that avoids the obstacle by passing over it. Once the trajectory is assigned and the motion of the manipulator starts, a second obstacle, dynamic, moves linearly along a path that intersects the kinematic chain of the robot. The effect of the second obstacle is perceptible at approximately $t = 0.8$ s, when the robot suddenly changes its configuration to avoid the collision between the obstacle end the link, without affecting the motion of the end-effector which remains on the planned trajectory.

3 VR Application

The algorithms presented in the previous section were preliminarily evaluated by simulation, testing different cases in order to fine-tune the control parameters. However, simulations are not enough when the aim is to test the effect of a real dynamic obstacle. For this purpose, the idea was to create a VR application in which a real obstacle is inserted in the simulation loop, while the robot is still completely virtual. The advantages of this approach are many: firstly, the algorithms can be modified and fine-tuned ensuring safety for the operators, secondly these changes can be made easily and quickly, reducing the development time of the real application (Fig. 4).

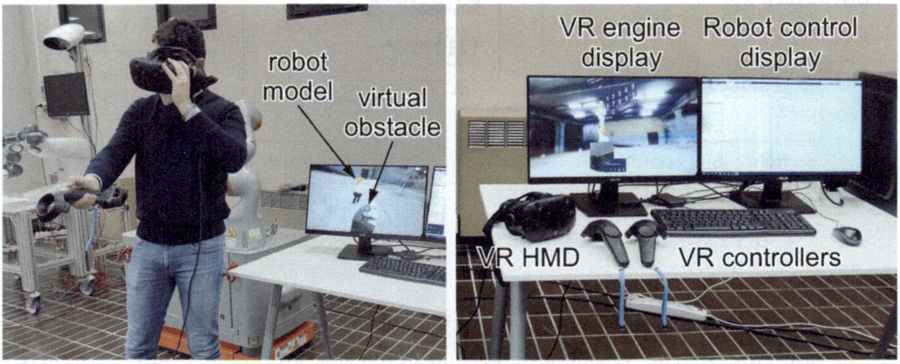

Fig. 4. VR system components

The control architecture used for the VR application is shown in Fig. 3. The core of the system is the VR engine, running on a standard PC, developed by the SteamVR development suite with Unreal Engine 4 used to model the kinematics of the KUKA robot. A classic HTC VR set is used to equip the operator: the HMD provides an immersive three-dimensional representation of the workspace, which in this case is simply the robot mounted on a stand; the controller, manually managed by the operator, is rigidly connected to a virtual obstacle, shaped like a sphere, which can be moved in the virtual space to interfere with the robot's motion. The control of the robot is executed by the same PC in a parallel thread developed in Matlab, with a frame rate set to a typical value for communication protocols with robots (e.g. 125 Hz or 250 Hz, depending on the manufacturer). Instead, the frame rate of the VR engine is set to 60 Hz, which is the conventional display refresh rate in VR. The communication between the two executed threads is realized by a standard TCP/IP protocol. A classic pick and place task is simulated, thus the robot controller sends joints rotations to the VR engine at each time step, similarly to what is done in real robotic systems with direct servoing by external controllers. Trajectory of the robot is updated in real time by the collision avoidance algorithm if the obstacle enters in a safety region of the manipulator. The position of the obstacle is known once the VR engine reads the coordinates of the controller hold by the hand of the operator and send this information to the robot controller.

In the example shown in Fig. 4 a single obstacle is moved by the operator through the controller; more complex scenarios can be tested using two controllers related to two different obstacles, while the avoidance of the operator's head can be obtained also considering the coordinates of the HMD device.

4 Conclusions

A VR application for the simulation of collision avoidance control for collaborative robots has been presented in this paper. The system allows to insert the real motion of one or more obstacles in the simulation loop of the robotic cell, guaranteeing the safety of the operators and simplifying the tuning of the algorithms. Future developments may concern the modeling of more realistic workstations with real industrial test cases, as well as the integration of external optical sensors or motion capture systems [1] to more accurately digitize the operator's movement in order to improve the perception capacity of the system.

Acknowledgement. The authors would like to thank Cadland Spa for their support in the development of this work.

Funding. This work was partly funded by the project URRA' ("usability of robots and reconfigurability of processes: enabling technologies and use cases"), based on the topics of User-Centered Manufacturing and Industry 4.0, which is part of the project EU ERDF, POR MARCHE Region FESR 2014/2020-AXIS 1-Specific Objective 2-ACTION 2.1,"HD3Flab-Human Digital Flexible Factory of the Future Laboratory", coordinated by the Polytechnic University of Marche.

References

1. Amorim, A., Guimares, D., Mendona, T., Neto, P., Costa, P., Moreira, A.P.: Robust human position estimation in cooperative robotic cells. Robot. Comput.-Integr. Manuf. **67**, 102035 (2021)
2. Arntz, A., Di Dia, A., Riebner, T., Eimler, S.C.: Machine learning concepts for dual-arm robots within virtual reality. In: 2021 IEEE International Conference on Artificial Intelligence and Virtual Reality (AIVR), pp. 168–172 (2021)
3. Brandstötter, M., et al.: Versatile collaborative robot applications through safety-rated modification limits. In: Berns, K., Görges, D. (eds.) RAAD 2019. AISC, vol. 980, pp. 438–446. Springer, Cham (2020). https://doi.org/10.1007/978-3-030-19648-6_50
4. Chiriatti, G., Palmieri, G., Scoccia, C., Palpacelli, M.C., Callegari, M.: Adaptive obstacle avoidance for a class of collaborative robots. Machines **9**(6), 113 (2021)
5. de Giorgio, A., Romero, M., Onori, M., Wang, L.: Human-machine collaboration in virtual reality for adaptive production engineering. Procedia Manuf. **11**, 1279–1287 (2017)
6. Dianatfar, M., Latokartano, J., Lanz, M.: Concept for virtual safety training system for human-robot collaboration. Procedia Manuf. **51**, 54–60 (2020). 30th International Conference on Flexible Automation and Intelligent Manufacturing (FAIM 2021)
7. Dianatfar, M., Latokartano, J., Lanz, M.: Review on existing VR/AR solutions in human-robot collaboration. Procedia CIRP **97**, 407–411 (2021)
8. Gammieri, L., Schumann, M., Pelliccia, L., Di Gironimo, G., Klimant, P.: Coupling of a redundant manipulator with a virtual reality environment to enhance human-robot cooperation. Procedia CIRP **62**, 618–623 (2017)

9. Nagy, B.G., et al.: Towards human-robot collaboration: an industry 4.0 VR platform with clouds under the hood. In: 2019 IEEE 27th International Conference on Network Protocols (ICNP), pp. 1–2 (2019)
10. Oyekan, J.O., et al.: The effectiveness of virtual environments in developing collaborative strategies between industrial robots and humans. Robot. Comput.-Integr. Manuf. **55**, 41–54 (2019)
11. Palmieri, G., Scoccia, C.: Motion planning and control of redundant manipulators for dynamical obstacle avoidance. Machines **9**(6), 121 (2021)
12. Paniti, I., Nacsa, J., Kovács, P., Szűr, D.: VR and depth camera based human-robot collision predictor system with 3-finger gripper assisted assembly device. In: 2020 23rd International Symposium on Measurement and Control in Robotics (ISMCR), pp. 1–5 (2020)
13. Papetti, A., Ciccarelli, M., Scoccia, C., Germani, M.: A multi-criteria method to design the collaboration between humans and robots. Procedia CIRP **104**, 939–944 (2021)
14. Safeea, M., Béarée, R., Neto, P.: Collision avoidance of redundant robotic manipulators using newton's method. J. Intell. Robot. Syst. Theory Appl. **99**(3–4), 673–681 (2020). https://doi.org/10.1007/s10846-020-01159-3
15. Scimmi, L.S., Melchiorre, M., Troise, M., Mauro, S., Pastorelli, S.: A practical and effective layout for a safe human-robot collaborative assembly task. Appl. Sci. **11**(4), 1763 (2021)
16. Scoccia, C., Ciccarelli, M., Palmieri, G., Callegari, M.: Design of a human-robot collaborative system: methodology and case study. In: Volume 7: 17th IEEE/ASME International Conference on Mechatronic and Embedded Systems and Applications (MESA). International Design Engineering Technical Conferences and Computers and Information in Engineering Conference (2021)
17. Scoccia, C., Palmieri, G., Palpacelli, M.C., Callegari, M.: A collision avoidance strategy for redundant manipulators in dynamically variable environments: on-line perturbations of off-line generated trajectories. Machines **9**(2), 30 (2021)
18. Shu, B., Sziebig, G., Pieskä, S.: Human-robot collaboration: task sharing through virtual reality. In: IECON 2018 - 44th Annual Conference of the IEEE Industrial Electronics Society, pp. 6040–6044 (2018)

Robots Are Actor-Networks: Awareness, Bottom-Up Ethics and Transforming Responsibility

Nicole Duller(✉)

KPK SEEROSE, University of Klagenfurt, Klagenfurt, Austria
nicole.duller@aau.at

Abstract. Robotic systems increasingly take part in many practices of the everyday life. Technological development and innovation continuously transform fields, including the realm of industrial robotics. However, new sets of possibilities challenge new forms of responsibilities. This paper discusses responsibility and ethics in robotics. Based on three expert talks, three qualitative expert interviews and five qualitative narrative interviews with robotic engineers, it gives three preliminary findings of the research project Collective Ethical Responsibility for Robotic Systems Engineering with Safety & Security (CERSE): 1. Robotics are networks (The term network, used in this paper, does not refer to computer or technological networks, but to the assemblage of interconnected actors [1, p. 228] as used in Actor-Network Theory (ANT).) formed by hybrid actors. 2. These networks distribute and (self-)govern responsibility. 3. Engineers call for pro-active reflection and communication to raise awareness. This paper supports a radical transformation of the notion of responsibility within the realms of robotics. In line with the complexity, interconnectedness, and fluidity of robotics, robot ethics must move towards more openness, inclusion, and transdisciplinary research.

Keywords: Robots · Robot ethics · AI ethics · Process-ethics · Responsibility · Governance · Actor-Network Theory

1 Introduction

The 21st century is characterized by a vast acceleration in both technological innovation and the integration of technologies into all areas of life. Thereby, new media and technologies not only take part in, but actively re-configure and transform societies. Robotization can be added to the notion of deep mediatization [2]. Robots are increasingly expected to "play a more complex and wider range of roles in society" [3, p. 6]. Robotic systems perform as an assemblage of parts that are capable to move, so there is a chance of intentional or non-intentional injury of humans [4, p. 1]. Research on moral

This work was supported by the Karl-Popper-Kolleg SEEROSE (Responsible, Safe, and Secure Robotic Systems Engineering) at the University of Klagenfurt. The author especially thanks Larissa Krainer, and Joan Ramon Rodriguez-Amat for their advisory support.

A. Müller and M. Brandstötter (Eds.): RAAD 2022, MMS 120, pp. 605–612, 2022.
https://doi.org/10.1007/978-3-031-04870-8_71

and legal responsibility, trust and human-robot-interactions, applications from love to war, Artificial Intelligence (AI) and the future of robot ethics is urgent [4].

In 2022 the term "forklift driver" can refer to both, human or non-human operators, like robotic shuttles or mobile manipulators, that are autonomous, half-autonomous or remotely steered robotic systems that e.g., contain of a platform of an automated guided vehicle and at least one serial manipulator, like a robotic arm [5, p. 284]. One engineer tells the story about a human forklift driver and his non-human robotic co-worker in a factory. To reduce the risk of human failure, that can result in vast financial loss, the company started to integrate more automation into the chain of transportation. Supposedly in fear of being replaced by his robotic colleague, the human forklift driver "accidentally" picked up the robot and let it drop. The engineer continues addressing the existential fear of this person, and urges for societal actions to counter these, such as the introduction of a machine tax [6]. Furthermore, facing the challenges of current digital revolutions and transformations, some engineers reclaim a general enhancement of the level of education, global norms and standards, and the restructuring of society, for example via the introduction of a basic income [6]. Another engineer, in a very popular robotic system, reports the arduous path of disclosing security breaches to obtain acknowledgment and mitigation of the risk [7]. Examples like these challenge our understanding of who/what performs actions, our traditional concepts of responsibility, and ethics.

This paper starts with a short overview, describes the research design, and reports the first findings of CERSE. CERSE's aim is to research the distribution, and governance of responsibility and ethics with robotic systems engineering with safety and security. These findings are based on first empirical data; six talks and interviews with experts, and five interviews with engineers, mainly from fields of industrial robotics. In conclusion, it supports the demand for updating the traditional concept of responsibility and suggests to radically transform its notion accordingly [8, p. 8].

2 Robots: Safety and Security, Ethics Codes and Responsibility

Robots "are software-and, sometimes, artificial intelligence-with sensors and moving parts in physical space, so their crashes can be literal" [4, p. 1]. It is common sense, that programmers should work as error-free as possible [4, p. 2], but safety and security in robotics is a complex issue of functional security, physical safety and cybersecurity. It is an ongoing effort to make and keep robotic systems safe and secure. Practical and legal regulations of physical safety and system security are steered by laws, norms, and standards, including e.g., the International Organization for Standardization (ISO) safety requirements for robots and robotic devices [9], and the International Electrotechnical Commission, IEC 62443 standard series on security [10]. Practices like threat modeling also allow to describe and identify possibilities of attacks and the harm they could inflict in systems [11, p. 2]. A view on robot ethics even poses a bigger challenge.

The current emergence of ethics guidelines [12] has still low implementation impact, and by some is seen as ineffective [13, p. 100]. Furthermore, internal commitments and ethics guidelines often marketed on companies' public appearances pose the risk of suggesting sufficient self-governance, calm critical voices, and discourage the establishment of legal frameworks [13, p. 100]. "AI Ethics – or ethics in general – lacks mechanism to reinforce its own normative claims" [13, p. 99]. A survey of 22 guidelines on AI ethics states that "virtually no guideline deals with the "hidden" social and ecological costs of AI systems" [13, p. 105] or highlights the issues of increasing "public-private partnerships and industry-funded research in the field of AI" [13, p. 106]. The logic of economics, bound to time and money, does not empower critical ethical thinking nor establishes an ethical education for engineers, or organizational structures that foster ethical reflection within corporate contexts on a larger scale [13, p. 108]. Rather than striving for global cooperation, the race for AI discourages dialogue between countries, companies or researchers [13, p. 107].

Responsibility is a concept closely linked to the ethics of technology. The paradigm shift towards applied ethics of technology within specific contexts [14, p. 270] inspired approaches, such as Responsible Research and Innovation (RRI) [15, p. 49], that promote the transition towards more flexibility, interconnectedness and communication. The rapid developments in robotics stimulate a machine and robot ethics growing field of engagement with ethical questions [4, 16]–[19] that align with general increases in standardization efforts. Researchers, and initiatives highlight the need of a supportive social structure of decision makers and politicians to acknowledge the benefits of ethical engineering [20, p. 139]. The *Vienna Manifesto on Digital Humanism* [21] e.g., calls for the establishment of effective regulations that are based on a broad public discourse, increasing transdisciplinary science, a more open engagement of scientists with society, a need for practitioners acknowledgement of their shared responsibility; and it calls for new curricula, and education on the impacts of technology, to add to the technical skills an ethical awareness [21, p. 2].

3 Problem Definition and Research Design: GT and ANT

In practice, the awareness and governance of ethical issues and responsibility often falls short, and is messy. Concrete bottom-up proactive considerations of ethical issues are often in tension with abstract expert issued top-down policies [22, p. 167]. The CERSE research project calls for the incorporation of more voices, including the ones of non-human actors and engineers, into the processes and ethics [23] of engineering. There are diverging opinions on whether, and if, to what extent, engineers are ethically responsible for their work [20, p. 71]. It is therefore critical to analyze the distribution of responsibility, the places and translations of ethical responsibility, and its governance. Research on these issues is key, especially as the concentration of power is distributed amongst few big corporations. This questions the levels of democracy, and the quality of policy and decision-making participation [22, p. 170].

This project engages with the following research questions (RQ): RQ 1 Amongst whom and what is responsibility distributed in robotic systems engineering? RQ 2 How is responsibility perceived, governed, and dealt with? Which ethical challenges arise? RQ 3 How can these findings be integrated within engineering robotic systems, and current technological and societal developments? The research design is open, to respond to the heterogeneity of the field. The mixed-methods design combines a triangulation of qualitative and quantitative data [24] within the inductive and flexible approaches of Grounded Theory (GT) [25] and ANT [26]. ANT assumes, that society as a "collective" consists of actor-networks, associations of humans and non-humans [26, p. 11], with the networks being the actor, instead of the single individual in the network [27, p. 46]. Data collection included an initial review of literature, six talks, and interviews with experts in robotics, and technology assessment and five qualitative narrative interviews [28] with engineers. In addition, the interview candidates were asked to visualize the statement: "Me & responsibility, safety & security within my work in robotics". By genders 7 interviews were to cis male, and 4 to cis female. At later stages, this research will include further forms of data relevant to the study.

4 Robots as Actor-Networks: Responsibility, Ethics, Awareness

One engineer visualizes some of the actors, such as science, technology, sensors, communication, vehicles, robots, safety, data, security, automation, and usability, and states that these are just some of many that perform engineering [29]. The complexity and interconnectedness of the actor-network applied to engineering enables a holistic approach towards the issues that arise with increasingly automated robotic systems. There is a need to shift focus towards the automatic systems' interactions and connections in regards to safety and security issues [29].

4.1 Distributed Responsibility and (Self-)Governance: "It's Complicated and It Depends", the Liminal Space of the Programmer and Bottom-Up Ethics

Complex networks forming robotic systems also distribute responsibility, and its (self-) governance. Thereby, responsibility is portrayed as a relational variable constructing specific hybrid networks: "Who is responsible if something happens? It's complicated and it depends. In some cases, it's easy. If you didn't adhere to the standards and this can be proven, it is the fault of the one who took the robot into production. But, also here you have some exceptions; it can be the end user, who actively decided to not follow or to not do a certification. It could also be the integrator, who didn't implement the stuff correctly, or it could be the certifying body, that made a mistake. The really interesting stuff starts, where autonomy comes into play; whenever the robot has some kind of freedom of decision, it's rather undecidable" [30]. Other actors that may intervene are insurance entities or the state, and to one engineer "hopefully" the engineer is not the one regarded responsible; if innovation would be punished in that sense, no one could expect for anything to be invented anymore [29].

This discussion on the distribution of responsibility places the programmer in a liminal space between the public perceptions and expectations, the existential dependencies, and the frequently unseen management element that governs the processes of engineering, and ethics. Programmers' scope of action is often limited, as one engineer puts it, the programmer decides incredibly little in regards to what she or he implements. Typically, the decision is made within project management, or product management [7]. This limits engineers' individual ability to engage with ethical questions in depth: "to be an ethical trail blazer, one needs to be very independent. Both economically and mentally. But in engineering, this is not a typical condition" [31].

Responsible disclosure is one example for the bottom-up (self-)governance of security awareness. To contribute to more safety and security, engineers should possibly communicate any issues upon detection. Yet, this is easier said than done. An engineer tells about one robotic system, based on a "catastrophically insecure" operating system [7]. The engineer disclosed the issue to management, who opted to actively ignore the problem, until the company finally started to engage with it. Before fixing the problem, responsibility for any incident occurring would have fallen on the integrator of the system. End users, in practice, often do not know the technical details, hence they are not aware of the risks, and implicitly assume the system to be secure. The absence of an official discussion or documentation about the security of the system jeopardizes its safety. As the engineer states, it is very hard to assign responsibility, even after the acknowledgement of the problem, but in this case, this should fall on the management of the manufacturer [7]. The aim of some engineers is to convince as many entities in the network as possible to act on pressing safety and security related issues; but such decisions are not governed by laws and standards yet [7]: "After years of beating the drums, people finally notice, that security is important. Not all of them, of course. There still are robot vendors who think that insecurity is a feature, but some of them are on the track of recognizing that." [30].

These endeavours are accompanied by a bottom-up vision towards an amplification of engineers' voices, and more self-governance of engineering communities to constructively engage with safety, security and ethics, aside from the regulatory pressure and norms. Therefore, besides the educational key requirements, adequate tools and methods are needed [7]. Self-regulatory measures play an important role: although present in many companies, ethical guidelines often don't play a prominent role in the engineers' actual work environment [31]. Instead of such top-down processes, a bottom-up approach originating from the engineers' views should be fostered, especially in big enterprises [32].

For technology development and testing, engineers, as another form of bottom-up ethics, voice the need to broaden the scope of engineering towards more inclusivity, and diversity: "Engineering often is not directed towards minorities" [32]. Monetary gain drives most of the innovation aspects and it seems hard to get funding for projects for people with disabilities or diseases [32]. This also calls for diversity during testing and the development of robotic systems: "because you don't only want to detect persons that are male and walking straight up, you also want to detect people in wheelchairs, children, people who are small, who are tall, whatever" [30]. Including diversity into developing technology throughout the whole process is crucial, as "this is especially then necessary when you use something like a depth camera, where we know that this camera has issues in detecting people that have darker skin color. We are also trying to develop, if we get the funding, guidelines on how your test data set should look like in order to reflect this diversity. However, there are no established standards on how to test for human presence, for example" [30].

4.2 Pro-active Reflection and Communication for More Awareness

Another key finding is the engineers' call for more discourse, and basic education on robotics and robot ethics for all, to contribute to broader awareness for current and future technologies. As one engineer puts it: "In general, in tech, there is a very interesting conflict between the AI and the public perception of what it can really do. I laugh every day, when people say that there is going to be a rise of the robots against people, because I know how utterly stupid robots are. This general misunderstanding and fear of this technology is really based on nothing, based on maybe science fiction. This is not really what's happening" [31], and many of these problems can be solved by education [31]. Engineers advocate for ethical education in all areas of the robot-network, and beyond: "besides the inclusion of ethics into curricula of engineers, it is more important to embed ethical and basic schooling on responsibility also into management curricula, and the overall educational system in general" [7].

5 Conclusion: Robot-Networks Radically Transforming Responsibility

Robotics are networks of hybrid actors in action. Navigation and management of responsibility and ethics extends along complex interconnected systems, including legal norms, corporate guidelines, personal standards, educational curricula, practices and cultures of engineering, time and money, among others (RQ1). Robot ethics are movements of fluid bottom-up (self-)governance and top-down governance of distributed responsibility (RQ2). This asks for a widespread pro-active reflection and communication that raises awareness; under the forms of discourses on the actual abilities of robots and general education on ethics, responsibility, safety and security in relation to robotics (RQ3). These first preliminary findings require a shift in focus. They insist on research that states, that what is needed are not necessarily more ethical guidelines nor corporate policies and statements on AI ethics, which were not "primarily responsible for pressuring tech companies and governments to set guardrails on the use of AI", but instead that it is required to engage the ones that provoked the actual change, "community groups, workers, journalists, and researchers" [33, p. 11].

These initial findings serve as a chance to approach robotic engineering by exploring the extension of their fluid interconnectedness. This enables a stronger theoretically developed understanding of engineering and ethics as performative actions. Methodologically this approach allows to trace the actors in whatever form they appear; and empirically data can focus on the nuanced differences, translations, and connections that form the network [34, p. 114]. This paper points at some of these flows: movements translating explicit top-down laws, guidelines and norms, as well as their absence, into implicit bottom-up endeavors, including the responsible disclosure of security breaches, or the engineers' visions on more diversity in development. Furthermore, it identifies a void that emerges, as of both, the top-down endeavors, and the bottom-up efforts not fully arrive at their intended destinations, yet.

To comprehend these movements, we have to shift attention on what connects them and sets them into motion. Researching robot ethics, from this perspective, broadens the understanding of the field and provides a more holistic view. Digital transformation might not need more ethics, but new forms of governance which root in a networked reality [27, p. 47]. This especially applies for robotics, as "technical artefacts are primarily seen as isolated entities that can be optimized by experts to find technical solutions for technical problems. What is often lacking is a consideration of the wider contexts and the comprehensive relationship networks in which technical systems are embedded" [13, p. 103] such as social and ecological ones. This brings new angles on concepts that traditionally mainly were attributed to humans only, like agency, autonomy and responsibility [8, p. 1], that is "moral, legal, political, economic, social responsibility, and various other forms of responsibility" [8, p. 7]. This is a radically transformed concept of responsibility, that helps engaging with current technological innovations. A redefinition of the subject of responsibility as a network of dynamic and distributed shared interaction is part of this redefinition [8, p. 13].

It is time to rethink robotics and to shift some of the research attention: Instead of over-focusing on who is responsible, these initial findings of the CERSE project call for a closer look into the networks of governance of robotics, and of distributed responsibility that challenge the traditional conceptualization of its notion and suggest a radical update. Networks are fluid, flexible, complex, and innovative [27, p. 199], therein lies both the challenge and the chance to consider robot ethics further.

References

1. Latour, B.: Eine neue Soziologie für eine neue Gesellschaft. Suhrkamp, Frankfurt (2010)
2. Hepp, A.: Deep Mediatization. Routledge, London (2020)
3. Lin, P.: Introduction to robot ethics. In: Lin, P., Abney, K., Bekey, G.A. (eds.) Robot Ethics: The Ethical and Social Implications of Robotics, pp. 3–15. MIT Press, Cambridge (2012)
4. Lin, P., Jenkins, R., Abney, K. (eds). Robot Ethics 2.0: From Autonomous Cars to Artifical Intelligence. Oxford University Press, New York, NY (2017)
5. Rathmair, M., Brandstötter, M., Dieber, B., Nativel, C., Hofbaur, M.: Sichere und zuverlässige mobile manipulation. Elektrotech. Inftech. **137**(6), 283–290, (2020). https://doi.org/10.1007/s00502-020-00819-4
6. E2: Interview with Robotic Systems Engineer 2 (2021)
7. Ex1: Expert Interview with Robotic Systems Engineer 1 (2021)
8. Coeckelbergh, M., Loh, J.: Transformations of responsibility in the age of automation: being answerable to human and non-human others. In: Beck, B., Kühler, M. (eds.) Technology, anthropology, and Dimensions of Responsibility, vol. 1, pp. 7–22. J.B. Metzler, Stuttgart (2020). https://doi.org/10.1007/978-3-476-04896-7_2
9. International Organization for Standardization, ISO 10218-1:2011 (2011)
10. International Electrotechnical Commission, IEC 62443 | IEC
11. Simonjan, J., Taurer, S., Dieber, B.: A generalized threat model for visual sensor networks. Sensors **20**(13), 3629 (2020). https://doi.org/10.3390/s20133629
12. EU High-Level Expert Group on Artificial Intelligence: Ethics guidelines for trustworthy AI (2019)
13. Hagendorff, T.: The ethics of AI ethics: an evaluation of guidelines. Mind. Mach. **30**(1), 99–120 (2020). https://doi.org/10.1007/s11023-020-09517-8
14. Grunwald, A.: Verantwortung und Technik: zum Wandel des Verantwortungsbegriffs in der Technikethik. In: Seibert-Fohr, A (ed.) Entgrenzte Verantwortung: zur Reichweite und Regulierung von Verantwortung in Wirtschaft, Medien, Technik und Umwelt, pp. 265–283. Springer, Berlin [Heidelberg] (2020)
15. Dignum, V.: Responsible Artificial Intelligence: How to Develop and Use AI in a Responsible Way. Springer, Cham (2019). https://doi.org/10.1007/978-3-030-30371-6

16. Funk, M.: What is Robot Ethics? ...And Can It Be Standardized? In: Norskov, M., Seibt, J., Quick, O.S. (eds.) Culturally Sustainable Social Robotics: Proceedings of Robophilosophy, Washington, 2020, pp. 469–480 (2020)
17. Lin, P., Abney, K., Bekey, G.A., (Eds.) Robot Ethics: The Ethical and Social Implications of Robotics. MIT Press, Cambridge, Massachusetts (2012)
18. Loh, J.: Maschinenethik und Roboterethik. In: Bendel, O. (ed.) Handbuch Maschinenethik, pp. 1–19. Springer, Wiesbaden (2018). https://doi.org/10.1007/978-3-658-17484-2_6-1
19. Veruggio, G., Operto, F.: Roboethics: social and ethical implications of robotics. In: Siciliano, B., Khatib, O. (eds.). Springer Handbook of Robotics, pp. 1499–1524. Springer, Heidelberg (2008). https://doi.org/10.1007/978-3-540-30301-5_65
20. Bowen, W.R.: Engineering Ethics: Challenges and Opportunities. Springer, Heidelberg (2004). https://doi.org/10.1007/978-3-319-04096-7
21. Vienna Manifesto on Digital Humanism – DIGHUM, May 2019
22. Coeckelbergh, M.: AI Ethics. The MIT Press, Cambridge (2020)
23. Krainer, L., Heintel, P.: Prozessethik: zur Organisation ethischer Entscheidungsprozesse, 1st edn. VS Verlag für Sozialwissenschaften, Wiesbaden (2010)
24. Flick, U.: Doing Triangulation and Mixed Methods, 1st edn. SAGE Publications, Thousand Oaks (2017)
25. Glaser, B., Strauss, A.: The Discovery of Grounded Theory: Strategies for Qualitative Research. Routledge, New Brunswick (2000)
26. Latour, B.: Wir sind nie modern gewesen - Versuch einer symmetrischen Anthropologie, 1. Suhrkamp, Frankfurt am Main (2008)
27. Belliger, A., Krieger, D.J; Essays zur digitalen Transformation, 1. Auflage. Bielefeld: transcript (2021)
28. Keunecke, S.: Qualitatives Interview. In: Mikos, L., Wegener, C. (eds.) Qualitative Medienforschung: ein Handbuch, pp. 254–267. UVK Verlag, Konstanz (2005)
29. E3: Interview with Robotic Systems Engineer 3 (2021)
30. ExT1: Expert Talk with Expert 1 (2021)
31. E4: Interview with Robotic Systems Engineer 4 (2021)
32. Ex2: Expert Interview with Robotic Systems Engineer 2 (2021)
33. Crawford, K., et al.: AI Now Report, AI Now Institute, New York (2019)
34. Wieser, M.: Das Netzwerk von Bruno Latour: die Akteur-Netwerk-Theorie zwischen Science & Technology Studies und poststrukturalistischer Soziologie. Transcript, Bielefeld (2012)

Correction to: The Use of Robots in Aquatic Biomonitoring with Special Focus on Biohybrid Entities

Wiktoria Rajewicz, Thomas Schmickl, and Ronald Thenius

Correction to:
Chapter "The Use of Robots in Aquatic Biomonitoring with Special Focus on Biohybrid Entities"
in: A. Müller and M. Brandstötter (Eds.):
***Advances in Service and Industrial Robotics*, MMS 120,**
https://doi.org/10.1007/978-3-031-04870-8_61

The original version of the Chapter "The Use of Robots in Aquatic Biomonitoring with Special Focus on Biohybrid Entities" was previously published as non-open access. It has now been changed to Open Access under the CC BY 4.0 license and the copyright holder updated to 'The Author(s)'. The book and the chapter have been updated with this change.

The updated version of this chapter can be found at
https://doi.org/10.1007/978-3-031-04870-8_61

A. Müller and M. Brandstötter (Eds.): RAAD 2022, MMS 120, p. C1, 2022.
https://doi.org/10.1007/978-3-031-04870-8_72

Author Index

A. Müller and M. Brandstötter (Eds.): RAAD 2022, MMS 120, pp. 613–615, 2022.
https://doi.org/10.1007/978-3-031-04870-8

The manufacturer's authorised representative in the EU is Springer Nature Customer Service Centre GmbH, Europaplatz 3, 69115 Heidelberg, Germany. If you have any concerns regarding our products, please contact ProductSafety@springernature.com

Printed and bound by CPI Group (UK) Ltd, Croydon, CR0 4YY
06/10/2026
02225252-0017